Lecture Notes in Computer Science   10389

Commenced Publication in 1973
Founding and Former Series Editors:
Gerhard Goos, Juris Hartmanis, and Jan van Leeuwen

More information about this series at http://www.springer.com/series/7407

Faith Ellen · Antonina Kolokolova
Jörg-Rüdiger Sack (Eds.)

Algorithms
and Data Structures

15th International Symposium, WADS 2017
St. John's, NL, Canada, July 31 – August 2, 2017
Proceedings

 Springer

Editors
Faith Ellen 🆔
University of Toronto
Toronto, ON
Canada

Jörg-Rüdiger Sack 🆔
Carleton University
Ottawa, ON
Canada

Antonina Kolokolova
Memorial University of Newfoundland
St. John's, NL
Canada

ISSN 0302-9743   ISSN 1611-3349   (electronic)
Lecture Notes in Computer Science
ISBN 978-3-319-62126-5   ISBN 978-3-319-62127-2   (eBook)
DOI 10.1007/978-3-319-62127-2

Library of Congress Control Number: 2017945725

LNCS Sublibrary: SL1 – Theoretical Computer Science and General Issues

Printed on acid-free paper

This Springer imprint is published by Springer Nature
The registered company is Springer International Publishing AG
The registered company address is: Gewerbestrasse 11, 6330 Cham, Switzerland

Preface

This volume contains the papers presented at the 15th International Algorithms and Data Structures Symposium (WADS 2017), which was held from July 31 to August 2, 2017, in St. John's, Newfoundland, Canada. WADS, which alternates with the Scandinavian Symposium and Workshops on Algorithm Theory, SWAT, is a forum for researchers in the area of design and analysis of algorithms and data structures.

In response to the call for papers, 109 papers were submitted. From these submissions, the Program Committee selected 49 papers for presentation at WADS 2017, using a combination of online discussion in EasyChair and a one-day video conference. In addition, invited lectures were given by Pankaj Agarwal (Duke University), Michael Saks (Rutgers University), and Virginia Vassilevska Williams (MIT).

Special issues of papers selected from WADS 2017 are planned for two journals, *Algorithmica* and *Computational Geometry: Theory and Applications*.

We gratefully acknowledge the support of the WADS 2017 sponsors: Memorial University of Newfoundland, The Fields Institute for Research in Mathematical Sciences, Elsevier, and Springer.

July 2017

Faith Ellen
Antonina Kolokolova
Jörg Rüdiger Sack

Organization

Conference Chair and Local Arrangements Chair

Antonina Kolokolova   Memorial University of Newfoundland, Canada

Program Committee Co-chairs

Faith Ellen	University of Toronto, Canada
Antonina Kolokolova	Memorial University of Newfoundland, Canada
Jörg-Rüdiger Sack	Carleton University, Canada

Steering Committee

Frank Dehne	Carleton University, Canada
Faith Ellen	University of Toronto, Canada
Ian Munro	University of Waterloo, Canada
Jörg-Rüdiger Sack	Carleton University, Canada
Roberto Tamassia	Brown University, USA

Program Committee

Ittai Abraham	VMware Research, Israel
Hee-Kap Ahn	Pohang University of Science and Technology, Korea
Nina Amenta	University of California, Davis, USA
Sayan Bhattacharya	University of Warwick, UK
Therese Biedl	University of Waterloo, Canada
Joan Boyar	University of Southern Denmark, Denmark
Martin Dietzfelbinger	Technische Universität Ilmenau, Germany
Stephane Durocher	University of Manitoba, Canada
Funda Ergun	Indiana University, USA
Martín Farach-Colton	Rutgers University, USA
Fedor Fomin	University of Bergen, Norway
Travis Gagie	Diego Portales University, Chile
Thore Husfeld	IT University of Copenhagen, Denmark and Lund University, Sweden
Michael Kerber	Graz University of Technology, Austria
David Kirkpatrick	University of British Columbia, Canada
Ramesh Krishnamurti	Simon Fraser University, Canada
Kasper Green Larsen	Aarhus University, Denmark
Vangelis Markakis	Athens University of Economics and Business, Greece
Aleksandar Nikolov	University of Toronto, Canada
Naomi Nishimura	University of Waterloo, Canada

Richard Peng	Georgia Institute of Technology, USA
Seth Pettie	University of Michigan, USA
Liam Roditty	Bar-Ilan University, Israel
Sandeep Sen	Indian Institute of Technology, Delhi, India
Monique Teillaud	Inria Nancy-Grand Est, France
Takeshi Tokuyama	Tohoku University, Japan
Virginia Vassilevska Williams	Massachusetts Institute of Technology, USA
Todd Wareham	Memorial University of Newfoundland, Canada
Qin Zhang	Indiana University Bloomington, USA

Additional Reviewers

Mohammad Ali Abam	Olivier Devillers
Josh Alman	Michael Dinitz
Spyros Angelopoulos	Yago Diez Donoso
Esther Arkin	Feodor Dragan
Martin Aumüller	Ran Duan
Franz Aurenhammer	Charilaos Efthymiou
Mauricio Ayala-Rincon	Eduard Eiben
Erfan Sadeqi Azer	Matthias Englert
Sang Won Bae	David Eppstein
Aritra Banik	Leah Epstein
Gill Barequet	William Evans
Djamal Belazzougui	Rolf Fagerberg
Robert Benkoczi	Lene Favrholdt
Binay Bhattacharya	Andreas Emil Feldmann
Arijit Bishnu	Yuval Filmus
Hans L. Bodlaender	Johannes Fischer
Piotr Borowiecki	Kyle Fox
Jean Cardinal	Mathew Francis
Paz Carmi	Jie Gao
Timothy M. Chan	Daya Gaur
Steven Chaplick	Ellen Gethner
Jiecao Chen	Arijit Ghosh
Man Kwun Chiu	Archontia Giannopoulou
Keerti Choudhary	Petr Golovach
Nachshon Cohen	Carsten Grimm
Vincent Cohen-Adda	Allan Grønlund
Alexander Conway	Martin Groß
Søren Dahlgaard	Joachim Gudmundsson
Peter Damaschke	Manoj Gupta
Syamantak Das	Shalmoli Gupta
Mark de Berg	Sariel Har-Peled
Jean-Lou De Carufel	Meng He

Martin Hoefer
Chien-Chung Huang
Thomas Hull
Ehsan Iranmanesh
Mark Jerrum
Hossein Jowhari
Stasys Jukna
Mark Keil
Kamyar Khodamoradi
Elena Khramtcova
Philipp Kindermann
Peter Kling
Chip Klostermeyer
Christian Knauer
Mathias Bæk Tejs Knudsen
Janne H. Korhonen
Matias Korman
Daniel Kral
Stefan Kratsch
Ravishankar Krishnaswamy
Janardhan Kulkarni
Amit Kumar
Woden Kusner
Jakub Łącki
Kim Skak Larsen
Samuli Leppänen
Jian Li
Pak Ching Li
Yi Li
Bingkai Lin
Daniel Lokshtanov
Aaron Lowe
Anna Lubiw
Bin Ma
Pasin Manurangsi
Euripides Markou
Tyler Mayer
Saeed Mehrabi
Moti Medina
Victor Milenkovic
Tillmann Miltzow
Gopinath Misra
Joseph Mitchell
Dieter Mitsche
Valia Mitsou
Reza Modarres

Debajyoti Mondal
Gila Morgenstern
Miguel A. Mosteiro
Wolfgang Mulzer
Ian Munro
Nabil Mustafa
Petra Mutzel
Jesper Nederlof
Stefan Neumann
Huy L. Nguyen
Patrick Nicholson
Jesper Sindahl Nielsen
Arnur Nigmetov
Nicolas Nisse
Jakob Nordström
Eunjin Oh
Yota Otachi
John Owens
Jeff M. Phillips
Rameshwar Pratap
Sharath Raghvendra
Benjamin Raichel
Rajiv Raman
Mathias Rav
Saurabh Ray
Jean-Florent Raymond
André van Renssen
Marcel Roeloffzen
Luis M.S. Russo
Yogish Sabharwal
Sushant Sachdeva
Vera Sacristan
Toshiki Saitoh
Paweł Schmidt
Hannah Schreiber
Oded Schwartz
Roy Schwartz
Akiyoshi Shioura
Rakesh Sinha
Michiel Smid
Shay Solomon
Bettina Speckmann
Abhinav Srivastav
Frank Staals
Thomas Steinke
Torstein Strømme

Jan Arne Telle
Sharma V. Thankachan
Csaba Toth
Meng-Tsung Tsai
Charalampos Tsourakakis
Ryuhei Uehara
Seeun William Umboh
Birgit Vogtenhuber
Erik Waingarten
Di Wang
Haitao Wang
Junxing Wang

Josh Wang
Sebastian Wild
Ryan Williams
Steve Wismath
Prudence Wong
Christian Wulff-Nilsen
Jie Xue
Sang Duk Yoon
Hamid Zarrabi-Zadeh
Haoyu Zhang
Yuan Zhou

Abstracts of Invited Lectures

Materials of Insured Lectures

Algorithms for Geometric Similarity: Recent Developments

Pankaj K. Agarwal

Department of Computer Science, Duke University, Durham, USA
pankaj@cs.duke.edu

Abstract. A basic problem in classifying, or searching for similar objects, in a large set of geometric objects is computing similarity between two objects. There has been extensive work on computing geometric similarity between two objects. In many applications, it is not sufficient to return a single similarity score. Instead, a map between two objects that identifies shared structures is needed.

This talk discusses some recent work on computing maps between two or more objects. The talk consists of three parts. The first part focuses on computing maps between two weighted point sets, say, distributions. The second part is devoted to computing maps between a pair of trajectories. The third part will briefly discuss computing Gromov-Hausdorff distance between two metric spaces.

This work is supported in part by NSF under grants CCF-15-13816, CCF-15-46392, and IIS-14-08846, by ARO grant W911NF-15-1-0408, and by grant 2012/229 from the U.S.-Israel Binational Science Foundation.

How efficiently can easy dynamic programs be approximated?

Michael Saks

Department of Mathematics, Rutgers University, New Brunswick, USA
saks@math.rutgers.edu

Abstract. In many of the simplest examples of dynamic programming, inputs of size n are processed by constructing an $n \times n$ matrix, where each entry is obtained by a simple function of a few entries above and to the left. This yields a simple $O(n^2)$ algorithm for such problems. These algorithms naturally arise, for example, in evaluating various distance measures between two strings, such as LCS (longest common subsequence) distance, Edit Distance, Frechet Distance, and Dynamic Time Warping Distance, and the i,j entry of the matrix gives the desired measure between the length i prefix of the first string, and the length j prefix of the second. With few exceptions (such as the Longest Increasing Subsequence (LIS) problem where the quadratic time algorithm has been improved to $O(n \log(n))$), these quadratic time dynamic programming algorithms remain essentially the fastest exact algorithms (except for $n^{o(1)}$ factor improvements). This phenomenon has been the focus of much recent research in *fine grain complexity*, and it has been shown that for many such problems, reducing the running time to $O(n^{2-\varepsilon})$ would contradict the Strong Exponential Time Hypothesis (e.g., Bringmann [7], Abboud, Backurs and Williams [2], Backurs and Indyk [6] and Bringmann and Kunnermann [8].)

If we are willing to accept a good approximation (rather than the exact answer), then there is much less evidence that quadratic complexity is needed. Bringmann [7] proved that the Strong Exponential Hypothesis implies that truly subquadratic algorithms cannot achieve approximation factors arbitrarily close to 1. Abboud and Backurs [1] provided complexity theoretic evidence that truly subquadratic deterministic algorithms cannot achieve approximation factors arbitrarily close to 1 for edit distance and LCS-distance.

If we allow randomized algorithms, it is quite possible that problems such as edit distance and LCS distance have constant factor approximation algorithms that are significantly faster than quadratic. Andoni, Krauthgamer and Onak [4] gave a nearly linear time algorithm that achieves a polylogarithmic approximation to edit distance. For certain special cases of LCS-distance, arbitrarily good *additive* εn *error* approximation algorithms are known that are substantially faster than the best exact algorithms. For the LIS Problem, Saks and Seshadhri [11] (following the work of Ailon, Chazelle, Comandur and Liu [3] and Parnas, Ron and Rubinfeld [10]) developed such an additive approximation

Supported by Simons Foundation Award 332622.

whose running time is only polylogarithmic in the length of the input. For the special case of LCS-distance between two permutations of $\{1, \ldots, n\}$ (Ulam Distance), Naumovitz, Saks and Seshadhri [9] (following earlier work of Andoni and Nguyen [5]) obtained such an additive approximation running in time $\tilde{O}(\sqrt{n})$.

References

1. Abboud, A., Backurs, A.: Towards hardness of approximation for polynomial time problems. In: Conference on Innovations in Theoretical Computer Science (ITCS) (2017)
2. Abboud, A., Backurs, A., Williams, V.V.: Quadratic-Time Hardness of LCS and other Sequence Similarity Measures. CoRR, abs/1501.07053 (2015)
3. Ailon, N., Chazelle, B., Comandur, S., Liu, D.: Estimating the distance to a monotone function. Random Struct. Algorithms **31**, 371–383 (2017)
4. Andoni, A., Krauthgamer, R., Onak, K.: Polylogarithmic approximation for edit distance and the asymmetric query complexity. In: IEEE Symposium on Foundations of Computer Science (FOCS), pp. 377–386 (2010)
5. Andoni, A., Nguyen, H.L.: Near-optimal sublinear time algorithms for Ulam distance. In: Proceedings of the 21st Symposium on Discrete Algorithms (SODA), pp. 76–86 (2010)
6. Backurs, A., Indyk, P.: Edit distance cannot be computed in strongly subquadratic time (unless SETH is false). In: Proceedings of the Forty-Seventh Annual ACM on Symposium on Theory of Computing (STOC), pp. 51–58 (2015)
7. Bringmann, K.: Why walking the dog takes time: Frechet distance has no strongly subquadratic algorithms unless SETH fails. CoRR, abs/1404.1448 (2014)
8. Bringmann, K., Kunnemann, M.: Quadratic conditional lower bounds for string problems and dynamic time warping. In: Proceedings of the 56th Annual IEEE Symposium on Foundations of Computer Science (FOCS), pp. 79–97 (2015)
9. Naumovitz, T., Saks, M., Seshadhri, C.: Accurate and nearly optimal sublinear approximations to Ulam distance. In: SIAM-ACM Symposium on Discrete Algorithms, pp. 2012–2031 (2017)
10. Parnas, M., Ron, D., Rubinfeld, R.: Tolerant property testing and distance approximation. J. Comput. Syst. Sci. **6**, 1012–1042 (2006)
11. Saks, M., Seshadhri, C.: Estimating the longest increasing sequence in polylogarithmic time. SIAM J. Comput. **46**(2), 774–823 (2017)

Fine-Grained Complexity of Problems in P

Virginia Vassilevska Williams

Computer Science and Artificial Intelligence Laboratory,
Massachusetts Institute of Technology, Cambridge, USA
virgi@mit.edu

Abstract. A central goal of algorithmic research is to determine how fast computational problems can be solved in the worst case. Theorems from complexity theory state that there are problems that, on inputs of size n, can be solved in $t(n)$ time but not in $t(n)^{1-\varepsilon}$ time for $\varepsilon > 0$. The main challenge is to determine where in this hierarchy various natural and important problems lie. Throughout the years, many ingenious algorithmic techniques have been developed and applied to obtain blazingly fast algorithms for many problems. Nevertheless, for many other central problems, the best known running times are essentially those of their classical algorithms from the 1950s and 1960s.

Unconditional lower bounds seem very difficult to obtain, and so practically all known time lower bounds are conditional. For years, the main tool for proving hardness of computational problems have been NP-hardness reductions, basing hardness on P $\neq$ NP. However, when we care about the exact running time (as opposed to merely polynomial vs non-polynomial), NP-hardness is not applicable, especially if the problem is already solvable in polynomial time. In recent years, a new theory has been developed, based on "fine-grained reductions" that focus on exact running times. Mimicking NP-hardness, the approach is to (1) select a key problem X that is conjectured to require essentially $t(n)$ time for some t, and (2) reduce X in a fine-grained way to many important problems. This approach has led to the discovery of many meaningful relationships between problems, and even sometimes to equivalence classes.

The main key problems used to base hardness on have been: the 3SUM problem, the CNF-SAT problem (based on the Strong Exponential Time Hypothesis (SETH)) and the All Pairs Shortest Paths Problem. Research on SETH-based lower bounds has flourished in particular in recent years showing that the classical algorithms are optimal for problems such as Approximate Diameter, Edit Distance, Frechet Distance, Longest Common Subsequence, many dynamic graph problems, etc.

In this talk I will give an overview of the current progress in this area of study, and will highlight some exciting new developments.

Supported by an NSF CAREER Award, NSF Grants CCF-1417238, CCF-1528078 and CCF-1514339, and BSF Grant BSF:2012338.

Contents

Covering Segments with Unit Squares

Ankush Acharyya, Subhas C. Nandy, Supantha Pandit, and Sasanka Roy

Indian Statistical Institute, Kolkata, India

Abstract. We study several variations of line segment covering problem with axis-parallel unit squares in the plane. Given a set S of n line segments, the objective is to find the minimum number of axis-parallel unit squares which cover at least one end-point of each segment. The variations depend on the orientation and length of the input segments. We prove some of these problems to be NP-complete, and give constant factor approximation algorithms for those problems. For the general version of the problem, where the segments are of arbitrary length and orientation, and the squares are given as input, we propose a factor 16 approximation result based on multilevel linear programming relaxation technique. This technique may be of independent interest for solving some other problems. We also show that our problems have connections with the problems studied by Arkin et al. [2] on conflict-free covering problem. Our NP-completeness results hold for more simplified types of objects than those of Arkin et al. [2].

Keywords: Segment cover, unit square, NP-hardness, linear programming, approximation algorithms, PTAS.

1 Introduction

In this paper, we study different interesting variations of line segment covering problem. Here, a set $S = \{s_1, s_2, \ldots, s_n\}$ of line segments in $\mathbb{R}^2$ is given. An axis-parallel square t is said to *cover* a line segment $s \in S$ if t contains at least one end-point of s. We deal with two classes of covering problem: i) *continuous*, and ii) *discrete*.

CONTINUOUS COVERING SEGMENTS BY UNIT SQUARES (*CCSUS*): Given a set S of segments in the plane, the goal is to find a smallest set T of unit squares which covers all the segments in S.

DISCRETE COVERING SEGMENTS BY UNIT SQUARES (*DCSUS*): Given a set S of segments and a set T of unit squares in the plane, the goal is to find a subset $T' \subseteq T$ of minimum cardinality which can cover all the segments in S.

The motivation of studying this problem comes from an applications to network security [9]. Here, a set of physical devices is deployed over a geographical area. These devices communicate with each other through physical links. The objective is to check the security of the network by placing minimum number of devices which can sense at least one end point of every link. This problem

© Springer International Publishing AG 2017
F. Ellen et al. (Eds.): WADS 2017, LNCS 10389, pp. 1–12, 2017.
DOI: 10.1007/978-3-319-62127-2_1

can be modelled as *line segment covering problem*, where the links can be interpreted as straight-line segments and the objects can be interpreted as unit squares. In [9], several other applications are also stated.

We study the following variations of covering problem for line segments which are classified depending upon their lengths and orientations.

CONTINUOUS COVERING

▶ *CCSUS-H1-US:* Horizontal unit segments inside a unit height strip.
▶ *CCSUS-H1:* Horizontal unit segments.
▶ *CCSUS-HV1:* Horizontal and vertical unit segments.
▶ *CCSUS-ARB:* Segments with arbitrary length and orientation.

DISCRETE COVERING

▶ *DCSUS-ARB:* Segments with arbitrary length and orientation.

We define some terminologies and definitions used in this paper. We use *segment* to denote a line segment, and *unit square* to denote an axis-parallel unit square. For a given non-vertical segment s, we define $l(s)$ and $r(s)$ to be its left and right end-points. For a vertical segment s, $l(s)$ and $r(s)$ are defined to be the end-points of s with highest and lowest y-coordinates respectively. The *center* of a square t is the point of intersection of its two diagonals. We use $t(a, b)$ to denote a square whose center is at the point a and whose side length is b.

Definition 1. *Two segments in S are said to be **independent** if no unit square can cover both the segments. A subset $S' \subseteq S$ is an **independent set** if every pair of segments in S' is independent. A subset $S' \subseteq S$ of segments is a **maximal independent set** if for any $s \in S \setminus S'$, $S' \cup \{s\}$ is not an independent set.*

Known results: Arkin et al. [2] studied a related problem the *conflict-free covering*. Given a set P of n color classes, where each color class contains exactly two points, the goal is to find a set of conflict-free objects of minimum cardinality which covers at least one point from each color class. An object is said to be *conflict-free* if it contains at most one point from each color class. Arkin et al. [1, 2] showed that, both discrete and continuous versions of conflict-free covering problem are NP-complete where the points are on a real line and objects are intervals of arbitrary length. These results are also valid for covering arbitrary length segments on a line with unit intervals. They provided 2- and 4-factor approximation algorithms for the continuous and discrete versions of conflict-free covering problem with arbitrary length intervals respectively. If the points of the same color class are either vertically or horizontally unit separated, then they proved that the continuous version of the conflict-free covering problem with axis-parallel unit squares is NP-complete, and proposed a factor 6 approximation algorithm. Finally, they remarked the existence of a polynomial time dynamic programming based algorithm for the continuous version of conflict-free covering problem with unit intervals where the points are on a real line and each pair of same color points is unit separated. Recently, Kobylkin [9] studied the problem of covering the edges of a given straight line embedding of a planar graph by minimum number of unit disks, where an edge is said to be covered by a disk if any point on that edge lies inside that disk. He proved NP-completeness results for some special graphs. A similar study is made in [10], where a set of line segments is given, the objective is to cover these segments with minimum number of unit disks, where the covering of a segment by a disk is defined as in [9]. For continuous version of the problem, they proposed a PTAS where the segments are non-intersecting. For the discrete version, they showed that the problem is APX-hard.

Our contributions: In Section 2, we show that the *CCSUS-H1* problem is NP-complete, and propose an $O(n \log n)$ time factor 3 approximation algorithm for the *CCSUS-HV1* problem. This improves the factor 6 approximation result of Arkin et al. [2] while keeping the time complexity unchanged. We also provide a PTAS for *CCSUS-HV1* problem. For the *CCSUS-ARB* problem, we give an $O(n \log n)$ time factor 6 approximation algorithm. In Section 3, give a polynomial time factor 16 approximation algorithm for the *DCSUS-ARB* problem.

2   Continuous covering

In this version, the segments are given, and the objective is to place minimum number of unit squares for covering at least one end-point of all the segments.

2.1   *CCSUS-H1-US* problem

Let S be a set of n horizontal unit segments inside a unit height horizontal strip. Start with an empty set T'. Sort the segments in S from left to right with respect to their right end-points. Repeat the following steps until all segments of S are processed. In each step, select the left-most segment s among the uncovered segments in S. Add a unit square $t \in T'$ which is inside the strip aligning its left boundary at $r(s)$. Mark all the segments that are covered by t as processed. Finally, return T' as the output. Thus, we have an $O(n \log n)$ time algorithm for the *CCSUS-H1-US* problem.

2.2   *CCSUS-H1* problem

We prove that *CCSUS-H1* is NP-complete by using a reduction from the rectilinear version of planar 3 SAT *(RPSAT(3))* problem [8]. We also propose an $O(n \log n)$-time 2-approximation algorithm for this problem.

RPSAT(3) [8]: Given a 3 SAT instance ϕ with n variables and m clauses, where the variables are positioned on a horizontal line and each clause containing 3 literals is formed with three vertical and one horizontal line segment. Each clause is connected with its three variables either from above or from below such that no two line segments corresponding to two different clauses intersect. The objective is to find a satisfying assignment of ϕ. See Figure 1(a) for an instance of *RPSAT(3)* problem. Here the solid (resp. dotted) *vertical segment* attached to the horizontal line of a clause represents that the corresponding variable appears as a positive (resp. negative) literal in that clause.

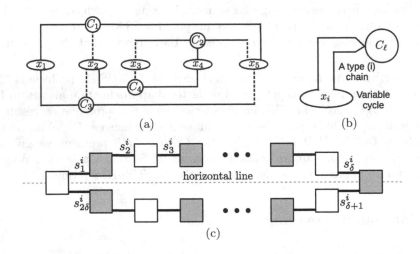

Fig. 1. (a) *RPSAT(3)* representation. (b) Connection of a cycle and a chain. (c) Cycle gadget for a variable x_i.

We now describe the construction of an instance I of *CCSUS-H1* problem from an instance ϕ of *RPSAT(3)* problem. Let $\{x_1, x_2, \ldots, x_n\}$ be n variables and $\{C_1, C_2, \ldots, C_m\}$ be m clauses of ϕ. We describe the construction for the clauses connecting to the variables from above. A similar construction can be done for the clauses connecting to the variables from below.

Let d be the maximum number of vertical segments connected to every variable from different clauses either from above or from below. Let, $\delta = 4d + 3$. Each variable gadget for x_i may consist of a **cycle** and at most $2d$ **chains**. The *cycle* consists of 2δ unit horizontal segments $\{s_1^i, s_2^i, \ldots, s_{2\delta}^i\}$ in two sides of a (dotted) horizontal line (see Figure 1(c)). The segments $\{s_1^i, s_2^i, \ldots, s_{\delta}^i\}$ are above the horizontal line and the segments $\{s_{\delta+1}^i, s_{\delta+2}^i, \ldots, s_{2\delta}^i\}$ are below the horizontal line. The chains correspond to the vertical segments connecting a variable x_i with the clauses containing it. There are three types of **chains**: (i) " Γ ", (ii) " $\daleth$ ", and (iii) " I " (see Figure 1(a)). The gadget corresponding to three types of chains are shown in Figures 2(a), 2(b), and 2(c) respectively. The chains are connected to the cycle, and together it forms a chain of **big-cycle** (see Figure 1(b)). It needs to be mentioned that the number of segments is not fixed for every chain, even for similar chains of different clauses. Note that, at the joining point (to construct a big-cycle) we slightly perturb two unit segments little upward.

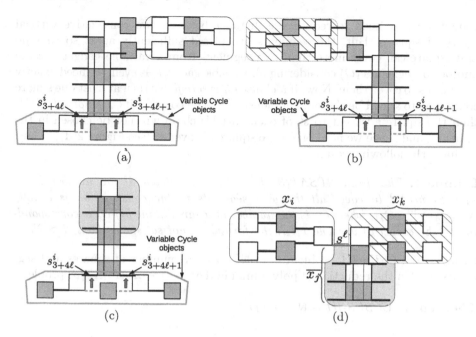

Fig. 2. Gadget for (a) type (1) chain, (b) type (ii) chain (c) type (iii) chain, and (d) Demonstration of clause-segment s^ℓ corresponding to the clause $C_\ell = (x_i \vee \overline{x_j} \vee x_k)$; shaded portions in (a), (b), (c) represent the connection of s^ℓ with the variables in C_ℓ.

Let $0, 1, 2, \ldots, \kappa$ ($\kappa \leq d$) be the left to right order of the vertical segments corresponding to the clauses which are connected to the gadget corresponding to the variable x_i. Consider the ℓ-th clause C_ℓ in this order. If x_i is a positive literal, then the segments $s^i_{3+4\ell}$ and $s^i_{3+4\ell+1}$ are perturbed (moved upward as shown using upward arrow in Figures 2(a), 2(b), 2(c)) to connect the corresponding chain of C_ℓ with the cycle of variable x_i. Otherwise, if x_i is a negative literal, then the segments $s^i_{3+4\ell+1}$ and $s^i_{3+4\ell+2}$ are perturbed.

Note that, the squares are not given as a part of the input. In Figures 1(c), 2(a), 2(b), and 2(c) a possible set of unit squares are also depicted. Each square can cover exactly two segments. Therefore, we have the following observation:

Observation 1 *Exactly half of the squares (either all empty or all shaded) can cover all the segments in the big-cycle corresponding to the variable x_i. This solution represents the truth value (empty for true and shaded for false) of the corresponding variable x_i.*

Further, for the clause C_ℓ, we take a single unit horizontal segment s^ℓ that connects the chain corresponding to three variables. This is referred to as a **clause-segment**. The placement of s^ℓ is shown in Figure 2(d). Note that, in order to maintain the alternating empty and shaded vertical layers in a variable gadget we may need to reduce the distance between two consecutive vertical layers of squares. But, the segments are placed sufficiently apart so that no unit square can cover more than two segments from a variable gadget. As the number of segments (Q) considering all variable gadgets, is even, we need exactly $\frac{Q}{2}$ squares to cover them. Now, if a clause C_ℓ is *satisfiable* then at least one square connected to s^ℓ will be chosen, and hence s^ℓ will be covered; if C_ℓ is *not satisfiable* then the square adjacent to s^ℓ of each variable chain will not be chosen in the solution, and hence we need one more square to cover s^ℓ (see Figure 2(d)). Thus, we have the following result.

Lemma 1. *The given RPSAT(3) formula is satisfiable if the number N of squares needed to cover all the unit segments in the construction is exactly $N_0 = \frac{1}{2}(\sum_{i=1}^{n} Q_i)$, where Q_i is the number of squares in the big-cycle corresponding to the gadget of variable x_i. If the formula is not satisfiable then $N > N_0$,*

Clearly, (i) CCSUS-H1 problem is in NP, and (ii) number of squares and segments used in the reduction is polynomial in the the size of RPSAT(3) problem.

Theorem 1. *CCSUS-H1 is NP-complete.*

Approximation algorithm Let S be a set of n unit horizontal segments on the plane. We first partition the entire region into a set of ℓ disjoint unit height horizontal strips $H_1, H_2, \ldots, H_\ell$. Let $S_i \subseteq S$ be the set of segments in the strip H_i, for $i = 1, \ldots, \ell$. Clearly, (i) $S_i \cap S_j = \emptyset$, for $i \neq j$, and (ii) any unit square cannot cover two segments, one from S_i and the other from S_j where $j - i \geq 2$, for $i = 1, \ldots, \ell - 2$, and $j = 3, \ldots, \ell$.

We calculate minimum number of unit squares covering S_i using the algorithm in Section 2.1. Let Q_i be the set of squares returned by our algorithm. Let $Q^{odd} = \{Q_1 \cup Q_3 \cup \ldots\}$ and $Q^{even} = \{Q_2 \cup Q_4 \cup \ldots\}$ be the optimum solutions for the segments in odd and even numbered strips respectively. We have $Q^{odd} \cap Q^{even} = \emptyset$, and we report $Q = Q_{odd} \cup Q_{even}$. Let OPT be a minimum sized set of unit squares covering S. Now, $|OPT| \geq \max(|Q_{odd}|, |Q_{even}|)$. Thus, $|Q| = |Q_{odd}| + |Q_{even}| \leq 2|OPT|$. Since $S_i \cap S_j = \emptyset$ for $i \neq j$, and computing Q_i needs $O(|S_i| \log |S_i|)$ time, the overall running time of the algorithm is $O(n \log n)$.

Theorem 2. *A 2-factor approximation solution for the CCSUS-H1 problem can be computed in $O(n \log n)$ time.*

2.3   CCSUS-HV1 problem

Here, we have both horizontal and vertical segments in S which are of unit length. An easy way to get a factor 4 approximation algorithm for this problem is as follows. Let $S = S_H \cup S_V$, where S_H and S_V are the sets of horizontal and vertical unit segments respectively. By Theorem 2, we already have a factor 2 approximation algorithm for covering the members in S_H (resp. S_V). If Q_H and Q_V be the

Fig. 3. (a) Placement of unit squares for a horizontal and vertical unit segment s.

set of squares returned by our algorithm for covering S_H and S_V respectively, and OPT_H and OPT_V be the optimum solution for S_H and S_V respectively, then $|Q_H| + |Q_V| \leq 2|OPT_H| + 2|OPT_V| \leq 4|OPT|$, where OPT be the overall optimum solution for $S_H \cup S_V$ (since $|OPT| \geq \max(|OPT_H|, |OPT_V|)$).

We now propose a factor 3 approximation algorithm for this problem using a sweep-line technique. During the execution of the algorithm, we maintain a set of segments LB such that no two members in LB can be covered by one unit square. Elements in LB form an independent set. For each segment in S we maintain a *flag* variable; its value is 1 or 0 depending on whether it is covered or not by the chosen set of squares corresponding to the members in LB. We also maintain a range tree $\mathcal{T}$ with the end-points of the members in S. Each element in $\mathcal{T}$ has a pointer to the corresponding element in S.

Let $OUTPUT$ be a set containing the chosen squares to cover the members in LB. Initially, we set $OUTPUT - \emptyset$. We sort the unit segments in S from top to bottom with respect to their $r(.)$ values. Next, we consider the elements in S in order. For each segment $s \in S$, if its *flag* bit is 0, then (i) if s is horizontal then we define three unit squares $\{t_1, t_2, t_3\}$ (see Figure 3(a)), and (ii) if s is vertical then we define two unit squares $\{t_1, t_2\}$ (see Figure 3(b)). Store s in LB and the generated squares are put in $OUTPUT$, and for each newly generated squares t, we search in the range tree $\mathcal{T}$ to identify the members in S covered by t. The *flag* bit of s and those members in S covered by t are set to 1. The process continues until the flag bit of all the members in S are set to 1.

Theorem 3. *The proposed algorithm produces a 3-factor approximation result for the CHSUS-HV1 problem in $O(n \log n)$ time using $O(n \log n)$ space.*

Polynomial time approximation scheme We propose a PTAS for the *CCSUS-HV1* problem using the **shifting strategy** of Hochbaum and Maass [7]. We are given a set S of n horizontal and vertical unit segments. Enclose the segments inside a integer length square box B; partition B into vertical strips of width 1, and also partition B into horizontal strips of height 1. We choose a constant $k \in \mathbb{Z}^+$, and define a **k-strip** consisting of at most k consecutive strips. Now, we define the concept of **shifts**. We have k different shifts in the vertical direction. Each vertical shift consists of some disjoint k-strips. In the i-th shift $(i = 0, 1, \ldots, k-1)$, the first k-strip consists of i unit vertical strips at extreme left, and then onwards each k-strip is formed with k consecutive unit vertical strips except the last k-strip which may contain lesser than k vertical strips. Similarly, k shifts are defined in horizontal direction. Now consider $\textbf{\textit{shift}}(\textbf{\textit{i, j}})$ as the i-th vertical shift and j-th horizontal shift. This splits the box B into rectangular $\textbf{\textit{cells}}$ of size at most $k \times k$. The following observation is important to analyze the complexity of our algorithm.

Observation 2 *There exists an optimal solution that contains squares such that two boundaries of each square is attached to an end-point(s) of some segments.*

Lemma 2. *Finding a feasible solution for each $shift(i, j)$ requires $O(n^{2k^2})$ time.*

In our algorithm, for each $shift(i, j)$ we calculate optimal solution in each cell and combine them to get a feasible solution. Finally, return the minimum among these k^2 feasible solutions.

Let OPT be an optimum set of unit squares covering S, and Q be a feasible solution returned by our algorithm described above. Following the proof technique of Lemma 2.1 of [7], we can prove the following theorem.

Theorem 4. $|Q| \leq (1 + \frac{1}{k})^2 |OPT|$ *and the running time of the above algorithm is $O(k^2 n^{2k^2})$.*

2.4   *CCSUS-ARB* problem

Here, we first design a 8-approximation algorithm for $CCSUS$-ARB problem as follows. Next, we improve the approximation factor to 6.

Observation 3 *Let s_1 and s_2 be two segments in S. If neither of the squares $t(l(s_1), 2)$ and $t(r(s_1), 2)$ covers s_2, then s_1 and s_2 are independent.*

Here also we start with an empty set $OUTPUT$ and LB, and each segment in S is attached with a flag bit. We maintain a range tree $\mathcal{T}$ with the end-points of the members in S. Each time, an arbitrary segment $s \in S$ with $flag(s) = 0$ is chosen, and inserted in LB. Its flag bit is set to 1. Insert four unit squares $\{t_1, t_2, t_3, t_4\}$ which fully cover the square $t(l(s), 2)$ and four unit squares $\{t_1', t_2', t_3', t_4'\}$ which fully cover the square $t(r(s), 2)$ in $OUTPUT$. Remove all the segments in S that are covered by $\{t_1, t_2, t_3, t_4, t_1', t_2', t_3', t_4'\}$ by performing range searching in $\mathcal{T}$ as stated in Section 2.3. The end-points of the deleted segments are also deleted from $\mathcal{T}$. This process is repeated until all the members in S are flagged. Finally, return the set $OUTPUT$. Observation 3 suggests the following result.

Lemma 3. *The above algorithm for CCSUS-ARB problem runs in $O(n \log n)$ time, and the produced solution is factor 8 approximation of the optimal solution.*

We now improve the approximation factor to 6 using a sweep-line technique introduced in Biniaz et al. [4]. We sort the segments in S with respect to their left end-points[1], and process the elements in S in order. When an element $s \in S$ is processed, if $flag(s) = 0$ then we put six squares, two of them covering the 1×2 rectangle $t_1 =$ right-half of $t(l(s), 2)$ and four of them covering the 2×2 square $t_2 = t(r(s), 2)$. We also identify the segments in S that are covered by t_1 and t_2, and their flag bit is set to 1. Thus, we have the following result:

Theorem 5. *The above algorithm for the CCSUS-ARB problem produces a factor 6 approximation result in $O(n \log n)$ time.*

3   Discrete covering: *DCSUS-ARB* problem

In this section, we give a 16 factor approximation algorithm for *DCSUS-ARB* problem. We use the following problem to solve the *DCSUS-ARB* problem.

DCPUS Problem: Given a point set P in $\mathbb{R}^2$ and a set T of unit squares, find a subset of T of minimum cardinality to cover all the points in P.

For an *ILP* z_w, the corresponding *LP* is denoted by $\overline{\mathsf{z}}_w$. We denote OPT_w^I and OPT_w^F as the value of the optimal solution for z_w and $\overline{\mathsf{z}}_w$ respectively.

Lemma 4. *If there exists an LP-based factor α approximation algorithm for the DCPUS problem, then there exists a LP-based factor 2α approximation result for the DCSUS-ARB problem.*

Proof. Let $T_1 \subseteq T$ (resp. $T_2 \subseteq T$) be the set of all squares which cover the left (resp. right) end-points of the segments in S. Consider binary variable x_i for each square $t_i \in T_1$, and y_j for each square $t_j \in T_2$. Create an *ILP* as follows.

$$\mathsf{z}_0 : \min \sum_{i \mid t_i \in T_1} x_i + \sum_{j \mid t_j \in T_2} y_j$$

$$\text{s.t.} \sum_{i \mid l(s_k) \in t_i \in T_1} x_i + \sum_{j \mid r(s_k) \in t_j \in T_2} y_j \geq 1 \ \forall \ k \mid s_k \in S; \ x_i, y_j \in \{0,1\} \ \forall \ i \mid t_i \in T_1 \ \& \ j \mid t_j \in T_2$$

In the solution of z_0, $x_i = 1$ or 0 (resp. $y_j = 1$ or 0) depending on whether the square $t_i \in T_1$ (resp. $t_j \in T_2$) is in an optimal solution or not. We solve the corresponding *LP*, $\overline{\mathsf{z}}_0$. Create two partitions: $S_1 \subseteq S$ with those segments s_k such that $\sum_{i \mid l(s_k) \in t_i \in T_1} x_i \geq 1/2$, and $S_2 \subseteq S$ with those segments s_ℓ for which $\sum_{j \mid r(s_\ell) \in t_j \in T_2} y_i \geq 1/2$. Now, consider the following two *ILP*'s z_1 and z_2.

$$\mathsf{z}_1 : \min \sum_{i \mid t_i \in T_1} x_i$$

$$\text{s.t.} \sum_{i \mid l(s_k) \in t_i \in T_1} x_i \geq 1, \ \forall \ k \mid s_k \in S_1$$

$$x_i \in \{0,1\} \ \forall i \mid t_i \in T_1$$

$$\mathsf{z}_2 : \min \sum_{j \mid t_j \in T_2} y_j$$

$$\text{s.t.} \sum_{j \mid r(s_k) \in t_j \in T_2} y_j \geq 1, \ \forall \ k \mid s_k \in S_2$$

$$y_j \in \{0,1\} \ \forall \ j \mid t_j \in T_2$$

---
[1] The left end-point of a vertical segment is defined as its top end-point.

By an analysis identical to Gaur et al. [6], we conclude that $OPT_1^F + OPT_2^F \leq 2OPT_0^F \leq 2OPT_0^I$. $\square$

Observe that, both the ILP's, Z_1 and Z_2 are the problems of covering points by unit squares. We consider the problem Z_1, and use a multilevel LP relaxation (see Theorem 6 stated below) to have a factor 8 approximation algorithm for the $DCPUS$ problem. It consists of 3 **steps** of LP relaxations. In each step, we use linear programming to partition the considered points into two disjoint subsets, and finally we reach to the **Restricted-Point-Cover** problem, as follows.

Restricted-Point-Cover: Given a point set P in $\mathbb{R}^2$ above the x-axis and a set $\mathcal{R}$ of unit width (x-span) rectangles such that bottom boundary of each rectangle in $\mathcal{R}$ coincides with x-axis, find a subset of $\mathcal{R}$ of minimum size to cover all the points in P.

Lemma 5. *For the standard ILP formulation of the Restricted-Point-Cover problem Z_{RPC}, if OPT_{RPC}^I and OPT_{RPC}^F are the optimum solutions of Z_{RPC} and its LP-relaxation respectively, then $OPT_{RPC}^I \leq 2OPT_{RPC}^F$.*

In Section 3.2 of [3], Bansal and Pruhs showed that $OPT_{RPC}^I \leq \alpha OPT_{RPC}^F$ for some positive constant α for a more generic version of this problem. In our simplified case $\alpha \leq 2$ for the proof of Lemma 5).

Chan et al. [5] proposed an $O(1)$-approximation algorithm for the $DCPUS$ problem using quasi-uniform sampling. Thus using Lemma 4, we have an $O(1)$ approximation for our $DCSUS$-ARB problem. However, the following theorem says that one can get a factor 8 approximation algorithm for the $DCPUS$ problem.

Theorem 6. *For the standard ILP formulation of the DCPUS problem Z_1 with a set of points P_1 and a set of unit squares T_1, if OPT_1^I and OPT_1^F are the optimum solutions of Z_1 and its LP-relaxation respectively, then $OPT_1^I \leq 8OPT_1^F$.*

Proof. It consists of three steps as follows:

Step 1: Divide the plane into unit strips by drawing horizontal lines. No unit square in T_1 can intersect more than one line. Partition T_1 into two sets T_{11} and T_{12}, where T_{11} (resp. T_{12}) consists of all squares which intersect even (resp. odd) indexed lines. Define binary variables x_i for $t_i \in T_{11}$ and y_j for $t_j \in T_{12}$. Then, Z_1 is equivalent to the following ILP.

$$Z_1 : \min \sum_{i | t_i \in T_{11}} x_i + \sum_{j | t_j \in T_{12}} y_j$$

$$\text{s.t.} \sum_{i | p \in t_i \in T_{11}} x_i + \sum_{j | p \in t_j \in T_{12}} y_j \geq 1 \ \forall \ p \in P_1; \ x_i, y_j \in \{0,1\} \ \forall \ i \mid t_i \in T_{11} \ \& \ j \mid t_j \in T_{12}$$

We solve $\overline{Z}_1$. Now, create two subsets $P_{11}, P_{12} \subseteq P_1$. P_{11} consists of those points $p \in P_1$ such that $\sum_{i | p \in t_i \in T_{11}} x_i \geq 1/2$, and P_{12} consists of those points $p \in P_1$ for which $\sum_{j | p \in t_j \in T_{12}} y_j \geq 1/2$. Again consider two ILP's, Z_{11} and Z_{12} as follows.

$$Z_{11} : \min \sum_{i|t_i \in T_{11}} x_i$$

s.t. $\sum_{i|p \in t_i \in T_{11}} x_i \geq 1, \ \forall \, p \in P_{11}$

$x_i \in \{0,1\} \quad \forall \, i \mid t_i \in T_{11}$

$$Z_{12} : \min \sum_{j|t_j \in T_{12}} y_j$$

s.t. $\sum_{j|p \in t_j \in T_{12}} y_j \geq 1, \ \forall \, p \in P_{12}$

$y_j \in \{0,1\} \quad \forall \, j \mid t_j \in T_{12}$

By an analysis identical to Gaur et al. [6], we have $OPT_{11}^F + OPT_{12}^F \leq 2 OPT_1^F$.

Step 2: In step 1, there are two ILP's, Z_{11} and Z_{12} corresponding to Z_1. Observe that, Z_{11} (resp. Z_{12}) is the problem of covering the points of $P_{11} \subseteq P_1$ (resp. $P_{12} \subseteq P_1$) by those unit squares which intersect even (resp. odd) indexed lines.

Now, focus our attention on Z_{11}. Since the subset of points covered by the squares intersected by two different even indexed lines ℓ_i and ℓ_j ($i \neq j$) are disjoint, we can split Z_{11} into different ILPs corresponding to each even indexed line, which can be independently solved.

Consider Z_{11}^{ξ} corresponding to the line ℓ_{ξ} (ξ is even). Let P_{11}^{ξ} be the set of points which are to be covered by the set of squares T_{11}^{ξ}, intersected by ℓ_{ξ}. We split P_{11}^{ξ} into disjoint sets P_{111}^{ξ} and P_{112}^{ξ}; P_{111}^{ξ} (resp. P_{112}^{ξ}) are the set of points which are above (resp. below) ℓ_{ξ}. The objective is to cover the members of P_{111}^{ξ} (resp. P_{112}^{ξ}) by minimum number of squares. Thus, the ILP Z_{11}^{ξ} can be written as,

$$Z_{11}^{\xi} : \min \sum_{i|t_i \in T_{11}^{\xi}} x_i$$

s.t. $\sum_{i|p \in t_i \in T_{11}^{\xi}} x_i \geq 1 \ \forall \, p \in P_{111}^{\xi}$, and $\sum_{i|p \in t_i \in T_{11}^{\xi}} x_i \geq 1 \ \forall \, p \in P_{112}^{\xi}$

$x_i \in \{0,1\} \quad \forall \, i \mid t_i \in T_{11}^{\xi}$

Again, since the sets P_{111}^{ξ} and P_{112}^{ξ} are disjoint, we may consider the following two ILP's, Z_{111}^{ξ} and Z_{112}^{ξ} as follows.

$$Z_{111}^{\xi} : \min \sum_{i|t_i \in T_{11}^{\xi}} x_i$$

s.t. $\sum_{i|p \in t_i \in T_{11}^{\xi}} x_i \geq 1 \ \forall \, p \in P_{111}^{\xi}$

$x_i \in \{0,1\} \quad \forall \, i \mid t_i \in T_{11}^{\xi}$

$$Z_{112}^{\xi} : \min \sum_{i|t_i \in T_{11}^{\xi}} x_i$$

s.t. $\sum_{i|p \in t_i \in T_{11}^{\xi}} x_i \geq 1 \ \forall \, p \in P_{112}^{\xi}$

$x_i \in \{0,1\} \quad \forall \, i \mid t_i \in T_{11}^{\xi}$

Let $\widetilde{x^*}$ be an optimal fractional solution of $\overline{Z}_{11}^{\xi}$. Clearly, $\widetilde{x^*}$ satisfies all the constraints in both $\overline{Z}_{111}^{\xi}$ and $\overline{Z}_{112}^{\xi}$. Also, it is observe that $OPT_{111}^{\xi F} \leq OPT_{11}^{\xi F}$ and $OPT_{112}^{\xi F} \leq OPT_{11}^{\xi F}$. Combining, we conclude that $OPT_{111}^{\xi F} + OPT_{112}^{\xi F} \leq 2 OPT_{11}^{\xi F}$. A similar equation can be shown for Z_{12}^{ξ} as follows: $OPT_{121}^{\xi F} + OPT_{122}^{\xi F} \leq 2 OPT_{12}^{\xi F}$.

Finally, we have the following four equations,

1. $\sum_{\xi \text{ even}} OPT_{111}^{\xi F} + \sum_{\xi \text{ even}} OPT_{112}^{\xi F} \leq 2 \sum_{\xi \text{ even}} OPT_{11}^{\xi F} \leq 2OPT_{11}^{F}$,
2. $\sum_{\xi \text{ odd}} OPT_{121}^{\xi F} + \sum_{\xi \text{ odd}} OPT_{122}^{\xi F} \leq 2 \sum_{\xi \text{ odd}} OPT_{12}^{\xi F} \leq 2OPT_{12}^{F}$,

Step 3: In this step we apply Lemma 5 independently on each of Z_{111}^{ξ}, Z_{112}^{ξ}, where ξ is even, and Z_{121}^{ξ}, Z_{122}^{ξ} where ξ is odd to get the following four equations.

(i) $OPT_{111}^{\xi I} \leq 2OPT_{111}^{\xi F}$,   (ii) $OPT_{112}^{\xi I} \leq 2OPT_{112}^{\xi F}$,

(iii) $OPT_{121}^{\xi I} \leq 2OPT_{121}^{\xi F}$,   (iv) $OPT_{122}^{\xi I} \leq 2OPT_{122}^{\xi F}$.

Now combining all the inequalities of Step 3, we have

$$\sum_{\xi \text{ even}} OPT_{111}^{\xi I} + \sum_{\xi \text{ even}} OPT_{112}^{\xi I} + \sum_{\xi \text{ odd}} OPT_{121}^{\xi I} + \sum_{\xi \text{ odd}} OPT_{122}^{\xi I}$$
$$\leq 2 \left(\sum_{\xi \text{ even}} OPT_{111}^{\xi F} + \sum_{\xi \text{ even}} OPT_{112}^{\xi F} + \sum_{\xi \text{ odd}} OPT_{121}^{\xi F} + \sum_{\xi \text{ odd}} OPT_{122}^{\xi F} \right)$$
$$\leq 8 \, OPT_0^I, \text{ by applying the inequalities in Steps 3, 2 and 1 in this order. } \square$$

Lemma 4 and Theorem 6 lead to the following theorem.

Theorem 7. *There exists a factor* 16 *approximation algorithm for DCSUS-ARB problem that runs in polynomial time.*

References

1. E. M. Arkin, A. Banik, P. Carmi, G. Citovsky, M. J. Katz, J. S. B. Mitchell, and M. Simakov. Choice is hard. In *ISAAC*, pages 318–328, 2015.
2. E. M. Arkin, A. Banik, P. Carmi, G. Citovsky, M. J. Katz, J. S. B. Mitchell, and M. Simakov. Conflict-free covering. In *CCCG*, 2015.
3. N. Bansal and K. Pruhs. The geometry of scheduling. *SIAM J. Comput.*, 43(5):1684–1698, 2014.
4. A. Biniaz, P. Liu, A. Maheshwari, and M. Smid. Approximation algorithms for the unit disk cover problem in 2D and 3D. *Computational Geometry*, 60:8–18, 2016.
5. T. M. Chan, E. Grant, J. Konemann and M. Sharpe. Weighted capacited, priority, and geometric set cover via improved quasi-uniform sampling In *SODA*, Pages 1576–1585, 2012.
6. D. R. Gaur, T. Ibaraki, and R. Krishnamurti. Constant ratio approximation algorithms for the rectangle stabbing problem and the rectilinear partitioning problem. *Journal of Algorithms*, 43(1):138–152, 2002.
7. D. S. Hochbaum and W. Maass. Approximation schemes for covering and packing problems in image processing and VLSI. *J. ACM*, 32(1):130–136, January 1985.
8. D. E. Knuth and A. Raghunathan. The problem of compatible representatives. *SIAM Journal on Discrete Mathematics*, 5(3):422–427, 1992.
9. K. Kobylkin. Computational complexity of guarding of proximity graphs. *CoRR*, abs/1605.00313, 2016.
10. R. R. Madireddy and A. Mudgal. Stabbing line segments with disks and related problems. In *CCCG*, 2016.

Replica Placement on Bounded Treewidth Graphs

Anshul Aggarwal[1], Venkatesan T. Chakaravarthy[2], Neelima Gupta[1], Yogish Sabharwal[2], Sachin Sharma[1], and Sonika Thakral[1*]

[1] University of Delhi, India   ngupta@cs.du.ac.in, sonika.ta@gmail.com
[2] IBM Research, India   {vechakra,ysabharwal}@in.ibm.com

Abstract. We consider the replica placement problem: given a graph and a set of clients, place replicas on a minimum set of nodes of the graph to serve all the clients; each client is associated with a request and maximum distance that it can travel to get served; there is a maximum limit (capacity) on the amount of request a replica can serve. The problem falls under the general framework of capacitated set cover. It admits an $O(\log n)$-approximation and it is NP-hard to approximate within a factor of $o(\log n)$. We study the problem in terms of the treewidth t of the graph and present an $O(t)$-approximation algorithm.

1 Introduction

We study a form of capacitated set cover problem [5] called *replica placement* (RP) that finds applications in settings such as data distribution by internet service providers (ISPs) and video on demand service delivery (e.g., [6,8]). In this problem, we are given a graph representing a network of servers and a set of clients. The clients are connected to the network by attaching each client to a specific server. The clients need access to a database. We wish to serve the clients by placing replicas (copies) of the database on a selected set of servers and clients. While the selected clients get served by the dedicated replicas (i.e., cached copies) placed on themselves, we serve the other clients by assigning them to the replicas on the servers. The assignments must be done taking into account Quality of Service (QoS) and capacity constraints. The QoS constraint stipulates a maximum distance between each client and the replica serving it. The clients may have different demands (the volume of database requests they make) and the capacity constraint specifies the maximum demand that a replica can handle. The objective is to minimize the number of replicas opened. The problem can be formally defined as follows.

    Problem Definition (RP): The input consists of a graph $G = (\mathcal{V}, E)$, a set of clients $\mathcal{A}$ and a capacity W. Each client a is attached to a node $u \in \mathcal{V}$, denoted $\text{att}(a)$. For each client $a \in \mathcal{A}$, the input specifies a request $r(a)$ and a distance $d_{\max}(a)$. For a client $a \in \mathcal{A}$ and a node $u \in \mathcal{V}$, let $d(a, u)$ denote the length of the shortest path between u and $\text{att}(a)$, the node to which a is attached

---

* Corresponding author

© Springer International Publishing AG 2017
F. Ellen et al. (Eds.): WADS 2017, LNCS 10389, pp. 13–24, 2017.
DOI: 10.1007/978-3-319-62127-2_2

- the length is measured by the number of edges and we take $d(a, u) = 0$, if $u = \texttt{att}(a)$. We say that a client $a \in \mathcal{A}$ can *access* a node $u \in \mathcal{V}$, if $d(a, u)$ is at most $d_{\max}(a)$. A feasible solution consists of two parts: (i) it identifies a subset of nodes $S \subseteq \mathcal{V}$ where a replica is placed at each node in S; (ii) for each client $a \in \mathcal{A}$, it either opens a dedicated replica at a itself for serving the client's request or assigns the request to the replica at some node $u \in S$ accessible to a. The solution must satisfy the constraint that for each node $u \in S$, the sum of requests assigned to the replica at u does not exceed W. The cost of the solution is the number of replicas opened, i.e., cardinality of S plus the number of dedicated replicas opened at the clients. The goal is to compute a solution of minimum cost. In order to ensure feasibility, without loss of generality, we assume $r(a) \leq W, \forall\, a \in \mathcal{A}$.   □

The RP problem falls under the framework of the capacitated set cover problem, the generalization of the classical set cover problem wherein each set is associated with a capacity specifying the number of elements it can cover. The latter problem is known to have an $O(\log n)$-approximation algorithm [5]. Using the above result, we can derive an $O(\log n)$-approximation algorithm for the RP problem as well. On the other hand, we can easily reduce the classical dominating set problem to RP: given a graph representing an instance of the dominating set problem, we create a new client for each vertex and attach it to the vertex; then, we set $d_{\max}(\cdot) = 1$ for all the clients and $W = \infty$. Since it is NP-hard to approximate the dominating set problem within a factor of $o(\log n)$ [7], by the above reduction, we get the same hardness result for the RP problem as well.

The RP problem is NP-hard even on the highly restricted special case where the graph is simply a path, as can be seen via the following reduction from the bin packing problem. Given K bins of capacity W and a set of items of sizes $s_1, s_2, \ldots, s_n$, for each item i, we create a client a with demand $r(a) = s_i$. We then construct a path of nodes of length K and attach all the clients to one end of the path and take W to be the capacity of the nodes.

Prior Results: Prior work has studied a variant of the RP problem where the network is a directed acyclic graph (DAG), and a client a can access a node u only if there is a directed path from a to u of the length at most $d_{\max}(a)$. Under this setting, Benoit et al. [3] considered the special case of rooted trees and presented a greedy algorithm with an approximation ratio of $O(\Delta)$, where Δ is the maximum degree of the tree. For the same problem, Arora et al. [2] (overlapping set of authors) devised a constant factor approximation algorithm via LP rounding.

Progress has been made on generalizing the above result to the case of bounded treewidth DAGs. Recall that treewidth [4] is a classical parameter used for measuring how close a given graph is to being a tree. For a DAG, the treewidth refers to the treewidth t of the underlying undirected graph. Notice that the reduction from the bin-packing problem shows that the problem is NP-hard even for trees (i.e., $t = 1$) and rules out the possibility of designing an exact algorithm running in time $n^{O(t)}$ (say via dynamic programming) or FPT algorithms with parameter t.

Arora et al. [1] made progress towards handling DAGs of bounded treewidth and designed an algorithm for the case of bounded-degree, bounded-treewidth graphs. Their algorithm achieves an approximation ratio of $O(\Delta + t)$, where Δ is the maximum degree and t is the treewidth of the DAG. Their result also extends for networks comprising of bounded-degree bounded-treewidth subgraphs connected in a tree like fashion.

Our Result and Discussion: We study the RP problem on undirected graphs of bounded treewidth. Our main result is an $O(t)$-approximation algorithm running in polynomial time (the polynomial is independent of t and the approximation guarantee). In contrast to prior work, the approximation ratio depends only on the treewidth and is independent of parameters such as the maximum degree.

Our algorithm is based on rounding solutions to a natural LP formulation, as in the case of prior work [2, 1]. However, the prior algorithms exploit the acyclic nature of the graphs and the bounded degree assumption to transform a given LP solution to a solution wherein each client is assigned to at most two replicas. In other words, they reduce the problem to a capacitated vertex cover setting, for which constant factor rounding algorithms are known [10].

The above reduction does not extend to the case of general bounded treewidth graphs. Our algorithm is based on an entirely different approach. We introduce the notion of "clustered solutions", wherein the partially open nodes are grouped into clusters and each client gets served only within a cluster. We show how to transform a given LP solution to a new solution in which a partially-open node participates in at most $(t + 1)$ clusters. This allows us to derive an overall approximation ratio $O(t)$. The notion of clustered solutions may be applicable in other capacitated set cover settings as well.

Other Related Work: The RP problem falls under the framework of the capacitated set cover problem (CSC), which admits an $O(\log n)$-approximation [5]. Two versions of the CSC problem have been studied: soft capacity and hard capacity settings. Our work falls under the more challenging hard capacity setting, wherein a set can be picked at most once. The capacitated versions of the vertex cover problem (e.g., [10]) and dominating set problem (e.g., [9]) have also been studied. Our result applies to the capacitated dominating problem with uniform capacities and yields $O(t)$-approximation.

Full Version: Due to space constraints, some of the proofs and details of analysis could not be included in this version. A full version of the paper is available as an Arxiv preprint (https://arxiv.org/abs/1705.00145).

2 Overview of the Algorithm

Our $O(t)$-approximation algorithm is based on rounding solution to a natural LP formulation. In this section, we present an outline of the algorithm highlighting its main features, deferring a detailed description to subsequent sections. We assume that the input includes a decomposition $\mathcal{T}$ of treewidth t of the input network $G = (\mathcal{V}, E)$.

LP Formulation: For each node $u \in \mathcal{V}$, we introduce a variable $y(u)$ to represent the extent to which a replica is opened at u and similarly, for each client $a \in \mathcal{A}$, we add a variable $y(a)$ to represent the extent to which a dedicated replica is opened at a itself. For each client $a \in \mathcal{A}$ and each node $u \in \mathcal{V}$ accessible to a, we use a variable $x(a, u)$ to represent the extent to which a is assigned to u. For a client $a \in \mathcal{A}$ and a node $u \in \mathcal{V}$, we use the shorthand "$a \sim u$" to mean that a can access u.

$$\min \quad \sum_{a \in \mathcal{A}} y(a) \quad + \quad \sum_{u \in \mathcal{V}} y(u)$$

$$y(a) + \sum_{u \in \mathcal{V} \,:\, a \sim u} x(a, u) \geq 1 \qquad \text{for all } a \in \mathcal{A} \tag{1}$$

$$\sum_{a \in \mathcal{A} \,:\, a \sim u} x(a, u) \cdot r(a) \leq y(u) \cdot W \qquad \text{for all } u \in \mathcal{V} \tag{2}$$

$$x(a, u) \leq y(u) \qquad \text{for all } a \in \mathcal{A} \text{ and } u \in \mathcal{V} \text{ with } a \sim u \tag{3}$$

$$0 \ \leq \ y(u), y(a) \leq 1 \qquad \text{for all } u \in \mathcal{V} \text{ and } a \in \mathcal{A} \tag{4}$$

Constraint (3) stipulates that a client a cannot be serviced at a node u for an amount exceeding the extent to which u is open. For an LP solution $\sigma = \langle x, y \rangle$, let $\text{cost}(\sigma)$ denote the objective value of σ.

The following simple notations will be useful in our discussion. With respect to an LP solution σ, we classify the nodes into three categories based on the extent to which they are open. A node u is said to be *fully-open*, if $y(u) = 1$; *partially-open*, if $0 < y(u) < 1$; and *fully-closed*, if $y(u) = 0$. A client a is said to be *assigned* to a node u, if $x(a, u) > 0$. For a set of nodes U, let $y(U)$ denote the extent to which the vertices in U are open, i.e., $y(U) = \sum_{u \in U} y(u)$.

Outline: The major part of the rounding procedure involves transforming a given LP solution $\sigma_{\text{in}} = \langle x_{\text{in}}, y_{\text{in}} \rangle$ into an *integrally open solution*: wherein which each node $u \in \mathcal{V}$ is either fully open or closed. Such a solution differs from an integral solution as a client may be assigned to multiple nodes (possibly to its own dedicated replica as well). We address the issue easily via a cycle cancellation procedure to get an integral solution.

The procedure for obtaining an integrally open solution works in two stages. First it transforms the input solution into a "clustered" solution, which is then transformed into an integrally open solution. The notion of clustered solution lies at the heart of the rounding algorithm. Intuitively, in a clustered solution, the set of partially open (and closed) nodes are partitioned into a collection of clusters $\mathcal{C}$ and the clients can be partitioned into a set of corresponding groups satisfying three useful properties, as discussed below.

Let $\sigma = \langle x, y \rangle$ be an LP solution. It will be convenient to express the three properties using the notion of linkage: we say that a node u is *linked* to a node v, if there exists a client a assigned to both u and v. For constants α and ℓ, the solution σ is said to be (α, ℓ)-*clustered*, if the set of partially-open nodes can be partitioned into a collection of clusters, $\mathcal{C} = \{C_1, C_2, \ldots, C_k\}$ (for some k), such that the the following properties are true:

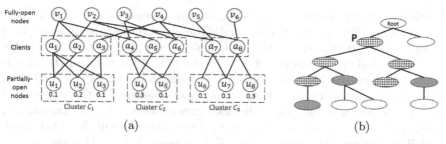

(a) (b)

Fig. 1. (a) Illustration for clustered solution. Three clusters are shown C_1, C_2 and C_3, open to an extent of 0.4, 0.4 and 0.5; the clusters are linked to the sets of fully-open nodes $\{v_1, v_2, v_4\}$, $\{v_1, v_2, v_3, v_4\}$, and $\{v_2, v_4, v_5, v_6\}$. The solution is $(0.5, 4)$-clustered. (b) Illustration for regions. The figure shows an example tree decomposition. The bags filled solidly represent already identified boundary bags. All checkered bags belong to the region headed by P.

- *Localization:* assignments from clients to the partially-open nodes is localized, i.e., two partially-open nodes are linked only if they belong to the same cluster.
- *Distributivity:* assignments from the clients to fully-open nodes are restricted, i.e., for any C_j, there are at most ℓ fully-open nodes that are linked to the nodes in C_j.
- *Bounded opening:* clusters are tiny, i.e., the total extent to which any cluster is open is at most α, i.e., $y(C_j) \leq \alpha$.

Figure 1 (a) provides an illustration. In the first stage of the rounding algorithm, we transform the input solution σ_{in} into an $(\alpha, t+1)$-clustered solution with the additional guarantee that the number of clusters is at most a constant factor of $\text{cost}(\sigma_{\text{in}})$, where $\alpha \in [0, 1/2]$ is a tunable parameter. The lemma below specifies the transformation performed by the first stage.

Lemma 1. *Fix any constant $\alpha \leq 1/2$. Any LP solution σ can be transformed into a $(\alpha, t+1)$-clustered solution σ' such that $\text{cost}(\sigma')$ is at most $2 + 6(t+1)\text{cost}(\sigma)/\alpha$. Furthermore, the number of clusters is at most $3 + 8 \cdot \text{cost}(\sigma)/\alpha$.*

At a high level, the lemma is proved by considering the tree decomposition $\mathcal{T}$ of the input graph $G = (V, E)$ and performing a bottom-up traversal that identifies a suitable set of boundary bags. We use these boundary bags to split the tree into a set of disjoint regions and create one cluster per region. We then fully open the nodes in the boundary bags and transfer assignments from the nodes that stay partially-open to these fully-open nodes. The transfer of assignments is performed in such a manner that clusters get localized and have distributivity of $(t+1)$. By carefully selecting the boundary bags, we shall enforce that each cluster is open to an extent of only α and that the number of clusters is also bounded. The proof is discussed in Section 3.

The goal of the second stage is to transform a $(1/4, t+1)$-clustered solution (obtained from Lemma 1) into an integrally open solution. At a high level, the localization property allows us to independently process each cluster $C \in \mathcal{C}$ and its corresponding group of clients A. The clients in A are assigned to a set of fully-open nodes, say F. For each node $u \in F$, we identify a suitable node $v \in C$ called the "consort" of $u \in C$ and fully open v. Then the idea is to transfer assignments from the non-consort nodes to the nodes in F and their consorts in such a manner that at the end, no client is assigned to the non-consort nodes. This allows us to fully close the non-consort nodes. The localization and bounded opening properties facilitate the above maneuver. On the other hand, the distributivity property ensures that F is at most $(t + 1)$. This means that we fully open at most $(t + 1)$ consorts per cluster. Thus, overall increase in cost is at most $(t + 1)|\mathcal{C}|$. Since $|\mathcal{C}|$ is guaranteed to be linear in $\mathrm{cost}(\sigma_{\mathrm{in}})$, we get an $O(t)$ approximation factor.

Lemma 2. *Let* $\sigma = \langle x, y \rangle$ *be a* $(1/4, t + 1)$-*clustered solution via a collection of clusters* $\mathcal{C}$. *The solution can be transformed into an integrally open solution* $\sigma' = \langle x', y' \rangle$ *such that* $\mathrm{cost}(\sigma') \leq 2 \cdot \mathrm{cost}(\sigma) + 2(t + 1)|\mathcal{C}|$.

Once we obtain an integrally open solution, it can be transformed to an integral solution by applying a cycle cancellation strategy, as given by the following lemma.

Lemma 3. *Any integrally open solution* $\sigma = \langle x, y \rangle$ *can be transformed to an integral solution* $\sigma' = \langle x', y' \rangle$ *such that* $\mathrm{cost}(\sigma') \leq 4 \cdot \mathrm{cost}(\sigma)$.

We can transform any input LP solution σ_{in} into an integral solution σ_{out} by applying the above three transformations leading to the following main result of the paper: the RP problem admits on $O(t)$-approximation poly-time algorithm.

3   Clustered Solutions: Proof of Lemma 1

The goal is to transform a given solution into an $(\alpha, t+1)$-clustered solution with the properties claimed in the lemma. The idea is to select a set of partially-open or closed nodes and open them fully, and then transfer assignments from the other partially-open nodes to them in such a manner that the partially-open nodes get partitioned into clusters satisfying the three properties of clustered solutions. An issue in executing the above plan is that the capacity at a newly opened node may be exceeded during the transfer. We circumvent the issue by first performing a pre-processing step called de-capacitation.

3.1   De-capacitation

Consider an LP solution $\sigma = \langle x, y \rangle$ and let u be a partially-open or closed node. The clients that can access u might have been assigned to other partially-open nodes under σ. We call the node u *de-capacitated*, if even when all the above assignments are transferred to u, the capacity at u is not exceeded; meaning,

$$\sum_{a \sim u} \sum_{v:\, a \sim v\, \wedge\, v \in \mathrm{PO}} x(a, v) < W,$$

For each partially-open node v (considered in an arbitrary order)
  For each client a that can access both u and v (considered in an arbitrary order)
    Compute capacity available at u: $\texttt{cap}(u) = W - \sum_{b \in \mathcal{A} \,:\, b \sim u} x(b, u) \cdot r(b)$
    If $\texttt{cap}(u) = 0$ exit the procedure
    $\delta = \min \left\{ x(a, v), \frac{\texttt{cap}(u)}{r(a)} \right\}$
    Increment $x(a, u)$ by δ and decrement $x(a, v)$ by δ.

Fig. 2. Pulling procedure for a given partially-open or closed node u.

where PO is the set of partially-open nodes under σ (including u). The solution σ is said to be *de-capacitated*, if all the partially-open and the closed nodes are de-capacitated.

The preprocessing step transforms the input solution into a de-capacitated solution by performing a pulling procedure on the partially-open and closed nodes. Given a partially-open or closed node u, the procedure transfers assignments from other partially-open nodes to u, as long as the capacity at u is not violated. The procedure is shown in Figure 2, which we make use of in other components of the algorithm as well.

Lemma 4. *Any LP solution $\sigma = \langle x, y \rangle$ can be transformed into a de-capacitated solution $\sigma' = \langle x', y' \rangle$ such that $\mathrm{cost}(\sigma') \leq 2 \cdot \mathrm{cost}(\sigma)$.*

Proof. We consider the partially-open and closed nodes, and process them in an arbitrary order, as follows. Let u be a partially-open or closed node. Hypothetically, consider applying the pulling procedure on u. The procedure may terminate in one of two ways: (i) it exits mid-way because of reaching the capacity limit; (ii) the process executes in its entirety. In the former case, we fully open u and perform the pulling procedure on u. In the latter case, the node u is de-capacitated and so, we leave it as partially-open or closed, without performing the pulling procedure. It is clear that the above method produces a de-capacitated solution σ'. We next analyze the cost of σ'. Let s be the number of partially-open or closed nodes converted to be fully-open. Apart from these conversions, the method does not alter the cost and so, $\mathrm{cost}(\sigma')$ is at most $s + \mathrm{cost}(\sigma)$. Let the total amount of requests be $r_{\mathrm{tot}} = \sum_{a \in \mathcal{A}} r(a)$. The extra cost s is at most $\lfloor r_{\mathrm{tot}}/W \rfloor$, since any newly opened node is filled to its capacity. Due to the capacity constraints, the input solution σ must also incur a cost of at least $\lfloor r_{\mathrm{tot}}/W \rfloor$. It follows that $\mathrm{cost}(\sigma')$ is at most $2 \cdot \mathrm{cost}(\sigma)$. $\qquad\square$

3.2 Clustering

Given Lemma 4, assume that we have a de-capacitated solution $\sigma = \langle x, y \rangle$. We next discuss how to transform σ into an $(\alpha, t+1)$-clustered solution. The transformation would perform a bottom-up traversal of the tree decomposition and identify a set of partially-open or closed nodes. It would then fully open them and perform the pulling procedure on these nodes. The advantage is that the above nodes are de-capacitated and so, the pulling procedure would run to its entirety (without having to exit mid-way because of reaching capacity limits). As a consequence, the linkage between the nodes gets restricted, leading to a clustered solution. Below we first describe the transformation and then present an analysis.

Transformation: Consider the given tree decomposition $\mathcal{T}$. We select an arbitrary bag of $\mathcal{T}$ and make it the root. A bag P is said to be an *ancestor* of a bag Q, if P lies on the path connecting Q and the root; in this case, Q is called a *descendant* of P. We consider P to be both an ancestor and descendant of itself. A node u may occur in multiple bags; among these bags the one closest to the root is called the *anchor* of u and it is denoted $\texttt{anchor}(u)$. A *region* in $\mathcal{T}$ refers to any set of contiguous bags (i.e., the set of bags induce a connected sub-tree).

In transforming σ into a clustered solution, we shall encounter three types of nodes and it will be convenient to color them as red, blue and brown. To start with, all the fully-open nodes are colored red and the remaining nodes (partially-open nodes and closed nodes) are colored blue. The idea is to carefully select a set of blue nodes, fully-open them and perform the pulling procedure on these nodes; these nodes are then colored brown. Thus, while the blue nodes are partially-open or closed, the red and the brown nodes are fully-open, with the brown and blue nodes being de-capacitated.

The transformation identifies two kinds of nodes to be colored brown: *helpers* and *boundary nodes*. We say that a red node $u \in \mathcal{V}$ is *proper*, if it has at least one neighbor $v \in \mathcal{V}$ which is a blue node. For each such proper red node u, we arbitrarily select one such blue neighbor $v \in \mathcal{V}$ and declare it to be the helper of u. Multiple red nodes are allowed to share the same helper. Once the helpers have been identified, we color them all brown. The boundary brown nodes are selected via a more involved bottom-up traversal of $\mathcal{T}$ that works by identifying a set $\mathcal{B}$ of bags, called the *boundary bags*. To start with, $\mathcal{B}$ is initialized to be the empty set. We arrange the bags in $\mathcal{T}$ in any bottom-up order (i.e., a bag gets listed only after all its children are listed) and then iteratively process each bag P as per the above order. Consider a bag P. We define the *region headed by P*, denoted $\texttt{Region}(P)$, to be the set of bags Q such that Q is a descendant of P, but not the descendant of any bag already in $\mathcal{B}$. See Figure 1 (b) for an illustration. A blue node u is said to be *active at P*, if it occurs in some bag included in $\texttt{Region}(P)$. Let $\texttt{active}(P)$ denote the set of blue nodes active at P. We declare P to be a boundary bag and add it to $\mathcal{B}$ under three scenarios: (i) P is the root bag. (ii) P is the anchor of some red node. (iii) the extent to which the nodes in $\texttt{active}(P)$ are open is at least α, i.e., $\sum_{u \in \texttt{active}(P)} y(u) \geq \alpha$. If P is identified as a boundary bag, then we select all the blue nodes appearing in the bag and change their color to be brown. Once the bottom-up traversal is completed, we have a set of brown nodes (helpers and boundary nodes). We consider these nodes in any arbitrary order, open them fully, and perform the pulling procedure on them. We take σ' to be the solution obtained by the above process. This completes the construction of σ'. We note that a node may change its color from blue to brown in the above process, and the new color is to be considered while determining the active sets thereafter. Notice that during the whole process of the above transformation, the solution continues to remain de-capacitated.

Analysis: We now show that σ' is an $(\alpha, t + 1)$-clustered solution. To start with, we have a set of red nodes that are fully-open and a set of blue nodes that are either partially-open or closed under σ. The red nodes do not change color during the transformation. On the other hand, each blue node u becomes active at some boundary bag P. If u occurs in the bag P, it changes its color to brown, otherwise it stays blue. Thus, the transformation partitions the set of originally blue nodes into a set of brown nodes and a set of nodes that stay blue. In the following discussion, we shall use the term 'blue' to refer to the nodes that stay blue. With respect to the solution σ', the red and brown nodes are fully-open, whereas the blue nodes are partially-open or closed.

Recall that with respect to σ', two nodes u and v are linked, if there is a client a assigned to both u and v. In order to prove the properties of $(\alpha, t + 1)$-clustering, we need to analyze the linkage information for the blue nodes. We first show that the blue nodes cannot be linked to brown nodes, by proving the following stronger observation.

Proposition 1. *If a client $a \in \mathcal{A}$ is assigned to a blue node u under σ', then a cannot access any brown node v.*

Proposition 1 rules out the possibility of a blue node u being linked to any brown node. Thus, u may be linked to a red node or another blue node. The following lemma establishes a crucial property on the connectivity in these two settings.

Lemma 5. *(a) If two blue nodes u and v are linked under σ', then there must exist a path connecting u and v consisting of only blue nodes. (b) If a blue node u is linked to a red node v under σ', then there must exist a path p connecting u and v such that barring v, the path consists of only blue nodes.*

The transformation outputs a set of boundary bags $\mathcal{B}$; let $\overline{\mathcal{B}}$ denote the set of non-boundary bags. If we treat the bags in $\mathcal{B}$ as cut-vertices and delete them from $\mathcal{T}$, the tree splits into a collection $\mathcal{R}$ of disjoint regions. Alternatively, these regions can be identified in the following manner. For each bag $P \in \mathcal{B}$ and each of its non-boundary child $Q \in \overline{\mathcal{B}}$, add the region headed by Q (Region(Q)) to the collection $\mathcal{R}$. Let the collection derived be $\mathcal{R} = \{R_1, R_2, \ldots, R_k\}$. It is easy to see that $\mathcal{R}$ partitions $\overline{\mathcal{B}}$ and that the regions in $\mathcal{R}$ are pairwise disconnected (not connected by edges of the tree decomposition).

In order to show that σ' is an $(\alpha, t + 1)$-clustered solution, let us suitably partition the set of blue nodes into a collection of clusters $\mathcal{C}$. For each region R_j, let C_j be the set of partially-open nodes that occur in some bag of R_j. We take $\mathcal{C}$ to be the collection $\{C_1, C_2, \ldots, C_k\}$. It can be verified that the collection $\mathcal{C}$ is a partitioning of the set of partially-open nodes. Based on Lemma 5 we can establish the following result.

Lemma 6. *The solution σ' is $(\alpha, t+1)$-clustered.*

We next analyze the cost of the solution $\sigma' = \langle x', y' \rangle$. Let Red, Blue and Brown denote the set of red, brown and blue nodes, respectively. We can see that in constructing σ' the brown nodes are the only new nodes opened fully and hence, $\text{cost}(\sigma') \leq \text{cost}(\sigma) + |\text{Brown}|$. We create a brown helper node for each red node and furthermore, for each boundary bag $P \in \mathcal{B}$, we convert all the blue nodes in P to be brown and the number of blue nodes per bag is at most $(t+1)$. Thus, the number of brown nodes is at most $|\text{Red}| + (t+1)|\mathcal{B}|$. A bag P is made into a boundary bag under one of the three scenarios: (i) P is the root bag; (ii) P is the anchor of some red node; (iii) the total extent to which the nodes in active(P) are open is at least α. The number of boundary bags of the first two types are $1 + |\text{Red}|$ and those of the third type can be $(1/\alpha)$ times the extent to which the blue nodes are open, which is in turn, at most $\text{cost}(\sigma)$. Using the above arguments, we can show that $|\mathcal{B}|$ is at most $2 + |\text{Red}| + \text{cost}(\sigma)/\alpha$ and $\text{cost}(\sigma')$ is at most $2 + 3(t+1)\text{cost}(\sigma)/\alpha$. The preprocessing step of de-capacitation incurs a 2-factor increase in cost. Taking this into account, we get the cost bound claimed in the statement of Lemma 1.

As mentioned earlier, an issue with the collection $\mathcal{C}$ is that it may have more clusters than the bound claimed in Lemma 1. The issue can be resolved as follows. Consider each boundary bag P. All the non-boundary children of P have a corresponding cluster in $\mathcal{C}$ and let $\mathcal{C}_P$ denote the collection of these clusters. We merge any two clusters C, C' from $\mathcal{C}_P$ having $y(C), y(C') \leq \alpha/2$. The process is stopped when we cannot find two such clusters.

It can be shown that the process of merging does not affect distributivity and the total number of clusters in the transformed solution is at most $3 + 4\text{cost}(\sigma)/\alpha$. The preprocessing step of de-capacitation incurs a 2-factor increase in cost. Taking this into account, we get the bound on number of clusters claimed in the statement of Lemma 1.

4 Integrally Open Solution: Proof of Lemma 2

Our goal is to transform a given $(1/4, t+1)$-clustered solution $\sigma = \langle x, y \rangle$ into an integrally open solution σ'. We classify the clients into two groups, *small* and *large*, based on the extent to which they are served by dedicated replicas: a client $a \in \mathcal{A}$ said to be *small*, if $y(a) \leq 1/2$, and it is said to be *large* otherwise. Let $\mathcal{A}_s$ and $\mathcal{A}_l$ denote the set of small and large clients, respectively.

We pre-process the solution σ by opening a dedicated replica at each large client a and removing its assignments to the nodes (set $y(a) = 1$ and set $x(a, u) = 0$ for all nodes u accessible to a). We see that the transformation at most doubles the cost and the solution remains $(1/4, t+1)$-clustered.

Consider the pre-processed solution σ. Let $\mathcal{C}$ denote the set of clusters (of the partially-open nodes) under σ. For each cluster $C \in \mathcal{C}$, we shall fully open a selected set of at most $2(t+1)$ nodes and fully close rest of the nodes in it.

We now describe the processing for a cluster $C \in \mathcal{C}$. Let $A \subseteq \mathcal{A}_s$ denote the set of clients assigned to the nodes in C. By the distributivity property, these clients are assigned to at most $(t+1)$ fully-open nodes, denoted $F = \{u_1, u_2, \ldots, u_{t+1}\}$. A client $a \in A$ may be assigned to multiple nodes from F. In our procedure, it would be convenient if each client is assigned to at most one node from F and we obtain such a structure using the following transformation.

Proposition 2. *Given a solution $\sigma = \langle x, y \rangle$, a set of fully-open nodes F and a set of clients A, we can obtain a solution $\sigma' = \langle x', y' \rangle$ such that each client $a \in A$ is assigned to at most one node from F. Furthermore, the transformation does not alter the other assignments, i.e., for any node $u \in V$ and any client $a \in \mathcal{A}$, if $u \notin F$ or $a \notin A$, then $x'(a, u) = x(a, u)$. Moreover, $\mathrm{cost}(\sigma') \leq \mathrm{cost}(\sigma) + |F|$.*

The above proposition is proved via a cycle cancellation procedure that transfers assignments amongst the nodes in F. The procedure can ensure that, except for at most $|F|$ clients, every other client $a \in \mathcal{A}$ is assigned to at most one node from F. We open dedicated replicas at the exceptional clients and this results in an cost increase of at most $|F|$.

The proposition does not alter the other assignments and so, its output solution is also $(1/4, t+1)$-clustered. Given the proposition and the pre-processing, we can assume that $\sigma = \langle x, y \rangle$ is $(1/4, t+1)$-clustered wherein each client $a \in A$ is assigned to at most one node from F and that $y(a) \leq 1/2$. For each node $u_i \in F$, let $A_i \subseteq A$ denote the set of clients assigned to the node u_i. The proposition guarantees that these sets are disjoint.

For a node v and a client a, let $\mathrm{load}(a, v)$ denote the amount of load imposed by a on v towards the capacity: $\mathrm{load}(a, v) = x(a, v)r(a)$. It will be convenient to define the notion over sets of clients and nodes. For a set of clients B and a set of nodes U, let $\mathrm{load}(B, U)$ denote the load imposed by the clients in B on the nodes U: $\mathrm{load}(B, U) = \sum_{a \in B, v \in U : a \sim v} x(a, v)r(a)$; when the sets are singletons, we shall omit the curly braces. Similarly, for a subset $C' \subseteq C$, let $\mathrm{load}(C') = \sum_{v \in C'} \mathrm{load}(v)$.

The intuition behind the remaining transformation is as follows. We shall identify a suitable set of nodes $L = \{v_1, v_2, \ldots, v_{t+1}\}$ from C, with v_i being called the *consort* of u_i in C, and fully open all these nodes. Then, we consider the non-consort nodes $C' = C - L$ and for each $i \leq t+1$, we transfer the load $\mathrm{load}(A_i, C')$ to the node u_i. As a result, no clients are assigned to the non-consort nodes any more and so, they can be fully closed. In order to execute the transfer, for each $i \leq t+1$, we create space in u_i by pushing a load equivalent to $\mathrm{load}(A_i, C')$ from u_i to its (fully-opened) consort v_i. The amount of load $\mathrm{load}(A_i, C')$ involved in the transfer is very small: the bounded opening property ensures that $y(C) < 1/4$ and thus, $\mathrm{load}(A_i, C') < W/4$. The fully-opened consort v_i has enough additional space to receive the load: $y(v_i) \leq 1/4$ and so, $\mathrm{load}(A, v_i) \leq W/4$, which means that if we fully open the consort, we get an additional space of $(3/4)W$. However, an important issue is that a consort v_i may not be accessible to all the clients in A_i. Therefore, we need to carefully choose the consorts in such a manner that each fully open node u_i has enough load accessible to the consort v_i that can be pushed to v_i. Towards this purpose, we define the notion of *pushable load*. For a node $u_i \in F$ and a node $v \in C$, let $\mathrm{pushable}(u_i, v)$ denote the amount of load on u_i that is accessible to v: $\texttt{pushable}(u_i, v) = \sum_{a \in A_i : a \sim v} x(a, u_i) r(a)$. We next show how to identify a suitable set of consorts such that the pushable load is more than the load that we wish to transfer.

Lemma 7. *We can find a set of nodes $L = \{v_1, v_2, \ldots, v_{t+1}\}$ such that for all $i \leq t+1$, $\texttt{pushable}(u_i, v) \geq \texttt{load}(A_i, C')$.*

We have shown that each node u_i has a load of at least $\texttt{load}(A_i, C')$ which can be pushed to its consort v_i. As observed earlier $\texttt{load}(A_i, C') < W/4$ and $\texttt{load}(A_i, v_i) \leq W/4$. Hence, when we fully open the consort, we get an additional space of $(3/4)W$, which is sufficient to receive the load from u_i.

Given the above discussion, we iteratively consider each cluster $C_j \in \mathcal{C}$ and perform the above transformation. This results in $(t+1)$ consorts from C_j being fully-opened and all the other nodes in C_j being fully closed. At the end of processing all the clusters, we get a solution in which each node either fully open or fully close. For each cluster C_j, we incur an extra cost of at most $(t+1)$ while applying Proposition 2, and an additional cost of $(t+1)$ for opening the consorts. Thus, the cost increases by at most $2(t+1)|\mathcal{C}|$.

References

1. S. Arora, V. Chakaravarthy, K. Gupta, N. Gupta, and Y. Sabharwal. Replica placement on directed acyclic graphs. In V. Raman and S. Suresh, editors, *Proceedings of the 34th International Conference on Foundation of Software Technology and Theoretical Computer Science (FSTTCS)*, pages 213–225, 2014.
2. S. Arora, V. Chakaravarthy, N. Gupta, K. Mukherjee, and Y. Sabharwal. Replica placement via capacitated vertex cover. In A. Seth and N. Vishnoi, editors, *Proceedings of the 33rd International Conference on Foundations of Software Technology and Theoretical Computer Science (FSTTCS)*, pages 263–274, 2013.
3. A. Benoit, H. Larchevêque, and P. Renaud-Goud. Optimal algorithms and approximation algorithms for replica placement with distance constraints in tree networks. In *Proceedings of the 26th IEEE International Parallel and Distributed Processing Symposium (IPDPS)*, pages 1022–1033, 2012.
4. H. Bodlaender and A. Koster. Combinatorial optimization on graphs of bounded treewidth. *Computer Journal*, 51(3):255–269, 2008.
5. J. Chuzhoy and J. Naor. Covering problems with hard capacities. *SIAM Journal of Computing*, 36(2):498–515, 2006.
6. I. Cidon, S. Kutten, and R. Soffer. Optimal allocation of electronic content. *Computer Networks*, 40:205–218, 2002.
7. U. Feige. A threshold of ln n for approximating set cover. *Journal of the ACM*, 45(4):634–652, 1998.
8. K. Kalpakis, K. Dasgupta, and O. Wolfson. Optimal placement of replicas in trees with read, write, and storage costs. *IEEE Transactions on Parallel and Distributed Systems*, 12:628–637, 2001.
9. M. Kao, H. Chen, and D. Lee. Capacitated domination: Problem complexity and approximation algorithms. *Algorithmica*, 72(1):1–43, 2015.
10. B. Saha and S. Khuller. Set cover revisited: Hypergraph cover with hard capacities. In A. Czumaj, K. Mehlhorn, A. Pitts, and R. Wattenhofer, editors, *Proceedings of the 39th International Colloquium on Automata, Languages, and Programming (ICALP)*, volume 7391 of *LNCS*, pages 762–773. Springer, 2012.

Fast Exact Algorithms for Survivable Network Design with Uniform Requirements *

Akanksha Agrawal[1], Pranabendu Misra[1], Fahad Panolan[1], and Saket Saurabh[1,2]

[1] Department of Informatics, University of Bergen, Norway.
{Akanksha.Agrawal|Pranabendu.Misra|Fahad.Panolan}@uib.no
[2] The Institute of Mathematical Sciences, HBNI, Chennai, India. saket@imsc.res.in

Abstract. We design exact algorithms for the following two problems in survivable network design: (i) designing a minimum cost network with a desired value of edge connectivity, which is called MINIMUM WEIGHT λ-CONNECTED SPANNING SUBGRAPH and (ii) augmenting a given network to a desired value of edge connectivity at a minimum cost which is called MINIMUM WEIGHT λ-CONNECTIVITY AUGMENTATION. Many well known problems such as MINIMUM SPANNING TREE, HAMILTONIAN CYCLE, MINIMUM 2-EDGE CONNECTED SPANNING SUBGRAPH and MINIMUM EQUIVALENT DIGRAPH reduce to these problems in polynomial time. It is easy to see that a minimum solution to these problems contains at most $2\lambda(n-1)$ edges. Using this fact one can design a brute-force algorithm which runs in time $2^{O(\lambda n(\log n + \log \lambda))}$. However no better algorithms were known. In this paper, we give the first single exponential time algorithm for these problems, i.e. running in time $2^{O(\lambda n)}$, for both undirected and directed networks. Our results are obtained via well known characterizations of λ-connected graphs, their connections to linear matroids and the recently developed technique of dynamic programming with representative sets.

1 Introduction

The survivable network design problem involves designing a cost effective communication network that can survive equipment failures. The failure may be caused by any number of things such as a hardware or software breakage, human error or a broken link between two network components. Designing a network which satisfies certain connectivity constraints, or augmenting a given network to a certain connectivity are important and well studied problems in network design. In terms of graph theory these problems correspond to finding a spanning subgraph of a graph which satisfies given connectivity constraints and, augmenting the given graph with additional edges so that it satisfies the given constraints, respectively. Designing a minimum cost network which connects all the nodes, is the well-known MINIMUM SPANNING TREE(MST) problem. However such a

---

* Supported by "Parameterized Approximation" ERC Starting Grant 306992 and "Rigorous Theory of Preprocessing" ERC Advanced Investigator Grant 267959.

F. Ellen et al. (Eds.): WADS 2017, LNCS 10389, pp. 25–36, 2017.
DOI: 10.1007/978-3-319-62127-2_3

network fails on the failure of a single link. This leads to the question of designing a minimum cost network which can survive one or more link failures. Such a
network must be λ-connected, in order to survive $\lambda - 1$ link failures (we use the
term λ-connected to represent λ-edge connected). This problem is NP-hard (for
$\lambda \geq 2$), and a 2-approximation algorithm is known [19]. In the special case when
the weights are 1 or ∞, i.e. we wish to find a minimum spanning λ-connected
subgraph, a $1 + \frac{2}{\lambda+1}$ approximation may be obtained in polynomial time [6]. The
above results also hold in the case of directed graphs. The case of $\lambda = 1$ for
digraphs, known as MINIMUM STRONG SPANNING SUBGRAPH(MSSS), is NP-
hard as it is a generalization of the HAMILTONIAN CYCLE. Further, the MINIMUM
EQUIVALENT GRAPH(MEG) problem reduces to it in polynomial time.

Adding a minimum number of edges to make the graph satisfy certain connectivity constrains is known as minimum augmentation problem. Minimum
augmentation find application in designing survivable networks [12, 16] and in
data security [14, 17]. Watanabe and Nakamura [25] gave a polynomial time algorithm for solving the λ-edge connectivity augmentation in an undirected graph,
where we want to add minimum number of edges to the graph to make it λ-edge
connected. Frank gave a polynomial time algorithm for the same problem in
directed graphs [11]. However in the weighted case, or when the augmenting set
must be a subset of a given set of links, the problem becomes NP-Hard problem.
Even the restricted case of augmenting the edge connectivity of a graph from
$\lambda - 1$ to λ remains NP-hard [1]. A 2-approximation may be obtained for these
problems, by choosing a suitable weight function and applying the algorithm of
[19]. We refer to [1, 4, 18, 20] for more details, other related problems and further
applications. A few results are also known in the frameworks of parameterized
complexity and exact exponential time algorithms. Marx and Végh gave an FPT
algorithm for computing a minimum cost set of at most k links, which augments
the connectivity of an undirected graph from $\lambda - 1$ to λ [22]. Basavaraju et al.
[2] improved the running time of this algorithm and, also gave an algorithm for
another variant of this problem. Bang-Jensen and Gutin [1, Chapter 12] obtain
an FPT algorithm for a variant of MSSS in unweighted graphs. The first exact
algorithms for MEG and MSSS, running in time $\mathcal{O}(2^{\mathcal{O}(m)} \cdot n^{\mathcal{O}(1)})$, where m is
the number of edges in the graph, were given in by Moyles and Thompson [23] in
1969. Only recently, Fomin et al. [10] gave the first single-exponential algorithm
for MEG and MSSS, i.e. with a running time of $2^{\mathcal{O}(n)}$. For the special case of
HAMILTONIAN CYCLE, a $\mathcal{O}(2^n)$ time algorithm is known [15, 3] for digraphs from
1960s. It was recently improved to $\mathcal{O}(1.657^n)$ for undirected graphs [5], and to
$\mathcal{O}(1.888^n)$ for bipartite digraphs [9] (but these are randomized algorithms). For
other results and more details we refer to Chapter 12 of [1].

In this paper we consider the problem of designing an exact algorithm for
finding a minimum weight spanning subgraph of a given λ-connected (di)graph.

MINIMUM WEIGHT λ-CONNECTED SPANNING SUBGRAPH
Input: A graph G (or digraph D), and a weight function w on the edges(or
the arcs).
Output: A minimum weight spanning λ-connected subgraph.

One can observe that such a subgraph contains at most $\lambda(n-1)$ edges ($2\lambda(n-1)$ arcs for digraphs). Hence a solution can be obtained by enumerating all possible subgraphs with at most these many edges and testing if it is λ-connected. However such an algorithm will take $2^{\mathcal{O}(\lambda n(\log n + \log \lambda))}$ time. One may try a more clever approach, by using the observation that we can enumerate all possible minimal λ-connected graphs in $2^{\mathcal{O}(\lambda n)}$ time. Then we test if any of these graph is isomorphic to a subgraph of the input graph. However, subgraph isomorphism requires $2^{\lambda n(\log n + \log \lambda)}$ unless the Exponential Time Hypothesis fails [7]. In this paper, we give the first single exponential algorithm for this problem that runs in time $2^{\mathcal{O}(\lambda n)}$. As a corollary, we also obtain single exponential time algorithm for the minimum weight connectivity augmentation problem.

MINIMUM WEIGHT λ-CONNECTIVITY AUGMENTATION
Input: A graph G (or a digraph D), a set of links $L \subseteq V \times V$ (ordered pairs in case of digraphs), and a weight function $w : L \to \mathbb{N}$.
Output: A minimum weight subset L' of L such that $G \cup L$ (or $D \cup L$) is λ-connected

---

Our Methods and Results. We extend the algorithm of Fomin et al. for finding a MINIMUM EQUIVALENT GRAPH [10], to solve MINIMUM WEIGHT λ- CONNECTED SUB-DIGRAPH, exploiting the structural properties of λ-connected (di)graphs. A digraph D is λ-connected if and only if for some $r \subset V(D)$, there is a collection $\mathbb{I}$ of λ arc disjoint in-branchings rooted at r and a collection $\mathbb{O}$ of λ arc disjoint out-branchings rooted at r. Then computing a $\mathbb{I}$ and a $\mathbb{O}$ with the largest possible intersection yields a minimum weight λ connected spanning sub-digraph. We show that the solution can be embedded in a linear matroid of rank $\mathcal{O}(\lambda n)$, and then compute the solution by a dynamic programming algorithm with representative sets over this matroid.

Theorem 1. *Let D be a λ-edge connected digraph on n vertices and $w : A(D) \to \mathbb{N}$. Then we can find a min-weight λ-edge connected subgraph of D in $2^{\mathcal{O}(\lambda n)}$ time.*

For the case of undirected graphs, no equivalent characterization is known. However, we obtain a characterization by converting the graph to a digraph with labels on the arcs, corresponding to the undirected edges. Then computing a solution that minimizes the number of labels used, gives the following theorem.

Theorem 2. *Let G be a λ-edge connected graph on n vertices and $w : E(G) \to \mathbb{N}$. Then we can find a min-weight λ-edge connected subgraph of G in $2^{\mathcal{O}(\lambda n)}$ time.*

For the problem of augmenting a network to a given connectivity requirement, at a minimum cost, we obtain the following results by applying the previous theorems with suitably chosen weight functions.

Theorem 3. *Let D be a digraph (or a graph) on n vertices, $L \subseteq V(D) \times V(D)$ be a collection of links with weight function $w : L \to \mathbb{N}$. For any integer λ, we can find a min-weight $L' \subseteq L$ such that $D' = (V(D), A(D) \cup L')$ is λ-edge connected, in time $2^{\mathcal{O}(\lambda n)}$.*

Preliminaries. Due to space constraints, standard definitions, notations related graphs and matroids have been omitted from this extended abstract. These preliminaries and other results on matriods and representative sets, may be found in [8, 10]. We only mention the following. We say that a family $\mathcal{S} = \{S_1, \ldots, S_t\}$ of subsets of a universe U is a *p-family* if each set in $\mathcal{S}$ has cardinality at most p. For two families $\mathcal{S}_1$ and $\mathcal{S}_2$ of a universe U, define $\mathcal{S}_1 \bullet \mathcal{S}_2 = \{S_i \cup S_j \mid S_i \in \mathcal{S}_1, S_j \in \mathcal{S}_2 \text{ and } S_i \cap S_j = \emptyset\}$. We use ω to denote the exponent of matrix multiplication.

Definition 1 (Min/Max q-Representative Family [8, 10, 21]). *Given a matroid $\mathcal{M} = (E, \mathcal{I})$, a p-family $\mathcal{B}$ of E and a non-negative weight function $w : \mathcal{B} \to \mathbb{N}$. We say that $\widehat{\mathcal{B}} \subseteq \mathcal{B}$ is a min (max) q-representative for $\mathcal{B}$ if for every set $Y \subseteq E$ of size at most q, if there is a set $X \in \mathcal{B}$, such that $X \cap Y = \emptyset$ and $X \cup Y \in \mathcal{I}$, then there is a set $\widehat{X} \in \widehat{\mathcal{B}}$ such that $\widehat{X} \cap Y = \emptyset$, $\widehat{X} \cup Y \in \mathcal{I}$ and $w(\widehat{X}) \leq w(X)$ $(w(\widehat{X}) \geq w(X))$. If $\widehat{\mathcal{B}} \subseteq \mathcal{B}$ is a min (max) q-representative for $\mathcal{B}$ then we denote it by $\widehat{\mathcal{B}} \subseteq_{minrep}^q \mathcal{B}$ $(\widehat{\mathcal{B}} \subseteq_{maxrep}^q \mathcal{B})$.*

Theorem 4 ([8, 10]). *Let $\mathcal{M} = (E, \mathcal{I})$ be a linear matroid of rank $k = p + q$, and matrix $A_\mathcal{M}$ be a representation of $\mathcal{M}$ over a field $\mathbb{F}$. Also, let $\mathcal{B} = \{B_1, B_2, \ldots, B_t\}$ be a p-family of independent sets in E and $w : \mathcal{B} \to \mathbb{N}$ be a non-negative weight function. Then, there exists $\widehat{\mathcal{B}} \subseteq_{minrep}^q \mathcal{B}$ $(\widehat{\mathcal{B}} \subseteq_{maxrep}^q \mathcal{B})$ of size at most $\binom{p+q}{p}$. Moreover, $\widehat{\mathcal{B}} \subseteq_{minrep}^q \mathcal{B}$ $(\widehat{\mathcal{B}} \subseteq_{maxrep}^q \mathcal{B})$ can be computed in at most $\mathcal{O}(\binom{p+q}{p} t p^\omega + t \binom{p+q}{p}^{\omega-1})$ operations over $\mathbb{F}$.*

2   Directed Graphs

In this section, we give a single exponential exact algorithm, that is of running time $2^{\mathcal{O}(\lambda n)}$, for computing a minimum weight spanning λ-connected subgraph of a λ connected n-vertex digraph. We first consider the unweighted version of the problem and it will be clear that the same algorithm works for weighted version as well. In a digraph D, we define $\mathsf{Out}_D(v) = \{(v, w) \in A(D)\}$ and $\mathsf{In}_D(v) = \{(u, v) \in A(D)\}$ to be the set of *out-edges* and *in-edges* of v, respectively. We begin with the following characterization of λ-connectivity in digraphs.

Lemma 1 ($*^3$). *Let D be a digraph. Then D is λ-connected if and only if for any $r \in V(D)$, there is a collection of λ arc disjoint in-branchings rooted at r, and a collection of λ arc disjoint out-branchings rooted at r.*

Let D be the input to our algorithm, which is a λ-connected digraph on n vertices. Let us fix a vertex $r \in V(D)$. By Lemma 1, any minimal λ-connected subgraph of D is a union of a collection $\mathbb{I}$ of λ arc disjoint in-branchings and a collection $\mathbb{O}$ of λ arc disjoint out-branchings which are all rooted at vertex r. The following lemma relates the size of such a minimal subgraph to the number of arcs which appear in both $\mathbb{I}$ and $\mathbb{O}$ and it follows easily from Lemma 1. Here,

---

^{3} Proof of the results marked ($*$) are omitted due to space constraints.

$A(\mathbb{I})$ denotes the set of arcs which are present in some $I \in \mathbb{I}$ and $A(\mathbb{O})$ denotes the set of arcs which are present in some $O \in \mathbb{O}$.

Lemma 2. *Let D be a λ-connected digraph, r be a vertex in $V(D)$ and $\ell \in [\lambda(n-2)]$. Then a subdigraph D' with at most $2\lambda(n-1) - \ell$ arcs, is a minimal λ-connected spanning subdigraph of D if and only if D' is a union of a collection $\mathbb{I}$ of arc disjoint in-branchings rooted at r, and a collection $\mathbb{O}$ of arc disjoint out-branchings rooted at r such that $|A(\mathbb{I}) \cap A(\mathbb{O})| \geq \ell$ (i.e. they have at least ℓ common arcs).*

By Lemma 2, a minimum λ connected subgraph of D is $\mathbb{I} \cup \mathbb{O}$, where $\mathbb{O} = \{O_1, O_2, \ldots O_\lambda\}$ is a collection of λ arc disjoint out-branchings rooted at r, $\mathbb{I} = \{I_1, I_2, \ldots I_\lambda\}$ is a collection of λ arc disjoint in-branchings rooted at r, and $A(\mathbb{O}) \cap A(\mathbb{I})$ is maximized. To explain the concept of the algorithm let us assume that the number of arcs in a minimum λ connected spanning subdigraph D' is $2\lambda(n-1) - \ell$ and let $A(D') = A(\mathbb{O}) \cup A(\mathbb{I})$, where $\mathbb{O} = \{O_1, O_2, \ldots O_\lambda\}$ is a collection of λ arc disjoint out-branchings rooted at r and $\mathbb{I} = \{I_1, I_2, \ldots I_\lambda\}$ is a collection of λ arc disjoint in-branchings rooted at r. Note that $|A(\mathbb{O}) \cap A(\mathbb{I})| = \ell$. The first step of our algorithm is to construct the set $A(\mathbb{O}) \cap A(\mathbb{I})$, and then, given the intersection, we can construct $\mathbb{O}$ and $\mathbb{I}$ in polynomial time. Observe that $A(\mathbb{O})$ and $A(\mathbb{I})$ can intersect in at most $\lambda(n-2)$ arcs. The main idea is to enumerate a subset of potential candidates for the intersection, via dynamic programming. But note that there could be as many as $n^{\mathcal{O}(\lambda n)}$ such candidates, and enumerating them all will violate the claimed running time. So we try a different approach. We first observe that the arcs in a solution, $\mathbb{O} \cup \mathbb{I}$, can be embedded into a linear matroid of rank $\mathcal{O}(\lambda n)$. Then we prove that, it is enough to keep a representative family of the partial solutions in the dynamic programming table. Since, the size of the representative family is bounded by $2^{\mathcal{O}(\lambda n)}$, our algorithm runs in the claimed running time.

Let us delve into the details of the algorithm. Let D_r^- be the digraph obtained from D after removing the arcs in $\mathsf{Out}_D(r)$. Similarly, let D_r^+ be the digraph obtained from D after removing the arcs in $\mathsf{In}_D(r)$. Observe that the arc sets of D_r^- and D_r^+ can be partitioned as follows. $A(D_r^-) = \biguplus_{v \in V(D_r^-)} \mathsf{Out}_{D_r^-}(v)$ and $A(D_r^+) = \biguplus_{v \in V(D_r^+)} \mathsf{In}_{D_r^+}(v)$. We construct a pair of matroids corresponding to each of the λ in-branching in $\mathbb{I}$ and each of the λ out-branching in $\mathbb{O}$. For each in-branching $I_i \in \mathbb{I}$, we have a matroid $\mathcal{M}_{I,1}^i = (E_{I,1}^i, \mathcal{I}_{I,1}^i)$ which is a graphic matroid in D and $E_{I,1}^i$ is a copy of the arc set of D. And similarly, for each out-branching $O_i \in \mathbb{O}$, we have a matroid $\mathcal{M}_{O,1}^i = (E_{O,1}^i, \mathcal{I}_{O,1}^i)$ which is a graphic matroid in D and $E_{O,1}^i$ is again a copy of the arc set of D. Note that the rank of these graphic matroids is $n - 1$. Next, for each I_i, we define matroid $\mathcal{M}_{I,2}^i = (E_{I,2}^i, \mathcal{I}_{I,2}^i)$ which is a partition matroid where $E_{I,2}^i$ is a copy of the arc set of D_r^- and $\mathcal{I}_{I,2}^i = \{X \mid X \subseteq E_{I,2}^i, |X \cap \mathsf{Out}_{D_r^-}(v)| \leq 1, \text{ for all } v \in V(D_r^-)\}$ [4]. Since $\mathsf{Out}_{D_r^-}(r) = \emptyset$ and $|V(D_r^-)| = n$, we have that the rank of

---

[4] We slightly abuse notation for the sake of clarity, as strictly speaking X and $\mathsf{Out}_{D_G^r}(v)$ are disjoint, since they are subsets of two different copies of the arc set.

these partition matroids, $\mathcal{M}_{I,2}^i$, $i \in [\lambda]$, is $n-1$. Similarly, for each O_i, we define $\mathcal{M}_{O,2}^i = (E_{O,2}^i, \mathcal{I}_{O,2}^i)$ as the partition matroid, where $E_{O,2}^i$ is a copy of the arc set of D_r^+ and $\mathcal{I}_{O,2}^i = \{X \mid X \subseteq E_{O,2}^i, |X \cap \ln_{D_r^+}(v)| \leq 1, \text{ for all } v \in V(D_r^+)\}$ Since $\ln_{D_r^+}(r) = \emptyset$ and $V(D_r^+) = n$, we have that the rank of these partition matroids, $\mathcal{M}_{O,2}^i$, $i \in [\lambda]$, is $n-1$. We define two uniform matroids $\mathcal{M}_I$ and $\mathcal{M}_O$ of rank $\lambda(n-1)$, corresponding to $\mathbb{I}$ and $\mathbb{O}$, on the ground sets E_I and E_O, respectively, where E_i and E_O are copies of the arc set of D. We define matroid $\mathcal{M} = (E, \mathcal{I})$ as the direct sum of $\mathcal{M}_I, \mathcal{M}_O, \mathcal{M}_{I,j}^i, \mathcal{M}_{O,j}^i$, for $i \in [\lambda]$ and $j \in \{1,2\}$. That is, $\mathcal{M} = \left(\bigoplus_{i \in [\lambda], j \in \{1,2\}} (\mathcal{M}_{I,j}^i \oplus \mathcal{M}_{O,j}^i) \right) \oplus \mathcal{M}_I \oplus \mathcal{M}_O$ Since the rank of $\mathcal{M}_{I,j}^i, \mathcal{M}_{O,j}^i$ where $i \in [\lambda]$ and $j \in \{1,2\}$, are $n-1$ each, and rank of $\mathcal{M}_I$ and $\mathcal{M}_O$ is $\lambda(n-1)$, we have that the rank of $\mathcal{M}$ is $6\lambda(n-1)$. We briefly discuss the representation of these matroids. The matroids $\mathcal{M}_{I,1}^i$, $\mathcal{M}_{O,1}^i$ for $i \in [\lambda]$ are graphic matroids, which are representable over any field of size at least 2. The matroids $\mathcal{M}_{I,2}^i, \mathcal{M}_{O,1}^i$ are partition matroids with partition size 1, and therefore they are representable over any field of at least 2 as well. Finally, the two uniform matroids, $\mathcal{M}_I$ and $\mathcal{M}_O$, are representable over any field with at least $|A(D)| + 1$ elements. Hence, at the start of our algorithm, we choose a representation of all these matroids over a field $\mathbb{F}$ of size at least $|A(D)| + 1$. So $\mathcal{M}$ is representable over any field of size at least $|A(D)| + 1$ (see [8, 10]).

For an arc $e \in A(D)$ not incident to r there are $4\lambda + 2$ copies of it in $\mathcal{M}$. Let $e_{J,j}^i$ denotes it's copy in $E_{J,j}^i$, where $i \in [\lambda]$, $j \in \{1,2\}$ and $J \in \{I,O\}$. An arc incident to r has only $3\lambda + 2$ copies in $\mathcal{M}$. For an arc $e \in \ln_D(r)$ we will denote its copies in $E_{I,1}^i, E_{O,1}^i, E_{I,2}^i$ by $e_{I,1}^i, e_{O,1}^i, e_{I,2}^i$, and similarly for an arc $e \in \text{Out}_D(r)$ we will denote its copies in $E_{I,1}^i, E_{O,1}^i, E_{O,2}^i$ by $e_{I,1}^i, e_{O,1}^i, e_{O,2}^i$. And finally, for any arc $e \in A(D)$, let e_I and e_O denote it's copies in E_I and E_O, respectively. For $e \in A(D) \setminus \text{Out}_D(r)$ and $i \in [\lambda]$, let $S_{I,e}^i = \{e_{I,1}^i, e_{I,2}^i\}$. Similarly for $e \in A(D) \setminus \ln_D(r), i \in [\lambda]$, let $S_{O,e}^i = \{e_{O,1}^i, e_{O,2}^i\}$. Let $S_e = (\cup_{i=1}^\lambda S_{I,e}^i) \bigcup (\cup_{j=1}^\lambda S_{O,e}^j) \bigcup \{e_I, e_O\}$. For $X \in \mathcal{I}$, let A_X denote the set of arcs $e \in A(D)$ such that $S_e \cap X \neq \emptyset$.

Observation 1 *(i) Let I be an in-branching in D rooted at r. Then for any $i \in [\lambda]$, $\{e_{I,1}^i \mid e \in A(I)\}$ is a basis in $\mathcal{M}_{I,1}^i$ and $\{e_{I,2}^i \mid e \in A(I)\}$ is a basis in $\mathcal{M}_{I,2}^i$. And conversely, let X and Y be basis of $\mathcal{M}_{I,1}^i$ and $\mathcal{M}_{I,2}^i$, respectively, such that $A_X = A_Y$. Then A_X is an in-branching rooted at r in D.*

(ii) Similarly, let O be an out-branching in D. Then for any $i \in [\lambda]$, $\{e_{O,1}^i \mid e \in A(O)\}$ is a basis in $\mathcal{M}_{O,1}^i$ and $\{e_{O,2}^i \mid e \in A(O)\}$ is a basis in $\mathcal{M}_{O,2}^i$. And conversely, let X and Y be basis of $\mathcal{M}_{O,1}^i$ and $\mathcal{M}_{O,2}^i$, respectively, such that $A_X = A_Y$. Then A_X is an out-branching rooted at r in D.

Observe that any arc $e \in A(D)$ can belong to at most one in-branching in $\mathbb{I}$ and at most one out-branching in $\mathbb{O}$, because both $\mathbb{I}$ and $\mathbb{O}$ are collection of arc disjoint subgraphs of D. Because of Observation 1, if we consider that each $I_i \in \mathbb{I}$ is embedded into $\mathcal{M}_{I,1}^i$ and $\mathcal{M}_{I,2}^i$ and each $O_i \in \mathbb{O}$ is embedded into $\mathcal{M}_{O,1}^i$ and $\mathcal{M}_{O,2}^i$, then we obtain an independent set Z' of rank $4\lambda(n-1)$ corresponding to $\mathbb{I} \cup \mathbb{O}$ in the matroid $\mathcal{M}$. Further, since the collection $\mathbb{I}$ is arc disjoint, $\{e_I \mid e \in A(\mathbb{I})\}$ is a basis of $\mathcal{M}_I$. And similarly, $\{e_O \mid e \in A(\mathbb{O})\}$ is a basis of $\mathcal{M}_O$. Therefore, $Z = Z' \cup \{e_I \mid e \in A(\mathbb{I})\} \cup \{e_O \mid e \in A(\mathbb{O})\}$ is a basis of $\mathcal{M}$. Now observe that, each arc in the intersection $\mathbb{I} \cap \mathbb{O}$ has six copies in the independent set Z. The remaining arcs in $\mathbb{I} \cup \mathbb{O}$ have only three copies each, and this includes any arc which is incident on r. Now, we define a function $\phi : \mathcal{I} \times A(D) \rightarrow \{0,1\}$, where for $W \in \mathcal{I}$ and $e \in A(D)$, $\phi(W, e) = 1$ if and only if exactly one of the following holds. Either, $W \cap S_e = \emptyset$. Or, $\{e_I, e_O\} \subseteq W$ and there exists $t, t' \in [\lambda]$ such that $S_{I,e}^t \subseteq W$ and $S_{O,e}^{t'} \subseteq W$. And for each $i \in [\lambda] \setminus \{t\}$ and $j \in [\lambda] \setminus \{t'\}$, $S_{I,e}^i \cap W = \emptyset$ and $S_{O,e}^j \cap W = \emptyset$. Using function ϕ we define the following collection of independent sets of $\mathcal{M}$. $\mathcal{B}^{6\ell} = \{W \mid W \in \mathcal{I}, |W| = 6\ell, \forall e \in A(D) \; \phi(W, e) = 1\}$ By the definitions of ϕ, $\mathbb{I}$ and $\mathbb{O}$, $\bigcup_{e \in A(\mathbb{O}) \cap A(\mathbb{I})} S_e$ is an independent set of $\mathcal{M}$, which is contained in $\mathcal{B}^{6\ell}$. In fact, for the optimal value of ℓ, the collection $\mathcal{B}^{6\ell}$ contains all possible candidates for the intersection of $\mathbb{O}'$ and $\mathbb{I}'$, where $\mathbb{O}'$ and $\mathbb{I}'$ are collections of arc disjoint in-branchings and arc disjoint out-branchings which form an optimum solution. Our goal is to find one such candidate partial solution from $\mathcal{B}^{6\ell}$. We are now ready to state the following lemma which shows that a representative family of $\mathcal{B}^{6\ell}$ always contains a candidate partial solution which can be extended to a complete solution.

Lemma 3. *Let D be a λ-connected digraph on n vertices, $r \in V(D)$ and $\ell \in [\lambda(n-2)]$. There exists a λ-connected spanning subdigraph D' of D with at most $2\lambda(n-1) - \ell$ arcs if and only if, there exists $\widehat{T} \in \widehat{\mathcal{B}}^{6\ell} \subseteq_{rep}^{n'-6\ell} \mathcal{B}^{6\ell}$, where $n' - 6\lambda(n-1)$, such that D has λ arc disjoint in-branchings containing $A_{\widehat{T}}$ and λ arc disjoint out-branchings containing $A_{\widehat{T}}$, which are all rooted at r.*

Proof. In the forward direction consider a λ-connected spanning subdigraph D' of D with at most $2\lambda(n-1) - \ell$ arcs. By Lemma 2, D' is union of a collection $\mathbb{I} = \{I_1, I_2, \ldots, I_\lambda\}$ of arc disjoint in-branchings rooted at r, and a collection $\mathbb{O} = \{O_1, O_2, \ldots, O_\lambda\}$ of arc disjoint out-branchings rooted at r such that $|A(\mathbb{I}) \cap A(\mathbb{O})| \geq \ell$. By Observation 1, for all $i \in [\lambda]$, $\{e_{I,1}^i \mid e \in A(I_i)\}$ is a basis in $\mathcal{M}_{I,1}^i$ and $\{e_{I,2}^i \mid e \in A(I_i)\}$ is a basis in $\mathcal{M}_{I,2}^i$. Similarly, by Observation 1, for all $i \in [\lambda]$, $\{e_{O,1}^i \mid e \in A(O_i)\}$ is a basis in $\mathcal{M}_{O,1}^i$ and $\{e_{O,2}^i \mid e \in A(O_i)\}$ is a basis in $\mathcal{M}_{O,2}^i$. Further $\{e_I \mid e \in A(\mathbb{I})\}$ and $\{e_O \mid e \in A(\mathbb{O})\}$ are bases of $\mathcal{M}_I$ and $\mathcal{M}_O$, respectively. Hence the set $Z_{D'} = \{e_{I,1}^i, e_{I,2}^i \mid e \in A(I_i), i \in [\lambda]\} \cup \{e_{O,1}^i, e_{O,2}^i \mid e \in A(O_i), i \in [\lambda]\} \cup \{e_I \mid e \in A(\mathbb{I})\} \cup \{e_O \mid e \in A(\mathbb{O})\}$ is an independent set in $\mathcal{M}$. Since $|Z_{D'}| = 6\lambda(n-1)$, $Z_{D'}$ is actually a basis in $\mathcal{M}$. Consider $T \subseteq A(\mathbb{I}) \cap A(\mathbb{O})$ with exactly ℓ arcs. Let $T' = \{e_{I,1}^i, e_{I,2}^i \mid e \in T \cap I_i, \text{ for some } i \in [\lambda]\} \cup \{e_{O,1}^i, e_{O,2}^i \mid e \in T \cap O_i, \text{ for some } i \in [\lambda]\} \cup \{e_I, e_O \mid e \in T\}$. Note that T' is a set of six copies of the ℓ arcs that are common to a pair of an in-branching in $\mathbb{I}$ and an out-branching in $\mathbb{O}$. Therefore, by the definition of $\mathcal{B}^{6\ell}$, $T' \in \mathcal{B}^{6\ell}$. Then, by the definition of representative family, there exists $\widehat{T} \in \widehat{\mathcal{B}}^{6\ell} \subseteq_{rep}^{n'-6\ell} \mathcal{B}^{6\ell}$, such that $\widehat{Z} = (Z_{D'} \setminus T') \cup \widehat{T}$ is an independent set in $\mathcal{M}$. Note that $|\widehat{Z}| = 6\lambda(n-1)$, and hence it is a basis in $\mathcal{M}$. Also note that $A_{\widehat{T}} \subseteq A_{\widehat{Z}}$.

Claim. 1 $(*)$ *(i) For any* $i \in [\lambda]$ *and* $e \in A(D)$, *either* $\{e_{I,1}^i, e_{I,2}^i\} \subseteq \widehat{Z}$ *or* $\{e_{I,1}^i, e_{I,2}^i\} \cap \widehat{Z} = \emptyset$. *And further for every* $e \in A(D)$ *such that* $e_{I,1}^i \in \widehat{Z}$ *for some* $i \in [\lambda]$, $\widehat{Z}$ *also contains* e_I. *Similarly, for any* $i \in [\lambda]$ *and* $e \in A(D)$, *either* $\{e_{O,1}^i, e_{O,2}^i\} \subseteq \widehat{Z}$ *or* $\{e_{O,1}^i, e_{O,2}^i\} \cap \widehat{Z} = \emptyset$, *and further, for every* $e \in A(D)$ *such that* $e_{O,1}^i \in \widehat{Z}$ *for some* $i \in [\lambda]$, $\widehat{Z}$ *also contains* e_O.

(ii) And for any $i, j \in [\lambda]$, $i \neq j$, *either* $\{e_{I,1}^i, e_{I,2}^i\} \cap \widehat{Z} = \emptyset$ *or* $\{e_{I,1}^j, e_{I,2}^j\} \cap \widehat{Z} = \emptyset$. *Similarly, for any* $i, j \in [\lambda]$, $i \neq j$, *either* $\{e_{O,1}^i, e_{O,2}^i\} \cap \widehat{Z} = \emptyset$ *or* $\{e_{O,1}^i, e_{O,2}^i\} \cap \widehat{Z} = \emptyset$.

Since $\widehat{Z}$ is a basis in $\mathcal{M}$, for any $i \in [\lambda], j \in \{1, 2\}$ and $k \in \{I, O\}$, we have that $\widehat{Z} \cap E_{k,j}^i$ is a basis in $\mathcal{M}_{k,j}^i$ (see [8, 10]). For each $i \in [\lambda]$, let $\widehat{X}_1^i = \widehat{Z} \cap E_{I,1}^i$ and $\widehat{X}_2^i = \widehat{Z} \cap E_{I,2}^i$. By Claim 1, $A_{\widehat{X}_1^i} = A_{\widehat{X}_2^i}$ and hence, by Observation 1, $\widehat{I}_i = A_{\widehat{X}_1^i}$ forms an in-branching rooted at r. Because of Claim 1, $\{\widehat{I}_i \mid i \in \lambda\}$ are pairwise arc disjoint as $\widehat{I}_i \cap \widehat{I}_j = \emptyset$ for every $i \neq j \in [\lambda]$. Further $A_{\widehat{T}}$ is covered in arc disjoint in-branchings $\{A_{I_{i,1}} \mid i \in \lambda\}$, as $\widehat{T} \cap E_{I,j}^i \subseteq \widehat{X}_j^i$ for $j \in \{1, 2\}$. By similar arguments we can show that there exist a collection $\{\widehat{O}_i \mid i \in [\lambda]\}$ of λ out-branchings rooted at r containing $A_{\widehat{T}}$. The reverse direction of the lemma follows from Lemma 2. $\qquad\square$

Lemma 4. *Let* D *be a* λ *connected digraph on* n *vertices and* $\ell \in [\lambda(n-2)]$. *In time* $2^{\mathcal{O}(\lambda n)}$ *we can compute* $\widehat{\mathcal{B}}^{6\ell} \subseteq_{rep}^{n'-6\ell} \mathcal{B}^{6\ell}$ *such that* $|\widehat{\mathcal{B}}^{6\ell}| \leq \binom{n'}{6\ell}$. *Here* $n' = 6\lambda(n-1)$.

Proof. We give an algorithm via dynamic programming. Let $\mathcal{D}$ be an array of size $\ell + 1$. For $i \in \{0, 1, \dots, \ell\}$ the entry $\mathcal{D}[i]$ will store the family $\widehat{\mathcal{B}}^{6i} \subseteq_{rep}^{n'-6i} \mathcal{B}^{6i}$. We will fill the entries in array $\mathcal{D}$ according to the increasing order of index i, i.e. from $0, 1, \dots, \ell$. For $i = 0$, we have $\widehat{\mathcal{B}}^0 = \{\emptyset\}$. Let $\mathcal{W} = \{\{e_I, e_O, e_{I,1}^i, e_{I,2}^i, e_{O,1}^j, e_{O,2}^j\} \mid i, j \in [\lambda], e \in A(D)\}$ and note that $|\mathcal{W}| = \lambda^2 m$, where $m = |A(D)|$. Given that we have filled all the entries $\mathcal{D}[i']$, where $i' < i + 1$, we fill the entry $\mathcal{D}[i+1]$ at step $i + 1$ as follows. Let $\mathcal{F}^{6(i+1)} = (\widehat{\mathcal{B}}^{6i} \bullet \mathcal{W}) \cap \mathcal{I}$.

Claim. 2 $(*)$ $\mathcal{F}^{6(i+1)} \subseteq_{rep}^{n'-6(i+1)} \widehat{\mathcal{B}}^{6(i+1)}$, *for all* $i \in \{0, 1, \dots \ell - 1\}$

Now the entry for $\mathcal{D}[i+1]$ is $\widehat{\mathcal{F}}^{6(i+1)}$ which is $n'-6(i+1)$ representative family for $\mathcal{F}^{6(i+1)}$, and it is computed as follows. By Theorem 4 we have that $|\widehat{\mathcal{B}}^{6i}| \leq \binom{n'}{6i}$, Hence it follows that $|\mathcal{F}^{6(i+1)}| \leq \lambda^2 m \binom{n'}{6i}$ and moreover, we can compute $\mathcal{F}^{6(i+1)}$ in time $\mathcal{O}(\lambda^2 mn\binom{n'}{6i})$. We use Theorem 4 to compute $\widehat{\mathcal{F}}^{6(i+1)} \subseteq_{rep}^{n'-6(i+1)} \mathcal{F}^{6(i+1)}$ of size at most $\binom{n'}{6(i+1)}$. This step can be done in time $\mathcal{O}(\binom{n'}{6(i+1)} t p^\omega + t\binom{n'}{6(i+1)}^{\omega-1})$, where $t = |\mathcal{F}^{6(i+1)}| = \lambda^2 m\binom{n'}{6i}$. We know from Claim 2 that $\mathcal{F}^{6(i+1)} \subseteq_{rep}^{n'-6(i+1)} \mathcal{B}^{6(i+1)}$. Therefore by the transitive property of representative sets [8, 10], we have $\widehat{\mathcal{B}}^{6(i+1)} = \widehat{\mathcal{F}}^{6(i+1)} \subseteq_{rep}^{n'-6(i+1)} \mathcal{B}^{6(i+1)}$. Finally, we assign the family $\widehat{\mathcal{B}}^{6(i+1)}$ to $\mathcal{D}[i+1]$. This completes the description of the algorithm and its correctness. Now, since $\ell \leq n'/6$, we can bound the total running time of this algorithm as $\mathcal{O}\big(\sum_{i=1}^{\ell} \big(i^{\omega}\binom{n'}{6(i+1)} + \binom{n'}{6(i+1)}^{\omega-1}\big)\lambda^2 m\binom{n'}{6i}\big) \leq 2^{\mathcal{O}(\lambda n)}$.   □

We have the following algorithm for computing $\mathbb{I}$ and $\mathbb{O}$ given $A(\mathbb{I}) \cap A(\mathbb{O})$. This algorithm extends a given set of arcs to an minimum weight collection of λ arc disjoint out-branchings. This is a simple corollary of [24, Theorem 53.10] and it also follows from the results of Gabow [13].

Lemma 5 (∗). *Let D be a digraph and w be a weight function on the arcs. For any subset X of arcs of D, a vertex r and an integer λ, we can find a minimum weight collection $\mathbb{O}$ of λ arc disjoint out-branchings rooted at r, such that $X \subseteq A(\mathbb{O})$, if it exists, in polynomial time.*

Theorem 5. *Let D be a λ edge connected digraph on n vertices. Then we can find a minimum λ edge connected subgraph of D in $2^{\mathcal{O}(\lambda n)}$ time.*

Proof. Let $n' = 6\lambda(n-1)$. We fix an arbitrary $r \in V(D)$ and for each choice of ℓ, the cardinality of $|A(\mathbb{I}) \cap A(\mathbb{O})|$, we attempt to construct a solution. By Lemma 3 we know that there exists a λ-connected spanning subdigraph D' of D with at most $2\lambda(n-1) - \ell$ arcs if and only if there exists $\widehat{T} \in \widehat{\mathcal{B}}^{6\ell} \subseteq_{rep}^{n'-6\ell} \mathcal{B}^{6\ell}$, where $n' = 6\lambda(n-1)$, such that D has a collection $\mathbb{I} = \{I_1, I_2, \ldots, I_\lambda\}$ of arc disjoint in-branchings rooted at r and a collection $\mathbb{O} = \{O_1, O_2, \ldots, O_\lambda\}$ of arc disjoint out-branchings rooted at r such that $A_{\widehat{T}} \subseteq A(\mathbb{I}) \cap A(\mathbb{O})$. Using Lemma 4 we compute $\widehat{\mathcal{B}}^{6\ell} \subseteq_{rep}^{n'-6\ell} \mathcal{B}^{6\ell}$ in time $2^{\mathcal{O}(\lambda n)}$, and for every $F \in \widehat{\mathcal{B}}^{6\ell}$ we check if A_F can be extended to a collection of λ arc disjoint out-branchings rooted at r and a collection of λ arc disjoint in-branchings rooted at r, using Lemma 5. Since $\ell \leq \lambda(n-2)$, the running time of the algorithm is bounded by $2^{\mathcal{O}(\lambda n)}$.   □

An similar algorithm can be obtained for the weighted version of the problem using the notion of weighted representative sets, thus proving Theorem 1.

3   Undirected Graphs

In this section, we give an algorithm for computing a minimum λ-connected subgraph of an undirected graph G. As before, we only consider the unweighted version of the problem. While there is no equivalent characterization of λ-connected graphs as there was in the case of digraphs, we show that we can obtain a characterization by converting the graph to a digraph with labels on the arcs. Then, as in the previous section, we embed the solutions in a linear matroid and compute them by a dynamic programming algorithm with representative families. Let D_G be the digraph with $V(D_G) = V(G)$ and for each edge $e = (u, v) \in E(G)$, we have two arcs $a_e = (u, v)$ and $a'_e = (v, u)$ in $A(D_G)$. We label the arcs a_e and a'_e by the edge e, which is called the *type* of these arcs. For $X \subseteq A(D_G)$ let $\mathsf{Typ}(X) = \{e \in E(G) \mid a_e \in X \text{ or } a'_e \in X\}$. The following two lemmata relate λ-connected subgraphs of G with collections of out-branchings in D_G.

Lemma 6 (∗). *Let G be an undirected graph and D_G be the digraph constructed from G as described above. Then G is λ-connected if and only if for any $r \in V(D_G)$, there are λ arc disjoint out-branchings rooted at r in D_G.*

By Lemma 6 we know that G is λ-connected if and only if for any $r \in V(D)$, there is a collection $\mathbb{O}$ of λ arc disjoint out-branchings rooted at r in D_G. Given a collection of out-branchings, we can obtain a λ-connected subgraph of G with at most $\lambda(n-1)$ edges. For an edge $e \in E(G)$ which is not incident on r, the two arcs corresponding to it in D_G may appear in two distinct out-branchings of $\mathbb{O}$, but for an edge e incident on r in G, only the corresponding outgoing arc of r may appear in $\mathbb{O}$. Since there are $\lambda(n-1)$ arcs in total that appear in $\mathbb{O}$ and at least λ of those are incident on r, the number of edges of G such that both the arcs corresponding to it appear in $\mathbb{O}$ is upper bounded by $\frac{\lambda(n-2)}{2}$. So any minimal λ-connected subgraph of G has $\lambda(n-1) - \ell$ edges where $\ell \in [\lfloor \frac{\lambda(n-2)}{2} \rfloor]$.

Lemma 7 (∗). *Let G be an undirected λ-connected graph on n vertices and $\ell \in [\lfloor \frac{\lambda(n-2)}{2} \rfloor]$. G has a λ-connected subgraph G' with at most $\lambda(n-1) - \ell$ edges if and only if for any $r \in V(D_G)$, $D_{G'}$ has λ arc disjoint out-branchings $\mathbb{O} = \{O_1, O_2, \ldots, O_\lambda\}$ rooted at r such that $|\mathsf{Typ}(A(\mathbb{O}))| \leq \lambda(n-1) - \ell$.*

By Lemma 7, a collection $\mathbb{O}$ of out-branchings rooted at some vertex r, that minimizes $|\mathsf{Typ}(A(\mathbb{O}))|$ corresponds to a minimum λ-connected subgraph of G. In the rest of this section, we design an algorithm that finds a collection of arc disjoint out-branchings $\mathbb{O}$ in D_G such that $|\mathsf{Typ}(A(\mathbb{O}))|$ is minimized. The first step of our algorithm is to compute the set of edges of G such that both the arcs corresponding to it appear in the collection $\mathbb{O}$, and then we can extend this to a full solution in polynomial time. Fix a vertex r. Let D_G^r denote the digraph obtained from D_G by removing the arcs in $\mathsf{In}_{D_G}(r)$. Observe that $A(D_G^r)$ can be partitioned as follows. $A(D_G^r) = \biguplus_{v \in V(D_G^r)} \mathsf{In}_{D_G^r}(v)$ We construct a pair of a graphic matroid and a partition matroid, corresponding to each of the λ out-branching that we want to find. For each $i \in [\lambda]$, we define a matroid $\mathcal{M}_1^i = (A_1^i, \mathcal{I}_1^i)$ which is a graphic matroid of D_G^r whose ground set A_1^i is a copy of the arc set $A(D_G^r)$. Similarly, for each $i \in [\lambda]$ we define matroid $\mathcal{M}_2^i = (A_2^i, \mathcal{I}_2^i)$, which is a partition matroid on the ground set A_2^i, which is a copy of the arc set $A(D_G^r)$, such that the following holds. $\mathcal{I}_2^i = \{I \mid I \subseteq A_2^i, |I \cap \mathsf{In}_{D_G^r}(v)| \leq 1,$ *for all* $v \in V(D_G^r)\}$ Next, let $\mathcal{M}_O$ be a uniform matroid of rank $\lambda(n-1)$ on the ground set A_O where A_O is also a copy of $A(D_G^r)$. Finally, we define the matroid $\mathcal{M} = (A_\mathcal{M}, \mathcal{I})$ as the direct sum of $\mathcal{M}_O$ and $\mathcal{M}_1^i, \mathcal{M}_2^i$, for $i \in [\lambda]$, i.e. $\mathcal{M} = \left(\bigoplus_{i \in [\lambda]} (\mathcal{M}_1^i \oplus \mathcal{M}_2^i) \right) \oplus \mathcal{M}_O$. Note that the rank of this matroid is $3\lambda(n-1)$ and it is representable over any field of size at least $|A(D_G^r)| + 1$. For an arc $a \in A(D_G^r)$, we denote its copies in A_1^i, A_2^i and A_O by a_1^i, a_2^i and a_O respectively. For a collection $\mathbb{O}$ of λ out-branchings in D_G^r, by $A(\mathbb{O})$ we denote the set of arcs which is present in some $O \in \mathbb{O}$. For $X \in \mathcal{I}$, by A_X we denote the set of arcs $a \in A(D_G^r)$ such that $X \cap \bigcup_{i=1}^\lambda \{a_1^i, a_2^i\} \neq \emptyset$. For $e \in E(G)$ and $i \in [\lambda]$, we let $S_e^i = \{(a_e)_1^i, (a_e)_2^i, (a_e')_1^i, (a_e')_2^i\}$ and $S_e = \{(a_e)_O, (a_e')_O\} \cup \left(\bigcup_{i=1}^\lambda S_e^i \right)$. We define a function $\psi : \mathcal{I} \times E(G) \to \{0, 1\}$, where for $W \in \mathcal{I}, e \in E(G), \psi(W, e) = 1$

if and only if exactly one of the following holds. Either $W \cap S_e = \emptyset$; or, there exists $t, t' \in [\lambda], t \neq t'$, such that, (i) $(a_e)_O, (a'_e)_O \in W$, (ii) $S_e^t \cap W = \{(a_e)_1^t, (a_e)_2^t\}$, (iii) $S_e^{t'} \cap W = \{(a'_e)_1^{t'}, (a'_e)_2^{t'}\}$, and (iv) $\forall i \in [\lambda] \setminus \{t, t'\}, S_e^i \cap W = \emptyset$. Now for each $\ell \in [\lfloor \lambda(n-2)/2 \rfloor]$, we define the following set. $\mathcal{B}^{6\ell} = \{W \mid W \in \mathcal{I}, |W| = 6\ell \text{ and } \forall e \in E(G), \psi(W, e) = 1\}$ Observe that for every $W \in \mathcal{B}^{6\ell}$, $|\mathsf{Typ}(A_W)| = \ell$ and, $a_e \in A_W$ if and only if $a'_e \in A_W$. Therefore, any set in this collection corresponds to a potential candidate for the subset of arcs which appear in exactly two out-branchings in $\mathbb{O}$. The following lemma, relates the computation of λ out-branchings minimizing types and representative sets.

Lemma 8 (∗). *Let G be a λ-connected undirected graph on n vertices, D_G its corresponding digraph and $\ell \in [\lfloor \frac{\lambda(n-2)}{2} \rfloor]$. Let $n' = 3\lambda(n-1)$. Then there exists a set $\mathbb{O}$ of out-branchings rooted at r, with $|\mathsf{Typ}(A(\mathbb{O}))| \leq \lambda(n-1) - \ell$ in D_G if and only if there exists $\widehat{T} \in \widehat{\mathcal{B}}^{6\ell} \subseteq_{rep}^{n'-6\ell} \mathcal{B}^{6\ell}$, such that D_G has a collection $\widehat{\mathbb{O}}$ of λ out-branchings rooted at r, $A_{\widehat{T}} \subseteq A(\widehat{\mathbb{O}})$ and $|\mathsf{Typ}(\widehat{\mathbb{O}})| \leq \lambda(n-1) - \ell$. Further, we can compute $\widehat{\mathcal{B}}^{6\ell} \subseteq_{rep}^{n'-6\ell} \mathcal{B}^{6\ell}$ such that $|\widehat{\mathcal{B}}^{6\ell}| \leq \binom{n'}{6\ell}$ in time $2^{\mathcal{O}(\lambda n)}$.*

Finally, Lemmata 5, 7 and 8 give us the following theorem. As before, this theorem can be extended to prove Theorem 2.

Theorem 6 (∗). *Let G be a λ edge connected graph on n vertices. Then we can find a minimum λ edge connected subgraph of G in $2^{\mathcal{O}(\lambda n)}$ time.*

4 Augmentation Problems

The algorithms for MINIMUM WEIGHT λ-CONNECTED SPANNING SUBGRAPH may be used to solve instances of MINIMUM WEIGHT λ-CONNECTIVITY AUGMENTATION as well. Given an instance (D, L, w, λ) of the augmentation problem, we construct an instance (D', w', λ) of MINIMUM WEIGHT λ-CONNECTED SPANNING SUBGRAPH, where $D' = D \cup L$ and w' is a weight function that gives a weight 0 to arcs in $A(D)$ and it is w for the arcs from L. It is easy to see that the solution returned by our algorithm contains a minimum weight augmenting set. A similar approach works for undirected graphs as well, proving Theorem 3.

References

1. Bang-Jensen, J., Gutin, G.Z.: Digraphs: theory, algorithms and applications. Springer Science & Business Media (2008)
2. Basavaraju, M., Fomin, F.V., Golovach, P., Misra, P., Ramanujan, M., Saurabh, S.: Parameterized algorithms to preserve connectivity. In: Automata, Languages, and Programming, pp. 800–811. Springer (2014)
3. Bellman, R.: Dynamic programming treatment of the travelling salesman problem. Journal of the ACM (JACM) 9(1), 61–63 (1962)
4. Berman, P., Dasgupta, B., Karpinski, M.: Approximating transitive reductions for directed networks. In: Proceedings of the 11th International Symposium on Algorithms and Data Structures. pp. 74–85. Springer-Verlag (2009)

5. Bjorklund, A.: Determinant sums for undirected hamiltonicity. SIAM Journal on Computing 43(1), 280–299 (2014)
6. Cheriyan, J., Thurimella, R.: Approximating minimum-size k-connected spanning subgraphs via matching. SIAM Journal on Computing 30(2), 528–560 (2000)
7. Cygan, M., Fomin, F.V., Golovnev, A., Kulikov, A.S., Mihajlin, I., Pachocki, J., Socala, A.: Tight bounds for graph homomorphism and subgraph isomorphism. In: Proceedings of the Twenty-Seventh Annual ACM-SIAM Symposium on Discrete Algorithms, SODA 2016, Arlington, VA, USA, January 10-12, 2016. pp. 1643–1649 (2016)
8. Cygan, M., Fomin, F.V., Kowalik, L., Lokshtanov, D., Marx, D., Pilipczuk, M., Pilipczuk, M., Saurabh, S.: Parameterized Algorithms. Springer (2015)
9. Cygan, M., Kratsch, S., Nederlof, J.: Fast hamiltonicity checking via bases of perfect matchings. In: Proceedings of the forty-fifth annual ACM Symposium on Theory of Computing. pp. 301–310. ACM (2013)
10. Fomin, F.V., Lokshtanov, D., Panolan, F., Saurabh, S.: Efficient computation of representative families with applications in parameterized and exact algorithms. J. ACM 63(4), 29:1–29:60 (Sep 2016), http://doi.acm.org/10.1145/2886094
11. Frank, A.: Augmenting graphs to meet edge-connectivity requirements. SIAM Journal on Discrete Mathematics 5(1), 25–53 (1992)
12. Frank, H., Chou, W.: Connectivity considerations in the design of survivable networks. Circuit Theory, IEEE Transactions on 17(4), 486–490 (1970)
13. Gabow, H.N.: A matroid approach to finding edge connectivity and packing arborescences. Journal of Computer and System Sciences 50(2), 259–273 (1995)
14. Gusfield, D.: A graph theoretic approach to statistical data security. SIAM Journal on Computing 17(3), 552–571 (1988)
15. Held, M., Karp, R.M.: A dynamic programming approach to sequencing problems. Journal of the Society for Industrial and Applied Mathematics 10(1), 196–210 (1962)
16. Jain, S., Gopal, K.: On network augmentation. Reliability, IEEE Transactions on 35(5), 541–543 (1986)
17. Kao, M.Y.: Data security equals graph connectivity. SIAM Journal on Discrete Mathematics 9(1), 87–100 (1996)
18. Khuller, S.: Approximation algorithms for finding highly connected subgraphs. Vertex 2, 2 (1997)
19. Khuller, S., Vishkin, U.: Biconnectivity approximations and graph carvings. Journal of the ACM (JACM) 41(2), 214–235 (1994)
20. Kortsarz, G., Nutov, Z.: Approximating minimum cost connectivity problems. In: Dagstuhl Seminar Proceedings. Schloss Dagstuhl-Leibniz-Zentrum für Informatik (2010)
21. Marx, D.: A parameterized view on matroid optimization problems. Theor. Comput. Sci. 410(44), 4471–4479 (2009), http://dx.doi.org/10.1016/j.tcs.2009.07.027
22. Marx, D., Végh, L.A.: Fixed-parameter algorithms for minimum-cost edge-connectivity augmentation. ACM Transactions on Algorithms (TALG) 11(4), 27 (2015)
23. Moyles, D.M., Thompson, G.L.: An algorithm for finding a minimum equivalent graph of a digraph. Journal of the ACM (JACM) 16(3), 455–460 (1969)
24. Schrijver, A.: Combinatorial optimization: polyhedra and efficiency, vol. 24. Springer Science & Business Media (2003)
25. Watanabe, T., Narita, T., Nakamura, A.: 3-edge-connectivity augmentation problems. In: Circuits and Systems, 1989., IEEE International Symposium on. pp. 335–338. IEEE (1989)

The Complexity of Tree Partitioning

Zhao An[1] Qilong Feng[1] Iyad Kanj[2] Ge Xia[3]

[1] School of Information Science and Engineering, Central South University, China.
anzhao1990@126.com, csufeng@csu.edu.cn
[2] School of Computing, DePaul University, Chicago, IL. ikanj@cs.depaul.edu
[3] Dept. of Computer Science, Lafayette College, Easton, PA. xiag@lafayette.edu

Abstract. Given a tree T on n vertices, and $k, b, s_1, \ldots, s_b \in \mathbb{N}$, the
Tree Partitioning problem asks if at most k edges can be removed
from T so that the resulting components can be grouped into b groups
such that the number of vertices in group i is s_i, for $i = 1, \ldots, b$. The
case when $s_1 = \cdots = s_b = n/b$, referred to as the Balanced Tree
Partitioning problem, was shown to be $\mathcal{NP}$-complete for trees of maximum degree at most 5, and the complexity of the problem for trees of
maximum degree 4 and 3 was posed as an open question. The parameterized complexity of Balanced Tree Partitioning was also posed as
an open question in another work.
In this paper, we answer both open questions negatively. We show that
Balanced Tree Partitioning (and hence, Tree Partitioning) is
$\mathcal{NP}$-complete for trees of maximum degree 3, thus closing the door on
the complexity of Balanced Tree Partitioning, as the simple case
when T is a path is in $\mathcal{P}$. In terms of the parameterized complexity of
the problems, we show that both Balanced Tree Partitioning and
Tree Partitioning are $W[1]$-complete. Finally, using a compact representation of the solution space for an instance of the problem, we present
a dynamic programming algorithm for Tree Partitioning (and hence,
for Balanced Tree Partitioning) that runs in subexponential-time
$2^{O(\sqrt{n})}$, adding a natural problem to the list of problems that can be
solved in subexponential time.

1 Introduction

Problem Definition and Motivation. We consider the Tree Partitioning problem defined as follows:

Tree Partitioning
Given: A tree T; $k, b, s_1, \ldots, s_b \in \mathbb{N}$
Parameter: k
Question: Does there exist a subset $E' \subseteq E(T)$ of at most k edges such that
the components of $T - E'$ can be grouped into b groups, where group i contains
s_i vertices, for $i = 1, \ldots, b$?

The special case of the problem when $s_1 = \cdots = s_b = |V(T)|/b$ is referred to
as Balanced Tree Partitioning.

© Springer International Publishing AG 2017
F. Ellen et al. (Eds.): WADS 2017, LNCS 10389, pp. 37–48, 2017.
DOI: 10.1007/978-3-319-62127-2_4

The two problems are special cases of the BALANCED GRAPH PARTITIONING problem, which has applications in the areas of parallel computing [3], computer vision [3], VLSI circuit design [4], route planning [8], and image processing [22]. The special case of BALANCED GRAPH PARTITIONING, corresponding to $b = 2$, is the well-known $\mathcal{NP}$-complete problem BISECTION [16]. The BALANCED GRAPH PARTITIONING problem has received a lot of attention from the area of approximation theory (for instance, see [2, 12, 21]). Moreover, the complexity and the approximability of the problem restricted to special graph classes, such as grids, trees, and bounded degree trees [12–14, 20], have been studied.

Our Results. We study the complexity and the parameterized complexity of TREE PARTITIONING and BALANCED TREE PARTITIONING, and design subexponential time algorithms for these problems. Our results are:

(A) We prove that BALANCED TREE PARTITIONING, and hence TREE PARTITIONING, is $\mathcal{NP}$-complete for trees with maximum degree at most 3. This answers an open question in [13] about the complexity of BALANCED TREE PARTITIONING for trees of maximum degree 4 and 3, after they had shown the $\mathcal{NP}$-completeness of the problem for trees of maximum degree at most 5. This also closes the door on the complexity of these problems on trees, as the simple case when the tree is a path is in $\mathcal{P}$.

(B) We prove that both TREE PARTITIONING and BALANCED TREE PARTITIONING are $W[1]$-hard. This answers an open question in [23]. We also prove the membership of the problems in the class $W[1]$, using the characterization of $W[1]$ given by Chen et al. [7].

(C) We present an exact subexponential-time algorithm for TREE PARTITIONING, and hence for BALANCED TREE PARTITIONING, that runs in time $2^{\mathcal{O}(\sqrt{n})}$, where n is the number of vertices in the tree.

For the lack of space, many details and proofs in this paper have been omitted, and can found in [1].

Related Work and Our Contributions. Feldmann and Foschini [13] studied BALANCED TREE PARTITIONING. They showed that the problem is $\mathcal{NP}$-complete for trees of maximum degree at most 5, and left the question about the complexity of the problem for maximum degree 4 and 3 open. Whereas the reduction used in the current paper to prove the $\mathcal{NP}$-hardness of BALANCED TREE PARTITIONING on trees of maximum degree at most 3 starts from the same problem (3-PARTITION) as in [13], and is inspired by their construction, the reduction in this paper is much more involved in terms of the gadgets employed and the correctness proofs.

Bevern et al. [23] showed that the parameterized complexity of BALANCED GRAPH PARTITIONING is $W[1]$-hard when parameterized by the combined parameters (k, μ), where k is (an upper bound on) the cut size, and μ is (an upper bound on) the number of resulting components after the cut. It was observed in [23], however, that the employed $\mathcal{FPT}$-reduction yields graphs of unbounded treewidth, which motivated the authors to ask about the parameterized complexity of the problem for graphs of bounded treewidth, and in particular for trees. We answer their question by showing that the problem is $W[1]$-complete.

Bevern *et al.* [23] also showed that BALANCED GRAPH PARTITIONING is $W[1]$-hard on forests by a reduction from the UNARY BIN PACKING problem, which was shown to be $W[1]$-hard in [18]. We note that the disconnectedness of the forest is crucial to their reduction, as they represent each number x in an instance of BIN PACKING as a separate path of x vertices. For BALANCED TREE PARTITIONING, in contrast to UNARY BIN PACKING (and hence, to BALANCED GRAPH PARTITIONING on forests), the difficulty is not in grouping the components into groups (bins) because enumerating all possible distributions of $k + 1$ components (resulting from cutting k edges) into $b \leq k + 1$ groups can be done in $\mathcal{FPT}$-time; the difficulty, however, stems from not knowing which tree edges to cut. The $\mathcal{FPT}$-reduction we use to show the $W[1]$-hardness is substantially different from both of those in [18, 23], even though we use the idea of non-averaging sets in our construction—a well-studied notion in the literature (e.g., see [5]), which was used for the $W[1]$-hardness result of UNARY BIN PACKING in [18].

Many results in the literature have shown that certain $\mathcal{NP}$-hard graph problems are solvable in subexponential time. Some of these rely on topological properties of the underlying graph that guarantee the existence of a balanced graph-separator of sub-linear size, which can then be exploited in a divide-and-conquer approach (*e.g.*, see [6, 9]). There are certain problems on restricted graph classes that resist such approaches due to the the problem specifications; designing subexponential-time algorithms for such problems usually requires exploiting certain properties of the solution itself, in addition to properties of the graph class (see [15, 19] for such recent results). In the case of TREE PARTITIONING and BALANCED TREE PARTITIONING, since every tree has a balanced separator consisting of a single vertex, yet the two problems remain $\mathcal{NP}$-hard on trees, clearly a divide-and-conquer approach based solely on balanced separators does not yield subexponential-time algorithms for these problems. To design subexponential-time algorithms for them, we rely on the observation that the number of possible partitions of an integer $n \in \mathbb{N}$ is subexponential in n; this allows for a "compact representation" of all solutions using a solution space of size $2^{\mathcal{O}(\sqrt{n})}$, enabling a dynamic programming approach that solves the problems within the same time upper bound.

Terminologies. We refer the reader to [10, 11] for more information about graph theory and parameterized complexity.

Let T be a rooted tree. For an edge $e = uv$ in T such that u is the parent of v, by the subtree of T *below* e we mean the subtree T_v of T rooted at v. For two edges e, e' in T, e is said to be *below* e' if e in an edge of the subtree of T below e'. A *nice binary tree* T is defined recursively as follows. If $|V(T)| \leq 1$ then T is a nice binary tree. If $V(T) > 1$, then T is nice if (1) each of the left-subtree and right-subtree of T is nice and (2) the sizes of the left-subtree and the right-subtree differ by at most 1. For any $n \in \mathbb{N}$, there is a nice binary tree of order

n. A *star* S is a tree consisting of a single vertex r, referred to as the root of
the star, attached to degree-1 vertices, referred to each as a *star-leaf*; we refer
to an edge between r and a leaf in S as a *star-edge*; we refer to a subtree of S
containing r as a *substar* of S.

A *solution* P to an instance $(T, k, b, s_1, \ldots, s_b)$ of TREE PARTITIONING is a
pair (E_P, λ_P), where E_P is a set of k edges in T, and λ_P is an assignment that
maps the connected components in $T - E_P$ into b groups so that the total number
of vertices assigned to group i is s_i, for $i \in [b]$. We call a connected component
in $T - E_P$ a *P-component*, and denote by C_P the set of all P-components.

By a *cut in a tree* T we mean the removal of an edge from T. A solution
$P = (E_P, \lambda_P)$ to $(T, k, b, s_1, \ldots, s_b)$ *cuts* an edge e in T if $e \in E_P$. Let T' be a
subtree of T such that P cuts at least one edge in T'. By a *lowest P-component*
in T' we mean a subtree T'' below an edge e of T' such that T'' is a P-component
(*i.e.*, P does not cut any edge below e in T').

The restriction of TREE PARTITIONING to instances in which $s_1 = \cdots = s_b = |T|/b$ is denoted BALANCED TREE PARTITIONING; an instance of BALANCED
TREE PARTITIONING is a triplet (T, k, b). The restriction of TREE PARTITION-
ING and BALANCED TREE PARTITIONING to trees of maximum degree at most 3
are denoted DEGREE-3 TREE PARTITIONING and BALANCED DEGREE-3 TREE
PARTITIONING, respectively. For $\ell \geq 1 \in \mathbb{N}$, we write $[\ell]$ for the set $\{1, \ldots, \ell\}$.

2 $\mathcal{NP}$-completeness

In this section, we show that BALANCED DEGREE-3 TREE PARTITIONING, and
hence DEGREE-3 TREE PARTITIONING, is $\mathcal{NP}$-complete. Without loss of generality, we will consider the version of BALANCED DEGREE-3 TREE PARTITIONING
in which we ask for a cut of size exactly k, as opposed to at most k; it is easy to
see that the two problems are polynomial-time reducible to one another.

To prove that BALANCED DEGREE-3 TREE PARTITIONING is $\mathcal{NP}$-hard, we
will show that the strong $\mathcal{NP}$-hard problem 3-PARTITION [16] is polynomial-
time reducible to it. Our reduction is inspired by the construction of Feldmann
and Foschini [13]. Whereas the construction in [13] uses gadgets such that each
consists of five chains joined at a vertex, the construction in this paper uses
gadgets consisting of nearly-complete binary trees, that are referred to as nice
binary trees. The idea behind using nice binary trees is that we can combine
them to construct a degree-3 tree in which the cuts must happen at specific
edges in order to produce components of certain sizes.

An instance of the 3-PARTITION problem consists of an integer $s > 0$ and
a collection $S = \langle a_1, \ldots, a_{3k} \rangle$ of $3k$ positive integers, where each a_i satisfies
$s/4 < a_i < s/2$, for $i \in [3k]$. The problem is to decide whether S can be
partitioned into k groups $S_1, \ldots, S_k$, each of cardinality 3, such that the sum
of the elements in each S_i is s, for $i \in [k]$. By pre-processing the instance of
3-PARTITION, we can assume that $\sum_{i=1}^{3k} a_i = k \cdot s$, and that s is divisible by 4.

For the reduction, we construct a degree-3 tree T as follows. For each $a_i \in S$,
we create a binary tree T_i, whose left subtree L_i is a nice binary tree of size a_i,

and whose right subtree R_i is a nice binary tree of size $s-2$. We denote by R_i^l and R_i^r the left and right subtrees of R_i, respectively. Let $H = (p_1, \ldots, p_{3k})$ be a path on $3k$ vertices. The tree T is constructed by adding an edge between each p_i in H and the root of T_i, for $i \in [3k]$. See Figure 1 for illustration. It is clear from the construction that T is a degree-3 tree of $4k \cdot s$ vertices, since each T_i has size $a_i + s - 1$ and P has $3k$ vertices. We will show that (S, s) is a yes-instance of 3-PARTITION if and only if the instance $I = (T, 6k - 1, b = 4k)$ is a yes-instance of BALANCED DEGREE-3 TREE PARTITIONING.

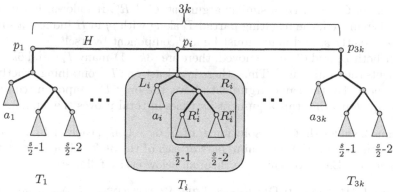

Fig. 1. Illustration of the construction of the tree T.

Note that the size of T is $4k \cdot s$, and hence, if the vertices in T can be grouped into $4k$ groups of equal size, then each group must contain s vertices. From the aforementioned statement, it follows that at least one cut is required in each tree T_i because the size of each T_i is $a_i + s - 1 > s$.

Suppose that the instance I has a solution P that cuts $6k - 1$ edges in T.

Lemma 1. *For $i \in [3k]$, R_i is not a P-component in C_P and T_i does not contain a lowest P-component of size less than $s/4$.*

Lemma 2. *For $i \in [3k]$, L_i is the only P-component contained in T_i, and the subtree of T induced by $(V(T_i) - V(L_i)) \cup \{p_i\}$ is a P-component of size s.*

Proof. Since $|T_i| > s$, any T_i must contain at least one P-component. Since C_P has $6k$ P-components, at least one of the $3k$ T_i's contains at most one P-component, because otherwise the P-components containing vertices in H are not accounted for. Therefore, at least one T_i contains exactly one P-component C, which must be a lowest P-component in T_i. By Lemma 1, $C \neq R_i$ and $|C| \geq s/4$, and hence C cannot be any proper subtree of L_i, R_i^l, or R_i^r. This leaves L_i, R_i^l, and R_i^l as the only possible choices for C.

Suppose that $C = R_i^l$. After removing C, the partial-T_i, denoted T_i^-, has size $s - 1 + a_i - (s/2 - 1) = s/2 + a_i$, and contains no P-components. Let D be the set of vertices that are not in T_i^-, and are in the same group as T_i^-. Observe that for any $j \neq i, j \in [3k]$, if a vertex in L_j is in D then all vertices in L_j are in D. This is true because, by Lemma 1, *all* vertices in L_j belong to the same P-component; otherwise L_j would have a lowest P-component of size less than $s/4$. This means that the P-component containing T_i^- has size $|T_i^-| + |D| \geq s/2 + a_i + a_j > s$. Therefore, D does not include any vertex in L_j. Similarly, D does not include any vertex in R_j^l or R_j^r. It follows that D consists only of vertices in H, the roots of the T_i's, and the roots of the R_i's, $i \in [3k]$. However, there are only $9k$ such vertices, which means $|D| \leq 9k$ and hence the P-component containing T_i^- has size $|T_i^-| + |D| \leq s/2 + a_i + 9k < s$. The last inequality is true because $a_i \leq s/2 - 18k$, for $i \in [k]$.

Therefore, $C \neq R_i^l$. By a similar argument, $C \neq R_i^r$. It follows that $C = L_i$. After L_i is removed, the resulting partial-T_i along with p_i in H induces a subtree C_i of size exactly s, and hence must be a P-component by itself.

After both L_i and C_i are removed, there are $(3k-1)$-many T_i's and $6k-2$ P-components remaining in T. Thus, there is at least one T_j containing exactly one P-component. By the same argument above, the only P-component contained in T_j is L_j. Repeating this argument $3k$ times in total proves the lemma.　　□

Lemma 2 shows that, in a solution P of $(T, 6k-1, 4k)$, each of $L_1, \ldots, L_{3k}$ is a P-component, and the remaining part of each of the $3k$ T_i's is a P-component whose size is s. Based on this, the theorem below easily follows:

Theorem 1. BALANCED DEGREE-3 TREE PARTITIONING *is $\mathcal{NP}$-complete.*

3　$W[1]$-completeness

To show that TREE PARTITIONING is $W[1]$-hard (membership in $W[1]$ is shown using a characterization of $W[1]$ given in [7]), we give a fixed-parameter tractable reduction ($\mathcal{FPT}$-reduction) from the k-MULTI-COLORED CLIQUE PROBLEM (k-MCC), which is $W[1]$-complete ([11]), and is defined as follows: Given a graph $M = (V(M), E(M))$ and a proper k-coloring of the vertices $f : V(M) \longrightarrow C$, where $C = \{1, 2, ..., k\}$ and each color class has the same cardinality, decide whether there exists a clique $Q \subseteq V(M)$ of size k such that, $\forall u, v \in Q$, $f(u) \neq f(v)$. For $i \in [k]$, we define $C_i = \{v \in M \mid f(v) = i\}$ to be the color class consisting of all vertices whose color is i. Let $n = |C_i|$, $i \in [k]$, and let $N = k \cdot n$. We label the vertices in C_i arbitrarily as $v_1^i, \ldots, v_n^i$. We first introduce some terminologies.

For a finite set $X \subseteq \mathbb{N}$ and $\ell \in \mathbb{Z}^+$, we say that X is ℓ-*non-averaging* if for any ℓ numbers $x_1, \ldots, x_\ell \in X$, and for any number $x \in X$, the following holds: if $x_1 + \cdots + x_\ell = \ell \cdot x$ then $x_1 = \cdots = x_\ell = x$.

Let $X = \{x_1, \ldots, x_n\}$ be a $(k-1)$-non-averaging set. It is known that we can construct such a set X such that each element $x_i \in X$, $i \in [n]$, is polynomial in n (for instance, see [5]). Jensen *et al.* [18] showed that a $(k-1)$-non-averaging set of cardinality n, in which each number is at most $k^2 n^2 \leq n^4$, can be constructed in polynomial time in n; we will assume that X is such a set. Let $k' = k + \binom{k}{2}$, and let $z = k'^2 n^5$. Choose $2k$ numbers $b_1, \ldots, b_k, c_1, \ldots, c_k \in \mathbb{N}$ such that $b_j = k'^{2j} \cdot z$ for $j \in [k]$, and $c_j = k'^{2(k+j)} \cdot z$ for $j \in [k]$. Observe that each number in the sequence $b_1, \ldots, b_k, c_1, \ldots, c_k$ is equal to the preceding number multiplied by k'^2, and that the smallest number b_1 in this sequence is $k'^2 \cdot z \geq k'^4 n^5$. For each $j, j' \in [k], j < j'$, we choose a number $c_j^{j'} = c_k \cdot k'^{2((j-1)k-j(j-1)/2+j'-j)}$. That is, each number in the sequence $c_1^2, \ldots, c_1^k, c_2^3, \ldots, c_2^k, \ldots, c_{k-1}^k$ is equal to the preceding one multiplied by k'^2, and the smallest number c_1^2 in this sequence is equal to $k'^2 \cdot c_k$.

We construct a tree T rooted at a vertex r as follows. For a vertex v_i^j, $i \in [n], j \in [k]$, we correspond a *vertex-gadget* (for vertex v_i^j) that is a star $S_{v_i^j}$ with $c_j - (k-1)b_j - (k-1)x_i - 1$ leaves, and hence with $c_j - (k-1)b_j - (k-1)x_i$ vertices; we label the root of the star $r_{v_i^j}$, and add the edge $r r_{v_i^j}$ to T. See Figure 2 for illustration. For each edge e in M between two vertices v_i^j and v_p^q, $i, p \in [n], j, q \in [k], j < q$, we create two stars $S'_{v_i^j}$ and $S'_{v_p^q}$, with $b_j + x_i - 1$ and $b_q + x_p - 1$ leaves, respectively, and of roots $r'_{v_i^j}$ and $r'_{v_p^q}$, respectively. We introduce a star S_e with root r_e and $c_j^q - 1$ leaves, and connect r_e to $r'_{v_i^j}$ and $r'_{v_p^q}$ to form a tree T_e with root r_e that we call an *edge-gadget* (for edge e). We connect r_e to r. See Figure 3 for illustration. Note that the number of vertices in T_e that are not in $S'_{v_i^j} \cup S'_{v_p^q}$ is exactly c_j^q. Finally, we create $k' + 1$ copies of a star S_{fix} consisting of $c_{k-1}^k + k' + 1$ many vertices, and connect the root r of T to the root of each of these copies. This completes the construction of T. Let $t = |T|$. We define the reduction from k-MULTI-COLORED CLIQUE to TREE PARTITIONING to be the map that takes an instance $I = (M, f)$ of k-MULTI-COLORED CLIQUE and produces the instance $I' = (T, k', b = k \cdot \binom{k}{2}, c_1, \ldots, c_k, c_1^2, \ldots, c_1^k, c_2^3, \ldots, c_2^k \ldots, c_{k-1}^k, t')$, where $k' = k + 3\binom{k}{2}$ and $t' = t - \sum_{j=1}^{k} c_j - \sum_{j,q \in [k], j < q} c_j^q$. Clearly, this reduction is an $\mathcal{FPT}$-reduction. Next, we describe the intuition behind this reduction.

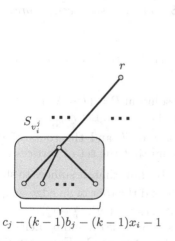
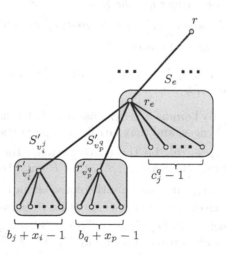

Fig. 2. Illustration of the vertex-gadget for v_i^j.

Fig. 3. Illustration of the edge-gadget for $e = v_i^j v_p^q$.

Each number c_j, $j \in [k]$, chosen above, will serve as a "signature" for class C_j, in the sense that it will ensure that in any solution to the instance, a vertex-gadget corresponding to a vertex in class C_j is "cut" and placed in the group of size c_j. Each number $c_j^{j'}$, $j, j' \in [k]$, $j < j'$, will serve as a "signature" for the class-pair $(C_j, C_{j'})$, in the sense it will ensure that in a solution exactly one edge-gadget corresponding to an edge e between classes C_j and C_j' is cut and the star S_e is placed in the group whose size is $c_j^{j'}$. Each number b_j, $j \in [k]$, will serve as a "signature" for any edge such that one of its endpoints is in C_j (i.e., a signature for an arbitrary vertex in C_j), ensuring that in a solution, $k - 1$ of these edges are cut. Finally, the choice of the x_i's, for $i \in [n]$, to be elements of a $(k - 1)$-non-averaging set, will ensure that all the edges cut that are incident to vertices in the same class C_j, $j \in [k]$, are incident to the same vertex in C_j.

Next, we prove the correctness of the reduction. One direction is easy:

Lemma 3. *If (M, f) is a yes-instance of k-MULTI-COLORED CLIQUE then I' is a yes-instance of TREE PARTITIONING.*

To prove the converse, let $P = (E_P, \lambda_P)$ be a solution to the instance $I' = (T, k', b = k + \binom{k}{2}, c_1, \ldots, c_k, c_1^2, \ldots, c_1^k, c_2^3, \ldots, c_2^k \ldots, c_{k-1}^k, t')$ of TREE PARTITIONING. Let G_j, $j \in [k]$, denote the group of size c_j, G_j^q, $j, q \in [k]$, $j < q$, denote the group of size c_j^q, and G_{rest} denote the group of size t'. We have:

Lemma 4. *There is a solution P that cuts exactly $k' = k + 3\binom{k}{2}$ edges from T as follows. For each $j \in [k]$, P cuts exactly one edge between the root r of T and the root of a vertex-gadget corresponding to a vertex in color class C_j; moreover, λ_P assigns the resulting vertex-gadget to group G_j. For each $j, q \in [k]$, $j < q$, P cuts exactly 3 edges from one edge-gadget T_e, corresponding to an edge e between a vertex v_i^j, $i \in [n]$, in color classes C_j, and a vertex v_p^q, $p \in [n]$, in color class C_q; those 3 edges are the edges rr_e, $r_e r'_{v_i^j}$, and $r_e r'_{v_p^q}$, where r_e is the root of star S_e in T_e, and $r'_{v_i^j}, r'_{v_p^q}$ are the roots of stars $S'_{v_i^j}, S'_{v_p^q}$ in T_e, respectively; moreover, λ_P assigns S_e to group G_j^q.*

Lemma 5. *If I' is a yes-instance of TREE PARTITIONING then (M, f) is a yes-instance of k-MCC.*

Proof. By Lemma 4, we can assume that I' has a solution $P = (E_P, \lambda_P)$ that cuts $k + 3\binom{k}{2}$ edges, and that satisfies the properties in the lemma. Let $rr_{v_{i_1}^1}, \ldots, rr_{v_{i_k}^k}$, $i_1, \ldots, i_k \in [n]$, be the edges between the root r of T and the roots of the vertex-gadgets $S_{v_{i_1}^1}, \ldots, S_{v_{i_1}^k}$ that P cuts. We claim that the set of vertices $Q = \{v_{i_1}^1, \ldots, v_{i_k}^k\}$ induce a multi-colored clique in M. To show that, it suffices to show that each of the $\binom{k}{2}$ edges rr_e cut by P, between r and the root of an edge-gadget T_e, where $e = v_i^j v_p^q$, $i, p \in [n], p, q \in [k], p < q$, satisfies that $v_i^j, v_p^q \in Q$.

Consider an arbitrary group G_j, $j \in [k]$. The size of G_j is c_j, and by Lemma 4, λ_P assigns the star $S_{v_{i_j}^j}$ of size $c_j - (k-1)b_j - (k-1)x_{i_j}$ to G_j. Each star S_e is assigned to some group G_p^q whose size is exactly $|S_e|$. Therefore, each group G_j

contains a vertex-gadget and some of the stars $S'_{v^{j'}_{i'}}$, $i' \in [n]$, $j' \in [k]$. Observe that group G_j, $j < k$, cannot contain a star $S'_{v^{j'}_{i'}}$ such that $j' > j$ because the size of such a star is at least $b_{j'} > k'^2 b_j$, and hence the size of such a star plus the size of $S_{v^j_{i_j}}$ would exceed the size of G_j. Since there are exactly $k-1$ stars, of the form $S'_{v^k_*}$ contained in edge-gadgets corresponding to edges incident to class C_k, it follows that all these stars must be assigned by λ_P to group G_k. Moreover, no other star $S'_{v^j_*}$, $j < k$, can be assigned to G_k, as the size of such a star would be at least $b_1 > (k-1)x_i$ for any $i \in [n]$; hence, G_k would contain vertex gadget $S_{v^k_{i_k}}$ of size $c_k - (k-1)b_k - (k-1)x_{i_k}$, plus $k-1$ stars $S'_{v^k_*}$ of total size greater than $(k-1)b_k$, plus a star of size at least $b_1 > (k-1)x_{i_k}$, and the size of G_k would exceed c_k.

Similarly, all the $k-1$ stars of the form $S'_{v^{k-1}_*}$ contained in edge-gadgets corresponding to edges incident to class C_{k-1} are assigned to group G_{k-1}, and following this argument, we obtain that for each $j \in [k]$, the $(k-1)$ stars of the form $S'_{v^j_*}$ must be assigned to group G_j. We claim that all these stars must correspond to the same vertex $v^j_{i_j}$. Observe that this will prove that Q is a clique, since it will imply that each vertex in Q is incident to exactly $k-1$ of the $\binom{k}{2}$ many edges between the color classes.

Let $S'_{v^j_{i'_1}}, \ldots, S'_{v^j_{i'_{k-1}}}$ be the $k-1$ stars placed in G_j. The sizes of these stars are $b_j + x_{i'_1}, \ldots, b_j + x_{i'_{k-1}}$, respectively. The size c_j of G_j is equal to the sum of the sizes of these $k-1$ stars, plus that of $S_{v^j_{i_j}}$. Therefore: $c_j = c_j - (k-1)b - (k-1)x_{i_j} + (k-1)b + x_{i'_1} + \cdots + x_{i'_{k-1}}$, and hence, $(k-1) \cdot x_{i_j} = x_{i'_1} + \cdots + x_{i'_{k-1}}$. Since the set X is $(k-1)$-non-averaging, it follows that $x_{i_j} = x_{i'_1} = \cdots = x_{i'_{k-1}}$, and hence, the $(k-1)$ stars $S'_{v^j_*}$ must correspond to vertex $v^j_{i_j}$. $\square$

Theorem 2. TREE PARTITIONING *and* BALANCED TREE PARTITIONING *are* W[1]*-complete.*

4 Subexponential-time Algorithms

Let $n \in \mathbb{Z}^+$. A *partition* of n is a collection X of positive integers such that $\sum_{x \in X} x = n$. Let $p(n)$ denote the total number of (distinct) partitions of n. It is well known that $p(n) = 2^{\mathcal{O}(\sqrt{n})}$ [17]. It follows that the total number of partitions of all integers n', where $0 < n' \leq n$, is $\sum_{0 < n' \leq n} p(n') = 2^{\mathcal{O}(\sqrt{n})}$.

Let L be a list of numbers in $\mathbb{N}$ that are not necessarily distinct. We denote by $L(i)$ the ith number in L, and by L_i the sublist of L consisting of the first i numbers. The *length* of L, denoted $|L|$, is the number of elements in L.

Let $(T, k, b, s_1, \ldots, s_b)$ be an instance of TREE PARTITIONING. Let $n = |T|$. Consider a partial assignment of $n' \leq n$ vertices of T to the b groups, with the possibility of some groups being empty. Since the groups are indistinguishable, such an assignment corresponds to a partition of the n' vertices into at most b

parts, and can be represented by a *sorted* list L of b numbers in $\mathbb{N}$ whose sum is n', where $L(i) \leq n'$ for $i \in [b]$, is the number of vertices assigned to group i; we call such a representation of the groups, under a partial assignment, a *size representation*, denoted as σ-representation. Note that the zeroes in a σ-representation appear at the beginning. Since each σ-representation corresponds uniquely to a partition of a number $n' \leq n$ prefixed by less than $b \leq n$ zeroes, it follows that the total number of σ-representations is $n \cdot 2^{\mathcal{O}(\sqrt{n})} = 2^{\mathcal{O}(\sqrt{n})}$.

Let X, Y, Z be three lists of the same length. We write $X = Y \Diamond Z$ if there is a list Y' obtained via a permutation of the numbers in Y, and a list Z' obtained via a permutation of the numbers in Z, such that $X(i) = Y'(i) + Z'(i)$, for every $i \in [\|X\|]$; that is, in the context when the lists are σ-representations, $X = Y \Diamond Z$ if each group-size in X can be obtained, in a one-to-one fashion, by adding a group-size in Y to a group-size in Z (including group-sizes zero). We have:

Proposition 1. *There is a subroutine* **Check-Realizability** *(X, Y, Z) that determines if $X = Y \Diamond Z$ in time $2^{\mathcal{O}(\sqrt{n})}$.*

Let $(T, k, b, s_1, \ldots, s_b)$ be an instance of TREE PARTITIONING. The key observation that leads to a subexponential-time algorithm is that the b groups are indistinguishable. Therefore, all assignments of the n vertices in T to the b groups can be compactly represented by lists of numbers, where each list corresponds to a partition of n into b parts. This simple, yet crucial, observation allows for a "compact representation" of all solutions using a solution space of size $2^{\mathcal{O}(\sqrt{n})}$.

Suppose that T is rooted at an arbitrary vertex r. The algorithm uses dynamic programming, starting from the leaves of T, and climbing T up to its root r. At each vertex v in T, we construct a table Γ_v that contains the following information. For each σ-representation X, for each $k' = 0, \ldots, n$, and for each $s \in [n]$, $\Gamma_v(k', X, s)$ is TRUE if and only if there is a cut C of k' edges in T_v such that the component P_v containing v in $T_v - C$ has size s (note that this component, so far, is still attached to the rest of the tree above v), and such that there is an assignment to the components in $T_v - C - P_v$ to the b groups whose σ-representation is X; otherwise, $\Gamma_v(k', X, s)$ is FALSE. If $\Gamma_v(k', X, s)$ is TRUE, we store a witness that realizes such a partial solution. To compute Γ_v, we consider the children of v one by one. After a child u_i of v is considered, we have computed a partial table Γ_i containing partial solutions up to child u_i; this is done by considering the two possibilities of whether or not the edge vu_i is in the cut C. Although the above may seem like we are enumerating all possibilities for the edges between v and its children to be cut or not, the crucial ingredient for this approach to achieve the desired running time is that the table Γ_v—at vertex v—can be computed based on the tables corresponding to the children of v in $2^{\mathcal{O}(\sqrt{n})}$ time. This analysis works similarly to iterative compression, as the table Γ_i is a compressed table, storing $2^{\mathcal{O}(\sqrt{n})}$ many entries.

Suppose that the algorithm is at vertex v whose children are $u_1, \ldots, u_d$, and that the tables $\Gamma_{u_1}, \ldots, \Gamma_{u_d}$ associated with $u_1, \ldots, u_d$, respectively, have been constructed. To compute Γ_v, we iterate through the edges $vu_1, \ldots, vu_d$. Let T_p, for $p \in [d]$, be the subtree of T rooted at v that is induced by the vertex-set $(\bigcup_{j=1}^{p} V(T_{u_j})) \cup \{v\}$. Consider edge vu_i, and assume inductively, that a table Γ_{i-1} has been computed (based on tables $\Gamma_{u_1}, \ldots, \Gamma_{u_{i-1}}$) that contains the following information. For each $k' = 0, \ldots, n$, for each $s \in [n]$, and for each σ-representation X, $\Gamma_{i-1}(k', X, s)$ is TRUE if and only if there is a cut C of k' edges in T_{i-1}, with s being the size of the component P_v containing v in $T_{i-1} - C$, and an assignment to the components in $T_{i-1} - C - P_v$ that realizes X. After considering vu_i, we will compute a table Γ_i such that, for each $k' = 0, \ldots, n$, for each $s \in [n]$, and for each σ-representation X, $\Gamma_i(k', X, s)$ is TRUE if and only if there is a cut C of k' edges in T_i, with s being the size of the component P_v containing v in $T_i - C$, and an assignment to the components in $T_i - C - P_v$ that realizes X. We explain how the Boolean value $\Gamma_i(k', X, s)$ is computed, and omit how the witness is stored. After we are done computing Γ_d, we set $\Gamma_v = \Gamma_d$.

To compute Γ_i, we compute two tables Γ_i^- and Γ_i^+, and set $\Gamma_i = \Gamma_i^- \cup \Gamma_i^-$. Table Γ_i^- contains the solutions that can be obtained by cutting edge uv_i, and Γ_i^+ contains those that can be obtained by not cutting edge vu_i.

1. To compute Γ_i^-, we enumerate each possible triplet (k', X, s), where $k' = 0, \ldots, n$, $s \in [n]$, and X is a σ-representation. Fix such a triplet (k', X, s). To compute $\Gamma_i^-(k', X, s)$, we iterate through every entry in Γ_{u_i} containing (k_{u_i}, Y, s_{u_i}) and every entry of Γ_{i-1} containing (k_{i-1}, Z, s_{i-1}) such that $k' = k_{u_i} + k_{i-1} + 1$ (because 1 more cut is introduced, corresponding to the edge vu_i), and $s = s_{i-1}$ because the component P_{u_i} containing u_i of size s_{u_i} becomes a separate component after vu_i is cut. Since P_{u_i} becomes a separate component, it will be placed into one of the groups, and hence, it contributes its size to one of the numbers in the σ-representation X. We enumerate each number in X as the number that P_{u_i} contributes to. For each number j in X satisfying $j \geq |P_{u_i}|$, we subtract $|P_{u_i}|$ from j in X to obtain a new σ-representation X' from X, and then call **Check-Realizability**(X', Y, Z); $\Gamma_i^-(k', X, s)$ is TRUE iff for some number j in X, **Check-Realizability**(X', Y, Z) returns TRUE.

2. To compute Γ_i^+, we enumerate each triplet (k', X, s), where $k' = 0, \ldots, n$, $s \in [n]$, and X is a σ-representation. Fix such a triplet (k', X, s). To compute $\Gamma_i^+(k', X, s)$, we iterate through every entry in Γ_{u_i} containing (k_{u_i}, Y, s_{u_i}), and every entry in Γ_{i-1} containing (k_{i-1}, Z, s_{i-1}), such that $k' = k_{u_i} + k_{i-1}$, and $s = s_{u_i} + s_{i-1}$ (because s_{u_i} is attached to v). We call **Check-Realizability**(X, Y, Z), and set $\Gamma_i^+(k', X, s)$ to TRUE iff **Check-Realizability**(X, Y, Z) returns TRUE.

Theorem 3. *The dynamic programming algorithm described above solves* TREE PARTITIONING *and* BALANCED TREE PARTITIONING *in time* $2^{\mathcal{O}(\sqrt{n})}$.

Bibliography

[1] Z. An, Q. Feng, I. Kanj, and G. Xia. The Complexity of Tree Partitioning. Available at http://arxiv.org/abs/1704.05896.

[2] K. Andreev and H. Räcke. Balanced graph partitioning. *Theory of Computing Systems*, 39(6):929–939, 2006.

[3] P. Arbenz, G. van Lenthe, U. Mennel, R. Müller, and M. Sala. Multi-level μ-finite element analysis for human bone structures. In *PARA 2006*, pages 240–250, 2006.

[4] S. Bhatt and F. Leighton. A framework for solving VLSI graph layout problems. *Journal of Computer and System Sciences*, 28(2):300–343, 1984.

[5] Á. Boscznay. On the lower estimation of non-averaging sets. *Acta Mathematica Hungariga*, 53(1-1):155–157, 1989.

[6] J. Chen, I. Kanj, L. Perkovic, E. Sedgwick, and G. Xia. Genus characterizes the complexity of certain graph problems: Some tight results. *Journal of Computer and System Sciences*, 73(6):892–907, 2007.

[7] Y. Chen, J. Flum, and M. Grohe. Machine-based methods in parameterized complexity theory. *Theoretical Computer Science*, 339(2-3):167–199, 2005.

[8] D. Delling, A. Goldberg, T. Pajor, and R. Werneck. Customizable route planning. In *SEA 2011*, pages 376–387, 2011.

[9] E. Demaine, F. Fomin, M. Hajiaghayi, and D. Thilikos. Subexponential parameterized algorithms on bounded-genus graphs and H-minor-free graphs. *J. ACM*, 52:866–893, 2005.

[10] Reinhard Diestel. *Graph Theory, 4th Edition*. Springer, 2012.

[11] R. Downey and M. Fellows. *Fundamentals of Parameterized Complexity*. Springer, New York, 2013.

[12] A. Feldmann. Balanced partitions of grids and related graphs, 2012. Ph.D. thesis, ETH, Zurich, Switzerland.

[13] A. Feldmann and L. Foschini. Balanced partitions of trees and applications. *Algorithmica*, 71(2):354–376, 2015.

[14] A. Feldmann and P. Widmayer. An $O(n^4)$ time algorithm to compute the bisection width of solid grid graphs. *Algorithmica*, 71(1):181–200, 2015.

[15] F. Fomin, S. Kolay, D. Lokshtanov, F. Panolan, and S. Saurabh. Subexponential algorithms for rectilinear steiner tree and arborescence problems. In *SoCG 2016*, pages 39:1–39:15, 2016.

[16] M. Garey and D. Johnson. *Computers and Intractability: A Guide to the Theory of NP-Completeness*. W. H. Freeman, New York, 1979.

[17] G. Hardy and S. Ramanujan. Asymptotic formulae in combinatory analysis. *Proceedings of the London Mathematical Society*, 17(2):75–115, 1918.

[18] K. Jansen, S. Kratsch, D. Marx, and I. Schlotter. Bin packing with fixed number of bins revisited. *Journal of Computer and System Sciences*, 79(1):39–49, 2013.

[19] P. Klein and D. Marx. A subexponential parameterized algorithm for subset TSP on planar graphs. In *SODA 2014*, pages 1812–1830, 2014.

[20] R. MacGregor. On partitioning a graph: a theoretical and empirical study, 1978. Ph.D. thesis, University of California at Berkeley, California, USA.

[21] H. Räcke and R. Stotz. Improved approximation algorithms for balanced partitioning problems. In *STACS 2016*, pages 58:1–58:14, 2016.

[22] J. Shi and J. Malik. Normalized cuts and image segmentation. *IEEE Transactions on Pattern Analysis and Machine Intelligence*, 22(8):888–905, 2000.

[23] R. van Bevern, A. Feldmann, M. Sorge, and O. Suchý. On the parameterized complexity of computing balanced partitions in graphs. *Theory of Computing Systems*, 57(1):1–35, 2015.

Effectiveness of Local Search for Art Gallery Problems

Sayan Bandyapadhyay[1]* and Aniket Basu Roy[2]

[1] Computer Science, University of Iowa, Iowa City, USA
[2] Computer Science and Automation, Indian Institute of Science, Bangalore, India
sayan-bandyapadhyay@uiowa.edu, aniket.basu@csa.iisc.ernet.in

Abstract. We study the variant of the art gallery problem where we
are given an orthogonal polygon P (possibly with holes) and we want to
guard it with the minimum number of sliding cameras. A sliding camera
travels back and forth along an orthogonal line segment s in P and a
point p in P is said to be visible to the segment s if the perpendicular
from p onto s lies in P. Our objective is to compute a set containing the
minimum number of sliding cameras (orthogonal segments) such that
every point in P is visible to some sliding camera. We study the following
two variants of this problem: *Minimum Sliding Cameras* problem, where
the cameras can slide along either horizontal or vertical segments in P,
and *Minimum Horizontal Sliding Cameras* problem, where the cameras
are restricted to slide along horizontal segments only. In this work, we
design local search PTASes for these two problems improving over the
existing constant factor approximation algorithms. We note that in the
first problem, the polygons are not allowed to contain holes. In fact, there
is a family of polygons with holes for which the performance of our local
search algorithm is arbitrarily bad.

1 Introduction

Local search is a popular technique for designing time and cost efficient approximation algorithms. It has a long history in combinatorial optimization and has
proved to be very effective for achieving near-optimum solutions. The use of this
technique in geometric approximation is relatively new, but has resulted in improved approximation for metric and geometric versions of many combinatorial
problems. In fact, this technique has been used in this domain to achieve several
breakthrough results.

In this article, we restrict ourselves to the works based on local search in
geometric approximation. One of the first results of local search for the problems
in metric space is a $3+\epsilon$ approximation algorithm for k-median due to Arya et al.
[1]. An arguably simplified analysis was later given by Gupta and Tangwongsan
[15]. Building on the work of Arya et al., Kanungo et al. [16] have designed a
$9 + \epsilon$ approximation algorithm for k-means. In a celebrated work Mustafa and

* The author has been supported by NSF under Grant CCF-1615845.

Ray [20] designed the first PTASes for the Hitting Set problem for a wide class of geometric range spaces. Around the same time Chan and Har-Peled [7] gave PTASes for Maximum Independent Set problem of geometric objects including fat objects and pseudodisks in the plane. Cohen-Addad and Mathieu [8] have shown the effectiveness of local search for achieving PTASes for facility location type problems. In a recent breakthrough, Cohen-Addad et al. [9] and Friggstad et al. [13] have independently designed the first PTAS for k-means which was a long standing open problem. Very recently, Govindarajan et al. [14] have obtained the first PTASes for the Set Cover and Dominating Set problems for non-piercing regions. See also [2, 3, 4, 5] for related work on local search.

In this paper, we consider art gallery problems and design PTASes using local search proving the effectiveness of this technique for a new family of problems.

1.1 Art Gallery Problems

In the classical *art gallery* problem, we are given a polygon P (possibly with holes), and the goal is to determine the number of guards (points) needed that can see all parts of P. A guard can see the portion of the polygon not obstructed by the polygon boundary, considering the boundary to be opaque. Over the years different variants of the problem have been studied based on the shape of the polygon, the type of guards employed, and the notion of visibility. The polygons can be of specific type like orthogonal[3], monotone etc. Guards can be stationary or mobile. In case of stationary guards, either one can place the guards anywhere inside the polygon (point guards) or restrict them only to the vertices of the polygon (vertex guards). The notion of mobile guards was introduced by Toussaint [21], where each guard can travel to and fro along a segment inside the polygon, and every point in the polygon must be seen by at least one guard at some point of time along its path. If the notion of visibility is altered, it gives rise to a lot of other variants like k-transmitters [22], multi-guarding [18] etc. One such variant is called sliding cameras, where the polygon is considered to be orthogonal. A sliding camera is a mobile guard that can move back and forth along an orthogonal (or axis-parallel) segment. In the following, we formally define this notion.

Definition 1 (Sliding Camera). Given an orthogonal polygon P, any orthogonal (horizontal or vertical) segment contained in P is called a *sliding camera*. If the corresponding segment is horizontal (resp. vertical), then the sliding camera is horizontal (resp. vertical). A sliding camera $\bar{s}$ can *guard* a point $p \in P$ if there is a point $q \in \bar{s}$ such that the segment $\overline{pq}$ is perpendicular to $\bar{s}$ and is contained in P. A set S of sliding cameras can *guard* the polygon P, if for every point $p \in P$, there is a sliding camera in S that can guard p.

In this article, we study the following two problems involving sliding cameras.

Definition 2 (Minimum Horizontal Sliding Cameras Problem (MHSC)). Given an orthogonal polygon P, compute a set S containing the minimum number of horizontal sliding cameras, that can guard P.

---

[3] A polygon is said to be *orthogonal* if all of its sides are axis-parallel.

Definition 3 (Minimum Sliding Cameras Problem (MSC)). Given an orthogonal polygon P, compute a set S containing the minimum number of sliding cameras, that can guard P.

The MSC and MHSC problems were introduced by Katz and Morgenstern [17]. They showed that if P is simple (has no holes), MHSC can be solved in polynomial time. For monotone P, they obtain a polynomial time 2 approximation for MSC. Later for this special case, de Berg et al. [10] gave a linear time exact algorithm. MHSC and MSC both are known to be NP-hard for polygons with holes [6, 11]. Recently, Biedl et al. [6] have given the first constant factor approximation algorithms for MHSC and MSC for polygons possibly with holes. Currently, this is the best known approximation factor for both problems.

1.2 Related Work

The art gallery problem is related to the 1.5D Terrain Guarding problem. In the latter problem, the input is a terrain T consisting of an x-monotone polygonal chain, and we need to guard the terrain T by choosing the minimum number of point guards on it. The discrete version of the problem comes with two finite sets $G, X \subseteq T$ and one needs to compute the minimum sized subset $G' \subseteq G$ such that G' guards X. A local search framework similar to the ones in [7, 20] was used by Krohn et al. [19] to give a PTAS for the discrete version of the problem. Building on this result, Friedrichs et al. [12] gave a PTAS for the continuous version. Also note that the collection of problems considered in the local search framework of [2] includes problems on guarding polygons and terrains with limited visibility and appropriate shallowness assumption.

1.3 Our Results and Techniques

In this article, we use a standard local search framework to obtain PTASes for the MHSC and MSC problems. In fact, we give PTASes for certain hitting set problems involving segments that yield the PTASes for MHSC and MSC. We note, that we get a PTAS for MHSC even in the case where the polygon contains holes. However, we get the PTAS for MSC only if the polygon is simple (has no holes). In fact, one can show that in case of MSC where the polygon contains holes, the performance of local search can be arbitrarily bad. Our main contribution is the design of two planar graph embedding schemes which are used to prove the planarity of certain graphs. In Section 2 we define some notations and describe the local search framework that we use. In Section 3 and 4 we use the framework for MHSC and MSC, respectively, and prove that the framework yields PTASes for both problems. Due to lack of space, many proofs are omitted which can be found in the full version of the paper.

2 Preliminaries

Consider the orthogonal polygon P in the MSC or MHSC problem. An orthogonal segment $s \in P$ is called *canonical* if (i) s is an extension of a side of P, and (ii) s is maximal in terms of length, i.e., there is no other segment in P that strictly contains it. We note that the endpoints of a canonical segment lie on the boundary of P. Also the number of canonical segments is at most the number of sides of P. It is not hard to see that there exists a canonical segment s' for any segment $s \in P$, such that s' can guard all the points in P guarded by s. Henceforth, for any solution S of MHSC or MSC, without loss of generality we assume that S consists of only canonical segments. Obviously, for MHSC, any solution consists of only horizontal canonical segments.

2.1   The Local Search Framework

Now we describe a local search framework similar to the ones in [7, 20] for designing and analyzing local search algorithms. In the following sections we will apply this framework to get near optimum approximations for MHSC and MSC. We note that this framework is applicable for discrete optimization problems, where given a set S, the goal is to find an optimum subset $S' \subseteq S$ that is feasible. Notably, not every subset $S' \subseteq S$ is a feasible solution. Moreover, there is an initial feasible solution that can be found in polynomial time, and given any $S' \subseteq S$, one can decide the feasibility of S' in polynomial time. We note that for MHSC (resp. MSC), S is the finite set of all horizontal (resp. horizontal and vertical) canonical segments, which is a feasible solution. Let $n = |S|$. Fix $\epsilon > 0$. As we deal with only minimization problems in this article, the local search algorithm we consider is the following.

Local Search Algorithm. Choose a parameter k. Start with some feasible solution A. At each iteration, search for $A' \subseteq A$ and $C \subseteq S \setminus A$ such that $|A'| \leq k$, $|C| < |A'|$, and $(A \setminus A') \cup C$ is a feasible solution. If such A' and C exist, update A to $(A \setminus A') \cup C$ and reiterate. Otherwise, return A.

As we shoot for a $(1 + \epsilon)$ approximation we set $k := c/\epsilon^2$ for some constant c. From the above discussion, it is not hard to see that the running time of the algorithm is polynomial. A feasible solution A is said to be *local optimum* if there is no $A' \subseteq A$ and $C \subseteq S \setminus A$ such that $|A'| \leq k$, $|C| < |A'|$, and $(A \setminus A') \cup C$ is a feasible solution. Note that the local search algorithm always returns a local optimum solution. Let $\mathcal{R}$ and $\mathcal{B}$ be an optimum solution and the local search solution, respectively. For simplicity, we assume that $\mathcal{R} \cap \mathcal{B} = \emptyset$. Otherwise, one can remove the common elements $\mathcal{R} \cap \mathcal{B}$ from both $\mathcal{R}$ and $\mathcal{B}$, and perform a similar analysis. As we remove the same number of elements from both $\mathcal{R}$ and $\mathcal{B}$ the approximation ratio of the original instance is at most the approximation ratio of the restricted one. Now we state the conditions required for the local search algorithm to be a PTAS.

Theorem 1. *[7, 20]. Consider a minimization problem Π. Suppose there exists a planar bipartite graph $H = (\mathcal{R} \cup \mathcal{B}, E)$, that satisfies the local exchange property: For any subset $\mathcal{B}' \subseteq \mathcal{B}$, $(\mathcal{B} \setminus \mathcal{B}') \cup N(\mathcal{B}')$ is a feasible solution. Then the Local Search Algorithm is a PTAS. Here $N(\mathcal{B}')$ is the set of neighbors of $\mathcal{B}'$ in H.*

3   The MHSC Problem

Instead of directly working with MHSC we consider the following *Orthogonal Segment Covering (OSC)* problem: Given a finite set $\mathcal{H}$ of non-intersecting horizontal segments and a finite set $\mathcal{V}$ of vertical segments in the plane, find a minimum cardinality set $S \subseteq \mathcal{H}$ such that any segment in $\mathcal{V}$ is intersected by S. From the work due to Biedl et al. [6] we have the following theorem.

Theorem 2. *[6] MHSC reduces to the Orthogonal Segment Covering problem. In particular, an α approximation for the Orthogonal Segment Covering problem yields an α approximation for MHSC.*

We note that the non-intersecting assumption of the horizontal segments comes from the fact that we consider only the canonical horizontal segments in MHSC to guard P which are non-intersecting. Biedl et al. [6] gave a constant approximation for OSC which implies a constant approximation for MHSC. We use the local search framework to obtain a PTAS for OSC; by Theorem 2 we obtain a PTAS for MHSC as well.

Let L be the local search solution and G be an optimum solution. We show that there exists a planar bipartite graph $H=(G \cup L, E)$ that satisfies the local exchange property; by Theorem 1 the local search algorithm is a PTAS for OSC. As mentioned before, we can assume WLOG that $L \cap G = \emptyset$. For any two intersecting vertical segment v and horizontal segment h, let $I(h,v)$ be the intersection point.

Construction of H. Initially the set of vertices of H consists of the intersection points of the segments in $L \cup G \cup \mathcal{V}$ and the endpoints of the segments in $L \cup G$. We first join the vertices on any horizontal segment using edges. In particular, for any segment $h \in L \cup G$, let $\{p_1, \ldots, p_t\}$ be the vertices in H corresponding to h in increasing order of their x-coordinates. For each $1 \le i \le t-1$, draw a horizontal edge between p_i and p_{i+1}. Then we join the "consecutive" horizontal segments. Formally, for any two segments $h_1 \in L$ and $h_2 \in G$, if h_1 and h_2 intersect a segment v of $\mathcal{V}$, and $\nexists h_3 \in L \cup G \setminus \{h_1, h_2\}$ such that $I(h_3, v)$ is on the segment joining $I(h_1, v)$ and $I(h_2, v)$, draw a vertical edge between $I(h_1, v)$ and $I(h_2, v)$. For any segment $h \in L \cup G$, contract all the vertices in H corresponding to h into a single vertex. The modified graph is the desired graph H.

It is obvious that before the contractions, H is a planar graph, as it is a subgraph of the graph corresponding to the arrangement of the segments in $L \cup G \cup \mathcal{V}$. Now note that the segments in $L \cup G$ are non-intersecting. Thus there is a 1-1 correspondence between the vertices in final H and the segments in $L \cup G$. Furthermore, any edge in H is between a vertex corresponding to a segment in L and a vertex corresponding to a segment in G. As planar graphs are closed under contraction of edges, H is planar.

Observation 4. $H = (G \cup L, E)$ *is a planar bipartite graph.*

Observation 5. *For any segments h_1, h_2, v such that $h_1 \in L$, $h_2 \in G$, $v \in \mathcal{V}$, if there is an edge in H between $I(h_1, v)$ and $I(h_2, v)$ before the contractions, then there is an edge corresponding to h_1 and h_2 in the final graph H.*

Lemma 3. $H = (G \cup L, E)$ *satisfies the local exchange property.*

Proof. It is sufficient to show that for any $v \in V$, there exist $h_1 \in L$ and $h_2 \in G$ such that h_1 and h_2 intersect v, and there is an edge between h_1 and h_2 in H. Consider any $v \in V$. Also consider the subset of segments $S \subseteq L \cup G$ that intersect v. Then S must contain two segments h_1, h_2 such that $h_1 \in L$, $h_2 \in G$ and $\nexists h_3$ such that $I(h_3, v)$ is on the segment joining $I(h_1, v)$ and $I(h_2, v)$. Thus there must be an edge in H before the contractions between $I(h_1, v)$ and $I(h_2, v)$. From Observation 5 it follows that there is an edge in H between h_1 and h_2 which completes the proof.   □

We conclude this section with the following theorem.

Theorem 4. *There exists a local search PTAS for the MHSC problem, where the input polygon may or may not contain holes.*

4   The MSC Problem

We use the local search framework to get a PTAS for MSC problem assuming P does not contain any holes. In fact, one can construct a family of polygons with holes for which our local search scheme performs very poorly (see full paper). We start the local search algorithm with the set of canonical segments. Now consider the local search solution L and an optimum solution $\mathcal{R}$. Then we have the following lemma which will be useful later.

Lemma 5. *Consider the local search solution L and an optimum solution $\mathcal{R}$. There is a local optimum solution $\mathcal{B}$ such that $|\mathcal{B}| = |L|$, and for every point $p \in P$, there exists segments $b \in \mathcal{B}$ and $r \in \mathcal{R}$ both of which guard p and either of the following is true.*

1. b *and* r *are both horizontal or both vertical.*
2. *One of* b *and* r *is horizontal and the other is vertical, and* $r \cap b \neq \emptyset$.

Henceforth, we consider the local optimum solution $\mathcal{B}$ in Lemma 5 and show that $|\mathcal{B}| \leq (1 + \epsilon)|\mathcal{R}|$. As $|L| = |\mathcal{B}|$ the local search algorithm is a PTAS. Like before, WLOG we can assume that $\mathcal{B} \cap \mathcal{R} = \emptyset$. We will show the existence of a bipartite planar graph $G(\mathcal{R} \cup \mathcal{B})$ that satisfies the local exchange property.

Denote the arrangement of a set of segments S by $\mathcal{A}(S)$. Consider the edges and the vertices of $\mathcal{A}(S)$. One can visualize these edges and vertices as a plane graph and define its connected components accordingly. Now consider any such component $\mathcal{C}$ and let S' be the subset of segments of S corresponding to $\mathcal{C}$. We define the closure of $\mathcal{C}$ (denoted by closure($\mathcal{C}$)) as the union of the points on the segments in S' and the points in the bounded cells of $\mathcal{A}(S')$. Define $\mathcal{H}(S)$ as the union of the closures of the components in $\mathcal{A}(S)$, i.e., $\mathcal{H}(S) := \cup_{\mathcal{C} \in \mathcal{A}(S)}$ closure($\mathcal{C}$). Also define $\partial\mathcal{H}(S)$ as the union of the boundaries of the closures in $\mathcal{H}(S)$ (see Figure 1b). Later, we will prove the following theorem.

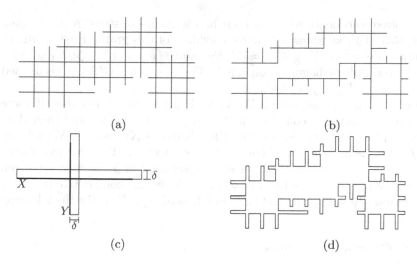

Fig. 1: *(a) A connected component of the arrangement. (b) The boundary of the closure of the component. (c) A horizontal segment X and a vertical segment Y (bolded). The two slabs generated by taking the Minkowski sum of the two segments with S_δ, a square of side length δ. (d) The fattened boundary of the closure.*

Theorem 6. *Given a set of canonical horizontal segments X and a set of canonical vertical segments Y, one can construct a planar graph $G(X \cup Y)$ satisfying the following properties.*

1. *The vertices in $G(X \cup Y)$ correspond to the segments in $X \cup Y$.*
2. *For each $X_i, X_j \in X$, such that there exists a vertical segment v contained in P that intersects X_i and X_j, and $\nexists X_k \in X$ that intersects v and lies between X_i and X_j, (X_i, X_j) is in $G(X \cup Y)$.*
3. *For each $Y_i, Y_j \in Y$, such that there exists a horizontal segment h contained in P that intersects Y_i and Y_j, and $\nexists Y_k \in Y$ that intersects h and lies between Y_i and Y_j, (Y_i, Y_j) is in $G(X \cup Y)$.*
4. *For each $X_i \in X$, and $Y_j \in Y$, such that $X_i \cap Y_j$ is on $\partial \mathcal{H}(X \cup Y)$, (X_i, Y_j) is in $G(X \cup Y)$.*

We construct the planar graph $G(\mathcal{R} \cup \mathcal{B})$ by applying Theorem 6 on the horizontal and vertical segments of $\mathcal{R} \cup \mathcal{B}$. We remove all the edges from $G(\mathcal{R} \cup \mathcal{B})$ that joins two segments of $\mathcal{R}$ or two segments of $\mathcal{B}$ making it bipartite for the partition $(\mathcal{R}, \mathcal{B})$. The following lemma completes our claim that the local search algorithm is a PTAS. The proof of the lemma partly follows from Lemma 5.

Lemma 7. $G(\mathcal{R} \cup \mathcal{B})$ *satisfies the local exchange property.*

4.1   Graph Construction Algorithm

In this section, we prove Theorem 6. Given a set of horizontal segments X and a set of vertical segments Y, we define a graph $G(X \cup Y)$ on $X \cup Y$. The construction is a bit involved and hence we describe it in a sequence of steps. After defining the graph $G(\mathcal{X} \cup \mathcal{Y})$ we show a planar embedding of this graph. First we define three graphs $G_{\mathcal{X}} = (\mathcal{X}, E_{\mathcal{X}})$, $G_{\mathcal{Y}} = (\mathcal{Y}, E_{\mathcal{Y}})$ and $G_{\mathcal{X},\mathcal{Y}} = (\mathcal{X} \cup \mathcal{Y}, E_{\mathcal{X},\mathcal{Y}})$ with respect to the simple orthogonal polygon P. The edge set of $G(\mathcal{X} \cup \mathcal{Y})$ is defined to be the union of $E_{\mathcal{X}}, E_{\mathcal{Y}}$, and $E_{\mathcal{X},\mathcal{Y}}$.

A pair of horizontal segments $X_i, X_j \in \mathcal{X}$ are called *consecutive* if there exists a vertical segment v contained in P that joins X_i and X_j, and there does not exist $X_k \in \mathcal{X}$ that intersects v and lies between X_i and X_j. We add an edge to $E_{\mathcal{X}}$ between each pair of consecutive segments in $\mathcal{X}$. If X_i and X_j are consecutive but there does not exist a vertical segment in $\mathcal{Y}$ intersecting them, then we call (X_i, X_j) a soft edge. Likewise, we define the consecutive segments in $\mathcal{Y}$, the graph $G_{\mathcal{Y}} = (\mathcal{Y}, E_{\mathcal{Y}})$ and the soft edges of $G_{\mathcal{Y}}$. We make the following claim.

Lemma 8. $G_{\mathcal{X}}$ *and* $G_{\mathcal{Y}}$ *are forests.*

We assume that the arrangement $\mathcal{A}(\mathcal{X} \cup \mathcal{Y})$ of the segments in $\mathcal{X} \cup \mathcal{Y}$ has only one connected component and later show how to remove this assumption. Then for each pair $X_i, X_j \in \mathcal{X}$, there is a chain of vertical and horizontal segments in $\mathcal{X} \cup \mathcal{Y}$ that connects X_i and X_j. Thus by definition $G_{\mathcal{X}}$ is also connected and hence is a tree. Similarly, $G_{\mathcal{Y}}$ is a tree. We refer to the closure of the component in $\mathcal{A}(\mathcal{X} \cup \mathcal{Y})$ by $\mathcal{H}$ and its boundary by $\partial \mathcal{H}$. Define $G_{\mathcal{X},\mathcal{Y}} = (\mathcal{X} \cup \mathcal{Y}, E_{\mathcal{X},\mathcal{Y}})$ to be the bipartite graph on $\mathcal{X}$ and $\mathcal{Y}$, such that $(X, Y) \in E_{\mathcal{X},\mathcal{Y}}$ iff $X \cap Y \cap \partial \mathcal{H} \neq \emptyset$.

We note that, by definition $G(\mathcal{X} \cup \mathcal{Y})$ satisfies the four properties mentioned in Theorem 6. Thus it is sufficient to show that $G(\mathcal{X} \cup \mathcal{Y})$ is planar, to which we turn next. First we use the embedding of $\mathcal{X} \cup \mathcal{Y}$ to get an intermediate plane graph H. Then we perform some planarity preserving operations on H to obtain $G(\mathcal{X} \cup \mathcal{Y})$. Hence we get a planar drawing of $G(\mathcal{X} \cup \mathcal{Y})$, thus proving $G(\mathcal{X} \cup \mathcal{Y})$ is planar.

To construct the plane graph H we use the given embedding of the segments in $\mathcal{X} \cup \mathcal{Y}$. We consider the endpoints of the segments in $\mathcal{X} \cup \mathcal{Y}$ and the intersection points of the segments in $\mathcal{X} \cup \mathcal{Y}$ to be the vertex points of H. We first join the vertices on any horizontal segment using edges. For any segment $h \in \mathcal{X} \cup \mathcal{Y}$, let $\{p_1, \ldots, p_t\}$ be the vertices in H corresponding to h in increasing order of their x-coordinates. For each $1 \leq i \leq t - 1$, draw a horizontal edge between p_i and p_{i+1}. Similarly, we join the vertices on any vertical segment using vertical edges and call them the edges corresponding to that vertcal segment. It is not hard to see that H is a plane graph. One can visualize H as a graph where each $X \in \mathcal{X}$ is being represented by its left endpoint, and each $Y \in \mathcal{Y}$ is being represented by the segment Y itself. Later we will contract the edges corresponding to $Y \in \mathcal{Y}$ to represent them using points in the plane. Note that the edges of $E_{\mathcal{Y}}$ are already present in H. Thus to get a drawing of $G(\mathcal{X} \cup \mathcal{Y})$ we need to draw the edges of $E_{\mathcal{X}}$ and $E_{\mathcal{X},\mathcal{Y}}$. In the remainder of the section, we will show the planarity preserving operations on H that leads to the drawing of $G(\mathcal{X} \cup \mathcal{Y})$. This will be done in two stages. In the first stage, we add some edges to H. In the second stage, we contract several edges.

First, we fix some notations. For every $X \in \mathcal{X}$, denote its left (resp. right) endpoint by $\ell(X)$ (resp. $r(X)$), and the set of vertical segments in $\mathcal{Y}$ intersecting it by $vert(X)$. Also, denote any non-empty intersection point $X \cap Y$ by $p(X, Y)$, where $X \in \mathcal{X}$ and $Y \in \mathcal{Y}$.

One can visualize $\partial \mathcal{H}$ as a self-intersecting curve. For simplicity we will disentangle it to get a simple closed curve. To this end, we consider the object $\mathcal{H}' = \mathcal{H} \oplus S_\delta$, which is the Minkowski-sum of $\mathcal{H}$ and a square S_δ with side length δ, where δ is an arbitrarily small positive quantity (see Figure 1c and 1d). This is basically "fattening" the boundary $\partial \mathcal{H}$ such that the boundary of $\mathcal{H}'$ is a simple closed curve. Also note that it fattens only towards the positive horizontal and vertical axes. Let $\partial \mathcal{H}'$ be the boundary of $\mathcal{H}'$. Now for every $p(X, Y) \in \partial \mathcal{H}$, there exists at least one point on $\partial \mathcal{H}'$ among the following points — $p(X, Y)$, $p(X, Y) + (0, \delta)$, $p(X, Y) + (\delta, 0)$, and $p(X, Y) + (\delta, \delta)$. Refer to that point as $p'(X, Y)$. If multiple of these points exist on $\partial \mathcal{H}'$, then choose one arbitrarily. For convenience we use $p'(X, Y)$ as a "proxy" for $p(X, Y)$ while adding edges to H whose one endpoint is $p(X, Y)$. At the end we will use the original points $p(X, Y)$. As the two points are arbitrarily close the latter conversion does not affect the planarity. Hereafter, for any reference to a point $p'(X, Y)$, we will assume that $p(X, Y) \in \partial \mathcal{H}$. Now note that the only edges that are needed to be drawn are the edges in $E_\mathcal{X}$ and $E_{\mathcal{X}, \mathcal{Y}}$. The endpoints of the edges in $E_\mathcal{X}$ will be the left endpoints of the segments in $\mathcal{X}$. Also the endpoints of the edges in $E_{\mathcal{X}, \mathcal{Y}}$ will be the left endpoints of the segments in $\mathcal{X}$ and the intersection points $p(X, Y)$ for $X \in \mathcal{X}$ and $Y \in \mathcal{Y}$ such that $p(X, Y) \in \partial \mathcal{H}$. Let $Q = \bigcup_X \ell(X) \cup \bigcup_{X,Y} p'(X, Y)$, the set of endpoints of the edges we will draw. Note that all the points of Q are on $\partial \mathcal{H}'$.

We consider $G_\mathcal{X}$ as a tree rooted at some segment $X_r \in \mathcal{X}$. Later we will use this tree structure to add edges to H in an inductive manner. We denote the parent of a segment X_i with respect to $G_\mathcal{X}$ by $parent(X_i)$. Let $r(X_r) = (x', y')$. We cut open the closed curve $\partial \mathcal{H}'$ at $(x', y' + \delta)$ such that the open curve is homeomorphic to an interval M, where $r(X_r)$ gets mapped before $\ell(X_r)$. For simplicity, we denote the mapped points by their original notations. Thus $r(X_r) < \ell(X_r)$ in M. Note that one can consider another homeomorphism to an interval (the reflection of M with respect to its middle point) so that $\ell(X_r) < r(X_r)$. One can visualize M in a way such that $\partial \mathcal{H}'$ is disconnected at the point $(x', y' + \delta)$ and straightened out to get M. The way we have defined M it has the following property: for any child X_i of X_r, if X_i lies above X_r, $r(X_r) < \ell(X_r) < \ell(X_i) < r(X_i)$ and if X_i lies below X_r, $r(X_r) < r(X_i) < \ell(X_i) < \ell(X_r)$. Observe, that the endpoints of M are not in Q and we already have an embedding of the points of Q on the interval M. We use this embedding of the points to draw the edges between them so that the edges do not cross each other.

For any edge $(a, b) \in E_\mathcal{X} \cup E_{\mathcal{X}, \mathcal{Y}}$, (a, b) is called an edge *corresponding* to a segment X of $G_\mathcal{X}$ if either a or b is X, and the other endpoint is not $parent(X)$. For any subtree, we call all the edges corresponding to its nodes as its *corresponding* edges. Now we proceed towards the inductive process of addition of edges. For that we need to define a concept called zone for every segment X_i of

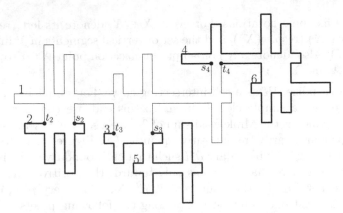

Fig. 2: *The bolded curves corresponding to the children* $(2, 3, 4)$ *of* 1 *are mapped to the corresponding zones in* M. s_i *and* t_i *are the respective left and right endpoint of the zone of* i *for* $i = 2, 3, 4$.

G_X, which is basically a sub-interval of M. We will ensure that the endpoints of the edges corresponding to X_i lies in its zone. The structure of $\partial \mathcal{H}'$ or M ensures that the zones have many useful properties, which we will exploit heavily.

For the root segment X_r, $zone(X_r)$ is defined to be the entire interval M. For any other $X_i \in \mathcal{X}$, let Y_ℓ and Y_r be the leftmost and the rightmost vertical segment in $vert(X_i) \cap vert(parent(X_i))$, respectively. If X_i lies above its parent, then $zone(X_i) = [p(X_i, Y_\ell), p(X_i, Y_r) + (\delta, 0)]$. Otherwise, $zone(X_i) = [p(X_i, Y_r) + (\delta, \delta), p(X_i, Y_\ell) + (0, \delta)]$. See Figure 2. Now we have the following observation.

Observation 6. *The following statements are true regarding zones.*

1. $zone(X_i) \subset zone(parent(X_i))$ *for every* $X_i \in \mathcal{X}$.
2. *For any segment* $X \in \mathcal{X}, Y \in \mathcal{Y}$, $p'(X, Y)$ *and* $\ell(X)$ *lies outside* $zone(X_j)$, *where* X_j *is a child of* X.
3. *The endpoints of the edges corresponding to a segment* $X \in \mathcal{X}$ *lie within the zone of* X. *Moreover, for all edge corresponding to the subtree rooted at* X, *both of the endpoints lie on the same side of* $\ell(X)$.
4. *The zones of the segments corresponding to the children of* X *are pairwise disjoint.*

In the following lemma, we show the first stage of our construction, where the edges are drawn without crossing.

Lemma 9. *Given the plane graph* H *as defined before, the edges of* $E_{\mathcal{X}}$ *and* $E_{\mathcal{X}, \mathcal{Y}}$ *can be drawn in a non-crossing manner.*

Proof. We note that the edges of $E_{\mathcal{X}} \cup E_{\mathcal{X}, \mathcal{Y}}$ are drawn outside of the closed curve $\partial \mathcal{H}'$. Thus they do not cross the edges of H as the latter edges lie inside $\partial \mathcal{H}'$. Consider the set Q of the endpoints of the edges to be drawn. We use the embedding of the points of Q on M for drawing edges between them. As M is homeomorphic to the open curve we obtain by cutting open $\partial\mathcal{H}'$ this lemma follows. Now we give an explicit construction of a drawing of the edges. In each step, we draw the edges corresponding to a subtree of $G_\mathcal{X}$ rooted at some node X, given the drawings corresponding to the subtrees rooted at the children of X. In the base case, we draw the edges corresponding to the leaves of $G_\mathcal{X}$. Note that each leaf is corresponding to a segment $X \in \mathcal{X}$. The only edges we draw corresponding to X are of the form $(\ell(X), p'(X,Y))$, and hence it is easy to see that these edges can be drawn without any crossing.

Now consider any non-leaf node X. By induction we already have a crossing-free drawing corresponding to the subtree rooted at any child of X. From statements (1) and (3) of Observation 6 it follows that the endpoints of the edges in this drawing must lie within the zone of the child. As zones of the children are disjoint by statement (4) of the same Observation, we get a crossing-free drawing with respect to the children of X. Thus the only additional edges we need to draw have $\ell(X)$ as an endpoint. The other endpoint could be $l(X_i)$ for a child X_i of X or $p'(X,Y)$ for $(X,Y) \in E_{\mathcal{X},\mathcal{Y}}$. Now $l(X)$ and $p'(X,Y)$ do not belong to the zone of any child of X by statement (4) of Observation 6. Also by statement (3) of the same observation, both of the endpoints of any existing edge corresponding to the subtree rooted at a child X_j of X lie on the same side of $l(X_j)$. Hence all the additional edges can be drawn without crossing any existing edge. $\qquad\square$

Now we begin the second stage of planarity preserving operations, i.e. the edge contractions. We contract $(p(X,Y), r(X))$ for every $X \in \mathcal{X}$ where Y is the rightmost vertical segment among the ones intersecting X. After this for every $Y \in \mathcal{Y}$ we contract every edge subdividing Y such that a single point remains at the end corresponding to Y. Thus for every Y, there exists a unique vertex point. Hence, by construction all the properties in Theorem 6 are satisfied.

One can extend the construction for the case when $\mathcal{A}(\mathcal{X} \cup \mathcal{Y})$ has more than one components. We conclude this section with the following theorem.

Theorem 10. *There exists a local search PTAS for the MSC problem, where the input polygon does not contain holes.*

Acknowledgments

We would like to thank the anonymous referees, Abhiruk Lahiri, and Kasturi Varadarajan for their valuable comments, which have helped us improve the presentation of the paper.

References

[1] V. Arya, N. Garg, R. Khandekar, A. Meyerson, K. Munagala, and V. Pandit. Local search heuristics for k-median and facility location problems. *SIAM J. Comput.*, 33(3):544–562, 2004.

[2] R. Aschner, M. J. Katz, G. Morgenstern, and Y. Yuditsky. Approximation schemes for covering and packing. In *WALCOM 2013.*, pages 89–100.

[3] P. Ashok, A. Basu Roy, and S. Govindarajan. Local search strikes again: PTAS for variants of geometric covering and packing. In *COCOON*, 2017.

[4] S. Bandyapadhyay and K. R. Varadarajan. On variants of k-means clustering. In *SoCG 2016,*, pages 14:1–14:15, 2016.

[5] V. V. S. P. Bhattiprolu and S. Har-Peled. Separating a voronoi diagram via local search. In *SoCG 2016,*, pages 18:1–18:16, 2016.

[6] T. C. Biedl, T. M. Chan, S. Lee, S. Mehrabi, F. Montecchiani, and H. Vosoughpour. On guarding orthogonal polygons with sliding cameras. *CoRR*, abs/1604.07099, 2016.

[7] T. M. Chan and S. Har-Peled. Approximation algorithms for maximum independent set of pseudo-disks. *DCG*, 48(2):373–392, 2012.

[8] V. Cohen-Addad and C. Mathieu. Effectiveness of local search for geometric optimization. In *SoCG 2015,*, pages 329–343, 2015.

[9] V. Cohen-Addad, P. N. Klein, and C. Mathieu. Local search yields approximation schemes for k-means and k-median in euclidean and minor-free metrics. In *FOCS 2016*, pages 353–364, 2016.

[10] M. de Berg, S. Durocher, and S. Mehrabi. Guarding monotone art galleries with sliding cameras in linear time. In *COCOA 2014*, pages 113–125, 2014.

[11] S. Durocher and S. Mehrabi. Guarding orthogonal art galleries using sliding cameras: Algorithmic and hardness results. In *MFCS*, pages 314–324, 2013.

[12] S. Friedrichs, M. Hemmer, J. King, and C. Schmidt. The continuous 1.5d terrain guarding problem: Discretization, optimal solutions, and PTAS. *JoCG*, 7(1):256–284, 2016.

[13] Z. Friggstad, M. Rezapour, and M. R. Salavatipour. Local search yields a PTAS for k-means in doubling metrics. In *FOCS 2016*, pages 365–374.

[14] S. Govindarajan, R. Raman, S. Ray, and A. Basu Roy. Packing and covering with non-piercing regions. In *ESA 2016*, pages 47:1–47:17, 2016.

[15] A. Gupta and K. Tangwongsan. Simpler analyses of local search algorithms for facility location. *CoRR*, abs/0809.2554, 2008.

[16] T. Kanungo, D. M. Mount, N. S. Netanyahu, C. D. Piatko, R. Silverman, and A. Y. Wu. A local search approximation algorithm for k-means clustering. *Comput. Geom.*, 28(2-3):89–112, 2004.

[17] M. J. Katz and G. Morgenstern. Guarding orthogonal art galleries with sliding cameras. *Int. J. Comput. Geometry Appl.*, 21(2):241–250, 2011.

[18] D. G. Kirkpatrick. An o(lg lg opt)-approximation algorithm for multi-guarding galleries. *DCG*, 53(2):327–343, 2015.

[19] E. Krohn, M. Gibson, G. Kanade, and K. Varadarajan. Guarding terrains via local search. *Journal of Computational Geometry*, 5(1):168–178, 2014.

[20] N. H. Mustafa and S. Ray. Improved results on geometric hitting set problems. *Discrete & Computational Geometry*, 44(4):883–895, 2010.

[21] J. O'Rourke. *Art Gallery Theorems and Algorithms.* Oxford University Press, Inc., New York, NY, USA, 1987. ISBN 0-19-503965-3.

[22] J. O'Rourke. Computational geometry column 52. *SIGACT News*, 43(1): 82–85, Mar. 2012. ISSN 0163-5700.

Parameterized Complexity of Geometric Covering Problems Having Conflicts [*]

Aritra Banik[1], Fahad Panolan[2], Venkatesh Raman[3], Vibha Sahlot[1], and Saket Saurabh[3]

[1] Indian Institute of Technology, Jodhpur, India.
{aritrabanik|sahlotvibha}@gmail.com
[2] Department of Informatics, University of Bergen, Norway.
fahad.panolan@ii.uib.no
[3] The Institute of Mathematical Sciences, HBNI, Chennai, India.
{vraman|saket}@imsc.res.in

Abstract. The input for the GEOMETRIC COVERAGE problem consists of a pair $\Sigma = (P, \mathcal{R})$, where P is a set of points in $\mathbb{R}^d$ and $\mathcal{R}$ is a set of subsets of P defined by the intersection of P with some geometric objects in $\mathbb{R}^d$. These coverage problems form special instances of the SET COVER problem which is notoriously hard in several paradigms including approximation and parameterized complexity. Motivated by what are called *choice problems* in geometry, we consider a variation of the GEOMETRIC COVERAGE problem where there are conflicts on the covering objects that precludes some objects from being part of the solution if some others are in the solution.

As our first contribution, we propose two natural models in which the conflict relations are given: (a) by a graph on the covering objects, and (b) by a representable matroid on the covering objects. We consider the parameterized complexity of the problem based on the structure of the conflict relation. Our main result is that as long as the conflict graph has bounded arboricity (that includes all the families of intersection graphs of low density objects in low dimensional Euclidean space), there is a parameterized reduction to the problem without conflicts on the covering objects. This is achieved through a randomization-derandomization trick. As a consequence, we have the following results when the conflict graph has bounded arboricity.

- If the GEOMETRIC COVERAGE problem is fixed parameter tractable (FPT), then so is the conflict free version.
- If the GEOMETRIC COVERAGE problem admits a factor α-approximation, then the conflict free version admits a factor α-approximation algorithm running in FPT time.

As a corollary to our main result we get a plethora of approximation algorithms running in FPT time. Our other results include an FPT algorithm and a W[1]-hardness proof for the conflict-free version of COVERING POINTS BY INTERVALS. The FPT algorithm is for the case when the conflicts are given by a representable matroid, and the W[1]-hardness result

---

[*] Supported by Parameterized Approximation, ERC Starting Grant 306992, and Rigorous Theory of Preprocessing, ERC Advanced Investigator Grant 267959.

F. Ellen et al. (Eds.): WADS 2017, LNCS 10389, pp. 61–72, 2017.
DOI: 10.1007/978-3-319-62127-2_6

is for all the families of conflict graphs for which the INDEPENDENT SET problem is W[1]-hard.

1   Introduction, Motivation, Model and Our Results

There are many real life geometric covering problems, for which there exist additional constrains that need to be enforced. In this paper, we attempt to address these problems and hope that this will initiate a new line of research directed at bridging the gap between theory and practice.

To define our model of covering with conflicts, we start by defining the classic covering problem. The input to a covering problem consists of a universe U of size n, a family $\mathcal{F}$ of size m of subsets of U and a positive integer k. Our objective is to check whether there exists a subfamily $\mathcal{F}' \subseteq \mathcal{F}$ of size at most k satisfying some desired properties. If $\mathcal{F}'$ is required to contain all the elements of U, then it corresponds to the classical SET COVER problem and $\mathcal{F}'$ is called a *set cover*. The SET COVER problem is part of Karp's 21 NP-complete problems [11].

We begin the development with a conflict free problem already studied, CONFLICT FREE INTERVAL COVERING, introduced in [1,2,3]. Let P be a set of points on the x-axis, and let $\mathcal{I} = \{I_1, \dots, I_m\}$ be a set of intervals on the x-axis. Furthermore, let $\mathcal{C} = \{C_1, C_2, \dots, C_\ell\}$ denote a set of color classes, where each color class C_i consists of a pair of intervals from $\mathcal{I}$. Moreover, for any pair of integers i,j ($1 \leq i < j \leq \ell$), $C_i \cap C_j = \emptyset$. We term $\mathcal{C}$ a *matching family*. For a set of intervals $Q \subseteq \mathcal{I}$, Q is *conflict free* if Q contains at most one interval from each color class, i.e. $\forall_{1 \leq i \leq \ell} |Q \cap C_i| \leq 1$. Finally, for an interval $I = [a, b]$ and a point c on x-axis, we say I *covers* p if and only if $a \leq c \leq b$. Now we are ready to define the problem formally.

RAINBOW COVERING
Input: A set of points P on the x-axis, a set of intervals $\mathcal{I} = \{I_1, \dots, I_m\}$ on the x-axis and a matching family $\mathcal{C} = \{C_1, C_2, \dots, C_\ell\}$.
Question: Does there exist a conflict free subset Q of intervals which covers all the points in P?

Our first goal is to define *a model* in which we can express much more generalized version of conflicts beyond the matching family of conflict graphs.

To define our model we revisit SET COVER, as the model is best defined in the most general setting. Recall that the input to a SET COVER consists of a universe U of size n, a family $\mathcal{F}$ of subsets of U of size m. A natural way to model conflict is by using graphs. Formally stated, we have a graph $CG_{\mathcal{F}}$, on the vertex set $\mathcal{F}$ and there is an edge between two sets $F_i, F_j \in \mathcal{F}$ if F_i and F_j are in conflict. We call $CG_{\mathcal{F}}$ a *conflict graph*. Observe that in the RAINBOW COVERING problem, the family $\mathcal{C}$ would corresponds to $CG_{\mathcal{C}}$ with degree at most one. That is, edges of $CG_{\mathcal{C}}$ form a matching. And the question of finding a conflict free subset Q of intervals covering all the points in P becomes a problem of finding a set Q of intervals that covers all the points in P and $CG_{\mathcal{C}}[Q]$ is an independent set. The set cover $\mathcal{F}'$ such that $CG_{\mathcal{F}}[\mathcal{F}']$ is an independent set will be called *conflict free set cover*.

Our Contributions. In this paper we study the following problems in "geo-metric settings" in the realm of Parameterized Complexity. For more details about parameterized complexity we refer to monographs [4].

GRAPHICAL CONFLICT FREE SET COVER (GRAPHICAL CF-SC)
Input: A universe U of size n, a family $\mathcal{F}$ of size m of subsets of U, a conflict graph $CG_{\mathcal{F}}$ and a positive integer k.
Parameter: k
Question: Does there exist a set cover $\mathcal{F}' \subseteq \mathcal{F}$ of size at most k such that $CG_{\mathcal{F}}[\mathcal{F}']$ is an independent set?

Let $(\mathcal{A}, \mathcal{B})$-SET COVER denote a restriction of SET COVER, where every instance $(U, \mathcal{F}, k)$ of SET COVER satisfies the property that $U \subseteq \mathcal{A}$ and $\mathcal{F} \subseteq \mathcal{B}$. For example in this setting, COVERING POINTS BY INTERVALS corresponds to $(\mathcal{A}, \mathcal{B})$-SET COVER where $\mathcal{A}$ is the set of points on x-axis and $\mathcal{B}$ is the set of intervals on x-axis. Given $(\mathcal{A}, \mathcal{B})$-SET COVER, the corresponding GRAPHICAL CF-SC corresponds to $(\mathcal{A}, \mathcal{B})$-GRAPHICAL CF-SC.

Observe that GRAPHICAL CF-SC becomes SET COVER if $CG_{\mathcal{F}}$ is an independent set. As the general SET COVER is hard in the parameterized framework, to design an FPT algorithm for GRAPHICAL CF-SC, it is important that the base SET COVER problem is FPT. This restricts us to $(\mathcal{A}, \mathcal{B})$-SET COVER which are either FPT or polynomial time solvable. If we are seeking FPT approximation algorithms then we can also restrict ourselves to $(\mathcal{A}, \mathcal{B})$-SET COVER which has either polynomial time approximation scheme (PTAS), constant factor approximation algorithm or FPT approximation algorithms, even if the problem is not in FPT. For example $(\mathcal{A}, \mathcal{B})$-SET COVER, where $\mathcal{A}$ is set of points in $\mathbb{R}^2$ and $\mathcal{B}$ is a set of unit discs in $\mathbb{R}^2$ is known to be W[1] hard [14] but admits a PTAS [10]. We will call $(\mathcal{A}, \mathcal{B})$-SET COVER *tractable* if it admits one of the following: a polynomial time algorithm, an FPT algorithm, an (E)PTAS, a constant factor approximation algorithm, an FPT approximation algorithm.

The next natural question is if we restrict ourselves to tractable $(\mathcal{A}, \mathcal{B})$-SET COVER, can an arbitrary conflict graph $CG_{\mathcal{F}}$ yield tractable algorithms for the conflict-free versions of $(\mathcal{A}, \mathcal{B})$-SET COVER? To formalize this question, let $\mathcal{G}$ denote a family of graphs. Then, the question is for which family of graphs $\mathcal{G}$, does $(\mathcal{A}, \mathcal{B})$-GRAPHICAL CF-SC admit an FPT algorithm or an FPT approximation algorithm when $CG_{\mathcal{F}}$ belongs to $\mathcal{G}$. For example, if $\mathcal{G}$ is the family of *cliques*, then even GRAPHICAL CF-SC trivially becomes polynomial time solvable when $CG_{\mathcal{F}}$ belongs to this family of cliques.

A problem that will be central to our study is the following. Let $\mathscr{P}$ and $\mathscr{I}$ denote a set of points and a set of intervals on the x-axis, respectively.

$(\mathscr{P}, \mathscr{I})$-GRAPHICAL CF-SC   **Parameter:** k
Input: A set of points $P \subseteq \mathscr{P}$, a set of intervals $\mathcal{I} = \{I_1, \ldots, I_m\} \subseteq \mathscr{I}$, a conflict graph $CG_{\mathcal{I}}$ and a positive integer k.
Question: Does there exist a conflict free set cover of size at most k?

In $(\mathscr{P}, \mathscr{I})$-GRAPHICAL CF-SC, when $CG_{\mathcal{I}}$ belongs to the family of matchings then the problem becomes PARAMETERIZED RAINBOW COVERING. This problem was studied in [1] and shown to be NP-complete. In fact, even if we do not care about the size of the conflict free set cover we seek, *just the decision version of a conflict free set cover set is the same as* RAINBOW COVERING, which is known to be NP-complete. Thus, seeking a conflict free set cover can transform a problem from being tractable to intractable.

In order to restrict the family of graphs to which a *conflict graph* belongs, we need to define the notion of *arboricity*. The arboricity of an undirected graph is the minimum number of forests into which its edges can be partitioned. A graph G is said to have *arboricity d* if the edges of G can be partitioned into at most d forests. Let $\mathcal{G}_d$ denote the family of graphs of arboricity d. This family includes the family of intersection graphs of low density objects in low dimensional Euclidean space as explained in [8,9]. Specifically, this includes planar graphs, graphs excluding a fixed graph as a minor, graphs of bounded expansion, and graphs of bounded degeneracy. Har-Peled and Quanrud [8,9] showed that low-density geometric objects form a subclass of the class of graphs that have polynomial expansion, which in turn, is contained in the class of graphs of bounded arboricity. Thus, our restriction of the family of conflict graphs to a family of graphs of bounded arboricity covers a large class of low-density geometric objects.

Theorem 1. *Let $(\mathcal{A}, \mathcal{B})$-SET COVER be tractable and let $\mathcal{G}_d$ be the family of graphs of arboricity d. Then, the corresponding $(\mathcal{A}, \mathcal{B})$-GRAPHICAL CF-SC is also tractable if $CG_{\mathcal{F}}$ belongs to $\mathcal{G}_d$. In particular we obtain following results when $CG_{\mathcal{F}}$ belongs to $\mathcal{G}_d$:*

- *If $(\mathcal{A}, \mathcal{B})$-SET COVER admits an FPT algorithm with running time $\tau(k) \cdot n^{\mathcal{O}(1)}$, then $(\mathcal{A}, \mathcal{B})$-GRAPHICAL CF-SC admits an FPT algorithm with running time $2^{\mathcal{O}(dk)} \cdot \tau(k) \cdot n^{\mathcal{O}(1)}$.*
- *If $(\mathcal{A}, \mathcal{B})$-SET COVER admits a factor α-approximation running in time $n^{\mathcal{O}(1)}$ then $(\mathcal{A}, \mathcal{B})$-GRAPHICAL CF-SC admits a factor α-FPT-approximation algorithm running in time $2^{\mathcal{O}(dk)} \cdot n^{\mathcal{O}(1)}$.*

The proof of Theorem 1 is essentially a black-box reduction to the non-conflict version of the problem. Thus, Theorem 1 covers a number of conflict-free version of many fundamental geometric coverage problems as illustrated in Table 1. In light of Theorem 1, it is natural to ask whether or not, these problems admit polynomial time *approximation* algorithms. Unfortunately, we cannot expect these problems to admit even a factor $o(n)$-approximation algorithm. This is because for most of these problems even deciding whether there exists a conflict free solution, *with no restriction on the size of the solution*, is NP-complete (for example RAINBOW COVERING is NP-complete [1]). Thus, having an $o(n)$-approximation algorithm would imply a polynomial time algorithm for the decision version of the problem, which we do not expect unless P=NP. Hence, the best we can expect for the $(\mathcal{A}, \mathcal{B})$-GRAPHICAL CF-SC problems is an FPT-approximation algorithm, as for many of them we can neither have an FPT algorithm, nor a polynomial time approximation algorithm.

We complement our algorithmic findings by a hardness reduction. Let $\mathcal{G}$ denote a family of graphs. Let $\mathcal{G}$-INDEPENDENT SET be the problem where the

$(\mathbb{R}^2, \mathcal{A})$-SC	Complexity of $(\mathbb{R}^2, \mathcal{A})$-SC	Complexity of $(\mathbb{R}^2, \mathcal{A})$-GRAPHICAL CF-SC
Disks/pseudo-disks	PTAS	α-FPT approx., $\forall \alpha > 1$
Fat triangles of same size	$\mathcal{O}(1)$	$\mathcal{O}(1)$-FPT approx.
Fat objects in $\mathbb{R}^2$	$\mathcal{O}(\log^* \mathsf{OPT})$	$\mathcal{O}(\log^* \mathsf{OPT})$-FPT approx.
$\mathcal{O}(1)$ density objects in $\mathbb{R}^2$	PTAS	α-FPT approx., $\forall \alpha > 1$
Objects with polylog density	QPTAS	$2^{\mathcal{O}(k)} n^{\mathcal{O}(\log^* n)}$ time approx., $\forall \alpha > 1$
Objects with density $\mathcal{O}(1)$ in $\mathbb{R}^d$	PTAS	α-FPT approx., $\forall \alpha > 1$
$(\mathcal{A}, \mathcal{B})$-SET COVER where every instance $(U, \mathcal{F})$ has VC dimension d	$\mathcal{O}(d \log(d\mathsf{OPT}))$	$\mathcal{O}(d \log(d\mathsf{OPT}))$-FPT approx.
POINT GUARD ART GALLERY	$\mathcal{O}(\log \mathsf{OPT})$	$\mathcal{O}(\log \mathsf{OPT})$-FPT approx.
TERRAIN GUARDING	PTAS	α-FPT approx., $\forall \alpha > 1$
$(\mathcal{P}, \mathcal{I})$-SET COVER	Polynomial Time	$2^{\mathcal{O}(dk)} \cdot n^{\mathcal{O}(1)}$-FPT algorithm

Table 1. Corollaries of Theorem 1. Here $(\mathbb{R}^2, \mathcal{A})$-SET COVER ($(\mathbb{R}^2, \mathcal{A})$-SC) is a geometric set cover problem where $\mathbb{R}^2$ is a set of points in the plane and the covering objects are specified in the first column. The conflict graph for all the problems is $\mathcal{G}_d$, family of graphs of arboricity d, for some constant d. For the definitions of density and fatness we refer to [8]. The entries in the second column give the approximation ratio of the $(\mathbb{R}^2, \mathcal{A})$-SC problem based on Theorem 1.

input is a graph $G \in \mathcal{G}$ and a positive integer k, and the objective is to decide whether there is a set S of size at least k such that $G[S]$ is an independent set.

Theorem 2. *Let $\mathcal{G}$ denote a family of graphs such that $\mathcal{G}$-INDEPENDENT SET is W[1]-hard. If $CG_\mathcal{I}$ belongs to $\mathcal{G}$, then $(\mathcal{P}, \mathcal{I})$-GRAPHICAL CF-SC does not admit an FPT algorithm, unless FPT =W[1].*

The proof of Theorem 2 is a Turing reduction based on (n, k)-*perfect hash families* [16] that takes time $2^{\mathcal{O}(k)} \cdot n^{\mathcal{O}(1)}$. In fact, for any fixed $\mathcal{A}$ and $\mathcal{B}$, one should be able to follow this proof and show W[1]-hardness for $(\mathcal{A}, \mathcal{B})$-GRAPHICAL CF-SC, where $CG_\mathcal{F}$ belongs to a graph family $\mathcal{G}$ for which $\mathcal{G}$-INDEPENDENT SET is W[1]-hard. Due to paucity of space the proof of Theorem 2 is deferred to the full version of the paper.

Theorem 1 captures those families of conflict graphs that are "everywhere sparse". However, the $(\mathcal{A}, \mathcal{B})$-GRAPHICAL CF-SC problem is also tractable if the conflict graphs belong to the family of cliques. When the conflict graph belongs to a "dense family" of graphs, we design a general theorem using matroid machinery.

Let $(U, \mathcal{F}, k)$ be an instance of SET COVER. In the matroidal model of representing conflicts, we are given a matroid $M = (E, \mathcal{J})$, where the ground set $E = \mathcal{F}$, and $\mathcal{J}$ is a family of subsets of $\mathcal{F}$ satisfying all the three properties of a matroid. In this paper we assume that $M = (E, \mathcal{J})$ is a *linear or representable matroid*, and the corresponding linear representation is given as part of the input. In the RAINBOW COVERING problem, let $\mathcal{Q}$ denote the family of conflict free subsets of intervals in $\mathcal{I}$. One can define a *partition matroid* on $\mathcal{F}$ such that $\mathcal{J} = \mathcal{Q}$. Thus, the question of finding a conflict free subset of intervals covering all the points in P becomes a problem of finding an independent set in $\mathcal{J}$ that covers all the points in P. The MATROIDAL CONFLICT FREE SET COVER problem (MATROIDAL CF-SC, in short) is defined similarly to GRAPHICAL CF-SC. In particular, the input consists of a linear matroid $M = (\mathcal{F}, \mathcal{J})$ over the ground set $\mathcal{F}$ such that the set cover $\mathcal{F}' \in \mathcal{J}$.

Theorem 3. $(\mathscr{P}, \mathscr{I})$-MATROIDAL CF-SC *is FPT for all representable matroids* $M = (\mathcal{I}, \mathcal{J})$ *defined over* $\mathcal{I}$. *In fact, given a linear representation, the algorithm runs in time* $2^{\omega k} \cdot (n + m)^{\mathcal{O}(1)}$. *Here,* ω *is the exponent in the running time of matrix multiplication.*

A graph is called a *cluster graph*, if all its connected components are cliques. Since cluster graphs can be captured by partition matroids, Theorem 3 implies that $(\mathscr{P}, \mathscr{I})$-MATROIDAL CF-SC is FPT if $CG_{\mathcal{F}}$ is a cluster graph.

Notations. For $t \in \mathbb{N}$, we use $[t]$ as a shorthand for $\{1, 2, \ldots, t\}$. A family of sets $\mathcal{A}$ is called a p-family, if the cardinality of all the sets in $\mathcal{A}$ is p. Given two families of sets $\mathcal{A}$ and $\mathcal{B}$, we define $\mathcal{A} \bullet \mathcal{B} = \{X \cup Y \mid X \in \mathcal{A} \text{ and } Y \in \mathcal{B} \text{ and } X \cap Y = \emptyset\}$. Given a graph G, $V(G)$ and $E(G)$ denote its vertex-set and edge-set, respectively. We borrow notations from the book of Diestel [5] for graph-related notations.

2 FPT Algorithms

In this section we prove Theorems 1 and Theorem 3. The Proof of Theorem 1 is based on a randomization scheme while the proof of Theorem 3 uses the idea of efficient computation of representative families [6].

2.1 FPT Algorithms for Graphical CF-SC

Our algorithm for Theorem 1 is essentially a randomized reduction from $(\mathcal{A}, \mathcal{B})$-GRAPHICAL CF-SC to $(\mathcal{A}, \mathcal{B})$-SET COVER, when the conflict graph has bounded arboricity. Towards this, we start with a forest decomposition of graphs of bounded arboricity and then apply a randomized process to obtain an instance of $(\mathcal{A}, \mathcal{B})$-SET COVER. However, to design a deterministic algorithm we use the construction of universal sets. For this, we will exploit the following definition and theorem.

Definition 1 ([16]). *An* (n, t)-*universal set* $\mathscr{F}$ *is a set of functions from* $\{1, \ldots, n\}$ *to* $\{0, 1\}$, *such that for every subset* $S \subseteq \{1, \ldots, n\}$, $|S| = t$, *the set* $\mathscr{F}|_S = \{f|_S \mid f \in \mathscr{F}\}$ *is equal to the set* 2^S *of all the functions from* S *to* $\{0, 1\}$.

Theorem 4 ([16]). *There is a deterministic algorithm with* $\mathcal{O}(2^t t^{\mathcal{O}(\log t)} n \log n)$ *run time that constructs an* (n, t)-*universal set* $\mathscr{F}$ *such that* $|\mathscr{F}| = 2^t t^{\mathcal{O}(\log t)} \log n$.

Now we are ready to give the proof of Theorem 1[4]

Proof (Proof of Theorem 1). Let $(U, \mathcal{F}, CG_{\mathcal{F}}, k)$ be an instance of $(\mathcal{A}, \mathcal{B})$-GRAPHICAL CF-SC, where $CG_{\mathcal{F}}$ belongs to $\mathcal{G}_d$. Our algorithm has the following phases.

---

[4] The idea used in the proof of Theorem 1 is inspired by a proof used in [13].

Decomposing $CG_{\mathcal{F}}$ into Forests. We apply the known polynomial time algorithm [7] to decompose the graph $CG_{\mathcal{F}}$ into $T_1, \ldots, T_d$ where T_i is a forest in $CG_{\mathcal{F}}$ and $\bigcup_{i=1}^{d} E(T_i) = E(CG_{\mathcal{F}})$. Let v_{root} be a special vertex such that v_{root} does not belong to $V(CG_{\mathcal{F}}) = \mathcal{F}$. Now for every T_i, and for every connected component of T_i, we pick an arbitrary vertex and connect it to v_{root}. Now if we look at the tree induced on $V(T_i) \cup \{v_{\mathsf{root}}\}$ then it is connected and we will denote this tree by T_i'. Furthermore, we will treat each T_i' as a tree rooted at v_{root}. This automatically defines *parent-child* relationship among the vertices of T_i'. This completes the partitioning of the edge set of $CG_{\mathcal{F}}$ into forests.

Step 1: Randomized event and probability of success. Independently color the vertices of $CG_{\mathcal{F}}$ into blue and green uniformly at random. That is, we color the vertices of $CG_{\mathcal{F}}$ blue and green with probability $\frac{1}{2}$. Furthermore, we color $\{v_{\mathsf{root}}\}$ to blue. Let $\mathcal{F}'$ be a conflict free set cover of size at most k. We consider the following event to be *good*.

Every vertex in $\mathcal{F}'$ is colored green and every *parent* of every vertex in $\mathcal{F}'$ in every tree T_i' is colored blue.

Let S_{parent} denote the set of parents of every vertex in $\mathcal{F}'$ in every tree T_i'. Since, we have at most d trees and the size of $\mathcal{F}'$ is upper bounded by k we have that $|S_{\mathsf{parent}}| \leq kd$. We say that $\mathcal{F}'$ (S_{parent}) is green (blue) to mean that every vertex in $\mathcal{F}'$ (S_{parent}) is colored green (blue). Thus,

$$\Pr[\text{good event happens}] = \Pr[\mathcal{F}' \text{ is green} \wedge S_{\mathsf{parent}} \text{ is blue}]$$

$$= \Pr[\mathcal{F}' \text{ is green}] \times \Pr[S_{\mathsf{parent}} \text{ is blue}] \geq \frac{1}{2^{k(d+1)}}.$$

The second equality follows from the following fact. The set $\mathcal{F}'$ is an independent set in $CG_{\mathcal{F}}$ and $S_{\mathsf{parent}} \subseteq N_{CG_{\mathcal{F}}}(\mathcal{F}') \cup \{v_{\mathsf{root}}\}$. Thus, these sets are pairwise disjoint and hence the events $\mathcal{F}'$ is colored green and S_{parent} is colored blue are independent.

Step 2: A cleaning process. Let $p = \frac{1}{2^{kd}}$. Now we apply a cleaning procedure so that we get a set Z such that $CG_{\mathcal{F}}[Z]$ is an independent set in $CG_{\mathcal{F}}$ and it contains $\mathcal{F}'$. Let $\mathcal{B}$ denote the set of vertices that have been colored blue. We start by deleting every vertex in $\mathcal{B}$. Now for every edge (f_1, f_2) in $CG_{\mathcal{F}}[V(CG_{\mathcal{F}}) \setminus \mathcal{B}]$, we do as follows. We know that (f_1, f_2) belongs to some tree T_i' and thus either f_1 is a *child* of f_2 or vice-versa. If f_1 is a child then we delete f_1, otherwise we delete f_2. Let the resulting set of vertices be Z. By construction Z is an independent set in $CG_{\mathcal{F}}$. Next we show that $\mathcal{F}' \subseteq Z$ with probability $p/2^k$. Clearly, with probability $\frac{1}{2^k}$ we know that no vertex of $\mathcal{F}'$ is colored blue and thus with probability $\frac{1}{2^k}$ we know that $\mathcal{F}' \subseteq V(CG_{\mathcal{F}}) \setminus \mathcal{B}$. Observe that with probability p, we have that all the parents of $\mathcal{F}'$ in any tree T_i' have been colored blue. Thus, a vertex $x \in V(CG_{\mathcal{F}}) \setminus \mathcal{B}$, colored green, can not belong to $\mathcal{F}'$, if it is a child of some vertex in some tree T_i' after deleting the vertices of $\mathcal{B}$. This is the reason when we delete a vertex from an edge (f_1, f_2), we delete the one which is a child in some tree T_i'. Thus, by deleting a vertex that is a child in an edge (f_1, f_2), we do not delete any vertex from $\mathcal{F}'$. This implies that with probability $\frac{1}{2^{k(d+1)}}$, we have that $\mathcal{F}' \subseteq Z$. This completes the proof.

Solving the problem. Let $\mathscr{Q}$ be a parameterized algorithm for $(\mathcal{A}, \mathcal{B})$-SET COVER running in time $\tau(k) \cdot n^{\mathcal{O}(1)}$. Recall that $(U, \mathcal{F}, CG_{\mathcal{F}}, k)$ is an instance of $(\mathcal{A}, \mathcal{B})$-GRAPHICAL CF-SC. Now to test whether there exists a conflict free set cover $\mathcal{F}'$ of size at most k, we run $\mathscr{Q}$ on (U, Z, k). If the algorithm return Yes, we return the same for $(\mathcal{A}, \mathcal{B})$-GRAPHICAL CF-SC. Else, we repeat the process by randomly finding another Z^* by following Steps 1 and 2 and then running the algorithm $\mathscr{Q}$ on the instance (U, Z^*, k) and returning the answer accordingly. We repeat the process $2^{k(d+1)}$ time. If we fail to detect whether $(U, \mathcal{F}, k, CG_{\mathcal{F}})$ is a Yes instance of $(\mathcal{A}, \mathcal{B})$-GRAPHICAL CF-SC in $2^{k(d+1)}$ rounds, then we return that the given instance is a No instance. Thus, if $(U, \mathcal{F}, k, CG_{\mathcal{F}})$ is No instance of $(\mathcal{A}, \mathcal{B})$-GRAPHICAL CF-SC, then we always return No. However, if $(U, \mathcal{F}, k, CG_{\mathcal{F}})$ is a Yes instance of $(\mathcal{A}, \mathcal{B})$-GRAPHICAL CF-SC then there exists a set $\mathcal{F}'$, that is a conflict free set cover of size at most k. The probability that we will not find a set Z containing $\mathcal{F}'$ in $q = 2^{k(d+1)}$ rounds is upper bounded by $\left(1 - \frac{1}{q}\right)^q \leq \frac{1}{e}$. Thus, the probability that we will find a set Z containing $\mathcal{F}'$ in q rounds is at least $1 - \frac{1}{e} \geq \frac{1}{2}$. Thus, if the given instance is a Yes instance then the algorithm succeeds with probability at least $\frac{1}{2}$. The running time of the algorithm is upper bounded by $\tau(k) \cdot 2^{k(d+1)} \cdot n^{\mathcal{O}(1)}$.

Derandomizing the algorithm. Now to design our deterministic algorithm all we will need to do is to replace the randomized coloring function with a deterministic coloring function that colors the vertices in $\mathcal{F}'$ green and all the vertices in S_{parent} to blue. To design such a coloring function we set $t = k(d+1)$, and use Theorem 4 to construct an (n, t)-universal set $\mathscr{F}$ such that $|\mathscr{F}| = 2^t t^{\mathcal{O}(\log(t))} \log n$. The algorithm to construct $\mathscr{F}$ takes $\mathcal{O}(2^t t^{\mathcal{O}(\log(t))} n \log n)$. Finally, to derandomize our algorithm, rather than randomly coloring vertices with $\{\mathsf{blue}, \mathsf{green}\}$, we go through each function f in the family $\mathscr{F}$ and view the vertices that have assigned 0 as blue and others as green. By the properties of (n, t)-universal set we know that there exists a function f that correctly colors the vertices in $\mathcal{F}'$ with 1 and every vertex in S_{parent} with 0. Thus, the set Z_f we will obtain by applying Step 2 will contain the set $\mathcal{F}'$. After this the correctness of the algorithm follows from the correctness of the algorithm $\mathscr{Q}$. Thus, the running time of the algorithm is upper bounded by $\tau(k) \cdot |\mathscr{F}| \cdot n^{\mathcal{O}(1)} = \tau(k) \cdot 2^{k(d+1)+o(kd)} \cdot n^{\mathcal{O}(1)}$. This completes the proof of the first part.

Let $\mathscr{S}$ be a factor α-approximation algorithm for $(\mathcal{A}, \mathcal{B})$-SET COVER running in time $n^{\mathcal{O}(1)}$. To obtain the desired FPT approximation algorithm with factor α, we do as follows. We only give the deterministic version of the algorithm based on the uses of universal sets. As before, let $(U, \mathcal{F}, CG_{\mathcal{F}}, k)$ be an instance of $(\mathcal{A}, \mathcal{B})$-GRAPHICAL CF-SC, where $CG_{\mathcal{F}}$, belongs to $\mathcal{G}_d$. We again set $t = k(d+1)$, and use Theorem 4 to construct an (n, t)-universal set $\mathscr{F}$ such that $|\mathscr{F}| = 2^t t^{\mathcal{O}(\log(t))} \log n$. The algorithm to construct $\mathscr{F}$ takes $\mathcal{O}(2^t t^{\mathcal{O}(\log(t))} n \log n)$. We go through each function f in the family $\mathscr{F}$ and view the vertices that have been assigned 0 as blue and others as green. If there exists a conflict free set cover $\mathcal{F}'$ of size at most k, then by the properties of (n, t)-universal set we know that there exists a function f that correctly color the vertices in $\mathcal{F}'$ with 1 and every vertex in S_{parent} with 0. Thus, the set Z_f we will obtain by applying Step 2, will contain the set $\mathcal{F}'$. Thus, to design the approximation algorithm, for every $f \in \mathcal{F}$, we first construct Z_f. And for each such Z_f we run $\mathcal{S}$ on (U, Z_f, k). This could either return that there is No solution, or returns a solution $\mathcal{F}'$ which is a factor α-approximation to the instance (U, Z_f, k). If for some $f \in \mathcal{F}$, $\mathcal{S}$ returns $\mathcal{F}'$ of size at most αk when run on (U, Z_f, k) then the algorithm returns $\mathcal{F}'$. In all other cases the algorithm returns that the given instance is a No instance. The correctness of the algorithm follows from the properties of universal sets and the correctness of the algorithm $\mathcal{S}$. The running time of the algorithm is upper bounded by: $|\mathcal{F}| \times$ Running time of $\mathcal{S} = 2^{k(d+1)+o(kd)} \cdot n^{\mathcal{O}(1)}$. This completes the proof. $\qquad\square$

2.2 FPT Algorithm for $(\mathscr{P}, \mathscr{I})$-Matroidal CF-SC

In this section we will design an FPT algorithm proving Theorem 3. Towards that we need to define some basic notions related to representative families and results regarding their fast and efficient computation. For definitions related to matroids and a broad overview of representative families we refer to [4, Chapter 12].

Definition 2 (q-Representative Family [15,4]). *Given a matroid $M = (E, \mathcal{J})$ and a family S of subsets of E, we say that a subfamily $\hat{S} \subseteq S$ is q-representative for S if the following holds: for every set $Y \subseteq E$ of size at most q, if there is a set $X \in S$ disjoint from Y with $X \cup Y \in \mathcal{J}$, then there is a set $\hat{X} \in \hat{S}$ disjoint from Y with $\hat{X} \cup Y \in \mathcal{J}$. If $\hat{S} \subseteq S$ is q-representative for S we write $\hat{S} \subseteq_{rep}^q S$.*

Lemma 1 ([6]). *Let $M = (E, \mathcal{J})$ be a matroid and S be a family of subsets of E. If $S' \subseteq_{rep}^q S$ and $\hat{S} \subseteq_{rep}^q S'$, then $\hat{S} \subseteq_{rep}^q S$.*

Lemma 2 ([12]). *Let $M = (E, \mathcal{J})$ be a linear matroid of rank n and let $S = \{S_1, \ldots, S_t\}$ be a p-family of independent sets. Let A be a $n \times |E|$ matrix representing M over a field $\mathbb{F}$, where $\mathbb{F} = \mathbb{F}_{p^\ell}$ or $\mathbb{F}$ is $\mathbb{Q}$. Then there is a deterministic algorithm computing $\hat{S} \subseteq_{rep}^q S$ of size $np\binom{p+q}{p}$ in $\mathcal{O}\left(\binom{p+q}{p} t p^3 n^2 + t \binom{p+q}{q}^{\omega-1} (pn)^{\omega-1}\right) + (n + |E|)^{\mathcal{O}(1)}$ operations over $\mathbb{F}$.*

Now we are ready to prove Theorem 3. Let $(P, \mathcal{I}, k, M = (\mathcal{I}, \mathcal{J}))$ be an instance of $(\mathscr{P}, \mathscr{I})$-MATROIDAL CF-SC, where P is a set of points on the x-axis, $\mathcal{I} = \{I_1, \ldots, I_m\}$ is a set of intervals on the x-axis and $M = (\mathcal{I}, \mathcal{J})$ is a matroid over the ground set $\mathcal{I}$. The objective is to find a set cover $S \subseteq \mathcal{I}$ of size at most k such that $S \in \mathcal{J}$.

To design our algorithm for $(\mathscr{P}, \mathscr{I})$-MATROIDAL CF-SC, we will use efficient computation of representative families applied on a dynamic programming algorithm. Let $P = \{p_1, \ldots, p_n\}$ denote the set of points sorted from left to right. Next we introduce the notion of family of partial solutions. Let

$$\mathcal{P}^i = \left\{ X \mid X \subseteq \mathcal{I}, X \in \mathcal{J}, |X| \leq k, X \text{ covers } p_1, \ldots, p_i \right\}$$

denote the family of subsets of intervals of size at most k that covers first i points and are independent in the matroid $M = (\mathcal{I}, \mathcal{J})$. Furthermore, for every $j \in [k]$, by $\mathcal{P}^{ij}$, we denote the subset of $\mathcal{P}^i$ containing sets of size *exactly* j. Thus,

$$\mathcal{P}^i = \biguplus_{j=1}^{k} \mathcal{P}^{ij}.$$

In this subsection whenever we talk about independent sets, these are independent sets of the matroid $M = (\mathcal{I}, \mathcal{J})$. Furthermore, we *assume that we are given*, A_M, *the linear representation of* M. Without loss of generality we can assume that A_M is a $n' \times |\mathcal{I}|$ matrix, where $n' \leq |\mathcal{I}|$.

Observe that $(P, \mathcal{I}, k, M = (\mathcal{I}, \mathcal{J}))$ is a Yes instance of $(\mathscr{P}, \mathscr{I})$-MATROIDAL CF-SC if and only if $\mathcal{P}^n$ is non-empty. This implies that $\mathcal{P}^n$ is non-empty if and only if $\widehat{\mathcal{P}}^n \subseteq_{rep}^0 \mathcal{P}^n$ is non-empty. We capture this into the following lemma.

Lemma 3. *Let* $(P, \mathcal{I}, k, M = (\mathcal{I}, \mathcal{J}))$ *be an instance of* $(\mathscr{P}, \mathscr{I})$-MATROIDAL CF-SC. *Then,* $(P, \mathcal{I}, k, M = (\mathcal{I}, \mathcal{J}))$ *is a* Yes *instance of* $(\mathscr{P}, \mathscr{I})$-MATROIDAL CF-SC *if and only if* $\mathcal{P}^n$ *is non-empty if and only if* $\widehat{\mathcal{P}}^n \subseteq_{rep}^0 \mathcal{P}^n$ *is non-empty.*

For an ease of presentation by $\mathcal{P}^0$, we denote the set $\{\emptyset\}$. The next lemma provides an efficient computation of the family $\widehat{\mathcal{P}}^i \subseteq_{rep}^{1\cdots k} \mathcal{P}^i$. In particular, for every $1 \leq i \leq n$, we compute

$$\widehat{\mathcal{P}}^i = \bigcup_{j=1}^{k} \left(\widehat{\mathcal{P}}^{ij} \subseteq_{rep}^{k-j} \mathcal{P}^{ij} \right).$$

Lemma 4. *Let* $(P, \mathcal{I}, k, M = (\mathcal{I}, \mathcal{J}))$ *be an instance of* $(\mathscr{P}, \mathscr{I})$-MATROIDAL CF-SC. *Then for every* $1 \leq i \leq n$, *a collection of families* $\widehat{\mathcal{P}}^i \subseteq_{rep}^{1\cdots k} \mathcal{P}^i$, *of size at most* $2^k \cdot |\mathcal{I}| \cdot k$ *can be found in time* $2^{\omega k} \cdot (n + |\mathcal{I}|)^{\mathcal{O}(1)}$.

Proof. We describe a dynamic programming based algorithm. Let $P = \{p_1, \ldots, p_n\}$ denote the set of points sorted from left to right and $\mathcal{D}$ be a $n + 1$-sized array indexed with $\{0, \ldots, n\}$. The entry $\mathcal{D}[i]$ will store a family $\widehat{\mathcal{P}}^i \subseteq_{rep}^{1\cdots k} \mathcal{P}^i$. We fill the entries in the matrix $\mathcal{D}$ in the increasing order of index. For $i = 0$, $\mathcal{D}[i] = \{\emptyset\}$. Let $i \in \{0, 1, \ldots, n\}$ and assume that we have filled all the entries until the row i (i.e, $\mathcal{D}[i]$ will contain a family $\widehat{\mathcal{P}}^i \subseteq_{rep}^{1\cdots k} \mathcal{P}^i$). For any interval $I \in \mathcal{I}$, let ℓ_I be the lowest index in $[n]$ such that p_{ℓ_I} is covered by I. Let $\mathcal{Z}_{i+1}$ denote the set of intervals $I \in \mathcal{I}$ that covers the point p_{i+1}. Now we compute

$$\mathcal{N}^{i+1} = \bigcup_{I \in \mathcal{Z}_{i+1}} (\mathcal{D}[\ell_I - 1] \bullet \{I\}) \cap \mathcal{J} \tag{1}$$

Notice that in the Equation 1, the union is taken over $I \in \mathcal{Z}_{i+1}$. Since for any $I \in \mathcal{Z}_{i+1}$, I covers p_{i+1}, the value $\ell_I - 1$ is strictly less than $i + 1$ and hence Equation 1 is well defined. Let $\mathcal{N}^{(i+1)j}$ denote the subset of $\mathcal{N}^{i+1}$ containing subsets of size exactly j.

Claim. $\mathcal{N}^{i+1} \subseteq_{rep}^{1\cdots k} \mathcal{P}^{i+1}$.

Proof. Let $S \in \mathcal{P}^{(i+1)j}$ and Y be a set of size $k - j$ (which is essentially an independent set of M) such that $S \cap Y = \emptyset$ and $S \cup Y \in \mathcal{J}$. We will show that there exists a set $\widehat{S} \in \mathcal{N}^{(i+1)}$ such that $\widehat{S} \cap Y = \emptyset$ and $\widehat{S} \cup Y \in \mathcal{J}$. This will imply the desired result.

Since S covers $\{p_1, \ldots, p_{i+1}\}$, there is an interval J in S which covers p_{i+1}. Since S covers $\{p_1, \ldots, p_{i+1}\}$ and J covers p_{i+1}, the set of intervals $S' = S \setminus \{J\}$ covers $\{p_1, \ldots, p_{i+1}\} \setminus \{p_{\ell_J}, \ldots, p_{i+1}\}$ and J covers $\{p_{\ell_J}, \ldots p_{i+1}\}$. Let $Y' = Y \cup \{J\}$. Notice that $S' \cup Y' = S \cup Y \in \mathcal{J}$, $|S'| = j - 1$, $|Y'| = k - j + 1$ and S' covers $\{p_1, \ldots, p_{i+1}\} \setminus \{p_{\ell_J}, \ldots, p_{i+1}\}$. This implies that $S' \in \mathcal{P}^{(\ell_J - 1)(j-1)}$ and by our assumption that $\mathcal{D}[\ell_J - 1]$ contain $\widehat{\mathcal{P}}^{(\ell_J - 1)(j-1)} \subseteq_{rep}^{k-j+1} \mathcal{P}^{(\ell_J - 1)(j-1)}$, we have that there exists $S^* \in \mathcal{D}[\ell_J - 1]$ such that $S^* \cap Y' = \emptyset$ and $S^* \cup Y' \in \mathcal{J}$. By Equation 1, $S^* \cup \{J\}$ in $\mathcal{N}^{i+1}$, because $S^* \cup \{J\} \in \mathcal{J}$. Now we set $\widehat{S} = S^* \cup \{J\}$. Observe that $\widehat{S} \cap Y = \emptyset$ and $\widehat{S} \cup Y \in \mathcal{J}$. This completes the proof of the claim. $\quad\square$

We fill the entry for $\mathcal{D}[i + 1]$ as follows.

$$\mathcal{D}[i + 1] = \bigcup_{j=1}^{k} \left(\widehat{\mathcal{N}}^{(i+1)j} \subseteq_{rep}^{k-j} \mathcal{N}^{(i+1)j} \right) \tag{2}$$

In Equation 2, for every $1 \leq j \leq k$, $\mathcal{N}^{(i+1)j}$ denote the subset of $\mathcal{N}^{(i+1)}$ containing sets of size *exactly* j and $\widehat{\mathcal{N}}^{(i+1)j}$ can be computed using Lemma 2. Lemma 1 and Claim 2.2 implies that $\mathcal{D}[i + 1] \subseteq_{rep}^{1 \cdots k} \mathcal{P}^{i+1}$.

Now we analyse the running time of the algorithm. Consider the time to compute $\mathcal{D}[i + 1]$. We already have computed the family corresponding to $\mathcal{D}[r]$ for all $r \in [i]$. By Lemma 2, for any $r \in [i]$ and $j \in [k]$, the subset of $\mathcal{D}[r]$ containing sets of size exactly j is upper bounded by $|\mathcal{I}| \cdot k \cdot \binom{k}{j}$. Hence, the cardinality of $\mathcal{N}^{(i+1)j}$ is upper bounded by $|\mathcal{I}|^2 \cdot n \cdot k \cdot \binom{k}{j}$. Thus, by Lemma 2, the time to compute $\widehat{\mathcal{N}}^{(i+1)j} \subseteq_{rep}^{k-j} \mathcal{N}^{(i+1)j}$ is bounded by $\left(\binom{k}{j}^2 + \binom{k}{j}^\omega \right) (n + |\mathcal{I}|)^{\mathcal{O}(1)} = \binom{k}{j}^\omega \cdot (n + |\mathcal{I}|)^{\mathcal{O}(1)}$ number of operation over the field in which A_M is given and $|\widehat{\mathcal{N}}^{(i+1)j}| \leq |\mathcal{I}| \cdot k \cdot \binom{k}{j}$. Hence the total running time to compute $\mathcal{D}[i + 1]$ for any $i + 1 \in [n]$ is

$$\sum_{j=1}^{k} \binom{k}{j}^\omega \cdot (n + |\mathcal{I}|)^{\mathcal{O}(1)}) = 2^{\omega k} \cdot (n + |\mathcal{I}|)^{\mathcal{O}(1)}.$$

By Lemma 2, the cardinality of $\mathcal{D}[i + 1]$ is bounded by,

$$|\mathcal{D}[i + 1]| = \sum_{j=1}^{k} |\widehat{\mathcal{N}}^{(i+1)j}| \leq \sum_{j=1}^{k} |\mathcal{I}| \cdot k \cdot \binom{k}{j} = 2^k |\mathcal{I}| \cdot k.$$

This completes the proof. $\quad\square$

Theorem 3 follows from Lemmata 3 and 4. Now we explain an application of Theorem 3. Consider the problem $(\mathscr{P}, \mathscr{I})$-GRAPHICAL CF-SC, where $CG_\mathcal{I}$ is a cluster graph. Let $(P, \mathcal{I}, CG_\mathcal{I}, k)$ be an instance of $(\mathscr{P}, \mathscr{I})$-GRAPHICAL CF-SC.

Let $C_1, \ldots C_t$ be the connected components of $CG_\mathcal{I}$, where each C_i is a clique for all $i \in [t]$. In any solution we are allowed to pick at most one vertex (an interval) from C_i for any $i \in [t]$. This information can be encoded using a partition matroid $M = (\mathcal{I} = V(C_1) \uplus \ldots \uplus V(C_t), \mathcal{J})$ where any subset $\mathcal{I}' \subseteq \mathcal{I}$ is independent in M if and only if $|\mathcal{I}' \cap V(C_i)| \leq 1$ for any $i \in [t]$. Moreover, a linear representation of a partition matroid can be found in polynomial time ([15, Proposition 3.5]). As a result, by applying Theorem 3 and Proposition 3.5 of [15], we get the following corollary.

Corollary 1. $(\mathscr{P}, \mathscr{I})$-GRAPHICAL CF-SC, when $CG_\mathcal{I}$ is a cluster graph, can be solved in time $2^{\omega k} \cdot (n + |\mathcal{I}|)^{\mathcal{O}(1)}$.

References

1. Arkin, E.M., Banik, A., Carmi, P., Citovsky, G., Katz, M.J., Mitchell, J.S.B., Simakov, M.: Choice is hard. In: ISAAC. pp. 318–328 (2015)
2. Arkin, E.M., Banik, A., Carmi, P., Citovsky, G., Katz, M.J., Mitchell, J.S.B., Simakov, M.: Conflict-free covering. In: CCCG (2015)
3. Banik, A., Panolan, F., Raman, V., Sahlot, V.: Fréchet distance between a line and avatar point set. FSTTCS pp. 32:1–32:14 (2016)
4. Cygan, M., Fomin, F.V., Kowalik, L., Lokshtanov, D., Marx, D., Pilipczuk, M., Pilipczuk, M., Saurabh, S.: Parameterized algorithms. Springer (2015)
5. Diestel, R.: Graph Theory, 4th Edition, Graduate texts in mathematics, vol. 173. Springer (2012)
6. Fomin, F.V., Lokshtanov, D., Panolan, F., Saurabh, S.: Efficient computation of representative families with applications in parameterized and exact algorithms. J. ACM 63(4), 29 (2016)
7. Gabow, H.N., Westermann, H.H.: Forests, frames, and games: Algorithms for matroid sums and applications. Algorithmica 7(5&6), 465–497 (1992)
8. Har-Peled, S., Quanrud, K.: Approximation algorithms for low-density graphs. CoRR abs/1501.00721 (2015)
9. Har-Peled, S., Quanrud, K.: Approximation algorithms for polynomial-expansion and low-density graphs. In: ESA. vol. 9294, pp. 717–728. Springer (2015)
10. Hunt III, H.B., Marathe, M.V., Radhakrishnan, V., Ravi, S.S., Rosenkrantz, D.J., Stearns, R.E.: NC-approximation schemes for NP- and PSPACE-hard problems for geometric graphs. J. Algorithms 26(2), 238–274 (1998)
11. Karp, R.M.: Reducibility among combinatorial problems. In: Proceedings of a symposium on the Complexity of Computer Computations. pp. 85–103 (1972)
12. Lokshtanov, D., Misra, P., Panolan, F., Saurabh, S.: Deterministic truncation of linear matroids. In: ICALP, Proceedings, Part I. vol. 9134, pp. 922–934. Springer (2015)
13. Lokshtanov, D., Panolan, F., Saurabh, S., Sharma, R., Zehavi, M.: Covering small independent sets and separators with applications to parameterized algorithms. ArXiv e-prints (May 2017)
14. Marx, D.: Parameterized complexity of independence and domination on geometric graphs. In: IPEC, pp. 154–165. Springer (2006)
15. Marx, D.: A parameterized view on matroid optimization problems. Theor. Comput. Sci. 410(44), 4471–4479 (2009)
16. Naor, M., Schulman, J.L., Srinivasan, A.: Splitters and near-optimal derandomization. In: FOCS. pp. 182–191 (1995)

Obedient Plane Drawings for Disk Intersection Graphs

Bahareh Banyassady[1,*], Michael Hoffmann[2],
Boris Klemz[1], Maarten Löffler[3,**], Tillmann Miltzow[4,***]

[1] Institute of Computer Science, Freie Universität Berlin, Germany
[2] Department of Computer Science, ETH Zürich, Switzerland
[3] Dept. of Computing and Information Sciences, Utrecht University, the Netherlands
[4] Université libre de Bruxelles (ULB), Brussels, Belgium

Abstract. Let $\mathcal{D}$ be a set of disks and G be the intersection graph of $\mathcal{D}$. A drawing of G is *obedient* to $\mathcal{D}$ if every vertex is placed in its corresponding disk. We show that deciding whether a set of unit disks $\mathcal{D}$ has an obedient plane straight-line drawing is $\mathcal{NP}$-hard regardless of whether a combinatorial embedding is prescribed or an arbitrary embedding is allowed. We thereby strengthen a result by Evans *et al.*, who show $\mathcal{NP}$-hardness for disks with arbitrary radii in the arbitrary embedding case. Our result for the arbitrary embedding case holds true even if G is *thinnish*, that is, removing all triangles from G leaves only disjoint paths. This contrasts another result by Evans *et al.* stating that the decision problem can be solved in linear time if $\mathcal{D}$ is a set of unit disks and G is *thin* , that is, (1) the (graph) distance between any two triangles is larger than 48 and (2) removal of all disks within (graph) distance 8 of a triangle leaves only *isolated* paths. A path in a disk intersection graph is isolated if for every pair A, B of disks that are adjacent along the path, the convex hull of $A \cup B$ is intersected only by disks adjacent to A or B. Our reduction can also guarantee the triangle separation property (1). This leaves only a small gap between tractability and $\mathcal{NP}$-hardness, tied to the path isolation property (2) in the neighborhood of triangles. It is therefore natural to study the impact of different restrictions on the structure of triangles. As a positive result, we show that an obedient plane straight-line drawing is always possible if all triangles in G are *light* and the disks are in general position (no three centers collinear). A triangle in a disk intersection graph is light if all its vertices have degree at most three or the common intersection of the three corresponding disks is empty. We also provide an efficient drawing algorithm for that scenario.

1 Introduction

Disk intersection graphs have been long studied in mathematics and computer science due to their wide range of applications in a variety of domains. They can be used to model molecular bonds as well as the structure of the cosmic web, interference in communication networks and social interactions in crowds. Finding planar realizations of

* supported by DFG project MU/3501-2
** supported by the Netherlands Organisation for Scientific Research (NWO) under project no. 639.021.123 and 614.001.504
*** supported by the ERC grant PARAMTIGHT: "Parameterized complexity and the search for tight complexity results", no. 280152

© Springer International Publishing AG 2017
F. Ellen et al. (Eds.): WADS 2017, LNCS 10389, pp. 73–84, 2017.
DOI: 10.1007/978-3-319-62127-2_7

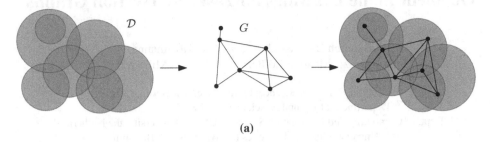

(a)

Fig. 1: A set $\mathcal{D}$ of disks, the induced graph G, and an obedient plane straight-line drawing of G.

disk intersection graphs is an important tool to model sensor networks so as to translate the network connectivity information into virtual coordinates [11]. The planar model can help with topology extraction [10, 15] and enable geometric routing schemes [5, 12].

Rapidly increasing data collection rates call for efficient algorithms to run simulations. But computational problems involving sets of disks are notoriously difficult to grasp: they often lie in a transition zone between tractable and intractable classes of shapes. In this paper, we investigate the computational complexity of one such problem: determining whether a disk intersection graph admits an obedient plane straight-line drawing.

Disk-obedient Drawings. A set $\mathcal{D}$ of disks in the plane induces a graph $G = (V, E)$, called the *disk intersection graph*, which has a vertex for every disk in $\mathcal{D}$ and an edge between two vertices whose (closed) disks intersect. A straight-line *drawing* of G is an injective map $\varphi : V \to \mathbb{R}^2$ so that for every edge $uv \in E$ the open line segment $\varphi(u)\varphi(v)$ is disjoint from $\varphi(V)$. A drawing is *plane* (also called an *embedding*) if the edges do not intersect except at common endpoints. A drawing is *obedient* if every vertex is contained in its corresponding disk. Disk-obedient drawings were introduced by Evans *et al.* [8], who prove that recognizing whether G admits an obedient plane straight-line drawing is $\mathcal{NP}$-hard. This decision problem is called PLANAR DISK OBEDIENCE RECOGNITION. The motivation to study disk-obedience stems from dealing with data uncertainty [7, 13]. The problem is strongly related to *Anchored* Planar Graph Drawing (AGD) (shown to be $\mathcal{NP}$-hard by Angelini *et al.* [2]): Given a planar graph G and an associated unit disk for each circle, produce a planar embedding of G such that each vertex is contained in its disk. Our problem is different from AGD since for us, G itself is defined by the disks. Keszegh *et al.* [9] also study a related problem of placing vertices of a disk intersection graph inside their respective disks such that the resulting drawing is C-oriented; however, the disks in [9] can only touch (their interiors are disjoint), ensuring the drawing is always a plane embedding.

Results and Overview. In this paper we show that several natural restrictions of PLANAR DISK OBEDIENCE RECOGNITION remain $\mathcal{NP}$-hard: even when the disks are unit disks (Section 2), and / or the combinatorial embedding of the graph is given (Section 4), the problem is still hard. In the former case, our result holds true even if removing all triangles leaves only disjoint paths and even if all triangles are far apart. This creates an interesting contrast to results by Evans et al. [8]. Since it is unclear whether the problem is in $\mathcal{NP}$ (see Section 5), our results indicate that the problem is indeed very hard, and it is probably difficult to attack the general problem using a combinatorial approach. On the positive side, we show that the problem can be solved efficiently if the degree of vertices belonging to ply-3 triangles (that is, the three corresponding disks have a common intersection) is bounded by three (Section 3). In this result no assumption concerning the uniformity of the disks' radii is needed. Due to space restrictions, some of the proofs are sketched or omitted in this short paper.

Notation. Throughout this paper disks are closed disks in the plane. Let $\mathcal{D}$ be a set of disks, let $G = (V, E)$ be the intersection graph of $\mathcal{D}$ and let Γ be a straight-line drawing of G. We use capital letters to denote disks and non-capital letters to denote the corresponding vertices, e.g., the vertex corresponding to disk $D \in \mathcal{D}$ is $d \in V$. We use the shorthand uv to denote the edge $\{u, v\} \in E$. Further, uvw refers to the triangle composed of the edges uv, vw and wu. We identify vertices and edges with their geometric representations, e.g., uv also refers to the line segment between the points representing u and v in Γ. We use int(D) to refer to the interior of a disk $D \in \mathcal{D}$. The *ply* of a point $p \in \mathbb{R}^2$ with respect to $\mathcal{D}$ is the cardinality $|\{D \in \mathcal{D} \mid p \in D\}|$. The *ply* of $\mathcal{D}$ is the maximum ply of any point $p \in \mathbb{R}^2$ with respect to $\mathcal{D}$.

Planar Montone 3-Satisfiability. Let $\varphi = (\mathcal{U}, \mathcal{C})$ be a 3-SATISFIABILITY (3SAT) formula where $\mathcal{U}$ denotes the set of variables and $\mathcal{C}$ denotes the set of clauses. We call the formula φ *monotone* if each clause $c \in \mathcal{C}$ is either *positive* or *negative*, that is, all literals of c are positive or all literals of c are negative. Note that this is not the standard notion of monotone Boolean formulas. Formula $\varphi = (\mathcal{U}, \mathcal{C})$ is *planar* if its *variable clause graph* $G_\varphi = (\mathcal{U} \uplus \mathcal{C}, E)$ is planar. The graph G_φ is bipartite and every edge in E is incident to both a *clause* vertex from $\mathcal{C}$ and a *variable* vertex from $\mathcal{U}$. The edge $\{c, u\}$ is contained in E if and only if a literal of variable $u \in \mathcal{U}$ occurs in $c \in \mathcal{C}$. In the decision problem PLANAR MONOTONE 3-SATISFIABILITY we are given a planar and monotone 3SAT formula φ together with a *monotone rectilinear representation* $\mathcal{R}$ of the variable clause graph of φ and we need to decide whether φ is satisfiable. The representation $\mathcal{R}$ is a contact representation on an integer grid, see Figure 2a. In $\mathcal{R}$ the variables are represented by horizontal line segments arranged on a line ℓ. The clauses are represented by E-shapes. All positive clauses are placed above ℓ and all negative clauses are placed below ℓ. PLANAR MONOTONE 3-SATISFIABILITY is $\mathcal{NP}$-complete [4].

2 Thinnish Unit Disk Intersection Graphs

Evans et al. [8] showed that PLANAR DISK OBEDIENCE RECOGNITION is $\mathcal{NP}$-hard. Further, they provide a polynomial time algorithm to recognize disk-obedience graphs for the case that the respective disk intersection graph is thin and unit. A disk intersection graph G is *thin* if (i) the graph distance between any two triangles of G is larger than 48 and (ii) removal of all disks within graph distance 8 of a triangle decomposes the graph into *isolated* paths. A path is *isolated* if for any pair of adjacent disks A and B of the path, the convex hull of $A \cup B$ is intersected only by disks adjacent to A or B.

In this section we strengthen the $\mathcal{NP}$-hardness result by Evans et al. by showing that PLANAR DISK OBEDIENCE RECOGNITION is $\mathcal{NP}$-hard even for *unit* disks. Further, we show that the path-isolation property of thin disk intersection graphs is essential to make the problem tractable. This is implied by the fact that our result holds even for disk intersection graphs that are *thinnish*, that is, removing all disks that belong to a

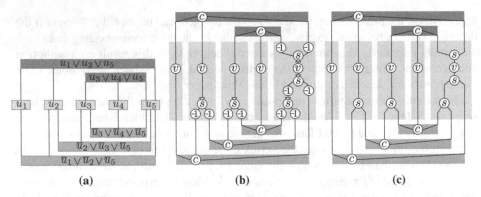

Fig. 2: (a) A monotone rectilinear representation. For the purposes of illustration, the line segments representing variables and clauses are thickened and labeled. (b) Gadget layout mimicking the monotone rectilinear representation from (a) in the proof of Theorem 1. The v-nodes are variable gadgets, the c-nodes are clause gadgets, the s-nodes are splitter gadgets and the (-1)-nodes are inverter gadgets. The input of the splitter gadgets are marked with an arrow. The black polygonal paths represent wires. (c) Gadget layout for the proof of Theorem 3

triangle decomposes the graph into disjoint paths (which are not necessarily isolated). Being thinnish does not impose any distance or other geometric constraint on the set of disks and its intersection graph. Nevertheless, our reduction also works if the distance between any two triangles is lower bounded by some value. In particular, this implies that spacial separation of triangles is not sufficient for tractability.

Theorem 1. PLANAR DISK OBEDIENCE RECOGNITION *is $\mathcal{NP}$-hard even under any combination of the following restrictions: (1) all disks have the same radius (unit disks); (2) the intersection graph of the disks is thinnish; (3) the graph distance between any two triangles is lower bounded by some value that is polynomial in the number of disks.*

Proof. We describe a polynomial-time reduction from PLANAR MONOTONE 3-SAT. Let $\varphi = (\mathcal{U}, \mathcal{C})$ be a planar monotone 3SAT formula where $\mathcal{U}$ is the set of variables and $\mathcal{C}$ is the set of clauses. On an intuitive level our reduction works as follows. We introduce five different types of gadgets. A variable gadget is created for each variable of $\mathcal{U}$. The gadget has two combinatorial states that are used to encode the truth state of its variable. Wire gadgets are used to propagate these states to other gadgets. In particular, we create a clause gadget for each clause $c \in \mathcal{C}$ and use wires to propagate the truth states of the variables occurring in c to the clause gadget of c. The purpose of the clause gadget is to enforces that at least one of the literals of c is satisfied. In order to appropriately connect the variables with the clauses we require two more gadgets. The splitter gadget splits a wire into two wires and the inverter gadget inverts the state transported along a wire. The gadgets are arranged according to the monotone rectilinear representation $\mathcal{R}$ for φ, see Figure 2b. We proceed by describing our gadgets in detail.

Variables, Wires and Inverters. For each variable vertex we create a *variable* gadget as depicted in Figure 3a. Note how on the left side of the obedient plane straight-line drawing of the disk intersection graph the vertices of the lower path belong to the

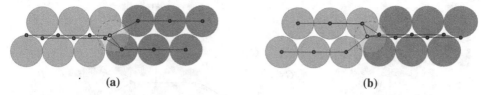

Fig. 3: The variable gadget has two possible states: narrow and wide. It can also be used as an inverter as either the wire to the left or to the right has to be in narrow position.

upper path of disks and vice versa. We say these paths are in a *narrow* position. On the other hand, the paths on the right side are in a *wide* position: the upper path in the obedient plane straight-line drawing of the disk intersection graph belongs to the upper path of disks. By flipping the combinatorial embedding of the subgraph induced by the edges incident to the triangle vertices of the gadget, we can allow the left paths to be in wide position and the right path to be in narrow position; see Figure 3b. However, observe that (i) without flipping the embedding it is not possible to switch between the narrow and the wide positions and (ii) exactly one side has to be in narrow and the other in wide position in order to avoid edge-crossings. We shall use these two states of the variable gadget to encode the truth state of the corresponding variable. The parallel paths of disks to either side of the triangle act as *wires* that propagate the state of the variable gadget. Figure 4a illustrates the information propagation and shows that it is not necessary that the disk centers of the two parallel paths are collinear. Thus, wires are very flexible structures that allow us to transmit the truth states of the variable gadgets to other gadgets in our construction. Observe that our variable gadget can also be used as an *inverter*. If the narrow state is propagated to the triangle from one side, the other side is forced to propagate the wide state and vice versa.

Clauses. For each clause vertex we create a *clause* gadget as depicted in Figure 4b. Each of the three sets of disks $\{R_1, R_2, ...\}$, $\{G_1, G_2, ...\}$ and $\{B_1, B_2, ...\}$ belong to one wire. Note that if a wire is in narrow position, the position of the vertices of the wire is essentially *unique* up to an arbitrarily small wiggle room. The clause gadget is designed such that if all three of its wires are in narrow position the disk intersection graph can not be drawn obediently. The reason for this is that the unique positions of the vertices of the triangle $r_1 g_1 b_1$ in the middle of the gadget enforce a crossing, see Figure 4b. However, if at least one of the incident wires is in wide position and, thus, at least one of u, v or w can be placed freely in its disk, then the gadget can be drawn obediently.

Splitters. The final gadget in our construction is the *splitter* gadget, see Figure 5a. It works as follows. The three sets of disks $\{R_1, R_2, ...\}$, $\{G_1, G_2, ...\}$ and $\{B_1, B_2, ...\}$ belong to three wires r, g, b respectively. We also created a disk O which almost completely overlaps with G_2. Without the disk O the gadget would contain a vertex of degree 3 that is not part of a triangle. The disk O artificially creates a triangle so that the resulting disk intersection graph is thinnish. We refer to the wire b as the *input* of the gadget and to the wires r and g as the *outputs*. If the input is narrow then both outputs have to be wide due to the unique positions of the vertices in the narrow wires. However,

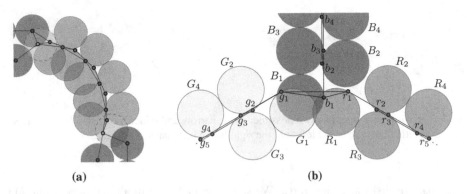

Fig. 4: (a) A curved wire that propagates the narrow position between the two disks with dashed boundary. (b) The clause gadget.

if the input is wide, then any of the outputs can have any combination of states. For instance, both can be narrow. For our reduction we shall always connect the two outputs to inverters so that if the input of the splitter is narrow, then the inverted outputs are also narrow.

Layout. Figure 2b illustrates how we layout and combine our gadgets. We mimic the monotone rectilinear representation $\mathcal{R}$ for φ. We place the variable gadgets and clause gadgets according to $\mathcal{R}$. Consider a variable $u \in \mathcal{U}$. From the variable gadget of u one wire w_t leads to the top; another wire w_b leads to the bottom. If u occurs as a literal only once in a positive clause, the top wire w_t leads directly to the corresponding clause gadget. Otherwise, it leads to the input of a splitter. As stated earlier, we connect the outputs of the splitter to inverters. We split the resulting wires recursively until we have created as many wires as there are literals of u in positive clauses. We call the created wires the *children* of w_t and w_t is their *origin*. Similarly, the wire w_b that leads from the variable gadget of u to the bottom is connected to the negative clauses in which u occurs and the resulting wires are the *children* of w_b and w_b is their *origin*. Further, we refer to u as the *variable* of w_t and w_b. Note that while in some of our gadgets we require very precise coordinates, the required precision does not depend on the input size. Thus, the construction can be carried out in polynomial time.

Correctness. It remains to argue that our reduction is correct. Recall that if the input of a splitter is narrow then the outputs are wide. Since we place inverters at the outputs of each splitter it follows that all children of a narrow wire w are also narrow. Conversely, if a wire connected to a clause gadget is wide then it is a child of a wide wire.

Assume there exists an obedient plane straight-line drawing of the disk intersection graph of the set of disks we created. We create a satisfying truth assignment for φ. In an obedient plane straight-line drawing, for each clause gadget c there is at least one wire w connected to c that is wide; otherwise there is a crossing in the subdrawing of the clause gadget. Consequently, the origin of w is wide as well. If c is positive, we set the variable of w to true and if c is negative we set the variable of w to false. Thus, we have created a truth assignment in which for each clause, there is at least one satisfied literal. Note that it is not possible that we set a variable to both true and false since a wire can

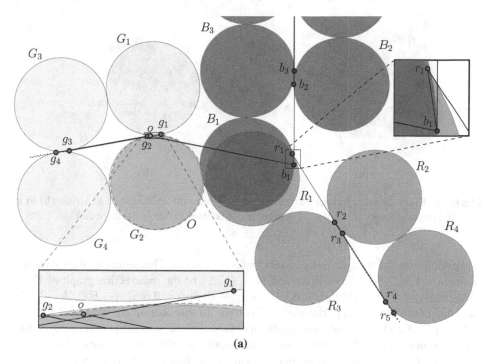

Fig. 5: The splitter gadget.

only be the origin of either positive of negative clauses due to the fact that in a monotone rectilinear representation all negative clauses are below and all positive clauses are above the variable line.

Finally, assume that φ is satisfiable. For each variable u we orient its variable gadget such that the wire that leads to the positive clauses is wide if and only if u is true. We draw the splitter gadgets such that all children of wide wires are wide. Since every clause has a satisfied literal, the corresponding clause gadget is connected to a wide wire and, thus, can be drawn without introducing crossing.

Spacial Separation. Note that the splitter, variable, clause and inverter gadgets each contain one triangle and recall that all gadgets are connected by wires, which do not contain triangles. Thus, it is straightforward to ensure a minimum distance between any two triangles by simply increasing the length of our wires accordingly.   □

3   Disk Intersection Graphs with Light Triangles

In Section 2 we give an $\mathcal{NP}$-hardness proof for a very restricted class of instances of PLANAR DISK OBEDIENCE RECOGNITION. A key ingredient of our reduction from 3SAT is a triangle in the middle of each variable gadget. The two truth states of the gadget are encoded by the combinatorial embedding of the subgraph induced by edges incident to the triangle vertices. It seems natural to study recognition of disk obedience graphs where the degree of vertices in triangles and, thus, the number of combinatorial

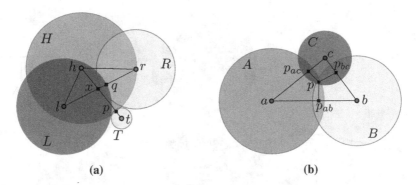

(a)   (b)

Fig. 6: (a) Every crossing between centered edges implies the existence of an arrow. (b) In a centered drawing, every point in a ply-3 triangle abc belongs to one of the three disks.

embeddings for the triangle-induced subgraphs is bounded. This motivates the following definition. Let $\mathcal{D}$ be a set of disks and $G = (V, E)$ be the intersection graph of $\mathcal{D}$. A triangle abc of G is called *light* if $\deg(a), \deg(b), \deg(c) \leq 3$ or if $A \cap B \cap C = \emptyset$. We say that $\mathcal{D}$ is *light* if every triangle of G is light. In this section we show that for any light set $\mathcal{D}$ of disks there always exists an obedient plane straight-line drawing of the intersection graph of $\mathcal{D}$. Note that we do not require the disks in $\mathcal{D}$ to have unit radius.

We begin by introducing some notations and by stating some helpful observations. A set of disks is *connected* if the union of all disks is connected. A set $\mathcal{D}$ of disks is said to be in *general position* if for any connected subset $\mathcal{D}' \subseteq \mathcal{D}$, $|\mathcal{D}'| = 3$ the disk centers of $\mathcal{D}'$ are non-collinear. Let $G = (V, E)$ be the intersection graph of a set of disks $\mathcal{D}$ and Γ be a straight-line drawing of G. A vertex $v \in V$ that is placed at its respective disk's center in Γ is called *centered*. An edge $e \in E$ between two centered vertices in Γ is called *centered*. The drawing Γ is called *centered* if all vertices in V are centered. An *arrow* (h, t, l, r) in a straight-line drawing Γ of a graph $G = (V, E)$ is a sequence of vertices $h, t, l, r \in V$ such that $ht, hl, hr, lr \in E$ and such that ht and lr cross in Γ, see Figure 6a. We refer to h, t, l, r as the arrow's *head, tail, left* and *right* vertex respectively.

Evans et al. [8] show that any set of disks with ply 2 in general position admits an obedient plane straight-line drawing. We observe that with some minor adaptations, their observation furthermore yields an explicit statement regarding the graph structure in non-plane centered drawings. We restate their proof together with our modifications to show that every crossing between centered edges implies the existence of an arrow. Furthermore, we strengthen this statement by showing that if the intersection point x of the two edges is contained in the interior of one of the corresponding disks, then there always is an arrow whose head's disk contains x in its interior.

Lemma 1. *Let $G = (V, E)$ be the disk intersection graph of a set of disks $\mathcal{D}$ in general position. Let Γ be a straight-line drawing of G. For any crossing in Γ between centered edges ab and cd there exists an arrow (h, t, l, r) where $H \cap L \cap R \cap lr \neq \emptyset$ and either (i) $ht = ab$ and $lr = cd$, or (ii) $ht = cd$ and $lr = ab$. Furthermore, if $x = ab \cap cd$ is contained in the interior of at least one of A, B, C, D, then $x \in \text{int}(H)$.*

Next, we observe that if $A \cap B \cap C \neq \emptyset$, then in a centered drawing any point in the triangle abc is contained in one of the disks A, B or C.

Observation 1. *Let $\mathcal{D} = \{A, B, C\}$ be a set of disks whose intersection graph G is the triangle abc and let Γ be a centered drawing of G. If the closed triangle abc contains a point $p \in A \cap B \cap C$ then any point q in the closed triangle abc is contained in $A \cup B \cup C$.*

Our final auxiliary result states that under certain conditions, the number of crossings along one edge in a centered drawing is at most one.

Lemma 2. *Let $\mathcal{D}$ be a light set of disks in general position and let $G = (V, E)$ be the intersection graph of $\mathcal{D}$. Let Γ be a straight-line drawing of G. Let (a, b, c, d) be an arrow in Γ where a, b, c, d are centered and $A \cap C \cap D \cap cd \neq \emptyset$. Then there exists no centered edge other than ab that crosses cd in Γ.*

Proof Sketch. According to the definition of an arrow, we know that $ab, cd, ac, ad \in E$, see Figure 7c (left). Assume that some centered edge $fg \neq ab$ crosses cd. According to Lemma 1, the intersection of fg and cd implies existence of an arrow $\mathcal{A}$ consisting of c, d and two other vertices f and g, at least one of which has to be different from a and b. Since acd is a light triangle in G and since $A \cap C \cap D \neq \emptyset$, the degree of a, c, d is bounded by 3. Due to $\deg(a) \leq 3$ the vertex a can not be connected to any vertex other than b, c, d, which means that a cannot be an endpoint of the edge fg. Note that b could be equal to f or g. We perform a case distinction regarding the containment of the edges bc and bd in E.

First, assume that $bc \in E$ and $bd \in E$, see Figure 7a (left). Because of the degree restrictions for c and d, neither of them can be adjacent to a vertex different from a, b, c, d. Thus, cd can not be the head-tail edge of $\mathcal{A}$ since the head of $\mathcal{A}$ is connected to f and g. If cd is the left-right edge of $\mathcal{A}$, then the head of $\mathcal{A}$ has to be b and without loss of generality f is equal to b. Due to Lemma 1 we know that $B \cap C \cap D \neq \emptyset$ and, thus, since bcd is light the degree of b is bounded by 3. However, this is a contradiction to the fact that b is connected to g.

Now, assume that $bc \in E$ and $bd \notin E$, see Figure 7b. Similar to the last case, the degree restriction for c implies that c can not be adjacent to any vertex other than a, b, d. This implies that c can not be the head of $\mathcal{A}$ and it can be the right or the left vertex of $\mathcal{A}$ only if b is the head of $\mathcal{A}$. In this case cd is the left-right edge of $\mathcal{A}$ and d has to be connected to the head b of $\mathcal{A}$, which contradicts $bd \notin E$. Thus, c can only be the tail of $\mathcal{A}$, and so d is the head. The head d is connected to f and g both of which have to be different from a and b due to $\deg(a) \leq 3$ and due to $bd \notin E$ respectively. This is a contradiction to $\deg(d) \leq 3$.

We sketch the final case. Assume $bc \notin E$ and $bd \notin E$, see Figure 7c (left). Due to degree restrictions we see that cd is the left-right edge of $\mathcal{A}$ and w.l.o.g. $g \neq b$ is the head. Head g can not be located inside triangle acd since this would imply an additional crossing with ab, which contradicts degree restrictions. If g is exterior to acd and f is interior, Observation 1 implies a contradiction to the degree bounds of a, c, d. If g and f are exterior to acd, fg has to cross ad or ac, which implies the existence of another arrow, again contradicting degree restrictions. $\qquad\square$

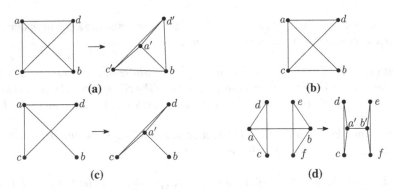

Fig. 7: Removing crossing by moving vertices.

Theorem 2. *Let $\mathcal{D}$ be a light set of disks in general position whose intersection graph is G. Then G has a plane straight-line drawing obedient to $\mathcal{D}$.*

Proof. We describe an iterative approach that transforms a centered drawing of G into a crossing-free drawing obedient to $\mathcal{D}$. In each *step* we change the position of precisely one, two or three vertices to remove one or two crossings. During the entire procedure, each vertex is moved at most once. We maintain the following *invariant*: After and before each step, all crossing edges are centered. We proceed by describing our algorithm. After that we show that the invariant is maintained and, thus, that the algorithm is correct.

Algorithm. Let ab and cd be two centered edges that cross in a point x. By Lemma 1 there exists an arrow consisting of a, b, c, d, w.l.o.g. (a, b, c, d), where $A \cap C \cap D \cap cd \neq \emptyset$, see Figure 7c (left). Note that this implies that the degree of a, c, d is bounded by 3 since $\mathcal{D}$ is light and since $A \cap C \cap D \neq \emptyset$. In order to remove the crossing we move some of a, b, c, d and we use a', b', c', d' to denote the new postions of these vertices.

We distinguish two cases. First assume that $x \in A \cap B \cap C \cap D$ and $x \notin \text{int}(A) \cup \text{int}(B) \cup \text{int}(C) \cup \text{int}(D)$, i.e. x is on the boundary of all four disks. In this case we set $a' = x$ and we move vertices c, d by a distance $\varepsilon \in \mathbb{R}^+$ in the direction given by the vector $\overrightarrow{ba}$, see Figure 7a. Value ε should be chosen small enough such that $c' \in C$, $d' \in D$ and $c'd'$ does not cross an edge ef, unless cd already crosses ef.

Now we consider the case that x is in the interior of at least one of A, B, C, D. By Lemma 1 we can assume without loss of generality that $x \in \text{int}(A)$. In order to remove the crossing we move a in the direction given by the vector $\overrightarrow{ab}$ for distance $|ax| + \varepsilon < |ab|$, see Figure 7c. The value $\varepsilon \in \mathbb{R}^+$ should be chosen small enough such that $a' \in A$ and such that there is no crossing between $a'c$ or $a'd$ and any other edge ef, unless ef also intersects cd. To shorten notation, we refer to this procedure as 'removing the crossing by moving a to x', although technically a' is close to x but $a' \neq x$. It remains to treat the special case that ab has a crossing with some additional edge $ef \neq cd$. In this case, in addition to moving a to x, we move b to $y = ab \cap ef$, see Figure 7d. In the following paragraphs we will see that there exists at most one such additional crossing edge ef and that e, f are distinct from c, d as illustrated.

Correctness. Clearly our invariant holds for the initial centered drawing. In order to show that our invariant is maintained it suffices to show that none of the moved vertices is incident to an edge that has a crossing. In our algorithm we considered two main cases. In the first case we moved vertices a, c and d, which formed a complete graph with vertex b. Note that all these vertices have exactly degree three due to the degree restrictions. Therefore none of them can be adjacent to a vertex different from a, b, c, d. In the second case we moved vertex a, which is adjacent to exactly b, c and d due the degree restriction on a. Therefore, in both cases the moved vertices can only be incident to edges with crossing if there exists some edge $ef \neq ab, cd$ that has a crossing with ab or cd. According to Lemma 2 there is no edge that crosses cd except for ab. Due to Lemma 1, if ef crosses ab, there exists an arrow $\mathcal{A}$ composed of the edge ef and ab. According to Lemma 2, ab can not be the left-right edge of $\mathcal{A}$. The degree of a is bounded by 3 and, thus, a has to be the tail, b has to be the head and e, f have to be the left and right vertex of $\mathcal{A}$.

We perform a case distinction regarding the equivalence of e, f and c, d. First, asumme that $\{e, f\} = \{c, d\}$ and without loss of generality $e = d$ and $f = c$. Then $\mathcal{A}$ consists of a, b, c, d. Next, assume that exactly one of e, f is equal to one of c, d and without loss of generality $e = d$ and $f \neq a, b, c, d$. Then d is adjacent to a, c, b, f, which contradicts the degree bound for d. Finally assume that both e, f are distinct from a, b, c, d. By Lemma 2, neither cd nor ef can have a crossing with any edge other than ab. Hence, the situation looks like the one illustrated in Figure 7d. In this case, in addition to moving a to x, we move b to $y = ab \cap ef$ as described above. Now, $a'b'$ does not have any crossing and, thus, a' is not incident to an edge with a crossing. For symmetric reasons, neither is b'. $\qquad\square$

4 Embedded Unit Disk Intersection Graphs

In Section 2 we proved that PLANAR DISK OBEDIENCE RECOGNITION is $\mathcal{NP}$-hard even for disk intersection graphs that are unit and thinnish. In the reduction from 3SAT used for the proof, the truth state of a variable gadget corresponds to the combinatorial embedding of the respective subgraph. The $\mathcal{NP}$-hardness proof by Evans et al. [8] also establishes a correspondence between truth states and combinatorial embeddings. This raises the question, whether $\mathcal{NP}$-hardness holds if a combinatorial embedding is prescribed. The following theorem answers this question in the affirmative. On a high level, the proof idea is a reduction from PLANAR MONOTONE 3-SATISFIABILITY similar to the one in the proof of Theorem 1. However, for the reduction in Theorem 3 we have to heavily rely on geometric arguments rather than combinatorial embeddings to encode the truth states of variable and wire gadgets.

Theorem 3. PLANAR DISK OBEDIENCE RECOGNITION *is $\mathcal{NP}$-hard even for embedded unit disk intersection graphs.*

5 Remarks and Open Problems

Other Shapes. The notion of obedient drawings naturally extends to other shapes. The reduction strategies used in the hardness proofs in this paper and the paper by Evans et al. [8] seem to apply for several other shapes as well, e.g., for unit squares. This raises the interesting question whether a more general statement can be made that captures all these hardness results at once.

NP-Membership. For many combinatorial problems, showing $\mathcal{NP}$-membership is an easy exercise. For disk-obedience the question turns out to be much more intricate. A naive idea to show NP-membership would be to guess the coordinates of all vertices. However, it is not obvious that there always exists a rational representation of bounded precision. Indeed, there are several geometric problems where this approach is known to fail. In some cases an explicit rational representation may require an exponential number of bits, in others optimal solutions may require irrational coordinates, see [1,3,14]. Many problems initially not known to lie in $\mathcal{NP}$ turned out to be $\exists\mathbb{R}$-complete. The complexity class $\exists\mathbb{R}$ captures all computational problems that are equivalent under polynomial time reductions to the satisfiability of arbitrary polynomial equations and inequalities over the reals, see [6,14]. We leave it as an open problem to determine the relation of disk obedient plane straight-line drawings with respect to $\mathcal{NP}$ and $\exists\mathbb{R}$.

Acknowledgments. This work was initiated during the *Fixed-Parameter Computational Geometry* Workshop at Lorentz Center, April 4–8, 2016. We thank the organizers and all participants for the productive and positive atmosphere.

References

1. Abrahamsen, M., Adamaszek, A., Miltzow, T.: Irrational guards are sometimes needed. arXiv preprint arXiv:1701.05475 (2017)
2. Angelini, P., Lozzo, G.D., Bartolomeo, M.D., Battista, G.D., Hong, S., Patrignani, M., Roselli, V.: Anchored drawings of planar graphs. In: Graph Drawing. Lecture Notes in Computer Science, vol. 8871, pp. 404–415. Springer (2014)
3. Bajaj, C.: The algebraic degree of geometric optimization problems. Discrete & Computational Geometry 3(2), 177–191 (1988)
4. de Berg, M., Khosravi, A.: Optimal binary space partitions for segments in the plane. Int. J. Comput. Geometry Appl. 22(3), 187–206 (2012)
5. Bose, P., Morin, P., Stojmenović, I., Urrutia, J.: Routing with guaranteed delivery in ad hoc wireless networks. Wireless networks 7(6), 609–616 (2001)
6. Cardinal, J.: Computational geometry column 62. SIGACT News 46(4), 69–78 (2015)
7. Evans, W., Kirkpatrick, D., Löffler, M., Staals, F.: Minimizing co-location potential of moving entities. SIAM J. Comput. 45(5), 1870–1893 (2016)
8. Evans, W.S., van Garderen, M., Löffler, M., Polishchuk, V.: Recognizing a DOG is hard, but not when it is thin and unit. In: FUN 2016. pp. 16:1–16:12 (2016)
9. Keszegh, B., Pach, J., Pálvölgyi, D.: Drawing planar graphs of bounded degree with few slopes. SIAM Journal on Discrete Mathematics 27(2), 1171–1183 (2013)
10. Kröller, A., Fekete, S.P., Pfisterer, D., Fischer, S.: Deterministic boundary recognition and topology extraction for large sensor networks. In: SoDA 2006. pp. 1000–1009 (2006)
11. Kuhn, F., Moscibroda, T., Wattenhofer, R.: Unit disk graph approximation. In: Proceedings of the 2004 joint workshop on Foundations of mobile computing. pp. 17–23. ACM (2004)
12. Kuhn, F., Wattenhofer, R., Zhang, Y., Zollinger, A.: Geometric ad-hoc routing: of theory and practice. In: PoDC. pp. 63–72 (2003)
13. Löffler, M.: Data Imprecision in Computational Geometry. Ph.D. thesis, Utrecht University (2009)
14. Matoušek, J.: Intersection graphs of segments and $\exists\mathbb{R}$. arXiv preprint arXiv:1406.2636 (2014)
15. Wang, Y., Gao, J., Mitchell, J.S.B.: Boundary recognition in sensor networks by topological methods. In: MobiCom 2006. pp. 122–133 (2006)

δ-Greedy t-spanner [*]

Gali Bar-On and Paz Carmi

Department of Computer Science,
Ben-Gurion University of the Negev, Israel

Abstract. We introduce a new geometric spanner, δ-*Greedy*, whose construction is based on a generalization of the known *Path-Greedy* and *Gap-Greedy* spanners. The δ-Greedy spanner combines the most desirable properties of geometric spanners both in theory and in practice. More specifically, it has the same theoretical and practical properties as the Path-Greedy spanner: a natural definition, small degree, linear number of edges, low weight, and strong $(1 + \varepsilon)$-spanner for every $\varepsilon > 0$. The δ-Greedy algorithm is an improvement over the Path-Greedy algorithm with respect to the number of shortest path queries and hence with respect to its construction time. We show how to construct such a spanner for a set of n points in the plane in $O(n^2 \log n)$ time.

The δ-Greedy spanner has an additional parameter, δ, which indicates how close it is to the Path-Greedy spanner on the account of the number of shortest path queries. For $\delta = t$ the output spanner is identical to the Path-Greedy spanner, while the number of shortest path queries is, in practice, linear.

Finally, we show that for a set of n points placed independently at random in a unit square the expected construction time of the δ-Greedy algorithm is $O(n \log n)$. Our analysis indicates that the δ-Greedy spanner gives the best results among the known spanners of expected $O(n \log n)$ time for random point sets. Moreover, analysis implies that by setting $\delta = t$, the δ-Greedy algorithm provides a spanner identical to the Path-Greedy spanner in expected $O(n \log n)$ time.

1 Introduction

Given a set P of points in the plane, a Euclidean t-spanner for P is an undirected graph G, where there is a t-spanning path in G between any two points in P. A path between points p and q is a t-spanning path if its length is at most t times the Euclidean distance between p and q (i.e., $t|pq|$).

The most known algorithm for computing t-spanner is probably the *Path-Greedy* spanner. Given a set P of n points in the plane, the Path-Greedy spanner algorithm creates a t-spanner for P as follows. It starts with a graph G having a vertex set P, an empty edge set E and $\binom{n}{2}$ pairs of distinct points sorted in a non-decreasing order of their distances. Then, it adds an edge between p and

---

[*] The research is partially supported by the Lynn and William Frankel Center for Computer Science.

F. Ellen et al. (Eds.): WADS 2017, LNCS 10389, pp. 85–96, 2017.
DOI: 10.1007/978-3-319-62127-2_8

q to the set E if the length of the shortest path between p and q in G is more than $t|pq|$, see Algorithm 1 for more details. It has been shown in [9, 8, 12, 11, 17, 21] that for every set of points, the Path-Greedy spanner has $O(n)$ edges, a bounded degree and total weight $O(wt(MST(P)))$, where $wt(MST(P))$ is the weight of a minimum spanning tree of P. The main weakness of the Path-Greedy algorithm is its time complexity – the naive implementation of the Path-Greedy algorithm runs in near-cubic time. By performing $\binom{n}{2}$ shortest path queries, where each query uses Dijkstra's shortest path algorithm, the time complexity of the entire algorithm reaches $O(n^3 \log n)$, where n is the number of points in P. Therefore, researchers in this field have been trying to improve the Path-Greedy algorithm time complexity. For example, the *Approximate-Greedy* algorithm generates a graph with the same theoretical properties as the Path-Greedy spanner in $O(n \log n)$ time [13, 19]. However, in practice there is no correlation between the expected and the unsatisfactory resulting spanner as shown in [15, 16]. Moreover, the algorithm is complicated and difficult to implement.

Another attempt to build a t-spanner more efficiently is introduced in [14, 15]. This algorithm uses a matrix to store the length of the shortest path between every two points. For each pair of points, it first checks the matrix to see if there is a t-spanning path between these points. In case the entry in the matrix for this pair indicates that there is no t-spanning path, it performs a shortest path query and updates the matrix. The authors in [15] have conjectured that the number of performed shortest path queries is linear. This has been shown to be wrong in [5], as the number of shortest path queries may be quadratic. In addition, Bose et al. [5] have shown how to compute the Path-Greedy spanner in $O(n^2 \log n)$ time. The main idea of their algorithm is to compute a partial shortest path and then extend it when needed. However, the drawback of this algorithm is that it is complex and difficult to implement. In [1], Alewijnse et al. compute the Path-Greedy spanner using linear space in $O(n^2 \log^2 n)$ time by utilizing the Path-Greedy properties with respect to the Well Separated Pair Decomposition (WSPD). In [2], Alewijnse et al. compute a t-spanner in $O(n \log^2 n \log^2 \log n)$ expected time by using bucketing for short edges and by using WSPD for long edges. Their algorithm is based on the assumption that the Path-Greedy spanner consists of mostly short edges.

Additional effort has been put in developing algorithms for computing t-spanner graphs, such as θ-Graph algorithm [10, 18], Sink spanner, Skip-List spanner [3], and WSPD-based spanners [6, 7]. However, none of these algorithms produces a t-spanner as good as the Path-Greedy spanner in all aspects: size, weight and maximum degree, see [15, 16].

Therefore, our goal is to develop a simple and efficient algorithm that achieves both the theoretical and practical properties of the Path-Greedy spanner. In this paper we introduce the δ-Greedy algorithm that constructs such a spanner for a set of n points in the plane in $O(n^2 \log n)$ time. Moreover, we show that for a set of n points placed independently at random in a unit square the expected running time of the δ-Greedy algorithm is $O(n \log n)$.

Algorithm 1 Path-Greedy(P, t)

Input: A set P of points in the plane and a constant $t > 1$
Output: A t-spanner $G(V, E)$ for P
1: sort the $\binom{n}{2}$ pairs of distinct points in non-decreasing order of their distances and store them in list L
2: $E \longleftarrow \emptyset$
3: **for** $(p, q) \in L$ consider pairs in increasing order **do**
4:    $\pi \longleftarrow$ length of the shortest path in G between p and q
5:    **if** $\pi > t|pq|$ **then**
6:       $E := E \cup |pq|$
7: **return** $G = (P, E)$

---

2 δ-Greedy

In this section we describe the δ-Greedy algorithm (Section 2.1) for a given set P of points in the plane, and two real numbers t and δ, such that $1 < \delta \leq t$. Then, in Section 2.2 we prove that the resulting graph is indeed a t-spanner with bounded degree. Throughout this section we assume that $\delta < t$ (for example, $\delta = t^{\frac{4}{5}}$ or $\delta = \frac{1+4t}{5}$), except in Lemma 4, where we consider the case that $\delta = t$.

2.1 Algorithm description

For each point $p \in P$ we maintain a collection of cones C_p with the property that for each point $q \in P$ that lies in C_p there is a t-spanning path between p and q in the current graph. The main idea of the δ-Greedy algorithm is to ensure that two cones of a constant angle with apexes at p and q are added to C_p and to C_q, respectively, each time the algorithm runs a shortest path query between points p and q.

The algorithm starts with a graph G having a vertex set P, an empty edge set, and an initially empty collection of cones C_p for each point $p \in P$. The algorithm considers all pairs of distinct points of P in a non-decreasing order of their distances. If $p \in C_q$ or $q \in C_p$, then there is already a t-spanning path that connects p and q in G, and there is no need to check this pair. Otherwise, let d be the length of the shortest path that connects p and q in G divided by $|pq|$. Let $c_p(\theta, q)$ denote the cone with apex at p of angle θ, such that the ray $\overrightarrow{pq}$ is its bisector. The decision whether to add the edge (p, q) to the edge set of G is made according to the value of d. If $d > \delta$, then we add the edge (p, q) to G, a cone $c_p(2\theta, q)$ to C_p, and a cone $c_q(2\theta, p)$ to C_q, where $\theta = \frac{\pi}{4} - \arcsin(\frac{1}{\sqrt{2} \cdot t})$. If $d \leq \delta$, then we do not add this edge to G, however, we add a cone $c_p(2\theta, q)$ to C_p and a cone $c_q(2\theta, p)$ to C_q, where $\theta = \frac{\pi}{4} - \arcsin(\frac{d}{\sqrt{2} \cdot t})$.

In Algorithm 2, we give the pseudo-code description of the δ-Greedy algorithm. In Figure 1, we illustrate a cone collection C_p of a point p and how it is modified during the three scenarios of the algorithm. The figure contains the point p, its collection C_p colored in gray, and three points v, u, and w, such that $|pv| < |pu| < |pw|$. Point v lies in C_p representing the first case, where the algorithm does not change the spanner and proceeds to the next pair without performing a shortest path query. The algorithm runs a shortest path query between p and u, since $u \notin C_p$ (for the purpose of illustration assume $p \notin C_u$). Figure 1(b) describes the second case of the algorithm, where the length of the shortest path between p and u is at most $\delta|pu|$. In this case the algorithm adds a cone to C_p without updating the spanner. Figure 1(c) describes the third case of the algorithm, where the length of the shortest path between p and w is more than $\delta|pw|$. In this case the algorithm adds a cone to C_p and the edge (p, w) to the spanner.

Algorithm 2 δ-Greedy

Input: A set P of points in the plane and two real numbers t and δ s.t. $1 < \delta \le t$
Output: A t-spanner for P
1: sort the $\binom{n}{2}$ pairs of distinct points in non-decreasing order of their distances (breaking ties arbitrarily) and store them in list L
2: $E \longleftarrow \emptyset$ /* E is the edge set */
3: $C_p \longleftarrow \emptyset \;\; \forall p \in P$ /* C_p is set of cones with apex at p */
4: $G \longleftarrow (P, E)$ /* G is the resulting t-spanner */
5: **for** $(p, q) \in L$ consider pairs in increasing order **do**
6:    **if** $(p \notin C_q)$ and $(q \notin C_p)$ **then**
7:       $d \longleftarrow$ length of the shortest path in G between p and q divided $|pq|$
8:       **if** $d > \delta$ **then**
9:          $E \longleftarrow E \cup \{(p, q)\}$
10:          $d \longleftarrow 1$
11:          $\theta \longleftarrow \frac{\pi}{4} - \arcsin(\frac{d}{\sqrt{2} \cdot t})$ /* $\frac{1}{\cos\theta - \sin\theta} = \frac{t}{d}$ */
12:          $c_p(2\theta, q) \longleftarrow$ cone of angle 2θ with apex at p and bisector $\overrightarrow{pq}$
13:          $c_q(2\theta, p) \longleftarrow$ cone of angle 2θ with apex at q and bisector $\overrightarrow{qp}$
14:          $C_p \longleftarrow C_p \cup c_p(2\theta, q)$
15:          $C_q \longleftarrow C_q \cup c_q(2\theta, p)$
16: **return** $G = (P, E)$

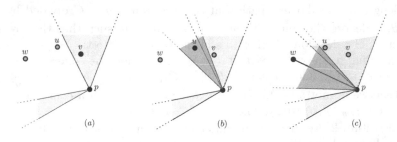

(a)   (b)   (c)

Fig. 1. Illustration of the three scenarios of the δ-Greedy algorithm. (a) $v \in C_p$; (b) $u \notin C_p$ and $d \le \delta$; (c) $w \notin C_p$ and $d > \delta$.

2.2 Algorithm analysis

In this section we analyze several properties of the δ-Greedy algorithm, including the spanning ratio and the degree of the resulting graph.

The following lemma is a generalization of Lemma 6.4.1. in [20].

Lemma 1. *Let* t, δ *and* θ *be real numbers, such that* $1 \leq \delta \leq t$ *and* $0 \leq \theta \leq \frac{\pi}{4}$. *Let* p, q, *and* r *be points in the plane, such that*

1. $p \neq r$,
2. $|pr| \leq |pq|$,
3. $\frac{1}{\cos\theta - \sin\theta} \leq \frac{t}{\delta}$, *where* θ *is the angle* $\angle rpq$ *(i.e.,* $\angle rpq = \theta \leq \frac{\pi}{4} - \arcsin(\frac{\delta}{\sqrt{2} \cdot t})$*).*

Then $\delta|pr| + t|rq| \leq t|pq|$.

Proof. Let r' be the orthogonal projection of r onto segment $\overline{pq}$. Then, $|rr'| = |pr| \sin\theta$, $|pr'| = |pr| \cos\theta$, and $|r'q| = |pq| - |pr'|$. Thus, $|r'q| = |pq| - |pr| \cos\theta$. By the triangle inequality

$$|rq| \leq |rr'| + |r'q|$$
$$\leq |pr| \sin\theta + |pq| - |pr| \cos\theta$$
$$= |pq| - |pr|(\cos\theta - \sin\theta).$$

We have, $\delta|pr| + t|rq| \leq \delta|pr| + t(|pq| - |pr|(\cos\theta - \sin\theta))$
$$= t|pq| - t|pr|(\cos\theta - \sin\theta) + \delta|pr|$$
$$\leq t|pq| - t|pr|(\cos\theta - \sin\theta) + t(\cos\theta - \sin\theta)|pr|$$
$$\leq t|pq|.$$

□

Lemma 2. *The number of shortest path queries performed by δ-Greedy algorithm for each point is* $O(\frac{1}{t/\delta - 1})$.

Proof. Clearly, the number of shortest path queries performed for each point is at most $n - 1$. Thus, we may assume that $t/\delta > 1 + 1/n$. Consider a point $p \in P$ and let (p, q) and (p, r) be two pairs of points that δ-Greedy algorithm has run shortest path queries for. Assume w.l.o.g. that the pair (p, r) has been considered before the pair (p, q), i.e., $|rp| \leq |pq|$. Let d be the length of the path computed by the shortest path query for (p, r) divide by $|pr|$. If $d \leq \delta$, then the cone added to the collection C_p has an angle of at least $\frac{\pi}{4} - \arcsin(\frac{\delta}{\sqrt{2} \cdot t})$. Otherwise, the algorithm adds the edge (p, r) to G and a new cone to the collection of cones C_p, where the angle of this cone is $\frac{\pi}{4} - \arcsin(\frac{1}{\sqrt{2} \cdot t})$. Thus, after the shortest path query performed for the pair (p, r), the collection C_p contains a cone $c_p(\theta, r)$, where θ is at least $\frac{\pi}{2} - 2\arcsin(\frac{1}{\sqrt{2} \cdot t})$. The δ-Greedy algorithm performs a shortest path query for (p, q) only if $p \notin C_q$ and $q \notin C_p$. Thus, the angle $\angle rpq$ is at least $\frac{\pi}{4} - \arcsin(\frac{\delta}{\sqrt{2} \cdot t})$, and we have at most $k = \frac{2\pi}{\theta}$ shortest path queries for a point.

Let us consider the case where $t > 1$ and $\frac{t}{\delta} \to 1$. The equation $\theta = \frac{\pi}{4} - \arcsin(\frac{\delta}{\sqrt{2} \cdot t})$ implies that $\frac{1}{\cos\theta - \sin\theta} = \frac{t}{\delta}$. Then, we have

$$\theta \to 0, \quad \frac{t}{\delta} \sim 1 + \theta, \text{ and } \theta \sim \frac{t}{\delta} - 1.$$

Thus, we have $k \sim \frac{2\pi}{\frac{t}{\delta} - 1} = O(\frac{1}{t/\delta - 1})$. □

Observation 1 For $\delta = t^{\frac{x-1}{x}}$, where $x > 1$ is a fixed integer, the number of shortest path queries performed by δ-Greedy algorithm for each point is $O(\frac{x}{t-1})$.

Proof. As in Lemma 2, let us consider the case where $t > 1$ and $\frac{t}{\delta} \to 1$. Then, we have

$$\theta \to 0, \quad \frac{t}{\delta} \sim 1 + \theta, \quad \frac{t}{t^{(\frac{x-1}{x})}} \sim 1 + \theta, \quad t^{(\frac{1}{x})} \sim 1 + \theta,$$

$$t \sim (1 + \theta)^x, \quad t \sim 1 + x \cdot \theta, \text{ and } \theta \sim \frac{t-1}{x}.$$

Thus, we have $k \sim \frac{2\pi x}{t-1} = O(\frac{x}{t-1})$.

□

Observation 2 The running time of δ-Greedy algorithm is $O(\frac{n^2 \log n}{(t/\delta - 1)^2})$.

Proof. The algorithm sorts the $\binom{n}{2}$ pairs of distinct points in non-decreasing order of their distances, this takes $O(n^2 \log n)$ time. A shortest path query is done by Dijkstra's shortest path algorithm on a graph with $O(\frac{n}{t/\delta - 1})$ edges and takes $O(\frac{n}{t/\delta - 1} + n \log n)$ time. By Lemma 2 each point performs $O(\frac{1}{t/\delta - 1})$ shortest path queries. Therefore, we have that the running time of δ-Greedy algorithm is $O((\frac{n}{t/\delta - 1})^2 \log n)$. □

Observation 3 The number of cones that each point has in its collection along the algorithm is constant depending on t and δ $(O(\frac{1}{t/\delta - 1}))$.

Proof. As shown in Lemma 2, the number of shortest path queries for each point is $O(\frac{1}{t/\delta - 1})$. The subsequent step of a shortest path query is the addition of two cones, meaning that for each point p the number of cones in the collection of cones C_p is $O(\frac{1}{t/\delta - 1})$. □

Corollary 1. The additional space for each point p for the collection C_p is constant.

Lemma 3. The output graph $G = (P, E)$ of δ-Greedy algorithm (Algorithm 2) is a t-spanner for P (for $1 < \delta < t$).

Proof. Let $G = (P, E)$ be the output graph of the δ-Greedy algorithm. To prove that G is a t-spanner for P we show that for every pair $(p, q) \in P$, there exists a t-spanning path between them in G. We prove the above statement by induction on the rank of the distance $|pq|$, i.e., the place of (p,q) in a non-decreasing distances order of all pairs of points in P.

Base case: Let (p,q) be the first pair in the ordered list (i.e., the closest pair). The edge (p,q) is added to E during the first iteration of the loop in step 9 of Algorithm 2, and thus there is a *t*-spanning path between p and q in G.

Induction hypothesis: For every pair (r,s) that appears before the pair (p,q) in the ordered list, there is a *t*-spanning path between r and s in G.

The inductive step: Consider the pair (p,q). We prove that there is a *t*-spanning path between p and q in G. If $p \notin C_q$ and $q \notin C_p$, we check whether there is a δ-spanning path in G between p and q. If there is a path which length is at most $\delta|pq|$, then $\delta|pq| \leq t|pq|$, meaning there is a *t*-spanning path between p and q in G. If there is no path of length of at most $\delta|pq|$, we add the edge (p,q) to G, which forms a *t*-spanning path.

Consider that $p \in C_q$ or $q \in C_p$, and assume w.l.o.g. that $q \in C_p$. Let (p,r) be the edge handled in Step 5 in Algorithm 2 when the cone containing q has been added to C_p (Step 12 in Algorithm 2). Notice that $|pr| \leq |pq|$. Step 7 of Algorithm 2 has computed the value d for the pair (p,r). In the algorithm there are two scenarios depending on the value of d.

The first scenario is when $d > \delta$, then the algorithm has added the edge (p,r) to G and a cone $c_p(\theta,r)$ to C_p, where $\theta = 2(\frac{\pi}{4} - \arcsin(\frac{1}{\sqrt{2} \cdot t}))$. Thus, the angle between (p,q) and (p,r) is less than $\theta/2$. Hence, $|rq| < |pq|$ and by the induction hypothesis there is a *t*-spanning path between r and q. Consider the shortest path between p and q that goes through the edge (p,r). The length of this path is at most $|pr| + t|rq|$. By Lemma 1, we have $|pr| + t|rq| \leq \delta|pr| + t|rq| \leq t|pq|$ for $\delta - 1$. Therefore, we have a *t*-spanning path between p and q.

The second scenario is when $d \leq \delta$, then the algorithm has added a cone $c_p(\theta,r)$ to C_p, where $\theta = 2(\frac{\pi}{4} - \arcsin(\frac{d}{\sqrt{2} \cdot t}))$. Thus, the angle between (p,q) and (p,r) is less than $\theta/2$. Hence, $|rq| < |pq|$ and by the induction hypothesis there is a *t*-spanning path between r and q. Consider the shortest path between p and q that goes through r. The length of this path is at most $d|pr| + t|rq|$. By Lemma 1, we have $d|pr| + t|rq| \leq t|pq|$. Therefore, we have a t-spanning path between p and q. □

Theorem 4. *The δ-Greedy algorithm computes a t-spanner for a set of points* P *with the same properties as the Path-Greedy t-spanner, such as degree and weight, in* $O((\frac{n}{t/\delta - 1})^2 \log n)$ *time.*

Proof. Clearly, the degree of the δ-Greedy is at most the degree of the Path-Greedy δ-spanner. The edges of the δ-Greedy spanner satisfy the δ-leap frog property, thus, the weight of the δ-Greedy is as Path-Greedy *t*-spanner. Hence, we can pick δ close to *t*, such that we will have the required bounds. □

Lemma 4. *If* $t = \delta$*, the result of the δ-Greedy algorithm is identical to the result of the Path-Greedy algorithm.*

Proof. Assume towards contradiction that for $t = \delta$ the resulting graph of the δ-Greedy algorithm, denoted as $G = (P, E)$, differs from the result of the Path-Greedy algorithm, denoted as $G' = (P, E')$. Assuming the same order of the sorted edges, let (p, q) be the first edge that is different in G and G'. Notice that δ-Greedy algorithm decides to add the edge (p, q) to G when there is no t-spanning path between p and q in G. Since until handling the edge (p, q) the graphs G and G' are identical, the Path-Greedy algorithm also decides to add the edge (p, q) to G'. Therefore, the only case we need to consider is $(p, q) \in E'$ and $(p, q) \notin E$. The δ-Greedy algorithm does not add an edge (p, q) to G in two scenarios: (i) there is a t-spanning path between p and q in the current graph G – which contradicts that the Path-Greedy algorithm adds the edge (p, q) to G'; (ii) $p \in C_q$ or $q \in C_p$ – the δ-Greedy algorithm does not perform a shortest path query between p and q. Assume w.l.o.g., $q \in C_p$, and let (p, r) be the edge considered in Step 5 in Algorithm 2 when the cone containing q has been added to C_p. The angle of the added cone is $\theta = \frac{\pi}{2} - 2 \arcsin(\frac{d}{\sqrt{2} \cdot t})$, where d is the length of the shortest path between p and r divided $|pr|$. Thus, we have $|pr| \leq |pq|$ and $\frac{1}{\cos \alpha - \sin \alpha} \leq \frac{t}{d}$, where $\alpha \leq \theta$ is the angle $\angle rpq$. Then, by Lemma 1, $\delta|pr| + t|rq| \leq t|pq|$, and since there is a path from p to r of length at most $\delta|pr|$, we have that there is t-spanning path between p and q in the current graph. This is in contradiction to the assumption that the Path-Greedy algorithm adds the edge (p, q) to E'. $\square$

3 δ-Greedy in Expected $O(n \log n)$ Time for Random Set

In this section we show how a small modification in the implementation improves the running time of the δ-Greedy algorithm. This improvement yields an expected $O(n \log n)$ time for random point sets. The first modification is to run the shortest path query between points p to q up to $\delta|pq|$. That is, running Dijkstras shortest path algorithm with source p and terminating as soon as the minimum key in the priority queue is larger than $\delta|pq|$.

Let P be a set of n points in the plane uniformly distributed in a unit square. To prove that δ-Greedy algorithm computes a spanner for P in expected $O(n \log n)$ time, we need to show that:

- each point runs a constant number of shortest path queries – follows from Lemma 2;
- the expected number of points visited in each query is constant – The fact that the points are randomly chosen uniformly in the unit square implies that the expected number of points at distance of at most r from point p is $\Theta(r^2 \cdot n)$. A shortest path query from a point p to a point q terminates as soon as the minimum key in the priority queue exceeds $\delta|pq|$, thus, it is expected to visit $O(n \cdot (\delta|pq|)^2)$ points.

By Lemma 2 the number of shortest path queries performed by the algorithm for a point p is $O(\frac{1}{t/\delta - 1})$. Each such query defines a cone with apex at p of angle $\Omega(t/\delta - 1)$, such that no other shortest path query from p will be performed to a point in this cone. By picking $k = \frac{1}{t/\delta - 1}$ and $r = \frac{k}{\sqrt{n}}$, we have that the expected number of points around each point in a distance of r is $\Theta(k^2) = \Theta(\frac{1}{(t/\delta - 1)^2})$.

Assume we partition the plane into k equal angle cones with apex at point p. The probability that there exists a cone that does not contain a point from the set of points of distance $\frac{k}{\sqrt{n}}$ is at most $k \cdot (1 - \frac{1}{k})^{k^2}$. Let Q be the set of points that p computed a shortest path query to, and let $q \in Q$ be the farthest point in Q from p. Then, the expected Euclidean distance between p and q is less than $\frac{k}{\sqrt{n}}$. Thus, the expected number of points visited by the entire set of shortest path queries from a point is $O(\frac{\delta^2 k^2}{t/\delta-1}) = O(\frac{\delta^2}{(t-\delta)^3})$;

- the next pair to be processed can be obtained in expected $O(\log n)$ time without sorting all pairs of distinct points – Even-though this is quite straight forward, for completeness we give a short description how this can be done. Divide the unit square to $n \times n$ grid cells of side length $1/n$. A hash table of size $3n$ is initialized, and for each non-empty grid cell (at most n such cells) we map the points in it to the hash table. In addition, we maintain a minimum heap H_p for each point $p \in P$ (initially empty), and one main minimum heap H that contains the top element of each H_p. Each heap H_p contains a subset of the pairs that include p.

For each point $p \in P$, all the cells of distance at most $\frac{k}{\sqrt{n}}$ from p are scanned (using the hash table) to find all the points in these cells, where k is a parameter that we fix later. All the points found in these cells are added to H_p according to their Euclidean distance from p.

The heap H holds the relevant pairs in an increasing order, therefore the pairs are extracted from the main heap H. After extracting the minimum pair in H that belongs to a point p, we add to H the next minimum in H_p. To insure the correctness of the heaps, when needed we increase the distance to the scanned cells. Observe that there may be a pair (p,q) such that $|pq| < |rw|$, where the pair (r,w) is the top pair in H. This can occur only when the pair (p,q) has not been added to H_p nor H_q, and this happens when $p \in C_q$ or $q \in C_p$. However, in this case we do not need to consider the pair (p,q).

Notice that the only cells that are not contained in C_p are scanned to add more pairs to H_p. Thus, points that are in C_p are ignored.

Therefore, the total expected running time of the algorithm is $O(\frac{\delta^2}{(t-\delta)^3} n \log n)$. Since both t and t/δ are constants bigger than one, the expected running time of the δ-Greedy algorithm is $O(n \log n)$.

A very nice outcome of δ-Greedy algorithm and its analysis can be seen when δ is equal to t. Assume that δ-Greedy algorithm (for $\delta = t$) has computed a shortest path query for two points p and q and the length of the received path is $d|pq|$. If the probability that $t/d > 1 + \varepsilon$ is low (e.g, less than $1/2$), for some constant $\varepsilon > 0$, then δ-Greedy algorithm computes the Path-Greedy spanner with linear number of shortest path queries. Thus δ-Greedy algorithm computes the Path-Greedy spanner for a point set uniformly distributed in a square in expected $O(n \log n)$ time.

Not surprisingly our experiments have shown that this probability is indeed low (less than $1/100$), since most of the shortest path queries are performed on pairs of points placed close to each other (with respect to Euclidean distance), and thus with a high probability their shortest path contains a constant number of points. Moreover, it seems that for a "real-life" input this probably is low. Thus, there is a very simple algorithm to compute the Path-Greedy spanner in expected $O(n^2 \log n)$ time for real-life inputs, based on the δ-Greedy algorithm

For real-life input we mean that our analysis suggests that in the current computers precision (Memory) one cannot create an instance of points set with more than 1000 points, where the Path-Greedy spanner based on the δ-Greedy algorithm has more than $O(n^2 \log n)$ constructing time.

4 Experimental Results

In this section we discuss the experimental results by considering the properties of the graphs generated by different algorithms and the number of shortest path queries performed during these algorithms. We have implemented the Path-Greedy, δ-Greedy, Gap-Greedy, θ-Graph, Path-Greedy on θ-Graph algorithms. The Path-Greedy on θ-graph t-spanner, first computes a θ-graph t'-spanner, where $t' < t$, and then runs the Path-Greedy t/t'-spanner on this t'-spanner. The shortest path queries criteria is used for an absolute running time comparison that is independent of the actual implementation. The known theoretical bounds for the algorithms can be found in Table 1.

The experiments were performed on a set of 8000 points, with different values of the parameter δ (between 1 and t). We have chosen to present the parameter δ for the values $t, t^{0.9}$ and $\sqrt{t}$. This values do not have special properties, they where chosen arbitrary to present the behavior of the spanner.

To avoid the effect of specific instances, we have run the algorithms several times and taken the average of the results. However, in all the cases the difference between the values is negligible. Table 2–4 show the results of our experiments for different values of t and δ. The columns of the weight (divided by $wt(MST)$) and the degree are rounded to integers, and the columns of the edges are rounded to one digit after the decimal point (in k).

The implementation details and the results analysis appear in [4].

Acknowledgments We would like to thank Rachel Saban for implementing the algorithms.

Algorithm	Edges	$\frac{Weight}{wt(MST)}$	Degree	Time
Path-Greedy	$O(\frac{n}{t-1})$	$O(1)$	$O(\frac{1}{t-1})$	$O(n^3 \log n)$
Gap-Greedy	$O(\frac{n}{t-1})$	$O(\log n)$	$O(\frac{1}{t-1})$	$O(n \log^2 n)$
θ-Graph	$O(\frac{n}{\theta})$	$O(n)$	$O(n)$	$O(\frac{n}{\theta} \log n)$
δ-Greedy	$O(\frac{n}{t/\delta-1})$	$O(1)$	$O(\frac{1}{t/\delta-1})$	$O(\frac{1}{t/\delta-1} \cdot n^2 \log n)$

Table 1. Theoretical bounds of different t-spanner algorithms

Algorithm	δ	Edges (in K)	Weight $\overline{wt(MST)}$	Degree	Shortest path queries (in K)
Path-Greedy	-	35.6	10	17	31996
δ-Greedy	1.1	35.6	10	17	254
δ-Greedy	1.0896	37.8	12	18	242
δ-Greedy	1.048	51.6	19	23	204
θ-Graph	-	376.6	454	149	-
Greedy on θ-Graph	1.0896	37.8	12	18	3005
Greedy on θ-Graph	1.048	52	19	23	693
Gap-Greedy	-	51.6	19	23	326

Table 2. Comparison between several *t*-spanner algorithms for $t = 1.1$

Algorithm	δ	Edges (in K)	Weight $\overline{wt(MST)}$	Degree	Shortest path queries (in K)
Path-Greedy	-	15.1	3	7	31996
δ-Greedy	1.5	15.1	3	7	82
δ-Greedy	1.44	16	3	8	77
δ-Greedy	1.224	22.5	5	11	63
θ-Graph	-	118.6	76	53	-
Greedy on θ-Graph	1.44	16	3	8	817
Greedy on θ-Graph	1.224	22.5	6	11	198
Gap-Greedy	-	22.6	5	11	95

Table 3. Comparison between several *t*-spanner algorithms for $t = 1.5$

Algorithm	δ	Edges (in K)	Weight $\overline{wt(MST)}$	Degree	Shortest path queries (in K)
Path-Greedy	-	11.4	2	5	31996
δ-Greedy	2	11.4	2	5	55
δ-Greedy	1.866	11.9	2	5	52
δ-Greedy	1.414	16.3	3	8	44
θ-Graph	-	85.3	48	42	-
Greedy on θ-Graph	1.866	11.9	3	6	493
Greedy on θ-Graph	1.414	16.5	3	8	129
Gap-Greedy	-	16	3	8	63

Table 4. Comparison between several *t*-spanner algorithms for $t = 2$

References

1. S. P. A. Alewijnse, Q. W. Bouts, A. P. ten Brink, and K. Buchin. Computing the greedy spanner in linear space. *Algorithmica*, 73(3):589–606, 2015.
2. S. P. A. Alewijnse, Q. W. Bouts, A. P. ten Brink, and K. Buchin. Distribution-sensitive construction of the greedy spanner. *Algorithmica*, pages 1–23, 2016.
3. S. Arya, D. M. Mount, and M. H. M. Smid. Randomized and deterministic algorithms for geometric spanners of small diameter. In *FOCS*, pages 703–712, 1994.
4. G. Bar-On and P. Carmi. δ-greedy t-spanner. *CoRR*, abs/1702.05900, 2017.
5. P. Bose, P. Carmi, M. Farshi, A. Maheshwari, and M. H. M. Smid. Computing the greedy spanner in near-quadratic time. In *SWAT*, pages 390–401, 2008.
6. P. B. Callahan. Optimal parallel all-nearest-neighbors using the well-separated pair decomposition. In *FOCS*, pages 332–340, 1993.
7. P. B. Callahan and S. R. Kosaraju. A decomposition of multi-dimensional point-sets with applications to k-nearest-neighbors and n-body potential fields. In *STOC*, pages 546–556, 1992.
8. B. Chandra. Constructing sparse spanners for most graphs in higher dimensions. *Inf. Process. Lett.*, 51(6):289–294, 1994.
9. B. Chandra, G. Das, G. Narasimhan, and J. Soares. New sparseness results on graph spanners. *Int. J. Comp. Geom. and Applic.*, 5:125–144, 1995.
10. K. L. Clarkson. Approximation algorithms for shortest path motion planning. In *STOC*, pages 56–65, 1987.
11. G. Das, P. J. Heffernan, and G. Narasimhan. Optimally sparse spanners in 3-dimensional Euclidean space. In *SoCG*, pages 53–62, 1993.
12. G. Das and G. Narasimhan. A fast algorithm for constructing sparse Euclidean spanners. *Int. J. Comp. Geom. and Applic.*, 7(4):297–315, 1997.
13. G. Das and G. Narasimhan. A fast algorithm for constructing sparse Euclidean spanners. *Int. J. Comp. Geom. and Applic.*, 7(4):297–315, 1997.
14. M. Farshi and J. Gudmundsson. Experimental study of geometric *t*-spanners. In *ESA*, pages 556–567, 2005.
15. M. Farshi and J. Gudmundsson. Experimental study of geometric t-spanners: A running time comparison. In *WEA*, pages 270–284, 2007.
16. M. Farshi and J. Gudmundsson. Experimental study of geometric *t*-spanners. *ACM Journal of Experimental Algorithmics*, 14, 2009.
17. J. Gudmundsson, C. Levcopoulos, and G. Narasimhan. Fast greedy algorithms for constructing sparse geometric spanners. *SIAM J. Comput.*, 31(5):1479–1500, 2002.
18. J. M. Keil. Approximating the complete Euclidean graph. In *SWAT*, pages 208–213, 1988.
19. C. Levcopoulos, G. Narasimhan, and M. H. M. Smid. Improved algorithms for constructing fault-tolerant spanners. *Algorithmica*, 32(1):144–156, 2002.
20. G. Narasimhan and M. Smid. *Geometric Spanner Networks*. Cambridge University Press, New York, NY, USA, 2007.
21. J. Soares. Approximating Euclidean distances by small degree graphs. *Discrete & Computational Geometry*, 11(2):213–233, 1994.

Dynamic Graph Coloring[*]

Luis Barba[1], Jean Cardinal[2], Matias Korman[3], Stefan Langerman[2],
André van Renssen[4,5], Marcel Roeloffzen[4,5], and Sander Verdonschot[6]

[1] Dept. of Computer Science, ETH Zürich, Switzerland. `luis.barba@inf.ethz.ch`
[2] Départment d'Informatique, Université Libre de Bruxelles, Brussels, Belgium.
`{jcardin,stefan.langerman}@ulb.ac.be`
[3] Tohoku University, Sendai, Japan. `mati@dais.is.tohoku.ac.jp`
[4] National Institute of Informatics, Tokyo, Japan. `{andre,marcel}@nii.ac.jp`
[5] JST, ERATO, Kawarabayashi Large Graph Project.
[6] School of Computer Science, Carleton University, Ottawa, Canada.
`sander@cg.scs.carleton.ca`

Abstract. In this paper we study the number of vertex recolorings that
an algorithm needs to perform in order to maintain a proper coloring of
a graph under insertion and deletion of vertices and edges. We present
two algorithms that achieve different trade-offs between the number
of recolorings and the number of colors used. For any $d > 0$, the first
algorithm maintains a proper $O(\mathcal{C}dN^{1/d})$-coloring while recoloring at most
$O(d)$ vertices per update, where $\mathcal{C}$ and N are the maximum chromatic
number and maximum number of vertices, respectively. The second
algorithm reverses the trade-off, maintaining an $O(\mathcal{C}d)$-coloring with
$O(dN^{1/d})$ recolorings per update. We also present a lower bound,
showing that any algorithm that maintains a c-coloring of a 2-colorable
graph on N vertices must recolor at least $\Omega(N^{\frac{2}{c(c-1)}})$ vertices per update,
for any constant $c \geq 2$.

1 Introduction

It is hard to underestimate the importance of the graph coloring problem in
computer science and combinatorics. The problem is certainly among the most
studied questions in those fields, and countless applications and variants have
been tackled since it was first posed for the special case of maps in the mid-
nineteenth century. Similarly, the maintenance of some structures in *dynamic
graphs* has been the subject of study of several volumes in the past couple
of decades [1,2,11,18,19,20]. In this setting, an algorithmic graph problem is
modelled in the dynamic environment as follows. There is an online sequence of
insertion and deletion of edges or vertices, and our goal is to maintain the solution
of the graph problem after each update. A trivial way to maintain this solution is
to run the best static algorithm for this problem after each update; however, this

[*] M. K. was partially supported by MEXT KAKENHI grant Nos. 12H00855, and
17K12635. A. v. R. and M. R. were supported by JST ERATO Grant Number
JPMJER1305, Japan. L. B. was supported by the ETH Postdoctoral Fellowship.
S. V. was partially supported by NSERC and the Carleton-Fields postdoctoral award.

© Springer International Publishing AG 2017
F. Ellen et al. (Eds.): WADS 2017, LNCS 10389, pp. 97–108, 2017.
DOI: 10.1007/978-3-319-62127-2_9

is clearly not optimal. A dynamic graph algorithm seeks to maintain some clever data structure for the underlying problem such that the time taken to update the solution is much smaller than that of the best static algorithm.

In this paper, we study the problem of maintaining a coloring in a dynamic graph undergoing insertions and deletions of both vertices and edges. At first sight, this may seem to be a hopeless task, since there exist near-linear lower bounds on the competitive factor of online graph coloring algorithms [9], a restricted case of the dynamic setting. In order to break through this barrier, we allow a "fair" number of *vertex recolorings* per update. We focus on the combinatorial aspect of the problem – the trade-off between the number of colors used versus the number of recolorings per update. We present a strong general lower bound and two simple algorithms that provide complementary trade-offs.

Definitions and Results. Let $\mathcal{C}$ be a positive integer. A $\mathcal{C}$-*coloring* of a graph G is a function that assigns a color in $\{1, \ldots, \mathcal{C}\}$ to each vertex of G. A $\mathcal{C}$-coloring is *proper* if no two adjacent vertices are assigned the same color. We say that G is $\mathcal{C}$-*colorable* if it admits a proper $\mathcal{C}$-coloring, and we call the smallest such $\mathcal{C}$ the *chromatic number* of G.

A *recoloring algorithm* is an algorithm that maintains a proper coloring of a simple graph while that graph undergoes a sequence of updates. Each update adds or removes either an edge or a vertex with a set of incident edges. We say that a recoloring algorithm is c-*competitive* if it uses at most $c \cdot \mathcal{C}_{max}$ colors, where $\mathcal{C}_{max}$ is the maximum chromatic number of the graph during the updates.

For example, an algorithm that computes the optimal coloring after every update is 1-competitive, but may recolor every vertex for every update. At the other extreme, we can give each vertex a unique color, resulting in a linear competitive factor for an algorithm that recolors at most 1 vertex per update. In this paper, we investigate intermediate solutions that use more than $\mathcal{C}$ colors but recolor a sublinear number of vertices per update. Note that we do not assume that the value $\mathcal{C}$ is known in advance, or at any point during the algorithm.

In Section 2, we present two complementary recoloring algorithms: an $O(dN^{1/d})$-competitive algorithm with an amortized $O(d)$ recolorings per update, and an $O(d)$-competitive algorithm with an amortized $O(dN^{1/d})$ recolorings per update, where d is a positive integer parameter and N is the maximum number of vertices in the graph during a sequence of updates. Interestingly, for $d = \Theta(\log N)$, both are $O(\log N)$-competitive with an amortized $O(\log N)$ vertex recolorings per update. Using standard techniques, the algorithms can be made sensitive to the current (instead of the maximum) number of vertices in the graph.

We provide lower bounds in Section 3. In particular, we show that for any recoloring algorithm A using c colors, there exists a specific 2-colorable graph on N vertices and a sequence of m edge insertions and deletions that forces A to perform at least $\Omega(m \cdot N^{\frac{2}{c(c-1)}})$ vertex recolorings. Thus, any x-competitive recoloring algorithm performs in average at least $\Omega(N^{\frac{1}{x(2x-1)}})$ recolorings per update.

To allow us to focus on the combinatorial aspects, we assume that we have access to an algorithm that, at any time, can color the current graph (or an induced subgraph) using few colors. Of course, finding an optimal coloring of an

n-vertex graph is NP-complete in general [13] and even NP-hard to approximate to within $n^{1-\epsilon}$ for any $\epsilon > 0$ [22]. Still, this assumption is not as strong as it sounds. Most practical instances can be colored efficiently [4], and for several important classes of graphs the problem is solvable or approximable in polynomial time, including bipartite graphs, planar graphs, k-degenerate graphs, and unit disk graphs [15].

Related results. *Dynamic graph coloring.* The problem of maintaining a coloring of a graph that evolves over time has been tackled before, but to our knowledge, only from the points of view of heuristics and experimental results. This includes for instance results from Preuveneers and Berbers [17], Ouerfelli and Bouziri [16], and Dutot et al. [7]. A related problem of maintaining a graph-coloring in an online fashion was studied by Borowiecki and Sidorowicz [3]. In that problem, vertices lose their color, and the algorithm is asked to recolor them.

Online graph coloring. The online version of the problem is closely related to our setting, except that most variants of the online problem only allow the coloring of new vertices, which then cannot be recolored later. Near-linear lower bounds on the best achievable competitive factor have been proven by Halldórsson and Szegedy more than two decades ago [9]. They show their bound holds even when the model is relaxed to allow a constant fraction of the vertices to change color over the whole sequence. This, however, does not contradict our results. We allow our algorithms to recolor all vertices at some point, but we bound only the number of recolorings *per update*. Algorithms for online coloring with competitive factor coming close, or equal to this lower bound have been proposed by Lovász et al. [14], Vishwanathan [21], and Halldórsson [8].

Dynamic graphs. Several techniques have been used for the maintenance of other structures in dynamic graphs, such as spanning trees, transitive closure, and shortest paths. Surveys by Demetrescu et al. [5,6] give a good overview of those. Recent progress on dynamic connectivity [12] and approximate single-source shortest paths [10] are witnesses of the current activity in this field.

2 Upper bound: Recoloring-algorithms

For the description of our algorithms we consider only inserting a vertex with its incident edges. Deletions cannot invalidate the coloring and edge insertions can be done by removing and adding one of the vertices with the appropriate edges.

Our algorithms partition the vertices into a set of *buckets*, each of which has its own distinct set of colors. All our algorithms guarantee that the subgraph induced by the vertices inside each bucket is properly colored and this implies that the entire graph is properly colored at all times.

The algorithms differ in the number of buckets they use and the size (maximum number of vertices) of each bucket. Typically, there is a sequence of buckets of increasing size, and one *reset bucket* that can contain arbitrarily many vertices and that holds vertices whose color has not changed for a while. Initially, the size of each bucket depends on the number of vertices in the input graph. As vertices are inserted and deleted, the current number of vertices changes. When certain

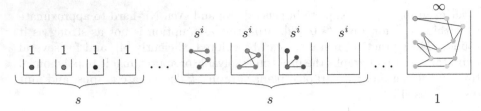

Fig. 1: The small-buckets algorithm uses d levels, each with s buckets of capacity s^i, where i is the level, $s = \lceil N_R^{1/d} \rceil$, and N_R is the number of vertices during the last reset.

buckets are full, we *reset* everything, to ensure that we can accommodate the new number of vertices. This involves emptying all buckets into the reset bucket, computing a proper coloring of the entire graph, and recomputing the sizes of the buckets in terms of the current number of vertices.

We refer to the number of vertices during the most recent reset as N_R, and we express the size of the buckets in $s = \lceil N_R^{1/d} \rceil$, where $d > 0$ is an integer parameter that allows us to achieve different trade-offs between the number of colors and number of recolorings used. Since $s = O(N^{1/d})$, where N is the maximum number of vertices thus far, we state our bounds in terms of N. Note that it is also possible to keep N_R within a constant factor of the current number of vertices by triggering a reset whenever the current number of vertices becomes too small or too large. We omit these details for the sake of simplicity.

2.1 Small-buckets algorithm

Our first algorithm, called the *small-buckets algorithm*, uses a lot of colors, but needs very few recolorings. In addition to the reset bucket, the algorithm uses ds buckets, grouped into d *levels* of s buckets each. All buckets on level i, for $0 \leq i < d$, have capacity s^i (see Fig. 1). Initially, the reset bucket contains all vertices, and all other buckets are empty. Throughout the execution of the algorithm, we ensure that every level always has at least one empty bucket. We call this the *space invariant*.

When a new vertex is inserted, we place it in any empty bucket on level 0. The space invariant guarantees the existence of this bucket. Since this bucket has a unique set of colors, assigning one of them to the new vertex establishes a proper coloring. Of course, if this was the last empty bucket on level 0, filling it violates the space invariant. In that case, we gather up all s vertices on this level, place them in the first empty bucket on level 1 (which has capacity s and must exist by the space invariant), and compute a new coloring of their induced graph using the set of colors of the new bucket. If this was the last free bucket on level 1, we move all its vertices to the next level and repeat this procedure. In general, if we filled the last free bucket on level i, we gather up all at most $s \cdot s^i = s^{i+1}$ vertices on this level, place them in an empty bucket on level $i + 1$ (which exists by the space invariant), and recolor their induced graph with the new colors. If we fill up the last level $(d - 1)$, we reset the structure, emptying each bucket into the reset bucket and recoloring the whole graph.

Theorem 2.1. *For any integer $d > 0$, the small-buckets algorithm is an $O(dN^{1/d})$-competitive recoloring algorithm that uses at most $O(d)$ amortized vertex recolorings per update.*

Proof. (sketch) The total number of colors is bounded by the maximum number of non-empty buckets $(1 + d(s - 1))$, multiplied by the maximum number of colors used by any bucket. Let $\mathcal{C}$ be the maximum chromatic number of the graph. Since any induced subgraph of a $\mathcal{C}$-colorable graph is also $\mathcal{C}$-colorable, each bucket requires at most $\mathcal{C}$ colors. Thus, the total number of colors is at most $(1 + d(s - 1))\mathcal{C}$, and the algorithm is $O(dN^{1/d})$-competitive.

To analyze the number of recolorings, we use a simple charging scheme that places coins in the buckets and pays one coin for each recoloring. Whenever we place a vertex in a bucket on level 0, we give $d + 2$ coins to that bucket. One of these coins is immediately used to pay for the vertex's new color, leaving $d + 1$ coins. In general, we can maintain the invariant that each non-empty bucket on level i has $s^i \cdot (d - i + 1)$ coins from which the result follows. □

2.2   Big-buckets algorithm

Our second algorithm, called the *big-buckets algorithm*, is similar to the small-buckets algorithm, except it merges all buckets on the same level into a single larger bucket. Specifically, the algorithm uses d buckets in addition

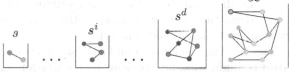

Fig. 2: Besides the reset bucket, the big-buckets algorithm uses d buckets, each with capacity s^{i+1}, where i is the bucket number.

to the reset bucket. These buckets are numbered sequentially from 0 to $d - 1$, with bucket i having capacity s^{i+1}, see Fig. 2. Since we use far fewer buckets, an upper bound on the total number of colors drops significantly, to $(d + 1)\mathcal{C}$. Of course, as we will see later, we pay for this in the number recolorings. Similar to the space invariant in the small-buckets algorithm, the big-buckets algorithm maintains the *high point invariant*: bucket i always contains at most $s^{i+1} - s^i$ vertices (its *high point*).

When a new vertex is inserted, we place it in the first bucket. Since this bucket may already contain other vertices, we recolor all its vertices, so that the subgraph induced by these vertices remains properly colored. This revalidates the coloring, but may violate the high point invariant. If we filled bucket i beyond its high point, we move all its vertices to bucket $i + 1$ and compute a new coloring for this bucket. We repeat this until the high point invariant is satisfied, or we fill bucket $d - 1$ past its high point. In the latter case we reset, adding all vertices to the reset bucket and computing a new coloring for the entire graph.

Theorem 2.2. *For any integer $d > 0$, the big-buckets algorithm is an $O(d)$-competitive recoloring algorithm that uses at most $O(dN^{1/d})$ amortized vertex recolorings per update.*

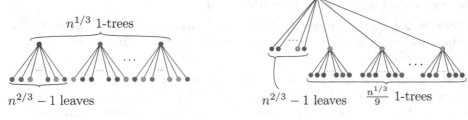

Fig. 3: (left) A 1-configuration is any forest that has many 1-trees as induced subgraphs. (right) A 2-tree is constructed by connecting the roots of many 1-trees.

3 Lower bound

In this section we prove a lower bound on the amortized number of recolorings for any algorithm that maintains a c-coloring of a 2-colorable graph, for any constant $c \geq 2$. We say that a vertex is c-*colored* if it has a color in $[c] = \{1, \dots, c\}$. For simplicity of description, we assume that a recoloring algorithm only recolors vertices when an edge is inserted and not when an edge is deleted, as edge deletions do not invalidate the coloring. This assumption causes no loss of generality, as we can delay the recolorings an algorithm would perform in response to an edge deletion until the next edge insertion.

The proof for the lower bound consists of several parts. We begin with a specific initial configuration and present a strategy for an adversary that constructs a large configuration with a specific colouring and then repeatedly performs costly operations in this configuration. In light of this strategy, a recoloring algorithm has a few choices: it can allow the configuration to be built and perform the recolorings required, it can destroy the configuration by recoloring parts of it instead of performing the operations, or it can prevent the configuration from being built in the first place by recoloring parts of the building blocks. We show that all these options require a large number of amortized recolorings.

3.1 Maintaining a 3-coloring

To make the general lower bound easier to understand, we first show that to maintain a 3-coloring, we need at least $\Omega(n^{1/3})$ recolorings on average per update.

Lemma 3.1. *For any sufficiently large n and any $m \geq 2n^{1/3}$, there exists a forest with n vertices, such that for any recoloring algorithm A, there exists a sequence of m updates that forces A to perform $\Omega(m \cdot n^{1/3})$ vertex recolorings to maintain a 3-coloring throughout this sequence.*

Proof. Let A be any recoloring algorithm that maintains a 3-coloring of a forest under updates. We use an adversarial strategy to choose a sequence of updates on a specific forest with n nodes that forces A to recolor "many" vertices. We start by describing the initial forest structure.

A *1-tree* is a rooted (star) tree with a distinguished vertex as its root and

$n^{2/3} - 1$ leaf nodes attached to it. Initially, our forest consists of $n^{1/3}$ pairwise disjoint 1-trees, which account for all n vertices in our forest. The sequence of updates we construct never performs a cut operation among the edges of a 1-tree. Thus, the forest remains a *1-configuration*: a forest of rooted trees with the $n^{1/3}$ independent 1-trees as induced subgraphs; see Fig. 3 (left). We require that the induced subtrees are not *upside down*, that is, the root of the 1-tree should be closer to the root of the full tree than its children. Intuitively, a 1-configuration is simply a collection of our initial 1-trees linked together into larger trees.

Let F be a 1-configuration. We assume that A has already chosen an initial 3-coloring of F. We assign a color to each 1-tree as follows. Since each 1-tree is properly 3-colored, the leaves cannot have the same color as the root. Thus, a 1-tree T always has at least $\frac{n^{2/3}-1}{2}$ leaves of some color C, and C is different from the color of the root. We assign the color C to T. In this way, each 1-tree is assigned one of the three colors. We say that a 1-tree with assigned color C becomes *invalid* if it has no children of color C left. Notice that to invalidate a 1-tree, algorithm A needs to recolor at least $\frac{n^{2/3}-1}{2}$ of its leaves. Since the coloring uses only three colors, there are at least $\frac{n^{1/3}}{3}$ 1-trees with the same assigned color, say X. In the remainder, we focus solely on these 1-trees.

A *2-tree* is a tree obtained by merging $\frac{n^{1/3}}{9}$ 1-trees with assigned color X, as follows. First, we cut the edge connecting the root of each 1-tree to its parent, if it has one. Next, we pick a distinguished 1-tree with root r, and connect the root of each of the other $\frac{n^{1/3}}{9} - 1$ 1-trees to r. In this way, we obtain a 2-tree whose root r has $n^{2/3} - 1$ leaf children from the 1-tree of r, and $\frac{n^{1/3}}{9} - 1$ new children that are the roots of other 1-trees; see Fig. 3 (right) for an illustration. This construction requires $\frac{n^{1/3}}{9} - 1$ edge insertions and at most $\frac{n^{1/3}}{9}$ edge deletions (if every 1-tree root had another parent in the 1-configuration).

We build 3 such 2-trees in total. This requires at most $6(\frac{n^{1/3}}{9}) = \frac{2n^{1/3}}{3}$ updates. If none of our 1-trees became invalid, then since our construction involves only 1-trees with the same assigned color X, no 2-tree can have a root with color X. Further, since the algorithm maintains a 3-coloring, there must be at least two 2-trees whose roots have the same color. We can now perform a *matching link*, by connecting the roots of these two trees by an edge (in general, we may need to perform a cut first). To maintain a 3-coloring after a matching link, A must recolor the root of one of the 2-trees and either recolor all its non-leaf children or invalidate a 1-tree. If no 1-tree has become invalidated, this requires at least $\frac{n^{1/3}}{9}$ recolorings, and we again have two 2-trees whose roots have the same color. Thus, we can perform another matching link between them. We keep doing this until we either performed $\frac{n^{1/3}}{6}$ matching links, or a 1-tree is invalidated.

Therefore, after at most $n^{1/3}$ updates ($\frac{2n^{1/3}}{3}$ for the construction of the 2-trees, and $\frac{n^{1/3}}{3}$ for the matching links), we either have an invalid 1-tree, in which case A recolored at least $\frac{n^{2/3}-1}{2}$ nodes, or we performed $\frac{n^{1/3}}{6}$ matching links, which forced at least $\frac{n^{1/3}}{6} \cdot \frac{n^{1/3}}{9} = \frac{n^{2/3}}{54}$ recolorings. In either case, we forced A to perform at least $\Omega(n^{2/3})$ vertex recolorings, using at most $n^{1/3}$ updates.

Since no edge of a 1-tree was cut, we still have a valid 1-configuration, where the process can be restarted. Consequently, for any $m \geq 2n^{1/3}$, there exists a sequence of m updates that starts with a 1-configuration and forces A to perform $\lfloor \frac{m}{n^{1/3}} \rfloor \Omega(n^{2/3}) = \Omega(m \cdot n^{1/3})$ vertex recolorings. $\qquad\square$

3.2 On k-trees

We are now ready to describe a general lower bound for any number of colors c. The general approach is the same as when using 3 colors: We construct trees of height up to $c + 1$, each exclud-

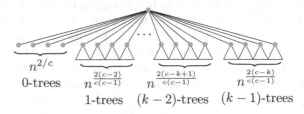

Fig. 4: A k-tree is constructed by connecting the roots of a large number of $(k-1)$-trees.

ing a different color for the root of the merged trees. By now connecting two such trees, we force the algorithm A to recolor the desired number of vertices.

A *0-tree* is a single node, and for each $1 \leq k \leq c$, a *k-tree* is a tree obtained recursively by merging $2 \cdot n^{\frac{2(c-k)}{c(c-1)}}$ $(k-1)$-trees as follows: Pick a $(k-1)$-tree and let r be its root. Then, for each of the $2 \cdot n^{\frac{2(c-k)}{c(c-1)}} - 1$ remaining $(k-1)$-trees, connect their root to r with an edge; see Fig. 4 for an illustration.

As a result, for each $0 \leq j \leq k - 1$, a k-tree T consists of a root r with $2 \cdot n^{\frac{2(c-j-1)}{c(c-1)}} - 1$ j-trees, called the *j-subtrees* of T, whose root hangs from r. The root of a j-subtree of T is called a *j-child* of T. By construction, r is also the root of a j-tree which we call the *core j-tree* of T.

Whenever a k-tree is constructed, it is assigned a color that is present among a "large" fraction of its $(k-1)$-children. Indeed, whenever a k-tree is assigned a color c_k, we guarantee that it has at least $\left\lceil \frac{2}{c} \cdot n^{\frac{2(c-k)}{c(c-1)}} \right\rceil$ $(k-1)$-children of color c_k. We describe later how to choose the color that is assigned to a k-tree.

We say that a k-tree that was assigned color c_k has a *color violation* if its root no longer has a $(k-1)$-child with color c_k. We say that a k-tree T becomes *invalid* if either (1) it has a color violation or (2) if a core j-tree of T has a color violation for some $1 \leq j < k$; otherwise we say that T is *valid*.

Observation 1 *To obtain a color violation in a k-tree constructed by the above procedure, A needs to recolor at least $\left\lceil \frac{2}{c} \cdot n^{\frac{2(c-k)}{c(c-1)}} \right\rceil$ vertices.*

Notice that a valid c-colored k-tree of color c_k cannot have a root with color c_k. Formally, color c_k is *blocked* for the root of a k-tree if this root has a child with color c_k. In particular, the color assigned to a k-tree and the colors assigned to its core j-trees for $1 \leq j \leq k - 1$ are blocked as long as the tree is valid.

3.3 On k-configurations

A 0-configuration is a set F_0 of c-colored nodes, where $|F_0| = T_0 = \alpha n$, for some sufficiently large constant α which will be specified later. For $1 \leq k < c$, a

k-*configuration* is a set F_k of T_k k-trees, where

$$T_k = \frac{\alpha}{(4c)^k} \cdot n^{1-\sum_{i=1}^{k} \frac{2(c-i)}{c(c-1)}}.$$

Note that the trees of a k-configuration may be part of m-trees for $m > k$. If at least $\frac{T_k}{2}$ k-trees in a k-configuration are valid, then the configuration is *valid*.

For our construction, we let the initial configuration F_0 be an arbitrary c-colored 0-configuration in which each vertex is c-colored. To construct a k-configuration F_k from a valid $(k-1)$-configuration F_{k-1}, consider the at least $\frac{T_{k-1}}{2}$ valid $(k-1)$-trees from F_{k-1}. Recall that the trees of F_{k-1} may be part of larger trees, but since we consider edge deletions as "free" operations we can separate the trees. Since each of these trees has a color assigned, among them at least $\frac{T_{k-1}}{2c}$ have the same color assigned to them. Let c_{k-1} denote this color.

Because each k-tree consists of $2 \cdot n^{\frac{2(c-k)}{c(c-1)}}$ $(k-1)$-trees, to obtain F_k we merge $\frac{T_{k-1}}{2c}$ $(k-1)$-trees of color c_{k-1} into T_k k-trees, where

$$T_k = \frac{T_{k-1}}{2c} \cdot \frac{1}{2 \cdot n^{\frac{2(c-k)}{c(c-1)}}} = \frac{\alpha}{(4c)^k} \cdot n^{1-\sum_{i=1}^{k} \frac{2(c-i)}{c(c-1)}}.$$

Once the k-configuration F_k is constructed, we perform a *color assignment* to each k-tree in F_k as follows: For a k-tree τ of F_k whose root has $2 \cdot n^{\frac{2(c-k)}{c(c-1)}} - 1$ c-colored $(k-1)$-children, we assign τ a color that is shared by at least $\left\lfloor \frac{2}{c} \cdot n^{\frac{2(c-k)}{c(c-1)}} - 1 \right\rfloor$ of these $(k-1)$-children. Therefore, τ has at least $\left\lfloor \frac{2}{c} \cdot n^{\frac{2(c-k)}{c(c-1)}} \right\rfloor$ children of its assigned color. After these color assignments, if each $(k-1)$-tree used is valid, then each of the T_k k-trees of F_k is also valid. Thus, F_k is a valid configuration. Moreover, for F_k to become invalid, A would need to invalidate at least $\frac{T_k}{2}$ of its k-trees. Since we use $(k-1)$-trees with the same assigned color to construct k-trees, we can conclude the following about the use of colors in any k-tree.

Lemma 3.2. *Let F_k be a valid k-configuration. For each $1 \leq j < k$, each core j-tree of a valid k-tree of F_k has color c_j assigned to it. Moreover, $c_i \neq c_j$ for each $1 \leq i < j < k$.*

We also provide bounds on the number of updates needed to construct a k-configuration.

Lemma 3.3. *Using $\Theta(\sum_{i=j}^{k} T_i) = \Theta(T_j)$ edge insertions, we can construct a k-configuration from a valid j-configuration.*

Proof. To merge $\frac{T_{k-1}}{2c}$ $(k-1)$-trees to into T_k k-trees, we need $\Theta(T_{k-1})$ edge insertions. Thus, in total, to construct a k-configuration from a j-configuration, we need $\Theta(\sum_{i=j}^{k} T_i) = \Theta(T_j)$ edge insertions. $\qquad\square$

3.4   Reset phase

Throughout the construction of a k-configuration, the recoloring-algorithm A may recolor several vertices which could lead to invalid subtrees in F_j for any $1 \leq j < k$. Because A may invalidate some trees from F_j while constructing F_k from F_{k-1}, one of two things can happen. If F_j is a valid j-configuration for each $1 \leq j \leq k$, then we continue and try to construct a $(k+1)$-configuration from F_k. Otherwise a *reset* is triggered as follows.

Let $1 \leq j < k$ be an integer such that F_i is a valid i-configuration for each $0 \leq i \leq j-1$, but F_j is not valid. Since F_j was a valid j-configuration with at least T_j valid j-trees when it was first constructed, we know that in the process of constructing F_k from F_j, at least $\frac{T_j}{2}$ j-trees where invalidated by A. We distinguish two ways in which a tree can be invalid:

(1) the tree has a color violation, but all its $j-1$-subtrees are valid and no core i-tree for $1 \leq i \leq j-1$ has a color violation; or

(2) A core i-tree has a color violation for $1 \leq i \leq j-1$, or the tree has a color violation and at least one of its $(j-1)$-subtrees is invalid.

In case (1) the algorithm A has to perform fewer recolorings, but the tree can be made valid again with a color reassignment, whereas in case (2) the j-tree has to be rebuild.

Let Y_0, Y_1 and Y_2 respectively be the set of j-trees of F_j that are either valid, or are invalid by case (1) or (2) respectively. Because at least $\frac{T_j}{2}$ j-trees were invalidated, we know that $|Y_1| + |Y_2| > \frac{T_j}{2}$. Moreover, for each tree in Y_1, A recolored at least $\frac{2}{c} \cdot n^{\frac{2(c-j)}{c(c-1)}} - 1$ vertices to create the color violation on this j-tree by Observation 1. For each tree in Y_2 however, A created a color violation in some i-tree for $i < j$. Therefore, for each tree in Y_2, by Observation 1, the number of vertices that A recolored is at least $\frac{2}{c} \cdot n^{\frac{2(c-i)}{c(c-1)}} - 1 > \frac{2}{c} \cdot n^{\frac{2(c-j+1)}{c(c-1)}} - 1$.

Case 1: $|Y_1| > |Y_2|$. Recall that each j-tree in Y_1 has only valid $(j-1)$-subtrees by the definition of Y_1. Therefore, each j-tree in Y_1 can be made valid again by performing a color assignment on it while performing no update. In this way, we obtain $|Y_0| + |Y_1| > \frac{T_j}{2}$ valid j-trees, i.e., F_j becomes a valid j-configuration contained in F_k. Notice that when a color assignment is performed on a j-tree, vertex recolorings previously performed on its $(j-1)$-children cannot be counted again towards invalidating this tree.

Since we have a valid j-configuration instead of a valid k-configuration, we "wasted" some edge insertions. We say that the insertion of each edge in F_k that is not an edge of F_j is a *wasted* edge insertion. By Lemma 3.3, to construct F_k from F_j we used $\Theta(T_j)$ edge insertions. That is, $\Theta(T_j)$ edge insertions became wasted. However, while we wasted $\Theta(T_j)$ edge insertions, we also forced A to perform $\Omega(|Y_1| \cdot n^{\frac{2(c-j)}{c(c-1)}}) = \Omega(T_j \cdot n^{\frac{2(c-j)}{c(c-1)}})$ vertex recolorings. Since $1 \leq j < k \leq c-1$, we know that $n^{\frac{2(c-j)}{c(c-1)}} \geq n^{\frac{2}{c(c-1)}}$. Therefore, we can charge A with $\Omega(n^{\frac{2}{c(c-1)}})$ vertex recolorings per wasted edge insertion. Finally, we remove each edge corresponding to a wasted edge insertion, i.e., we remove all the edges used to construct F_k

from F_j. Since we assumed that A performs no recoloring on edge deletions, we are left with a valid j-configuration F_j.

Case 2: $|Y_2| > |Y_1|$. In this case $|Y_2| > \frac{T_j}{4}$. Recall that F_{j-1} is a valid $(j-1)$-configuration by our choice of j. In this case, we say that the insertion of each edge in F_k that is not an edge of F_{j-1} is a *wasted* edge insertion. By Lemma 3.3, we constructed F_k from F_{j-1} using $\Theta(T_{j-1})$ wasted edge insertions. However, while we wasted $\Theta(T_{j-1})$ edge insertions, we also forced A to perform $\Omega(|Y_2| \cdot n^{\frac{2(c-j+1)}{c(c-1)}}) = \Omega(T_j \cdot n^{\frac{2(c-j+1)}{c(c-1)}})$ vertex recolorings. That is, we can charge A with $\Omega(\frac{T_j}{T_{j-1}} \cdot n^{\frac{2(c-j+1)}{c(c-1)}})$ vertex recolorings per wasted edge insertions. Since $\frac{T_{j-1}}{T_j} = 4c \cdot n^{\frac{2(c-j)}{c(c-1)}}$, we conclude that A was charged $\Omega(n^{\frac{2}{c(c-1)}})$ vertex recolorings per wasted edge insertion. Finally, we remove each edge corresponding to a wasted edge insertion, i.e., we go back to the valid $(j-1)$-configuration F_{j-1} as before.

Regardless of the case, we know that during a reset consisting of a sequence of h wasted edge insertions, we charged A with the recoloring of $\Omega(h \cdot n^{\frac{2}{c(c-1)}})$ vertices. Notice that each edge insertion is counted as wasted at most once as the edge that it corresponds to is deleted during the reset phase. A vertex recoloring may be counted more than once. However, a vertex recoloring on a vertex v can count towards invalidating any of the trees it belongs to. Recall though that v belongs to at most one i-tree for each $0 \leq i \leq c$. Moreover, two things can happen during a reset phase that count the recoloring of v towards the invalidation of a j-tree containing it: either (1) a color assignment is performed on this j-tree or (2) this j-tree is destroyed by removing its edges corresponding to wasted edge insertions. In the former case, we know that v needs to be recolored again in order to contribute to invalidating this j-tree. In the latter case, the tree is destroyed and hence, the recoloring of v cannot be counted again towards invalidating it. Therefore, the recoloring of a vertex can be counted towards invalidating any j-tree at most c times throughout the entire construction. Since c is assumed to be a constant, we obtain the following result.

Lemma 3.4. *After a reset phase in which h edge insertions become wasted, we can charge A with $\Omega(h \cdot n^{\frac{2}{c(c-1)}})$ vertex recolorings. Moreover, A will be charged at most $O(1)$ times for each recoloring.*

If A stops triggering resets, then at some point we reach a $(c-1)$-configuration with at least $T_{c-1} \geq 2(c+1)$ trees and $c+1$ valid ones. By linking together two such trees with the same color we can force algorithm A to trigger a reset.

Theorem 3.5. *Let c be a constant. For any sufficiently large integers n and α depending only on c, and any $m = \Omega(n)$ sufficiently large, there exists a forest F with αn vertices, such that for any recoloring algorithm A, there exists a sequence of m updates that forces A to perform $\Omega(m \cdot n^{\frac{2}{c(c-1)}})$ vertex recolorings to maintain a c-coloring of F.*

References

1. S. Baswana, M. Gupta, and S. Sen. Fully dynamic maximal matching in $O(\log n)$ update time. *SIAM J. on Comp.*, 44(1):88–113, 2015.
2. S. Baswana, S. Khurana, and S. Sarkar. Fully dynamic randomized algorithms for graph spanners. *ACM Trans. on Alg.*, 8(4):35, 2012.
3. P. Borowiecki and E. Sidorowicz. Dynamic coloring of graphs. *Fundamenta Informaticae*, 114(2):105–128, 2012.
4. O. Coudert. Exact coloring of real-life graphs is easy. In *Proc. 34th Design Autom. Conf.*, pages 121–126. ACM, 1997.
5. C. Demetrescu, D. Eppstein, Z. Galil, and G. F. Italiano. Dynamic graph algorithms. In M. J. Atallah and M. Blanton, editors, *Algorithms and Theory of Computation Handbook*. Chapman & Hall/CRC, 2010.
6. C. Demetrescu, I. Finocchi, and P. Italiano. Dynamic graphs. In D. Mehta and S. Sahni, editors, *Handbook on Data Structures and Applications*, Computer and Information Science. CRC Press, 2005.
7. A. Dutot, F. Guinand, D. Olivier, and Y. Pigné. On the decentralized dynamic graph-coloring problem. In *Proc. Worksh. Compl. Sys. and Self-Org. Mod.*, 2007.
8. M. M. Halldórsson. Parallel and on-line graph coloring. *J. Alg.*, 23(2):265–280, 1997.
9. M. M. Halldrsson and M. Szegedy. Lower bounds for on-line graph coloring. *Theo. Comp. Sci.*, 130(1):163 – 174, 1994.
10. M. Henzinger, S. Krinninger, and D. Nanongkai. A subquadratic-time algorithm for decremental single-source shortest paths. In *Proc. 25th ACM-SIAM Symp. on Discr. Alg.*, pages 1053–1072, 2014.
11. J. Holm, K. De Lichtenberg, and M. Thorup. Poly-logarithmic deterministic fully-dynamic algorithms for connectivity, minimum spanning tree, 2-edge, and biconnectivity. *J. ACM*, 48(4):723–760, 2001.
12. B. M. Kapron, V. King, and B. Mountjoy. Dynamic graph connectivity in polylogarithmic worst case time. In *Proc. 24th ACM-SIAM Symp. on Discr. Alg.*, pages 1131–1142, 2013.
13. R. M. Karp. Reducibility among combinatorial problems. In *Complexity of computer computations*, pages 85–103. Plenum, New York, 1972.
14. L. Lovász, M. E. Saks, and W. T. Trotter. An on-line graph coloring algorithm with sublinear performance ratio. *Discr. Math.*, 75(1-3):319–325, 1989.
15. M. V. Marathe, H. Breu, H. B. Hunt, III, S. S. Ravi, and D. J. Rosenkrantz. Simple heuristics for unit disk graphs. *Networks*, 25(2):59–68, 1995.
16. L. Ouerfelli and H. Bouziri. Greedy algorithms for dynamic graph coloring. In *Proc. Int. Conf. on Comm., Comp. and Control App.*, pages 1–5, 2011.
17. D. Preuveneers and Y. Berbers. ACODYGRA: an agent algorithm for coloring dynamic graphs. In *Symb. Num. Alg. Sci. Comp.*, pages 381–390, 2004.
18. L. Roditty and U. Zwick. Improved dynamic reachability algorithms for directed graphs. In *Proc. 43rd IEEE Sym. Found. Comp. Sci.*, pages 679–688, 2002.
19. L. Roditty and U. Zwick. Dynamic approximate all-pairs shortest paths in undirected graphs. In *Proc. 45th IEEE Sym. Found. Comp. Sci.*, pages 499–508, 2004.
20. M. Thorup. Fully-dynamic min-cut. *Combinatorica*, 27(1):91–127, 2007.
21. S. Vishwanathan. Randomized online graph coloring. *J. Alg.*, 13(4):657–669, 1992.
22. D. Zuckerman. Linear degree extractors and the inapproximability of max clique and chromatic number. *Theory Comp.*, 3:103–128, 2007.

Universal Hinge Patterns for Folding Strips Efficiently into Any Grid Polyhedron

Nadia M. Benbernou[1] *, Erik D. Demaine[2],
Martin L. Demaine[2], and Anna Lubiw[3]

[1] Google Inc., nbenbern@gmail.com.
[2] MIT Computer Science and Artificial Intelligence Laboratory,
32 Vassar St., Cambridge, MA 02139, USA, {edemaine,mdemaine}@mit.edu
[3] David R. Cheriton School of Computer Science, University of Waterloo,
Waterloo, Ontario N2L 3G1, Canada, alubiw@uwaterloo.ca

Abstract. We present two universal hinge patterns that enable a strip of material to fold into any connected surface made up of unit squares on the 3D cube grid—for example, the surface of any polycube. The folding is efficient: for target surfaces topologically equivalent to a sphere, the strip needs to have only twice the target surface area, and the folding stacks at most two layers of material anywhere. These geometric results offer a new way to build programmable matter that is substantially more efficient than what is possible with a square $N \times N$ sheet of material, which can fold into all polycubes only of surface area $O(N)$ and may stack $\Theta(N^2)$ layers at one point. We also show how our strip foldings can be executed by a rigid motion without collisions (albeit assuming zero thickness), which is not possible in general with 2D sheet folding.

To achieve these results, we develop new approximation algorithms for milling the surface of a grid polyhedron, which simultaneously give a 2-approximation in tour length and an 8/3-approximation in the number of turns. Both length and turns consume area when folding a strip, so we build on past approximation algorithms for these two objectives from 2D milling.

1 Introduction

In *computational origami design*, the goal is generally to develop an algorithm that, given a desired shape or property, produces a crease pattern that folds into an origami with that shape or property. Examples include folding any shape [9], folding approximately any shape while being watertight [10], and optimally folding a shape whose projection is a desired metric tree [14,15]. In all of these results, every different shape or tree results in a completely different crease pattern; two shapes rarely share many (or even any) creases.

The idea of a *universal hinge pattern* [6] is that a finite set of *hinges* (possible creases) suffice to make exponentially many different shapes. The main result along these lines is that an $N \times N$ "box-pleat" grid suffices to make any polycube

---

* Work performed while at MIT.

© Springer International Publishing AG 2017
F. Ellen et al. (Eds.): WADS 2017, LNCS 10389, pp. 109–120, 2017.
DOI: 10.1007/978-3-319-62127-2_10

made of $O(N)$ cubes [6]. The *box-pleat grid* is a square grid plus alternating diagonals in the squares, also known as the "tetrakis tiling". For each target polycube, a subset of the hinges in the grid serve as the crease pattern for that shape. Polycubes form a *universal* set of shapes in that they can arbitrarily closely approximate (in the Hausdorff sense) any desired volume.

The motivation for universal hinge patterns is the implementation of *programmable matter*—material whose shape can be externally programmed. One approach to programmable matter, developed by an MIT–Harvard collaboration, is a *self-folding sheet*—a sheet of material that can fold itself into several different origami designs, without manipulation by a human origamist [12,1]. For practicality, the sheet must consist of a fixed pattern of hinges, each with an embedded actuator that can be programmed to fold or not. Thus for the programmable matter to be able to form a universal set of shapes, we need a universal hinge pattern.

The box-pleated polycube result [6], however, has some practical limitations that prevent direct application to programmable matter. Specifically, using a sheet of area $\Theta(N^2)$ to fold N cubes means that all but a $\Theta(1/N)$ fraction of the surface area is wasted. Unfortunately, this reduction in surface area is necessary for a roughly square sheet, as folding a $1 \times 1 \times N$ tube requires a sheet of diameter $\Omega(N)$. Furthermore, a polycube made from N cubes can have surface area as low as $\Theta(N^{2/3})$, resulting in further wastage of surface area in the worst case. Given the factor-$\Omega(N)$ reduction in surface area, an average of $\Omega(N)$ layers of material come together on the polycube surface. Indeed, the current approach can have up to $\Theta(N^2)$ layers coming together at a single point [6]. Real-world robotic materials have significant thickness, given the embedded actuation and electronics, meaning that only a few overlapping layers are really practical [12].

Our results: strip folding. In this paper, we introduce two new universal hinge patterns that avoid these inefficiencies, by using sheets of material that are long only in one dimension ("strips"). Specifically, Fig. 1 shows the two hinge patterns: the *canonical strip* is a $1 \times N$ strip with hinges at integer grid lines and same-oriented diagonals, while the *zig-zag strip* is an N-square zig-zag with hinges at just integer grid lines. We show in Section 2 that any *grid surface*— any connected surface made up of unit squares on the 3D cube grid—can be folded from either strip. The strip length only needs to be a constant factor larger than the surface area, and the number of layers is at most a constant throughout the folding. Most of our analysis concerns (genus-0) *grid polyhedra*, that is, when the surface is topologically equivalent to a sphere (a manifold without boundary, so that every edge is incident to exactly two grid squares, and without handles, unlike a torus). We show in Section 4 that a grid polyhedron of

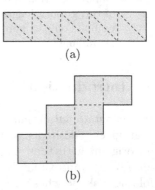

Fig. 1. Two universal hinge patterns in strips. (a) A canonical strip of length 5. (b) A zig-zag strip of length 6. The dashed lines are hinges.

surface area N can be folded from a canonical strip of length $2N$ with at most two layers everywhere, or from a zig-zag strip of length $4N$ with at most four layers everywhere.

The improved surface efficiency and reduced layering of these strip results seem more practical for programmable matter. In addition, the panels of either strip (the facets delineated by hinges) are connected acyclically into a path, making them potentially easier to control. One potential drawback is that the reduced connectivity makes for a flimsier device; this issue can be mitigated by adding tabs to the edges of the strips to make full two-dimensional contacts across seams and thereby increase strength.

We also show in Section 5 an important practical result for our strip foldings: under a small assumption about feature size, we give an algorithm for actually folding the strip into the desired shape, while keeping the panels rigid (flat) and avoiding self-intersection throughout the motion. Such a rigid folding process is important given current fabrication materials, which put flexibility only in the creases between panels [12]. An important limitation, however, is that we assume zero thickness of the material, which would need to be avoided before this method becomes practical.

Our approach is also related to the 1D chain robots of [7], but based on thin material instead of thick solid chains. Most notably, working with thin material enables us to use a few overlapping layers to make any desired surface without scaling, and still with high efficiency. Essentially, folding long thin strips of sheet material is like a fusion between 1D chains of [7] and the square sheet folding of [6,12,1].

Milling tours. At the core of our efficient strip foldings are efficient approximation algorithms for *milling* a grid polyhedron. Motivated by rapid-fabrication CNC milling/cutting tools, milling problems are typically stated in terms of a 2D region called a "pocket" and a cutting tool called a "cutter", with the goal being to find a path or tour for the cutter that covers the entire pocket. In our situation, the "pocket" is the surface of the grid polyhedron, and the "cutter" is a unit square constrained to move from one grid square of the surface to an (intrinsically) adjacent grid square.

The typical goals in milling problems are to minimize the length of the tour [3] or to minimize the number of turns in the tour [2]. Both versions are known to be strongly NP-hard, even when the pocket is an integral orthogonal polygon and the cutter is a unit square. We conjecture that the problem remains strongly NP-hard when the pocket is a grid polyhedron, but this is not obvious.

In our situation, both length and number of turns are important, as both influence the required length of a strip to cover the surface. Thus we develop one algorithm that simultaneously approximates both measures. Such results have also been achieved for 2D pockets [2]; our results are the first we know for surfaces in 3D. Specifically, we develop in Section 3 an approximation algorithm for computing a milling tour of a given grid polyhedron, achieving both a 2-approximation in length and an 8/3-approximation in number of turns.

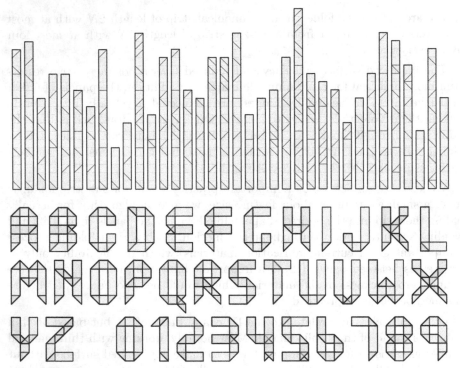

Fig. 2. Strip folding of individual letters typeface, A–Z and 0–9: unfolded font (top) and folded font (bottom), where the face incident to the bottom edge remains face-up.

Fonts. To illustrate the power of strip folding, we designed a typeface, representing each letter of the alphabet by a folding of a $1 \times x$ strip for some x, as shown in Fig. 2. The individual-letters typeface consists of two fonts: the unfolded font is a puzzle to figure out each letter, while the folded font is easy to read. These crease patterns adhere to an integer grid with orthogonal and/or diagonal creases, but are not necessarily subpatterns of the canonical hinge pattern. This extra flexibility gives us control to produce folded half-squares as desired, increasing the font's fidelity.

We have developed a web app that visualizes the font.[4] Currently in development is the ability to chain letters together into one long strip folding; Fig. 9 at the end of the paper shows one example.

The full version of this paper [5] contains details omitted from this extended abstract.

2   Universality

In this section, we prove that both the canonical strip and zig-zag strip of Fig. 1, of sufficient length, can fold into any grid surface. We begin with milling tours

---

[4] http://erikdemaine.org/fonts/strip/

Fig. 3. (a) Left and (b) right turn with a canonical strip.

which provide an abstract plan for routing the strip, and then turn to the details of how to manipulate each type of strip.

Dual graph. Recall that a *grid surface* consists of one or more *grid squares*—that is, squares of the 3D cube grid—glued edge-to-edge to form a connected surface (ignoring vertex connections). Define the *dual graph* to have a dual vertex for each such grid square, and any two grid squares sharing an edge define a dual edge between the two corresponding dual vertices. Our assumption of the grid surface being connected is equivalent to the dual graph being connected.

Milling tours. A *milling tour* is a (not necessarily simple) spanning cycle in the dual graph, that is, a cycle that visits every dual vertex at least once (but possibly more than once). Equivalently, we can think of a milling tour as the path traced by the center of a moving square that must cover the entire surface while remaining on the surface, and return to its starting point. Milling tours always exist: for example, we can double a spanning tree of the dual graph to obtain a milling tour of length less than double the given surface area.

At each grid square, we can characterize a milling tour as going straight, turning, or U-turning—intrinsically on the surface—according to which two sides of the grid square the tour enters and exits. If the sides are opposite, the tour is *straight*; if the sides are incident, the tour *turns*; and if the sides are the same, the tour *U-turns*. Intuitively, we can imagine unfolding the surface and developing the tour into the plane, and measuring the resulting planar turn angle at the center of the grid square.

Strip folding. To prove universality, it suffices to show that a canonical strip or zig-zag strip can follow any milling tour and thus make any grid polyhedron. In particular, it suffices to show how the strip can go straight, turn left, turn right, and U-turn. Then, in 3D, the strip would be further folded at each traversed edge of the grid surface, to stay on the surface. Indeed, U-turns can be viewed as folding onto the opposite side of the same surface, and thus are intrinsically equivalent to going straight; hence we can focus on going straight and making left/right turns.

Canonical strip. Fig. 3 shows how a canonical strip can turn left or right; it goes straight without any folding. Each turn adds 1 to the length of the strip, and adds 2 layers to part of the grid square where the turn is made. Therefore a milling tour of length L with t turns of a grid surface can be followed by a canonical strip of length $L + t$. Furthermore, if the milling tour visits each grid square at most c times, then the strip folding has at most $3c$ layers covering any point of the surface.

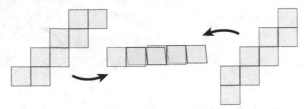

Fig. 4. Going straight with a zig-zag strip requires at most two unit squares per grid square. Left and right crease patterns show two different parities along the strip.

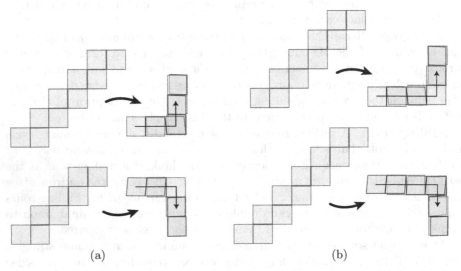

Fig. 5. Turning with a zig-zag strip has two cases because of parity. (a) Turning left at an odd position requires three grid squares, whereas turning right requires one grid square. (b) Turning left at an even position requires one grid square, whereas turning right requires three grid squares.

Zig-zag strip. Fig. 4 shows how to fold a zig-zag strip in order to go straight. In this straight portion, each square of the surface is covered by two squares of the strip. Fig. 5 shows left and right turns. Observe that turns require either one or three squares of the strip. Therefore a milling tour of length L with t turns can be followed by a zig-zag strip of length at most $2L + t$. Furthermore, if the milling tour visits each grid square at most c times, then the strip folding has at most $3c$ layers covering any point of the surface.

Proposition 1. *Every grid surface of area N can be folded from a canonical strip of length $4N$, with at most eight layers stacked anywhere, and from a zig-zag strip of length $6N$, with at most twelve layers stacked anywhere.*

The goal in the rest of this paper is to achieve better bounds for grid polyhedra, using more carefully chosen milling tours.

3   Milling Tour Approximation

This section presents a constant-factor approximation algorithm for milling a (genus-0) grid polyhedron P with respect to both length and turns. Specifically, our algorithm is a 2-approximation in length and an 8/3-approximation in turns. Our milling tours also have special properties that make them more amenable to strip folding.

Our approach is to reduce the milling problem to vertex cover in a tripartite graph. Then it follows that our algorithm is a 2α-approximation in turns, where α is the best approximation factor for vertex cover in tripartite graphs. The best known bounds on α are $34/33 \leq \alpha \leq 4/3$. Clementi et al. [8] proved that minimum vertex cover in tripartite graphs is not approximable within a factor smaller than $34/33 = 1.\overline{03}$ unless P = NP. Theorem 1 of [13] implies a 4/3-approximation for minimum weighted vertex cover for tripartite graphs (assuming we are given the 3-partition of the vertex set, which we know in our case). Thus we use $\alpha = 4/3$ below. An improved approximation ratio α would improve our approximation ratios, but may also affect the stated running times, which currently assume use of [13].

3.1   Bands

The basis for our approximation algorithms is the notion of "bands" for a grid polyhedron P. Let x_{min} and x_{max} respectively be the minimum and maximum x coordinates of P; define $y_{min}, y_{max}, z_{min}, z_{max}$ analogously. These minima and maxima have integer values because the vertices of P lie on the integer grid. Define the ith x-slab $S_x(i)$ to be the slab bounded by parallel planes $x = x_{min} + i$ and $x = x_{min} + i + 1$, for each $i \in \{0, 1, \ldots, x_{max} - x_{min} - 1\}$. The intersection of P with the ith x-slab $S_x(i)$ (assuming i is in the specified range) is either a single *band* (i.e., a simple cycle of grid squares in that slab), or a collection of such bands, which we refer to as x-bands. Define y-bands and z-bands analogously.

Two bands *overlap* if there is a grid square contained in both bands. Each grid square of P is contained in precisely two bands (e.g., if a grid square's outward normal were in the $+z$-direction, then it would be contained in one x-band and one y-band). Two bands B_1 and B_2 are *adjacent* if they do not overlap, and a grid square of B_1 shares an edge with a grid square of B_2. A *band cover* for P is a collection of x-, y-, and z-bands that collectively cover the entire surface of P. The *size* of a band cover is the number of its bands.

3.2   Cover Bands

The starting point for the milling approximation algorithm is to find an approximately minimum band cover, as the minimum band cover is a lower bound on the number of turns in any milling tour:

Proposition 2. [2, Lemma 4.9] *The size of a minimum band cover of a grid polyhedron P is a lower bound on the number of turns in any milling tour of P.*

Next we describe how to find a near-optimal band cover. Consider the graph G_P with one vertex per band of a grid polyhedron P, connecting two vertices by an edge if their corresponding bands overlap. It turns out that an (approximately minimum) vertex cover in G_P will give us an (approximately minimum) band cover in P:

Proposition 3. *A vertex cover for G_P induces a band cover of the same size and vice versa.*

Because the bands fall into three classes (x-, y-, and z-), with no overlapping bands within a single class, G_P is tripartite. Hence we can use an α-approximation algorithm for vertex cover in tripartite graphs to find an α-approximate vertex cover in G_P and thus an α-approximate band cover of P.

3.3 Connected Bands

Our next goal will be to efficiently tour the bands in the cover. Given a band cover S for a grid polyhedron P, define the *band graph* G_S to be the subgraph of G_P induced by the subset of vertices corresponding to S. We will construct a tour of the bands S based on a spanning tree of G_S. Our first step is thus to show that G_S is connected (Lemma 5 below). We do so by showing that adjacent bands (as defined in Section 3.1) are in the same connected component of G_S, using the following lemma of Genc [11]:

Lemma 4. [11][5] *For any band B in a grid polyhedron P, let N_b be the bands of P overlapping B. (Equivalently, N_b is the set of neighbors of B in G_P). Then the subgraph of G_P induced by N_B is connected.*

Lemma 5. *If S is a band cover for a grid polyhedron P, then the graph G_S is connected.*

3.4 Band Tour

Now we can present our algorithm for transforming a band cover into an efficient milling tour.

Theorem 6. *Let P be a grid polyhedron with N grid squares. In $O(N^2 \log N)$ time, we can find a milling tour of P that is a 2-approximation in length and an 8/3-approximation (or more generally, a 2α-approximation) in turns.*

We now state some additional properties of any milling tour produced by the approximation algorithm of Theorem 6, which will be useful for later applications to strip folding in Section 4.

---

[5] Genc [11] uses somewhat different terminology to state this lemma: "straight cycles in the dual graph" are our bands, and "crossing" is our overlapping. The induced subgraph is also defined directly, instead of as an induced subgraph of G_P.

Proposition 7. *Let P be a grid polyhedron, and consider a milling tour of P obtained from the approximation algorithm of Theorem 6. Then the following properties hold:*

1. *A grid square of P is either visited once, in which case it is visited by a straight part of the tour; or it is visited twice, by two straight junctions or by two turn junctions.*
2. *In the case of a turn junction, the length of the milling tour between the two visits to the grid square (counting only one of the two visits to the grid square in the length measurement) is even.*
3. *The tour can be modified to alternate between left and right turns (without changing its length or the number of turns).*

3.5 Polynomial Time

The algorithm described above is polynomial in the surface area N of the grid polyhedron, or equivalently, polynomial in the number of unit cubes making up the polyomino solid. For our application to strip folding, this is polynomial in the length of the strip, and thus sufficient for most purposes. On the other hand, polyhedra are often encoded as a collection of n vertices and faces, with faces possibly much larger than a unit square. In this case, the algorithm runs in *pseudopolynomial* time.

Although we do not detail the approach here, our algorithm can be modified to run in polynomial time. To achieve this result, we can no longer afford to deal with unit bands directly, because their number is polynomially related to the number N of grid squares, not the number n of vertices. To achieve polynomial time in n, we concisely encode the output milling tour using "fat" bands rather than unit bands, which can then be easily decoded into a tour of unit bands. By making each band as wide as possible, their number is polynomially related to n instead of N. Details of an $O(n^3 \log n)$-time milling approximation algorithm (with the same approximation bounds as above) can be found in [4].

4 Strip Foldings of Grid Polyhedra

In this section, we show how we can use the milling tours from Section 3 to fold a canonical strip or zig-zag strip efficiently into a given (genus-0) grid polyhedron. For both strip types, define the *length* of a strip to be the number of grid squares it contains; refer to Fig. 1. For a strip folding of a polyhedron P, define the *number of layers covering a point q* on P to be the number of interior points of the strip that map to q in the folding, excluding *crease points*, that is, points lying on a hinge that gets folded by a nonzero angle. (This definition may undercount the number of layers along one-dimensional creases, but counts correctly at the remaining two-dimensional subset of P.) We will give bounds on the length of the strip and also on the maximum number of layers of the strip covering any point of the polyhedron.

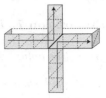

The main idea for canonical strips is that Properties (1) and (3) of Proposition 7 allow us to make turns as shown in Fig. 6, so that we do not waste an extra square of the strip per turn.

Fig. 6. A turn junction for a canonical strip.

Theorem 8. *Let P be a grid polyhedron, and let N be the number of grid squares of P. Then P can be covered by a folding of a canonical strip of length $2N$, and with at most two layers covering any point of P.*

For zig-zag strips, we instead use Properties (1) and (2) of Proposition 7:

Theorem 9. *Let P be a grid polyhedron, and let N be the number of grid squares of P. Then P can be covered by a folding of a zig-zag strip of length $4N$, and with at most four layers covering any point of P.*

By coloring the two sides of the zig-zag strip differently, we can also bicolor the surface of P in any pattern we wish, as long as each grid square is assigned a uniform color. We do not prove this result formally here, but mention that the bounds in length would become somewhat worse, and the rigid motions presented in Section 5 do not work in this setting.

5   Rigid Motion Avoiding Collision

So far we have focused on just specifying a final folded state for the strip, and ignored the issue of how to continuously move the strip into that folded state. In this section, we show how to achieve this using a *rigid folding motion*, that is, a continuous motion that keeps all polygonal faces between hinges rigid/planar, bending only at the hinges, and avoids collisions throughout the motion. Rigid folding motions are important for applications to real-world programmable matter made out of stiff material except at the hinges. Our approach may still suffer from practical issues, as it requires a large (temporary) accumulation of many layers in an accordion form.

We prove that, if the grid polyhedron P has feature size at least 2, then we can construct a rigid motion of the strip folding without collision. (By *feature size at least* 2, we mean that every exterior voxel of P is contained in some empty $2 \times 2 \times 2$ box.) If the grid polyhedron we wish to fold has feature size 1, then one solution is to scale the polyhedron by a factor of 2, and then the results here apply.

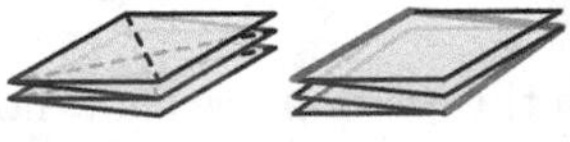

Fig. 7. Accordion for (a) canonical strip and (b) zig-zag strip, with hinges drawn thick for increased visibility.

Our approach is to keep the yet-unused portion of the strip folded up into an *accordion* and then to unfold only what is needed for the current move: straight, left, or right. Fig. 7 shows the accordion state for the canonical strip and for the zig-zag strip. We will perform the strip folding in such a way that the strip never penetrates the interior of the polyhedron P, and it never weaves under

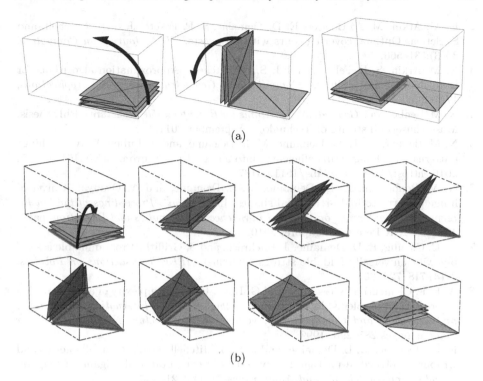

Fig. 8. Canonical strip, face-up annon. (a) Straight. (b) Turn where e_{23} is flat.

previous portions of the strip. Thus, we could wrap the strip around P's surface even if P's interior were already a filled solid. This restriction helps us think about folding the strip locally, because some of P's surface may have already been folded (and it thus should not be penetrated) by earlier parts of the strip.

It suffices to show, regardless of the local geometry of the polyhedron at the grid square where the milling tour either goes straight or turns, and regardless of whether the accordion faces up or down relative to the grid square it is covering, that we can maneuver the accordion in a way that allows us to unroll as many squares as necessary to perform the milling-tour move. Fig. 8 shows two key cases for unrolling part of the accordion of a canonical strip. See [5] for details.

Acknowledgments. We thank ByoungKwon An and Daniela Rus for several helpful discussions about programmable matter that motivated this work.

References

1. B. An, N. Benbernou, E. D. Demaine, and D. Rus. Planning to fold multiple objects from a single self-folding sheet. *Robotica*, 29(1):87–102, 2011. Special issue on Robotic Self-X Systems.

2. E. M. Arkin, M. A. Bender, E. D. Demaine, S. P. Fekete, J. S. B. Mitchell, and S. Sethia. Optimal covering tours with turn costs. *SIAM Journal on Computing*, 35(3):531–566, 2005.
3. E. M. Arkin, S. P. Fekete, and J. S. B. Mitchell. Approximation algorithms for lawn mowing and milling. *Computational Geometry: Theory and Applications*, 17(1–2):25–50, 2000.
4. N. M. Benbernou. *Geometric Algorithms for Reconfigurable Structures*. PhD thesis, Massachusetts Institute of Technology, September 2011.
5. N. M. Benbernou, E. D. Demaine, M. L. Demaine, and A. Lubiw. Universal hinge patterns for folding strips efficiently into any grid polyhedron. arXiv:1611.03187, 2016. https://arXiv.org/abs/1611.03187.
6. N. M. Benbernou, E. D. Demaine, M. L. Demaine, and A. Ovadya. Universal hinge patterns to fold orthogonal shapes. In *Origami[5]: Proceedings of the 5th International Conference on Origami in Science, Mathematics and Education*, pages 405–420. A K Peters, Singapore, 2010.
7. K. C. Cheung, E. D. Demaine, J. Bachrach, and S. Griffith. Programmable assembly with universally foldable strings (moteins). *IEEE Transactions on Robotics*, 27(4):718–729, 2011.
8. A. E. F. Clementi, P. Crescenzi, and G. Rossi. On the complexity of approximating colored-graph problems. In *Proceedings of the 5th Annual International Conference on Computing and Combinatorics*, volume 1627 of *Lecture Notes in Computer Science*, pages 281–290, 1999.
9. E. D. Demaine, M. L. Demaine, and J. S. B. Mitchell. Folding flat silhouettes and wrapping polyhedral packages: New results in computational origami. *Computational Geometry: Theory and Applications*, 16(1):3–21, 2000.
10. E. D. Demaine and T. Tachi. Origamizer: A practical algorithm for folding any polyhedron. Manuscript, 2017.
11. B. Genc. *Reconstruction of Orthogonal Polyhedra*. PhD thesis, University of Waterloo, 2008.
12. E. Hawkes, B. An, N. M. Benbernou, H. Tanaka, S. Kim, E. D. Demaine, D. Rus, and R. J. Wood. Programmable matter by folding. *Proceedings of the National Academy of Sciences of the United States of America*, 107(28):12441–12445, 2010.
13. D. S. Hochbaum. Efficient bounds for the stable set, vertex cover and set packing problems. *Discrete Applied Mathematics*, 6(3):243–254, 1983.
14. R. J. Lang. A computational algorithm for origami design. In *Proceedings of the 12th Annual ACM Symposium on Computational Geometry*, pages 98–105, Philadelphia, PA, May 1996.
15. R. J. Lang and E. D. Demaine. Facet ordering and crease assignment in uniaxial bases. In *Origami[4]: Proceedings of the 4th International Conference on Origami in Science, Mathematics, and Education*, pages 189–205, Pasadena, California, September 2006. A K Peters.

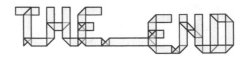

Fig. 9. An example of joining together a few letters from our typeface in Fig. 2. Unfolding (bottom) not to scale with folding (top).

An optimal XP algorithm for Hamiltonian Cycle on graphs of bounded clique-width

Benjamin Bergougnoux[1], Mamadou Moustapha Kanté[1], and O-joung Kwon[2] *

[1] Université Clermont Auvergne, LIMOS, CNRS, Aubière, France
[2] Logic and Semantics, TU Berlin, Berlin, Germany
benjamin.bergougnoux@uca.fr, mamadou.kante@uca.fr, ojoungkwon@gmail.com

Abstract. For MSO_2-expressible problems like EDGE DOMINATING SET or HAMILTONIAN CYCLE, it was open for a long time whether there is an algorithm which given a clique-width k-expression of an n-vertex graph runs in time $f(k) \cdot n^{\mathcal{O}(1)}$ for some function f. Recently, Fomin et al. (*SIAM. J. Computing*, 2014) presented several lower bounds; for instance, there are no $f(k) \cdot n^{o(k)}$-time algorithms for EDGE DOMINATING SET and for HAMILTONIAN CYCLE unless the Exponential Time Hypothesis (ETH) fails. They also provided an algorithm running in time $n^{\mathcal{O}(k)}$ for EDGE DOMINATING SET, but left open whether HAMILTONIAN CYCLE can be solved in time $n^{\mathcal{O}(k)}$.

In this paper, we prove that HAMILTONIAN CYCLE can be solved in time $n^{\mathcal{O}(k)}$. This improves the naive algorithm that runs in time $n^{\mathcal{O}(k^2)}$ by Espelage et al. (WG 2001). We present a general technique of representative sets using two-edge colored multigraphs on k vertices. The essential idea behind is that for a two-edge colored multigraph, the existence of an Eulerian trail that uses edges with different colors alternatively can be determined by two information: the number of colored edges incident with each vertex, and the connectedness of the multigraph. With this idea, we avoid the bottleneck of the naive algorithm, which stores all the possible multigraphs on k vertices with at most n edges. We can apply this technique to other problems such as q-CYCLE COVERING or DIRECTED HAMILTONIAN CYCLE as well.

1 Introduction

Tree-width is one of the graph width parameters that plays an important role in graph algorithms. Various problems which are NP-hard on general graphs, have been shown to be solvable in polynomial time on graphs of bounded tree-width [1,2]. A celebrated algorithmic meta-theorem by Courcelle [3] states that every graph property expressible in monadic second-order logic which allows

---

* B. Bergougnoux and M.M. Kanté are supported by French Agency for Research under the GraphEN project (ANR-15-CE-0009). O. Kwon is supported by the European Research Council (ERC) under the European Union's Horizon 2020 research and innovation programme (ERC consolidator grant DISTRUCT, agreement No. 648527).

© Springer International Publishing AG 2017
F. Ellen et al. (Eds.): WADS 2017, LNCS 10389, pp. 121–132, 2017.
DOI: 10.1007/978-3-319-62127-2_11

quantifications over edge and vertex sets (MSO_2) can be decided in linear time on graphs of bounded tree-width. MINIMUM DOMINATING SET, q-COLORING, and HAMILTONIAN CYCLE problems are such graph problems.

Courcelle and Olariu [5] defined the notion of *clique-width* of graphs, whose modeling power is strictly stronger than tree-width. The motivation of clique-width came from the observation that many algorithmic problems are tractable on classes of graphs that can be recursively decomposed along vertex partitions (A, B) where the number of neighbourhood types between A and B is small. We formally define clique-width in Section 2. Courcelle, Makowsky, and Rotics [4] extended the meta-theorem on graphs of bounded tree-width [3] to graphs of bounded clique-width, at a cost of a smaller set of problems, namely, the class of problems expressible in MSO_1, which allows quantifications on vertex sets only. Some of the known examples of graph problems that are MSO_2-definable, but not MSO_1-definable are MAX-CUT, EDGE DOMINATING SET, and HAMILTONIAN CYCLE problems.

A natural question is whether such problems allow an algorithm with running time $f(k) \cdot n^{\mathcal{O}(1)}$ for some function f, when a *clique-width k-expression* of an n-vertex input graph is given. This question has been carefully answered by Fomin et al. [8,9]. In particular, they showed that for MAX-CUT and EDGE DOMINATING SET, there is no $f(k) \cdot n^{o(k)}$-time algorithm unless the Exponential Time Hypothesis (ETH) fails, and proposed for both problems algorithms with running time $n^{\mathcal{O}(k)}$. They proved that HAMILTONIAN CYCLE also cannot be solved in time $f(k) \cdot n^{o(k)}$, unless ETH fails, but left open the question of finding an algorithm running in time $n^{\mathcal{O}(k)}$. Until now, the best algorithm is the one by Espelage, Gurski, and Wanke [7] which runs in time $n^{\mathcal{O}(k^2)}$.

Our Contribution and Approach. In this paper, we prove that HAMILTONIAN CYCLE can be solved in time $n^{\mathcal{O}(k)}$, thereby resolving the open problem in [9]. A *Hamiltonian cycle* in a graph is a cycle containing all vertices of the graph. The HAMILTONIAN CYCLE problem asks whether given a graph G, G contains a Hamiltonian cycle or not.

Theorem 1. *Given an n-vertex graph and its clique-width k-expression, one can solve* HAMILTONIAN CYCLE *in time* $n^{\mathcal{O}(k)}$.

A k-labeled graph is a graph whose vertices are labeled by integers in $\{1, \ldots, k\}$. Clique-width k-expressions are expressions which allow to recursively construct a graph with the following graph operations: (1) creating a graph with a single vertex labeled i, (2) taking the disjoint union of two k-labeled graphs, (3) adding all edges between vertices labeled i and vertices labeled j, for some $i \neq j$, (4) renaming all vertices labeled i into j. The clique-width of a graph is the minimum k such that it can be constructed using labels in $\{1, \ldots, k\}$. One observes that if a graph contains a Hamiltonian cycle C, then each k-labeled graph H introduced in the clique-width k-expression admits a partition of its vertex set into pairwise vertex-disjoint paths, *path-partitions*, which is the intersection of H and C. A natural approach is to enumerate all path-partitions. Furthermore, since the adjacency relations between this k-labeled graph and the remaining part only depends on the labels, it is sufficient to store for each pair of labels (i, j), the number of paths whose end vertices are labeled by i and j. As the number of paths of a path-partition is bounded by n, there are at most $n^{\mathcal{O}(k^2)}$ possibilities. This is the basic idea of the XP algorithm developed by Espelage, Gurski, and Wanke [7].

The essential idea of our algorithm is to introduce an equivalence relation between two path-partitions. Given a path-partition $\mathcal{P}$ that is a restriction of a Hamiltonian cycle C, we consider the maximal paths in $C - \bigcup_{P \in \mathcal{P}} E(P)$ as another path-partition $\mathcal{Q}$. As depicted in Figure 1, we can construct a multigraph associated with $\mathcal{P}$ and $\mathcal{Q}$ on the vertex set $\{v_1, \ldots, v_k\}$, by adding a red edge $v_i v_j$ (an undashed edge) if there is a path in $\mathcal{P}$ with end vertices labeled by i and j, and by adding a blue edge $v_i v_j$ (a dashed edge) if there is a path in $\mathcal{Q}$ with end vertices labeled by i and j. A crucial observation is that this multigraph admits an Eulerian trail where red edges and blue edges are alternatively used. This is indeed a characterisation of the fact that two such path-partitions can be joined into a Hamiltonian cycle. To determine the existence of such an Eulerian trail, it is sufficient to know the degree of each vertex and the connected components of the corresponding multigraphs of the two path-partitions. This motivates an equivalence relation between path-partitions. As a byproduct, we can keep in each equivalence class a representative and since the number of equivalence classes is bounded by $2^{k \log_2 k} \cdot n^k$, we can turn the naive algorithm into an $n^{\mathcal{O}(k)}$-time algorithm (there are at most $2^{k \log_2 k}$ partitions of k-elements set). A more detailed explanation of our algorithm is provided in Section 3.

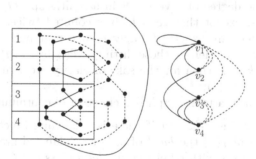

Fig. 1. The restriction of a Hamiltonian cycle to a k-labeled graph. The complement part can be considered as another set of paths.

The paper is organized as follows. Section 2 contains the necessary preliminaries and required notions. Section 3 is devoted to the overview of the algorithm and the proof of the existence of Eulerian trails in two-edge colored multigraphs. We introduce in Section 4 the equivalence relation between multigraphs on the vertex set $\{v_1, \ldots, v_k\}$, and introduce operations related to the update of path-partitions in clique-width k-expressions, and prove that they preserve the equivalence relation. We define the notion of representatives and give the algorithm in Section 5. We conclude with more applications of our representatives in Section 6. Some proofs are omitted (statements with $\star$) because of space constraints.

2   Preliminaries

The size of a set V is denoted by $|V|$, and we write $[V]^2$ to denote the set of all subsets of V of size 2. We denote by $\mathbb{N}$ the set of non-negative integers. We essentially follow [6] for our graph terminology, but we deal only with finite graphs. The vertex set of a graph G is denoted by $V(G)$ and its edge set by $E(G) \subseteq [V(G)]^2$. We write xy to denote an edge $\{x, y\}$. Let G be a graph. For $X \subseteq V(G)$, we denote by $G[X]$ the subgraph of G induced by X, and for $F \subseteq E(G)$, we write $G - F$ for the subgraph $(V(G), E(G) \setminus F)$. The *degree* of a vertex x, denoted by $\deg_G(x)$, is the number of edges incident with x. A *cut-edge* in a connected graph is an edge e such that $G - \{e\}$ is disconnected. For two sets $A, B \subseteq V(G)$, A is *complete* to B if for every $v \in A$ and $w \in B$, $vw \in E(G)$.

A graph is *non-trivial* if it contains an edge, otherwise it is said *trivial*. A *walk* of a graph is an alternating sequence of vertices and edges, starting and ending at some vertices, where for every consecutive pair of a vertex x and an edge e, x is incident with e. A *trail* of a graph is a walk where each edge is used at most once. A trail is *closed* if its first and last vertices are the same.

A *multigraph* is essentially a graph, but we allow two edges to be incident with the same set of vertices. Formally, a *multigraph* G is a pair $(V(G), E(G))$ of disjoint sets, also called sets of vertices and edges, respectively, together with a map $\text{mult}_G : E(G) \to V(G) \cup [V(G)]^2$, which maps every edge to one or two vertices, still called its end vertices. Note that we admit loops in multigraphs, while we do not in our definition of graphs. If there is $e \in E(G)$ such that $\text{mult}_G(e) = \{x, y\}$ (or $\text{mult}_G(e) = \{x\}$), we use the term *multiedge* to refer to $\{x, y\}$ (or $\{x\}$). The degree of a vertex x in a multigraph G, is defined analogously as in graphs, except that each loop is counted twice, and similarly for other notions. If there are exactly k edges e such that $\text{mult}_G(e) = \{x, y\}$ (or $\text{mult}_G(e) = \{x\}$), then we denote these distinct edges by $\{x, y\}_1, \ldots, \{x, y\}_k$ (or $\{x\}_1, \ldots, \{x\}_k$); if $k = 1$, then for the sake of clarity, we write $\{x, y\}$ (or $\{x\}$) instead of $\{x, y\}_1$ (or $\{x\}_1$).

An *Eulerian trail* in a multigraph is a closed trail containing all edges.

Clique-width. A graph is k-*labeled* if there is a labeling function $f : V(G) \to \{1, \ldots, k\}$, and we call $f(v)$ the *label* of v. For a k-labeled graph G, we simply call the set of all vertices with label i as the *label class i* of G.

The *clique-width* of a graph G is the minimum number of labels needed to construct G using the following four operations:

1. Creation of a new vertex v with label i (denoted by $i(v)$).
2. Disjoint union of two labeled graphs G and H (denoted by $G \oplus H$).
3. Joining by an edge each vertex with label i to each vertex with label j ($i \neq j$, denoted by $\eta_{i,j}$).
4. Renaming label i to j (denoted by $\rho_{i \to j}$).

Such an expression is called a *clique-width k-expression* or simply a k-*expression* if it uses at most k distinct labels. We can naturally represent this expression as a tree-structure. Such trees are known as *syntactic trees* associated with k-expressions.

A clique-width k-expression is called *irredundant* if whenever $\eta_{i,j}$ is applied, the constructed graph contains no prior edges between vertices with label i and vertices with label j. Courcelle and Olariu [5] proved that given a clique-width k-expression, it can be transformed into an irredundant k-expression in linear time. Therefore, we can assume that a given clique-width expression is irredundant.

Path-partition. For a graph G, a set $\mathcal{P} = \{P_1, \ldots, P_m\}$ of vertex-disjoint paths in G is called a *path-partition* of G if $\bigcup_{1 \leq i \leq m} V(P_i) = V(G)$. A path-partition $\mathcal{P}$ is *k-labeled* if the end vertices of each path in $\mathcal{P}$ are labeled by some integer in $\{1, \ldots, k\}$. For a path-partition $\mathcal{P}$ of a k-labeled graph G, the labeling of $\mathcal{P}$ *induced by the labeling of G* consists in assigning to each end vertex of a path in $\mathcal{P}$ its label in G. Lastly, for a k-labeled path-partition $\mathcal{P}$ of a graph, we define the auxiliary multigraph $Aux(\mathcal{P})$ with vertex set $\{v_1, \ldots, v_k\}$ and edge set $\bigcup_{i,j \in \{1,\ldots,k\}} \{\{v_i, v_j\}_1, \ldots, \{v_i, v_j\}_{\ell_{ij}}\}$ where ℓ_{ij} is the number of paths in $\mathcal{P}$ with end vertices labeled i and j respectively.

3   Overview of the algorithm

In an irredundant clique-width k-expression ϕ defining a given graph G, G is recursively constructed using k-labeled graphs. Such k-labeled graphs H arising in the k-expression defining G are subgraphs[3] of G and satisfy the following properties: (1) for two vertices $v, w \in V(H)$ with same labels in H, $N_G(v) \cap (V(G) \setminus V(H)) = N_G(w) \cap (V(G) \setminus V(H))$, and (2) for some label class i, say L_i, and label class j, say L_j, in H with $i \neq j$, if there exist $v \in L_i$, $w \in L_j$ with $vw \in E(G) \setminus E(H)$, then every vertex of L_i is adjacent to every vertex of L_j in G, and there are no edges between L_i and L_j in H. The former statement is because when we add vw, all vertices in each set L_i or L_j have the same label, and the latter statement is because of the irredundancy of ϕ. Given such a k-labeled subgraph H of G, for any Hamiltonian cycle C, the restriction of C to H induces a k-labeled path-partition of H. Because of the two properties (1) and (2), it is sufficient to store the end vertices of each path. This is naturally represented as a multigraph on k vertices, motivating the definition of the auxiliary multigraphs associated with k-labeled path-partitions. Our algorithm will be based on the following characterization of equivalent path-partitions.

Proposition 1 ($\star$). *Let $\mathcal{P}_1, \mathcal{P}_2$ be k-labeled path-partitions of H whose labelings are induced by the labeling of H such that $(\deg_{Aux(\mathcal{P}_1)}(v_1), \ldots, \deg_{Aux(\mathcal{P}_1)}(v_k)) = (\deg_{Aux(\mathcal{P}_2)}(v_1), \ldots, \deg_{Aux(\mathcal{P}_2)}(v_k))$, and $\{V(C) \mid C$ is a component of $Aux(\mathcal{P}_1)\} = \{V(C) \mid C$ is a component of $Aux(\mathcal{P}_2)\}$. Then G has a Hamiltonian cycle C_1 containing each path in $\mathcal{P}_1$ as a subpath and such that every edge in $C_1 - (\bigcup_{P \in \mathcal{P}_1} E(P))$ is contained in $E(G) \setminus E(H)$ if and only if G has a Hamiltonian cycle C_2 containing each path in $\mathcal{P}_2$ as a subpath and such that every edge in $C_2 - (\bigcup_{P \in \mathcal{P}_2} E(P))$ is contained in $E(G) \setminus E(H)$.*

---

[3] Disregarding the labels.

Suppose there are two such path-partitions $\mathcal{P}_1$ and $\mathcal{P}_2$ where there is a cycle C_1 satisfying the conditions of Proposition 1. Let $\mathcal{Q}$ be the subpaths of C_1 connecting two consecutive paths in $\mathcal{P}_1$. See Figure 1 for an illustration. Note that if a path in $\mathcal{Q}$ is an edge, then it is an edge between two label classes and contained in $E(G) \setminus E(H)$. By the property (2), these two label classes are complete in G.

We label each end vertex of a path in $\mathcal{Q}$ as the label of H, We consider the auxiliary multigraph $Aux(\mathcal{P}_1)$ and the auxiliary multigraph $Aux(\mathcal{Q})$ by considering $\mathcal{Q}$ as a path-partition of the underlying graph on $\bigcup_{Q \in \mathcal{Q}} V(Q)$. We obtain a multigraph F from the disjoint union of $Aux(\mathcal{P}_1)$ and $Aux(\mathcal{Q})$ by identifying each v_i. Following the Hamiltonian cycle C, one easily checks that there is an Eulerian trail which alternates between edges in $Aux(\mathcal{P}_1)$ and edges in $Aux(\mathcal{Q})$.

We will prove that if we replace $Aux(\mathcal{P}_1)$ with $Aux(\mathcal{P}_2)$ in F, then the new graph also admits an Eulerian trail, because of the given conditions in Proposition 1. To see this, we observe the following, which is a strengthening of Euler's theorem on Eulerian trails. It is well known that a connected graph contains an Eulerian trail if and only if every vertex has even degree. Moreover, when edges are colored by two colors, say red and blue, and each vertex is incident with the same number of edges for both colors, then we can find an Eulerian trail where the two colors appear alternatively. We call such an Eulerian trail a *red-blue alternating Eulerian trail*. For a multigraph G colored by red and blue and $v \in V(G)$, let $\mathrm{rdeg}_G(v)$ denote the number of red edges incident with v, and let $\mathrm{bdeg}_G(v)$ denote the number of blue edges incident with v.

Lemma 1 ($\star$). *Let G be a connected multigraph whose edges are colored by red and blue. Then G has a red-blue alternating Eulerian trail if and only if for every vertex v, $\mathrm{bdeg}_G(v) = \mathrm{rdeg}_G(v)$.*

Indeed, when we replace $Aux(\mathcal{P}_1)$ with $Aux(\mathcal{P}_2)$ in F, the set of components does not change (thus consists of one non-trivial component), and each vertex is incident with same number of red and blue edges, and by Lemma 1, the resulting graph has an Eulerian trail. We will show that one can construct a Hamiltonian cycle of G from paths of $\mathcal{P}_2$ using the properties (1) and (2).

Motivated by Proposition 1, we define in Section 4 an equivalence relation between two sets of multigraphs on the same vertex set $\{v_1, \ldots, v_k\}$. We further define operations on those multigraphs, corresponding to procedures of updating path-partitions, and prove that the equivalence between two sets is preserved under such operations. These results will form the skeleton of the main algorithm.

4 An equivalence relation between families of k-vertex multigraphs

For two multigraphs G and H on the same vertex set $\{v_1, \ldots, v_k\}$ and with disjoint edge sets, we denote by $G//H$ the multigraph with vertex set $\{v_1, \ldots, v_k\}$ and edge set $E(G) \cup E(H)$.

For families $\mathcal{F}, \mathcal{F}_1, \mathcal{F}_2$ of multigraphs on the vertex set $\{v_1, \ldots, v_k\}$ and two distinct integers $i, j \in \{1, \ldots, k\}$, we define the following operations:

1. $\mathcal{F} + (i,j)$ is the set of all multigraphs F' where F' can be obtained from a multigraph F in $\mathcal{F}$ as follows[4]: choose two distinct edges $\{v_i, v'_i\}_t$ and $\{v_j, v'_j\}_s$, and let F' be the multigraph on vertex set $\{v_1, \ldots, v_k\}$ and edge set $(E(F) \setminus \{\{v_i, v'_i\}_t, \{v_j, v'_j\}_s\}) \cup \{e\}$ with $e \notin E(F)$ mapped to $\{v'_i, v'_j\}$.

2. $\mathcal{F} + t(i,j)$ is the set constructed from $\mathcal{F}$ by doing the operation $+(i,j)$ t times.

3. $\mathcal{F}|_{i \to j}$ is the set of all multigraphs F where F can be obtained from a multigraph in $\mathcal{F}$ by replacing every edge with an end vertex v_i by an edge with an end vertex v_j.

4. $\mathcal{F}_1 \uplus \mathcal{F}_2 := \{F_1 /\!/ F_2 : F_1 \in \mathcal{F}_1, F_2 \in \mathcal{F}_2\}$.

Let $\mathcal{F}_1$ and $\mathcal{F}_2$ be two families of multigraphs on the vertex set $\{v_1, \ldots, v_k\}$. We write $\mathcal{F}_1 \lesssim \mathcal{F}_2$ if for every multigraph H on the vertex set $\{v_1, \ldots, v_k\}$,

- whenever there exists $G_2 \in \mathcal{F}_2$ such that $(\deg_{G_2}(v_1), \ldots, \deg_{G_2}(v_k)) = (\deg_H(v_1), \ldots, \deg_H(v_k))$ and $G_2 /\!/ H$ has at most one non-trivial component, there exists $G_1 \in \mathcal{F}_1$ such that $(\deg_{G_1}(v_1), \ldots, \deg_{G_1}(v_k)) = (\deg_H(v_1), \ldots, \deg_H(v_k))$ and $G_1 /\!/ H$ has at most one non-trivial component.

We say that $\mathcal{F}_1$ is *equivalent* to $\mathcal{F}_2$, written $\mathcal{F}_1 \equiv \mathcal{F}_2$, if $\mathcal{F}_1 \lesssim \mathcal{F}_2$ and $\mathcal{F}_2 \lesssim \mathcal{F}_1$.

We prove that the equivalence between two families is preserved by the operation $+(i,j)$.

Proposition 2. *Let $\mathcal{F}_1$ and $\mathcal{F}_2$ be two families of multigraphs on the vertex set $\{v_1, \ldots, v_k\}$. If $\mathcal{F}_1 \equiv \mathcal{F}_2$, then $\mathcal{F}_1 + (i,j) \equiv \mathcal{F}_2 + (i,j)$.*

Proof. Suppose $\mathcal{F}_1 \equiv \mathcal{F}_2$. It is sufficient to prove that $\mathcal{F}_1 + (i,j) \lesssim \mathcal{F}_2 + (i,j)$. For this, suppose there exist a graph H on $\{v_1, \ldots, v_k\}$ and $G_2 \in \mathcal{F}_2 + (i,j)$ such that $(\deg_{G_2}(v_1), \ldots, \deg_{G_2}(v_k)) = (\deg_H(v_1), \ldots, \deg_H(v_k))$ and $G_2 /\!/ H$ has at most one non-trivial component. Since $G_2 \in \mathcal{F}_2 + (i,j)$, there exist $F_2 \in \mathcal{F}_2$, edges $\{v_i, v'_i\}_t$, $\{v_j, v'_j\}_s$ in F_2 such that $G_2 = (V(F_2), E(F_2) \setminus \{\{v_i, v'_i\}_t, \{v_j, v'_j\}_s\} \cup \{e\})$ with $e \notin E(F_2)$ mapped to $\{v'_i, v'_j\}$ in G_2. Let $H' := (V(H), E(H) \cup \{e'\})$ with $e' \notin E(H)$ mapped to $\{v_i, v_j\}$ in H. We claim that

- $(\deg_{F_2}(v_1), \ldots, \deg_{F_2}(v_k)) = (\deg_{H'}(v_1), \ldots, \deg_{H'}(v_k))$ and
- $F_2 /\!/ H'$ has at most one non-trivial component.

By the construction of G_2 from F_2, for every $v_\ell \in V(F_2) \setminus \{v_i, v_j\}$, v_ℓ has the same degree in F_2 and G_2, and the degrees of v_i and v_j in G_2 are one less than the degrees in F_2, respectively. Since the degrees of v_i and v_j in H' are one more than the degrees in H, we have $(\deg_{F_2}(v_1), \ldots, \deg_{F_2}(v_k)) = (\deg_{H'}(v_1), \ldots, \deg_{H'}(v_k))$. Assume now that $F_2 /\!/ H'$ has at least two non-trivial components. First observe that the four vertices v_i, v_j, v'_i, v'_j are in the same non-trivial component C of $F_2 /\!/ H'$, and $\{v'_i, v'_j\}$ are in a same non-trivial component of $G_2 /\!/ H$. If C' is another non-trivial component of $F_2 /\!/ H'$, then it does not intersect $\{v_i, v'_i, v_j, v'_j\}$, that is, C' is non-trivial component in $G_2 /\!/ H$ that does not intersect the one containing $\{v'_i, v'_j\}$, yielding a contradiction.

---

[4] We allow v'_i (or v'_j) to be equal to v_i (or v_j).

Since $\mathcal{F}_1 \equiv \mathcal{F}_2$, there exists $F_1 \in \mathcal{F}_1$ such that $(\deg_{F_1}(v_1), \ldots, \deg_{F_1}(v_k)) = (\deg_{H'}(v_1), \ldots, \deg_{H'}(v_k))$ and $F_1//H'$ has at most one non-trivial component. By Lemma 1, $F_1//H'$ contains an Eulerian trail where edges in F_1 and edges in H' are alternatively used. Let $\{v_i, v_i''\}_{t'}$ and $\{v_j, v_j''\}_{s'}$ be the edges where $\{v_i, v_i''\}_{t'}, e', \{v_j, v_j''\}_{s'}$ appear in the Eulerian trail in this order (recall that e' is mapped to $\{v_i, v_j\}$). Clearly, if we remove the edges $\{v_i, v_i''\}_{t'}, e', \{v_j, v_j''\}_{s'}$ and add an edge f mapped to $\{v_i'', v_j''\}$ in $F_1//H'$, then the obtained multigraph K still admits an alternating Eulerian trail. Let $G_1 = (V(F_1), E(F_1) \setminus \{\{v_i, v_i''\}_{t'}, \{v_j, v_j''\}_{s'}\} \cup \{f\})$ with $f \notin E(F_1)$ mapped to $\{v_i'', v_j''\}$ in G_1. One easily checks that $K = G_1//H$, and since K has an Eulerian trail where edges in G_1 and edges in H are alternatively used, by Lemma 1, $(\deg_{G_1}(v_1), \ldots, \deg_{G_1}(v_k)) = (\deg_H(v_1), \ldots, \deg_H(v_k))$ and $G_1//H$ has at most one non-trivial component. Because $G_1 \in \mathcal{F}_1 + (i, j)$, we can thus conclude that $\mathcal{F}_1 + (i, j) \lesssim \mathcal{F}_2 + (i, j)$. $\square$

We prove a similar property for the other operations.

Proposition 3 ($\star$). *Let $\mathcal{F}_1$ and $\mathcal{F}_2$ be families of multigraphs on the vertex set $\{v_1, \ldots, v_k\}$ and let $i, j \in \{1, \ldots, k\}$ be two distinct integers. If $\mathcal{F}_1 \equiv \mathcal{F}_2$, then $\mathcal{F}_1|_{i \to j} \equiv \mathcal{F}_2|_{i \to j}$.*

Proposition 4 ($\star$). *Let $\mathcal{F}_1, \mathcal{F}_2, \mathcal{F}_3$ be families of multigraphs on the vertex set $\{v_1, \ldots, v_k\}$. If $\mathcal{F}_1 \equiv \mathcal{F}_2$, then $\mathcal{F}_1 \uplus \mathcal{F}_3 \equiv \mathcal{F}_2 \uplus \mathcal{F}_3$.*

5   Hamiltonian Cycle problem

We prove the main result of this paper. We recall the statement.

Theorem 2. *Given a graph G and its clique-width k-expression, one can solve* HAMILTONIAN CYCLE *in time $n^{\mathcal{O}(k)}$.*

We now define formally our notion of representatives based only on the degree sequence and connected components of auxiliary multigraphs associated with k-labeled path-partitions.

Definition 1 (Representatives by auxiliary multigraphs). *Let G and H be multigraphs on vertex set $\{v_1, \ldots, v_k\}$. We write $G \simeq H$ whenever $(\deg_G(v_1), \ldots, \deg_G(v_k)) = (\deg_H(v_1), \ldots, \deg_H(v_k))$ and $\{V(C) \mid C$ is a component of $G\}$ is equal to $\{V(C) \mid C$ is a component of $H\}$. One easily checks that $\simeq$ is an equivalence relation on any set $\mathcal{F}$ of multigraphs on vertex set $\{v_1, \ldots, v_k\}$.*

For a family $\mathcal{F}$ of multigraphs on vertex-set $\{v_1, \ldots, v_k\}$, let $\mathrm{reduce}(\mathcal{F})$ be the operation which takes in each equivalence class of $\mathcal{F}/\simeq$ a representative.

The following is a rephrasing of Proposition 1.

Proposition 5 ($\star$). *Let G be a graph with its irredundant clique-width k-expression ϕ, and let t be a node in the syntactic tree. Let G_t be the k-labeled graph constructed at t, and let $\mathcal{P}_1$ and $\mathcal{P}_2$ be k-labeled path-partitions of G_t whose labelings are induced by the labeling of G_t. If $Aux(\mathcal{P}_1) \simeq Aux(\mathcal{P}_2)$, then the following are equivalent.*

1. G has a Hamiltonian cycle C_1 containing each path in $\mathcal{P}_1$ as a subpath and such that every edge in $C_1 - (\bigcup_{P \in \mathcal{P}_1} E(P))$ is contained in $E(G) \setminus E(G_t)$.
2. G has a Hamiltonian cycle C_2 containing each path in $\mathcal{P}_2$ as a subpath and such that every edge in $C_2 - (\bigcup_{P \in \mathcal{P}_2} E(P))$ is contained in $E(G) \setminus E(G_t)$.

Proof (Sketch of Proof). Suppose G has a Hamiltonian cycle C_1 satisfying 1. Let $\mathcal{Q}$ be the set of all maximal paths in $C_1 - (\bigcup_{P \in \mathcal{P}_1} E(P))$, and let $H :=$ $G[\bigcup_{Q \in \mathcal{Q}} V(Q)]$. We consider $\mathcal{Q}$ as the path-partition of H where the end vertices of paths in $\mathcal{Q}$ are labeled by their labels in G_t. Since C_1 is a Hamiltonian cycle, $Aux(\mathcal{P}_1)//Aux(\mathcal{Q})$ has an Eulerian trail where edges in $Aux(\mathcal{P}_1)$ and edges in $Aux(\mathcal{Q})$ are alternatively used. Since $Aux(\mathcal{P}_2) \simeq Aux(\mathcal{P}_1)$, by Lemma 1, $Aux(\mathcal{P}_2)//Aux(\mathcal{Q})$ admits an Eulerian trail where the edges in $Aux(\mathcal{P}_2)$ and the edges in $Aux(\mathcal{Q})$ are alternatively used.

To construct a Hamiltonian cycle C_2 satisfying 2, let $e_1, e_2, \ldots, e_{2m}$ be the sequence of the edges in an Eulerian trail of $Aux(\mathcal{P}_2)//Aux(\mathcal{Q})$ where edges in $Aux(\mathcal{P}_2)$ and edges in $Aux(\mathcal{Q})$ are alternatively used such that $e_1 \in E(Aux(\mathcal{P}_2))$. If e_{2i} corresponds to a path of length 1 in $\mathcal{Q}$ between two label classes, then we add a direct edge between end vertices of paths in $\mathcal{P}$ corresponding to e_{2i-1} and e_{2i+1}. If e_{2i} corresponds to a path of length ≥ 2 in $\mathcal{Q}$, then we add this path to the subgraph and connect to the paths in $\mathcal{P}$ corresponding to e_{2i-1} and e_{2i+1}. In particular, the former procedure is possible because ϕ is irredundant, and thus two label classes are complete in G. In this way, we can construct a Hamiltonian cycle C_2 satisfying 2. $\square$

Proposition 0 tells us that if $\mathscr{F}$ is the set of possible k-labeled path-partitions at a node t of the syntactic tree, it is enough to store reduce($\{Aux(\mathcal{P}) \mid \mathcal{P} \in \mathscr{F}\}$).

Proposition 6 ($\star$). *Let $\mathcal{F}$ be a family of graphs on the vertex set $\{v_1, \ldots, v_k\}$. Then $\mathcal{F} \equiv$ reduce($\mathcal{F}$).*

Proof (Proof of Theorem 1). We assume that G has at least 3 vertices, otherwise we can automatically say it is a No-instance. Since every k-expression can be transformed into an irredundant k-expression in linear time, we may assume that G is given with an irredundant k-expression. Let ϕ be the given irredundant k-expression defining G, and T be the syntactic tree of ϕ. For every node t of T, let G_t be the subgraph of G defined at node t, and for each $i \in \{1, \ldots, k\}$, let $G_t[i]$ be the subgraph of G_t induced by the vertices with label i.

For each node t and each vector $(a_1, \ldots, a_k) \in \{0, 1, \ldots, n\}^k$, let $c[t, (a_1, \ldots, a_k)]$ be the set of all multigraphs F on the vertex set $\{v_1, \ldots, v_k\}$ where

- $F = Aux(\mathcal{P})$ for some k-labeled path-partition $\mathcal{P}$ of G_t,
- for each $i \in \{1, \ldots, k\}$, a_i is the degree of v_i in F.

Instead of computing the whole set $c[t, (a_1, \ldots, a_k)]$, we will compute a subset $r[t, (a_1, \ldots, a_k)]$ of $c[t, (a_1, \ldots, a_k)]$ of size $2^{\mathcal{O}(k \log k)}$ such that $r[t, (a_1, \ldots, a_k)] \equiv c[t, (a_1, \ldots, a_k)]$.

We explain how to decide whether G has a Hamiltonian cycle. Let t_{root} be the root node of T, and let $t_{lastjoin}$ be the node taking the disjoint union of two graphs and closest to the root node. We can observe that G has a Hamiltonian cycle if and only if there are some node t between t_{root} and $t_{lastjoin}$ with child t' and a path-partition $\mathcal{P}$ of $G_{t'}$ such that t is a join node labeled by $\eta_{i,j}$, and $\deg_{Aux(\mathcal{P})}(v_i) = \deg_{Aux(\mathcal{P})}(v_j) > 0$ and $\deg_{Aux(\mathcal{P})}(v_{i'}) = 0$ for all $i' \in \{1, \ldots, k\} \setminus \{i, j\}$. This is equivalent to that $c[t', (a_1, \ldots, a_k)] \neq \emptyset$ for some vector $(a_1, \ldots, a_k)$ where $a_i = a_j > 0$ and $a_{i'} = 0$ for all $i' \in \{1, \ldots, k\} \setminus \{i, j\}$. Therefore, if there is a Hamiltonian cycle, then $r[t', (a_1, \ldots, a_k)] \neq \emptyset$ for such a tuple of t, t', and $(a_1, \ldots, a_k)$, and we can correctly say that G has a Hamiltonian cycle, and otherwise, there are no such tuples, and we can correctly say that G has no Hamiltonian cycles.

Now, we explain how to recursively generate $r[t, (a_1, \ldots, a_k)]$.

1. (Creation of a vertex v with label i)
   If $a_i = 2$ and $a_j = 0$ for all $j \neq i$, then $c[t, (a_1, \ldots, a_k)]$ consists of one graph on the vertex set $\{v_1, \ldots, v_k\}$ with a loop incident with v_i, and otherwise, it is an empty set. So, we add the graph $(\{v_1, \ldots, v_k\}, \{v_i v_i\})$ to $r[t, (a_1, \ldots, a_k)]$ when $a_i = 2$ and $a_j = 0$ for all $j \neq i$, and set $r[t, (a_1, \ldots, a_k)] := \emptyset$ otherwise.

2. (Disjoint union node with two children t_1 and t_2)
   Since every path-partition of G_t is obtained by taking the disjoint union of a path-partition of G_{t_1} and a path-partition of G_{t_2}, we have

   $c[t, (a_1, \ldots, a_k)]$

   $$:= \bigcup_{(a_1^1, \ldots, a_k^1) + (a_1^2, \ldots, a_k^2) = (a_1, \ldots, a_k)} c[t_1, (a_1^1, \ldots, a_k^1)] \uplus c[t_2, (a_1^2, \ldots, a_k^2)].$$

   We assign

   $r[t, (a_1, \ldots, a_k)]$

   $$:= \mathrm{reduce}\left(\bigcup_{(a_1^1, \ldots, a_k^1) + (a_1^2, \ldots, a_k^2) = (a_1, \ldots, a_k)} r[t_1, (a_1^1, \ldots, a_k^1)] \uplus r[t_2, (a_1^2, \ldots, a_k^2)] \right).$$

3. (Join node with the child t' such that each vertex with label i is joined to each vertex with label j)
   Note that every path-partition of G_t is obtained from a path-partition of $G_{t'}$ by adding some edges between end vertices of label i and end vertices of label j. We can observe that when we add an edge between an end vertex v of a path P_1 with label i, and an end vertex w of a path P_2 with label j, these two paths P_1 and P_2 will be unified into a path whose end vertices are end vertices of P_1 and P_2 other than v and w. Thus, it corresponds to the operation $+(i, j)$ on auxiliary multigraphs. We observe that

   $$c[t, (a_1, \ldots, a_k)] := \bigcup_{\substack{a_i' - a_i = a_j' - a_j = \ell \geq 0 \\ a_x' = a_x \text{ for } x \neq i, j}} (c[t', (a_1', \ldots, a_k')] + \ell(i, j)).$$

We take all possible vectors $(a'_1, \ldots, a'_k)$ where $a'_i - a_i = a'_j - a_j \geq 0$, and for all $t \in \{1, \ldots, k\} \setminus \{i, j\}$, $a'_t = a_t$. Assume $\ell = a'_i - a_i$. For each $\ell \in \{0, 1, \ldots, n\}$, we assign

$$r_\ell := \text{reduce}(\cdots \text{reduce}(\text{reduce}(r[t', (a'_1, \ldots, a'_k)] + (i, j)) + (i, j)) \cdots + (i, j)),$$

where we repeat ℓ times, and assign

$$r[t, (a_1, \ldots, a_k)] := \text{reduce}(r_0 \cup r_1 \cup \cdots \cup r_n).$$

4. (Renaming node with a child t' such that the label of each vertex with label i is changed to j)
   Every path-partition of G_t is also a path-partition of $G_{t'}$, and vice versa. Since just labelings of vertices are changed, we can observe that if $a_i \neq 0$, then $c[t, (a_1, \ldots, a_k)]$ is the empty set, and otherwise, we have

$$c[t, (a_1, \ldots, a_k)] := \bigcup_{\substack{a_x = a'_x \text{ for all } x \neq i, j \\ a'_i + a'_j = a_j}} c[t', (a'_1, \ldots, a'_k)]|_{i \to j}.$$

If $a_i \neq 0$, then we assign the empty set to $r[t, (a_1, \ldots, a_k)]$, and otherwise, we assign

$$r[t, (a_1, \ldots, a_k)] := \text{reduce} \left(\bigcup_{\substack{a_x = a'_x \text{ for all } x \neq i, j \\ a'_i + a'_j = a_j}} r[t', (a'_1, \ldots, a'_k)]|_{i \to j} \right).$$

One can prove by induction that $r[t, (a_1, \ldots, a_k)] \equiv c[t, (a_1, \ldots, a_k)]$ for each t and $(a_1, \ldots, a_k)$. Therefore, we can correctly decide whether G has a Hamiltonian cycle or not using sets $r[t, (a_1, \ldots, a_k)]$.

Running time. Each constructed set $r[t, (a_1, \ldots, a_k)]$ consists of at most $2^{\mathcal{O}(k \log k)}$ graphs, as we keep at most one graph for each partition of $\{v_1, \ldots, v_k\}$ after the reduce operation. For the node taking the disjoint union of two graphs, for a fixed vector $(a_1, \ldots, a_k)$, there are $n^{\mathcal{O}(k)}$ ways to take two vectors A_1 and A_2 such that $A_1 + A_2 = (a_1, \ldots, a_k)$. So, we can update $r[\cdot, \cdot]$ in time $n^{\mathcal{O}(k)} \cdot 2^{\mathcal{O}(k \log k)}$. For the node joining edges between two classes, the value ℓ can be taken from 0 to n. Since each operation $+(i, j)$ take $k^2 \cdot 2^{\mathcal{O}(k \log k)}$ time, we can update $r[\cdot, \cdot]$ in time $n^2 \cdot 2^{\mathcal{O}(k \log k)}$. Clearly, we can update $r[\cdot, \cdot]$ in time $n \cdot 2^{\mathcal{O}(k \log k)}$ for the relabeling nodes. Therefore, we can solve HAMILTONIAN CYCLE for G in time $n^{\mathcal{O}(k)}$.   □

6   More applications

Let q be a positive integer. The q-CYCLE COVERING problem asks for a given graph G whether there is a set of at most q pairwise vertex-disjoint cycles in

G whose union contains all vertices of G. Definitely, 1-CYCLE COVERING is the HAMILTONIAN CYCLE problem. In the q-CYCLE COVERING problem, we relax the definition of path-partitions so that it may contain at most q cycles, and we keep the number of cycles in the path-partition. Also, we define its auxiliary multigraph $Aux(\mathcal{P})$ using those remaining paths. One can easily check that two such modified path-partitions $\mathcal{P}_1$ and $\mathcal{P}_2$ are equivalent for q-CYCLE COVERING if they contain the same number of cycles and $Aux(\mathcal{P}_1) \simeq Aux(\mathcal{P}_2)$.

The second application is for DIRECTED HAMILTONIAN CYCLE. Clique-width was also considered for directed graphs by Courcelle and Olariu [5]. The clique-width operations for directed graphs are the same as for the undirected graphs, except the one that add edges between two label classes, defined as follows: (3*) Adding an arc (u, v) for each vertex u with label i to each vertex v with label j ($i \neq j$, denoted by $\alpha_{i,j}$). The *clique-width* of a directed graph G is the minimum number of labels needed to construct G using these operations. In this case, we use directed auxiliary multigraphs. Similar to Lemma 1, we can show that

Lemma 2. *Let G be a connected directed multigraph whose arcs are colored by red and blue. Then the following are equivalent.*

1. *For every vertex v, the number of blue edges leaving v is the same as the number of red edges entering v, and the number of red edges leaving v is the same as the number of blue edges entering v.*
2. *G has a red-blue alternating Eulerian directed trail.*

Using Lemma 2, we can proceed same as Theorem 1.

References

1. Arnborg, S., Proskurowski, A.: Linear time algorithms for NP-hard problems restricted to partial k-trees. Discrete Appl. Math. 23, 11–24 (1989)
2. Bodlaender, H.L.: A partial k-arboretum of graphs with bounded treewidth. Theor. Comput. Sci. 209, 1–45 (1998)
3. Courcelle, B.: The monadic second-order logic of graphs I: Recognizable sets of finite graphs. Information and Computation 85, 12–75 (1990)
4. Courcelle, B., Makowsky, J.A., Rotics, U.: Linear time solvable optimization problems on graphs of bounded clique width. Theor. Comput. Sci. 33, 125–150 (2000)
5. Courcelle, B., Olariu, S.: Upper bounds to the clique width of graphs. Discrete Appl. Math. 101(1-3), 77–114 (2000)
6. Diestel, R.: Graph Theory. No. 173 in Graduate Texts in Mathematics, Springer, third edn. (2005)
7. Espelage, W., Gurski, F., Wanke, E.: How to solve NP-hard graph problems on clique-width bounded graphs in polynomial time. In: Graph-theoretic concepts in computer science (Boltenhagen, 2001), Lecture Notes in Comput. Sci., vol. 2204, pp. 117–128. Springer, Berlin (2001)
8. Fomin, F.V., Golovach, P.A., Lokshtanov, D., Saurabh, S.: Intractability of clique-width parameterizations. SIAM J. Comput. 39(5), 1941–1956 (2010)
9. Fomin, F.V., Golovach, P.A., Lokshtanov, D., Saurabh, S.: Almost optimal lower bounds for problems parameterized by clique-width. SIAM J. Comput. 43(5), 1541–1563 (2014)

Improved Algorithms for Computing
k-Sink on Dynamic Flow Path Networks

Binay Bhattacharya[1]*, Mordecai J. Golin[2]**, Yuya Higashikawa[3],***,†,
Tsunehiko Kameda[1], and Naoki Katoh[4],***

[1] School of Computing Science, Simon Fraser University, Burnaby, Canada
[2] Dept. of Computer Science, Hong Kong Univ. of Science and Technology, China
[3] Dept. of Information and System Engineering, Chuo University, Tokyo, Japan
[4] School of Science and Technology, Kwansei Gakuin University, Hyogo, Japan

Abstract. We address the problem of locating k sinks on dynamic flow path networks with n vertices in such a way that the evacuation completion time to them is minimized. Our two algorithms run in $O(n \log n + k^2 \log^4 n)$ and $O(n \log^3 n)$ time, respectively. When all edges have the same capacity, we also present two algorithms which run in $O(n + k^2 \log^2 n)$ time and $O(n \log n)$ time, respectively. These algorithms together improve upon the previously most efficient algorithms, which have time complexities $O(kn \log^2 n)$ [1] and $O(kn)$ [11], in the general and uniform edge capacity cases, respectively. The above results are achieved by organizing relevant data for subpaths in a strategic way during preprocessing, and the final results are obtained by extracting/merging them in an efficient manner.

1 Introduction

Ford and Fulkerson [5] introduced the concept of *dynamic flow* which models movement of commodities in a network. In this model, each vertex is assigned some initial amount of supply, each edge has a capacity, which limits the rate of commodity flow into it per unit time, and the transit time to traverse it. One variant of the dynamic flow problem is the *quickest transshipment* problem, where the source vertices have specified supplies and sink vertices have specified demands. The problem is to send exactly the right amount of commodity out of sources into sinks in minimum overall time. Hoppe and Tardos [12] provided a polynomial time algorithm for this problem in the case where the transit times are integral. However, the complexity of their algorithm is very high. Finding a practical polynomial time solution to this problem is still open. The reader is referred to a recent paper by Skutella [18] on dynamic flows.

This paper discusses a related problem, called the *evacuation problem* [8, 14], in which the supplies (i.e., evacuees) are discrete, and the sinks and their demands are not specified. In fact, the locations of the sinks are the output of the problem. Many disasters, such as earthquakes, nuclear plant accidents, volcanic

---

* Partially supported by a Discovery Grant from NSERC of Canada
** Partially supported by Hong Kong RGC GRF grant 16208415
*** Supported by JSPS KAKENHI Grant-in-Aid for Young Scientists (B) (17K12641)
† Supported by JST CREST (JPMJCR1402)

© Springer International Publishing AG 2017
F. Ellen et al. (Eds.): WADS 2017, LNCS 10389, pp. 133–144, 2017.
DOI: 10.1007/978-3-319-62127-2_12

eruptions, flooding, have struck in recent years in many parts of the world, and it is recognized that orderly evacuation planning is urgently needed.

A *k-sink* is a set of k sinks such that the evacuation completion time to sinks is minimized, and our objective is to find a k-sink on a dynamic flow path network. *Congestion* is said to occur when an evacuee cannot move at the maximum speed constrained only by transit time. Thus, when the capacities of the edges are sufficiently large, no congestion occurs and each evacuee can follow the shortest path to its nearest sink at the maximum speed. This is equivalent to the classical k-center problem in networks, which is known to be NP-hard even on bipartite planar graphs of maximum degree 4 [17]. To the best of our knowledge the most general polynomially solvable case for general k is where the underlying graphs are cacti or partial t-trees with constant t. Congestion could occur if vertex capacities are limited, in which case edges may get clogged and congestion backs up. Our results are valid regardless of whether vertex capacities (the number of evacuees that they can accommodate) are limited or not.

Mamada et al. [15] solved the 1-sink problem for the dynamic flow tree networks in $O(n \log^2 n)$ time under the condition that only a vertex can be a sink, where n is the number of vertices. When edge capacities are uniform, we have presented $O(n \log n)$ time algorithms with a more relaxed condition that the sink can be on an edge, as well as on a vertex [3, 10]. Dealing with congestion is non-trivial even in path networks. On dynamic flow path networks with uniform edge capacities, it is straightforward to compute the 1-sink in linear time, as shown by Cheng et al. [4]. Arumugam et al. [1] showed that the k-sink problem for dynamic flow path networks can be solved in $O(kn \log^2 n)$ time, and when the edge capacities are uniform Higashikawa et al. [11] showed that it can be solved in $O(kn)$ time.

In this paper we present two algorithms for the k-sink problem on dynamic flow path networks with general edge capacities. A path network can model an airplane aisle, a hall way in a building, a street, a highway, etc., to name a few. Unlike the previous algorithm for the k-sink problem [1] which uses dynamic programming, our algorithms adopt Megiddo's *parametric search* [16] and the *sorted matrices* introduced by Frederickson and Johnson [6, 7]. Together, they outperform all other known algorithms, and they are the first sub-quadratic algorithms for any value of k. These improvements were made possible by our method of merging evacuation times of subpaths stored in a hierarchical data structure. We also present two algorithms for the dynamic flow path networks with uniform edge capacities.

This paper is organized as follows. In the next section, we define our model and the terms that are used throughout the paper. Sec. 3 introduces a new data structure, named the *capacities and upper envelopes tree*, which plays a central role in the rest of the paper. In Sec. 4 we identify two important tasks that form building blocks of our algorithms, and also discuss a feasibility test. Sec. 5 presents several algorithms for uniform and general edge capacities. Finally, Sec. 6 concludes the paper.

2   Preliminaries

2.1   Definitions

Let $P = (V, E)$ be a path network, whose vertices $v_1, v_2, \ldots, v_n$ are arranged from left to right in this order. For $i = 1, 2, \ldots, n$, vertex v_i has an integral weight w_i (> 0), representing the number of evacuees, and each edge $e_i = (v_i, v_{i+1})$ has a fixed non-negative length l_i and an integral *capacity* c_i, which is the upper limit on the number of evacuees who can enter an edge per unit time. We assume that a sink has infinite capacity, so that the evacuees coming from the left and right of a sink do not interfere with each other. An evacuation starts at the same time from all the vertices, and all the evacuees from a vertex evacuate to the same sink. This is called "confluent flow" in the parlance of the network flow theory. This constraint is desirable in evacuation in order to avoid confusion among the evacuees at a vertex as to which way they should move.

By $x \in P$, we mean that point x lies on either an edge or a vertex of P. For two points $a, b \in P$, $a \prec b$ or $b \succ a$ means that a lies to the left of b. Let $d(a, b)$ denote the distance (sum of the edge lengths) between a and b. If a and/or b lies on an edge, we use the prorated distance. The transit time for a unit distance is denoted by τ, so that it takes $d(a, b)\tau$ time to travel from a to b, and τ is independent of the edge. Let $c(a, b)$ denote the minimum capacity of the edges on the subpath of P between a and b. The point that is arbitrarily close to v_i on its left (resp. right) side is denoted by v_i^- (resp. v_i^+). Let $P[a, b]$ denote the subpath of P between a and b satisfying $a \prec b$. If a, b or both are excluded, we denote them by $P(a, b]$, $P[a, b)$ or $P(a, b)$, respectively. Let $V[a, b]$ (resp. $V(a, b]$, $V[a, b)$ or $V(a, b)$) denotes the set of vertices on $P[a, b]$ (resp. $P(a, b]$, $P[a, b)$ or $P(a, b)$). We introduce a weight array $W[\cdot]$, defined by

$$W[i] \triangleq \sum_{v_j \in V[v_1, v_i]} w_j, \text{ for } i = 1, 2, \ldots, n, \tag{1}$$

and let $W[v_i, v_j] \triangleq W[j] - W[i-1]$ for $i \le j$.

2.2   Completion time functions

In our model, a set of k sinks accepts evacuees from k disjoint subpaths of P. We thus need to be able to compute the completion time for each such subpath $P[v_i, v_j]$. For simplicity, from now on, we assume that the optimal k sinks are on edges, not on vertices. Small modifications will be necessary if we allow some sinks to be on vertices. We define the *completion time from left* (*L-time* for short) to $x \succ v_j$ of vertex v_p on $P[v_i, v_j]$ to be the evacuation completion time to x for the evacuees on the vertices on $P[v_i, v_p]$, assuming that they all arrive at x continuously at a uniform rate $c(v_p, x)$. We similarly define the *completion time from right* (*R-time* for short) to $x \prec v_i$ of vertex v_p on $P[v_i, v_j]$ to be the evacuation completion time to x for all the evacuees on the vertices on $P[v_p, v_j]$, arriving at x continuously at a uniform rate $c(x, v_p)$. For any vertex $v_p \in V[v_i, v_j]$, its L-time and R-time are given mathematically as

$$\theta_L^{[i,j]}(x, v_p) \triangleq d(v_p, x)\tau + W[v_i, v_p]/c(v_p, x) \text{ for } x \succ v_j, \tag{2}$$

$$\theta_R^{[i,j]}(x, v_p) \triangleq d(x, v_p)\tau + W[v_p, v_j]/c(x, v_p) \text{ for } x \prec v_i, \tag{3}$$

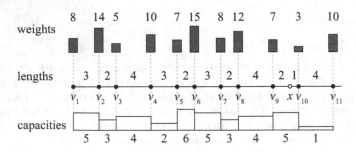

Fig. 1. An example path $P = (V, E)$ with $x \in e_9 = (v_9, v_{10})$.

respectively. For convenience we sometimes refer to the first (resp. second) term in the righthand side of (2) and (3) as the *distance time* (resp. *weight time*). Note that the distance time is linear in the distance to x. Consider an arbitrary subpath $P[v_i, v_j]$, where $i \leq j$.

Fig. 1 shows an example, where vertices $v_1, v_2, v_3, \ldots$ (represented by black circles) have weights $8, 14, 5, \ldots$, and edges $e_1, e_2, e_3, \ldots$ have lengths $3, 2, 4, \ldots$ and capacities $5, 3, 4, \ldots$ Point x is located on $e_9 = (v_9, v_{10})$ so that $d(v_9, x) = 2$ (represented by a white circle). Assuming $\tau = 1$, let us compute the L-time to x of vertex v_5 on $P[v_2, v_7]$. From $d(v_5, x) = 13$, $W[v_2, v_5] = 36$ and $c(v_5, x) = 3$, we obtain $\theta_L^{[2,7]}(x, v_5) = 25$.

To be more precise, the weight time should be $\lceil W[v_i, v_p]/c(v_p, x) \rceil$ and $\lceil W[v_p, v_j]/c(x, v_p) \rceil$ in (2) and (3), respectively, since the evacuees are discrete entities. Although only small modifications are necessary to get exact solutions as shown in [4], we use (2) and (3) for simplicity.

Lemma 1. [9] *Let s be the sink for a subpath $P[v_i, v_j]$ of a path network P. The evacuation completion time to s ($v_i \preceq v_h \prec s \prec v_{h+1} \preceq v_j$) for the evacuees on $P[v_i, v_j]$ is given by*

$$\Theta^{[i,j]}(s) \triangleq \max \left\{ \max_{v \in V[v_i, s]} \{\theta_L^{[i,h]}(s, v)\}, \max_{v' \in V(s, v_j]} \{\theta_R^{[h+1,j]}(s, v')\} \right\}. \quad (4)$$

Referring to (2) and (3), the vertex $v_p \in V[v_i, v_j]$ that maximizes $\theta_L^{[i,j]}(v_j^+, v_p)$ (resp. $\theta_R^{[i,j]}(v_i^-, v_p)$) is called the *L-critical* vertex (resp. *R-critical* vertex) of $P[v_i, v_j]$, and is denoted by $c_L^{[i,j]}$ (resp. $c_R^{[i,j]}$). Note that (v_j^+, v_p) (resp. (v_i^-, v_p)) is used instead of (v_j, v_p) (resp. (v_i, v_p)), and that we have $d(v_p, v_j^+) = d(v_p, v_j)$ and $c(v_p, v_j^+) = \min\{c(v_p, v_j), c_j\}$ (resp. $d(v_i^-, v_p) = d(v_i, v_p)$ and $c(v_i^-, v_p) = \min\{c(v_i, v_p), c_{i-1}\}$).

Using the example in Fig. 1 again, let us find the L-critical vertex of $P[v_2, v_7]$. We first compute $\theta_L^{[2,7]}(v_7^+, v_p)$ for $p = 2, \ldots, 7$: $\theta_L^{[2,7]}(v_7^+, v_2) = 14 + 14/2 = 21$, $\theta_L^{[2,7]}(v_7^+, v_3) = 12 + 19/2 = 21.5$, $\theta_L^{[2,7]}(v_7^+, v_4) = 8 + 29/2 = 22.5$, $\theta_L^{[2,7]}(v_7^+, v_5) = 5 + 36/3 = 17$, $\theta_L^{[2,7]}(v_7^+, v_6) = 3 + 51/3 = 20$, and $\theta_L^{[2,7]}(v_7^+, v_7) = 0 + 59/3 \approx 19.7$. Comparing these values, we obtain $c_L^{[2,7]} = v_4$.

Proposition 1. *Critical vertex* $v_p = c_L^{[i,j]}$ *(resp.* $v_p = c_R^{[i,j]}$*) maximizes* $\theta_L^{[i,j]}(x, v_p)$
(resp. $\theta_R^{[i,j]}(x, v_p)$*) for any point* $x \in (v_j, v_{j+1}]$ *(resp.* $x \in [v_{i-1}, v_i)$*).*

3   Data structures

A problem instance is said to be *t-feasible* if there are k sinks such that every
evacuee can reach a sink within time t. In our algorithms, we want to perform
t-feasibility tests for many different values of completion time t. Therefore, it is
worthwhile to spend some time during preprocessing to construct data structures
which facilitate these tests.

3.1   Capacities and upper envelopes (CUE) tree

We want to design a data structure with which critical vertices $c_L^{[i,j]}$ and $c_R^{[i,j]}$ can
be found efficiently for an arbitrary pair (i, j) with $1 \le i \le j \le n$. To this end we
introduce the *capacities and upper envelopes tree* (*CUE tree*, for short), denoted
by $\mathcal{T}$, with root ρ, whose leaves are the vertices of P arranged from left to right.
It is a balanced tree with height $O(\log n)$. In balancing, the vertex weights are
not considered. For a node[5] u of $\mathcal{T}$, let $\mathcal{T}(u)$ denote the subtree rooted at u,
and let $l(u)$ (resp. $r(u)$) denote the index of the leftmost (resp. rightmost) vertex
on P that belongs to $\mathcal{T}(u)$. See Fig. 2. Let u_l, u_r and u_p denote the left child
of u, the right child of u, and the parent of u, respectively. We say that node

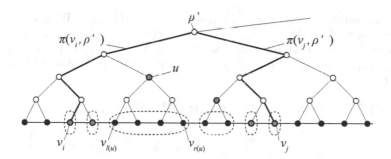

Fig. 2. Illustration of a part of CUE tree $\mathcal{T}$. The small gray disks represent nodes of
$N[v_i, v_j]$ and dashed circles enclose subpaths in $\mathcal{P}[v_i, v_j]$.

u *spans* subpath $P[v_{l(u)}, v_{r(u)}]$. At node u, we store $l(u)$, $r(u)$ and the capacity
$c(v_{l(u)}, v_{r(u)})$ among others. This information at every node can be computed
bottom up in $O(n)$ time by performing heap-like operations.

For two nodes u, u' of $\mathcal{T}$, let $\pi(u, u')$ denote the path from u to u' along
edges of $\mathcal{T}$. Suppose that for an index pair (i, j) with $1 \le i \le j \le n$, node ρ' is

---

[5] We use the term "node" here to distinguish it from the vertices on the path. A
  vertex, being a leaf of $\mathcal{T}$, is considered a node, but an interior node of $\mathcal{T}$ is not a
  vertex.

the lowest common ancestor of v_i and v_j in $\mathcal{T}$. Consider every node of $\mathcal{T}$ that is the right child of a node on $\pi(v_i, \rho')$ or the left child of a node on $\pi(v_j, \rho')$, but which itself is not on $\pi(v_i, \rho')$ or $\pi(v_j, \rho')$. Let $N[v_i, v_j]$ denote the set of such nodes plus v_i and v_j. Then clearly $N[v_i, v_j]$ consists of $O(\log n)$ nodes. Let $\mathcal{P}[v_i, v_j]$ denote the set of $O(\log n)$ subpaths spanned by nodes of $N[v_i, v_j]$.

In order to determine $c_L^{[i,j]}$ for a given pair (i, j), we need to compute

$$\max_{v_p \in V[v_i, v_j]} \{d(v_p, v_j)\tau + W[v_i, v_p]/c(v_p, v_{j+1})\}. \tag{5}$$

To facilitate such a computation for an arbitrary pair (i, j), at each node u, we precompute and store two upper envelope functions associated with subpath $\mathcal{P}[v_{l(u)}, v_{r(u)}]$. Then for $u \in N[v_i, v_j]$ that spans v_p, we have $W[v_i, v_p] = W[v_i, v_{l(u)-1}] + W[v_{l(u)}, v_p]$ and $c(v_p, v_{j+1}) = \min\{c(v_p, v_{r(u)+1}), c(v_{r(u)+1}, v_{j+1})\}$. Since (i, j), hence $W[v_i, v_{l(u)-1}]$ and $c(v_{r(u)+1}, v_{j+1})$, is not known during preprocessing, we replace these values with variables W and c, respectively, and express the two upper envelopes stored at u as functions of $W = W[v_i, v_{l(u)-1}]$ and $c = c(v_{r(u)+1}, v_{j+1})$, respectively. We can now break (5) down into a number of formulea, one for each $u \in N[v_i, v_j]$, which is given by

$$\max_{v_p \in V[v_{l(u)}, v_{r(u)}]} \{d(v_p, v_{r(u)})\tau + (W + W[v_{l(u)}, v_p]) / \min\{c(v_p, v_{r(u)+1}), c\}\}. \tag{6}$$

Using the concrete values of W and c, we can evaluate (5) by finding the maximum of the $|N[v_i, v_j]| = O(\log n)$ values, computed by (6).

Now we want to compute (6) efficiently for "arbitrary" W and c, but of course we have $W = W[v_i, v_{l(u)-1}]$ and $c = c(v_{r(u)+1}, v_j)$ in mind for some i and j. Consider two extreme cases, where for any p with $v_p \in V[v_{l(u)}, v_{r(u)}]$ (i) $c > c(v_p, v_{r(u)+1})$, and (ii) $c \le c(v_p, v_{r(u)+1})$, respectively. In Case (i), we have from (6)

$$\Theta_{L,1}^u(W) \triangleq \max_{v_p \in V[v_{l(u)}, v_{r(u)}]} \{d(v_p, v_{r(u)})\tau + (W + W[v_{l(u)}, v_p])/c(v_p, v_{r(u)+1})\} \tag{7}$$

$$= \max_{v_p \in V[v_{l(u)}, v_{r(u)}]} \{\theta_L^{[l(u), r(u)]}(v_{r(u)}^+, v_p) + W/c(v_p, v_{r(u)+1})\}. \tag{8}$$

Note that $c(v_p, v_{r(u)+1})$ gets smaller as v_p moves to the left. From (7) it is seen that $\Theta_{L,1}^u(W)$ is the upper envelope of linear functions of W and each coefficient of W is positive, which means that $\Theta_{L,1}^u(W)$ is piecewise linear, continuous, and increasing in W. Thus it can be encoded as a sequence of bending points. In Case (ii), we have from (6)

$$\Theta_{L,2}^u(c) = \max_{v_p \in V[v_{l(u)}, v_{r(u)}]} \{d(v_p, v_{r(u)})\tau + W[v_{l(u)}, v_p]/c\}. \tag{9}$$

Note that (9) was obtained from (6) by removing the term W/c, which does not depend on v_p. If we plot $\Theta_{L,2}^u(c)$ vs. $(1/c)$ as a graph, it is also piecewise linear, continuous, and increasing in $(1/c)$, and can be encoded as a sequence of bending points.

At node u we store both $\Theta^u_{L,1}(W)$ and $\Theta^u_{L,2}(c)$ in encoded form with bending points, which can be computed in $O(r(u) - l(u))$ time. Similarly, in order to determine $c^{[i,j]}_R$ for an arbitrary pair (i,j), we store two functions which are symmetric to $\Theta^u_{L,1}(W)$ and $\Theta^u_{L,2}(c)$, respectively, named $\Theta^u_{R,1}(W)$ and $\Theta^u_{R,2}(c)$, in linear time. We can now prove the following lemma.

Lemma 2. *Given a dynamic flow path network with n vertices, CUE tree $\mathcal{T}$ with associated data can be constructed in $O(n \log n)$ time and $O(n \log n)$ space.*

3.2 Using CUE tree

Suppose we want to find the L-critical vertex $c^{[i,j]}_L$ for subpath $P[v_i, v_j]$. We work on $P[v_{l(u)}, v_{r(u)}] \in \mathcal{P}[v_i, v_j]$ for each node $u \in N[v_i, v_j]$. Each such subpath provides a *candidate* for $c^{[i,j]}_L$. Clearly, among those candidates, the one that has the largest L-time is $c^{[i,j]}_L$.

Let us first compute $c = c(v_{r(u)+1}, v_{j+1})$. For this purpose, we trace the path $\pi(v_{r(u)+1}, v_{j+1})$ in $\mathcal{T}$ and, at each node $u' \in N[v_{r(u)+1}, v_{j+1}] \setminus \{v_{j+1}\}$, retrieve $c(v_{l(u')}, v_{r(u')})$ and $c_{r(u')+1}$. Taking the minimum of the retrieved capacities, we obtain $c(v_{r(u)+1}, v_{j+1})$, which costs $O(\log n)$ time.

Using binary search, we then find the largest index q $(l(u) \leq q \leq r(u))$, if any, such that $c(v_q, v_{r(u)+1}) < c = c(v_{r(u)+1}, v_{j+1})$ holds. Note that $c(v_q, v_{r(u)+1})$ is monotonically non-increasing as q decreases. To find q we trace the path $\pi(v_{r(u)}, u)$ in $\mathcal{T}$ as follows. Set $c_{\min}$ to $c_{r(u)+1}$ and u' to $v_{r(u)}$. If $c_{\min} < c$, q is determined as $r(u)$. Otherwise update u' to u'_p. While $u' \neq u$ and

$$\min\{c_{\min}, c(v_{l(u'_l)}, v_{r(u'_l)}), c_{r(u'_l)+1}\} \geq c, \tag{10}$$

update $c_{\min}$ to the L.H.S. of (10), and u' to u'_p. If $u' = u$ and (10) holds, such q does not exist. If (10) stops holding at some node u', then update u' to u'_l. While

$$\min\{c_{\min}, c(v_{l(u'_r)}, v_{r(u'_r)}), c_{r(u'_r)+1}\} \geq c, \tag{11}$$

update $c_{\min}$ to the L.H.S. of (11) and u' to u'_r. If (11) stops holding at some node u', then update u' to u'_r. This way we will eventually reach v_q, if it exists, in $O(\log n)$ time. If q exists, we partition $P[v_{l(u)}, v_{r(u)}]$ into two subpaths $P[v_{l(u)}, v_q]$ and $P[v_{q+1}, v_{r(u)}]$. Letting $V_1 = V[v_{l(u)}, v_q]$, $V_2 = V[v_{q+1}, v_{r(u)}]$, and $W = W[v_i, v_{l(u)-1}]$, we define

$$\tilde{\Theta}^u_{L,1}(W) = \max_{v_p \in V_1} \left\{ \theta^{[l(u),r(u)]}_L(v^+_{r(u)}, v_p) + W/c(v_p, v_{r(u)+1}) \right\}, \tag{12}$$

$$\tilde{\Theta}^u_{L,2}(c) = \max_{v_p \in V_2} \left\{ d(v_p, v_{r(u)})\tau + W[v_{l(u)}, v_p]/c \right\}. \tag{13}$$

Note that the range of maximization $v_p \in V_1$ in (12) (resp. $v_p \in V_2$ in (13)) is limited compared with (8) (resp. (9)). If q does not exist, we set $\tilde{\Theta}^u_{L,1}(W) = 0$ and $V_2 = V[v_{l(u)}, v_{r(u)}]$. It is clear that

$$\max \left\{ \tilde{\Theta}^u_{L,1}(W), \tilde{\Theta}^u_{L,2}(c) + W/c \right\} \tag{14}$$

is equal to (6), and its maximizing vertex corresponds to a candidate from $P[v_{l(u)}, v_{r(u)}]$ for the L-critical vertex of $P[v_i, v_j]$.

Let v_1^* (resp. v_2^*) be a vertex in $V_1 \cup V_2 = V[v_{l(u)}, v_{r(u)}]$ which maximizes the bracketed term in (8) (resp. (9)). Once $W = W[v_i, v_{l(u)-1}]$ and $c = c(v_{r(u)+1}, v_{j+1})$ are given, we can obtain v_1^* and v_2^* by binary search on the bending points of $\tilde{\Theta}_{L,1}^u(W)$ and $\tilde{\Theta}_{L,2}^u(c)$, respectively, which can be done in $O(\log n)$ time. We can now prove the following lemma.

Lemma 3.

(a) If $v_2^* \in V_1$, we have

$$\tilde{\Theta}_{L,1}^u(W) > \tilde{\Theta}_{L,2}^u(c) + W/c. \tag{15}$$

(b) If $v_1^* \in V_2$, we have

$$\tilde{\Theta}_{L,1}^u(W) \leq \tilde{\Theta}_{L,2}^u(c) + W/c. \tag{16}$$

(c) $v_1^* \in V_2$ and $v_2^* \in V_1$ cannot happen at the same time.

If $v_1^* \in V_1$ and $v_2^* \in V_2$, clearly v_1^* and v_2^* also achieve the maxima in (12) and (13), respectively. Therefore, if $\tilde{\Theta}_{L,1}^u(W) > \tilde{\Theta}_{L,2}^u(c) + W/c$ (resp. $\tilde{\Theta}_{L,1}^u(W) \leq \tilde{\Theta}_{L,2}^u(c) + W/c$), v_1^* (resp. v_2^*) is a candidate critical vertex from $P[v_{l(u)}, v_{r(u)}]$. Otherwise, by Lemma 3, $v_1^*, v_2^* \in V_1$ or $v_1^*, v_2^* \in V_2$ holds. Also by Lemma 3, if $v_1^*, v_2^* \in V_1$ (resp. $v_1^*, v_2^* \in V_2$), v_1^* (resp. v_2^*) is a candidate critical vertex. Based on the above arguments, we can prove the following lemma.

Lemma 4. *Suppose that CUE tree $\mathcal{T}$ is available. Consider subpath $P[v_i, v_j]$ with $1 \leq i < j \leq n$.*

(a) *For each node $u \in N[v_i, v_j]$, candidates from $P[v_{l(u)}, v_{r(u)}]$ for L-critical and R-critical vertices of $P[v_i, v_j]$ can be computed in $O(\log n)$ time.*

(b) *The L-critical and R-critical vertices for $P[v_i, v_j]$ can be computed in $O(\log^2 n)$ time.*

4   Building blocks

There are two useful tasks that we can call upon repeatedly. Given the starting vertex v_a, the first task is to find the rightmost vertex v_d such that all the evacuees on $V[v_a, v_d]$ can evacuate to a sink within time t. The second task is to find the cost of the 1-sink on a given subpath $P[v_i, v_j]$. To perform these tasks, we start with more basic procedures.

4.1   Basic algorithms

To implement the first task, note that for a given index $h > a$, there are $O(\log n)$ nodes in $N[v_a, v_h]$. For each such node u, we want to test where a sink s should be placed: to the left of $v_{l(u)}$, to the right of $v_{r(u)}$, or between $v_{l(u)}$ and $v_{r(u)}$.

Here is an algorithm for the first task.

Algorithm 1 1-Sink(t, v_a)

1. *Compute an integer b by binary search over h with $a \leq h \leq n$ such that the L-time of $c_L^{[a,b]}$ to v_b^+ does not exceed t but the L-time of $c_L^{[a,b+1]}$ to v_{b+1}^+ exceeds t.*
2. *Solve $\theta_L^{[a,b]}(v_b^+, c_L^{[a,b]}) + x\tau = t$, and place a sink $s \in (v_b, v_{b+1}]$ satisfying $d(v_b, s) = x$.*
3. *If $s \in (v_b, v_{b+1})$, set c to $b+1$. If $s = v_{b+1}$, set c to $b+2$. Compute an integer d by binary search over h with $c \leq h \leq n$ such that the R-time of $c_R^{[c,d]}$ to s does not exceed t but the R-time of $c_R^{[c,d+1]}$ to s exceeds t.*

Lemma 5. *If CUE tree $\mathcal{T}$ is available, 1-Sink(t, v_a) runs in $O(\log^3 n)$ time.*

Proof. In Step 1, for a fixed h, finding $c_L^{[a,h]}$ and computing the L-time of $c_L^{[a,h]}$ to v_h^+ take $O(\log^2 n)$ time by Lemma 4. Clearly, we repeat this computation $O(\log n)$ times, thus Step 1 takes $O(\log^3 n)$ time. Step 2 takes $O(1)$ time and Step 3 takes $O(\log^3 n)$ time similarly to Step 1. Summarizing these, we complete the proof. $\square$

Here is an algorithm for the second task.

Algorithm 2 Local-Cost(v_i, v_j)

1. *Let u be the node where the two paths $\pi(v_i, \rho)$ and $\pi(v_j, \rho)$ meet.*
2. *If the L-time of $c_L^{[i, r(u_l)]}$ and the R-time of $c_R^{[l(u_r), j]}$ have the same value at some point x on the edge $(v_{r(u_l)}, v_{l(u_r)})$, then output x as the 1-sink.*
3. *If the L-time of $c_L^{[i, r(u_l)]}$ is higher (resp. lower) than the R-time of $c_R^{[l(u_r), j]}$ at every point on edge $(v_{r(u_l)}, v_{l(u_r)})$, then let $u = u_l$ (resp. $u = u_r$) and repeat Step 2, using the new u_l and u_r.*

We have the following lemma.

Lemma 6. *If CUE tree $\mathcal{T}$ is available, Local-Cost(v_i, v_j) finds a 1-sink on subpath $\Gamma[v_i, v_j]$ in $O(\log^3 n)$ time.*

4.2 t-feasibility test

We carry out 1-Sink(t, v) repeatedly, starting from the left end of P, i.e., v_1. Clearly, the problem instance is t-feasible if and only if the rightmost vertex v_n belongs to the l-th isolated subpath, where $l \leq k$.

Lemma 7. *Given a dynamic flow path network, if CUE tree $\mathcal{T}$ is available, we can test its t-feasibility in $O(\min\{n \log^2 n, k \log^3 n\})$ time.*

Proof. Starting at the leftmost vertex v_1 of P, invoke 1-Sink(t, v_1), which isolates the first subpath in $O(\log^3 n)$ time by Lemma 5, and remove it from P. We repeat this at most $k - 1$ more times on the remaining subpath, spending $O(k \log^3 n)$ time.

On the other hand, when each 1-Sink(t, v_a) is executed, suppose we compute the L-time of $c_L^{[a,h]}$ to v_h^+ for $h = a, a+1, \ldots$ one by one at Step 1, and similarly the R-time of $c_R^{[c,h]}$ to s for $h = c, c+1, \ldots$ one by one, instead of binary search. Then, the computations of L-time and R-time are invoked at most n times during a t-feasibility test. Since each computation of L-time or R-time takes $O(\log^2 n)$ time by Lemma 4, the total time is $O(n \log^2 n)$ in this way. □

4.3 Uniform edge capacity case

The problem is much simplified if the edges have the same capacity. In particular, we can compute the critical vertex of a subpath resulting from concatenating two subpaths in constant time. At each node u of T bottom up, we compute and record the L- and R-critical vertices of $P[v_{l(u)}, v_{r(u)}]$ with respect to $v_{r(u)}^+$ and their costs, based on the following lemma.

Lemma 8. [11] *For a node u of CUE tree T, let $v_{l(u_l)} = v_h$, $v_{r(u_l)} = v_i$, $v_{l(u_r)} = v_{i+1}$, and $v_{r(u_r)} = v_j$, and assume that the critical vertices, $c_L^{[h,j]}$, $c_R^{[h,j]}$, $c_L^{[i+1,j]}$, and $c_R^{[i+1,j]}$ have already been computed.*

(a) The L-critical vertex $c_L^{[h,j]}$ is either $c_L^{[h,i]}$ or $c_L^{[i+1,j]}$.
(b) The R-critical vertex $c_R^{[h,j]}$ is either $c_R^{[h,i]}$ or $c_R^{[i+1,j]}$.

The following two lemmas provide counterparts to Lemmas 2 and 4, respectively.

Lemma 9. *Given a dynamic flow path network with n vertices and uniform edge capacities, CUE tree T with associated data can be constructed in $O(n)$ time and $O(n)$ space.*

Lemma 10. *Suppose that CUE tree T is available. For any i and j ($1 \le i < j \le n$), we can comput the L-critical and R-critical vertices for $P[v_i, v_j]$ in $O(\log n)$ time.*

Similarly to Lemma 7, we can prove the following lemma.

Lemma 11. *Given a dynamic flow path network with uniform edge capacities, if CUE tree T is available, we can test its t-feasibility in $O(\min\{n, k \log n\})$ time.*

5 Optimization

5.1 Parametric search approach

Lemma 12. [1] *If t-feasibility test can be tested in $\alpha(t)$ time, then the k-sink can be found in $O(k\alpha(t) \log n)$ time, excluding the preprocessing time.*

By Lemma 2 it takes $O(n \log n)$ time to construct T with weight and capacity data, and $\alpha(t) = O(k \log^3 n)$ by Lemma 7. Lemma 12 thus implies

Theorem 1. *Given a dynamic flow path network with n vertices, we can find an optimal k-sink in $O(n \log n + k^2 \log^4 n)$ time.*

Applying Megiddo's main theorem in [16] to Lemma 11, we obtain

Theorem 2. *Given a dynamic flow path network with n vertices and uniform edge capacities, we can find an optimal k-sink in $O(n + k^2 \log^2 n)$ time.*

5.2   Sorted matrix approach

Let $OPT(l, r)$ denote the evacuation time for the optimal 1-sink on subpath $P[v_l, v_r]$. Define an $n \times n$ matrix A whose entry (i, j) entry is given by

$$A[i, j] = \begin{cases} OPT(n - i + 1, j) & \text{if } n - i + 1 \le j \\ 0 & \text{otherwise.} \end{cases} \tag{17}$$

It is clear that matrix A includes $OPT(l, r)$ for every pair of integers (l, r) such that $1 \le l \le r \le n$. There exists a pair (l, r) such that $OPT(l, r)$ is the evacuation time for the optimal k-sink on the whole path. Then the k-sink location problem can be formulated as: "Find the smallest $A[i, j]$ such that the given problem instance is $A[i, j]$-feasible." Note that we do not actually compute all the elements of $A[\ ]$, but element $A[i, j]$ is computed on demand as needed.

A matrix is called a *sorted matrix* if each row and column of it is sorted in the nondecreasing order. In [6, 7], Frederickson and Johnson show how to search for an element in a sorted matrix. The following lemma is implicit in their papers.

Lemma 13. *Suppose that $A[i, j]$ can be computed in $g(n)$ time, and feasibility can be tested in $f(n)$ time with $h(n)$ preprocessing time. Then we can solve the k-sink problem in $O(h(n) + ng(n) + f(n) \log n)$ time.*

We have $h(n) = O(n \log n)$ by Lemma 2, $g(n) = O(\log^3 n)$ by Lemma 6, and $f(n) = O(n \log^2 n)$ by Lemma 7. Lemma 13 thus implies

Theorem 3. *Given a dynamic path network with n vertices and general edge capacities, we can find an optimal k-sink in $O(n \log^3 n)$ time.*

In the uniform edge capacity case, we have $h(n) = O(n)$ by Lemma 9, $g(n) = O(\log n)$ by Lemma 10, and $f(n) = O(n)$ by Lemma 11. Lemma 13 thus implies

Theorem 4. *Given a dynamic path network with n vertices and uniform edge capacities, we can find the k-sink in $O(n \log n)$ time.*

6   Conclusion and discussion

We have presented more efficient algorithms than the existing ones to solve the k-sink problem on dynamic flow path networks. Due to lack of space, we could not present all the proofs. All our results are valid if the model is changed slightly, so that the weights and edge capacities are not restricted to be integers. Then it becomes confluent transshipment problem.

For dynamic flow tree networks with uniform edge capacities, it is known that computing evacuation time to a vertex can be transformed to that on a path network [13]. We believe that our method is applicable to each "spine," which is a path in the spine decomposition of a tree [2], and we think we may be able to solve the k-sink problem on dynamic flow tree networks more efficiently. This is work in progress.

References

1. G. P. Arumugam, J. Augustine, M. J. Golin, and P. Srikanthan. A polynomial time algorithm for minimax-regret evacuation on a dynamic path. *arXiv:1404,5448v1*, 2014.
2. R. Benkoczi, B. Bhattacharya, M. Chrobak, L. Larmore, and W. Rytter. Faster algorithms for *k*-median problems in trees. *Mathematical Foundations of Computer Science, Springer-Verlag*, LNCS 2747:218–227, 2003.
3. Binay Bhattacharya and Tsunehiko Kameda. Improved algorithms for computing minmax regret sinks on path and tree networks. *Theoretical Computer Science*, 607:411–425, Nov. 2015.
4. S. W. Cheng, Y. Higashikawa, N. Katoh, G. Ni, B. Su, and Y. Xu. Minimax regret 1-sink location problem in dynamic path networks. In *Proc. Annual Conf. on Theory and Applications of Models of Computation (T-H.H. Chan, L.C. Lau, and L. Trevisan, Eds.), Springer-Verlag*, volume LNCS 7876, pages 121–132, 2013.
5. L. R. Ford and D. R. Fulkerson. Constructing maximal dynamic flows from static flows. *Operations research*, 6(3):419–433, 1958.
6. G. N. Frederickson. Optimal algorithms for tree partitioning. In *Proc. 2nd ACM-SIAM Symp. Discrete Algorithms*, pages 168–177, 1991.
7. G. N. Frederickson and D. B. Johnson. Finding *k*th paths and *p*-centers by generating and searching good data structures. *J. Algorithms*, 4:61–80, 1983.
8. H.W. Hamacher and S.A. Tjandra. Mathematical modeling of evacuation problems: a state of the art. *in: Pedestrian and Evacuation Dynamics, Springer Verlag,*, pages 227–266, 2002.
9. Y. Higashikawa. *Studies on the space exploration and the sink location under incomplete information towards applications to evacuation planning.* PhD thesis, Kyoto University, Japan, 2014.
10. Y. Higashikawa, M. J. Golin, and N. Katoh. Minimax regret sink location problem in dynamic tree networks with uniform capacity. *J. of Graph Algorithms and Applications*, 18.4:539–555, 2014.
11. Y. Higashikawa, M. J. Golin, and N. Katoh. Multiple sink location problems in dynamic path networks. *Theoretical Computer Science*, 607:2–15, 2015.
12. B. Hoppe and É. Tardos. The quickest transshipment problem. *Mathematics of Operations Research*, 25(1):36–62, 2000.
13. N. Kamiyama, N. Katoh, and A. Takizawa. An efficient algorithm for evacuation problem in dynamic network flows with uniform arc capacity. *IEICE Transactions*, 89-D(8):2372–2379, 2006.
14. S. Mamada, K. Makino, and S. Fujishige. Optimal sink location problem for dynamic flows in a tree network. *IEICE Trans. Fundamentals*, E85-A:1020–1025, 2002.
15. S. Mamada, T. Uno, K. Makino, and S. Fujishige. An $O(n \log^2 n)$ algorithm for a sink location problem in dynamic tree networks. *Discrete Applied Mathematics*, 154:2387–2401, 2006.
16. N. Megiddo. Combinatorial optimization with rational objective functions. *Math. Oper. Res.*, 4:414–424, 1979.
17. N. Megiddo and A. Tamir. New results on the complexity of *p*-center problems. *SIAM J. Comput.*, 12:751–758, 1983.
18. M. Skutella. An introduction to network flows over time. In *Research Trends in Combinatorial Optimization*, pages 451–482. Springer, 2009.

A 2-Approximation for the Height of Maximal Outerplanar Graph Drawings *

Therese Biedl and Philippe Demontigny

David R. Cheriton School of Computer Science, University of Waterloo, Waterloo, Ontario N2L 1A2, Canada.
biedl@uwaterloo.ca, phdemontigny@gmail.com

Abstract In this paper, we study planar drawings of maximal outerplanar graphs with the objective of achieving small height. (We do not necessarily preserve a given planar embedding.) A recent paper gave an algorithm for such drawings that is within a factor of 4 of the optimum height. In this paper, we substantially improve the approximation factor to become 2. The main ingredient is to define a new parameter of outerplanar graphs (the *umbrella depth*, obtained by recursively splitting the graph into graphs called *umbrellas*). We argue that the height of any poly-line drawing must be at least the umbrella depth, and then devise an algorithm that achieves height at most twice the umbrella depth.

1 Introduction

Graph drawing is the art of creating a picture of a graph that is visually appealing. In this paper, we are interested in drawings of so-called *outerplanar graphs*, i.e., graphs that can be drawn in the plane such that no two edges have a point in common (except at common endpoints) and all vertices are incident to the outerface. All drawings are required to be planar, i.e., to have no crossing. The drawing model used is that of flat visibility representations where vertices are horizontal segments and edges are horizontal or vertical segments, but any such drawing can be transformed into a poly-line drawing (or even a straight-line drawing if the width is of no concern) without adding height [6].

Every planar graph with n vertices has a straight-line drawing in an $n \times n$-grid [19,9]. Minimizing the area is NP-complete [17], even for outerplanar graphs [7]. In this paper, we focus on minimizing just one direction of a drawing (we use the height; minimizing the width is equivalent after rotation). It is not known whether minimizing the height of a planar drawing is NP-hard (the closest related result concerns minimizing the height if edges must connect adjacent rows [16]). Given the height H, testing whether a planar drawing of height H exists is fixed parameter tractable in H [12], but the run-time is exceedingly large in H. As such, approximation algorithms for the height of planar drawings are of interest.

---

* TB supported by NSERC. Part of this work appeared as PD's Master's thesis [10].

© Springer International Publishing AG 2017   145
F. Ellen et al. (Eds.): WADS 2017, LNCS 10389, pp. 145–156, 2017.
DOI: 10.1007/978-3-319-62127-2_13

It is known that any graph G with a planar drawing of height H has $pw(G) \leq H$ [13], where $pw(G)$ is the so-called pathwidth of G. This makes the pathwidth a useful parameter for approximating the height of a planar graph drawing. For a tree T, Suderman gave an algorithm to draw T with height at most $\lceil \frac{3}{2} pw(T) \rceil$ [20], making this an asymptotic $\frac{3}{2}$-approximation algorithm. It was discovered later that optimum-height drawings can be found efficiently for trees [18]. Approximation-algorithms for the height or width of order-preserving and/or upward tree drawing have also been investigated [1,2,8].

For outerplanar graphs, the first author gave two results that will be improved upon in this paper. In particular, every maximal outerplanar graph has a drawing of height at most $3 \log n - 1$ [3], or alternatively of height $4pw(G) - 3$ [5]. Note that the second result gives a 4-approximation on the height of drawing outerplanar graphs, and improving this "4" is the main objective of this paper. A number of results for drawing outerplanar graphs have been developed since paper [3]. In particular, any outerplanar graph with maximum degree Δ admits a planar straight-line drawing with area $O(\Delta n^{1.48})$ [15], or with area $O(\Delta n \log n)$ [14]. The former bound was improved to $O(n^{1.48})$ area [11]. Also, every so-called balanced outerplanar graph can be drawn in an $O(\sqrt{n}) \times O(\sqrt{n})$-grid [11].

In this paper, we present a 2-approximation algorithm for the height of planar drawings of maximal outerplanar graphs. The key ingredient is to define the so-called *umbrella depth* $ud(G)$ in Section 3. In Section 4, we show that any outerplanar graph G has a planar drawing of height at most $2ud(G) + 1$. This algorithm is a relatively minor modification of the one in [5], albeit described differently. The bulk of the work for proving a better approximation factor hence lies in proving a better lower bound, which we do in Section 5: Any maximal outerplanar graph G with a planar drawing of height H has $ud(G) \leq H - 1$.

2   Preliminaries

Throughout this paper, we assume that G is a simple graph with $n \geq 3$ vertices that is *maximal outerplanar*. Thus, G has a *standard planar embedding* in which all vertices are in the *outer face* (the infinite connected region outside the drawing) and form an n-cycle, and all *interior faces* are triangles. We call an edge (u, v) of G a *cutting edge* if $G - \{u, v\}$ is disconnected, and a *non-cutting edge* otherwise. In a maximal outerplanar graph, any cutting edge (u, v) has exactly two *cut-components*, i.e., there are two maximal outerplanar subgraphs G_1, G_2 of G such that $G_1 \cap G_2 = \{u, v\}$ and $G_1 \cup G_2 = G$.

The *dual tree* T of G is the weak dual graph of G in the standard embedding, i.e., T has a vertex for each interior face of G, and an edge between two vertices iff their corresponding faces in G share an edge. An *outerplanar path* P is a maximal outerplanar graph whose dual tree is a path. P *connects edges e and e'* if e is incident to the first face and e' is incident to the last face of the path that is the dual tree of P. An outerplanar path P with $n = 3$ is a triangle and connects any pair of its edges. Since any two interior faces are connected by a path in T, any two edges e, e' of G are connected by some outerplanar path.

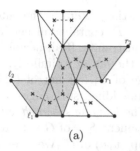 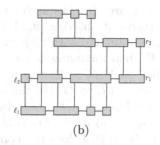

(a)   (b)

Figure 1: (a) A straight-line drawing in the standard embedding, including the dual tree (dashed edges) and an outerplanar path (shaded) connecting (ℓ_1, ℓ_2) with (r_1, r_2). (b) A flat visibility representation. Both drawings have height 4.

Graph drawing: A *drawing* of a graph assigns to each vertex a point or an axis-aligned box, and to each edge a polygonal curve connecting its endpoints. We only consider *planar drawings* where none of the points, boxes, or curves intersect unless the corresponding elements do in the graph. In this paper, a planar drawing is not required to reflect a graph's given planar embedding. We require that all defining features (points, endpoints of segments, bends) are placed at points with integer y-coordinates. A *layer* (or *row*) is a horizontal line with integer y-coordinate that intersects elements of the drawing, and the *height* is the number of layers.

In a *flat visibility representation* vertices are horizontal line segments, and edges are vertical or horizontal straight-line segments. (For ease of reading, we draw vertices as boxes of small height in our illustrations.) In a *poly-line drawing* vertices are points and edges are polygonal curves, while in a *straight-line drawing* vertices are points and edges are line segments. In this paper, we only study planar flat visibility representations, but simply speak of a *planar drawing*, because it is known that any planar flat visibility representation can be converted into a planar straight-line drawing of the same height and vice versa [6].

3   Umbrellas, bonnets and systems thereof

In this section, we introduce a method of splitting maximal outerplanar graphs into systems of special outerplanar graphs called *umbrellas* and *bonnets*.

Definition 1. *Let G be a maximal outerplanar graph, let U be a subgraph of G with $n \geq 3$, and let (u, v) be a non-cutting edge of G. We say that U is an umbrella with cap (u, v) if*

1. *U contains all neighbours of u and v,*
2. *there exists a non-empty outerplanar path $P \subseteq U$ (the handle) that connects (u, v) to some non-cutting edge of G, and*
3. *any vertex of U is either in P or a neighbour of u or v.*

See also Figure 2(a). For such an umbrella U, the *fan at u* is the outerplanar path that starts at an edge (u, x) of the handle P, contains all neighbours of u, and that is minimal with respect to these constraints. If all neighbours of u belong to P, then the fan at u is empty. Define the *fan at v* similarly, using v.

Any edge (a, b) of U that is a cutting edge of G, but not of U, is called an *anchor-edge* of U in G. (In the standard embedding, such edges are on the outerface of U but not on the outerface of G.) The *hanging subgraph with respect to anchor-edge (a, b) of U in G* is the cut-component $S_{a,b}$ of G with respect to cutting-edge (a, b) that does not contain the cap (u, v) of U. We often omit "of U in G" when umbrella and super-graph are clear from the context.

Definition 2. *Let G be a maximal outerplanar graph with $n \geq 3$, and let (u, v) be a non-cutting edge of G. An* umbrella system *$\mathcal{U}$ on G with root-edge (u, v) is a collection $\mathcal{U} = \mathcal{U}_0 \cup \mathcal{U}_1 \cup \cdots \cup \mathcal{U}_k$ of subgraphs of G for some $k \geq 0$ that satisfy the following:*

1. *$\mathcal{U}_0$ contains only one subgraph U_0 (the* root umbrella*), which is an umbrella with cap (u, v).*
2. *U_0 has k anchor-edges. We denote them by (u_i, v_i) for $i = 1, \ldots, k$, and let S_i be the hanging subgraph with respect to (u_i, v_i).*
3. *For $i = 1, \ldots, k$, $\mathcal{U}_i$ (the* hanging umbrella system*) is an umbrella system of S_i with root-edge (u_i, v_i).*

The depth *of such an umbrella system is defined recursively to be $d(\mathcal{U}) := 1 + \max_{1 \leq i \leq k} d(\mathcal{U}_i)$; in particular $d(\mathcal{U}) = 1$ if $k = 0$.*

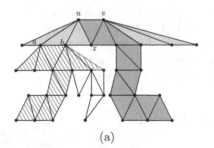 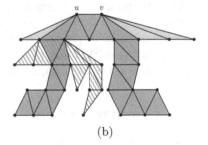

| (a) | (b) |

Figure 2: (a) An umbrella system of depth 3. The root umbrella is shaded, with its handle darker shaded. (b) The same graph has a bonnet system of depth 2, with the root bonnet shaded and its ribbon darker shaded.

See also Figure 2(a). A graph may have many different umbrella systems with the same root-edge. Define $ud(G; u, v)$ (the *(rooted) umbrella depth* of G) to be the minimum depth over all umbrella systems with root-edge (u, v). Note that the umbrella depth depends on the choice of the root-edge; define the *free umbrella depth* $ud(G) := ud^{free}(G)$ to be the minimum umbrella depth over all possible root-edges. (One can show that the free umbrella depth is at most one unit less than the rooted umbrella depth for any choice of root-edge; see [10].)

Bonnets: A *bonnet* is a generalization of an umbrella that allows two handles, as long as they go to different sides of the interior face at (u, v). Thus, condition (2) of the definition of an umbrella gets replaced by

2'. There exists a non-empty outerplanar path $P \subseteq U$ (the *ribbon*) that connects two non-cutting edges and contains u, v and their common neighbour.

Other than that, bonnets are defined exactly like umbrellas. See also Figure 2(b). We define *bonnet system*, *root bonnet*, etc., exactly as for an umbrella system, except that "bonnet" is substituted for "umbrella" everywhere. Let $bd(G; u, v)$ (the *rooted bonnet-depth of G*) be the minimum possible depth of a bonnet system with root-edge (u, v), and let $bd^{free}(G) = bd(G)$ be the minimum bonnet-depth over all choices of root-edge. Since any umbrella is a bonnet, we have $bd(G) \leq ud(G)$.

By definition the root bonnet U_0 must contain *all* edges incident to the ends u, v of the root-edge. If follows that no edge incident to u or v can be an anchor-edge of U_0, else the hanging subgraph at it would contain further neighbours of u (resp. v). We note this trivial but useful fact for future reference:

Observation 1 *In a bonnet system with root-edge (u, v), no edge incident to u or v is an anchor-edge of the root bonnet.*

4   From Bonnet System to Drawing

In this section, we show that any outerplanar graph G has a flat visibility representation of height at most $2ud(G) + 1$. We actually show a slightly stronger bound, namely a height of $2bd(G) + 1 \leq 2ud(G) + 1$. So fix a bonnet system of G of depth $bd(G)$ with root-edge (u, v). For merging purposes, we want to draw (u, v) in a special way: It *spans* the top layer, which means that u touches the top left corner of the drawing, and v touches the top right corner, or vice versa (see for example Figure 3(d)). We first explain how to draw the root bonnet U_0.

Lemma 1. *Let U_0 be the root bonnet of a bonnet system with root-edge (u, v). Then there exists a flat visibility representation Γ of U_0 on three layers such that*

1. *(u, v) spans the top layer of Γ.*
2. *Any anchor-edge of U_0 is drawn horizontally in the middle or bottom layer.*

Proof. As a first step, we draw the ribbon P of U_0 on 2 layers in such a way that (u, v) and all anchor-edges are drawn horizontally; see Figure 3(a) for an illustration. (This part is identical to [5].) To do this, consider the standard embedding of P in which the dual tree is a path, say it consists of faces $f_1, \ldots, f_k$. We draw $k + 1$ vertical edges between two layers, with the goal that the region between two consecutive ones belong to $f_1, \ldots, f_k$ in this order. Place u and v as segments in the top layer, and with an x-range such that they touch all the regions of faces that u and v are incident to. Similarly create segments for all other vertices. The placement for the vertices is uniquely determined by the standard planar embedding, except for the vertices incident to f_1 and f_k. We place those vertices such that the leftmost/rightmost vertical edge is not an anchor-edge. To see that this is possible, recall that P connects two non-cutting edges e_1, e_2 of G that are incident to f_1 and f_k. If $e_1 \neq (u, v)$, then choose the layer for the vertices of f_1 such that e_1 is drawn vertically. If $e_1 = (u, v)$, then one of its ends (say u) is the degree-2 vertex on f_1 and drawn in the top-left corner. The other edge e' incident to u is not an anchor-edge of U by Observation 1, and we draw e' vertically. So the leftmost vertical edge is either a non-cutting edge (hence not an anchor-edge) or edge e' (which is not an anchor-edge). We proceed similarly at f_k so that the rightmost vertical edge is not an anchor-edge. Finally all other vertical edges are cutting edges of U_0 and hence not anchor-edges.

The drawing of P obtained in this first step has (u, v) in the top layer. As a second step, we now *release* (u, v) as in [5]. This operation adds a layer above the drawing, moves (u, v) into it, and re-routes edges at u and v by expanding vertical ones and turning horizontal ones into vertical ones. In the result, (u, v) spans the top layer. See Figure 3(b) for an illustration and [5] for details.

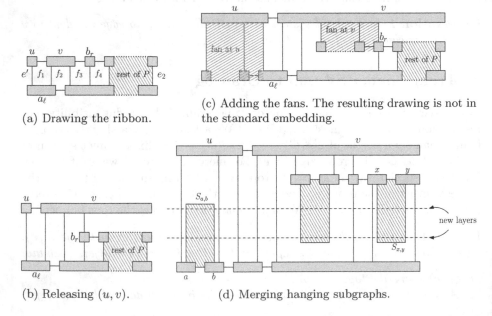

(a) Drawing the ribbon.

(c) Adding the fans. The resulting drawing is not in the standard embedding.

(b) Releasing (u, v).

(d) Merging hanging subgraphs.

Figure 3: From bonnet system to drawing.

As the third and final step, we add the fans. Consider the fan at v, and let (v, b_r) be the edge that it has in common with the ribbon P. Assume first that (v, b_r) was drawn horizontally after the first step, see Figure 3(a). After releasing (u, v) therefore no edge at b_r attaches on its left, see Figure 3(b). Into this space we insert, after adding columns, the remaining vertices of the fan at v, in order in which they appear around v in the standard embedding. See Figure 3(c)).

Else, (v, b_r) was drawn vertically after the first step. (Figure 3(c) does not illustrate this case for v, but illustrates the corresponding case for u.) Since the drawing of the first step is in the standard embedding, and (v, b_r) is on the outerface of the ribbon, therefore (v, b_r) must be the rightmost vertical edge. We can then simply place the vertices of the fan to the right of b_r and extend v.

The fan at u is placed in a symmetric fashion. It remains to show that all anchor-edges are horizontal and in the bottom two layers. We ensured that this is the case in the first step. Releasing (u, v) adds more vertical edges, but all of them are incident to u or v and not anchor-edges by Observation 1. Likewise, all vertical edges added when inserting the fans are incident to u or v. The only horizontal edge in the top layer is (u, v), which is not an anchor-edge.   □

Now we explain how to merge hanging subgraphs.

Theorem 1. *Any maximal outerplanar graph G has a planar flat visibility representation of height at most $2bd^{free}(G) + 1$.*

Proof. We show by induction that any graph with a bonnet system $\mathcal{U}$ of depth H has a drawing Γ of height $2H + 1$ where the root-edge (u, v) spans the top layer. This proves the theorem when using a bonnet system $\mathcal{U}$ of depth $bd^{free}(G)$.

Let U_0 be the root bonnet of the bonnet system, and draw U_0 on 3 layers using Lemma 1. Thus (u, v) spans the top and any anchor-edge (a, b) of U_0 is drawn as a horizontal edge in the bottom two layers of Γ_0. If $H = 1$ then there are no hanging subgraphs and we are done. Else add $2H - 2$ layers to Γ_0 between the middle and bottom layers. For each anchor-edge (a, b) of U_0, the hanging subgraph $S_{a,b}$ of U_0 has a bonnet system of depth at most $H - 1$ with root-edge (a, b). By induction $S_{a,b}$ has a drawing Γ_1 on at most $2H - 1$ layers with (a, b) spanning the top layer.

If (a, b) is in the bottom layer of Γ_0, then we can rotate (and reflect, if necessary) Γ_1 so that (a, b) is in the bottom layer of Γ_1 and the left-to-right order of a and b in Γ_1 is the same as their left-to-right order in Γ_0. This updated drawing of Γ_1 can then be inserted in the space between (a, b) in Γ_0. This fits because Γ_1 has height at most $2H - 1$, and in the insertion process we can re-use the layer spanned by (a, b). If (a, b) is in the middle layer of U_0, then we can reflect Γ_1 (if necessary) so that (a, b) has the same left-to-right order in Γ_1 as in Γ_0. This updated drawing of Γ_1 can then be inserted in the space between (a, b) in Γ_0. See Figure 3(d). Since we added $2H - 2$ layers to a drawing of height 3, the total height of the final drawing is $2H + 1$ as desired.   □

Our proof is algorithmic, and finds a drawing, given a bonnet system, in linear time. One can also show (see [10]) that the rooted bonnet depth, and an associated bonnet system, can be found in linear time using dynamic programming in the dual tree. The free bonnet depth can be found in quadratic time by trying all root-edges, but one can argue [10] that this will save at most one unit of depth and hence barely seems worth the extra run-time.

Comparison to [5]: The algorithm in [5] has only two small differences. The main one is that it does not do the "third step" when drawing the root bonnet, thus it draws the ribbon but not the fans. Thus in the induction step our algorithm always draws at least as much as the one in [5]. Secondly, [5] uses a special construction if $pw(G) = 1$ to save a constant number of levels. This could easily be done for our algorithm as well in the case where $pw(G) = 1$ but $bd(G) = 2$. As such, our construction never has worse height (and frequently it is better).

Comparison to [3]: One can argue that $bd(G) \leq \log(n+1)$ (see [10]). Since [3] uses $3 \log n - 1$ levels while ours uses $2bd(G) + 1 \leq 2 \log(n+1) + 1$ levels, the upper bound on the height is better for $n \geq 9$.

5   From Drawing to Umbrella System

The previous section argued that given an umbrella system (or even more generally, a bonnet system) of depth H, we can find a drawing of height at most $2H - 1$. To show that this is within a factor of 2 of the optimum, we show in this section that any drawing of height H gives rise to an umbrella system of depth at most $H - 1$. (Any umbrella system is also a bonnet system, so it also has a bonnet system of depth at most $H - 1$.)

We first briefly sketch the idea. We assume that we have a flat visibility representation, and further, for some non-cutting edge (u, v) we have an "escape path", i.e., a poly-line to the outerface that does not intersect the drawing. Now find an outerplanar path that connects the leftmost vertical edge (x, y) of the drawing with (u, v). This becomes the handle of an umbrella U with cap (u, v). One can now argue that any hanging subgraph of U is drawn with height at most $H - 1$, and furthermore, has an escape path from its anchor-edge. The claim then holds by induction.

We first clarify some definitions illustrated in Figure 4(a). Let Γ be a flat visibility representation, and let B_Γ be a minimum-height bounding box of Γ. A vertex $w \in G$ has a *right escape path* in Γ if there exists a polyline inside B_Γ from w to a point on the right side of B_Γ that is vertex-disjoint from Γ except at w, and for which all bends are on layers. We say that (r_1, r_2) is a *right-free edge* of Γ if it is vertical, and any layer intersected by (r_1, r_2) is empty, except for vertices r_1, r_2, to the right of the edge. In particular, for both r_1 and r_2 the rightward ray on its layer is an escape path. Define *left escape paths* and *left-free edges* symmetrically; an *escape path* is a left escape path or a right escape path.

Observe that in any flat visibility representation any leftmost vertical edge (v, w) is left-free. (Such vertical edges exist, presuming the graph has minimum degree 2, since the leftmost vertex in each layer has at most one incident horizontal edge.) For in any layer spanned by (v, w), no vertical edge is farther left by choice of (v, w), and no vertex can be farther left, else the incident vertical edge of the leftmost of them would be farther left. So (v, w) is left-free.

For the proof of the lower bound, we use as handle an outerplanar path connecting to a left-free edge. Recall that the definition of handle requires that it connects to a non-cutting edge, so we need a left-free edge that is not a cutting

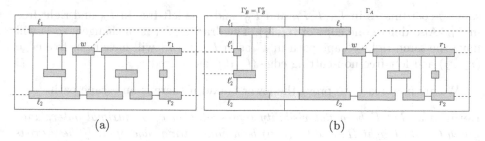

$$\Gamma_B' = \Gamma_B''$$ $$\Gamma_A$$

(a) (b)

Figure 4: w has a right escape path, (ℓ_1, ℓ_2) is left-free and (r_1, r_2) is right-free. After flipping the cutting component at (ℓ_1, ℓ_2), the non-cutting edge (ℓ_1', ℓ_2') becomes left-free.

edge. This does not exist in all drawings (see e.g. Figure 4(a)), but as we show now, we can modify the drawing without increasing the height such that such an edge exists. To be able to apply it later, we must also show that this modification does not destroy a given escape path.

Lemma 2. *Let Γ be a flat visibility representation of a maximal outerplanar graph G.*

1. *Let (r_1, r_2) be a right-free edge of Γ, and let w be a vertex that has a right escape path. Then there exists a drawing Γ' in which w has a right escape path, (r_1, r_2) is a right-free edge, and there exists a left-free edge that is not a cutting edge of G.*
2. *Let (ℓ_1, ℓ_2) be a left-free edge of Γ, and let w be a vertex that has a left escape path. Then there exists a drawing Γ' in which w has a left escape path, (ℓ_1, ℓ_2) is a left-free edge, and there exists a right-free edge that is not a cutting edge of G.*

In either case, the y-coordinates of all vertices in Γ are unchanged in Γ', and in particular both drawings have the same height.

Proof. We prove the claim by induction on n and show only the first claim (the other is symmetric). Let (ℓ_1, ℓ_2) be the leftmost vertical edge of Γ; this is left-free as argued above. If (ℓ_1, ℓ_2) is not a cutting edge of G, then we are done with $\Gamma' = \Gamma$. This holds in particular if $n = 3$ because then G has no cutting edge.

So assume $n \geq 4$ and (ℓ_1, ℓ_2) is a cutting edge of G. Let A and B be the cut-components of (ℓ_1, ℓ_2), named such that $w \in A$. Let Γ_A [resp. Γ_B] be the drawing of A [B] induced by Γ. Edge (ℓ_1, ℓ_2) is left-free for both Γ_A and Γ_B. Reflect Γ_B horizontally (this makes (ℓ_1, ℓ_2) right-free) to obtain Γ_B'. By induction, we can create a drawing Γ_B'' from Γ_B' in which (ℓ_1, ℓ_2) is right-free and there is a left-free edge (ℓ_1', ℓ_2') that is not a cutting edge of B. We have $(\ell_1', \ell_2') \neq (\ell_1, \ell_2)$, because the common neighbour of ℓ_1, ℓ_2 in B forces a vertex or edge to reside to the left of the right-free edge (ℓ_1, ℓ_2). So (ℓ_1', ℓ_2') is not a cutting edge of G either.

As in Figure 4(b), create a new drawing that places Γ_B'' to the left of Γ_A and extends ℓ_1 and ℓ_2 to join the two copies; this is possible since (ℓ_1, ℓ_2) has the same y-coordinates in $\Gamma_A, \Gamma, \Gamma_B$ and Γ_B'', and it is left-free in Γ_A and right-free in Γ_B''. Also delete one copy of (ℓ_1, ℓ_2). The drawing Γ_A is unchanged, so w will have the same right escape path in Γ' as in Γ, and Γ' will have right-free edge (r_1, r_2) and left-free non-cutting edge (ℓ_1', ℓ_2'), as desired.   □

We are now ready to prove the lower bound if there is an escape path.

Lemma 3. *Let Γ be a flat visibility representation of a maximal outerplanar graph G with height H, and let (u, v) be a non-cutting edge of G. If there exists an escape path from u or v in Γ, then G has an umbrella system with root-edge (u, v) and depth at most $H - 1$.*

Proof. We proceed by induction on H. Assume without loss of generality that there exists a right escape path from v (all other cases are symmetric). Using Lemma 2, we can modify Γ without increasing the height so that v has a right escape path, and there is a left-free edge (ℓ_1, ℓ_2) in Γ that is a not a cutting edge of G. Let P be the outerplanar path that connects edge (ℓ_1, ℓ_2) and (u, v). Let U_0 be the union of P, the neighbors of u, and the neighbors of v; we use U_0 as the root umbrella of an umbrella system.

We now must argue that all hanging subgraphs of U_0 are drawn with height at most $H - 1$ and have escape paths from their anchor-edges; we can then find umbrella systems for them by induction and combining them with U_0 gives the umbrella system for G as desired. To prove the height-bound, define "dividing paths" as follows. The outerface of U_0 in the standard embedding contains (ℓ_1, ℓ_2) (since it is not a cutting edge) as well as v. Let P_1 and P_2 be the two paths from ℓ_1 and ℓ_2 to v along this outerface in the standard embedding. Define the *dividing path* Π_i (for $i = 1, 2$) to be the poly-line in Γ that consists of the leftward ray from ℓ_i, the drawing of the path P_i (i.e., the vertical segments of its edges and parts of the horizontal segments of its vertices), and the right escape path from v. See Figure 5.

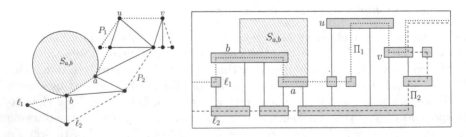

Figure 5: Extracting dividing paths from a flat visibility representation. P_1/Π_1 is dotted while P_2/Π_2 is dashed.

Now consider any hanging subgraph $S_{a,b}$ of U_0 with anchor-edge (a, b). No edge incident to v is an anchor-edge, and neither is (ℓ_1, ℓ_2), since it is not a cutting edge. So (a, b) is an edge of P_1 or P_2 (say P_1) that is not incident to v.

Therefore (a, b) (and with it $S_{a,b}$) is vertex-disjoint from P_2. It follows that the drawing Γ_S of $S_{a,b}$ induced by Γ is disjoint from the dividing path Π_2. Since Π_2 connects a point on the left boundary with a point on the right boundary, therefore Γ_S must be entirely above or entirely below Π_2, say it is above. Since Π_2 has all bends at points with integral y-coordinate, therefore the bottom layer of Γ is not available for Γ_S, and Γ_S has height at most $H - 1$ as desired.

Recall that (a, b) belongs to P_1 and is not incident to v. After possible renaming of a and b, we may assume that b is closer to ℓ_1 along P_1 than a. Then the sub-path of P_1 from b to ℓ_1 is interior-disjoint from $S_{a,b}$. The part of Π_1 corresponding to this path is a left escape path from b that resides within the top $H - 1$ layers, because it does not contain v and hence is disjoint from Π_2. We can hence apply induction to $S_{a,b}$ to obtain an umbrella system of depth at most $H - 2$ with root-edge (a, b). Repeating this for all hanging subgraphs, and combining the resulting umbrella systems with U_0, gives the result. $\qquad\square$

Theorem 2. *Let G be a maximal outerplanar graph. If G has a flat visibility representation Γ of height H, then $ud^{free}(G) \leq H - 1$.*

Proof. Using Lemma 2, we can convert Γ into a drawing Γ' of the same height in which some edge (u, v) is a right-free non-cutting edge. This implies that there is a right escape path from v, and by Lemma 3 we can find an umbrella system of G with root-edge (u, v) and depth $H - 1$. So $ud^{free}(G) \leq ud(G; u, v) \leq H - 1$. $\quad\square$

6   Conclusions and Future Work

We presented an algorithm for drawing maximal outerplanar graphs that is a 2-approximation for the optimal height. To this end, we introduced the umbrella depth as a new graph parameter for maximal outerplanar graphs, and used as key result that any drawing of height H implies an umbrella-depth of at least $H - 1$. Our result improves the previous best result, which was based on the pathwidth and gave a 4-approximation. We close with some open problems:

- Our result only holds for maximal outerplanar graphs. Can the algorithm be modified so that it becomes a 2-approximation for all outerplanar graphs? Clearly one could apply the algorithm after adding edges to make the graph maximal, but which edges should be added to keep the umbrella depth small?
- The algorithm from Section 4 creates a drawing that does not place all vertices on the outerface. Can we create an algorithm that approximates the optimal height in the standard planar embedding?
- What is the width achieved by the algorithm from Section 4 if we enforce integral x-coordinates? Any visibility representation can be modified without changing the height so that the width is at most $m+n$, where m is the number of edges and n is the number of vertices [6]. Thus the width is $O(n)$, but what is the constant?

Finally, can we determine the optimal height for maximal outerplanar graphs in polynomial time? This question is of interest both if (as in our algorithm) the embedding can be changed, or if the drawing must be in the standard embedding.

References

1. Md. J. Alam, Md. A.H. Samee, M. Rabbi, , and Md. S. Rahman. Minimum-layer upward drawings of trees. *J. Graph Algorithms Appl.*, 14(2):245–267, 2010.
2. J. Batzill and T. Biedl. Order-preserving drawings of trees with approximately optimal height (and small width), 2016. CoRR 1606.02233 [cs.CG]. In submission.
3. T. Biedl. Drawing outer-planar graphs in $O(n \log n)$ area. In S. Kobourov and M. Goodrich, editors, *Graph Drawing (GD'01)*, volume 2528 of *LNCS*, pages 54–65. Springer-Verlag, 2002. Full version included in [4].
4. T. Biedl. Small drawings of outerplanar graphs, series-parallel graphs, and other planar graphs. *Discrete and Computational Geometry*, 45(1):141–160, 2011.
5. T. Biedl. A 4-approximation algorithm for the height of drawing 2-connected outerplanar graphs. In T. Erlebach and G. Persiano, editors, *Workshop on Approximation and Online Algorithms (WAOA'12)*, volume 7846 of *LNCS*, pages 272–285. Springer-Verlag, 2013.
6. T. Biedl. Height-preserving transformations of planar graph drawings. In C. Duncan and A. Symvonis, editors, *Graph Drawing (GD'14)*, volume 8871 of *LNCS*, pages 380–391. Springer, 2014.
7. T. Biedl. On area-optimal planar grid-drawings. In J. Esparza, P. Fraigniaud, T. Husfeldt, and E. Koutsoupias, editors, *International Colloquium on Automata, Languages and Programming (ICALP '14)*, volume 8572 of *LNCS*, pages 198–210. Springer-Verlag, 2014.
8. T. Biedl. Ideal tree-drawings of approximately optimal width (and small height). *Journal of Graph Algorithms and Applications*, 21(4):631–648, 2017.
9. H. de Frayseix, J. Pach, , and R. Pollack. How to draw a planar graph on a grid. *Combinatorica*, 10:41–51, 1990.
10. P. Demontigny. A 2-approximation for the height of maximal outerplanar graphs. Master's thesis, University of Waterloo, 2016. See also CoRR report 1702.01719.
11. G. Di Battista and F. Frati. Small area drawings of outerplanar graphs. *Algorithmica*, 54(1):25–53, 2009.
12. V. Dujmovic, M. Fellows, M. Kitching, G. Liotta, C. McCartin, N. Nishimura, P. Ragde, F. Rosamond, S. Whitesides, , and D. Wood. On the parameterized complexity of layered graph drawing. *Algorithmica*, 52:267–292, 2008.
13. S. Felsner, G. Liotta, , and S. Wismath. Straight-line drawings on restricted integer grids in two and three dimensions. *J. Graph Alg. Appl*, 7(4):335–362, 2003.
14. F. Frati. Straight-line drawings of outerplanar graphs in $O(dn \log n)$ area. *Comput. Geom.*, 45(9):524–533, 2012.
15. A. Garg and A. Rusu. Area-efficient planar straight-line drawings of outerplanar graphs. *Discrete Applied Mathematics*, 155(9):1116–1140, 2007.
16. L.S. Heath and A.L. Rosenberg. Laying out graphs using queues. *SIAM Journal on Computing*, 21(5):927–958, 1992.
17. M. Krug and D. Wagner. Minimizing the area for planar straight-line grid drawings. In S. Hong, T. Nishizeki, and W. Quan, editors, *Graph Drawing (GD'07)*, volume 4875 of *LNCS*, pages 207–212. Springer-Verlag, 2007.
18. D. Mondal, Md. J. Alam, , and Md. S. Rahman. Minimum-layer drawings of trees. In N. Katoh and A. Kumar, editors, *Algorithms and Computations (WALCOM 2011)*, volume 6552 of *LNCS*, pages 221–232. Springer, 2011.
19. W. Schnyder. Embedding planar graphs on the grid. In *ACM-SIAM Symposium on Discrete Algorithms (SODA '90)*, pages 138–148, 1990.
20. M. Suderman. Pathwidth and layered drawings of trees. *Intl. J. Comp. Geom. Appl*, 14(3):203–225, 2004.

Splitting B_2-VPG Graphs into Outer-string and Co-comparability Graphs*

Therese Biedl and Martin Derka

David R. Cheriton School of Computer Science,
University of Waterloo, Waterloo, ON, N2L 3G1, Canada
{biedl,mderka}@uwaterloo.ca

Abstract. A B_2-VPG representation of a graph is an intersection representation that consists of orthogonal curves with at most 2 bends. In this paper, we show that the curves of such a representation can be partitioned into $O(\log n)$ groups that represent outer-string graphs or $O(\log^3 n)$ groups that represent permutation graphs. This leads to better approximation algorithms for hereditary graph problems, such as independent set, clique and clique cover, on B_2-VPG graphs.

1 Introduction

An *intersection representation* of a graph is a way of portraying a graph using geometric objects. In such a representation, every object corresponds to a vertex in the graph, and there is an edge between vertices u and v if and only if their two objects **u** and **v** intersect. One example are the *string graphs*, where the objects are (open) curves in the plane with no intersections that are overlaps or touch points. An *outer-string* representation is one where all the curves are inside a polygon P and touch the boundary of P at least once. A string representation is called a *1-string representation* if any two strings intersect at most once. It is called a B_k-*VPG-representation*[1] (for some $k \geq 0$) if every curve is an orthogonal curve with at most k bends. We naturally use the term *outer-string graph* for graphs that have an outer-string representation, and similarly for other types of intersecting objects.

Our contribution: This paper is concerned with partitioning string graphs (and other classes of intersection graphs) into subgraphs that have nice properties, such as being outer-string graphs or permutation graphs (defined formally below). We can then use such a partition to obtain approximation algorithms for some graph problems, such as weighted independent set, clique, clique cover and colouring. More specifically, "partitioning" in this paper usually means a *vertex partition*, i.e., we split the vertices of the graph as $V = V_1 \cup \cdots \cup V_k$ such that the subgraph induced by each V_i has nice properties. In one case we also do

---

* T.B. was supported by NSERC; M.D. was supported by Vanier CGS.
[1] Vertex intersection graphs of k-bend Paths in a Grid; see [8] and the references therein.

© Springer International Publishing AG 2017
F. Ellen et al. (Eds.): WADS 2017, LNCS 10389, pp. 157–168, 2017.
DOI: 10.1007/978-3-319-62127-2_14

an *edge-partition* where we partition $E = E_1 \cup E_2$ and then work on the two subgraphs $G_i = (V, E_i)$, for $i = 1, 2$.

Our paper was inspired by a paper by Lahiri et al. [8] in 2014. They gave an algorithm to approximate the maximum (unweighted) independent set in a B_1-VPG graph within a factor of $4 \log^2 n$ (log in this paper denotes $\log_2$). We greatly expand on their approach as follows. First, rather than solving maximum independent set directly, we instead split such a graph into subgraphs. This allows us to approximate not just independent set, but more generally any hereditary graph problem that is solvable in such graphs.

Secondly, rather than using co-comparability graphs for splitting as Lahiri et al. did, we use outer-string graphs. This allows us to stop the splitting earlier, reducing the approximation factor from $4 \log^2 n$ to $2 \log n$, and to give an algorithm for *weighted* independent set (wIS).

Finally, we allow much more general shapes. For splitting into outer-string graphs, we can allow any shape that can be described as the union of one vertical and any number of horizontal segments (we call such intersection graphs single-vertical). Our results imply a $2 \log n$-approximation algorithm for wIS in such graphs, which include B_1-VPG graphs, and a $4 \log n$-approximation for wIS in B_2-VPG graphs.

In the second part of the paper, we consider splitting the graph such that the resulting subgraphs are co-comparability graphs. This type of problem was first considered by Keil and Stewart [7], who showed that so-called subtree filament graphs can be vertex-partitioned into $O(\log n)$ co-comparability graphs. The work of Lahiri et al. [8] can be seen as proving that every B_1-VPG graph can be vertex-partitioned into $O(\log^2 n)$ co-comparability graphs. We focus here on the bigger class of B_2-VPG-graphs, and show that they can be vertex-partitioned into $O(\log^3 n)$ co-comparability graphs. Moreover, these co-comparability graphs have poset dimension 3, and if the B_2-VPG representation was 1-string, then they are permutation graphs. This leads to better approximation algorithms for clique, colouring and clique cover for B_2-VPG graphs.

2 Decomposing into outer-string graphs

We argue in this section how to split a graph into outer-string graphs if it has an intersection representation of a special form. A *single-vertical object* is a connected set $S \subset \mathbb{R}^2$ of the form $S = s_0 \cup s_1 \cup \cdots \cup s_k$, where s_0 is a vertical segment and $s_1, \ldots, s_k$ are horizontal segments, for some finite k. Given a number of single-vertical objects $S_1, \ldots, S_n$, we define the intersection graph of it in the usual way, by defining one vertex per object and adding an edge whenever objects have at least one point in common (contacts are considered intersections). We call such a representation a *single-vertical representation* and the graph a *single-vertical intersection graph*. The *x-coordinate* of one single-vertical object is defined to be the x-coordinate of the (unique) vertical segment. We consider a horizontal segment to be a single-vertical object as well, by attaching a zero-length vertical segment at one of its endpoints.

Theorem 1. *Let G be a single-vertical intersection graph. Then the vertices of G can be partitioned into at most $\max\{1, 2\log n\}$ sets such that the subgraph induced by each is an outer-string graph.*[2]

Our proof of Theorem 1 uses a splitting technique implicit in the recursive approximation algorithm of Lahiri et al. [8]. Let R be a single-vertical representation on G and S be an ordered list of the x-coordinates of all the objects in R. We define the *median* m of R to be a value such that at most $\frac{|S|}{2}$ x-coordinates in S are smaller than m and at most $\frac{|S|}{2}$ x-coordinates in S are bigger than m. (m may or may not be the x-coordinate of at least one object.) Now split R into three sets: The *middle* set M of objects that intersect the vertical line **m** with x-coordinate m; the *left* set L of objects whose x-coordinates are smaller than m and that do not belong to M, and the *right* set R of objects whose x-coordinates are bigger than m and that do not belong to M. Split M further into $M_L = \{\,c \mid$ the x-coordinate of c is less than $m\}$ and $M_R = M \setminus M_L$.

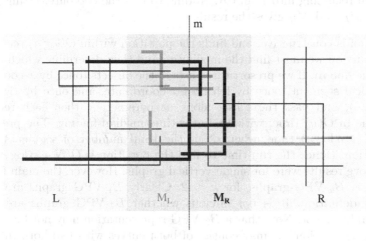

Fig. 1: The split of a representation into L, $M = M_L \cup M_R$ and R.

Lemma 2. *The subgraph induced by the objects in M_L is outer-string.*

Proof. All the objects in M_L intersect curve **m**. Since all the x-coordinates of those objects are smaller than m, all the intersections of the objects occur left of **m**. If an object is not a curve, one can replace it by a closed curve that traces around the part of the object that is left of **m** and that intersects the same set of objects (possibly repeatedly). Doing so for every object, one obtains a representation that induces the same graph as M_L and where all curves are in the left half-space of **m** and intersect **m**, hence an outer-string representation. $\qquad\square$

---

[2] This bound is not tight; a more careful analysis shows that we get at most $\max\{1, 2\lceil\log n\rceil - 2\}$ graphs.

A similar proof shows that the graph induced by objects in M_R is an outer-string graph. Now we can prove our main result:

Proof (of Theorem 1). Let G be a graph with a single-vertical representation. We proceed by induction on the number of vertices n in G. If $n \leq 2$, then the graph is outer-string and we are done, so assume $n \geq 3$, which implies that $\log n \geq \frac{3}{2}$. By Lemma 2, both M_L and M_R individually induce an outer-string graph. Applying induction, we get at most

$$\max\{1, 2\log|L|\} \leq \max\{1, 2\log(n/2)\} = \max\{1, 2\log n - 2\} = 2\log n - 2$$

outer-string subgraphs for L, and similarly at most $2\log n - 2$ outer-string subgraphs for R. Since the objects in L and R are separated by the vertical line $\mathbf{m}$, there are no edges between the corresponding vertices. Thus any outer-string subgraph defined by L can be combined with any outer-string subgraph defined by R to give one outer-string graph. We hence obtain $2\log n - 2$ outer-string graphs from recursing into L and R. Adding to this the two outer-string graphs defined by M_L and M_R gives the result.   □

Our proof is constructive, and finds the partition within $O(\log n)$ recursions. In each recursion we must find the median m and then determine which objects intersect the line $\mathbf{m}$. If we presort three lists of the objects (once by x-coordinate of the vertical segment, once by leftmost x-coordinate, and once by rightmost x-coordinate), and pass these lists along as parameters, then each recursion can be done in $O(n)$ time, without linear-time median-finding. The presorting takes $O(N + n\log n)$ time, where N is the total number of segments in the representation. Hence the run-time to find the partition is $O(N + n\log n)$.

The above results were for single-vertical graphs. However, the main focus of this paper is B_k-VPG-graphs, for $k \leq 2$. Clearly B_1-VPG graphs are single-vertical by definition. It is not obvious whether B_2-VPG graphs are single-vertical graphs or not. Note that a B_2-VPG representation may not be a single-vertical representation—it may consist of both curves with two horizontal segments and curves with two vertical segments (so no rotation of the representation can give a single-vertical representation). However, we can still handle them by doubling the number of graphs into which we split.

Lemma 3. *Let G be a B_2-VPG graph. Then the vertices of G can be partitioned into 2 sets such that the subgraph induced by each is a single-vertical B_2-VPG graph.*

Proof. Fix a B_2-VPG-representation of G. Let V_v be the vertices that have at most one vertical segment in their curve, and V_h be the remaining vertices. Since every curve has at most three segments, and all curves in V_h have at least two vertical segments, each of them has at most one horizontal segment. Clearly V_v induces a single-vertical B_2-VPG graph, and after rotating all curves by 90° V_h *also* induces a single-vertical B_2-VPG graph.   □

Combining this with Theorem 1, we immediately obtain:

Corollary 4. *Let G be a B_2-VPG graph. Then the vertices of G can be partitioned into at most $\max\{1, 4 \log n\}$ sets such that the subgraph induced by each is an outer-string graph.*

3 Decomposing into co-comparability graphs

We now show that by doing further splits, we can actually decompose B_2-VPG graphs into so-called co-comparability graphs of poset dimension 3 (defined formally below). While we require more subgraphs for such a split, the advantage is that numerous problems are polynomial for such co-comparability graphs, while for outer-string we know of no problem other than weighted independent set that is poly-time solvable.

We first give an outline of the approach. Given a B_2-VPG-graph, we first use Lemma 3 to split it into two single-vertical B_2-VPG-graphs. Given a single-vertical B_2-VPG-graph, we next use a technique much like the one of Theorem 1 to split it into $\log n$ single-vertical B_2-VPG-graphs that are "centered" in some sense. Any such graph can easily be edge-partitioned into two B_1-VPG-graphs that are "grounded" in some sense. We then apply the technique of Theorem 1 again (but in the other direction) to split a grounded B_1-VPG-graph into $\log n$ B_1-VPG-graphs that are "cornered" in some sense. The latter graphs can be shown to be permutation graphs. This gives the result after arguing that the edge-partition can be un-done at the cost of combining permutation graphs into co-comparability graphs.

We assume for this section that the B_2-VPG representation is in general position in the sense that no two horizontal or vertical segments overlap each other. Since curves do not overlap or touch, this is not a restriction for B_2-VPG representations.

3.1 Co-comparability graphs

We start by defining the graph classes that we use in this section only. A graph G with vertices $\{1, \dots, n\}$ is called a *permutation graph* if there exist two permutations π_1, π_2 of $\{1, \dots, n\}$ such that (i, j) is an edge of G if and only if π_1 lists i, j in the opposite order as π_2 does. Put differently, if we place $\pi_1(1), \dots, \pi_1(n)$ at points along a horizontal line, and $\pi_2(1), \dots, \pi_2(n)$ at points along a parallel horizontal line, and use the line segment $(\pi_1(i), \pi_2(i))$ to represent vertex i, then the graph is the intersection graph of these segments.

A *co-comparability graph* G is a graph whose complement can be directed in an acyclic transitive fashion. Rather than defining these terms, we describe here only the restricted type of co-comparability graphs that we are interested in. A graph G with vertices $\{1, \dots, n\}$ is called a *co-comparability graph of poset dimension k* if there exist k permutations $\pi_1, \dots, \pi_k$ such that (i, j) is an edge if and only if there are two permutations that list i and j in opposite order. (See Golumbic et al. [5] for more on these characterizations.) Note that a permutation graph is a co-comparability graph of poset dimension 2.

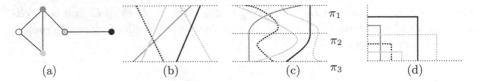

(a)　　　　　　　　(b)　　　　　　　(c)　π_3　　　　(d)

Fig. 2: (a) A graph that has simultaneously (b) a permutation representation; (c) a co-comparability representation of poset dimension 3; and (d) a cornered B_1-VPG graph.

3.2　Cornered B_1-VPG graphs

A B_1-VPG-representation is called *cornered* if there exists a horizontal and a vertical ray emanating from the same point such that any curve of the representation intersects both rays outside their common end. See Fig. 2(d) for an example.

Lemma 5. *If G has a cornered B_1-VPG-representation, say with respect to rays r_1 and r_2, then G is a permutation graph. Further, the two permutations defining G are exactly the two orders in which vertex-curves intersect r_1 and r_2.*

Proof. Since the curves have only one bend, the intersections with r_1 and r_2 determine the curve of each vertex. In particular, two curves intersect if and only if the two orders along r_1 and r_2 are *not* the same, which is to say, if their orders are different in the two permutations of the vertices defined by the orders along the rays. Hence using these orders shows that G is a permutation graph. □

3.3　From grounded to cornered

We call a B_1-VPG representation *grounded* if there exists a horizontal line segment ℓ_H that intersects all curves, and has all horizontal segments of all curves above it. See also Fig. 3 and [1] for more properties of graphs that have a grounded representation. We now show how to split a grounded B_1-VPG-representation into cornered ones. It will be important later that not only can we do such a split, but we know how the curves intersect ℓ_H afterwards. More precisely, the curves in the resulting representations may not be identical to the ones we started with, but they are modified only in such a way that the intersections points of curves along ℓ_H is unchanged.

Lemma 6. *Let R be a B_1-VPG-representation that is grounded with respect to segment ℓ_H. Then R can be partitioned into at most $\max\{1, 2\log n\}$ sets $R_1, \ldots, R_K$ such that each set R_i is cornered after upward translation and segment-extension of some of its curves.*

Proof. A single curve with one bend is always cornered, so the claim is easily shown for $n \leq 4$ where $\max\{1, 2\log n\} \geq n$. For $n \geq 5$, it will be helpful to split R first into two sets, those curves of the form $\ulcorner$ and those that form $\urcorner$ (no other shapes can exist in a grounded B_1-VPG-representation). The result follows if we show that each of them can be split into $\log n$ many cornered B_1-VPG-representations.

So assume that R consists of only $\ulcorner$'s. We apply essentially the same idea as in Theorem 1. Let again $\mathbf{m}$ be the vertical line along the median of x-coordinates of vertical segments of curves. Let M be all those curves that intersect $\mathbf{m}$. Since curves are $\ulcorner$'s, any curve in M intersects ℓ_H to the left of $\mathbf{m}$, and intersects $\mathbf{m}$ above ℓ_H. Hence taking the two rays along ℓ_H and $\mathbf{m}$ emanating from their common point shows that M is cornered.

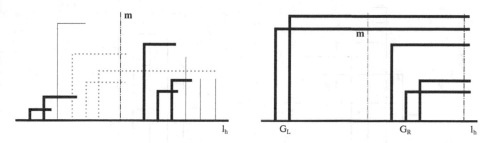

Fig. 3: An illustration for the proof of Lemma 6. (left) Splitting a cornered B_1-VPG graph. (right) Combining a graph G_L (solid black) from L with a graph G_R (solid blue) from R so that the result is a cornered B_1-VPG graph.

We then recurse both in the subgraph L of vertices entirely left of $\mathbf{m}$ and the subgraph R of vertices entirely right of $\mathbf{m}$. Each of them is split recursively into at most $\max\{1, \log(n/2)\} = \log n - 1$ subgraphs that are cornered. We must now argue how to combine two such subgraphs G_L and G_R (of vertices from L and R) such that they are cornered while modifying curves only in the permitted way.

Translate curves of G_L upward such that the lowest horizontal segment of G_L is above the highest horizontal segment of G_R. Extend the vertical segments of G_L so that they again intersect ℓ_H. Extend horizontal segments of both G_L and G_R rightward until they all intersect one vertical line segment. The resulting representation satisfies all conditions.

Since we obtain at most $\log n - 1$ such cornered representations from the curves in $R \cup L$, we can add M to it and the result follows.   □

Corollary 7. *Let G be a graph with a grounded B_1-VPG representation. Then the vertices of G can be partitioned into at most $\max\{1, 2\log n\}$ sets such that the subgraph induced by each is a permutation graph.*

3.4 From centered to grounded

We now switch to VPG-representations with 2 bends, but currently only allow those with a single vertical segment per curve. So let R be a single-vertical B_2-

VPG-representation. We call R *centered* if there exists a horizontal line segment ℓ_H that intersects the vertical segment of each curve. Given such a representation, we can cut each curve apart at the intersection point with ℓ_H. Then the parts above ℓ_H form a grounded B_1-VPG-representation, and the parts below form (after a $180°$ rotation) also a grounded B_1-VPG-representation. Note that this split corresponds to splitting the edges into $E = E_1 \cup E_2$, depending on whether the intersection for each edge occurs above or below ℓ_H. If curves may intersect repeatedly, then an edge may be in both sets. See Fig. 4 for an example. With this, we can now split into co-comparability graphs.

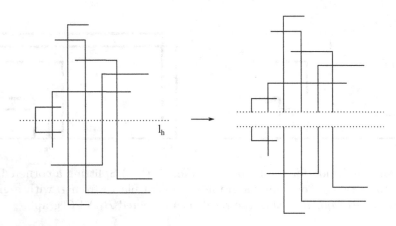

Fig. 4: Splitting a centered single-vertical B_2-VPG-representation into two grounded B_1-VPG-representations.

Lemma 8. *Let G be a graph with a single-vertical centered B_2-VPG representation. Then the vertices of G can be partitioned into at most $\max\{1, 4\log^2 n\}$ sets such that the subgraph induced by each is a co-comparability graph of poset dimension 3.*

Proof. The claim clearly holds for $n \leq 4$, so assume $n \geq 5$. Let ℓ_H be the horizontal segment along which the representation is centered. Split the edges into E_1 and E_2 as above, and let R_1 and R_2 be the resulting grounded B_1-VPG-representations, which have the same order of vertical intersections along ℓ_H. Split R_1 into $K \leq 2\log n$ sets of curves $R_1^1, \ldots, R_1^K$, each of which forms a cornered B_1-VPG-representation that uses the same order of intersections along ℓ_H. Similarly split R_2 into $K' \leq 2\log n$ sets $R_2^1, \ldots, R_2^{K'}$ of cornered B_1-VPG-representations.

   Now define $R_{i,j}$ to consist of all those curves r where the part of r above ℓ_H belongs to R_1^i and the part below belongs to R_2^j. This gives $K \cdot K' \leq 4\log^2 n$ sets of curves. Consider one such set $R_{i,j}$. The parts of curves in $R_{i,j}$ that were above ℓ_H are cornered at ℓ_H and some vertical upward ray, hence define a permutation π_1 along the vertical ray and π_2 along ℓ_H. Similarly the parts of curves below ℓ_H define two permutations, say π_2' along ℓ_H and π_3 along some vertical downward ray. But the split into cornered B_1-VPG-representation ensured that the intersections along ℓ_H did not changed, so $\pi_2 = \pi_2'$. The three permutations π_1, π_2, π_3 together hence define a co-comparability graph of poset dimension 3 as desired.   □

We can do slightly better if the representation is additionally 1-string.

Corollary 9. *Let G be a graph with a single-vertical centered 1-string B_2-VPG representation. Then the vertices of G can be partitioned into at most $\max\{1, 4\log^2 n\}$ sets such that the subgraph induced by each is a permutation graph.*

Proof. The split is exactly the same as in Lemma 8. Consider one of the subgraphs G_i and the permutations π_1, π_2, π_3 that came with it, where π_2 is the permutation of curves along the centering line ℓ_H. We claim that G_i is a permutation graph, using π_1, π_3 as the two permutations. Clearly if (u, v) is not an edge of G_i, then all of π_1, π_2, π_3 list u and v in the same order. If (u, v) is an edge of G_i, then two of π_1, π_2, π_3 list u, v in opposite order. We claim that π_1 and π_3 list u, v in opposite order. For if not, say u comes before v in both π_1 and π_3, then (to represent edge (u, v)) we must have u after v in π_2. But then the curves of u and v intersect both above and below ℓ_H, contradicting that we have a 1-string representation. So the two permutations π_1, π_3 define graph G_i.   □

3.5 Making single-vertical B_2-VPG-representations centered

Lemma 10. *Let G be a graph with a single-vertical B_2-VPG representation. Then the vertices of G can be partitioned into at most $\max\{1, \log n\}$ sets such that the subgraph induced by each has a single-vertical centered B_2-VPG-representation.*

Proof. The approach is quite similar to the one in Theorem 1, but uses a horizontal split and a different median. The claim is easy to show for $n \leq 3$, so assume $n \geq 4$. Recall that there are n vertical segments, hence $2n$ endpoints of such segments. Let m be a value such that at most n of these endpoints each are below and above m, and let $\mathbf{m}$ be the horizontal line with y-coordinate m.

Let M be the curves that are intersected by $\mathbf{m}$; clearly they form a single-vertical centered B_2-VPG-representation. Let B be all those curves whose vertical segment (and hence the entire curve) is completely below $\mathbf{m}$. Each such curve contributes two endpoints of vertical segments, hence $|B| \leq n/2$ by choice of m. Recursively split B into at most $\max\{1, \log(n/2)\} = \log n - 1$ sets, and likewise split the curves U above $\mathbf{m}$ into at most $\log n - 1$ sets.

Each chosen subset G_B of B is centered, as is each chosen subset G_U of U. Since G_B uses curves below $\mathbf{m}$ while G_U uses curves above, there are no crossings between these curves. We can hence translate the curves of G_B such they are centered with the same horizontal line as G_U. Therefore $G_B \cup G_U$ has a centered single-vertical B_2-VPG-representation. Repeating this for all of $R \cup U$ gives $\log n - 1$ centered single-vertical B_2-VPG-graphs, to which we can add the one defined by M.   □

3.6 Putting it all together

We summarize with our main result about splits into co-comparability graphs:

Theorem 11. *Let G be a B_2-VPG-graph. Then the vertices of G can be partitioned into at most $\max\{1, 8\log^3 n\}$ sets such that the subgraph induced by each is co-comparability graph of poset dimension 3. If G is a 1-string B_2-VPG graph, then the subgraphs are permutation graphs.*

Proof. The claim is trivial for $n \leq 3$ since then $n \leq 8\log^3 n$, so assume $n \geq 4$. Fix a B_2-VPG-representation R, and split it into two single-vertical B_2-VPG-representations (Lemma 3). Split each of them into $\log n$ single-vertical centered B_2-VPG-representations using Lemma 10, for a total of at most $2\log n$ sets of curves. Split each of them into $4\log^2 n$ co-comparability graphs (or permutation graphs if the representation was 1-string) using Lemma 8 or Corollary 9. □

We can do better for B_1-VPG-graphs. The subgraphs obtained in the result below are the same ones that were used implicitly in the $4\log^2 n$-approximation algorithm given by Lahiri et al.[8].

Theorem 12. *Let G be a B_1-VPG-graph. Then the vertices of G can be partitioned into at most $\max\{1, 4\log^2 n\}$ sets such that the subgraph induced by each is a permutation graph.*

Proof. The claim is trivial if $n = 1$, so assume $n > 1$. Fix a B_1-VPG-representation R, and split it into $\log n$ single-vertical centered B_1-VPG-representations using Lemma 10. Split each of them into two centered B_1-VPG-representations, one of those curves with the horizontal segment above the centering line, and one with the rest. Each of the resulting $2\log n$ centered B_1-VPG-representations is grounded (possibly after a 180° rotation) and can be split into $2\log n$ permutation graphs using Corollary 7, for a total of $4\log^2 n$ permutation graphs. □

4 Applications

We now show how Theorem 1 and 11 can be used for improved approximation algorithms for B_2-VPG-graphs. The techniques used here are virtually the same as the one by Keil and Stewart [7] and require two things. First, the problem considered needs to be solvable on the special graphs class (such as outer-string graph or co-comparability graph or permutation graph) that we use. Second, the problem must be *hereditary* in the sense that a solution in a graph implies a solution in an induced subgraph, and solutions in induced subgraphs can be used to obtain a decent solution in the original graph.

We demonstrate this in detail using weighted independent set, which Keil et al. showed to be polynomial-time solvable in outer-string graphs [6]. Recall that this is the problem, given a graph with vertex-weights, of finding a subset I of vertices that has no edges between them. The objective is to maximize $w(I) := \sum_{v \in I} w(v)$, where $w(v)$ denotes the weight of vertex v. The run-time to solve weighted independent set in outer-string graphs is $O(N^3)$, where N is the number of segments in the given outer-string representation.

Theorem 13. *There exists a* $(2 \log n)$*-approximation algorithm for weighted independent set on single-vertical graphs with run-time* $O(N^3)$, *where* N *is the total number of segments used among all single-vertical objects.*

Proof. If $n = 1$, then the unique vertex is the maximum weight independent set. Else, use Theorem 1 to partition the vertices of the given graph G into at most $2 \log n$ sets, each of which induces an outer-string graph, and find the largest weighted independent set in each applying the algorithm of Keil et al. If G_i had an outer-string representation with N_i segments in total, then this takes time $O(\sum N_i^3)$ time. Note that if a single-vertical object consisted of one vertical and ℓ horizontal segments, then we can trace around it with a curve with $O(\ell)$ segments. Hence all curves together have $O(N)$ segments and the total run-time is $O(N^3)$.

Let I_i^* be the maximum-weight independent set in G_i, and return as set I the set in $I_1^*, \ldots, I_k^*$ that has the maximum weight. To argue the approximation-factor, let I^* be the maximum-weight independent set of G, and define I_i to be all those elements of I^* that belong to R_i, for $i = 1, \ldots, k$. Clearly I_i is an independent set of G_i, and so $w(I_i) \leq w(I_i^*)$. But on the other hand $\max_i w(I_i) \geq w(I^*)/k$ since we split I^* into k sets. Therefore $w(I) = \max_i w(I_i^*) \geq w(I^*)/k$, and so $w(I)$ is within a factor of $k \leq 2 \log n$ of the optimum. $\square$

We note here that the best polynomial algorithm for independent set in general string graphs achieves an approximation factor of $O(n^\varepsilon)$, under the assumption that any two strings cross each other at most a constant number of times [3]. This algorithm only works for unweighted independent set; we are not aware of any approximation results for weighted independent set in arbitrary string graphs.

Because B_2-VPG-graphs can be vertex-split into two single-vertical B_2-VPG-representations, and the total number of segments used is $O(n)$, we also get:

Corollary 14. *There exists a* $(4 \log n)$*-approximation algorithm for weighted independent set on* B_2*-VPG-graphs with run-time* $O(n^3)$.

Another hereditary problem is *colouring*: Find the minimum number k such that we can assign numbers in $\{1, \ldots, k\}$ to vertices such that no two adjacent vertices receive the same number. Fox and Pach [3] pointed out that if we have a c-approximation algorithm for Independent Set, then we can use it to obtain an $O(c \log n)$-approximation algorithm for colouring. Therefore our result also immediately implies an $O(\log^2 n)$-approximation algorithm for colouring in single-vertical graphs and B_2-VPG-graphs.

Another hereditary problem is *weighted clique*: Find the maximum-weight subset of vertices such that any two of them are adjacent. (This is independent set in the complement graph.) Clique is NP-hard in outer-string graphs even in its unweighted version [2]. For this reason, we use the split into co-comparability graphs instead; weighted clique can be solved in quadratic time in co-comparability graphs (because weighted independent set is linear-time solvable in comparability graphs [4]). Weighted clique is also linear-time solvable on permutation graphs [4]. We therefore have:

Theorem 15. *There exists an* $(8 \log^3 n)$-*approximation algorithm for weighted clique on* B_2-*VPG-graphs with run-time* $O(n^2)$. *The run-time becomes* $O(n)$ *if the graph is a 1-string* B_2-*VPG graph, and the approximation factor becomes* $4 \log^2 n$ *if the graph is a* B_1-*VPG-graph.*

In a similar manner, we can get poly-time $(8 \log^3 n)$-approximation algorithms for any hereditary problem that is solvable on co-comparability graphs. This includes clique cover, maximum k-colourable subgraph, and maximum h-coverable subgraph. See [7] for the definition of these problems, and the argument that they are hereditary.

5   Conclusions

We presented a technique for decomposing single-vertical graphs into outer-string subgraphs, B_2-VPG-graphs into co-comparability graphs, and 1-string B_2-VPG-graphs into permutation graphs. We then used these results to obtain approximation algorithms for hereditary problems, such as weighted independent set.

As for open problems, we are very interested in approximation algorithms for B_k-VPG graphs, where k is a constant. Also, if curves are not required to be orthogonal, but have few bends, are there approximation algorithms better than those for arbitrary string graphs?

References

1. Jean Cardinal, Stefan Felsner, Tillmann Miltzow, Casey Tompkins, and Birgit Vogtenhuber. Intersection graphs of rays and grounded segments. Technical Report 1612.03638 [cs.DM], ArXiV, 2016.
2. Sergio Cabello, Jean Cardinal and Stefan Langerman. The Clique Problem in Ray Intersection Graphs. Discrete & Computational Geometry 50(3), 771–783, 2013.
3. Jacob Fox and János Pach. Computing the independence number of intersection graphs. In Dana Randall, editor, *Proceedings of the Twenty-Second Annual ACM-SIAM Symposium on Discrete Algorithms, SODA 2011, San Francisco, California, USA, January 23-25, 2011*, pages 1161–1165. SIAM, 2011.
4. M. C. Golumbic. *Algorithmic graph theory and perfect graphs*. Academic Press, New York, 1st edition, 1980.
5. Martin Charles Golumbic, Doron Rotem, and Jorge Urrutia. Comparability graphs and intersection graphs. *Discrete Mathematics*, 43(1):37–46, 1983.
6. J. Mark Keil, Joseph S. B. Mitchell, Dinabandhu Pradhan, and Martin Vatshelle. An algorithm for the maximum weight independent set problem on outerstring graphs. *Comput. Geom.*, 60:19–25, 2017.
7. J. Mark Keil and Lorna Stewart. Approximating the minimum clique cover and other hard problems in subtree filament graphs. *Discrete Applied Mathematics*, 154(14):1983–1995, 2006.
8. Abhiruk Lahiri, Joydeep Mukherjee, and C. R. Subramanian. Maximum independent set on B_1-VPG graphs. In Zaixin Lu, Donghyun Kim, Weili Wu, Wei Li, and Ding-Zhu Du, editors, *Combinatorial Optimization and Applications (COCOA 2015)*, volume 9486 of *Lecture Notes in Computer Science*, pages 633–646. Springer, 2015.

A Deterministic Algorithm
for Online Steiner Tree Leasing*

Marcin Bienkowski[1], Artur Kraska[1], and Paweł Schmidt[1]

Institute of Computer Science, University of Wrocław, Poland

Abstract. We study the Online Steiner Tree Leasing (OSTL) problem, defined in a weighted undirected graph with a distinguished root node r. There is a known set $\mathcal{L}$ of available lease types, where each type $\ell \in \mathcal{L}$ is characterized by its duration D_ℓ and cost factor C_ℓ. As an input, an online algorithm is given a sequence of terminals and has to connect them to the root r using leased edges. An edge of length d can be leased using lease type ℓ for cost $C_\ell \cdot d$ and remains valid for time D_ℓ.
The OSTL problem contains the online Steiner tree and the single-source rent-or-buy problems as specific subcases. We present the first deterministic online algorithm for OSTL, whose competitive ratio is $O(|\mathcal{L}| \cdot \log k)$, where k is the number of different terminals in the input. The currently best randomized algorithm attains the ratio of $O(\log |\mathcal{L}| \cdot \log n)$, where $n \geq k$ is the number of nodes in the graph.

Keywords: Steiner tree • Leasing • Competitive analysis • Online algorithms

1   Introduction

The traditional network design focuses on graph optimization problems, in which an algorithm purchases bandwidth on links to maintain certain graph properties, such as connectivity or throughput. A standard feature of most considered models is the permanence of bandwidth allocations. For example, in the Steiner tree problem [23], the goal is to buy a subset of edges connecting a given set of terminals to the chosen root node r. Even the online flavor of this problem [4,5,16,18,22] has this feature: the terminals arrive in online manner, and an algorithm irrevocably buys additional links, so that the terminals seen so far are connected to the root r. In this setting, each purchase is everlasting, i.e., the problem should be rather termed *incremental* Steiner tree.

Rent-or-Buy Variants. A well-studied modification of the online Steiner tree scenario is to relax the need of upfront commitment and additionally allow an algorithm to *rent* edges (at a fraction of the edge purchase price). Each terminal must be connected to the root r using either type of edges, but rented edges are

* Partially supported by the Polish National Science Centre grant 2016/22/E/ST6/00499.

© Springer International Publishing AG 2017
F. Ellen et al. (Eds.): WADS 2017, LNCS 10389, pp. 169–180, 2017.
DOI: 10.1007/978-3-319-62127-2_15

valid only for a single terminal and cannot be reused by subsequent ones. This variant is called *online single-source rent-or-buy* [3,10,11,12,15,17,21] and is also equivalent to the *file replication* problem.

Note that the rent-or-buy variant is still incremental: bought edges persist in the graph till the end and — if a request to connect a specific terminal appears sufficiently many times in the input — any reasonable algorithm finally buys a path connecting this terminal to the root r.

Leasing Variants. Many markets give an option of temporary leasing of resources. In particular, the advent of digital services in cloud computing changed the business model from buying physical servers to leasing virtual ones. This allowed companies to adapt quickly to varying requirements of their customers [7]. Furthermore, the software-defined networking enabled similar mechanisms on the network level, allowing companies to lease network links on the fly [13]. Typically, possible leases have different lengths and costs, and obey economies of scale, e.g., leasing a link for a week is more expensive than leasing it for a day, but not more than seven times. One can view rent-or-buy variants as an extreme case of leasing, where only two leases are available: a lifetime one (buying) and a lease of infinitesimal duration (renting).

These trends motivate the study of algorithmic leasing variants of popular network design mechanisms [1,2,6,19,20]. Note that from the algorithmic standpoint, leasing variants have a truly online nature: leases have finite duration and expire after some time. An online algorithm has to adapt itself to varying access patterns, e.g., by acquiring longer leases in response to increased demand.

1.1   The Model

In this paper, we study the Online Steiner Tree Leasing (OSTL) problem introduced by Meyerson [19]. The problem is defined in a weighted undirected graph G with a distinguished root node $r \in V(G)$. For each pair of nodes u and v, by $d_G(u, v)$ we denote the length (with respect to edge weights) of the shortest path between u and v. There is a known set $\mathcal{L}$ of available lease types, where each type $\ell \in \mathcal{L}$ is characterized by its duration D_ℓ and cost ratio C_ℓ. We denote the number of leases by $L = |\mathcal{L}|$.

An input to the problem consists of a sequence σ of requests, each being a terminal (a node of G), arriving sequentially in an online manner. We treat the root r also as a terminal. We assume that each request arrives at a different time (arrival times are real non-negative numbers). In response to a requested terminal σ_t, which appears at time t, an online algorithm has to connect σ_t to r using leased edges in G, leasing additional ones if necessary. (We note that a terminal may occur multiple times in sequence σ.) If an algorithm acquires a lease type $\ell \in \mathcal{L}$ of an edge $e = (u, v)$, it pays $C_\ell \cdot d_G(u, v)$. The edge leased at time t remains available for the period $[t, t + D_\ell)$; afterwards the lease expires.

To recap, an input instance $\mathcal{I}$ is a tuple $(G, d_G, r, \mathcal{L}; \sigma)$, where G, d_G, r, and $\mathcal{L}$ are known a priori, and σ is presented in an online fashion to an algorithm. For any algorithm A, $A(\mathcal{I})$ is the cost of A on input $\mathcal{I}$ and is subject to minimization.

The goal is to minimize the competitive ratio, defined as a worst-case ratio of the cost of an online algorithm to the cost of the optimal offline solution (denoted OPT) on the same input.

1.2 Previous Results

The OSTL problem on a single edge is known as the *parking permit problem* for which optimally competitive algorithms were given by Meyerson [19]: a deterministic $O(L)$-competitive one and a randomized $O(\log L)$-competitive one. (Note that the rent-or-buy variant on a single edge is equivalent to the classic ski rental problem with a trivially achievable constant competitive ratio.)

The randomized algorithm can be extended to trees [19]. As any n-node graph can be approximated by a random tree with expected distortion of $O(\log n)$ [14], this approach yields a randomized $O(\log L \cdot \log n)$-competitive solution for graphs.

Better or non-randomized algorithms were known only for specific variants of the OSTL. In particular, for a rent-or-buy variant (recall that it corresponds to a special 2-lease variant, where the cheaper lease suffices only for serving a single request, and the more expensive lease lasts forever) Awerbuch, Azar and Bartal gave a randomized $O(\log k)$-competitive algorithm and $O(\log^2 k)$-deterministic one [9]. The latter result was improved to $O(\log k)$ only recently by Umboh [21]. In these results, k denotes the number of different terminals in an input.

1.3 Our Contribution

In this paper, we present the first deterministic online algorithm for the OSTL problem. Our algorithm is $O(L \cdot \log k)$-competitive. It outperforms the randomized $O(\log L \cdot \log n)$-competitive solution by Meyerson [19] when k (the number of different terminals in the input) is small.

While the result might not be optimal, neither $O(L)$ nor $O(\log k)$ can be beaten by a deterministic solution: $\Omega(L)$ bound follows by the lower bound on the parking permit problem (which is equivalent to the OSTL on a single edge) and $\Omega(\log k)$ bound follows by the online Steiner tree problem (which is a specific case of the OSTL with a single lease of the infinite duration).

In our solution (presented in Sect. 4), a path that connects a requested terminal to an already existing Steiner tree is chosen greedily. However, we still have to decide which lease type to use for such path. To this end, we check how many requests were "recently" served in a "neighborhood" of the currently requested terminal; once certain thresholds are met, more expensive leases are acquired. While such approach is natural, the main difficulty stems from the dynamics of the leased edges. Namely, while in the rent-or-buy scenario the Steiner tree maintained by an algorithm may only grow, in the leasing variant it may also shrink as edge leases expire. As a result, the already aggregated serving cost may cease to be sufficient to cover a more expensive lease for the new connection. Coping with this issue is the main challenge we tackle in this paper.

We use a recent analysis technique by Umboh [21]: the online algorithm is run on a graph G, but its cost is compared to the cost of OPT run on a tree T.

This tree T is a hierarchically separated tree (HST), whose leaves are requested terminals and whose distances dominate graph distances. By showing that our algorithm is $O(L)$-competitive against OPT on T, for *any* choice of T, we obtain that it is $O(L \cdot \log k)$-competitive against OPT on the original graph G. The details of this reduction are presented in Sect. 2.

We emphasize that the competitive ratio of our algorithm is a function of the number of different terminals, k, and not the number of nodes in the graph, n, as it is the case for the randomized algorithm of [19]. In fact, our algorithm and its analysis work without changes also in any (infinite) metric space, e.g., on the Euclidean plane; in the paper, we use the graph terminology for simplicity.

1.4 Related Work

Other network design problems were also studied in leasing context. In particular, a randomized $O(\log L \cdot \log n)$-competitive algorithm was given for the Online Steiner Forest Leasing by Meyerson [19]. Deterministic algorithms for this problem are known only for the rent-or-buy subcase, for which an optimal competitive ratio of $O(\log k)$ was achieved by Umboh [21]. Other problems include the facility location [1,20] and the set cover [2].

The leasing setting was also applied to *offline* scenarios of the problems above by Anthony and Gupta [6], who showed an interesting reduction between leasing and stochastic optimization variants.

2   HST Embeddings

In this section, we show how to use hierarchically separated trees (HSTs) for the analysis of an algorithm for the OSTL problem. Unlike many online constructions for network design problems (see, e.g., [8,19]), here HSTs are not used for an algorithm construction. Moreover, in our analysis, an HST will approximate not the whole graph, but only the subgraph spanned by terminals.

Definition 1 (Dominating HST embedding of terminals). *Fix any input instance* $\mathcal{I} = (G, d_G, r, \mathcal{L}; \sigma)$ *of the OSTL problem. Let* $X \subseteq V(G)$ *be the set of terminals requested in* σ *(including the root* r*). Assume that the minimum distance between any pair of nodes from* X *is at least* 1.[1] *A dominating HST embedding of terminals of* $\mathcal{I}$ *is a rooted tree* T *with pairwise distances given by metric* d_T, *satisfying the following properties.*

1. *The leaves of* T *are exactly the nodes of* X *and they are on the same level.*
2. *The distance from any leaf of* T *to its parent is* 1.
3. *The edge lengths increase by a factor of* 2 *on any leaf-to-root path.*
4. d_T *dominates* d_G, *i.e.,* $d_T(u, v) \geq d_G(u, v)$ *for any pair of nodes* $u, v \in X$.

---

[1] For analysis, we may always scale the instance, so that this property holds.

Fix now any instance $\mathcal{I} = (G, d_G, r, \mathcal{L}; \sigma)$ of the OSTL and let (T, d_T) be any dominating HST embedding of terminals of $\mathcal{I}$. Let $\mathcal{I}_T = (T, d_T, r, \mathcal{L}; \sigma)$ be the instance $\mathcal{I}$, where graph G was replaced by tree T with distances given by d_T.

While estimating $\text{OPT}(\mathcal{I})$ directly may be quite involved, lower-bounding $\text{OPT}(\mathcal{I}_T)$ is much easier. In particular, for each request σ_t there is a unique path in T connecting σ_t with r. As d_T dominates d_G, it is also feasible to compare the cost of an online algorithm on $\mathcal{I}$ to $\text{OPT}(\mathcal{I}_T)$. Finally, it is possible to relate $\text{OPT}(\mathcal{I}_T)$ to $\text{OPT}(\mathcal{I})$ as stated in the following lemma, due to Umboh [21].

Lemma 2. *Let $\mathcal{I} = (G, d_G, r, \mathcal{L}; \sigma)$ be an instance of the OSTL problem and let X be the set of terminals of $\mathcal{I}$. There exists a dominating HST embedding (T^*, d_{T^*}) of terminals X, such that $\text{OPT}(\mathcal{I}_{T^*}) \leq O(\log |X|) \cdot \text{OPT}(\mathcal{I})$, where $\mathcal{I}_{T^*} = (T^*, d_{T^*}, r, \mathcal{L}; \sigma)$.*

Proof. Fix any dominating HST embedding (T, d_T) of terminals X. The solution $\text{OPT}(\mathcal{I})$ is a schedule that leases particular edges of G at particular times. Let $\text{OFF}(\mathcal{I}_T)$ be an offline solution that, for any leased edge $e = (u, v)$ in $\text{OPT}(\mathcal{I})$, leases all edges on the unique path in T from u to v, using the same lease type. While it is not necessary for the proof, it is worth observing that, by the domination property (cf. Definition 1), $\text{OPT}(\mathcal{I}) \leq \text{OFF}(\mathcal{I}_T)$.

By the FRT approximation [14], there exists a probability distribution $\mathcal{D}$ over dominating HST embeddings (T, d_T) of X, such that $\mathbf{E}_{T \sim \mathcal{D}}[d_T(u, v)] \leq O(\log |X|) \cdot d_G(u, v)$ for all $u, v \in X$. This relation summed over all edges (u, v) used in the solution of $\text{OPT}(\mathcal{I})$ yields that

$$\mathbf{E}_{T \sim \mathcal{D}}[\text{OFF}(\mathcal{I}_T)] \leq O(\log |X|) \cdot \text{OPT}(\mathcal{I}) .$$

By the average argument, there exists a dominating HST embedding (T^*, d_{T^*}), such that $\text{OFF}(\mathcal{I}_{T^*}) \leq O(\log |X|) \cdot \text{OPT}(\mathcal{I})$, and the proof follows by observing that $\text{OPT}(\mathcal{I}_{T^*})$ is at most $\text{OFF}(\mathcal{I}_{T^*})$. □

The lemma can be generalized to any network design problem whose objective function is a linear combination of edge lengths. In Sect. 4, we will construct an algorithm for the OSTL which is $O(L)$-competitive against the cost of OPT on *any* HST embedding. By Lemma 2, this algorithm is $O(L \cdot \log k)$-competitive.

3   Interval Model

In this section, we make several assumptions on the available leases. At the expense of a constant increase of the competitive ratio, they will make the construction of our algorithm easier. Similar assumptions were also made for the parking permit problem [19].

Definition 3. *In the interval model, the following conditions hold for the input instance.*

- *Costs factors and durations of all leases are powers of two.*

- *Lease types are sorted both by their costs and durations, i.e., if $\ell' < \ell$, then $D_{\ell'} < D_\ell$ and $C_{\ell'} < C_\ell$.*
- *Fix any lease type ℓ and let $J_\ell^m = [m \cdot D_\ell, (m+1) \cdot D_\ell)$ for any $m \in \mathbb{N}$. For any time t and any edge, there is a unique lease of type ℓ that can be acquired by an algorithm: it is the lease for period J_ℓ^m containing t.*

The last property of the interval model means that, unlike the standard leasing model outlined in Sect. 1.1, if an algorithm leases an edge at a time t using a lease type $\ell \in \mathcal{L}$, such transaction may occur within the lease duration. Hence, the acquired lease may expire earlier than at time $t + D_\ell$. We also define $J_\ell[t]$ to be the period J_ℓ^m containing time t.

Observation 4. *In the interval model, when lease of type ℓ expires, all leases of smaller types expire as well.*

Lemma 5. *Any (online or offline) algorithm for the original leasing model can be transformed into an algorithm for the interval model (and back) without changing its cost by more than a constant factor.*

The lemma above follows by standard rounding arguments (its proof is omitted; see [19] for a similar argument). Hence, if an algorithm is R-competitive for the interval model, it is $O(R)$-competitive for the original leasing model. Therefore, we will assume the interval model in the remaining part of the paper.

4   Algorithm Construction

We present our algorithm ACCUMULATE-AND-LEASE-GREEDILY (ALG). For simplicity of the description, we assume that a given graph G is complete (with the metric given by d_G). Such assumption is without loss of generality, as leasing the edge (u, v) can be always replaced by leasing a shortest path connecting u and v.

We will say that an edge e is ℓ-*leased* at time t, if an algorithm leased e for period $J_\ell[t]$ using lease type ℓ. Additionally, a request σ_t is ℓ-*leased* if at time t an algorithm ℓ-leases an edge $e = (\sigma_t, u)$ for some u.

By $F_\ell[t]$ and F_ℓ^m we denote the set of all requests that arrived during $J_\ell[t]$ and J_ℓ^m, respectively. Furthermore, $T_{\geq \ell}[t]$ denotes the set of requests that are connected, at time t, to the root r using edges of lease types at least ℓ.

High-level idea. In the execution of ALG, at any time t, the set of all currently leased edges will be a single (possibly empty) tree, called the Steiner tree of ALG. Furthermore, on any path from the root r that consists of leased edges, the closer we are to the root, the longer leases we have. In effect, $T_{\geq \ell}[t]$ always forms a tree. Moreover, when leases expire, the set of leased edges shrinks, but it remains connected.

When a request σ_t arrives, we check whether we can afford a lease ℓ for σ_t, starting from the longest (and the most expensive) available lease: We compute the distance d from σ_t to $T_{\geq \ell}[t]$. Then, we check if there were "sufficiently many" requests served "recently" in a "small" (compared to d) neighborhood of σ_t. If so, then we connect σ_t to $T_{\geq \ell}[t]$ using lease type ℓ.

Algorithm 1 ACCUMULATE-AND-LEASE-GREEDILY for the OSTL problem

---
1: **while** request σ_t arrives **do**
2:     **for** $\ell \leftarrow L \ldots 1$ **do**
3:         **if** σ_t is not connected to the root r with a path of leased edges **then**
4:             /* *Check whether we can afford lease ℓ for σ_t* */
5:             let x_ℓ be the node of $T_{\geq\ell}[t]$ closest to σ_t
6:             $j \leftarrow \lceil \log d_G(\sigma_t, x_\ell) \rceil$
7:             $N_\ell^t \leftarrow \{f \in F_\ell[t] : d_G(\sigma_t, f) \leq 2^{j-2}$ and $\mathrm{class}(f) = j\}$
8:             **if** $|N_\ell^t| \cdot C_1 \geq C_\ell$ or $\ell = 1$ **then**
9:                 ℓ-lease the edge (σ_t, x_ℓ)
10:                 $\mathrm{class}(\sigma_t) \leftarrow j$

Algorithm description. More precisely, fix any time t when a request σ_t is presented to ALG. ALG checks, for each lease type ℓ starting from the most expensive (the L-th one), what the cost of connecting σ_t to the tree $T_{\geq\ell}[t]$ would be. That is, among all the nodes of $T_{\geq\ell}[t]$, it finds the node x_ℓ closest to σ_t. If we ℓ-lease the edge (σ_t, x_ℓ) at the cost $C_\ell \cdot d_G(\sigma_t, x_\ell)$, then σ_t becomes connected to the root r via a path of leased edges (of lease type at least ℓ). We round the distance $d_G(\sigma_t, x_\ell)$ up to the smallest power of two, denoted 2^j. Then, we look at the set N_ℓ^t (cf. Line 7 in Algorithm 1) of requests that

- arrived at any time in $J_\ell[t]$,
- are at distance at most $2^j/4$ from σ_t,
- are of class j (i.e., upon their arrival, ALG connected them to its Steiner tree, using an edge of length from $(2^{j-1}, 2^j]$, i.e., roughly $d_G(\sigma_t, x_\ell)$).

Note that the terminals of N_ℓ^t are not necessarily connected to the Steiner tree of ALG at time t. If the number of requests in N_ℓ^t is at least C_ℓ/C_1, then ALG ℓ-leases the edge (σ_t, x_ℓ) and sets the class of σ_t to j. Otherwise, ALG proceeds to cheaper lease types. If no lease type ℓ satisfies the condition $|N_\ell^t| \geq C_\ell/C_1$, ALG eventually 1-leases the edge (σ_t, x_1). Note that ALG leases exactly one edge for each terminal that is not connected to the tree at the time of its arrival.

Pseudocode of ALG is given in Algorithm 1. We recall the property of ALG stated earlier in its informal description.

Observation 6. *For any time t and lease type $\ell \in \mathcal{L}$, $T_{\geq\ell}[t]$ is a single tree.*

5 Analysis

Throughout this section, we fix an input instance $\mathcal{I} = (G, d_G, r, \mathcal{L}; \sigma)$ and a corresponding "tree instance" $\mathcal{I}_T = (T, d_T, r, \mathcal{L}; \sigma)$, where (T, d_T) is a dominating HST embedding of terminals of $\mathcal{I}$ (cf. Sect. 2).

Without loss of generality, we assume that when a request σ_t arrives, it is not yet connected to the Steiner tree of ALG (otherwise σ_t would be ignored by ALG). If σ_t was ℓ-leased, then we call N_ℓ^t (computed in Line 7 of Algorithm 1) its *neighbor set*. We denote the set of all requests of class j by W_j. Additionally, let W_j^ℓ consist of all requests from W_j that were ℓ-leased.

A brief idea of the proof is as follows. Suppose each request $\sigma_t \in W_j$ receives a credit of $C_1 \cdot 2^j$ when it arrives. If σ_t was ℓ-leased, the actual cost paid by ALG was $C_\ell \cdot 2^j$. While the latter amount can be much larger for an individual request σ_t, in Sect. 5.1, we show that, for any fixed lease type ℓ, the total cost paid for ℓ-leased edges is bounded by the sum of all requests' credits. In Sect. 5.2, we exploit properties of dominating HST embeddings to show how all credits can be charged (up to constant factors) to the leasing costs OPT pays for particular edges of the tree T. Altogether, this will show that $\text{ALG}(\mathcal{I}) \leq O(L) \cdot \text{OPT}(\mathcal{I}_T)$. Along with Lemma 2, this will bound the competitive ratio of ALG (see Sect. 5.3).

5.1 Upper Bound on ALG

The core of this section is Lemma 8, which essentially states that for any lease type ℓ, all requests' credits can cover all leases of type ℓ. Before proceeding to its proof, we first show the following structural property.

Lemma 7. *Fix a class j, a lease type ℓ, and a pair of distinct requests $\sigma_s, \sigma_t \in W_j^\ell$. Their neighbor sets, N_ℓ^t and N_ℓ^s, are disjoint.*

Proof. Without loss of generality, $s < t$. We will prove the lemma by contradiction. Assume there exists a request $\sigma_u \in N_\ell^s \cap N_\ell^t$.

By the definition of neighbor sets, $\sigma_u \in F_\ell[s] \cap F_\ell[t]$. In the interval model, there are only two possibilities: either periods $J_\ell[s]$ and $J_\ell[t]$ are equal or they are disjoint. As in the latter case the corresponding sets $F_\ell[s]$ and $F_\ell[t]$ would be disjoint as well, it holds that $J_\ell[s] = J_\ell[t]$.

As the leases of type ℓ that ALG bought for σ_t and σ_s started and expired at the same time, σ_s was in the tree $T_{\geq \ell}[t]$ when σ_t arrived. Thus, the distance between σ_t and the tree $T_{\geq \ell}[t]$ was at most $d_G(\sigma_t, \sigma_s)$. From the triangle inequality and diameters of sets N_ℓ^s and N_ℓ^t, it follows that $d_G(\sigma_t, \sigma_s) \leq d_G(\sigma_t, \sigma_u) + d_G(\sigma_u, \sigma_s) \leq 2^{j-2} + 2^{j-2} = 2^{j-1}$. Hence, the request σ_t would be of class $j - 1$ or lower, which would contradict its choice. $\square$

Lemma 8. *For any class j and a lease type $\ell \in \mathcal{L}$, $C_\ell \cdot |W_j^\ell| \leq C_1 \cdot |W_j|$.*

Proof. The lemma follows trivially for $\ell = 1$, and therefore we assume that $\ell \geq 2$.

We look at any request $\sigma_t \in W_j^\ell$ and its neighbor set N_ℓ^t. As σ_t is of class j, N_ℓ^t contains requests only of class j, i.e., $N_\ell^t \subseteq W_j$. By Lemma 7, the neighbor sets of all requests from W_j^ℓ are disjoint, and hence $\sum_{\sigma_t \in W_j^\ell} |N_\ell^t| \leq |W_j|$.

As ALG ℓ-leases request σ_t, its neighbor set N_ℓ^t contains at least C_ℓ/C_1 requests. Therefore, $|W_j| \geq \sum_{\sigma_t \in W_j^\ell} |N_\ell^t| \geq \sum_{\sigma_t \in W_j^\ell} C_\ell/C_1 = |W_j^\ell| \cdot C_\ell/C_1$. $\square$

Lemma 9. *For any input $\mathcal{I}$, it holds that $\text{ALG}(\mathcal{I}) \leq L \cdot \sum_j |W_j| \cdot C_1 \cdot 2^j$.*

Proof. The cost of serving any request $\sigma_t \in W_j^\ell$ is at most $C_\ell \cdot 2^j$. Using Lemma 8, we obtain $\text{ALG}(\mathcal{I}) \leq \sum_{\ell \in \mathcal{L}} \sum_j |W_j^\ell| \cdot C_\ell \cdot 2^j \leq L \cdot \sum_j |W_j| \cdot C_1 \cdot 2^j$. $\square$

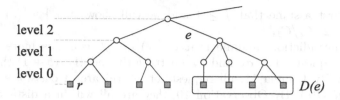

Fig. 1. An example of an HST embedding. Square nodes (leaves of the HST) represent terminals from an input sequence (including root r). Edge e is a 2-level edge (of length 2^2). $D(e)$ is the set of leaves below edge e.

5.2 Lower Bound on OPT

In this part, we bound the sum of all requests' credits by $O(1) \cdot \text{OPT}(\mathcal{I}_T)$.

We number edge levels of T starting from bottom ones and counting from 0. That is, a j-level edge is of length 2^j. Moreover, for an edge e, we denote the set of all leaves below it by $D(e)$, see Fig. 1. We denote the set of all j-level edges by E_j. The next observation follows immediately by Definition 1.

Observation 10. *Fix any j-level edge e. For any two leaves $u, v \in D(e)$, it holds that $d_G(u, v) \le d_T(u, v) \le 2^{j+1}$.*

Lemma 11. *For any request σ_t of class $j \ge 3$, there exists an edge $e \in E_{j-3}$, such that $\sigma_t \in D(e)$ and e lies on the unique path from σ_t to the root r in T.*

Proof. When ALG ℓ leases request σ_t and assigns class j to it, the distance between σ_t and the tree $T_{\ge \ell}[t]$ is larger than 2^{j-1}, and thus $d_G(\sigma_t, r) > 2^{j-1}$. By Observation 10, the unique path in T between r and σ_t must cross two $(j-2)$-level edges, and hence also two $(j-3)$-level edges. We pick e to be the $(j-3)$-level edge that is closer to σ_t (see Fig. 1). □

The lemma above implicitly creates a mapping φ from the set of all requests of class $j \ge 3$ to edges in tree T: a request of class j (i.e., connected by ALG with an edge of length at most 2^j) is mapped to a tree edge of level $j-3$ (of length 2^{j-3}). Note that a request of class j could be mapped to an edge in E_{j-2}. However, the next lemma requires that all the requests mapped to an edge e are close to each other. We extend φ to include also the requests of classes $j \le 2$, by mapping any such request σ_t to the edge $e \in E_0$ adjacent to σ_t in tree T. In these terms, $\varphi^{-1}(e)$ is a set of requests mapped to e. For an edge e of level $j \ge 1$, $\varphi^{-1}(e) = D(e) \cap W_{j+3}$, and for $e \in E_0$, we have $\varphi^{-1}(e) = D(e) \cap \bigcup_{j=0}^{3} W_j$.

Let $\text{OPT}(e)$ be the total leasing cost of e in the optimal solution for $\mathcal{I}_T$. Our goal now is to show that the sum of credits of requests in $\varphi^{-1}(e)$ is at most $O(\text{OPT}(e))$. To do so, we first prove a general bound on the amount of credit that holds for all possible periods J_ℓ^m. Later on, we will apply it to periods J_ℓ^m when OPT leased edge e.

Lemma 12. *Fix a lease type $\ell > 1$ and a j-level edge e of T. Then, for any $m \in \mathbb{N}$, $|\varphi^{-1}(e) \cap F_\ell^m| \le 8 \cdot C_\ell / C_1$.*

Proof. We first assume that $j \geq 1$ and we will show that $|\varphi^{-1}(e) \cap F_\ell^m| \leq C_\ell/C_1 + 1 \leq 2 \cdot C_\ell/C_1$.

For a contradiction, assume that $\varphi^{-1}(e) \cap F_\ell^m$ contains more than $b = C_\ell/C_1 + 1$ requests. Let σ_s and σ_t be the b-th and the $(b+1)$-th of them, respectively. By Lemma 11, all requests of $\varphi^{-1}(e)$ are of class $j+3$ and are contained in $D(e)$. By Observation 10, they are all within a distance of 2^{j+1} from σ_s in the graph. Therefore, N_ℓ^s, the neighbor set of σ_s considered by ALG, contains all previous requests from $\varphi^{-1}(e) \cap F_\ell^m$ (there are $b - 1 = C_\ell/C_1$ many of them). The condition at Line 8 of Algorithm 1 is thus fulfilled, and therefore ALG buys a lease of type at least ℓ for σ_s.

In effect, when σ_t arrives, σ_s is in $T_{\geq \ell}[t]$. Hence, the distance from σ_t to the tree $T_{\geq \ell}[t]$ in the graph is at most $d_G(\sigma_t, \sigma_s) \leq d_T(\sigma_t, \sigma_s) \leq 2^{j+1}$. Therefore, the class of σ_t is at most $j+1$, which contradicts the choice of σ_t.

The analysis above can be extended to any 0-level edge e. Because $D(e)$ for $e \in E_0$ contains exactly one terminal, all requests from $\varphi^{-1}(e) \cap F_\ell^m$ are always contained in the appropriate neighbor set. This implies that $|\varphi^{-1}(e) \cap F_\ell^m \cap W_i| \leq 2 \cdot C_\ell/C_1$ for any class $i \in \{0,1,2,3\}$. As $\varphi^{-1}(e) = D(e) \cap \bigcup_{i=0}^{3} W_i$, we obtain $|\varphi^{-1}(e) \cap F_\ell^m| \leq 4 \cdot 2 \cdot C_\ell/C_1$. $\qquad\square$

Lemma 13. *Fix a j-level edge e of T. Then, $|\varphi^{-1}(e)| \cdot C_1 \cdot 2^j \leq 8 \cdot \text{OPT}(e)$.*

Proof. By Lemma 11, for each request σ_t in $\varphi^{-1}(e)$, OPT has to have edge e leased at time t, as e lies on the only path between σ_t and the root r (see also Fig. 1).

Let $P(e)$ be the set of all pairs (ℓ, m), such that OPT ℓ-leases e for period J_ℓ^m. That is, $\text{OPT}(e) = \sum_{(\ell,m) \in P(e)} C_\ell \cdot 2^j$. In the optimal solution J_ℓ^m periods are pairwise disjoint for all pairs (ℓ, m) in $P(e)$, and hence so are sets F_ℓ^m. Thus,

$$|\varphi^{-1}(e)| \cdot C_1 \cdot 2^j = \sum_{(\ell,m) \in P(e)} |\varphi^{-1}(e) \cap F_\ell^m| \cdot C_1 \cdot 2^j$$

$$\leq \sum_{(\ell,m) \in P(e)} 8 \cdot C_\ell \cdot 2^j = 8 \cdot \text{OPT}(e) \ ,$$

where the inequality follows by Lemma 12. $\qquad\square$

Lemma 14. *For any input $\mathcal{I}$ and any dominating HST embedding (T, d_T) of terminals of $\mathcal{I}$, it holds that $\sum_j |W_j| \cdot C_1 \cdot 2^j \leq O(1) \cdot \text{OPT}(\mathcal{I}_T)$.*

Proof. Fix any level $j \geq 1$. Recall that all requests of class $j+3$ (and only them) are mapped by φ to edges from E_j. Hence, we obtain

$$|W_{j+3}| = \sum_{e \in E_j} |\varphi^{-1}(e)| \ . \tag{1}$$

On the other hand, all requests of class $j \in \{0,1,2,3\}$ (and only them) are mapped by φ to edges from E_0. Therefore, $\sum_{j \leq 3} |W_j| = \sum_{e \in E_0} |\varphi^{-1}(e)|$, and consequently

$$\sum_{j \leq 3} |W_j| \cdot 2^j \leq 8 \sum_{e \in E_0} |\varphi^{-1}(e)| \ . \tag{2}$$

We use (1) and (2) to bound $\sum_j |W_j| \cdot 2^j$:

$$\sum_{j \geq 0} |W_j| \cdot C_1 \cdot 2^j \leq \sum_{e \in E_0} 8 \cdot |\varphi^{-1}(e)| \cdot C_1 + \sum_{j \geq 1} \sum_{e \in E_j} 2^{j+3} \cdot |\varphi^{-1}(e)| \cdot C_1$$

$$= \sum_{j \geq 0} \sum_{e \in E_j} 8 \cdot 2^j \cdot |\varphi^{-1}(e)| \cdot C_1$$

$$\leq \sum_{j \geq 0} \sum_{e \in E_j} 8 \cdot 8 \cdot \text{OPT}(e) = O(1) \cdot \text{OPT}(\mathcal{I}_T) \ .$$

The second inequality is a consequence of Lemma 13. □

5.3 The Competitive Ratio

Theorem 15. ACCUMULATE-AND-LEASE-GREEDILY *is* $O(L \cdot \log k)$-*competitive.*

Proof. Fix an instance $\mathcal{I} = (G, d_G, r, \mathcal{L}; \sigma)$. By Lemma 2, there exists a dominating HST embedding (T, d_T), such that $\text{OPT}(\mathcal{I}_T) \leq O(\log k) \cdot \text{OPT}(\mathcal{I})$, where $\mathcal{I}_T = (T, d_T, r, \mathcal{L}; \sigma)$. By Lemma 9, the total cost of ALG is at most L times the sum of all requests' credits, $\sum_j |W_j| \cdot C_1 \cdot 2^j$. By Lemma 14, the latter amount is at most $O(1) \cdot \text{OPT}(\mathcal{I}_T)$, and hence $\text{ALG}(\mathcal{I}) \leq O(L) \cdot \text{OPT}(\mathcal{I}_T) \leq O(L \cdot \log k) \cdot \text{OPT}(\mathcal{I})$, which concludes the proof. □

6 Conclusions

We showed that the technique of analyzing greedy algorithms using HSTs can be also applied to the leasing variant of the online Steiner tree (the OSTL problem). A natural research direction is to employ it for other leasing variants of graph problems, such as Steiner forest or facility location.

Closing the gap between the current upper and lower bounds for the deterministic algorithms solving the OSTL problem ($O(L \cdot \log k)$ and $\Omega(L + \log k)$, respectively) is an intriguing open problem. In particular, it seems that improving the competitive ratio requires a very careful interplay between path-choosing and lease-upgrade routines. We remark that analogous gaps exist also for randomized algorithms for the OSTL problem and for a leasing variant of the facility location problem [20].

References

1. Abshoff, S., Kling, P., Markarian, C., Meyer auf der Heide, F., Pietrzyk, P.: Towards the price of leasing online. Journal of Combinatorial Optimization pp. 1–20 (2015)
2. Abshoff, S., Markarian, C., Meyer auf der Heide, F.: Randomized online algorithms for set cover leasing problems. In: Proc. 8th Int. Conf. on Combinatorial Optimization and Applications (COCOA). pp. 25–34 (2014)
3. Albers, S., Koga, H.: New on-line algorithms for the page replication problem. Journal of Algorithms 27(1), 75–96 (1998)

4. Alon, N., Azar, Y.: On-line Steiner trees in the Euclidean plane. In: Proc. 8th ACM Symp. on Computational Geometry (SoCG). pp. 337–343 (1992)
5. Angelopoulos, S.: On the competitiveness of the online asymmetric and Euclidean Steiner tree problems. In: Proc. 7th Workshop on Approximation and Online Algorithms (WAOA). pp. 1–12 (2009)
6. Anthony, B.M., Gupta, A.: Infrastructure leasing problems. In: Proc. 12th Int. Conf. on Integer Programming and Combinatorial Optimization (IPCO). pp. 424–438 (2007)
7. Armbrust, M., Fox, A., Griffith, R., Joseph, A.D., Katz, R.H., Konwinski, A., Lee, G., Patterson, D.A., Rabkin, A., Stoica, I., Zaharia, M.: Above the clouds: A Berkeley view of cloud computing. Tech. Rep. UCB/EECS-2009-28, EECS Department, University of California, Berkeley (2009), https://www2.eecs.berkeley.edu/Pubs/TechRpts/2009/EECS-2009-28.pdf
8. Awerbuch, B., Azar, Y.: Buy-at-bulk network design. In: Proc. 38th IEEE Symp. on Foundations of Computer Science (FOCS). pp. 542–547 (1997)
9. Awerbuch, B., Azar, Y., Bartal, Y.: On-line generalized Steiner problem. Theoretical Computer Science 324(2–3), 313–324 (2004)
10. Awerbuch, B., Bartal, Y., Fiat, A.: Competitive distributed file allocation. In: Proc. 25th ACM Symp. on Theory of Computing (STOC). pp. 164–173 (1993)
11. Bartal, Y., Fiat, A., Rabani, Y.: Competitive algorithms for distributed data management. Journal of Computer and System Sciences 51(3), 341–358 (1995)
12. Black, D.L., Sleator, D.D.: Competitive algorithms for replication and migration problems. Tech. Rep. CMU-CS-89-201, Department of Computer Science, Carnegie-Mellon University (1989)
13. Chowdhury, N.M.K., Boutaba, R.: A survey of network virtualization. Computer Networks 54(5), 862–876 (2010)
14. Fakcharoenphol, J., Rao, S., Talwar, K.: A tight bound on approximating arbitrary metrics by tree metrics. Journal of Computer and System Sciences 69(3), 485–497 (2004)
15. Fleischer, R., Głazek, W., Seiden, S.S.: New results for online page replication. Theoretical Computer Science 324(2–3), 219–251 (2004)
16. Imase, M., Waxman, B.M.: Dynamic Steiner tree problem. SIAM Journal on Discrete Mathematics 4(3), 369–384 (1991)
17. Lund, C., Reingold, N., Westbrook, J., Yan, D.C.K.: Competitive on-line algorithms for distributed data management. SIAM Journal on Computing 28(3), 1086–1111 (1999)
18. Matsubayashi, A.: Non-greedy online Steiner trees on outerplanar graphs. In: Proc. 14th Workshop on Approximation and Online Algorithms (WAOA). pp. 129–141 (2016)
19. Meyerson, A.: The parking permit problem. In: Proc. 46th IEEE Symp. on Foundations of Computer Science (FOCS). pp. 274–284 (2005)
20. Nagarajan, C., Williamson, D.P.: Offline and online facility leasing. Discrete Optimization 10(4), 361–370 (2013)
21. Umboh, S.: Online network design algorithms via hierarchical decompositions. In: Proc. 26th ACM-SIAM Symp. on Discrete Algorithms (SODA). pp. 1373–1387 (2015)
22. Westbrook, J., Yan, D.C.K.: The performance of greedy algorithms for the on-line Steiner tree and related problems. Mathematical Systems Theory 28(5), 451–468 (1995)
23. Wu, W., Huang, Y.: Steiner trees. In: Encyclopedia of Algorithms, pp. 2102–2107 (2016)

The I/O Complexity of Strassen's Matrix Multiplication with Recomputation*

Gianfranco Bilardi[1] and Lorenzo De Stefani[2]

[1] Department of Information Engineering, University of Padova,
Via Gradenigo 6B/Padova, Italy
bilardi@dei.unipd.it
[2] Department of Computer Science, Brown University,
115 Waterman Street/Providence, United States of America
lorenzo@cs.brown.edu

Abstract. A tight $\Omega((n/\sqrt{M})^{\log_2 7} M)$ lower bound is derived on the I/O complexity of Strassen's algorithm to multiply two $n \times n$ matrices, in a two-level storage hierarchy with M words of fast memory. A proof technique is introduced, which exploits the Grigoriev's flow of the matrix multiplication function as well as some combinatorial properties of the Strassen computational directed acyclic graph (CDAG). Applications to parallel computation are also developed. The result generalizes a similar bound previously obtained under the constraint of no-recomputation, that is, that intermediate results cannot be computed more than once.

1 Introduction

Data movement is increasingly playing a major role in the performance of computing systems, in terms of both time and energy. This technological trend [1] is destined to continue, since the very fundamental physical limitations on minimum device size and on maximum message speed lead to inherent costs when moving data, whether across the levels of a hierarchical memory system or between processing elements of a parallel system [2]. The communication requirements of algorithms have been the target of considerable research in the last four decades; however, obtaining significant lower bounds based on such requirements remains an important and challenging task.

In this paper, we focus on the I/O complexity of Strassen's matrix multiplication algorithm. Matrix multiplication is a pervasive primitive utilized in many applications. Strassen [3] showed that two $n \times n$ matrices can be multiplied with $O(n^\omega)$ operations, where $\omega = \log_2 7 \approx 2.8074$, hence with asymptotically fewer than the n^3 arithmetic operations required by the straightforward implementation of the definition of matrix multiplication. This result has motivated a number of efforts which have lead to increasingly faster algorithms, at least asymptotically, with the current record being at $\omega < 2.3728639$ [4].

* This work was supported, in part, by MIUR of Italy under project AMANDA 2012C4E3KT 004 and by the University of Padova under projects CPDA121378/12, and CPDA152255/15.

© Springer International Publishing AG 2017
F. Ellen et al. (Eds.): WADS 2017, LNCS 10389, pp. 181–192, 2017.
DOI: 10.1007/978-3-319-62127-2_16

Previous and Related Work: I/O complexity has been introduced in the seminal work by Hong and Kung [5]; it is essentially the number of data transfers between the two levels of a memory hierarchy with a fast memory of M words and a slow memory with an unbounded number of words. Hong and Kung presented techniques to develop lower bounds to the I/O complexity of computations modeled by *computational directed acyclic graphs* (CDAGs). The resulting lower bounds apply to all the schedules of the given CDAG, including those with recomputation, that is, where some vertices of the CDAG are evaluated multiple times. Among other results, they established an $\Omega\left(n^3/\sqrt{M}\right)$ lower bound to the I/O complexity of the definition-based matrix multiplication algorithm, which matched a known upper bound [6]. The techniques of [5] have also been extended to obtain tight communication bounds for the definition-based matrix multiplication in some parallel settings [7–9] and for the special case of *"sparse matrix multiplication"* [10]. Ballard et al. generalized the results on matrix multiplication of Hong and Kung [5] in [11,12] by using the approach proposed in [8] based on the Loomis-Whitney geometric theorem [13,14]. The same papers present tight I/O complexity bounds for various classical linear algebra algorithms, for problems such as LU/Cholesky/LDLT/QR factorization and eigenvalues and singular values computation.

It is natural to wonder what is the impact of Strassen's reduction of the number of arithmetic operations on the number of data transfers. In an important contribution, Ballard et al. [15], obtained an $\Omega((n/\sqrt{M})^{\log_2 7}M)$ I/O lower bound for Strassen's algorithm, using the *"edge expansion approach"*. The authors extend their technique to a class of *"Strassen-like"* fast multiplication algorithms and to fast recursive multiplication algorithms for rectangular matrices [16]. This result was later generalized to a broader class of *"Strassen-like"* algorithms by Scott et. al [17] using the *"path routing"* technique. In [18] (Chap. 4.5), De Stefani presented an alternative technique for obtaining I/O lower bounds for a large class of Strassen-like algorithms characterized by a recursive structure. This result combines the concept of Grigoriev's flow of a function and the *"dichotomy width"* technique [19]; it generalizes previous results and simplifies the analysis.

A parallel, *"communication avoiding"* implementation of Strassen's algorithm whose performance matches the known lower bound [15,17], was proposed by Ballard et al. [20]. A communication efficient algorithm for the special case of sparse matrices based on Strassen's algorithm was presented in [21].

On the impact of recomputation: The edge expansion technique of [15], the path routing technique of [17], and the *"closed dichotomy width"* technique of [19] all yield I/O lower bounds that apply only to computational schedules for which no intermediate result is ever computed more than once (*nr-computations*). While it is of interest to know what is the I/O complexity achievable by nr-computations, it is also important to investigate what can be achieved with recomputation. In fact, for some CDAGs, recomputing intermediate values reduces the space and/or the I/O complexity of an algorithm [22]. In [23], it is shown that some algorithms admit a *portable schedule* (i.e., a schedule which achieves optimal performance across memory hierarchies with different access costs) only if recomputation is allowed. Recomputation can also enhance the performance of simulations among networks (see [24] and references therein) and plays a key role in the design of efficient area-universal VLSI architectures with constant slowdown [25]. A number of lower bound techniques that allow for recomputation have been presented in the literature, including the "*S-partition* technique" [5], the "*S-span* technique" [22], and the "*S-covering* technique" [26] which merges and extends aspects from both [5] and [22]. However, none of these have been previously applied to fast matrix multiplication algorithms.

Our results: We extend the $\Omega((n/\sqrt{M})^{\log_2 7} M)$ I/O complexity lower bound for Strassen's algorithm to schedules with recomputation. A matching upper bound is known, and obtained without recomputation; hence, we can conclude that, for Strassen's algorithm, recomputation does not help in reducing I/O complexity if not, possibly, by a constant factor. Our proof technique is of independent interest, since it exploits to a significant extent the "*divide and conquer*" nature exhibited by many algorithms. We follow the dominator set approach pioneered by Hong and Kung in [5]. However, we focus the dominator analysis only on a select set of target vertices, specifically the outputs of the sub-CDAGs of Strassen's CDAG that correspond to sub-problems of a suitable size (i.e., chosen as a function of the fast memory capacity M). Any dominator set of a set of target vertices can be partitioned into two subsets, one internal and one external to the sub-CDAGs. The analysis of the internal component can be carried out based only on the fact that the sub-CDAGs compute matrix products, irrespective of the algorithm (in our case, Strassen's) by which the products are computed. To achieve this independence of the algorithm, we resort on the concept of Grigoriev's flow of a function [27] and on a lower bound to such flow established by Savage [28] for matrix multiplication..

In order to obtain our *general* lower bound for the I/O complexity, we then build on this result combining it with the analysis of the external component of the dominator, which requires instead rather elaborate arguments that are specific to Strassen's CDAG. The paper is organized as follows: In the first part of Sect. 2, we provide the details of our model and of several theoretical notions needed in our analysis. In the second part of Sect. 2, we analyze the relation between the Grigoriev's flow of a function and the size of the dominator sets of subsets of output vertices of a CDAG. In Sect. 3, we present the I/O complexity lower bound for Strassen's algorithm when recomputation is allowed. Extensions of the result to a parallel model are also discussed.

2 I/O Complexity, Dominator Sets and Grigoriev's Flow

We consider algorithms which compute the product $C = AB$ of $n \times n$ matrices A, B with entries from a ring $\mathcal{R}$. We focus on algorithms whose execution, for any given n, can be modeled as a *computational directed acyclic graph* (CDAG) $G = (V, E)$, where each vertex $v \in V$ represents either an input value or the result of a unit time operation (i.e., an intermediate result or one of the output values), while the directed edges in E represent data dependences. A *directed*

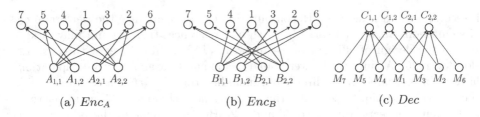

Fig. 1: Basic building blocks of Strassen's CDAG. Enc_A and Enc_B are isomorphic.

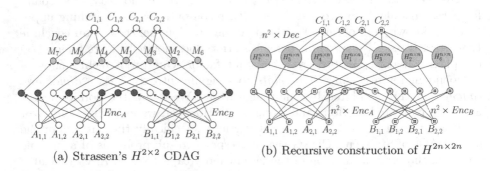

(a) Strassen's $H^{2\times 2}$ CDAG   (b) Recursive construction of $H^{2n\times 2n}$

Fig. 2: Black vertices represent combinations of the input values from the factor matrices A and B which are used as input values for the sub-problems M_i; Grey vertices represent the output of the seven sub-problems which are used to compute the output values of the product matrix C.

path connecting vertices $u, v \in V$ is an ordered sequence of vertices for which u and v are respectively the first and last vertex such that there is in E a (directed) edge pointing from each vertex in the sequence to its successor. We say that $G' = (V', E')$ is a *sub-CDAG* of $G = (V, E)$ if $V' \subseteq V$ and $E' \subseteq E \cap (V' \times V')$.

Properties of Strassen's CDAG: Consider Strassen's algorithm [3] when used to compute $C = AB$, where A and B are $n \times n$ matrices with entries from the ring $\mathcal{R}$. Let $H^{n\times n}$ denote the corresponding CDAG. For $n \geq 2$, $H^{n\times n}$ can be obtained by using a recursive construction which mirrors the recursive structure of the algorithm. The base of the construction is the $H^{2\times 2}$ CDAG which corresponds to the multiplication of two 2×2 matrices using Strassen's algorithm (Fig. 2a). $H^{2n\times 2n}$ can then be constructed by composing seven copies of $H^{n\times n}$, each corresponding to one of the seven sub-products generated by the algorithm (see Fig. 2b): n^2 disjoint copies of CDAG Enc_A (resp., Enc_B) are used to connect the input vertices of $H^{2n\times 2n}$, which correspond to the values of the input matrix A (resp., B) to the appropriate input vertices of the seven sub-CDAGs $H_i^{n\times n}$; the output vertices of the sub-CDAGs $H_i^{n\times n}$ (which correspond to the outputs of the seven sub-products) are connected to the appropriate output vertices of the entire $H^{2n\times 2n}$ CDAG using n^2 copies of the decoder sub-CDAG *Dec*.

We will exploit the following recursive structure of Strassen's CDAG:

Lemma 1. *Let $H^{n \times n}$ denote the CDAG of Strassen's algorithm for input matrices of size $n \times n$. For $0 \leq i \leq \log n - 1$, there are exactly 7^i disjoint sub-CDAGs $H^{n/2^i \times n/2^i}$.*

We will also capitalize on the existence of vertex-disjoint paths connecting the *"global"* input vertices of $H^{n \times n}$ to the *"local"* input vertices of the sub-CDAGs $H^{n/2^i \times n/2^i}$ for $0 \leq i \leq \log n - 1$, with the help of the following lemma.

Lemma 2. *Given an encoder CDAG, for any subset Y of its output vertices, there exists a subset X of its input vertices, with $\min\{|Y|, 1 + \lceil (|Y| - 1)/2 \rceil\} \leq |X| \leq |Y|$, such that there exist $|X|$ vertex-disjoint paths connecting the vertices in X to vertices in Y.*

We refer the reader to the extended on-line version of this paper [29] for a detailed presentation of Strassen's algorithm and for the proofs of Lemmas 1 and 2.

Model: We assume that sequential computations are executed on a system with a two-level memory hierarchy consisting of a fast memory or *cache* of size M, measured in words, and a *slow memory* of unlimited size. A memory word can store at most one value from $\mathcal{R}$. An operation can be executed only if all its operands are in cache. Data can be moved from the slow memory to the cache by *read* operations, and, in the other direction, by *write* operations. Read and write operations are also called *I/O operations*. We assume the input data to be stored in slow memory at the beginning of the computation. The evaluation of a CDAG in this model can be analyzed by means of the *"red-blue pebble game"* [5]. The number of I/O operations executed when evaluating a CDAG depends on the *"computational schedule,"* that is, on the order in which vertices are evaluated and on which values are kept in/discarded from cache. The *I/O complexity $IO_G(M)$* of a CDAG G is defined as the minimum number of I/O operations over all possible computational schedules.

We also consider a parallel model where P processors, each with a local memory of size M, are connected by a network. We assume that the input is initially distributed among the processors, thus requiring that $MP \geq 2n^2$. Processors can exchange point-to-point messages among each other. For this model, we derive lower bounds to the number of words that must be either sent or received by at least one processor during the CDAG evaluation.

Grigoriev's flow and dominator sets: The concept of *dominator set* was originally introduced in [5]. We use the following, slightly different, definition:

Definition 1 (Dominator set). *Given a CDAG $G = (V, E)$, let $I \subset V$ denote the set of input vertices. A set $D \subseteq V$ is a dominator set for $V' \subseteq V$ with respect to $I' \subseteq I$ if every path from a vertex in I' to a vertex in V' contains at least a vertex of D. When $I' = I$, D is simply referred as "a dominator set for $V' \subseteq V$".*

The *"flow of a function"* was introduced by Grigoriev [27]. We use a revised formulation by Savage [28]. The flow is an inherent property of a function, not of a specific algorithm by which the function may be computed.

Definition 2 (Grigoriev's flow). *A function $f : \mathcal{R}^p \to \mathcal{R}^q$ has a $w(u, v)$ Grigoriev's flow if for all subsets X_1 and Y_1, of its p input and q output variables,* with $|X_1| \geq u$ and $|Y_1| \geq v$, there is a sub-function h of f obtained by making some assignment to variables of f not in X_1 and discarding output variables not in Y_1, such that h has at least $|\mathcal{R}|^{w(u,v)}$ points in the image of its domain.

A lower bound on the Grigoriev's flow for the square matrix multiplication function $f_{n \times n} : \mathcal{R}^{2n^2} \to \mathcal{R}^{n^2}$ over the ring $\mathcal{R}$ was presented in [28] (Thm. 10.5.1).

Lemma 3 (Grigoriev's flow of $f_{n \times n} : \mathcal{R}^{2n^2} \to \mathcal{R}^{n^2}$ [28]). $f_{n \times n} : \mathcal{R}^{2n^2} \to \mathcal{R}^{n^2}$ has a $w_{n \times n}(u, v)$ Grigoriev's flow, where:

$$w_{n \times n}(u, v) \geq \frac{1}{2} \left(v - \frac{(2n^2 - u)^2}{4n^2} \right), \text{for } 0 \leq u \leq 2n^2, \ 0 \leq v \leq n^2. \quad (1)$$

The "*flow of a function*" measures the amount of information that suitable subsets of outputs encode about suitable subsets of inputs. Such information must be encoded by any dominator of those outputs, thus implying the following lower bound on the size of dominators.

Lemma 4. Let $G = (V, E)$ be a CDAG computing $f : \mathcal{R}^p \to \mathcal{R}^q$ with Grigoriev's flow $w_f(u, v)$. Let I (resp., O) denote the set of input (resp., output) vertices of G. Any dominator set D for any subset $O' \subseteq O$ with respect to any subset $I' \subseteq I$ satisfies $|D| \geq w_f(|I'|, |O'|)$.

Proof. Given $I' \subseteq I$ and $O' \subseteq O$, suppose the values of the input variables in $I \setminus I'$ to be fixed. Let D be a dominator set for $O' \subseteq O$ with respect to $I' \subseteq I$. The lemma follows combining statements (i) and (ii):
(i) By Definition 2, there exists an assignment of the input variables in I', such that the output variables in O' can assume $|\mathcal{R}|^{w_f(|I'|,|O'|)}$ distinct values.
(ii) Since all paths I' to O' intercept D, the values of the outputs in O' are determined by the inputs in $I \setminus I'$, which are fixed, and by the values of the vertices in D; hence, the outputs in O' can take at most $|\mathcal{R}|^{|D|}$ distinct values. □

We let $G^{n \times n}$ denote the CDAG corresponding to the execution of an *unspecified* algorithm for the square matrix multiplication function.

Lemma 5. Given $G^{n \times n}$, let $O' \subseteq O$ be a subset of its output vertices O. For any subset D of the vertices of $G^{n \times n}$ with $|O'| \geq 2|D|$, there exists a set $I' \subseteq I$ of the input vertices I with cardinality $|I'| \geq 2n\sqrt{|O'| - 2|D|}$, such that all vertices in I' are connected to some vertex in O' by directed paths with no vertex in D.

Proof. Lemma 5 follows by applying the results in Lemmas 3 and 4 to the CDAG $G^{n \times n}$. Let $I'' \subseteq I$ denote the set of all input vertices of $G^{n \times n}$, such that all paths connecting these vertices to the output vertices in O' include at least a vertex in D (i.e., I'' is the largest subset of I with respect to whom D is a dominator set for O'). From Lemmas 3 and 4 the following must hold:

$$|D| \geq w_{n \times n} \geq \frac{1}{2} \left(|O'| - \frac{(2n^2 - |I''|)^2}{4n^2} \right). \quad (2)$$

Let $I' = I \setminus I''$. By the definition of I'', the vertices in I' are exactly those that are connected to vertices in O' by directed paths with no vertex in D. Since $|I| = 2n^2$, from (2) we have $|I'|^2 \geq 4n^2 (|O'| - 2|D|)$. □

3 Lower Bounds for Schedules with Recomputation

Without recomputation, once an input value is loaded in memory or an intermediate result is computed, it must be kept in memory (either cache or slow) until the result of each operation which uses it has been evaluated. This is exploited by the "*dichotomy width* technique" [19], the "*boundary flow* technique" [30], and those yielding I/O lower bounds for Strassen's algorithm [16–18]. With recomputation, intermediate results can instead be deleted from *all* memory and recomputed starting from the global input values. This considerably complicates the analysis of the I/O cost (see [11] for an extensive discussion). In this section, we present a technique which addresses these complications. First, we obtain a lower bound for the minimum size of the dominator set of subset of vertices corresponding to the output values of the $(n/(2\sqrt{M}))^{\log_2 7}$ Strassen's sub-problems with input size $2\sqrt{M} \times 2\sqrt{M}$. In turn, this dominator bound yields an asymptotically tight I/O lower bound both in the sequential and the parallel model.

For $1 \leq M \leq n^2/4$, with M a power of four, we focus on the subset $\mathcal{Y}$ of the input vertices and the subset $\mathcal{Z}$ of the output vertices of the $(n/(2\sqrt{M}))^{\log_2 7}$ sub-CDAGs $H^{2\sqrt{M} \times 2\sqrt{M}}$ of $H^{n \times n}$. Further, we let $\mathcal{X}$ be the set of the "*global input vertices*" of $H^{n \times n}$ which correspond to the entries of matrices A and B.

Lemma 6. *Given $H^{n \times n}$, let Q be a set of internal (i.e., not input) vertices of its $\left(n/(2\sqrt{M})\right)^{\log_2 7}$ sub-CDAGs $H^{2\sqrt{M} \times 2\sqrt{M}}$. For any $Z \subseteq \mathcal{Z}$ with $|Z| \geq 2|Q|$ there exist $X \subseteq \mathcal{X}$ and $Y \subseteq \mathcal{Y}$ with $|X| = |Y| \geq 4\sqrt{M(|Z| - 2|Q|)}$ such that, (a) there are $|X| = |Y|$ vertex-disjoint paths from X to Y, and (b) each vertex in Y is connected to some vertex in Z by a directed path with no vertex in Q.*

Proof. For a fixed M, we proceed by induction on $n = 2\sqrt{M}, 4\sqrt{M}, \ldots$ In the base case, $H^{n \times n} = H^{2\sqrt{M} \times 2\sqrt{M}}$, and the sets $\mathcal{Y}$ and $\mathcal{X}$ coincide. The statement is a consequence of Lemma 5 as $H^{2\sqrt{M} \times 2\sqrt{M}}$ is a $G^{2\sqrt{M} \times 2\sqrt{M}}$ CDAG.

Assuming now inductively that the statement holds for $H^{n \times n}$, with $n \geq 2\sqrt{M}$, we shall show it also holds for $H^{2n \times 2n}$. Let $H_1^{n \times n}, H_2^{n \times n}, \ldots, H_7^{n \times n}$ denote the seven sub-CDAGs of $H^{2n \times 2n}$, each corresponding to one of the seven sub-products generated by the first recursive step of Strassen's algorithm.

Let Z_i, $\mathcal{Y}_i$ and Q_i respectively denote the subsets of Z, $\mathcal{Y}$ and Q in $H_i^{n \times n}$. Since, from Lemma 1, the seven sub-CDAGs $H_i^{n \times n}$ are mutually vertex-disjoint, clearly $Z_1, Z_2, \ldots, Z_7$ partition Z, $\mathcal{Y}_1, \mathcal{Y}_2, \ldots, \mathcal{Y}_7$ partition $\mathcal{Y}$ and $Q_1, Q_2, \ldots, Q_7$ partition Q. This implies $\sum_{i=1}^{7} |Z_i| = |Z|$, and $\sum_{i=1}^{7} |Q_i| = |Q|$. Letting $\delta_i = \max\{0, |Z_i| - 2|Q_i|\}$, we have $\delta = \sum_{i=1}^{7} \delta_i \geq |Z| - 2|Q|$.

Applying the inductive hypothesis to each $H_i^{n \times n}$, we have that there is a subset $Y_i \subseteq \mathcal{Y}_i$ with $|Y_i| \geq 4\sqrt{M\delta_i}$ such that vertices of Y_i are connected to vertices in Z_i via paths with no vertex in Q_i. In the sequel, the set Y referred to in the statement will be identified as a suitable subset of $\cup_{i=1}^{7} Y_i$ so that property (b) will be automatically satisfied. Towards property (a), we observe by the inductive hypothesis that vertices in Y_i can be connected to a subset K_i of the input vertices of $H_i^{n \times n}$ with $|K_i| = |Y_i|$ using vertex-disjoint paths. Since the sub-CDAGs $H_i^{n \times n}$ are vertex-disjoint, so are the paths connecting vertices in Y_i to vertices in K_i. It remains to show that at least $4\sqrt{M}(|Z| - 2|Q|)$ of these paths can be extended to $\mathcal{X}$ while maintaining them vertex-disjoint.

In Strassen's CDAG $H^{2n \times 2n}$ (Sect. 2), vertices in $\mathcal{X}$ corresponding to input matrix A (resp., B) are connected to vertices in $K_1, K_2, \ldots, K_7$ by means of n^2 encoding sub-CDAGs Enc_A (resp., Enc_B). None of these $2n^2$ encoding sub-CDAGs share any input or output vertices. No two output vertices of the same encoder sub-CDAG belong to the same sub-CDAG $H_i^{n \times n}$. This fact ensures that for a single sub-CDAG $H_i^{n \times n}$ it is possible to connect all the vertices in K_i to a subset of the vertices in $\mathcal{X}$ via vertex-disjoint paths.

For each of the $2n^2$ encoder sub-CDAGs, let us consider the vector $\mathbf{y}_j \in \{0,1\}^7$ such that $\mathbf{y}_j[i] = 1$ iff the corresponding i-th output vertex (respectively according to the numbering indicated in Fig. 1a or Fig. 1b) is in K_i. Therefore, $|\mathbf{y}_j|$ equals the number of output vertices of the j-th encoder sub-CDAG which are in K. From Lemma 2, for each encoder sub-CDAG there exists a subset $X_j \in \mathcal{X}$ of the input vertices of the j-th encoder sub-CDAG for which it is possible to connect each vertex in X_j to a distinct output vertex of the j-th encoder sub-CDAG using vertex-disjoint paths, each constituted by a singular edge with $\min\{|\mathbf{y}_j|, 1 + \lceil(|\mathbf{y}_j| - 1)/2\rceil\} \leq |X_j| \leq |\mathbf{y}_j|$. Therefore, the number of vertex-disjoint paths connecting vertices in $\mathcal{X}$ to vertices in $\cup_{i=1}^{7} K_i$ is at least $\sum_{j=1}^{2n^2} \min\{|\mathbf{y}_j|, 1 + \lceil(|\mathbf{y}_j| - 1)/2\rceil\}$ under the constraint that $\sum_{j=1}^{2n^2} \mathbf{y}_j[i] = 4\sqrt{M}\delta_i$. Let us assume, w.l.o.g., that $\delta_1 \geq \delta_2 \geq \ldots \geq \delta_7$. As previously stated, it is possible to connect all vertices in K_1 to vertices in $\mathcal{X}$ through vertex-disjoint paths. Consider now all possible dispositions of the vertices in $\cup_{i=2}^{7} K_i$ over the outputs of the $2n^2$ encoder sub-CDAGs. Recall that the output vertices of an encoder sub-CDAG belong each to a different $H^{n \times n}$ sub-CDAG. From Lemma 2, we have that for each encoder, there exists a subset $X_j \subset X$ of the input vertices of the j-th encoder sub-CDAG with $|X_j| \geq \min\Big\{|\mathbf{y}_j|, 1 + \lceil(|\mathbf{y}_j| - 1)/2\rceil\Big\} \geq \mathbf{y}_j[1] + \Big(\sum_{i=2}^{7} \mathbf{y}_j[i]\Big)/2$, for which it is possible to connect all vertices in X_j to $|X_j|$ distinct output vertices of the j-th encoder sub-CDAG which are in $\cup_{i=1}^{7} K_i$ using $|X_j|$, thus using vertex-disjoint paths. As all the Enc sub-CDAGs are vertex-disjoint, we can add their contributions so that the number of vertex-disjoint paths connecting vertices in $\mathcal{X}$ to vertices in $\cup_{i=1}^{7} K_i$ is at least $|K_1| + \frac{1}{2}\sum_{i=2}^{7} |K_i| = 4\sqrt{M}\left(\sqrt{\delta_1} + \frac{1}{2}\sum_{i=2}^{7}\sqrt{\delta_i}\right)$. Squaring this quantity leads to:

$$\left(4\sqrt{M}\left(\sqrt{\delta_1} + \frac{1}{2}\sum_{i=2}^{7}\sqrt{\delta_i}\right)\right)^2 = 16M\left(\delta_1 + \sqrt{\delta_1}\sum_{i=2}^{7}\sqrt{\delta_i} + \left(\frac{1}{2}\sum_{i=2}^{7}\sqrt{\delta_i}\right)^2\right),$$

since, by assumption, $\delta_1 \geq \ldots \delta_7$, we have: $\sqrt{\delta_1}\sqrt{\delta_i} \geq \delta_i$ for $i = 2, \ldots, 7$. Thus:

$$\left(4\sqrt{M}\left(\sqrt{\delta_1} + \frac{1}{2}\sum_{i=2}^{7}\sqrt{\delta_i}\right)\right)^2 \geq 16M\sum_{i=1}^{7}\delta_i \geq \left(4\sqrt{M\left(|Z| - 2|Q|\right)}\right)^2.$$

Thus, there are at least $4\sqrt{M\left(|Z| - 2|Q|\right)}$ vertex-disjoint paths connecting vertices in $\mathcal{X}$ to vertices in $\cup_{i=2}^{7}K_i$ as desired. $\qquad\square$

Lemma 7. *For $1 \leq M \leq n^2/4$, and for any subset $Z \subseteq \mathcal{Z}$ in $H^{n \times n}$ with $|Z| = 4M$, any dominator set D of Z satisfies $|D| \geq |Z|/2 = 2M$.*

Proof. Suppose for contradiction that D is a dominator set for Z in $H^{n \times n}$ such that $|D| \leq 2M - 1$. Let $D' \subseteq D$ be the subset of the vertices of D composed by vertices which are *not* internal to the sub-CDAGs $H^{2\sqrt{M} \times 2\sqrt{M}}$. From Lemma 6, with $Q = D \setminus D'$, there exist $X \subseteq \mathcal{X}$ and $Y \subseteq \mathcal{Y}$ with $|X| = |Y| \geq 4\sqrt{M(|Z| - 2(|D| - |D'|))}$ such that vertices in X are connected to vertices in Y by vertex-disjoint paths. Hence, each vertex in D' can be on at most one of these paths. Thus, there exists $X' \subseteq X$ and $Y' \subseteq Y$ with $|X'| = |Y'| \geq \nu = 4\sqrt{M(|Z| - 2(|D| - |D'|))} - |D'|$ paths from X' to Y' with no vertex in D'. From Lemma 6, we also have that all vertices in Y, and, hence, in Y', are connected to some vertex in Z by a path with no vertex in $D \setminus D'$. Thus, there are at least ν paths connecting vertices in $X' \subseteq \mathcal{X}$ to vertices in Z with no vertex in D. We shall now show that the contradiction assumption $|D| \leq 2M - 1$ implies $\nu > 0$:

$$\left(4\sqrt{M(|Z| - 2(|D| - |D'|))}\right)^2 = 16M\left(|Z| - 2\left(|D| - |D'|\right)\right),$$
$$= 16M\left(|Z| - 2|D|\right) + 32M|D'|.$$

By $|D| \leq 2M - 1$, we have $|Z| - 2|D| > 4M - 2(M - 1) > 0$. Furthermore, from $D' \subseteq D$, we have $32M > 2M - 1 > |D| \geq |D'|$. Therefore:

$$(\nu + |D'|)^2 = \left(4\sqrt{M(|Z| - 2(|D| - |D'|))}\right)^2 > |D'|^2. \tag{3}$$

Again, $|D| \leq 2M - 1$ implies $M\left(|Z| - 2\left(|D| - |D'|\right)\right) > 0$. Hence, we can take the square root on both sides of (3) and conclude that $\nu > 0$. Therefore, for $|D| \leq 2M - 1$ there are at least $\nu > 0$ paths connecting a global input vertex to a vertex in Z with no vertex in D, contradicting the assumption that D is a dominator of Z. $\qquad\square$

Lemma 7 provides us with the tools required to obtain our main result.

Theorem 1 (Lower bound I/O complexity Strassen's algorithm). *The I/O-complexity of Strassen's algorithm to multiply two matrices $A, B \in \mathcal{R}^{n \times n}$, on a sequential machine with cache of size $M \leq n^2$, satisfies:*

$$IO_{H^{n \times n}}(M) \geq \frac{1}{7}\left(\frac{n}{\sqrt{M}}\right)^{\log_2 7} M. \tag{4}$$

On P processors, each with a local memory of size $M \leq n^2$, the I/O complexity satisfies:

$$IO_{H^{n \times n}}(P, M) \geq \frac{1}{7} \left(\frac{n}{\sqrt{M}} \right)^{\log_2 7} \frac{M}{P}. \tag{5}$$

Proof. We start by proving (4). Let $n = 2^a$ and $\sqrt{M} = 2^b$ for some $a, b \in \mathbb{N}$. At least $3n^2 \geq 3M$ I/O operations are necessary in order to read the $2n^2$ input values from slow memory to the cache and to write the n^2 output values to the slow memory. The bound in (4) is therefore verified if $n \leq 2\sqrt{M}$.

For $n \geq 4\sqrt{M}$, let $\mathcal{Z}$ denote the set of output vertices of the $\left(n/(2\sqrt{M}) \right)^{\log_2 7}$ sub-CDAGs $H^{2\sqrt{M} \times 2\sqrt{M}}$ of $H^{n \times n}$. Let $\mathcal{C}$ be any computation schedule for the sequential execution of Strassen's algorithm using a cache of size M. We partition $\mathcal{C}$ into segments $\mathcal{C}_1, \mathcal{C}_2, \dots$ such that during each $\mathcal{C}_i$ exactly $4M$ distinct vertices in $\mathcal{Z}$ (denoted as Z_i) are evaluated for the *first time*. Since $|\mathcal{Z}| = 4M \left(n/(2\sqrt{M}) \right)^{\log 7}$, there are $\left(n/(2\sqrt{M}) \right)^{\log 7}$ such segments. Below we show that the number q_i of I/O operations executed during each $\mathcal{C}_i$ satisfies $q_i \geq M$, from which (4) follows.

To bound q_i, consider the set D_i of vertices of $H^{n \times n}$ corresponding to the at most M values stored in the cache at the beginning of $\mathcal{C}_i$ and to the at most q_i values loaded into the cache from the slow memory during $\mathcal{C}_i$ by means of a *read* I/O operation. Clearly, $|D_i| \leq M + q_i$. In order for the $4M$ values from Z_i to be computed during $\mathcal{C}_i$ there cannot be any path connecting any vertex in Z_i to any input vertex of $H^{n \times n}$ which does not have at least one vertex in D_i; that is, D_i has to be a *dominator set* of Z_i. We recall that $|Z_i| = 4M$ and, from Lemma 7, we have that any subset of $4M$ elements of $\mathcal{Z}$ has dominator size at least $2M$, whence $M + q_i \geq |D_i| \geq 2M$, which implies $q_i \geq M$ as stated above.

The proof for the bound for the parallel model in (5), follows a similar strategy: At least one of the P processors being used, denoted as P^*, must compute at least $|\mathcal{Z}|/P = 4M \left(n/(2\sqrt{M}) \right)^{\log 7}/P$ values corresponding to vertices in $\mathcal{Z}$. The bound follows by applying the same argument discussed for the sequential case to the computation executed by P^* (details the extended on-line version [29]). $\square$

Ballard et al. [20] presented a version of Strassen's algorithm whose I/O cost matches the lower bound of Theorem 1 to within a constant factor. Therefore, our bound is asymptotically tight, and the algorithm in [20] is asymptotically I/O optimal. Since in this algorithm no intermediate result is recomputed, recomputation can lead at most to a constant factor reduction of the I/O complexity.

The lower bound of Theorem 1 generalizes to $\Omega((n/\sqrt{M})^{\log_2 7} \frac{M}{B})$ in the *External Memory Model* introduced by Aggarwal and Vitter [31], where $B \geq 1$ values can be moved between cache and consecutive slow memory locations with a single I/O operation.

4 Conclusion

This work has contributed to the characterization of the I/O complexity of Strassen's algorithm by establishing asymptotically tight lower bounds that hold even when recomputation is allowed. Our technique exploits the recursive nature of the CDAG, which makes it promising for the analysis of other recursive algorithms, e.g., for fast rectangular matrix multiplication [32].

The relationship we have exploited between dominator size and Grigoriev's flow points at connections between I/O complexity, (pebbling) space-time tradeoffs [28], and VLSI area-time tradeoffs [33]; these connections deserve further attention.

Some CDAGs for which non-trivial I/O complexity lower bounds are known only in the case of no recomputations are described in [19]. These CDAGs are of interest in the *"limiting technology"* model, defined by fundamental limitations on device size and message speed, as they allow for speedups super-linear in the number of processors. Whether such speedups hold even when recomputation is allowed remains an open question, which our new technique might help answer.

While we know that recomputation may reduce the I/O complexity of some CDAGs, we are far from a characterization of those CDAGs for which recomputation is effective. This broad goal remains a challenge for any attempt toward a general theory of the communication requirements of computations.

References

1. Patterson, C.A., Snir, M., Graham, S.L.: Getting Up to Speed:: The Future of Supercomputing. National Academies Press (2005)
2. Bilardi, G., Preparata, F.P.: Horizons of parallel computation. Journal of Parallel and Distributed Computing **27**(2) (1995) 172–182
3. Strassen, V.: Gaussian elimination is not optimal. Numerische Mathematik **13**(4) (1969) 354–356
4. Le Gall, F.: Powers of tensors and fast matrix multiplication. In: Proc. ACM ISSAC, ACM (2014) 296–303
5. Hong, J., Kung, H.: I/o complexity: The red-blue pebble game. In: Proc. ACM STOC, ACM (1981) 326–333
6. Cannon, L.E.: A cellular computer to implement the Kalman filter algorithm. Technical report, DTIC Document (1969)
7. Ballard, G., Demmel, J., Holtz, O., Lipshitz, B., Schwartz, O.: Brief announcement: strong scaling of matrix multiplication algorithms and memory-independent communication lower bounds. In: Proc. ACM SPAA, ACM (2012) 77–79
8. Irony, D., Toledo, S., Tiskin, A.: Communication lower bounds for distributed-memory matrix multiplication. Journal of Parallel and Distributed Computing **64**(9) (2004) 1017–1026
9. Scquizzato, M., Silvestri, F.: Communication lower bounds for distributed-memory computations. arXiv preprint arXiv:1307.1805 (2013)
10. Pagh, R., Stöckel, M.: The input/output complexity of sparse matrix multiplication. In: Proc. ESA, Springer (2014) 750–761
11. Ballard, G., Demmel, J., Holtz, O., Schwartz, O.: Minimizing communication in numerical linear algebra. SIAM Journal on Matrix Analysis and Applications **32**(3) (2011) 866–901

12. Ballard, G., Demmel, J., Holtz, O., Schwartz, O.: Communication-optimal parallel and sequential Cholesky decomposition. SIAM Journal on Scientific Computing **32**(6) (2010) 3495–3523
13. Loomis, L.H., Whitney, H.: An inequality related to the isoperimetric inequality. Bull. Amer. Math. Soc. **55**(10) (10 1949) 961–962
14. V. A. Zalgaller, A. B. Sossinsky, Y.D.B. The American Mathematical Monthly **96**(6) (1989) 544–546
15. Ballard, G., Demmel, J., Holtz, O., Schwartz, O.: Graph expansion and communication costs of fast matrix multiplication. JACM **59**(6) (2012) 32
16. Ballard, G., Demmel, J., Holtz, O., Lipshitz, B., Schwartz, O.: Graph expansion analysis for communication costs of fast rectangular matrix multiplication. In: Design and Analysis of Algorithms. Springer (2012) 13–36
17. Scott, J., Holtz, O., Schwartz, O.: Matrix multiplication I/O complexity by Path Routing. In: Proc. ACM SPAA. (2015) 35–45
18. De Stefani, L.: On space constrained computations. PhD thesis, University of Padova (2016)
19. Bilardi, G., Preparata, F.: Processor-time trade offs under bounded speed message propagation. Lower Bounds. Theory of Computing Systems **32**(5) (1999) 531–559
20. Ballard, G., Demmel, J., H., O., Lipshitz, B., Schwartz, O.: Communication-optimal parallel algorithm for Strassen's matrix multiplication. In: Proc. ACM SPAA. (2012) 193–204
21. Jacob, R., Stóckel, M.: Fast output-sensitive matrix multiplication. In: Proc. ESA. Springer (2015) 766–778
22. Savage, J.E.: Extending the Hong-Kung model to memory hierarchies. In: Computing and Combinatorics. Springer (1995) 270–281
23. Bilardi, G., Peserico, E.: A characterization of temporal locality and its portability across memory hierarchies. In: Automata, Languages and Programming. Springer (2001) 128–139
24. Koch, R.R., Leighton, F.T., Maggs, B.M., Rao, S.B., Rosenberg, A.L., Schwabe, E.J.: Work-preserving emulations of fixed-connection networks. JACM **44**(1) (1997) 104–147
25. Bhatt, S.N., Bilardi, G., Pucci, G.: Area-time tradeoffs for universal VLSI circuits. Theoret. Comput. Sci. **408**(2-3) (2008) 143–150
26. Bilardi, G., Pietracaprina, A., D'Alberto, P.: On the space and access complexity of computation DAGs. In: Graph-Theoretic Concepts in Computer Science, Springer (2000) 47–58
27. Grigor'ev, D.Y.: Application of separability and independence notions for proving lower bounds of circuit complexity. Zapiski Nauchnykh Seminarov POMI **60** (1976) 38–48
28. Savage, J.E.: Models of Computation: Exploring the Power of Computing. 1st edn. Addison-Wesley Longman Publishing Co., Inc., Boston, MA, USA (1997)
29. Bilardi, G., Stefani, L.D.: The i/o complexity of strassen's matrix multiplication with recomputation. arXiv preprint arXiv:1605.02224 (2016)
30. Ranjan, D., Savage, J.E., Zubair, M.: Upper and lower I/O bounds for pebbling r-pyramids. Journal of Discrete Algorithms **14** (2012) 2–12
31. Aggarwal, A., Vitter, Jeffrey, S.: The input/output complexity of sorting and related problems. Commun. ACM **31**(9) (September 1988) 1116–1127
32. Le Gall, F.: Faster algorithms for rectangular matrix multiplication. In: Proc. IEEE FOCS, IEEE (2012) 514–523
33. Thompson, C.: Area-time complexity for VLSI. In: Proc. ACM STOC, ACM (1979) 81–88

Maximum Plane Trees in Multipartite
Geometric Graphs

Ahmad Biniaz[1], Prosenjit Bose[1], Kimberly Crosbie[1], Jean-Lou De Carufel[2],
David Eppstein[3], Anil Maheshwari[1], and Michiel Smid[1]

Abstract. A geometric graph is a graph whose vertices are points in the
plane and whose edges are straight-line segments between the points. A
plane spanning tree in a geometric graph is a spanning tree that is non-
crossing. Let R and B be two disjoint sets of points in the plane where
the points of R are colored red and the points of B are colored blue, and
let $n = |R \cup B|$. A bichromatic plane spanning tree is a plane spanning
tree in the complete bipartite geometric graph with bipartition (R, B).
In this paper we consider the maximum bichromatic plane spanning tree
problem, which is the problem of computing a bichromatic plane spanning tree of maximum total edge length.

1. For the maximum bichromatic plane spanning tree problem, we
   present an approximation algorithm with ratio $1/4$ that runs in
   $O(n \log n)$ time.
2. We also consider the multicolored version of this problem where the
   input points are colored with $k > 2$ colors. We present an approximation algorithm that computes a plane spanning tree in a complete
   k-partite geometric graph, and whose ratio is $1/6$ if $k = 3$, and $1/8$
   if $k \geqslant 4$.
3. We also revisit the special case of the problem where $k = n$, i.e., the
   problem of computing a maximum plane spanning tree in a complete
   geometric graph. For this problem, we present an approximation
   algorithm with ratio 0.503; this is an extension of the algorithm
   presented by Dumitrescu and Tóth (2010) whose ratio is 0.502.

1 Introduction

Let P be a set of n points in the plane in general position, i.e., no three points
are collinear. Let $K(P)$ be the *complete geometric graph* with vertex set P. It
is well known that the standard minimum spanning tree (MinST) problem in
$K(P)$ can be solved in $\Theta(n \log n)$ time. Also, any minimum spanning tree in
$K(P)$ is *plane*, i.e., its edges do not cross each other. The *maximum spanning
tree* (MaxST) problem is the problem of computing a spanning tree in $K(P)$
whose total edge length is maximum. Monma *et al.* [5] showed that this problem

---

[1] Carleton University, Ottawa. Supported by NSERC. ahmad.biniaz@gmail.com, kim-
berlycrosbie@cmail.carleton.ca, {jit, anil, michiel}@scs.carleton.ca
[2] University of Ottawa, Ottawa. Supported by NSERC. jdecaruf@uottawa.ca
[3] University of California, Irvine. Supported by NSF grant CCF-1228639. epp-
stein@ics.uci.edu

© Springer International Publishing AG 2017
F. Ellen et al. (Eds.): WADS 2017, LNCS 10389, pp. 193–204, 2017.
DOI: 10.1007/978-3-319-62127-2_17

can be solved in $\Theta(n \log n)$ time. However, a MaxST is not necessarily plane. Alon *et al.* [1] started the *maximum plane spanning tree* (MaxPST) problem, which is the problem of computing a plane spanning tree in $K(P)$ whose total edge length is maximum. It is not known whether or not this problem is NP-hard. They presented an approximation algorithm with ratio 0.5 for this problem. This approximation ratio was improved to 0.502 by Dumitrescu and Tóth [4].

Let R and B be two disjoint sets of points in the plane such that $R \cup B$ is in general position, and let $n = |R \cup B|$. Suppose that the points of R are colored red and the points of B are colored blue. Let $K(R, B)$ be the *complete bipartite geometric graph* with bipartition (R, B). The *minimum bichromatic spanning tree* (MinBST) problem is to compute a minimum spanning tree in $K(R, B)$. The *maximum bichromatic spanning tree* (MaxBST) problem is to compute a spanning tree in $K(R, B)$ whose total edge length is maximum. Recently, Biniaz *et al.* [2] showed that both the MinBST and the MaxBST problems can be solved in $\Theta(n \log n)$ time. We note that none of MinBST and MaxBST is necessarily plane; they might have crossing edges. Borgelt *et al.* [3] studied the problem of computing a minimum bichromatic plane spanning tree, which we refer to as the MinBPST problem. They showed that this problem is NP-hard, and also presented a polynomial-time approximation algorithm with approximation ratio of $O(\sqrt{n})$. In this paper we study the problem of computing a maximum bichromatic plane spanning tree, which we refer to as the MaxBPST problem. See Figure 1.

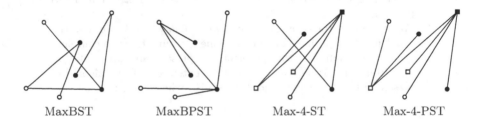

MaxBST   MaxBPST   Max-4-ST   Max-4-PST

Fig. 1. Maximum spanning trees.

A natural extension of the MinBST and the MaxBST problems is to have more than two colors. In this multicolored version, the input points are colored by $k > 2$ colors, and we are looking for a minimum (resp. maximum) spanning tree in which the two endpoints of every edge have distinct colors. In other words, we look for a minimum (resp. maximum) spanning tree in a complete k-partite geometric graph. We refer to these problems as the Min-k-ST and the Max-k-ST problems, respectively. Biniaz *et al.* [2] showed that both these problems can be solved in $O(n \log n \log k)$ time. Notice that the MinST and the MaxST problems are special cases of the Min-k-ST and the Max-k-ST problems in which every input point has a unique color, i.e., $k = n$. In this paper we also study the problem of computing a plane Max-k-ST, which we refer to as the Max-k-PST problem. See Figure 1.

1.1 Our contributions

In this paper we study the maximum plane spanning tree problems. In Section 3 we present an approximation algorithm with ratio 1/4 for the MaxBPST problem. We study the Max-k-PST problem in Section 4. For this problem, we present an approximation algorithm whose ratio is 1/6 if $k = 3$, and 1/8 if $k \geqslant 4$. In Section 5 we consider the MaxPST problem, where we modify the algorithm presented by Dumitrescu and Tóth [4] for this problem; this modification improves the approximation ratio to 0.503. All the presented approximation algorithms run in $O(n \log n)$ time, where n is the number of input points.

2 Preliminaries

For any two points p and q in the plane, we refer to the line segment between p and q as pq, and to the Euclidean distance between p and q as $|pq|$. The *lune* between p and q, which we denote by lune(p, q), is the intersection of the two disks of radius $|pq|$ that are centered at p and q.

For a point set P, the *diameter* of P is the maximum Euclidean distance between any two points of P. A pair of points that realizes the diameter of P is referred to as a *diametral pair*.

Let G be a geometric graph with colored vertices. We denote by $L(G)$ the total Euclidean length of the edges of G. A *star* is a tree with one internal node, which we refer to as the *center* of the star. For a color c, a c-*star* in G is a star whose center is colored c and the colors of its leaves are different from c.

3 The MaxBPST problem

In this section we consider the MaxBPST problem. Recall that in this problem we are given two sets R and B of red and blue points in the plane, respectively, and we are looking for a maximum plane spanning tree in $K(R, B)$. Let $n = |R \cup B|$. We present an approximation algorithm with ratio 1/4 for this problem that runs in $O(n \log n)$ time.

We will show that the length of the longest star in $K(R, B)$ is at least 1/4 times the length of an optimal MaxBPST. In fact, we present an algorithm that returns such a star. Moreover, we show that this estimate is the best possible for the length of a longest star. The longest star can easily be augmented to form a plane spanning tree as follows. The longest star has exactly one point of one color as its center, and all points of other color as its leaves. The edges of this star can be extended to partition the plane into convex cones, possibly except one cone; we split this cone into two convex cones by adding its bisector. Then, we connect all the remaining points in each cone to one of the leaves that is on the boundary of that cone.

If $|R| = 1$ or $|B| = 1$, then the problem is trivial. Assume $|R| \geqslant 2$ and $|B| \geqslant 2$. Our algorithm first computes a diametral pair (a, b) in R and a diametral pair (p, q) in B. Then, returns the longest star S_x in $K(R, B)$ that is centered at a point $x \in \{a, b, p, q\}$. Since diametral pairs of R and B can be computed in $O(n \log n)$ time, the running time of the algorithm follows. In the rest of this section we will show that the longest of these stars satisfies the approximation ratio.

Let T^* be an optimal MaxBPST and let L^* denote the length of T^*. We make an arbitrary point the root of T^* and partition the edges of T^* into two sets as follows. Let E_R^* be the set of edges (u, v) in T^* where u is a red point and u is the parent of v. Let E_B^* be the set of edges (u, v) where u is a blue point and u is the parent of v. The edges of E_R^* (resp. E_B^*) form a forest in which each component is a red-star (resp. blue-star). Let L_R^* and L_B^* denote the total lengths of the edges of E_R^* and E_B^*, respectively. Without loss of generality assume that $L_R^* \leqslant L_B^*$. Then,

$$L^* = L_B^* + L_R^* \leqslant 2L_B^*. \tag{1}$$

We will show that, in this case, the longest of S_p and S_q is a desired tree. To that end, let F_B be the set of edges that is obtained by connecting every red point to its farthest blue point. Notice that the edges of F_B form a forest in which every component is a blue-star. Moreover, observe that

$$L_B^* \leqslant L(F_B). \tag{2}$$

Lemma 1. $L(F_B) \leqslant \frac{2}{\sqrt{3}} \cdot (L(S_p) + L(S_q))$.

Proof. Let C_p and C_q be the two disks of radius $|pq|$ that are centered at p and q, respectively. Since (p, q) is a diameter of B, all blue points lie in lune(p, q); see Figure 2(a).

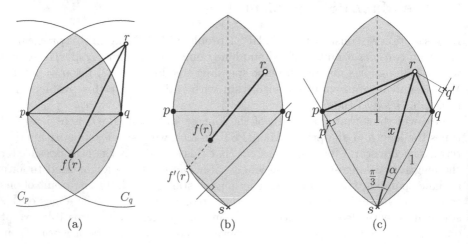

Fig. 2. Illustration of Lemmas 1 and 2: r is a red point, and p, q, $f(r)$ are blue points.

For any red point $r \in R$, let $f(r)$ denote its neighbor in F_B. Recall that $f(r)$ is the farthest blue point to r, and note that $f(r)$ is in lune(p, q). See Figure 2(a).

We are going to show that $|rf(r)| \leqslant \frac{2}{\sqrt{3}}(|rp| + |rq|)$. Depending on whether or not $r \in \text{lune}(p, q)$ we consider the following two cases.

- $r \notin \text{lune}(p, q)$. Thus, we have that $r \notin C_p$ or $r \notin C_q$. Without loss of generality assume $r \notin C_p$; see Figure 2(a). By the triangle inequality we have $|rf(r)| \leqslant |rq| + |qf(r)|$. Since pq is a diameter of B, we have $|qf(r)| \leqslant |pq|$. In addition, since $r \notin C_p$, we have $|pq| \leqslant |rp|$. By combining these inequalities, we get

$$|rf(r)| \leqslant |rq| + |qf(r)| \leqslant |rq| + |pq| \leqslant |rq| + |rp|.$$

- $r \in \text{lune}(p, q)$. Without loss of generality assume that pq has unit length, pq is horizontal, r is above pq, and r is closer to q than to p. If $f(r)$ is on or above pq, then $|rf(r)|$ is smaller than $|pq|$, and hence smaller than $|rp| + |rq|$. Assume $f(r)$ is below pq. Let s be the intersection point of the boundaries of C_p and C_q that is below pq as in Figure 2(b).

Claim 1. $|rf(r)| \leqslant |rs|$.

Proof of Claim 1. Let $f'(r)$ be the intersection point of the ray that is emanating from r and passing through $f(r)$ with the boundary of $\text{lune}(p, q)$. Note that $|rf(r)| \leqslant |rf'(r)|$. If $f'(r)$ is on the boundary of C_q, then the perpendicular bisector of the segment $sf'(r)$ passes through q. In this case r is in the same side of this perpendicular as $f'(r)$, and thus, $|rf'(r)| \leqslant |rs|$; see Figure 2(b). Similarly, if $f'(r)$ is on the boundary of C_p, then both r and $f'(r)$ are on a same side of the perpendicular bisector of $sf'(r)$ which passes through p. This proves the claim.

Extend the line segment sp from the endpoint p. Let p' be the point on the extended line that is closest to r. Define q' similarly. Note that $|rp'| \leqslant |rp|$ and $|rq'| \leqslant |rq|$. Based on this and Claim 1, in order to show that $|rf(r)| \leqslant \frac{2}{\sqrt{3}}(|rp| + |rq|)$, it suffices to show that $|rs| \leqslant \frac{2}{\sqrt{3}}(|rp'| + |rq'|)$. Let $\alpha = \angle rsq$, and note that $0 \leqslant \alpha \leqslant \frac{\pi}{6}$; see Figure 2(c). Since the triangles $\triangle rsp'$ and $\triangle rsq'$ are right-angled and $\triangle spq$ is equilateral, we have $|rq'| = |rs| \cdot \sin \alpha$ and $|rp'| = |rs| \cdot \sin(\pi/3 - \alpha)$. Thus,

$$|rq'| + |rp'| = |rs| \cdot (\sin \alpha + \sin(\pi/3 - \alpha)) \geqslant \frac{\sqrt{3}}{2}|rs|,$$

where the inequality is valid because $\sin \alpha + \sin(\pi/3 - \alpha)$ is at least $\sqrt{3}/2$ for all $0 \leqslant \alpha \leqslant \frac{\pi}{6}$. This implies that $|rs| \leqslant \frac{2}{\sqrt{3}}(|rp'| + |rq'|)$.

Since in both previous cases $|rf(r)| \leqslant \frac{2}{\sqrt{3}}(|rp| + |rq|)$, we have

$$L(F_B) = \sum_{r \in R} |rf(r)| \leqslant \sum_{r \in R} \frac{2}{\sqrt{3}}(|rp| + |rq|) = \frac{2}{\sqrt{3}}(L(S_p) + L(S_q)).$$

$\square$

Combining Inequalities (1), (2), and Lemma 1, we get

$$L^* \leqslant 2L_B^* \leqslant 2L(F_B) \leqslant \frac{4}{\sqrt{3}} \cdot (L(S_p) + L(S_q)).$$

Therefore, the length of the longest of S_p and S_q is at least $\frac{\sqrt{3}}{8} \approx 0.215$ times L^*. In the following lemma we improve the bound of Lemma 1 by proving that $L(F_B) \leqslant L(S_p) + L(S_q)$; this improves the approximation ratio to $1/4$. However, the proof of this lemma is algebraic.

Lemma 2. $L(F_B) \leqslant L(S_p) + L(S_q)$.

Proof. For any red point $r \in R$, let $f(r)$ denote its neighbor in F_B. In order to prove this lemma, as we have seen in the proof of Lemma 1, it suffices to show that $|rf(r)| \leqslant |rp| + |rq|$. Define lune$(p,q)$ as in the proof of Lemma 1. As we have seen there, if $r \notin$ lune(p,q), then $|rf(r)| \leqslant |rp| + |rq|$. Assume $r \in$ lune(p,q). Without loss of generality assume that pq has unit length, pq is horizontal, r is above pq, and r is closer to q than to p; see Figure 2(c). Define the point s as in the proof of Lemma 1. By Claim 1 in the proof of Lemma 1, in order to show that $|rf(r)| \leqslant |rp| + |rq|$, it suffices to show that $|rs| \leqslant |rp| + |rq|$. Let $x = |rs|$ and $\alpha = \angle rsq$; see Figure 2(c). Note that $\sqrt{3}/2 \leqslant x \leqslant \sqrt{3}$ and $0 \leqslant \alpha \leqslant \frac{\pi}{6}$. By the cosine rule we have

$$|rp| = \sqrt{1 + x^2 - 2x\cos(\pi/3 - \alpha)} \quad \text{and} \quad |rq| = \sqrt{1 + x^2 - 2x\cos\alpha}.$$

Define

$$f(x,\alpha) = \sqrt{1 + x^2 - 2x\cos(\pi/3 - \alpha)} + \sqrt{1 + x^2 - 2x\cos\alpha} - x. \quad (3)$$

Then, $|rp| + |rq| - |rs| = f(x,\alpha)$. In the full version of the paper we show that $f(x,\alpha) \geqslant 0$ for all $\sqrt{3}/2 \leqslant x \leqslant \sqrt{3}$ and $0 \leqslant \alpha \leqslant \frac{\pi}{6}$. This implies that $|rs| \leqslant |rp| + |rq|$. $\qquad\square$

To this end, we have proved the following theorem:

Theorem 1. *Let R and B be two disjoint sets of points in the plane such that $R \cup B$ is in general position, and let $n = |R \cup B|$. One can compute, in $O(n \log n)$ time, a plane spanning tree in $K(R,B)$ whose length is at least $1/4$ times the length of a maximum plane spanning tree.*

3.1 A matching upper bound

In this section we show that the above estimate is best possible for the length of the longest star in $K(R,B)$. Consider a set R of $n/2$ red points and a set B of $n/2$ blue points that are equally distributed in two circles of arbitrary very small radius with their centers at distance 1; see Figure 3. The bichromatic plane spanning tree/path that is shown in this figure, has $n-2$ edges of unit length and one small edge. Any star in $K(R,B)$ has $n/4$ edges of unit length, plus $n/4$ edges of very small length. Thus, the length of the longest star, in this example, is about $1/4$ times the length of an optimal MaxBPST (at the limit).

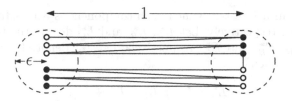

Fig. 3. Illustration of the upper bound.

4 The Max-k-PST problem

In the multicolored version of the maximum plane spanning tree problem, the input points are colored by more than two colors, and we want the two endpoints of every edge in the tree to have distinct colors. Formally, we are given a set P of n points in the plane in general position that is partitioned into subsets $P_1, \ldots, P_k$, with $k \geqslant 3$. For each $c \in \{1, \ldots, k\}$, assume the points of P_c are colored c. Let $K(P_1, \ldots, P_k)$ be the complete multipartite geometric graph on P, which has edges between every point of each set in the partition to all points of the other sets. The Max-k-PST problem is the problem of computing a maximum plane spanning tree in $K(P_1, \ldots, P_k)$. The standard MaxPST problem can be interpreted as an instance of this multicolored version in which $k = n$, i.e., each point has a unique color. In this section, we present an approximation algorithm, for the Max-k-PST problem, whose ratio is $1/6$ if $k = 3$, and $1/8$ if $k \geqslant 4$.

We will show that the length of the longest star in $K(P_1, \ldots, P_k)$ is at least $1/8$ (resp. $1/6$) times the length of an optimal Max-k-PST if $k \geqslant 4$ (resp. $k = 3$). In fact, we present an $O(n \log n)$-time algorithm that returns such a star. The algorithm is as follows. Compute a *bichromatic diameter* (p, q) of P, i.e., two points of different colors that have the maximum distance. It can easily be verified, by contradiction, that the MaxST in $K(P)$, which can be computed in $O(n \log n)$ time [5], contains a bichromatic diameter of P. Notice that the length of any edge in $K(P_1, \ldots, P_k)$ is at most $|pq|$. Without loss of generality assume that $p \in P_i$ and $q \in P_j$. Notice that all points of $P \setminus (P_i \cup P_j)$ lie in lune(p, q), because, otherwise (p, q) cannot be a bichromatic diameter of P. To simplify the notation, we write R, B, and G, for P_i, P_j, and $P \setminus (P_i \cup P_j)$, respectively. Moreover, we assume that the points of R, B, and G are colored red, blue, and green, respectively. Compute a diametral pair (r, r') in R, a diametral pair (b, b') in B, and a diametral pair (g, g') in G. Return the longest star S_x in $K(P_1, \ldots, P_k)$ that is centered at a point $x \in \{p, q, r, r', b, b', g, g'\}$. We show that the length of S_x is at least $1/8$ times the length of an optimal tree.

Let T^* be an optimal Max-k-PST, and let L^* denote the length of T^*. We make an arbitrary point the root of T^* and partition the edges of T^* into four sets as follows. Let E_R^* be the set of edges (u, v) in T^* where u is a red point and u is the parent of v. Let E_B^* be the set of edges (u, v) where u is a blue point and u is the parent of v. Let E_G^* be the set of edges (u, v) where u is a green point, v is not a green point, and u is the parent of v. Let E^* be the set of edges (u, v) where both u and v are green points. The edges of each of E_R^*,

E_B^*, and E_G^* form a forest in which each component is a red-star, a blue-star, and a green-star, respectively. Let L_R^*, L_B^*, and L_G^* denote the total lengths of the edges of E_R^*, E_B^*, and E_G^*, respectively. Let L_E^* denote the total length of the edges in E^*. Then,

$$L^* = L_R^* + L_B^* + L_G^* + L_E^*.$$

We consider the following two cases depending on where or not L_E^* is larger than $\max\{L_R^*, L_B^*, L_G^*\}$.

- $L_E^* \geqslant \max\{L_R^*, L_B^*, L_G^*\}$. In this case $L_E^* \geqslant \frac{1}{4}L^*$. The number of edges in E^* is at most $n(G) - 1$, where $n(G)$ is the number of points in G. Recall that the length of every edge in E^* is at most $|pq|$. Thus $L_E^* \leqslant (n(G) - 1) \cdot |pq|$. Each of the stars S_p and S_q has an edge to every point of G. Thus $L(S_p) + L(S_q) \geqslant n(G) \cdot |pq|$. Therefore,

$$\frac{1}{4}L^* \leqslant L_E^* \leqslant (n(G) - 1) \cdot |pq| < n(G) \cdot |pq| \leqslant L(S_p) + L(S_q),$$

  which implies that the longest of S_p and S_q has length at least $\frac{1}{8}L^*$.
- $L_E^* < \max\{L_R^*, L_B^*, L_G^*\}$. Without loss of generality assume that $L_B^* = \max\{L_R^*, L_B^*, L_G^*\}$. Thus, $L_B^* \geqslant \frac{1}{4}L^*$. Let F_B be the set of edges that is obtained by connecting every point of $R \cup G$ to its farthest blue point. Notice that the edges of F_B form a forest in which every component is a blue-star. Observe that $L_B^* \leqslant L(F_B)$. Moreover, by Lemma 2 we have $L(F_B) \leqslant L(S_b) + L(S_{b'})$. Therefore,

$$\frac{1}{4}L^* \leqslant L_B^* \leqslant L(F_B) \leqslant L(S_b) + L(S_{b'}),$$

  which implies that the longest of S_b and $S_{b'}$ has length at least $\frac{1}{8}L^*$.

The point set $\{p, q, r, r', b, b', g, g'\}$ can be computed in $O(n \log n)$ time, and thus, the running time of the algorithm follows.

Note 1. If $k = 3$, then E^* is empty, and thus, the longest star S_x with $x \in \{r, r', b, b', g, g'\}$ has length at least $\frac{1}{6}L^*$.

Note 2. If the diameter pair of P is monochromatic, then we get the ratio of $1/6$. Assume both points of a diametral pair (p, q) of P belong to P_i. Let $R = P_i$ and $G = R \setminus P_i$. Then, B is empty and $L^* = L_R^* + L_G^* + L_E^*$. Moreover, we have $r = p$ and $r' = q$. If $L_G^* \geqslant \frac{1}{3}L^*$, then the longest of S_g and $S_{g'}$ has length at least $\frac{1}{6}L^*$. If $L_G^* < \frac{1}{3}L^*$, then $L_R^* \leqslant L(S_r) + L(S_{r'})$ and $L_E^* \leqslant L(S_r) + L(S_{r'})$. Thus, the longest of S_r and $S_{r'}$ has length at least $\frac{1}{6}L^*$.

Theorem 2. *Let P be a set of n points in the plane in general position that is partitioned into subsets $P_1, \ldots, P_k$, with $k \geqslant 3$. One can compute, in $O(n \log n)$ time, a plane spanning tree in $K(P_1, \ldots, P_k)$ whose length is at least $1/8$ (resp. $1/6$) times the length of a maximum plane spanning tree if $k \geqslant 4$ (resp. $k = 3$).*

5 The MaxPST problem

In this section we study the MaxPST problem, a special case of the Max-k-PST problem where every input point has a unique color. Formally, given a set P of points in the plane in general position and we want to compute a maximum plane spanning tree in $K(P)$. We revisit this problem which was first studied by Alon, Rajagopalan and Suri (1993), and then by Dumitrescu and Tóth (2010). Alon *et al.* [1] presented an approximation algorithm with ratio 1/2 for this problem. In fact they proved that the length of the longest star in $K(P)$ is at least 0.5 times the length of an optimal tree; this bound is tight for a longest star. Dumitrescu and Tóth [4] improved the approximation ratio to 0.502. They proved that the length of the longest double-star in $K(P)$ (a double-star is a tree with two internal nodes) is at least 0.502 times the length of an optimal solution. They left as an open problem a more precise analysis of the approximation ratio of their algorithm. In this section we modify the algorithm of Dumitrescu and Tóth [4], and slightly improve the approximation ratio to 0.503. We will describe their algorithm briefly, and provide detail on the parts that we will modify. The algorithm outputs the longest of five plane trees S_a, S_b, S_h, E_a, E_b, that are describe below.

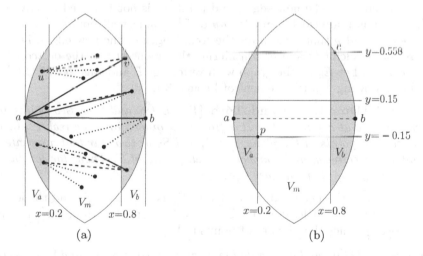

(a)   (b)

Fig. 4. (a) The plane tree E_a. The solid, dashed, and dotted edges belong to E_a^1, E_a^2, and E_a^3 respectively. (b) The distance between p and c is less than 0.948.

Let $n = |P|$, and let L^* denote the length of an optimal MaxPST in $K(P)$. Compute a diametral pair (a, b) of P. Without loss of generality assume that ab is a horizontal, $|ab| = 1$, $a = (0,0)$, and $b = (1,0)$. Since (a, b) is a diametral pair, the length of any edge in $K(P)$ is at most 1. Thus, $L^* \leqslant n - 1$. Moreover, all points of P are in lune(a, b). See Figure 4. Let $h = (x_h, y_h)$ be a point in P with the largest value of $|y|$ (absolute value of the y-coordinate). Without loss of generality assume that $y_h \geqslant 0$. Define S_a, S_b, and S_h as three spanning stars that are centered at a, b, and h respectively. We compute plane trees E_a and E_b as follows; this computation is different from the one that is presented in [4].

Set $w = 0.6$. Let V_a be the vertical strip between the lines $x = 0$ and $x = \frac{1-w}{2} = 0.2$, V_m be the vertical strip between the lines $x = 0.2$ and $x = \frac{1+w}{2} = 0.8$, and V_b be the vertical strip between the lines $x = 0.8$ and $x = 1$. Note that $a \in V_a$ and $b \in V_b$. See Figure 4(a). We describe how to construct E_a; the construction of E_b is analogous. Connect a to each point in V_b and let E_a^1 denotes the set of edges of the resulting star (the solid edges in Figure 4(a)). The edge ab has length 1 and every other edge has length at least $\frac{1+w}{2}$. The edges of E_a^1 partition V_a into convex regions. Each of these regions is bounded by at least one edge of E_a^1 from above or below. Take any region $\mathcal{A}$ of the partition of V_a. Let av be an edge of E_a^1 that bounds $\mathcal{A}$ either from above or blow. Note that $v \in V_b$. Connect all points that are in $\mathcal{A}$ (excluding a) to v. Let E_a^2 be the set of all such edges after considering all regions (the dashed edges in Figure 4(a)). Since any edge in E_a^2 connects a point in V_a to a point in V_b, each edge in E_a^2 has length at least w. The edges of $E_a^1 \cup E_a^2$ partition V_m into convex regions. Each of these regions is bounded by at least one edge of $E_a^1 \cup E_a^2$. Take any region $\mathcal{M}$ of the partition of V_m. Consider the following two cases: (a) If $\mathcal{M}$ is bounded by an edge uv of E_a^2, then connect all points that are in $\mathcal{M}$ to one of u and v that maximizes the total length of the new edges, and (b) if $\mathcal{M}$ is not bounded by any edge of E_a^2, then it is bounded by an edge av of E_a^1, connect all points that are in $\mathcal{M}$ to one of a and v that maximizes the total length of the new edges. Let E_a^3 be the set of all such edges after considering all regions of V_m (the dotted edges in Figure 4(a)). Let E_a be the graph with vertex set P and edge set $E_a^1 \cup E_b^2 \cup E_a^3$.

The following is a restatement of Lemma 3 in [4].

Lemma 3 (Dumitrescu and Tóth [4]). *Let a and b be two points in the plane that are at distance at least α from each other, for some real $\alpha > 0$. Let S be a set of m points in the plane. Let S_a and S_b be two star that are centered at a and b, respectively, and connected to all points of S. Then, the length of the longest of S_a and S_b is at least $\frac{\alpha \cdot m}{2}$.*

Recall that the length of each edge in E_a^1 is at least $\frac{1+w}{2}$ and the length of each edge in E_b^2 is at least w. By Lemma 3 the total length of the edges in E_a^3 is at least $\frac{w}{2}$ times the number of points in V_m.

Lemma 4. *Let n_a and n_b denote the number of points in V_a and V_b respectively. Then $L(E_a) \geqslant \frac{n_b}{2} + \frac{w}{2}(n + n_a) + \frac{1-3w}{2}$, $L(E_b) \geqslant \frac{n_a}{2} + \frac{w}{2}(n + n_b) + \frac{1-3w}{2}$, and consequently $L(E_a) + L(E_b) \geqslant \frac{n_a + n_b}{2} + \frac{w}{2}(2n + n_a + n_b) + 1 - 3w$.*

Proof. Since E_a^1, E_a^2, and E_a^3 are pairwise disjoint, we have $L(E_a) = L(E_a^1) + L(E_a^2) + L(E_a^3)$. E_a^1 contains n_b edges (including ab) of length at least $\frac{1+w}{2}$ with the length of ab is 1. Thus, $L(E_a^1) \geqslant \frac{1+w}{2}(n_b - 1) + 1$. E_a^2 has $n_a - 1$ edges of length at least w. Thus, $L(E_a^2) \geqslant w(n_a - 1)$. V_m contains $n - n_a - n_b$ points, and thus, the length of E_a^3 is at least $\frac{w}{2}(n - n_a - n_b)$. Therefore,

$$L(E_a) \geqslant \frac{1+w}{2}(n_b - 1) + 1 + w(n_a - 1) + \frac{w}{2}(n - n_a - n_b) = \frac{n_b}{2} + \frac{w}{2}(n + n_a) + \frac{1-3w}{2}.$$

The estimation of $L(E_b)$ is analogous. $\qquad\qquad\qquad\qquad\qquad\square$

The following lemma summarizes Lemmas 3, 4, 5, and 7 in [4]. Let $P = \{p_1, \ldots, p_n\}$, where $p_i = (x_i, y_i)$. Let $d_{max}(p_i)$ denote the maximum distance from p_i to other points in P.

Lemma 5 (Dumitrescu and Tóth [4]). *Let δ and t be two constants where $0 \leqslant \delta \leqslant t \leqslant 1$. Then*

1. *$L(S_a) + L(S_b) \geqslant n$.*
2. *$L^* \leqslant \sum_{i=1}^{n} d_{max}(p_i)$.*
3. *if $\sum_{i=1}^{n} |y_i| \geqslant \delta n$, then $L(S_a) + L(S_b) \geqslant 2n\sqrt{\frac{1}{4} + \delta^2}$.*
4. *if $\sum_{i=1}^{n} |y_i| \leqslant \delta n$ and $y_h \geqslant t$, then $L(S_h) \geqslant (t - \delta)n$.*

Set $\delta = 0.055$, $t = 0.558$, and $z = 0.49$; these constants are different from the ones that are chosen in [4].

Lemma 6. *Assume that $|y_h| \leqslant t$. Let $p_i = (x_i, y_i)$ be a point of P in V_m with $|y_i| \leqslant 0.15$. Then $d_{max}(p_i) \leqslant 0.948$.*

Proof. The maximum distance is attained when $p_i = (0.2, -0.15)$ or $p_i = (0.8, -0.15)$. Because of symmetry we assume that $p_i = (0.2, -0.15)$. Let $c = (x_c, y_c)$ be the rightmost intersection point of the line $y = t$ and lune(a, b). See Figure 4(b). The furthest point from p_i in the allowed region is c. Note that $y_c = t$ and $x_c = \sqrt{1 - t^2} < 0.83$. Thus, $x_c - x_i < 0.63$. Therefore $d_{max}(p_i) \leqslant |p_i c| < \sqrt{0.63^2 + (0.15 + t)^2} < 0.948$. $\qquad\square$

To prove the approximation ratio, we show that the length of the longest of S_a, S_b, S_h, E_a, E_b is at least 0.503 times L^*. In order to do that, we recall the following four cases that are considered in [4].

1. If $\sum_{i=1}^{n} |y_i| \geqslant \delta n$, then the output of the algorithm is not shorter than the longest of S_a and S_b. By Lemma 5, the approximation ratio is at least

$$\frac{L(S_a) + L(S_b)}{2L^*} \geqslant \sqrt{\frac{1}{4} + \delta^2} \geqslant 0.503.$$

2. If $\sum_{i=1}^{n} |y_i| < \delta n$ and $y_h \geqslant t$, then the output of the algorithm is not shorter than S_h. By Lemma 5, the approximation ratio is at least

$$\frac{L(S_h)}{L^*} \geqslant t - \delta = 0.503.$$

3. If $\sum_{i=1}^{n} |y_i| < \delta n$, $y_h < t$, and $n_a + n_b \geqslant (1 - z)n$, then the output of the algorithm is not shorter than the longest of E_a and E_b. As a consequence of Lemma 4, the approximation ratio is at least

$$\frac{L(E_a) + L(E_b)}{2L^*} \geqslant \frac{n_a + n_b}{2 \cdot 2n} + \frac{w(2n + n_a + n_b)}{2 \cdot 2n} + \frac{1 - 3w}{2n}$$

$$\geqslant \frac{(1 - z)(1 + w)}{4} + \frac{w}{2} + \frac{1 - 3w}{2n} \geqslant 0.503,$$

where the last inequality is valid for all $n \geqslant 400$. Based on the above calculations and the details that are provided in the proof of Lemma 4, it turns out that ratio of 0.502 that is claimed in [4] is valid when $n \geqslant 100$.

4. If $\sum_{i=1}^{n} |y_i| < \delta n$, $y_h < t$, and $n_a + n_b < (1 - z)n$, then the output of the algorithm is not shorter than the longest of S_a and S_b. There are at least $zn = .49n$ points in V_m. At most $\frac{11}{30}$ points have $|y_i| \geqslant 0.15$ because otherwise we have $\sum_{i=1}^{n} |y_i| \geqslant 0.15 \cdot \frac{11n}{30} = 0.055n = \delta n$, which is a contradiction. Thus, at least $\frac{49n}{100} - \frac{11n}{30} = \frac{37n}{300}$ points in V_m have $|y_i| \leqslant 0.15$. By Lemma 5 and Lemma 6 we have

$$L^* \leqslant \frac{263n}{300} + 0.948 \cdot \frac{37n}{300} < 0.994n.$$

The approximation ratio is at least

$$\frac{L(S_a) + L(S_b)}{2L^*} \geqslant \frac{n}{2 \cdot 0.994n} > 0.503.$$

Theorem 3. *Let P be a set of n points in the plane in general position. One can compute, in $O(n \log n)$ time, a plane spanning tree in $K(P)$ whose length is at least 0.503 times the length of a maximum plane spanning tree.*

6 Concluding Remarks

In this paper we presented constant factor approximation algorithms for the problem of computing a maximum plane tree in a multipartite geometric graph. It is not known whether or not this problem is NP-hard. A natural open problem is to improve any of the presented approximation ratios. Specifically, when the number of sets in the partition is more than two, we conjecture that the length of the longest star is at least $1/3$ times the length of a maximum spanning tree.

References

1. N. Alon, S. Rajagopalan, and S. Suri. Long non-crossing configurations in the plane. *Fundamenta Informaticae*, 22(4):385–394, 1995. Also in *Proceedings of the 9th ACM Symposium on Computational Geometry* (SoCG), 257–263, 1993.
2. A. Biniaz, P. Bose, D. Eppstein, A. Maheshwari, P. Morin, and M. Smid. Spanning trees in multipartite geometric graphs. *CoRR*, abs/1611.01661, 2016. Also submitted to *Algorithmica*.
3. M. G. Borgelt, M. J. van Kreveld, M. Löffler, J. Luo, D. Merrick, R. I. Silveira, and M. Vahedi. Planar bichromatic minimum spanning trees. *Journal of Discrete Algorithms*, 7(4):469–478, 2009.
4. A. Dumitrescu and C. D. Tóth. Long non-crossing configurations in the plane. *Discrete & Computational Geometry*, 44(4):727–752, 2010. Also in *Proceedings of the 27th International Symposium on Theoretical Aspects of Computer Science* (STACS), 311–322, 2010.
5. C. L. Monma, M. Paterson, S. Suri, and F. F. Yao. Computing Euclidean maximum spanning trees. *Algorithmica*, 5(3):407–419, 1990.

Local Routing in Spanners Based on WSPDs [*]

Prosenjit Bose[1], Jean-Lou De Carufel[2], Vida Dujmović[2], and Frédérik Paradis[2]

[1] School of Computer Science, Carleton University
Ottawa, Canada
jit@scs.carleton.ca
[2] School of Electrical Engineering and Computer Science, University of Ottawa
Ottawa, Canada
{jdecaruf,vida.dujmovic,fpara058}@uottawa.ca

Abstract. The well-separated pair decomposition (WSPD) of the complete Euclidean graph defined on points in $\mathbb{R}^2$ (Callahan and Kosaraju [JACM, 42 (1): 67-90, 1995]) is a technique for partitioning the edges of the complete graph based on length into a linear number of sets. Among the many different applications of WSPDs, Callahan and Kosaraju proved that the sparse subgraph that results by selecting an arbitrary edge from each set (called WSPD-spanner) is a $1 + 8/(s-4)$-spanner, where $s > 4$ is the separation ratio used for partitioning the edges.

Although competitive local-routing strategies exist for various spanners such as Yao-graphs, Θ-graphs, and variants of Delaunay graphs, few local-routing strategies are known for any WSPD-spanner. Our main contribution is a local-routing algorithm with a near-optimal competitive routing ratio of $1 + O(1/s)$ on a WSPD-spanner. Specifically, we present a 2-local and a 1-local routing algorithm on a WSPD-spanner with competitive routing ratios of $1 + 6/(s-2) + 4/s$ and $1 + 6/(s-2) + 6/s + 4/(s^2 - 2s) + 8/s^2$, respectively.

1 Introduction

A fundamental problem in networking is the routing of a message from one vertex to another in a graph. Because network resources are limited, it is often desirable that routing algorithms use as little memory as possible. At one extreme in this direction are *local* routing algorithms in which the routing algorithm must choose the next vertex to forward a message to based solely on knowledge of the destination vertex, the current vertex and some information about all vertices directly connected to the current vertex. When local routing is not possible, it is still desirable that a routing algorithm use little memory.

In many settings it is natural to model a network as a *geometric graph*, that is, a graph whose vertices are points and each edge is a line-segment whose weight is the Euclidean distance between its two endpoints. For example, geometric routing algorithms are important in wireless sensor networks (see [11] for a survey of the area) since routing strategies can take advantage of the fact that nodes in

---

[*] Research supported in part by NSERC and OGS.

© Springer International Publishing AG 2017
F. Ellen et al. (Eds.): WADS 2017, LNCS 10389, pp. 205–216, 2017.
DOI: 10.1007/978-3-319-62127-2_18

these networks have physical locations that can be used to help guide a packet to its destination.

A geometric routing strategy is said to be *competitive* if the length of the path found by the routing strategy is not more than a constant times the Euclidean distance between its endpoints. This constant is called the *routing ratio*. In order to find a competitive path between any two vertices of a graph, such a path must first exist. Graphs that meet this criterion are called (geometric) spanners. Formally, given a geometric graph G, the distance, $d_G(u, v)$, between two vertices u and v in G is the sum of the weights of the edges in the shortest path between u and v in G. G is a *t-spanner* if for all pairs of vertices u and v in G, $d_G(u, v) \leq t \cdot |uv|$ for $t \geq 1$. Here $|uv|$ denotes the Euclidean distance between u and v. The smallest value t for which G is a t-spanner is the *spanning ratio* or *stretch factor* of G. A family of graphs that are t-spanners, for some fixed constant t, are often referred to as simply *spanners*. Spanners have been extensively studied—for a detailed overview of results on geometric spanners, see [12].

Geometric spanners tend to fall into three categories: (i) Long-known geometric graphs that happen to be spanners, such as Delaunay triangulations; (ii) cone-based constructions, such as Keil's θ-graphs [10]; and (iii) well-separated pair decomposition (WSPD) based constructions introduced by Callaghan and Kosaraju [6] in the early '90s. Note that graphs in the first category have fixed worst-case spanning ratios bounded away from 1. Constructions in the second and third categories are designed for a given parameter. They can achieve spanning ratios arbitrarily close to 1 by choosing arbitrarily small values for this parameter. Significant work has gone into finding competitive local and low-memory routing algorithms for graphs in the first category, including Delaunay graphs (classical–, L_1–, L_∞–, and TD–Delaunay triangulations) [7,4,1,3]. In most cases, proving tight results about graphs in this category is difficult. For example, even the exact worst-case spanning ratio of the Delaunay triangulation is unknown, despite over 30 years of study [8,10,2,13].

For the second category—cone-based spanners—competitive local routing algorithms are usually trivial. These spanners are designed so that greedy choices produce paths of low stretch. Still, for certain cone-based spanners, there have been some refined results on competitive routing algorithms that produce exceptionally low competitive ratios. For example, Bose *et al* [3] present a routing algorithm for the Half-θ_6-graph with a competitive ratio of 2.887.

In this paper, we consider routing algorithms for the third category: WSPD-based spanners. Intuitively, a WSPD of a pointset is a partition of the edges of the complete geometric graph (on that pointset) such that all edges in the same partition are approximately of equal length.[3] Since its introduction by Callahan and Kosaraju [6] in early the '90s, the WSPD and WSPD-based spanners have found a plethora of applications in solving distance problems [12]. The main difficulty about local routing in these spanners stems from the fact that WSPD-spanners are based on WSPDs which are built globally and capture

---

[3] See the next section for the formal definition.

global distance properties of the given pointset. As such WSPD-spanners pose a challenge in designing local routing strategies.

Given a pointset and a separation ratio s, a WSPD with separation ratio s is (typically) not unique. Callahan and Kosaraju's original construction of a WSPD is based on fair split-trees and it computes a WSPD containing a linear number of edge partitions [6]. From this WSPD, we show how to construct a WSPD-spanner that facilitates local routing by selecting a well-chosen edge from each partition rather than picking an arbitrary edge (Section 2). As a side benefit, our WSPD-spanner has a slightly improved spanning ratio, $1+4/s+4/(s-2)$, over the original one, $1+8/(s-4)$. This improvement stems from the additional properties our well-chosen edges have. On this WSPD-spanner, we present a 2-local and a 1-local routing algorithm with competitive routing ratios of $1 + 4/s + 6/(s - 2)$ and $1 + 6/(s - 2) + 6/s + 4/(s^2 - 2s) + 8/s^2$, respectively (Sections 3 and 4). A routing algorithm on a graph G is k-local if each vertex v of G stores information about vertices that are at hop distance at most k from v. Ideally, one would like the routing ratio to be identical to the spanning ratio, however, this is rarely the case when routing locally since an adversary can often force an algorithm to stray from the actual shortest path. Finally, we prove a lower bound of $1+8/s$ on the spanning ratio of our WSPD-spanner, thereby proving the near-optimality of the spanning ratio of our WSPD-spanner and the near-optimality of the routing ratio of both our routing algorithms.

WSPDs have been used before as an aid to routing in unit-disk graphs by Kaplan et al. [9]. Their scheme applies to our setting when the unit distance is the diameter of the point set. Their routing scheme requires a header of $O(\log n \log D)$ bits, where D is the diameter, and routing tables of total size $O(nc^{-5} \log^2 n \log^2 D)$ bits. Their routing ratio is $1 + \epsilon$ where $\epsilon = (\alpha/s) \log |pq|$ with $\alpha \geq 192$. Our routing scheme does not use a header and requires routing tables of total size $O(s^2 nB)$ bits, where B is the maximum number of bits to store a bounding box. Our routing ratio is $1 + \epsilon$ where $\epsilon = 6/(s - 2) + 4/s$ for our 2-local algorithm and $\epsilon = 6/(s - 2) + 6/s + 4/(s^2 - 2s) + 8/s^2$ for our 1-local algorithm. The main advantage of our scheme is that we do not require a header.

2 Construction of t-Spanners Using WSPDs

In this section, we explain how to construct a WSPD-spanner on which our results are based. We also prove some useful geometric lemmas concerning these spanners. We begin with some definitions. Two point sets A and B are *well-separated* with separation ratio s if there exist two circles with the same radius, ρ, one containing A and the other containing B such that the minimum distance between the two circles is $s \cdot \rho$. A *well-separated pair decomposition* (WSPD) of a point set $S \subseteq \mathbb{R}^2$ with separation ratio s is a set of pairs of subsets of S: $\{\{A_1, B_1\}, \ldots, \{A_m, B_m\}\}$ such that for each pair $\{A_i, B_i\}$, $1 \leq i \leq m$, A_i and B_i are well-separated with separation ratio s and for any distinct points p and q in S, there is a unique pair $\{A_i, B_i\}$, $1 \leq i \leq m$, such that $p \in A_i$ and $q \in B_i$.

Callahan and Kosaraju's [6] classical construction of WSPDs uses the fair-split tree. We denote the bounding box of a point set A, the smallest axis-aligned rectangle enclosing A, by $R(A)$. The fair split tree is defined as follows. Take the bounding box $R(S)$ of the point set S and make it the root u of the split tree. Then, split $R(S)$ on its longest axis and make the bounding boxes of two resulting point sets the children of u. Repeat for each child until the leaves are the points of S. Callahan and Kosaraju's [6] classical construction of a spanner given a WSPD proceeds as follows: for each pair $\{A, B\}$, select an arbitrary point $a \in A$ as a *representative* of the set A and an arbitrary point $b \in B$ as a representative of the set B and add the edge ab to the graph. We call any such spanner a *WSPD-spanner*. Callahan and Kosaraju [5] proved that any WSPD-spanner has a spanning ratio of at most $1 + 8/(s - 4)$, where s is the separation ratio of the WSPD. At the heart of their proof is the following lemma.

Lemma 1 (Callahan and Kosaraju [5]). *Let $\{A, B\}$ be a well-separated pair with respect to the separation ratio $s > 0$. Let $p, p', p'' \in A$ and $q, q' \in B$. Then, $|p'p''| \leq (2/s)|pq|$ and $|p'q'| \leq (1 + 4/s)|pq|$.*

Any WSPD-spanner that is built using the WSPD resulting from Callahan and Kosaraju's fair-split tree, we call *ASW-Spanner* (standing for "Arbitrary representative Split tree based WSPD-spanner"). To facilitate our routing algorithm, rather than selecting an arbitrary point as representative of a set in a pair, we choose the rightmost point in each set as its representative. If there is more than one candidate, we choose the topmost point among the rightmost ones. We refer to an ASW-Spanner constructed in this way as an *RSW-Spanner* (standing for "Rightmost representative Split tree based WSPD-spanner"). It is on RSW-Spanners that we prove our routing results. Unless stated otherwise, for the remainder of this paper, we focus on RSW-Spanners.

By exploiting properties of the fair-split tree, we prove that the spanning ratio of RSW-Spanners is at most $1 + 4/(s - 2) + 4/s$ which is a slight improvement over the spanning ratio of $1 + 8/(s - 4)$, shown for any WSPD-spanner. When using a fair-split tree to construct a WSPD, each set A from a well-separated pair corresponds to a unique node u in the fair-split tree. The points in A are precisely the points in the subtree rooted at u. The following lemmas are consequences of the definition of fair-split trees, the construction of the WSPD, and the choice of representatives in an RSW-Spanner.

Lemma 2. *For any two nodes u and v in the fair split tree, their corresponding bounding boxes are either disjoint or one is a subset of the other.*

Lemma 3. *In an RSW-Spanner, consider two sets P and Q, each from a pair of the WSPD. Let p be a representative of P. If $Q \subset P$ and $p \in Q$, then p is also the representative of Q.*

Lemma 4. *In an RSW-Spanner, let $\{A, B\}$ be a pair in the WSPD, $a, x \in A$ be two points such that a is the representative of A, and $x \neq a$. There is a well-separated pair $\{C, D\}$ such that: (1) $a \in C$, (2) $x \in D$, (3) a is the representative of C, (4) C is a proper subset of A and (5) D is a proper subset of A.*

Lemma 5. *In an RSW-Spanner, let A be a set in a pair from the WSPD and let $a, x \in A$ be two points such that a is the representative of A and $x \neq a$. Let C be a set in a pair from the WSPD such that $A \subset C$ and a point $c \in C$ is the representative of C. If xc is an edge, then ac is an edge.*

Lemma 6. *In an RSW-Spanner, let $\{A, B\}$ and $\{C, D\}$ be two distinct pairs from the WSPD, such that $A \subset C$. Let c be the representative of C. Let x be any point in B and let $\{E, F\}$ be the unique pair from the WSPD separating $c \in E$ from $x \in F$. Then, c is the representative of E.*

Algorithm 1 finds a path between p and q in any WSPD-spanner and is inspired from the proof of Theorem 9.2.1 by Narasimhan and Smid in [12].

Algorithm 1 FINDPATH(p, q)

Precondition: $p \neq q$
  Let $\{A, B\}$ be the unique pair in the WSPD separating $p \in A$ from $q \in B$.
  Let a and b be the representatives of A and B.
  return FINDPATHREC(p, a, A), FINDPATHREC(q, b, B)

Algorithm 2 FINDPATHREC(v, w, E)

Precondition: $v, w \in E$ and w is the representative of E
  if $v = w$ **then return** v
  else
      Let $\{C, D\}$ be the pair in the WSPD separating $v \subset C$ from $w \in D$.
      Let c and d be the representatives of C and D, respectively.
      return FINDPATHREC(v, c, C), FINDPATHREC(w, d, D)

Lemma 7. *Let $p, q \in S$. Consider a call to FINDPATH(p, q) in a ASW-Spanner of S. Consider the call to FINDPATHREC(v, w, E) at recursion depth $k \geq 1$. For any two points $e, f \in E$, $|ef| \leq (2/s)^k |pq|$.*

Theorem 1. *The spanning ratio t of an RSW-Spanner is at most $4/(s-2) + 4/s + 1$.*

Proof. We find an upper bound on the spanning ratio by analyzing the path found by FINDPATH. Consider any of the calls to FINDPATHREC in FINDPATH(p, q). Since w is the representative of E, by Lemma 4, we know that w is the representative of D and, thus, $w = d$. This means that for each level $k \geq 1$, the call to FINDPATHREC(w, d, D) returns immediately. In other words, for all $k \geq 1$, there is exactly one edge of level k. Notice that $v, w \in E$ according to the preconditions of FINDPATHREC(v, w, E). Therefore, by Lemma 2, c and d are also both in E. From Lemma 7, we get $|cd| \leq (2/s)^k |pq|$. The fact that there is exactly one edge of level k allows us to get only $(2/s)^k$ as the sum of the length of all edges at level k. Then, if we sum up the length of all edges $[c, d]$ from level 1 to a maximum depth m, we find $\sum_{i=1}^{m} (2/s)^i |pq| \leq \sum_{i=1}^{\infty} (2/s)^i |pq| = (2/(s-2))|pq|$.

Let $\{A, B\}$ be the pair separating $p \in A$ and $q \in B$. Let $a \in A$ and $b \in B$ be the representatives of A and B, respectively. From Lemma 1, we have that

$|ab| \leq (1 + 4/s)|pq|$. To bound the path found by FINDPATH(p, q), we take the length of the path found by the call to FINDPATHREC(p, a, A), add the length of the edge $[a, b]$, and add the length of the path found by the call to FINDPATHREC(q, b, B). Thus, the path found in FINDPATH(p, q) has a length of at most $2 \cdot (2/(s - 2))|pq| + (1 + 4/s)|pq| = (4/(s - 2) + 4/s + 1)|pq|$.   □

Theorem 2. *For any $s > 0$, there exist an RSW-Spanner with a spanning ratio arbitrarily close to $1 + 8/s$.*

Proof. Let $0 < \epsilon < \pi$ be a real number. Let $S = \{p, p', q, q'\}$ be a point set such that $p = (\cos(\pi/2 + \epsilon), \sin(\pi/2 + \epsilon))$, $p' = (\cos(-\pi/2 + \epsilon), \sin(-\pi/2 + \epsilon))$, $q = (\cos(-\pi/2-\epsilon), \sin(-\pi/2-\epsilon)+s+2)$ and $q' = (\cos(\pi/2-\epsilon), \sin(\pi/2-\epsilon)+s+2)$. Let $A = \{p, p'\}$ and $B = \{q, q'\}$. By construction, there is a pair $\{A, B\}$ in the WSPD. Again by construction, p' is the representative of $R(A)$ and q' is the representative of $R(B)$. Hence, the only path between p and q is $pp'q'q$. We can show that the spanning ratio of the path between p and q and, therefore, the spanning ratio of the graph approaches $\lim_{\epsilon \to 0}(|pp'|+|p'q'|+|q'q|)/|pq| = 1+8/s$ as ϵ approaches 0.   □

3 2-Local-Routing Algorithm

In this section, we present our competitive 2-local routing algorithm on RSW-Spanners. We begin with some notation. For two distinct points $t, u \in S$, with $\{T, U\}$ being the unique pair in the WSPD with $t \in T$ and $u \in U$, let $B_{tu}(t) := R(T)$, and let $B_{tu}(u) := R(U)$. Let u^* be the representative of U, then notice that u^* is the representative of $B_{tu^*}(u^*) = B_{tu}(u) = R(U)$. Therefore, u^* has an edge to the representative of $B_{tu^*}(t)$.

In order to describe our 2-local routing algorithm, we need to describe the precise information available at each vertex of the RSW-Spanner Let c be the current vertex of the routing path. Let d be any neighbor of c. Let e be any neighbor of d. We assume that the following information is available at c: (1) $B_{cd}(c)$ and $B_{cd}(d)$; (2) $B_{de}(d)$ and $B_{de}(e)$. Notice that we know $B_{de}(d)$ and $B_{de}(e)$ even though the current point is c. This information makes our routing algorithm 2-local. In Section 4, we will modify our algorithm so that it does not need to know $B_{de}(e)$. This will lead to a 1-local routing algorithm with a slightly larger routing ratio. We want to find a path between two points $p \in S$ and $q \in S$. Let $\{A, B\}$ be the unique pair in the WSPD separating p from q. Let a and b be the representatives of A and B. The goal for our algorithm is to find competitive path from p to a, take the edge ab, and then, find a competitive path b from q.

To find a competitive path from p to a, we use the following strategy. Let v be the current point on the path from p to a produced by our algorithm (at the beginning $v = p$). Here is how our algorithm selects the next edge. For each vertex v' adjacent to v, let $B_{v'}$ be the largest bounding box such that v' is the representative of $B_{v'}$. The next edge chosen by our algorithm is the edge vw such that the size of B_w is maximized, B_w contains p, and B_w is contained in $B_{pq}(p) = R(A)$. Note that we do not know yet whether B_w is contained in $B_{pq}(p)$.

However, we will prove, in Section 3.1, that this is the case. Upon reaching a, we take the edge ab. To find a competitive path from b to q, we notice that b already has an edge to the representative of a bounding box smaller than $R(B)$ containing q. Thus, the algorithm simply takes that edge. Then, we repeat this procedure until we find q. Our algorithm is summarized in Algorithm 3. Note that sizeof$(B_{v'})$ denotes the area of $B_{v'}$ and $\mathcal{N}(v)$ denotes the set of all neighbours of v.

Algorithm 3 FINDPATHTWOLOCAL(v, p, q)

Input: the current point v, the source p and the destination q.
Output: The next point w on the path.
1: Consider all bounding boxes of every neighbor v' of v.
2: **if** v' is the representative of $B_{vq}(q)$ **then** $w \leftarrow v'$ // Reducing step
3: **else**// Enlarging step
4:     **if** v' is the representative of $B_{pq}(p)$ **then** $w \leftarrow v'$
5:     **else**
6:         Let $B_{v'}$ is the largest bounding box that v' is the representative of.
7:         Let $\mathcal{V} = \{v' \in \mathcal{N}(v) \mid p \in B_{v'},\ v'$ is not the representative of $B_{v'q}(v')\}$
8:         $w \leftarrow \mathrm{argmax}_{v' \in \mathcal{V}}\ \mathrm{sizeof}(B_{v'})$
9: **return** w

As said previously, Algorithm 3 is 2-local. Indeed, the information used is the bounding boxes of the current point v, every neighbor v' of v, and every neighbor v'' of every neighbor of v. However, it is not obvious how the bounding boxes of the v'''s are used. To know whether v' is the representative of $B_{pq}(p)$ or $B_{v'q}(v')$, we need to know if v' is the representative of a bounding box which is separated from another bounding box containing q.

3.1 Correctness: 2-Local

In this section, we prove the correctness of Algorithm 3 (see Theorem 3). For the rest of this paper, we denote by $P_t(p, q)$ the path from p to q with spanning ratio t, found by the FINDPATH algorithm, and, we denote by $P(p, q)$ the path from p to q found by Algorithm 3. The following lemma is used to prove the correctness of Algorithm 3 and to establish an upper bound on the routing ratio of Algorithm 3 (see Theorem 4). By using the fact that B_v is the largest bounding box that v is the representative of and supposing that $p \in B_v$, it establishes a relation between B_v and the bounding boxes containing p of the vertices of $P_t(p, q)$. This relation allows us to find the conclusion of Lemma 8.

Lemma 8. *Consider any RSW-Spanner. Let v be a point inside $B_{pq}(p)$ that is not the representative of $B_{pq}(p)$ such that p is in B_v. Among all edges of $P_t(p, q)$, let de be the edge such that $B_{de}(d)$ is the smallest bounding box containing p that is larger than B_v. Then, there is an edge between v and d.*

Proof. We first argue that the edge de is well-defined. Since v is inside $B_{pq}(p)$ but is not the representative of $B_{pq}(p)$, we know that B_v is smaller than and inside $B_{pq}(p)$ by Lemma 3 and 2, respectively. This implies that the set of edges $[\alpha, \beta]$ from $P_t(p, q)$ such that p is in $B_{\alpha\beta}(\alpha)$ and B_v is smaller than $B_{\alpha\beta}(\alpha)$ is non-empty. Indeed, the edge ab from $B_{pq}(p)$ to $B_{pq}(q)$ is in that set since $B_{ab}(a) = B_{pq}(p)$, and B_v is smaller than and inside $B_{pq}(p)$. Therefore, the edge de is well-defined.

Let c be the point before d in $P_t(p, q)$. Since d is in $P_t(p, q)$, then d is representative of $B_{de}(d)$. Therefore, by Lemma 4, we know that d is the representative of the bounding box separating d from p. Then, c is the representative of $B_{cd}(c)$ such that p is in $B_{cd}(c)$.

Because $B_{de}(d)$ is the smallest bounding box containing p larger than B_v, $c \in B_{cd}(c) \subseteq B_v$. If $v = c$ is the representative of $B_{cd}(c)$, then v has an edge to d. Otherwise, $B_{cd}(c) \subset B_v$ and we apply Lemma 5 in the following way. We have that c is in B_v, $B_v \subset B_{de}(d)$, d is the representative of $B_{de}(d)$ and there is an edge between c and d. Therefore, there is an edge from v to d by Lemma 5.   □

Recall that $\{A, B\}$ is the pair separating $p \in A$ from $q \in B$ and that $a \in A$ and $b \in B$ are the representatives of A and B, respectively.

Lemma 9. *The Enlarging step of Algorithm 3 finds a path in an RSW-Spanner from $p \in A$ to $a \in A$.*

Proof. In this proof, we use some notation introduced in Section 3. Recall that, in Algorithm 3, we define $\mathcal{V} = \{v' \in \mathcal{N}(v) \mid p \in B_{v'}, v'$ is not the representative of $B_{v'q}(v')\}$. We prove that each edge vw taken in the Enlarging step of Algorithm 3 leads to a bounding box B_w that is larger than B_v but not larger than $B_{pq}(q)$. Thus, Algorithm 3 finds a path from p to the representative of $B_{pq}(q)$ (i.e. the representative a of A). Suppose that the current point v is inside but is not the representative of $B_{pq}(p)$. From Lemma 8, we get that v has an edge to a point of $P_t(p, q)$ that has a bounding box larger than B_v. This proves that there is always a choice of edges in the Enlarging step such that B_w is larger than B_v.

Now, we prove that the next edge vw is chosen such that w is inside $B_{pq}(p)$. We prove this by contradiction. Suppose Algorithm 3 takes the edge vw where w is outside of $B_{pq}(p)$. Therefore, w must be the representatives of $B_{pq}(p)$ or must be in $\mathcal{V}$. Since w is outside of $B_{pq}(p)$, it cannot be the representative of $B_{pq}(p)$. Thus, it must be in $\mathcal{V}$. Since p is in B_w and w is outside of $B_{pq}(p)$, we have that B_w is larger than $B_{pq}(p)$. Since the representative of $B_{pq}(p)$ has an edge to a bounding box containing q, from Lemma 6, we also get that w has an edge to a bounding box containing q which contradicts the definition of $\mathcal{V}$.

Because $v = p$ is inside $B_{pq}(p)$ in the first call of Algorithm 3, we then get that each edge vw taken in the Enlarging step of Algorithm 3 leads to a bounding box B_w that is larger than B_v but not larger than $B_{pq}(q)$.   □

Once a is found, then Algorithm 3 follows the edge ab since a has an edge leading to a bounding box containing q. We get the following lemma from the fact that the Reducing step of Algorithm 3 follows exactly what the recursive algorithm FINDPATH does. The following lemma and theorem follow.

Lemma 10. *The Reducing step of Algorithm 3 finds a path in an RSW-Spanner from $b \in B$ to $q \in B$.*

Theorem 3. *Algorithm 3 finds a path in an RSW-Spanner from p to q.*

3.2 Routing Ratio: 2-Local

In this section, we find an upper bound on the routing ratio of Algorithm 3. We get the following lemma from the fact that the Reducing step of Algorithm 3 follows exactly what the recursive algorithm FINDPATH does.

Lemma 11. *When the Reducing step of Algorithm 3 is executed on an RSW-Spanner, the length of the part of the path from $b \in B$ to $q \in B$ is at most $(2/(s-2))|pq|$.*

Lemma 12. *When the Enlarging step of Algorithm 3 is executed on an RSW-Spanner, the length of the part of the path from $p \in A$ to $a \in A$ is at most $(4/(s-2))|pq|$.*

Proof. For the purpose of this proof, we consider the edges of $P(p,q)$ as directed edges. Thus, if uv is an edge in $P(p,q)$, then u precedes v in $P(p,q)$. We say that u is the *source* of the edge and that v is the *target* of the edge.

Let cde be a subpath of $P_t(p,q)$ such that $c, d \in A$ and the edge cd is at level i in the analysis of Theorem 1. Consider the set T_i of edges vw such that vw is an edge of $P(p,q)$ and the target w is in $B_{de}(d)$ but not in $B_{cd}(c)$. We claim that the sum of the lengths of the edges in T_i is at most $2(2/s)^i|pq|$, i.e. $\sum_{vw \in T_i} |vw| \le 2(2/s)^i|pq|$. Then, using the analysis of Theorem 1, if we sum up the lengths of all edges vw from level 1 to a maximum depth m, we get that the length of the path is at most $\sum_{i=1}^{m} \sum_{vw \in T_i} |vw| \le \sum_{i=1}^{m} 2(2/s)^i|pq| \le \sum_{i=1}^{\infty} 2(2/s)^i|pq| = (4/(s-2))|pq|$.

We now prove our claim. If T_i is empty, then the sum is zero. Otherwise, let an edge $w_{j-1}w_j \in T_i$, where w_j is the i-th edge in $P(p,q)$. From Lemma 7, we get $|w_{j-1}w_j| \le (2/s)^i|pq|$. We consider two cases: either (1) w_j is the representative of $B_{de}(d)$ or (2) it is not.

1. Consider the edge $w_{j-2}w_{j-1}$. We consider two subcases: either (a) w_{j-1} is in $B_{cd}(c)$ or (b) it is not.
   (a) In this case, only $w_{j-1}w_j$ has its target in $B_{de}(d)$ and $|w_{j-1}w_j| \le (2/s)^i|pq| \le 2(2/s)^i|pq|$. Notice that, in this case, w_{j-1} is the representative of $B_{cd}(c)$ (thus $w_{j-1} = c$) because $w_{j-1} = c$ can only belong to one pair separating it from $w_j = d$.
   (b) In this case, the edge $w_{j-2}w_{j-1}$ falls in case 2a, i.e. w_{j-2} is in $B_{cd}(c)$ but not the representative of $B_{cd}(c)$, and w_{j-1} is in $B_{de}(d)$ but not in $B_{cd}(c)$. Therefore, the sum of the lengths of all edges having their target in $B_{de}(d)$ is $|w_{j-2}w_{j-1}| + |w_{j-1}w_j| \le 2(2/s)^i|pq|$. See Figure 1a.

2. From Lemma 8, we get that w_{j-1} is in $B_{cd}(c)$ but not the representative of $B_{cd}(c)$. Otherwise, there would be an edge from w_{j-1} to d. Since w_{j-1} is in $B_{cd}(c)$, there is no other edge $w_{k-1}w_k$ of $P(p,q)$ preceding $w_{j-1}w_j$, where w_k is in $B_{de}(d)$ but not in $B_{cd}(c)$.

   Now, consider the edge $w_j w_{j+1}$. We consider two subcases: either (a) w_{j+1} is the representative of $B_{de}(d)$ or (b) it is not.

   (a) From Lemma 7, we get $|w_j w_{j+1}| \leq (2/s)^i |pq|$. Therefore, the sum of the lengths of all edges having their target in $B_{de}(d)$ is $|w_{j-1}w_j| + |w_j w_{j+1}| \leq 2(2/s)^i |pq|$. See Figure 1a.

   (b) From Lemma 8, we get that w_j has an edge to d. Because w_{j+1} is not the representative of $B_{de}(d)$, w_{j+1} must be outside of $B_{de}(d)$. Therefore, only $w_{j-1}w_j$ has its target in $B_{de}(d)$ and not in $B_{cd}(c)$ and $|w_{j-1}w_j| \leq (2/s)^i |pq| \leq 2(2/s)^i |pq|$. See Figure 1b.

These cases cover all possibilities of edges in T_i.   □

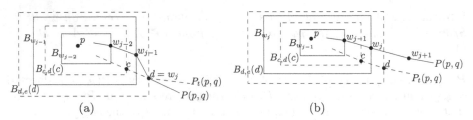

(a)   (b)

Fig. 1. (a) An illustration of the case 1b (and 2a.)(b) Illustration of the case 2b) of Lemma 12.

Theorem 4 combines Lemmas 11 and 12.

Theorem 4. *For a given RSW-Spanner, the routing ratio of Algorithm 3 is at most* $6/(s-2) + 4/s + 1$.

4 Improvement - 1-Local Routing Algorithm

An important aspect of routing algorithms is how much information each vertex needs to store. In this section, we present an algorithm that is slightly different from Algorithm 3. The main difference is that it is 1-local instead of 2-local. Let c be the current point, d be any neighbor of c and e be any neighbor of d. For Algorithm 3, we assumed that the following information was available at c: (1) $B_{cd}(c)$ and $B_{cd}(d)$; (2) $B_{de}(d)$ and $B_{de}(e)$. In this section, we explain how to design a routing algorithm that does not need to know $B_{de}(e)$. As a result, this increases the upper bound on the routing ratio by $8/s^2 + 2/s + 4/(s(s-2))$.

In our modified algorithm, only the Enlarging step differs from Algorithm 3 since the Reducing step in Algorithm 3 is already 1-local. To find a competitive path from p to a, we use the following modified strategy. Let v be the current vertex on the path from p to a produced by our algorithm (at the beginning $v = p$). Here is how our new algorithm selects the next edge. For each vertex v' adjacent to v, we modify the definition of $B_{v'}$. In this section, $B_{v'}$ is the largest bounding box such that v' is the representative of $B_{v'}$ and the distance between the enclosing circle $C_{v'}$ of $B_{v'}$ and q is at least s times the radius $\rho_{v'}$ of $C_{v'}$. The next edge chosen by our algorithm is the edge vw such that the size of B_w is maximized and B_w contains p. The strategy to a competitive path from b to q stays the same. Algorithm 4 below outlines the modified algorithm. Notice

---

Algorithm 4 FINDPATHONELOCAL(v, p, q)

Input: the current point v, the source p and the destination q.
Output: The next point w on the path.
 1: Consider all bounding boxes of every neighbor v' of v.
 2: **if** v' is the representative of $B_{vq}(q)$ **then** $w \leftarrow v'$ // Reducing step
 3: **else**// Enlarging step
 4:     $\forall v' \in \mathcal{N}(v)$, let $B_{v'}$ be the largest bounding box that v' is the representative of such that $|C_{v'}q| \geq s\rho_{v'}$.
 5:     Let $\mathcal{V} = \{v' \in \mathcal{N}(v) \mid p \in B_{v'}\}$
 6:     $w \leftarrow \text{argmax}_{v' \in \mathcal{V}} \text{sizeof}(B_{v'})$
 7: **return** w

---

that this new algorithm does not guarantee that the path stays inside $B_{pq}(p)$. However, as shown in the proof of Lemma 9, the first edge that goes outside of $B_{pq}(p)$ has an edge to the representative of a bounding box containing q. Thus, Algorithm 4 is entering the Reducing step right after this edge is taken. Before we prove an upper bound on the routing ratio of Algorithm 4 (see Theorem 5), we need the following lemma.

Lemma 13. *Consider any RSW-Spanner. In Algorithm 4, the diameter of the last enclosing circle in the Enlarging step is at most* $(2/s)|pq|$.

Theorem 5. *For a given RSW-Spanner, the routing ratio of Algorithm 4 is at most* $8/s^2 + 6/s + 4/(s(s-2)) + 6/(s-2) + 1$.

Proof. Consider the analysis of Lemma 12. Let cde be a subpath of $P_t(p,q)$ such that $c, d \in A$ and the edge cd is at level i in the analysis of Theorem 1. Recall that T_i is the set of edges vw such that vw is an edge of $P(p,q)$ and the target w is in $B_{de}(d)$ but not in $B_{cd}(c)$.

From Lemma 9, Algorithm 3 guarantees that the edge ab is taken while Algorithm 4 does not. However, as said above, the first edge that goes outside of $B_{pq}(p)$ has an edge to a bounding box containing q. This edge is the edge $w_j w_{j+1}$ in Case 2b, where the level i is 1 and $B_{de}(d) = B_{pq}(p)$. From Lemma 13, we know that $|w_j w_{j+1}| \leq (2/s)|pq|$. Since T_1 in Case 2b only contains the edge $w_{j-1}w_j$, adding the edge $w_j w_{j+1}$ to T_1 gives that the sum of the lengths of the edges of T_1 is at most $2(2/s)|pq|$. Thus, if we sum up the lengths of all edges vw from level 1 to a maximum depth m, we get that the length of the path found in the Enlarging step is still at most $\sum_{i=1}^{m} \sum_{vw \in T_i} |vw| \leq \sum_{i=1}^{m} 2(2/s)^i |pq| \leq \sum_{i=1}^{\infty} 2(2/s)^i |pq| = 4/(s-2)|pq|$.

Suppose we want to find a path between the point w_{j+1} and q with Algorithm 4. Let us rename w_{j+1} as u. Since u already has an edge to a bounding box containing q, Algorithm 4 enters immediately the Reducing step at u. This part of the algorithm does not differ from Algorithm 3. Therefore, the routing ratio is the same for the path from u to q with both algorithms. Because the Enlarging step does not occur, we get that the length of $P(u,q)$ is at most $(1 + 4/s + 2/(s-2))|uq|$.

Since the diameter of the enclosing circle of B_u is at most $(2/s)|pq|$ from Lemma 13 and p is in B_u, we have $|up| \leq (2/s)|pq|$. By the triangle inequality, we get that $|uq| \leq |up| + |pq| \leq (2/s)|pq| + |pq| = (1 + 2/s)|pq|$.

Let P_{pu} be the subpath from p to u of $P(p,q)$ and P_{uq} be the subpath from u to q of $P(p,q)$. We then get that the length of the path is at most $|P_{pu}| + |P_{uq}|$, which in turn is at most $(4/(s-2))|pq| + (1 + 4/s + 2/(s-2))|uq| \leq (4/(s-2))|pq| + (1 + 4/s + 2/(s-2))(1 + 2/s)|pq|$, which simplifies to $(8/s^2 + 6/s + 4/(s(s-2)) + 6/(s-2) + 1)|pq|$.   □

References

1. Bonichon, N., Bose, P., De Carufel, J.-L., Perkovic, L., van Renssen, A.: Upper and lower bounds for online routing on Delaunay triangulations. In: ESA. pp. 203–214 (2015)
2. Bose, P., Devroye, L., Löffler, M., Snoeyink, J., Verma, V.: Almost all Delaunay triangulations have stretch factor greater than pi/2. Comput. Geom. 44(2), 121–127 (2011)
3. Bose, P., Fagerberg, R., van Renssen, A., Verdonschot, S.: Optimal local routing on Delaunay triangulations defined by empty equilateral triangles. SIAM J. Comput. 44(6), 1626–1649 (2015)
4. Bose, P., Morin, P.: Online routing in triangulations. SIAM J. Comput. 33(4), 937–951 (2004)
5. Callahan, P.B., Kosaraju, S.R.: Faster algorithms for some geometric graph problems in higher dimensions. In: SODA. pp. 291–300 (1993)
6. Callahan, P.B., Kosaraju, S.R.: A decomposition of multidimensional point sets with applications to k-nearest-neighbors and n-body potential fields. J. ACM 42(1), 67–90 (1995)
7. Chew, P.: There is a planar graph almost as good as the complete graph. In: SOCG. pp. 169–177 (1986)
8. Dobkin, D.P., Friedman, S.J., Supowit, K.J.: Delaunay graphs are almost as good as complete graphs. Discrete Comput. Geom. 5, 399–407 (1990)
9. Kaplan, H., Mulzer, W., Roditty, L., Seiferth, P.: Routing in unit disk graphs. In: LATIN. pp. 536–548 (2016)
10. Keil, J.M., Gutwin, C.A.: Classes of graphs which approximate the complete euclidean graph. Discrete Comput. Geom. 7, 13–28 (1992)
11. Misra, S., Woungang, I., Misra, S.C.: Guide to Wireless Sensor Networks. Springer Publishing Company, Incorporated, 1st edn. (2009)
12. Narasimhan, G., Smid, M.H.M.: Geometric spanner networks. Cambridge University Press (2007)
13. Xia, G.: The stretch factor of the Delaunay triangulation is less than 1.998. SIAM J. Comput. 42(4), 1620–1659 (2013)

Relaxing the Irrevocability Requirement for Online Graph Algorithms*

Joan Boyar[1], Lene M. Favrholdt[1], Michal Kotrbčík[2], and Kim S. Larsen[1]

[1] University of Southern Denmark, Odense, Denmark,
{joan,lenem,kslarsen}@imada.sdu.dk
[2] The University of Queensland, Brisbane, Australia, m.kotrbcik@uq.edu.au

Abstract. Online graph problems are considered in models where the irrevocability requirement is relaxed. Motivated by practical examples where, for example, there is a cost associated with building a facility and no extra cost associated with doing it later, we consider the Late Accept model, where a request can be accepted at a later point, but any acceptance is irrevocable. Similarly, we also consider a Late Reject model, where an accepted request can later be rejected, but any rejection is irrevocable (this is sometimes called preemption). Finally, we consider the Late Accept/Reject model, where late accepts and rejects are both allowed, but any late reject is irrevocable. For Independent Set, the Late Accept/Reject model is necessary to obtain a constant competitive ratio, but for Vertex Cover the Late Accept model is sufficient and for Minimum Spanning Forest the Late Reject model is sufficient. The Matching problem has a competitive ratio of 2, but in the Late Accept/Reject model, its competitive ratio is $\frac{3}{2}$.

1 Introduction

For an online problem, the input is a sequence of requests. For each request, the algorithm has to make some decision without any knowledge about possible future requests. Often (part of) the decision is whether to accept or reject the request and the decision is usually assumed to be irrevocable. However, many online problems have applications for which total irrevocability is not inherent or realistic. Furthermore, when analyzing the quality of online algorithms, relaxations of the irrevocability constraint often result in dramatically different results, especially for graph problems. This has already been realized and several papers study various relaxations of the irrevocability requirement. In this paper we initiate a systematic study of the nature of irrevocability and of the implications for the performance of the algorithms. Our aim is to understand whether it is the absence of knowledge of the future or the irrevocability restrictions on the manipulation of the solution set that makes an online problem difficult.

We consider graph problems and focus on four classical problems, *Independent Set*, *Matching*, *Vertex Cover*, and *Minimum Spanning Forest*. Independent

---

* Supported in part by the Danish Council for Independent Research, Natural Sciences, grant DFF-1323-00247, and the Villum Foundation, grant VKR023219.

F. Ellen et al. (Eds.): WADS 2017, LNCS 10389, pp. 217–228, 2017.
DOI: 10.1007/978-3-319-62127-2_19

217

Set and Vertex Cover are studied in the vertex arrival model. In this model, vertices arrive one by one together with all the edges between the newly arrived vertex and previous vertices. Matching and Minimum Spanning Forest are studied in the edge arrival model, but the results hold in the vertex arrival model as well. In the edge arrival model, edges arrive one by one, and if a vertex incident with the newly-arrived edge was not seen previously, it is also revealed.

Relaxed irrevocability

For the four problems considered in this paper, the online decision is whether to accept or reject the current request. In the *standard* model of online problems, this decision is irrevocable and has to be made without any knowledge about possible future requests. We relax the irrevocability requirement by allowing the algorithm to perform two additional operations, namely *late accept* and *late reject*. Late accept allows the algorithm to accept not only the current request but also requests that arrived earlier. Thus, late accept relaxes irrevocability by not forcing the algorithm to discard the items that are not used immediately. Late reject allows the algorithm to remove items from the solution being constructed, relaxing the irrevocability of the decision to accept an item. When the algorithm is allowed to perform late accept or late reject, but not both, we speak of a *Late Accept model* and *Late Reject model*, respectively. Note that, in these two models, the late operations are irrevocable. We also consider the situation where the algorithm is allowed to perform *both* late accepts and late rejects, focusing on the *Late Accept/Reject model*, where any item can be late-accepted and late-rejected, but once it is late-rejected, this decision is irrevocable. In other words, if the algorithm performs both late accept and late reject on a single item, the late accept has to precede the late reject.

We believe that the Late Accept, Late Reject, and Late Accept/Reject models are appropriate modeling tools corresponding to many natural settings. Matching, for example, in the context of online gaming or chats, functions in the Late Accept model. Indeed, the users are in the pool until assigned, allowing the late accept, but once the users are paired, the connection should not be broken by the operator. Note that the matching problem is a maximization problem. For minimization problems, accepting a request may correspond to establishing a resource at some cost. Often there is no natural reason to require the establishment to happen at a specific time. Late acceptance was considered for the dominating set problem in [2], which also contains further feasible practical applications and additional rationale behind the model.

When the knapsack problem is studied in the Late Reject model, items are usually called *removable*; see for example [16, 13, 12, 4, 14]. For most other problems, late rejection is usually called *preemption* and has been studied in variants of many online problems, for example call control [1, 9], maximum coverage [23, 22], and weighted matching problems [6, 7]. Preemption was also previously considered for one of the problems we consider here, independent set, in [19], but in a model where advice is used, presenting lower bounds on the amount of advice necessary to achieve given competitive ratios in a stated range.

Online Vertex Cover was studied in [5], where they considered the possibility of swapping some of the accepted vertices for other vertices at the very end, at some cost depending on the number of vertices involved. A similar concept is studied in, for example, [15, 21, 10, 11] for online Steiner tree problems, MST, and TSP. Here, replacing an accepted edge with another is allowed, and the objective is to minimize the number of times this occurs while obtaining a good competitive ratio. The problem is said to allow *rearrangements* or *recourse*. TSP has also been studied [17] in a model where the actual acceptances and rejections (rejections carry a cost) are made at any time.

Competitive analysis

For each graph problem, we study online algorithms in the standard, Late Accept, Late Reject, and Late Accept/Reject models using the standard tool of competitive analysis [24, 18], where the performance of an online algorithm is compared to the optimum algorithm OPT via the competitive ratio. For any algorithm (online or offline), A, we let $A(\sigma)$ denote the value of the objective function when A is applied to the input sequence σ.

For minimization problems, we say that an algorithm, ALG, is *c-competitive*, if there exists a constant α such that, for all inputs σ, $\text{ALG}(\sigma) \leq c \cdot \text{OPT}(\sigma) + \alpha$. Similarly, for maximization problems, ALG is *c*-competitive, if there exists a constant α such that, for all inputs σ, $\text{OPT}(\sigma) \leq c \cdot \text{ALG}(\sigma) + \alpha$. In both cases, if the inequality holds for $\alpha = 0$, the algorithm is *strictly c*-competitive. The *(strict) competitive ratio* of ALG is the infimum over all c such that ALG is (strictly) *c*-competitive. The competitive ratio of a problem P is the infimum over the competitive ratio of all online algorithms for the problem. For all combinations of the problem and the model, we obtain matching lower and upper bounds on the competitive ratio.

For ease of notation for our results, we adopt the following conventions to express that a problem essentially has competitive ratio n, i.e., it is true up to an additive constant. We say that a problem has competitive ratio $n - \Theta(1)$ if (i) for any algorithm, there is a constant $b > 0$ such that the strict competitive ratio is at least $n - b$, and (ii) for any constant b, there is a strictly $(n - b)$-competitive algorithm for graphs with at least $b + 1$ vertices. Similarly, we say that a problem has competitive ratio $n/\Theta(1)$ if (i) for any algorithm, there is a constant $b > 0$ such that the strict competitive ratio is at least n/b, and (ii) for any constant b, there is an n/b-competitive algorithm for graphs with at least b vertices. This notation is used in Theorems 3 and 12. For all other results, the upper bounds hold for the strict competitive ratio. For convenience, when stating results containing both an upper bound on the strict competitive ratio and a lower bound on the competitive ratio, we use the term "competitive ratio" even though the result holds for the strict competitive ratio as well.

Our results

The paper shows that for some problems the Late Accept model allows for algorithms with significantly better competitive ratios, while for others it is the Late Reject model which does. For other problems, the Late Accept/Reject model is necessary to get these improvements. See Table 1. Note that only deterministic algorithms are considered, not randomized algorithms.

Our results on Minimum Spanning Forest follow from previous results. Thus, they are mainly included to give an example where late rejects bring down the competitive ratio dramatically. The technical highlights of the paper are the results for Independent Set in the Late Accept/Reject model, where, in Theorems 4 and 5, we prove matching lower and upper bounds of $3\sqrt{3}/2$ on the competitive ratio.

Table 1. Competitive ratios of the four problems in each of the four models. W is the ratio of the largest weight to the smallest.

Problem	Standard	Late Accept	Late Reject	Late Accept/Reject
Independent Set	$n-1$	$\frac{n}{\Theta(1)}$	$\lceil \frac{n}{2} \rceil$	$\frac{3\sqrt{3}}{2} \approx 2.598$
Matching	2	2	2	$\frac{3}{2}$
Vertex Cover	$n-1$	2	$n - \Theta(1)$	2
Min. Spanning Forest	W	W	1	1

We consider only undirected graphs $G = (V, E)$. Throughout the paper, G will denote the graph under consideration, and V and E will denote its vertex and edge set, respectively. Moreover, $n = |V|$ will always denote the number of vertices in G. We use uv for the undirected edge connecting vertices u and v, so vu denotes the same edge.

The missing proofs all appear in the full paper [3].

2 Independent Set

An *independent set* for a graph $G = (V, E)$ is a subset $I \subseteq V$ such that no two vertices in I are connected by an edge. For the problem called Independent Set, the objective is to find an independent set of maximum cardinality. We consider online Independent Set in the vertex arrival model.

In the standard model, no online algorithm can be better than $(n-1)$-competitive, since the adversary can give independent vertices, until the algorithm accepts a vertex, v, and then only give vertices adjacent to v. On the other hand, the greedy algorithm is $(n-1)$-competitive; if it ever rejects a vertex, the graph contains at least one edge and hence, OPT accepts at most $n-1$ vertices.

Theorem 1. *For Independent Set in the standard model, the strict competitive ratio is $n - 1$.*

In the Late Reject model, the greedy algorithm becomes $n/2$-competitive with the following modification. If a new vertex is adjacent to exactly one accepted vertex, v, then v is rejected and the new vertex is accepted. If the algorithm accepts only one vertex, the graph has a path containing all n vertices and OPT can accept at most $\lceil \frac{n}{2} \rceil$ vertices. For the lower bound, the adversary can give a bipartite graph by always connecting the new vertex to the only vertex (if any) accepted by the algorithm.

Theorem 2. *For Independent Set in the Late Reject model, the strict competitive ratio is $\lceil n/2 \rceil$.*

Allowing late accepts helps further, but not enough to obtain a finite (constant) competitive ratio. For a given positive constant c, an algorithm which does not accept any vertex until the presented graph has an independent set of size at least c, and then accepts any such set, is n/c-competitive. For the lower bound, if the adversary starts the input sequence with isolated vertices, any algorithm with a bounded competitive ratio will have to accept a vertex, v, at some point. From this point on, the adversary can give vertices with only v as a neighbor. If v was the cth vertex in the input, the algorithm can accept at most the first c vertices.

Theorem 3. *For Independent Set in the Late Accept model, the competitive ratio is $n/\Theta(1)$.*

The following two theorems show that, in the Late Accept/Reject model, the optimal competitive ratio for Independent set is $3\sqrt{3}/2$. The upper bound comes from a variant of the greedy algorithm, Algorithm 1, rejecting a set of vertices if it can be replaced by a set at least $\sqrt{3}$ as large. The algorithmic idea is natural and has been used before (with other parameters than $\sqrt{3}$) in [22, 23], for example. Thus, the challenge lies in deciding the parameter and proving the resulting competitive ratio. Pseudocode for Algorithm 1 is given below.

For Algorithm 1, we introduce the following notation. Let S be the current set of vertices that have been accepted and not late-rejected. Let R be the set of vertices that have been late-rejected, and let P denote the set $V - (R \cup S)$ of vertices that have not been accepted (and, hence, not late-rejected).

For a set U of vertices, let $N(U) = \cup_{v \in U} N(v)$, where $N(v)$ is the neighborhood of a vertex v (not including v). We call a set, T, of vertices *admissible* if all the following conditions are satisfied:

1) T is an independent set;
2) $T \subseteq P$;
3) $|T| \geq \sqrt{3} |N(T) \cap S|$.

Algorithm 1: Algorithm for Independent Set in the Late Accept/Reject model.

Result: Independent set S

1 $S = \emptyset$
2 **while** *a vertex v is presented* **do**
3   **if** $S \cup \{v\}$ *is independent* **then**
4          $S = S \cup \{v\}$
5   **else**
6   **while** *there exists an admissible set* **do**
7   Let T be an admissible set minimizing $|S \cap N(T)|$
8              $S = (S - N(T)) \cup T$

For the analysis of Algorithm 1, we partition S into the set, A, of vertices accepted in line 4 and the set, B, of vertices accepted in line 8. We let O be the independent set constructed by OPT. For any set, U, of vertices, we let $U^+ = U \cap O$ and $U^- = U - O$. Thus, $O = P^+ \cup S^+ \cup R^+ = P^+ \cup A^+ \cup B^+ \cup R^+$.

The following lemma follows from the fact that each time a set, X, of vertices is moved from S to R, a set at least $\sqrt{3}$ times as large as X is added to B.

Lemma 1. *When Algorithm 1 terminates, $|B| \geq (\sqrt{3} - 1)|R|$.*

The next lemma follows from the facts that when the algorithm terminates, P^+ is not admissible and $P^+ \cup S^+$ is independent, since $P^+ \cup S^+ \subseteq O$.

Lemma 2. *When Algorithm 1 terminates, $|P^+| < \sqrt{3}|S^-|$.*

Lemma 3. *When Algorithm 1 terminates, $|B^-| + |R^-| \geq \sqrt{3}|R^+|$.*

Proof. Consider a set, T, added to B in line 8. Let $Q = N(T) \cap S$. We prove that

$$|T^-| \geq \sqrt{3}|Q^+| \tag{1}$$

If $|Q^+| = 0$, this is trivially true. Thus, we can assume that Q^- is a proper subset of Q. Since T is admissible, it follows that

$$|T| \geq \sqrt{3}|Q| \tag{2}$$

Note that $(S - Q^-) \cup T^+$ is independent, since $(S - Q) \cup T$ is independent and there are no edges between Q^+ and T^+. Since the algorithm chooses T such that $|Q|$ is minimized, this means that

$$|T^+| < \sqrt{3}|Q^-| \tag{3}$$

Subtracting Ineq. (3) from Ineq. (2), we obtain Ineq. (1).

Let $T_1, T_2, \ldots, T_k$ be all the admissible sets that are chosen in line 8 during the run of the algorithm, and let $Q_1, Q_2, \ldots, Q_k$ be the corresponding sets that are removed from S. Then, $\cup_{i=1}^{k}T_i \subseteq B \cup R$, and thus, $\cup_{i=1}^{k}T_i^- \subseteq B^- \cup R^-$. Furthermore, $R = \cup_{i=1}^{k}Q_i$. Hence,

$$|B^-| + |R^-| \geq \sum_{i=1}^{k}|T_i^-| \geq \sum_{i=1}^{k}\sqrt{3}|Q_i^+| = \sqrt{3}|R^+|,$$

where the second inequality follows from Ineq. (1).   □

Using $(\sqrt{3}+1)(|B^+|+|R^+|) = \sqrt{3}(|B^+|+|R^+|) + |B| + |R| - (|B^-|+|R^-|)$, we obtain the following lemma via simple calculations using Lemmas 1 and 3.

Lemma 4. *When Algorithm 1 terminates,* $|B^+| + |R^+| \leq \frac{\sqrt{3}}{\sqrt{3}+1}|B^+| + \frac{\sqrt{3}}{2}|B|$.

The upper bound now follows from simple calculations using Lemmas 2 and 4:

Theorem 4. *For Independent Set in the Late Accept/Reject model, Algorithm 1 is strictly $3\sqrt{3}/2$-competitive.*

We prove a matching lower bound:

Theorem 5. *For Independent Set in the Late Accept/Reject model, the competitive ratio is at least $3\sqrt{3}/2$.*

Proof (Sketch of a proof.). Assume that ALG is strictly c-competitive for some $c > 1$. We first show that c is at least $3\sqrt{3}/2$ and then lift the strictness restriction. Assume for the sake of contradiction that $c < 3\sqrt{3}/2$.

Incrementally, we construct an input consisting of a collection of bags, where each bag is an independent set. Whenever a new vertex v belonging to some bag B is given, we make it adjacent to every vertex not in B, except vertices that have been late-rejected by ALG. Thus, if ALG accepts v, it cannot hold any vertex in any other bag. This implies that the currently accepted vertices of ALG always form a subset of a single bag, which we refer to as ALG's bag, and this is the crucial invariant in the proof. We say that ALG *switches* when it rejects the vertices of its current bag and accepts vertices of a different bag.

For the incremental construction, the first bag is special in the sense that ALG cannot switch to another bag. We discuss later when we decide to create the second bag, but all we will need is that the first bag is large enough. From the point where we have created a second bag, ALG has the option of switching. Whenever ALG switches to a bag, B', we start the next bag, B''. All that this means is that the vertices we give from this point on and until ALG switches bag again belong to B'', and ALG never holds vertices in the newest bag.

Now we argue that as long as we keep giving vertices, ALG will repeatedly have to switch bag in order to be c-competitive. Choose some $\varepsilon > 0$, let B be ALG's bag, B' be the new bag, and s be the number of vertices which are not adjacent to any vertices in B'. If ALG has accepted a vertices of B after $(c + \varepsilon)a - s$ vertices of the new bag B' have been given, ALG has to accept at least one additional vertex to be c-competitive, since at this point OPT could accept all of the vertices in B' and s additional vertices. Since B' is the new bag,

B has reached its final size, so eventually ALG will have to switch to a different bag.

For the proof, we keep track of relevant parts of the behavior of ALG using a tree structure. The first bag is the root of the tree. Recall that whenever ALG switches to a bag, say X, we start a new bag Y. In our tree structure we make Y a child of X.

Since ALG is c-competitive and always holds vertices only from a single bag B, the number a of vertices held in B satisfies $a \geq |B|/c$. Since, by assumption, $c < 3$, it follows that ALG can accept and then reject disjoint sets of vertices of B at most twice, or equivalently, that each bag in the tree has at most two children. As we proved above, ALG will have to keep switching bags, so if we keep giving vertices, this will eventually lead to leaves arbitrarily far from the root.

Consider a bag B_m that ALG holds after a "long enough" sequence has been presented. Label the bags from the root to ALG's bag by $B_1, \ldots, B_m$, where B_{i+1} is a child of B_i for each $i = 1, \ldots, m-1$. Let a_j, $1 \leq j < m$, be the number of vertices of B_j held by ALG immediately before it rejected already accepted vertices from B_j for the first time and let a_m be the number of vertices currently accepted in B_m. Let $n_j = |B_j|$, $1 \leq j \leq m$.

Furthermore, for each j, if j is even, let $s_j = a_2 + a_4 + \cdots + a_j$, and if j is odd, let $s_j = a_1 + a_3 + \cdots + a_j$. Note that our choice of adjacencies between bags implies that OPT can hold at least s_j vertices in bags $B_1, B_2, \ldots, B_j$.

Thus, just before ALG rejects the vertices in B_{j-1} (just before the n_jth vertex of B_j is given), we must have $ca_{j-1} \geq n_j - 1 + s_{j-2}$, by the assumption that ALG is c-competitive. We want to introduce the arbitrarily small ε chosen above and eliminate the "-1" in this inequality: Since OPT can always hold the a_1 vertices from the root bag, $ca_j \geq a_1$ must hold for all j. Since $a_1 \geq (n_1 - 1)/c$, we get that $a_j \geq (n_1 - 1)/c^2$. Thus, at the beginning of the input sequence, we can keep giving vertices for the first bag, making n_1 large enough such that a_j becomes large enough that $\varepsilon a_{j-1} \geq 1$. This establishes $(c + \varepsilon)a_{j-1} \geq n_j + s_{j-2}$. Trivially, $n_j \geq a_j$, so

$$(c + \varepsilon)a_{j-1} - s_{j-2} \geq a_j. \tag{4}$$

Next, we want to show that for any $1 \leq c < 3\sqrt{3}/2$, there exists an m such that

$$s_m > ca_m, \tag{5}$$

contradicting the assumption that ALG was c-competitive. To accomplish this, we repeatedly strengthen Ineq. (5) by replacing a_j with the bound from Ineq. (4), eventually arriving at an inequality which can be proven to hold, and then this will imply all the strengthened inequalities and, finally, Ineq. (5).   □

3 Matching

A *matching* in a graph $G = (V, E)$ is a subset of E consisting of pairwise non-incident edges. For the problem called Matching, the objective is to find a matching of maximum cardinality. We study online Matching in the edge arrival model, but note that the results hold in the vertex arrival model as well: For the upper bounds, an algorithm in the vertex arrival model can process the edges incident to the current vertex in any order. For the lower bounds, all adversarial sequences used in this section consist of paths, and hence, exactly the same input can be given in the vertex arrival model.

It is well known and easy to prove that the greedy algorithm which adds an edge to the matching whenever possible is 2-competitive and this is optimal in the standard model. The first published proof of this is perhaps in the classical paper of Korte and Hausmann [20].

For late accept, we can use the same adversarial sequence as for the standard model: First a number of isolated edges are presented. Then, for each edge, uv, accepted by the algorithm, two edges, xu and vy, are presented.

Theorem 6. *For Matching in the Late Accept model, the competitive ratio is 2.*

Late rejects do not improve the competitive ratio either. This can be seen from a sequence starting with a number of isolated edges. For each accepted edge, uv, the adversary presents an edge vx. If the algorithm late-rejects uv (and thus accepts vx), an edge xy is presented. Otherwise, an edge zu is presented.

Theorem 7. *For Matching in the Late Reject model, the competitive ratio is 2.*

In the Late Accept/Reject model, the competitive ratio is 3/2. Again, the adversarial sequence starts with a number of isolated edges. If an edge uv is accepted at any point, the adversary presents edges xu and vy. If uv is then late-rejected, edges $x'x$ and yy' are presented.

Theorem 8. *For Matching in the Late Accept/Reject model, the competitive ratio is at least 3/2.*

To prove a matching upper bound, we give an algorithm, Algorithm 2, which is strictly $\frac{3}{2}$-competitive in the Late Accept/Reject model.

Recall that for a matching M, a path $P = e_1, \ldots, e_k$ is *alternating* with respect to M, if for all $i \in \{1, \ldots, k\}$, e_i belongs to M if and only if i is even. Moreover, an alternating path P is called *augmenting* if neither endpoint of P is incident to a matched edge. Note that the symmetric difference of a matching M and an augmenting path with respect to M is a matching of size larger than M. We focus on local changes, called short augmentations in [26].

Algorithm 2: Algorithm for maximal matching in the Late Accept/Reject model.

  Result: Matching M

1 $M = \emptyset$

2 **while** *an edge e is presented* **do**

3   **if** $M \cup \{e\}$ *is a matching* **then**

4          $M = M \cup \{e\}$

5   **if** *there is an augmenting path $xuvy$ of length 3* **then**

6          $M = (M \cup \{ux, vy\}) - \{uv\}$

The fact that Algorithm 2 is a Late Accept/Reject algorithm follows from the observation that no matched vertex ever becomes unmatched again. For the upper bound, we use that if a maximal matching M does not admit augmenting paths of length 3, then $3|M| \geq 2|OPT|$. This fact is easy to prove and can be found as Lemma 2 of [8], for example.

Theorem 9. *For Matching in the Late Accept/Reject model, Algorithm 2 is strictly 3/2-competitive.*

4 Vertex Cover

A *vertex cover* for a graph $G = (V, E)$ is a subset $C \subseteq V$ such that for any edge, $uv \in E$, $\{u, v\} \cap C \neq \emptyset$. For the problem called Vertex Cover, the objective is to find a vertex cover of minimum cardinality. We study online Vertex Cover in the vertex arrival model.

In the standard model, no online algorithm can be better than $(n - 1)$-competitive: The adversary can present isolated vertices until some vertex, v, is rejected, and then vertices that are adjacent only to v. On the other hand, this competitive ratio is obtained by the algorithm that accepts only the second endpoint of each uncovered edge.

Theorem 10. *For Vertex Cover in the standard model, the strict competitive ratio is $n - 1$.*

Late accept changes the situation dramatically, since then the 2-approximation algorithm adding both endpoints of each uncovered edge can be used. Adding late rejects does not change the situation further; if the algorithm ever late-rejects a vertex, v, the adversary can add arbitrarily many neighbors of v.

Theorem 11. *For Vertex Cover in the Late Accept model and the Late Accept/Reject model, the competitive ratio is 2.*

In the Late Reject model, the competitive ratio is $n - \Theta(1)$. For the lower bound, the adversary can give isolated vertices until some vertex, v, is rejected, and then arbitrarily many neighbors of v. The upper bound is obtained by a family ALG_b of algorithms accepting the first $b + 1$ vertices and then rejecting the vertices not part of an optimal vertex cover of the graph seen so far.

Theorem 12. *For Vertex Cover in the Late Reject model, the competitive ratio is $n - \Theta(1)$.*

5 Minimum Spanning Forest

A *spanning forest* for a graph $G = (V, E)$ is a subset $T \subseteq E$ which forms a spanning tree on each of the connected components of G. Given a weight function $w \colon E \to \mathbb{R}^+$, the objective of the Minimum Spanning Forest problem is to find a spanning forest of minimum total weight. We let W denote the ratio between the largest and the smallest weight of any edge in the graph.

We study online Minimum Spanning Forest in the edge arrival model, but the results also hold in the vertex arrival model: For the upper bounds, an algorithm in the vertex arrival model can process the edges incident to the current vertex in any order. In the lower bound sequences presented here, all edges from a new vertex to all previous vertices are presented together in an arbitrary order.

Even in the standard model, the competitive ratio cannot be higher than W, since all spanning forests contain the same number of edges. A matching lower bound follows from the sequence consisting of a tree of edges of weight W and then a vertex with edges of weight 1 to all previous vertices. Since an online algorithm, even in the Late Accept model model, does not know when the input ends, it must always have a forest spanning all the vertices seen so far:

Theorem 13. *For Minimum Spanning Forest in the standard model or the Late Accept model, the competitive ratio is W.*

On the other hand, in the Late Reject model, the greedy online algorithm mentioned by Tarjan in [25] can be used: Each new edge is accepted, and if this results in a cycle, the heaviest edge in the cycle is (late-)rejected. Since the Late Reject model leads to an optimal spanning tree, any model allowing that possibility inherits the result.

Theorem 14. *For Minimum Spanning Forest in the Late Reject model or the Late Accept/Reject model, the competitive ratio is 1.*

Future Work

Since we prove tight results for all combinations of problems and models considered, we leave no immediate open problems. However, one could reasonably consider late operations a resource to be used sparingly, as for the rearrangements in [15, 21, 10, 11], for example. Thus, an interesting continuation of our work would be a study of trade-offs between the number of late operations employed and the quality of the solution (in terms of competitiveness). Obviously, one could also investigate other online problems and further model variations.

References

1. Y. Bartal, A. Fiat, and S. Leonardi. Lower bounds for on-line graph problems with application to on-line circuit and optical routing. In *28th STOC*, pages 531–540. ACM, 1996.
2. J. Boyar, S.J. Eidenbenz, L.M. Favrholdt, M. Kotrbčík, and K.S. Larsen. Online dominating set. In *15th SWAT*, volume 53 of *LIPIcs*, pages 21:1–21:15. Schloss Dagstuhl – Leibniz-Zentrum für Informatik GmbH, 2016.
3. J. Boyar, L.M. Favrholdt, M. Kotrbčík, and K.S. Larsen. Relaxing the irrevocability requirements for online graph algorithms. Technical Report arXiv:1704.08835 [cs.DS], arXiv, 2017.

4. M. Cygan, Ł. Jeż, and J. Sgall. Online knapsack revisited. *Theor. Comput. Syst.*, 58(1):153–190, 2016.
5. M. Demange and V.Th. Paschos. On-line vertex-covering. *Theor. Comput. Sci.*, 332:83–108, 2005.
6. L. Epstein, A. Levin, J. Mestre, and D. Segev. Improved approximation guarantees for weighted matching in the semi-streaming model. *SIAM J. Discrete Math.*, 25(3):1251–1265, 2011.
7. L. Epstein, A. Levin, D. Segev, and O. Weimann. Improved bounds for online preemptive matching. In *30th STACS*, volume 20 of *LIPIcs*, pages 389–399. Schloss Dagstuhl – Leibniz-Zentrum für Informatik GmbH, 2013.
8. J. Feigenbaum, S. Kannan, A. McGregor, S. Suri, and J. Zhang. On graph problems in a semi-streaming model. *Theor. Comput. Sci.*, 348(2–3):207–216, 2005.
9. J.A. Garay, I.S. Gopal, S. Kutten, Y. Mansour, and M. Yung. Efficient on-line call control algorithms. *J. Algorithm.*, 23(1):180–194, 1997.
10. A. Gu, A. Gupta, and A. Kumar. The power of deferral: Maintaining a constant-competitive steiner tree online. *SIAM J. Comput.*, 45(1):1–28, 2016.
11. A. Gupta and A. Kumar. Online steiner tree with deletions. In *25th SODA*, pages 455–467, 2014.
12. X. Han, Y. Kawase, and K. Makino. Randomized algorithms for online knapsack problems. *Theor. Comput. Sci.*, 562:395–405, 2015.
13. X. Han, Y. Kawase, K. Makino, and H. Guo. Online removable knapsack problem under convex function. *Theor. Comput. Sci.*, 540:62–69, 2014.
14. X. Han and K. Makino. Online minimization knapsack problem. *Theor. Comput. Sci.*, 609:185–196, 2016.
15. M. Imase and B.M. Waxman. Dynamic steiner tree problem. *SIAM J. Discrete Math.*, 4(3):369–384, 1991.
16. K. Iwama and S. Taketomi. Removable online knapsack problems. In *29th ICALP*, volume 2380 of *LNCS*, pages 293–305. Springer, 2002.
17. P. Jaillet and X. Lu. Online traveling salesman problems with rejection options. *Networks*, 64:84–95, 2014.
18. A.R. Karlin, M.S. Manasse, L. Rudolph, and D.D. Sleator. Competitive snoopy caching. *Algorithmica*, 3:79–119, 1988.
19. D. Komm, R. Královič, R. Královič, and C. Kudahl. Advice complexity of the online induced subgraph problem. In *41st MFCS*, volume 58 of *LIPIcs*, pages 59:1–59:13. Schloss Dagstuhl - Leibniz-Zentrum für Informatik, 2016.
20. B. Korte and D. Hausmann. An analysis of the greedy heuristic for independence systems. *Ann. Discrete Math.*, 2:65–74, 1978.
21. N. Megow, M. Skutella, J. Verschae, and A. Wiese. The power of recourse for online MST and TSP. *SIAM J. Comput.*, 45(3):859–880, 2016.
22. D. Rawitz and A. Rosén. Online Budgeted Maximum Coverage. In *24th ESA*, volume 57 of *LIIPCcs*, pages 73:1–73:17. Schloss Dagstuhl – Leibniz-Zentrum für Informatik GmbH, 2016.
23. B. Saha and L. Getoor. On maximum coverage in the streaming model & application to multi-topic blog-watch. In *9th SDM*, pages 697–708. SIAM, 2009.
24. D.D. Sleator and R.E. Tarjan. Amortized efficiency of list update and paging rules. *Communications of the ACM*, 28(2):202–208, 1985.
25. R.E. Tarjan. *Data Structures and Network Algorithms*, volume 44 of *CBMS-NSF regional conference series in applied mathematics*. SIAM, 1983.
26. D.E.D. Vinkemeier and S. Hougardy. A linear-time approximation algorithm for weighted matchings in graphs. *ACM T. Algorithms*, 1(1):107–122, 2005.

Approximating Small Balanced Vertex Separators in Almost Linear Time

Sebastian Brandt[✉] and Roger Wattenhofer

ETH Zürich, Zürich, Switzerland,
brandts@ethz.ch,
wattenhofer@ethz.ch

Abstract. For a graph G with n vertices and m edges, we give a randomized Las Vegas algorithm that approximates a small balanced vertex separator of G in almost linear time. More precisely, we show the following, for any $\frac{2}{3} \leq \alpha < 1$ and any $0 < \varepsilon < 1 - \alpha$: If G contains an α-separator of size K, then our algorithm finds an $(\alpha + \varepsilon)$-separator of size $\mathcal{O}(\varepsilon^{-1} K^2 \log^{1+o(1)} n)$ in time $\mathcal{O}(\varepsilon^{-1} K^3 m \log^{2+o(1)} n)$ w.h.p. In particular, if $K \in \mathcal{O}(\text{polylog } n)$, then we obtain an $(\alpha + \varepsilon)$-separator of size $\mathcal{O}(\varepsilon^{-1} \text{polylog } n)$ in time $\mathcal{O}(\varepsilon^{-1} m \text{ polylog } n)$ w.h.p. The presented algorithm does not require knowledge of K.

Due to space restrictions, no proofs are included in this version of the paper; the full version with a lot of additional material can be found at http://disco.ethz.ch/publications/wads2017-vertexsep.pdf.

1 Introduction

Motivation In order to solve a large computational problem, the problem is typically divided into smaller parts, and each part is solved on a single processor, in parallel. Some problems can be chopped into pieces in a straightforward way, e.g., using MapReduce or Spark. Other computational problems cannot be partitioned easily. Such difficult problems can frequently be represented as graphs: Each vertex represents some piece of work whereas an edge between two vertices denotes a relation between the two pieces, i.e., change at one vertex will directly affect the other (and possibly vice versa). There are dozens of software packages for distributed graph processing, e.g., Google's Pregel or PowerGraph.

In order to use multiple processors, the input graph has to be partitioned into multiple components, ideally of similar size. Then the vertices of a component are simulated on a single processor whereas edges between two vertices in different components are handled by the two processors responsible for the two components by exchanging messages. A natural objective of designing such a partition is to reduce the inter-processor communication as it is the expensive part in terms of runtime.

We argue that an input graph should be partitioned by means of a balanced vertex separator (and not a balanced edge cut), since vertex separators are often more efficient. For a simple example, consider a star graph, a tree where one single root is connected to all leaves. We want to partition the star for two

© Springer International Publishing AG 2017
F. Ellen et al. (Eds.): WADS 2017, LNCS 10389, pp. 229–240, 2017.
DOI: 10.1007/978-3-319-62127-2_20

processors. A star graph does not feature a small balanced edge cut, whereas the root is a perfectly good vertex separator. The root is simply replicated on both processors, and communication is reduced to the exchange between the two copies of the root vertex. In general, the computation and communication overhead of a vertex separator is asymptotically never worse than that of a balanced edge cut, whereas in some cases (such as the star graph) it can be a factor of n better, where n is the number of vertices in the graph.

In the last decades, algorithms research has made a lot of progress regarding balanced vertex separators, cf. [2, 6–9, 14]. To a large extent, these works focus on the fundamental case of dividing the input graph into two parts. For the remainder of this paper, we will also exclusively consider vertex separators that cut the input graph into two similar-sized pieces. Even though the algorithms given in the works cited above only need polynomial time, this is often too slow for practical purposes, as partitioning the input graph is the only non-parallel part of the whole process. What is needed is a "quick and dirty" way to compute a balanced vertex separator, i.e., an algorithm that (apart from a polylogarithmic factor) only reads the input once. So far, to the best of our knowledge, it is not known how to compute a balanced vertex separator for general graphs quickly.

Our goal is to *find* a reasonably small balanced vertex separator if there *exists* a small balanced vertex separator, e.g., of polylogarithmic size, and we want to achieve this in almost linear time. For a graph with n vertices and m edges, we show the following, for any $\frac{2}{3} \le \alpha < 1$ and any $0 < \varepsilon < 1 - \alpha$: If the graph contains an α-balanced vertex separator of size K, then our randomized Las Vegas algorithm finds an $(\alpha + \varepsilon)$-balanced vertex separator of size $\mathcal{O}(\varepsilon^{-1} K^2 \log^{1+o(1)} n)$ in time $\mathcal{O}(\varepsilon^{-1} K^3 m \log^{2+o(1)} n)$ w.h.p. Of course, this result can also be used for other practical applications related to balanced vertex separators, e.g., for determining quickly if a network has serious bottlenecks and locating them in the affirmative case. If no fixed K is considered, by successive doubling we can quickly reach a size K for which an α-separator exists, yielding only an additional small constant factor for the time complexity. In particular, using this technique, our algorithm does not require knowledge of K. If, on the other hand, the input graph does not contain a small separator, our algorithm will report the lack thereof. Note that graphs without small vertex separators may not be amenable to distributed graph processing in the first place, and one may wonder whether parallelism can speed up processing such graphs at all.

Related Work As discussed above, finding a balanced edge separator does not yield a balanced vertex separator with a similar approximation guarantee in general. Since there is an abundance of results regarding edge separators, we will only mention them if they are also related to vertex separators.

Let $G = (V, E)$ be a graph with n vertices and m edges. An α-separator of G is a triple (A, S, B) of disjoint subsets of V s.t. $V = A \cup S \cup B$, there are no edges between A and B, and $\max\{|A|, |B|\} \le \alpha |V|$. Its size is $|S|$.

The problem of finding an α-separator of minimum size is NP-hard, as shown by Bui and Jones [4]. Hence, one main focus of research in the context of balanced vertex separators has been to find approximation algorithms, cf., e.g., [2,6,7]. In their seminal paper [14], Leighton and Rao gave an $\mathcal{O}(\log n)$-approximation for balanced edge separator, incurring only an arbitrarily small loss in the balance. As they showed, their result extends to the case of directed edge separators and thereby to vertex separators. Feige, Hajiaghayi and Lee [8] proved that a $\frac{3}{4}$-separator of size $K \log^{\frac{1}{2}} K$ can be found in polynomial time if the input graph contains a $\frac{2}{3}$-separator of size K. Subsequently, Feige and Mahdian [9] showed, for any $\frac{2}{3} \le \alpha < 1$, how to find an α-separator of size K if such a separator exists, except when there is an $(\alpha + \varepsilon)$-separator of smaller size in which case they find the latter. Their runtime is polynomial if $K \in \mathcal{O}(\log n)$, for fixed ε.

As shown by Marx [17], the problem of finding an α-separator of minimum size is even $W[1]$-hard. In their work [9] mentioned above, Feige and Mahdian solve this issue by showing that the problem becomes fixed parameter tractable if the balance requirement is relaxed, obtaining a runtime of $n^{\mathcal{O}(1)} 2^{\mathcal{O}(K)}$ which is polynomial for $K \in \mathcal{O}(\log n)$. We show that if we relax the requirements on balance *and* size of the separator, then we can achieve an almost linear runtime.

The techniques used in the works above, e.g., linear or semidefinite programming, focus on achieving as good approximation ratios as possible while having polynomial time complexity. By applying their primal-dual approach for semidefinite programs [3] to the problem of approximating minimum balanced separators, Arora and Kale achieved a runtime of $\tilde{\mathcal{O}}(m^{\frac{3}{2}} + n^{2+\varepsilon})$ (resp. $\tilde{\mathcal{O}}(m^{\frac{3}{2}})$), for obtaining an approximation ratio of $\mathcal{O}(\log^{\frac{1}{2}} n)$ (resp. $\mathcal{O}(\log n)$). Although achieved in the context of directed edge separators, the given runtimes and approximation ratios apply directly to our problem of undirected vertex separators.

A different line of research consists in searching for primarily fast algorithms that yield separators of not necessarily near-optimal size. For graph classes with certain restrictions, there are a number of results obtaining good runtimes, often at the expense of the separator size depending polynomially on n. Gilbert, Hutchinson and Tarjan [10] gave a linear-time algorithm for finding a $\frac{2}{3}$-separator of size $\mathcal{O}((gn)^{1/2})$ where g is the genus of the given graph, thereby extending the famous planar separator theorem by Lipton and Tarjan [15]. The same linear runtime was achieved independently by Djidjev [5].

A further extension to graphs excluding certain minors was given by Alon, Seymour and Thomas [1]. They showed how to find, for a graph containing no minor K_j for some fixed integer j, a $\frac{2}{3}$-separator of size $\mathcal{O}(n^{\frac{1}{2}})$ in time $\mathcal{O}(n^{\frac{3}{2}})$. Reed and Wood [19] gave an algorithm which solves the same problem in linear time except that the separators are of somewhat larger size $\mathcal{O}(n^{\frac{2}{3}})$. Furthermore, they showed how to trade runtime for separator size in a parametrized way, bounded by those two results. Kawarabayashi and Reed [13] improved the runtime for finding a separator of size $\mathcal{O}(n^{\frac{1}{2}})$ to $\mathcal{O}(n^{1+\varepsilon})$, for any $\varepsilon > 0$, additionally improving the dependency of the separator size on the number j of vertices of the excluded minor. Unfortunately, the runtime depends heavily on j, making the algorithm infeasible in practice. Wulff-Nilsen [20] gave an algorithm which depends only polynomially on j, at a slight expense of runtime and separator size. Moreover, he showed how to find, for constant $c < 1$ and fixed

j, a separator of size $\mathcal{O}(n^c)$ in linear time. We are not aware of any results for general graphs (regarding balanced vertex separators) that focus on achieving a near-linear time complexity.

As mentioned earlier, recently various software packages to handle large graphs have been introduced, e.g., Pregel [16] or PowerGraph [11]. Some of them include simple heuristics to partition the input graph into pieces. Power-Graph, for instance, merely removes vertices with large degrees until the graph falls into small enough pieces. In practice, this seems to work well on power-law graphs, which include many interesting application areas such as, e.g., social networks. We believe that our work will help to find a theoretical foundation for this practical problem while also providing an implementable solution.

Our Approach In the following, we give an overview of our approach without providing formal accuracy. Exact definitions will follow in the next section. Our approach is based on maximum s-t-flows. By the very nature of flows, it is likely that such an approach can only find a near-optimally sized balanced vertex separator *quickly* if the considered graph actually contains a reasonably small balanced vertex separator. As explained before, this restricted problem is still very important in practice, thus we deem the presented approach to be a worthwhile endeavour while having the advantage of (conceptual) simplicity.

Assume we are given a graph G containing a small vertex separator and we have vertices s and t "on different sides" of the separator. Then, by Menger's Theorem (cf. [18]), the maximum number of pairwise vertex-disjoint s-t-paths is also small. We start by computing a set of maximum cardinality of pairwise vertex-disjoint s-t-paths. By using the Ford-Fulkerson algorithm (cf. [12]), this can be done in almost linear time as such a path collection corresponds to a maximum s-t-flow in an unweighted directed graph obtained from G by a simple transformation. From this collection of k paths we extract s-t-vertex cuts of the same cardinality k by taking one vertex from each path. These vertices have to be chosen carefully in order to actually separate s and t, but the existence of the s-t-vertex cuts is ensured, again, by Menger's Theorem. Using binary search, we determine two of the "best-balanced" of all these s-t-cuts, one closer to s and one closer to t. If one of these two cuts is sufficiently balanced, then we have found the desired small balanced vertex separator. Otherwise, consider the connected components cut off by the two s-t-cuts. We contract the two connected components containing s and t into new vertices s' and t', respectively.

All s'-t'-vertex cuts in the newly obtained graph are also s-t-cuts in G and additionally better-balanced than the above two s-t-cuts. We will prove that the maximum number of pairwise vertex-disjoint s'-t'-paths is larger than k (and therefore the same is true for the cardinality of any s'-t'-cut corresponding to such a path collection). We iterate the above process of finding vertex-disjoint paths, extracting some of the best-balanced s-t-cuts and contracting vertex sets, until we obtain s-t-cuts whose cardinality is equal to some predetermined value K (or observe that no such cut of cardinality K exists).

Consider an α-separator of size at most K separating s and t, where $\frac{2}{3} \leq \alpha < 1$. If the iterative process described above does not yield cuts whose (combined) balance is at least as good as α, then, as we will prove, at least one vertex of the α-separator must have been involved in one of the performed contractions. Thus, by iterating the whole process at most K times (with newly chosen s, t in each iteration), we obtain a balanced vertex separator (by collecting the relevant cuts obtained in the process). We will show that if G contains a small balanced vertex separator, then the obtained balanced vertex separator is also small.

Up to now, we assumed that we can find vertices s and t "on different sides" of a balanced separator. But because of the balance of the separator, this is actually the case with a large enough probability. By choosing s and t uniformly at random, applying the iterative process described above and then iterating the whole procedure on the largest obtained connected component, we obtain an almost linear runtime for finding a reasonably small balanced vertex separator, provided the given graph contains a small balanced vertex separator.

Due to space restrictions, all proofs (which show many of the intricacies of the presented work) are deferred to the full version, which can be found at http://disco.ethz.ch/publications/wads2017-vertexsep.pdf.

2   Conventions and Basic Definitions

In this work, we consider simple, undirected, connected graphs $G = (V, E)$ without self-loops, with n nodes and m edges. We call a triple (A, S, B) of pairwise disjoint subsets of V a *vertex separator* of G if $V = A \sqcup S \sqcup B$ and there is no $\{u, v\} \subset E$ s.t. $u \in A, v \in B$. We call $|S|$ the *size* of (A, S, B). Let $0 < \alpha < 1$. If $\max\{|A|, |B|\} \leq \alpha|V|$, then we call (A, S, B) α-*balanced* or, equivalently, an α-separator. Let $s, t \in V$ s.t. $s \in A$, $t \in B$. Then we call (A, S, B) an s-t-*vertex separator*. Let $s, t \in V, s \neq t$ and let $\{f_1, ..., f_k\}, k \in \mathbb{N}_{>0}$ be a set of s-t-paths in G. Then we say that $f_1, ..., f_k$ are *pairwise vertex-disjoint* if there are no vertices except s and t that appear in more than one of these paths. For all subsets $X \subseteq V$, we denote the induced subgraph of G whose vertex set is X by $G[X]$.

We will often consider two special vertices s, t. For the remainder of this work, we will assume that s and t are different, non-adjacent vertices if not specified otherwise. Moreover, for convenience, we will be not too technical regarding the distinction between sets and tuples.

In 1927, Karl Menger [18] stated the following famous theorem:

Theorem 1. *The maximum number of pairwise vertex-disjoint s-t-paths in a graph G is equal to the minimum number of vertices v, $s \neq v \neq t$, which have to be removed from G in order that there is no s-t-path in the resulting graph.*

Consider a set of maximum cardinality of pairwise vertex-disjoint s-t-paths. By Menger's Theorem, we can disconnect s from t by removing one vertex from each of those paths. Of course, if we choose these vertices arbitrarily (but still one per path), then it is not ensured that there is no s-t-path left. We call such a set of arbitrarily chosen vertices a *slice* whereas we call it a *cut* if its removal results in a disconnection of s from t. In more formal terms:

Definition 2. *Let G be a graph and $s, t \in V$. Let $\{f_1, ..., f_k\}$ be a set of pairwise vertex-disjoint s-t-paths in G. Then we call a tuple $(w_1, ..., w_k)$ a* slice *(w.r.t. $(f_1, ..., f_k)$) if $w_i \in f_i$, $s \neq w_i \neq t$ for all $1 \leq i \leq k$. Let X be an arbitrary subset of V s.t. $s, t \notin X$. If there is no s-t-path in $G[V \backslash X]$, then we say that X separates s and t. We call a slice that separates s and t a* cut.

Following Marx [17], the set of slices (w.r.t some fixed set of s-t-paths) can be partially ordered by their relative "closeness" to s. The following definition adapts the definition of the "dominance relation" given in [17] to our setting.

Definition 3. *Let $\{f_1, ..., f_k\}$ be a set of pairwise vertex-disjoint s-t-paths in G. Let $U = (u_1, ... u_k)$ and $W = (w_1, ..., w_k)$ be slices w.r.t. $(f_1, ..., f_k)$ s.t., for all $1 \leq i \leq k$, u_i is a predecessor of w_i in f_i or $u_i = w_i$. Then we say that U is* closer *to s than W and write $U \preceq W$. If additionally $u_i \neq w_i$ for some $1 \leq i \leq k$, then we say that U is* strictly *closer to s than W and write $U \prec W$. Analogously, we say that W is (strictly) closer to t than U. For convenience, we define the above analogously for the tuples $(s, s, ..., s)$ and $(t, t, ..., t)$. Thus we can, e.g., say that $(s, s, ..., s)$ is closer to s than any slice.*

The removal of a cut decomposes G into at least two connected components as s and t are not connected anymore. The component containing s and the component containing t are of special interest to us since we will develop a method to make them larger (by choosing "better" cuts) which, in turn, aids in finding cuts (or, more precisely, s-t-vertex separators) of "better balance".

Definition 4. *Let U be an arbitrary cut. We define $V_s(U)$ as the vertex set of the connected component of $G[V \backslash U]$ containing s, $V_t(U)$ as the vertex set of the connected component of $G[V \backslash U]$ containing t and $V_r(U)$ as the union of the vertex sets of the remaining connected components of $G[V \backslash U]$, i.e., those containing neither s nor t (so $V_r(U)$ may be empty).*

3   Closest Cuts

Consider a slice U. Among all cuts that are closer to t (resp. s) than U, we would like to single the "closest one" out. Our partial order "$\preceq$" provides a very intuitive way to do so, resulting in Definition 5. A proof for the uniqueness of such a cut is given in the full version. Lemma 6 shows that closest cuts can be computed in linear time. For the remainder of this paper, let $\{f_1, ..., f_k\}$ be a set of maximum cardinality of pairwise vertex-disjoint s-t-paths.

Definition 5. *Let U be a slice w.r.t. $(f_1, ..., f_k)$. Let X be a cut s.t. $U \preceq X$ and there is no cut X' satisfying $U \preceq X' \prec X$. Then we define $U^+ := X$. If there exists no X as described above, then set $U^+ := (t, t, ..., t)$. Analogously, let Y be a cut s.t. $Y \preceq U$ and there is no cut Y' satisfying $Y \prec Y' \preceq U$. Then we define $U^- := Y$. If there exists no Y as described above, then set $U^- := (s, s, ..., s)$.*

Lemma 6. *We can compute U^+ and U^- in time $\mathcal{O}(m)$.*

Algorithm 1 Find Innermost s-sided Cut of Minimum Size

Initialization: Given weights $g(s), g(t) \in \mathbb{N}_{>0}$ and the maximum number of pairwise vertex-disjoint s-t-paths $f_1, ..., f_k$, let $v_{ij}, 0 \le j \le \ell_i$, be the jth vertex of the path f_i where ℓ_i is the length of f_i and s is considered to be the 0th vertex of every f_i. Set $w_i := v_{i1}$ for all $1 \le i \le k$ and valid := **false**.

```
 1: for i = 1 to k do
 2:     c := 0                              // indexes the start of path f_i
 3:     d := ℓ_i                            // indexes the end of path f_i
 4:     while d ≠ c + 1 do
 5:         e := ⌈(c+d)/2⌉                  // binary search on f_i
 6:         W := (w_1, ..., w_{i-1}, v_{ie}, w_{i+1}, ..., w_k)  // get new slice by moving vertex on f_i
 7:         if W^+ ≠ (t, t, ..., t) and |V_s(W^+)| + g(s) ≤ |V_r(W^+) ∪ V_t(W^+)| + g(t) then
 8:             c := e      // W is suitable, continue binary search in direction towards t
 9:             valid := true                          // suitable cut found
10:         else
11:             d := e  // W is not suitable, continue binary search in direction towards s
12:         end if
13:     end while
14:     w_i := v_{ic}         // fix best vertex found on f_i, continue with next path
15: end for
16: if valid then
17:     return (w_1, ..., w_k)
18: else
19:     return (s, s, ..., s)
20: end if
```

---

4   An Algorithm for Finding Good Cuts of Bounded Size

Consider a graph G which contains an s-t-vertex separator of size at most K, for some fixed s, t. The goal of this section is to find a good cut of size at most K. Intuitively, a cut U is "good" if the connected components obtained by removing U from G can be divided into two groups in a balanced way. Unfortunately, it is not easy to find such a cut quickly. Thus, we relax our notion of "good" slightly and say that a cut U, that does not admit a balanced partition of the connected components, is still good if it satisfies the following property: For any s-t-vertex separator (A^*, S^*, B^*) of size at most K with a better balance than U, S^* is not contained in any of the connected components obtained by removing U. The idea is to iterate the process of finding a good cut on the largest component obtained by removing the (previous) good cut and to benefit from the fact that the size of (A^*, S^*, B^*) restricted to this component decreases by at least 1 in each iteration. The details of this idea will be discussed in Section 5.

The first step in order to design an algorithm that finds a good cut is to develop a method for finding a cut U (if it exists) s.t. $|V_s(U)| \le |V_r(U) \cup V_t(U)|$ and all cuts which are strictly closer to t than U violate that property. This can be done efficiently using binary search, as given by Algorithm 1. Essentially, this algorithm moves a vertex of some initial slice closer to t along an s-t-path, thereby obtaining a new slice, and checks if the cut closest to this slice in direction towards t still satisfies the above inequality. In the affirmative case, it iterates starting from this new slice, otherwise it goes back and tries another vertex. Since, later on, we will have to deal with graphs which are the result of a series of contractions, we design Algorithm 1 in a rather general way where weights are assigned to s and t. Regarding output and runtime, we prove the following, using Lemma 6 in the process:

Lemma 7. *Let $g(s), g(t)$ be positive integers. If there is a cut X s.t. $|V_s(X)| + g(s) \leq |V_r(X) \cup V_t(X)| + g(t)$, then Algorithm 1 returns such a cut X with the additional property that $|V_s(X')| + g(s) > |V_r(X') \cup V_t(X')| + g(t)$ for all cuts X' satisfying $X \prec X'$. If there is no such cut, then the algorithm returns the tuple $(s, s, ..., s)$. In both cases Algorithm 1 terminates in time $\mathcal{O}(km \log n)$.*

For reasons of symmetry, Algorithm 1 and Lemma 7 also work if s and t are reversed. The respective versions are given in the full paper. We will denote the "reversed" algorithm and lemma by Algorithm 1', resp. Lemma 7'.

With the tools gathered above, we are now able to design and analyze an algorithm (Algorithm 2) which finds a pair of tuples that contains a good cut. As we will perform contractions on a given graph G in the process, we give a short overview of the technical details. The contraction of a subset U of V transforms G into a graph H where $V(H) := (V \backslash U) \cup \{u\}$ and $E(H)$ contains an edge $\{u, w\}$ for each edge $\{v, w\} \in E$ satisfying $v \in U$, $w \in V \backslash U$ while all edges in G between vertices in $V \backslash U$ remain edges in H. We call u the contraction of U.

Essentially, Algorithm 2 uses Algorithm 1 and Algorithm 1' as subroutines in order to find two cuts that cut a preferably large part containing s, resp. t, off. Then it contracts these two parts into new nodes s and t and iterates on the obtained graph. We show in the following that the number of pairwise vertex-disjoint s-t-paths grows in each iteration and that the performed contractions ensure that s and t remain non-adjacent. This enables us to bound the runtime of Algorithm 2 (in the process using Ford and Fulkerson's maximum flow algorithm). For the proofs of both lemmas, we need Lemmas 7 and 7'.

Lemma 8. *Consider an iteration of the while loop in Algorithm 2 where (not necessarily non-trivial) contractions are performed. Then, at the end of the iteration, there is no edge $\{s, t\} \in E(H)$ and there are (at least) $k + 1$ pairwise vertex-disjoint s-t-paths in H.*

Lemma 9. *Algorithm 2 terminates in time $\mathcal{O}(K^2 m \log n)$.*

What is left to show is that the returned pair contains indeed a good cut ("good" as described in the introduction of Section 4). Theorem 10 takes care of that, using again Lemmas 7 and 7' in the proof. For ease of presentation, define, for any subset U of V, $L_U := \max\{|V(C)| \mid C$ is a connected component of $G[V \backslash U]\}$.

Algorithm 2 Find Innermost Cut of Bounded Size

Initialization: Given a positive integer K and two vertices $s, t \in V$, set $g(s) := g(t) := 1$, $k := 0$, $H := G$ and $S := T := \{\}$.

1: **while** the max. number of pairwise vertex-disjoint s-t-paths in H is at most K **do**
2:    find max. number of pairwise vertex-disjoint s-t-paths $f_1, ..., f_k$ in H, update k
3:    execute Algorithm 1, let U be output $//$ U cuts "large" part containing s off
4:    execute Algorithm 1', let W be output $//$ W cuts "large" part containing t off
5:    **if** $U \neq (s, s, ..., s)$ **then**
6:       $M_s := V_s(U) \cup U$ $//$ collect vertex set which is to be contracted
7:       $S := U$ $//$ U is best "s-sided" cut we found so far
8:    **else**
9:       $M_s := \{s\}$ $//$ no s-sided cut found, so nothing to contract
10:   **end if**
11:   **if** $W \neq (t, t, ..., t)$ **then**
12:      $M_t := V_t(W) \cup W$ $//$ collect vertex set which is to be contracted
13:      $T := W$ $//$ W is best "t-sided" cut we found so far
14:   **else**
15:      $M_t := \{t\}$ $//$ no t-sided cut found, so nothing to contract
16:   **end if**
17:   **if** $M_s \cap M_t \neq \emptyset$ or there is an edge from M_s to M_t in H **then**
18:      **break** $//$ contracting M_s and M_t impossible/problematic for later iterations
19:   **else**
20:      contract M_s and denote the contraction by s (and update H accordingly)
21:      contract M_t and denote the contraction by t (and update H accordingly)
22:      replace parallel edges of H by a single edge
23:      $g(s) := g(s) + |M_s| - 1$ $//$ update total number of vertices contracted into s
24:      $g(t) := g(t) + |M_t| - 1$ $//$ update total number of vertices contracted into t
25:   **end if**
26: **end while**
27: **return** the pair (S, T)

Theorem 10. *Let K be some positive integer and let s, t be two vertices of G. Let (S, T) be the pair returned by Algorithm 2 and suppose that there is an s-t-vertex separator (A^*, S^*, B^*) of G of size $K' \leq K$. Now if there is a connected component C of $G[V \backslash (S \cup T)]$ s.t. $S^* \subseteq V(C)$, then one of the following holds:*

    (i) $L_S \leq L_{S^*}$ *(ii)* $L_T \leq L_{S^*}$ *(iii)* $L_S \leq \frac{1}{2}|V|$ *(iv)* $L_T \leq \frac{1}{2}|V|$

5 Approximating Small Balanced Vertex Separators Quickly

In this section, we finally design an algorithm that uses Algorithm 2 as a subroutine in order to find a small and balanced vertex separator. The idea of Algorithm 3 is based on Algorithm 2 in conjunction with Theorem 10. Let $\frac{2}{3} \leq \alpha < 1$, and assume that there is a small α-separator (A^*, S^*, B^*) of G. By removing vertices of the given graph G, we obtain different connected components of which we then choose the largest one and iterate on this component. The goal is to reduce thereby the size of the largest component to approximately $\alpha|V|$ at most while removing only a small number of vertices in the process.

In more detail, Algorithm 3 chooses two vertices s, t in each iteration and then uses Algorithm 2 to find a small number of vertices which it then removes. There is a "large enough" probability that s and t are on different sides of the separator whose existence we assumed above since this separator is α-balanced. If s and t are on different sides, then Theorem 10 ensures that the size of the largest resulting connected component is at most $\alpha|V|$ or that the above separator contains strictly fewer vertices in the "separating set" S^* when restricted to this component. By iterating, we can reduce the number of vertices in this "separating set" to 0 (if the latter case occurs repeatedly) and then the balance of the above separator ensures that the largest component is of size at most $\alpha|V|$.

Since the balance of the separator may decrease radically by restricting it to the largest obtained component, so does the probability of s and t being on different sides of the separator. Thus, in order to obtain a good runtime, we stop when we achieve a balance close to α. Lemma 12 formalizes some of the above considerations. It enables us to prove Corollary 13 which gives an upper bound for the number of so-called successful iterations, an intuitive concept defined in the following. Using Corollary 13, we prove Lemma 14 which provides an upper bound for the runtime of Algorithm 3 in terms of the number of iterations. Note that the proof of Lemma 12 relies heavily on Theorem 10.

Definition 11. *Let (A^*, S^*, B^*) be a vertex separator. Then we call an iteration of the while loop of Algorithm 3 unsuccessful (w.r.t. (A^*, S^*, B^*)) if, for the vertices s, t chosen in that iteration, we have $s, t \in A^*$ or $s, t \in B^*$. If this is not the case, then we call the iteration successful.*

Lemma 12. *Let $\frac{2}{3} \leq \alpha < 1$ and let (A^*, S^*, B^*) be an α-separator of G of size at most K. Consider a successful iteration of the while loop of Algorithm 3 w.r.t. (A^*, S^*, B^*). Let H_0 and H_1 denote the graph H at the beginning, resp. the end, of this iteration. Then $|V(H_1) \cap S^*| \leq |V(H_0) \cap S^*| - 1$ or $|V(H_1)| \leq \alpha|V|$.*

Corollary 13. *Let $\frac{2}{3} \leq \alpha < 1$ and let (A^*, S^*, B^*) be an α-separator of G of size at most K. Then the number of successful iterations w.r.t. (A^*, S^*, B^*) is at most K in any execution of Algorithm 3.*

Lemma 14. *Let $\frac{2}{3} \leq \alpha < 1$. If there exists an α-separator of size at most K, then Algorithm 3 terminates after $\mathcal{O}(\varepsilon^{-1} K \log^{1+o(1)} n)$ iterations w.h.p.*

Using Lemma 9 and Lemma 14, we are able to prove our main result. The given bound on the separator size follows from the fact that the cardinality of the vertex sets S and T returned by Algorithm 2 never exceeds K.

Theorem 15. *Let $\frac{2}{3} \leq \alpha < 1$ and $0 < \varepsilon < 1 - \alpha$. If G contains an α-separator of size at most K, then Algorithm 3 finds an $(\alpha + \varepsilon)$-separator of size $\mathcal{O}(\varepsilon^{-1} K^2 \log^{1+o(1)} n)$ in time $\mathcal{O}(\varepsilon^{-1} K^3 m \log^{2+o(1)} n)$ w.h.p.*

Algorithm 3 Find Small Balanced Vertex Separator

Initialization: Given a graph G, a positive integer K, some $\frac{2}{3} \leq \alpha < 1$ and some $0 < \varepsilon < 1 - \alpha$, set $H := G$ and $S' := \{\}$.

1: **while** $|V(H)| > (\alpha + \varepsilon)|V|$ **do**
2:     choose two vertices s, t in H uniformly at random // can be identical/adjacent
3:     **if** $s = t$ or s and t are adjacent **then**
4:         find a largest connected component C of $H[V(H)\backslash(\{s\} \cup \{t\})]$ // remove s, t
5:         $H := C$ // continue on the resulting largest connected component
6:         $S' := S' \cup \{s\} \cup \{t\}$ // remember the removed vertices
7:     **else**
8:         execute Algorithm 2 (with input H, K, s, t) and denote the output by (S, T)
9:         compute a largest connected component C of $H[V(H)\backslash(S \cup T \cup \{s\} \cup \{t\})]$
10:        $H := C$ // continue on the resulting largest connected component
11:        $S' := S' \cup S \cup T \cup \{s\} \cup \{t\}$ // remember the removed vertices
12:    **end if**
13: **end while**
14: order the connected components $C_1, C_2, \ldots$ of $G[V\backslash S']$ s.t. $|V(C_1)| \geq |V(C_2)| \geq \ldots$
15: $A := V(C_1)$ // start collecting (vertex sets of) components
16: add the vertex sets $V(C_2), V(C_3), \ldots$ successively to A as long as the resulting A
    satisfies $|A| \leq \alpha|V|$ // note that $V(C_1)$ could already contain $(\alpha + \varepsilon)|V|$ vertices
17: $B := V(G)\backslash(A \cup S')$ // collect the "other side" of the vertex separator
18: **return** (A, S', B)

By Theorem 15, we can find a reasonably small $(\alpha + \varepsilon)$-separator in almost linear time, provided that G contains a small α-separator. In particular, we obtain:

- If $K \in \mathcal{O}(\text{polylog}\, n)$, then Algorithm 3 finds an $(\alpha + \varepsilon)$-separator of size $\mathcal{O}(\varepsilon^{-1} \text{polylog}\, n)$ in time $\mathcal{O}(\varepsilon^{-1} m\, \text{polylog}\, n)$ w.h.p.
- If $K \in \mathcal{O}(\log n)$, then Algorithm 3 finds an $(\alpha + \varepsilon)$-separator that has size $\mathcal{O}(\varepsilon^{-1} \log^{3+o(1)} n)$ in time $\mathcal{O}(\varepsilon^{-1} m \log^{5+o(1)} n)$ w.h.p.

Throughout this work, we supposed that we have a fixed K which gives us an upper bound for the size of the α-separator whose existence we assume. If we do not want to consider some specific K, but rather find some (ideally small) K for which Algorithm 3 returns a separator as specified in Theorem 15, we can do this by successively doubling K, each time executing Algorithm 3. The obtained total runtime is asymptotically the same as the runtime of Algorithm 3.

We consider our approach to be a first step in a new direction. We are confident that future work building on this approach can improve the presented theoretical bounds significantly. One reason (of many) for this is the following: One factor of K in the runtime is due to the possibility that, in each of K successful iterations, the number of vertices in S^* contained in the largest connected component potentially decreases by only 1 (see the proof of Corollary 26, full version). We expect that graphs exhibiting such an incremental decrease must have very specific structures that can be exploited. Moreover, from a practical (and entirely informal) standpoint, we note that the hidden constants are fairly small and that all three factors of K in the runtime should be significantly lower than K on average.

References

1. Noga Alon, Paul D. Seymour, and Robin Thomas. A separator theorem for graphs with an excluded minor and its applications. *STOC 1990*.
2. Eyal Amir, Robert Krauthgamer, and Satish Rao. Constant factor approximation of vertex-cuts in planar graphs. *STOC 2003*.
3. Sanjeev Arora and Satyen Kale. A combinatorial, primal-dual approach to semidefinite programs. *STOC 2007*.
4. Thang Nguyen Bui and Curt Jones. Finding good approximate vertex and edge partitions is NP-hard. *Information Processing Letters*, 42(3):153–159, 1992.
5. H. N. Djidjev. A linear algorithm for partitioning graphs of fixed genus. *Serdica*, 11(4):369–387, 1985.
6. Guy Even, Joseph Naor, Satish Rao, and Baruch Schieber. Divide-and-conquer approximation algorithms via spreading metrics. *FOCS 1995*.
7. Guy Even, Joseph Naor, Satish Rao, and Baruch Schieber. Fast approximate graph partitioning algorithms. *SODA 1997*.
8. Uriel Feige, Mohammad Taghi Hajiaghayi, and James R. Lee. Improved approximation algorithms for minimum-weight vertex separators. *STOC 2005*.
9. Uriel Feige and Mohammad Mahdian. Finding small balanced separators. *STOC 2006*.
10. John R. Gilbert, Joan P. Hutchinson, and Robert Endre Tarjan. A separator theorem for graphs of bounded genus. *Journal of Algorithms*, 5(3):391–407, 1984.
11. Joseph E. Gonzalez, Yucheng Low, Haijie Gu, Danny Bickson, and Carlos Guestrin. Powergraph: Distributed graph-parallel computation on natural graphs. *OSDI 2012*.
12. L. R. Ford Jr. and D. R. Fulkerson. Maximal flow through a network. *Canadian Journal of Mathematics*, 8:399–404, 1956.
13. Ken-ichi Kawarabayashi and Bruce A. Reed. A separator theorem in minor-closed classes. *FOCS 2010*.
14. Frank Thomson Leighton and Satish Rao. Multicommodity max-flow min-cut theorems and their use in designing approximation algorithms. *Journal of the ACM*, 46(6):787–832, 1999.
15. Richard J. Lipton and Robert E. Tarjan. A separator theorem for planar graphs. *SIAM Journal on Applied Mathematics*, 36(2):177–189, 1979.
16. Grzegorz Malewicz, Matthew H. Austern, Aart J. C. Bik, James C. Dehnert, Ilan Horn, Naty Leiser, and Grzegorz Czajkowski. Pregel: a system for large-scale graph processing. *SIGMOD 2010*.
17. Dániel Marx. Parameterized graph separation problems. *Theoretical Computer Science*, 351(3):394–406, 2006.
18. Karl Menger. Zur allgemeinen Kurventheorie. *Fundamenta Mathematicae*, 10(1):96–115, 1927.
19. Bruce A. Reed and David R. Wood. A linear-time algorithm to find a separator in a graph excluding a minor. *ACM Transactions on Algorithms*, 5(4), 2009.
20. Christian Wulff-Nilsen. Separator theorems for minor-free and shallow minor-free graphs with applications. *FOCS 2011*.

Balanced Line Separators of Unit Disk Graphs[*]

Paz Carmi[1], Man Kwun Chiu[2,3], Matthew J. Katz[1], Matias Korman[4],
Yoshio Okamoto[5], André van Renssen[2,3], Marcel Roeloffzen[2,3],
Taichi Shiitada[5], and Shakhar Smorodinsky[1]

[1] Ben-Gurion University of the Negev, Beer-Sheva, Israel
[2] National Institute of Informatics, Tokyo, Japan
[3] JST, ERATO, Kawarabayashi Large Graph Project, Japan
[4] Tohoku University, Sendai, Japan
[5] The University of Electro-Communications, Tokyo, Japan

Abstract. We prove a geometric version of the graph separator theorem
for the unit disk intersection graph: for any set of n unit disks in the plane
there exists a line ℓ such that ℓ intersects at most $O(\sqrt{(m+n)} \log n)$
disks and each of the halfplanes determined by ℓ contains at most $2n/3$
unit disks from the set, where m is the number of intersecting pairs of
disks. We also show that an axis-parallel line intersecting $O(\sqrt{m+n})$
disks exists, but each halfplane may contain up to $4n/5$ disks. We give
an almost tight lower bound (up to sublogarithmic factors) for our approach, and also show that no line-separator of sublinear size in n exists
when we look at disks of arbitrary radii, even when $m = 0$. Proofs are
constructive and suggest simple algorithms that run in linear time. Experimental evaluation has also been conducted, which shows that for
random instances our method outperforms the method by Fox and Pach
(whose separator has size $O(\sqrt{m})$).

1 Introduction

Balanced separators in graphs are a fundamental tool and used in many divide-
and-conquer-type algorithms as well as for proving theorems by induction. Given
an undirected graph $G = (V, E)$ with V as its vertex set and E as its edge set,
and a non-negative real number $\alpha \in [1/2, 1]$, we say that a subset $S \subseteq V$ is an
α-separator if the vertex set of $G \setminus S$ can be partitioned into two sets A and
B, each of size at most $\alpha|V|$ such that there is no edge between A and B. The
parameter α determines how balanced the two sets A and B are in terms of size.

[*] Chiu, van Renssen and Roeloffzen were supported by JST ERATO Grant Num-
ber JPMJER1305, Japan. Korman was supported in part by KAKENHI Nos.
12H00855 and 17K12635. Katz was partially supported by grant 1884/16 from the
Israel Science Foundation. Okamoto was partially supported by KAKENHI Grant
Numbers JP24106005, JP24220003 and JP15K00009, JST CREST Grant Number
JPMJCR1402, and Kayamori Foundation for Informational Science Advancement.
Smorodinsky's research was partially supported by Grant 635/16 from the Israel
Science Foundation.

© Springer International Publishing AG 2017
F. Ellen et al. (Eds.): WADS 2017, LNCS 10389, pp. 241–252, 2017.
DOI: 10.1007/978-3-319-62127-2_21

For a balanced separator to be useful we want both the size $|S|$ of the separator and $\alpha \geq 1/2$ to be small.

Much work has been done to prove the existence of separators with certain properties in general sparse graphs. For example, the well-known Lipton–Tarjan planar separator theorem [12] states that for any n-vertex planar graph, there exists a $2/3$-separator of size $O(\sqrt{n})$. Similar theorems have been proven for bounded-genus graphs [7], minor-free graphs [2], low-density graphs, and graphs with polynomial expansion [17, 8]. Note that graphs in each of these graph classes contain only $O(n)$ edges, where n is the number of vertices.

These separator results are for abstract planar graphs and are very general. Our focus of interest is geometric graphs, which often encode additional information other than an adjacency matrix. Even though one can use the separator tools in geometric graphs, often the additional information is lost in the process. As such, a portion of the literature has focused on the search of balanced separators that also preserve the geometric properties of the geometric graph.

Among several others, we highlight the work of Miller $et\ al.$ [16], and Smith and Wormald [18]. They considered intersection graphs of n balls in $\mathbb{R}^d$ and proved that if every point in d-dimensional space is covered by at most k of the given balls, then there exists a $(d+1)/(d+2)$-separator of size $O(k^{1/d}n^{1-1/d})$ (and such a separator can be found in deterministic linear time [4]). More interestingly, the separator itself and the two sets it creates have very nice properties; they show that there exists a $(d-1)$-dimensional sphere that intersects at most $O(k^{1/d}n^{1-1/d})$ balls and contains at most $(d+1)n/(d+2)$ balls in its interior and at most $(d+1)n/(d+2)$ balls in its exterior. In this case, the sphere acts as the separator (properly speaking, the balls that intersect the sphere), whereas the two sets A and B are the balls that are inside and outside the separator sphere, respectively. Note that the graph induced by the set A consists of the intersection graph of the balls inside the separator (similarly, B for the balls outside the separator and S for the balls intersecting the sphere).

We emphasize that, even though the size of the separator is larger than the one from Lipton–Tarjan (specially for high values of d), the main advantage is that the three subgraphs it creates are geometric graphs of the same family (intersection graphs of balls in $\mathbb{R}^d$). The bound on the separator size does not hold up well when k is large, even for $d = 2$: if $\sqrt{n}$ disks overlap at a single point and the other disks form a path we have $k = \sqrt{n}$ and $m = \Theta(n)$, where m is the number of edges in the intersection graph. Hence, the separator has size $O(\sqrt{kn}) = O(m^{3/4})$.

Fox and Pach [5] gave another separator result that follows the same spirit: the intersection graph of a set of Jordan curves in the plane has a $2/3$-separator of size $O(\sqrt{m})$ if every pair of curves intersects at a constant number of points.[6] A set of disks in $\mathbb{R}^2$ satisfies this condition, and thus the theorem applies to disk graphs. Their proof can be turned into a polynomial-time algorithm. However, we need to construct the arrangement of disks, which takes $O(n^2 2^{\alpha(n)})$ time,

---

[6] Without restriction on the number of intersection points for every pair of curves, the bound of $O(\sqrt{m} \log m)$ can be achieved [14].

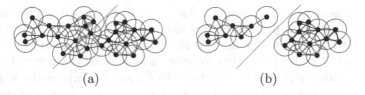

(a)   (b)

Fig. 1. An example of a line separator of a unit disk graph. (a) A family of unit disks (blue) and a line (red). (b) Removing the disks intersected by the red line leaves a disconnected graph.

where $\alpha(n)$ is the inverse Ackermann function [3], and in practice an efficient implementation is non-trivial.

From a geometric perspective these two results show that, given a set of unit disks in the plane, we can always find a closed curve in the plane (a circle [16, 18] and a Jordan curve [5], respectively) to partition the set. The disks intersected by the curve are those in the separator, and the two disjoint sets are the disks inside and outside the curve, respectively.

Results and Paper Organization. In this paper we continue the idea of geometric separators and show that a balanced separator always exists, even if we constrain the separator to be a line (see Fig. 1). Given a set of n unit disks with m pairwise distinct intersections, we show that a line 2/3-separator of size $O(\sqrt{(m+n)\log n})$ can be found in expected $O(n)$ time, and that an axis-parallel line 4/5-separator of size $O(\sqrt{m+n})$ can be found in deterministic $O(n)$ time.

Comparing our results with the previous work, our algorithm matches or improves in four ways, see also Table 1. (*i*) simplicity of the shape: circle [16, 18] vs. Jordan curve [5] vs. our line, (*ii*) balance of the sets A and B: 3/4 [16, 18] vs. 2/3 for both [5] and us, (*iii*) size of the separator: $O(m^{3/4})$ [16,18] vs. $O(\sqrt{m})$ [5] vs. our $\tilde{O}(\sqrt{m})$.[7] Finally, (*iv*) our algorithms are easy to implement and asymptotically faster: $O(n)$ [16, 18] vs. $\tilde{O}(n^2)$ [5] vs. our $O(n)$.

Table 1. Comparison of our results with other geometric separator results.

result	sep. shape	balance	sep. size	run-time	observations
[16], [18]	circle	3/4	$O(m^{3/4})$	$O(n)$	arbitrary disks
[5]	Jordan curve	2/3	$O(\sqrt{m})$	$\tilde{O}(n^2)$	pseudodisks
Thm. 1	line	2/3	$\tilde{O}(\sqrt{m})$	$O(n)$	unit disks
Thm. 2	axis-parallel line	4/5	$O(\sqrt{m})$	$O(n)$	unit disks
[1], [13], [9]	line	1/2	$O(\sqrt{n\log n})$	$O(n)$	**disjoint** unit disks
[9]	line	$1-\alpha$	$O(\sqrt{n(1-2\alpha)})$	$O(n)$	**disjoint** unit disks
[13]	axis-parallel line	9/10	$O(\sqrt{n})$	$O(n)$	**disjoint** unit disks

---

[7] The $\tilde{O}(\cdot)$ notation suppresses sublogarithmic factors. In particular, we note that our separator is slightly larger than the Fox-Pach separator.

We emphasize that our results focus on *unit* disk graphs, while the other results hold for disk graphs of arbitrary radii, too. Indeed, if we want to separate disks of arbitrary radii with a line, we show that the separator's size may be as large as $\Omega(n)$. We also prove that for unit disks our algorithm may fail to find a line 2/3-separator of size better than $O(\sqrt{m}\log(n/\sqrt{m}))$ in the worst case; the exact statement can be found in Section 3. In this sense, the size of our separators is asymptotically almost tight. In Section 4, experimental results are presented. We evaluate the performance of our algorithm, compare it with the method by Fox and Pach [5] in terms of the size of the produced separators for random instances, and conclude that our algorithm outperforms theirs.

Other Related Work. In a different context, line separators of *pairwise disjoint* unit disks have also been studied. Since the disks are pairwise disjoint, the intersection graph is trivially empty and can be easily separated. Instead, the focus is now to find a closed curve that intersects few disks, such that the two connected components it defines contain roughly the same number of disks.

Alon *et al.* [1] proved that for a given set $\mathcal{D}$ of n pairwise disjoint unit disks,[8] there exists a slope a such that every line with slope a intersects $O(\sqrt{n}\log n)$ unit disks of $\mathcal{D}$. In particular, the halving line of that slope will be a nice separator (each halfplane will have at most $n/2 - O(\sqrt{n\log n})$ disks fully contained in). Their proof is probabilistic, which can be turned into an expected $O(n)$-time randomized algorithm [13]. A deterministic $O(n)$-time algorithm was afterwards given by Hoffmann *et al.* [9], who also showed how to find a line ℓ that intersects at most $O(\sqrt{n/(1-2\alpha)})$ unit disks and each halfplane contains at most $(1-\alpha)n$ disks (for any $0 < \alpha < 1/2$). Löffler and Mulzer [13] proved that there exists an axis-parallel line ℓ such that ℓ intersects $O(\sqrt{n})$ disks, and each halfplane contains at most $9n/10$ unit disks. For comparison purposes, these three results are also shown in Table 1.

Preliminaries. In this paper, all disks are assumed to be closed (i.e., the boundaries are part of the disks), and a *unit disk* has radius one (thus diameter two). For a set S of n points in $\mathbb{R}^2$, there always exists a point $p \in \mathbb{R}^2$ such that every halfplane containing p contains at least $n/3$ points from S. Such a point p is called a *centerpoint* of S, and can be found in $O(n)$ time [11]. Let ℓ be a line through a centerpoint of S. Then, each of the two closed halfplanes bounded by ℓ contains at least $n/3$ points of S, which in turn means that each of the two open halfplanes bounded by ℓ contains at most $2n/3$ points of S each. Here, a halfplane H (closed or open) *contains* a point p if $p \in H$. We also say a halfplane H *contains* a disk D if $D \subseteq H$. Due to lack of space the proofs of some claims are deferred to the full version.

---

[8] The result extends to pairwise disjoint fat objects that are convex and of similar area (see Theorem 4.1 of [1]). For the sake of conciseness we only talk about unit disks.

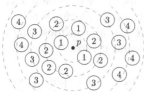 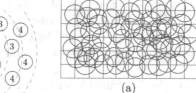 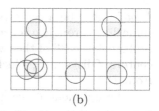

| (a) | (b) |

Fig. 2. Proof of Lemma 1. The number i in each disk means that it intersects the circle of radius $2i$ centered at p.

Fig. 3. Proof of Lemma 2. (a) A grid of $\sqrt{2} \times \sqrt{2}$ squares is laid over the family of unit disks. (b) Disks associated to the same cell intersect, but disks associated to different cells of the same color do not intersect. If j is the index for color yellow, then $n_j = 7$, $l_{j0} = 4$, and $l_{j1} = 1$ in this example.

2 Upper Bounds

Let $\mathcal{D}$ be a set of n unit disks in the plane. We first consider the case where the disks in $\mathcal{D}$ are pairwise disjoint. The results from this case will also be used for the more general case where the disks in $\mathcal{D}$ are not necessarily disjoint.

Lemma 1. *Let $\mathcal{D}$ be a set of n pairwise disjoint unit disks in the plane and let p be any point in the plane. Then the expected number of disks intersected by a random line through p is $O(\sqrt{n})$.*

We note that the lemma has a flavor similar to a theorem by Alon *et al.* [1].

Proof. Let C_i be the circle of radius $2i$ centered at p, for $i = 0, 1, \ldots$. Then each disk in $\mathcal{D}$ is intersected by at most two of these circles—if a disk is intersected by two circles the intersection takes place on the boundary of the disk. Let $\mathcal{D}_i \subseteq \mathcal{D}$ be the set of disks that have non-empty intersection with C_i, for $i = 0, 1, \ldots$. See Fig. 2. Note that $\sum_i |\mathcal{D}_i| \leq 2n$.

Given a random line ℓ through p, the number k_i of disks of $\mathcal{D}_i$ that are intersected by ℓ is at most four, due to disjointness, and its expectation is $O(|\mathcal{D}_i|/i)$. Therefore, by the linearity of expectation, the expected number of disks of $\mathcal{D}$ intersected by ℓ is bounded by

$$\mathbb{E}\left[\sum_{i \geq 0} k_i\right] = \mathbb{E}\left[\sum_{i:\, i \leq \sqrt{n}} k_i\right] + \mathbb{E}\left[\sum_{i:\, i > \sqrt{n}} k_i\right]$$

$$\leq 4\sqrt{n} + \sum_{i:\, i > \sqrt{n}} O(|\mathcal{D}_i|/\sqrt{n}) = O(\sqrt{n}) + O(n/\sqrt{n}) = O(\sqrt{n}). \qquad \square$$

Corollary 1. *Let $\mathcal{D}$ be a set of n pairwise disjoint unit disks in the plane. Then, there exists a line ℓ that intersects $O(\sqrt{n})$ disks of $\mathcal{D}$ such that each of the two open halfplanes bounded by ℓ contains at most $2n/3$ disks of $\mathcal{D}$. Moreover, such a line can be found in $O(n)$ time with probability at least $3/4$.*

Proof. Let p be a centerpoint of the set of centers of disks in $\mathcal{D}$. By Lemma 1, some line through p must intersect at most $O(\sqrt{n})$ disks from $\mathcal{D}$. Since p is a centerpoint, each of the two open halfplanes bounded by ℓ contains at most $2n/3$ centers of disks in $\mathcal{D}$, which means that the halfplane contains at most $2n/3$ disks from $\mathcal{D}$.

The argument above suggests the following algorithm: first we compute a centerpoint p of the centers of a given set of disks, and then choose a line through p uniformly at random. By Lemma 1 and Markov's inequality, the probability that the random line intersects more than $c\sqrt{n}$ disks is at most $1/4$ for some constant c. Thus, a desired line can be found with probability at least $3/4$. The running time is linear in the number of disks since a centerpoint can be found in linear time [11].   □

In the statement, the exact value of $3/4$ for the lower bound to the success probability is not important (any positive probability will suffice).

We now consider the general case where the disks are not necessarily disjoint.

Lemma 2. *Let $\mathcal{D}$ be a set of n unit disks in the plane with m intersecting pairs, and let p be any point in the plane. Then the expected number of disks intersected by a random line through p is $O(\sqrt{(m+n)\log n})$.*

Proof. Consider a grid of $\sqrt{2} \times \sqrt{2}$ squares. Each grid cell is treated as right-open and top-open so that it is of the form of $[x, x+\sqrt{2}) \times [y, y+\sqrt{2})$. Associate each disk in $\mathcal{D}$ with the grid cell containing its center, see Fig. 3(a).

Observe that one can color the grid cells with nine colors for every 3×3 block of grid cells, so that no two disks that are associated with different grid cells of the same color intersect, see Fig. 3(b). Consider one of the colors j, with $1 \leq j \leq 9$, and let $\mathbb{C}_j$ be the collection of subsets of $\mathcal{D}$ associated with the grid cells of this color:

$$\mathbb{C}_j = \{\mathcal{C} \subseteq \mathcal{D} \mid \text{the center of disks in } \mathcal{C} \text{ lie in the same grid cell of color } j\}.$$

Then, $\mathbb{C}_j$ has the following two properties: (i) each subset $\mathcal{C} \in \mathbb{C}_j$ in the same grid cell is a clique, i.e., any two disks in the subset intersect each other; (ii) any two disks from two different subsets in $\mathbb{C}_j$ are pairwise disjoint. Let $n_j = \sum_{\mathcal{C} \in \mathbb{C}_j} |\mathcal{C}|$ denote the number of disks in $\mathcal{D}$ associated to grid cells of color j.

We divide the cliques in $\mathbb{C}_j$ into $O(\log n_j)$ buckets $\mathbb{B}_{j0}, \mathbb{B}_{j1}, \ldots$, where $\mathbb{B}_{ji}$ consists of all cliques of $\mathbb{C}_j$ whose size is in the range $[2^i, 2^{i+1})$. Set $l_{ji} = |\mathbb{B}_{ji}|$, for $i = 0, 1, \ldots$. Then, the sum x_{ji} of the sizes of the cliques in $\mathbb{B}_{ji}$ is in the range $l_{ji}2^i \leq x_{ji} < l_{ji}2^{i+1}$. We also know that $\sum_i x_{ji} = n_j$. Let $m_{ji} = \sum_{\mathcal{C} \in \mathbb{B}_{ji}} |\mathcal{C}|(|\mathcal{C}|-1)/2$. Then, $l_{ji}(2^{2i-1} - 2^{i-1}) \leq m_{ji} < l_{ji}(2^{2i+1} - 2^i)$ and $\sum_i m_{ji} \leq \sum_j \sum_i m_{ji} \leq m$.

We first compute the expected number of disks in cliques of $\mathbb{B}_{ji}$ intersected by a random line through p. Since the union of the disks in each clique is contained in a disk of radius 2 such that they are disjoint, a random line through p intersects only $O(\sqrt{l_{ji}})$ cliques of $\mathbb{B}_{ji}$ by Lemma 1, and therefore only $O(\sqrt{l_{ji}}2^{i+1})$ disks of $\mathcal{D}$. By the definition of $\mathbb{B}_{ji}$, we have $l_{ji}2^{2i-1} \leq m_{ji} + l_{ji}2^{i-1} \leq m_{ji} + x_{ji}/2$. Thus, a random line through p intersects $O(\sqrt{m_{ji} + x_{ji}})$ disks in expectation.

Then, by the linearity of expectation, we sum the numbers for all $j = 1, \ldots, 9$ and $i = 0, \ldots, \log n_j$:

$$\sum_j \sum_i O(\sqrt{m_{ji} + x_{ji}}) \leq O\left(\sqrt{\sum_j \sum_i (m_{ji} + x_{ji})} \sqrt{\sum_j \sum_i 1}\right)$$

$$\leq O(\sqrt{(m + n)\log n}),$$

where the first inequality follows from the Cauchy-Schwarz inequality.   □

In the same way as Corollary 1 follows from Lemma 1, the following theorem follows from Lemma 2.

Theorem 1. *Let $\mathcal{D}$ be a set of n unit disks in the plane with m intersecting pairs. Then, there exists a line ℓ that intersects $O(\sqrt{(m + n)\log n})$ disks of $\mathcal{D}$ such that each of the two open halfplanes bounded by ℓ contains at most $2n/3$ disks of $\mathcal{D}$. Moreover, such a line can be found in $O(n)$ time with probability at least $3/4$.*

As before, the exact value of $3/4$ for the success probability is not important.

2.1   Axis-Parallel Separators

In this section we show an alternative, more restricted separator. Specifically, we show that a line separator that intersects fewer disks ($O(\sqrt{m + n})$) exists, even if we restrict ourselves to axis-parallel lines. However, this comes at the cost that the balancing parameter is worsened: on each side of the line we can certify only that there are at most $4n/5$ disks. The theorem below and its proof have a flavor similar to Löffler and Mulzer [13].

Theorem 2. *Let $\mathcal{D}$ be a set of n unit disks in the plane with m intersecting pairs. Then, there exists an axis-parallel line ℓ that intersects $O(\sqrt{m + n})$ disks of $\mathcal{D}$ such that each of the two open halfplanes bounded by ℓ contains at most $4n/5$ disks of $\mathcal{D}$. Moreover, such a line can be found in $O(n)$ time.*

Proof (sketch). Let P be the set of disk centers and assume that there are no two points (centers) in P with the same x-coordinate or the same y-coordinate. Let ℓ_d (resp., ℓ_u) be a horizontal line such that there are exactly $n/5$ points of P below it (resp., above it). Let H be the distance between ℓ_d and ℓ_u. Similarly, let ℓ_l (resp., ℓ_r) be a vertical line such that there are exactly $n/5$ points of P to its left (resp., to its right). Let V be the distance between ℓ_l and ℓ_r. From the above definitions, it follows that any horizontal line ℓ between ℓ_d and ℓ_u, and any vertical line ℓ between ℓ_l and ℓ_r have at most $4n/5$ disks of $\mathcal{D}$ on each of the two open halfplanes bounded by ℓ. We will show that one of these lines intersects $O(\sqrt{m + n})$ disks of $\mathcal{D}$.

Consider the rectangle $\mathcal{R}$ defined by ℓ_d, ℓ_u, ℓ_l and ℓ_r. Now let h_i be the horizontal line above ℓ_d whose distance from ℓ_d is exactly i, for $i = 1, \ldots, \lceil H-1 \rceil$, and similarly, let v_i be the vertical line to the right of ℓ_l whose distance from ℓ_l is exactly i, for $i = 1, \ldots, \lceil V-1 \rceil$. We may assume that at least one of H or V is at least 2; otherwise it is easy. Then, we prove that one (or more) of the lines h_i or one (or more) of the lines v_i intersects at most $\lambda \sqrt{m} + n/10$ disks, for some constant λ.   □

3 Almost Tightness of Our Approach

We now show that the approach used in Theorem 1 cannot be drastically improved. Specifically, we present a family $\mathcal{D}$ of n unit disks with a centerpoint p of the centers of those unit disks such that any line that passes through p will intersect many disks. Although this example can be constructed for any number of disks $n > 0$ and any desired number of intersecting pairs $m > n$, the details are a bit tedious. Instead, given the desired values n and m, we find $n' \approx n$ and $m' \approx m$ that satisfy the properties. This greatly simplifies the proof and, asymptotically speaking, the bounds are unaffected.

Theorem 3. *For any $n, m \in \mathbb{N}$ such that $9n \leq m \leq \lfloor n^2/6 \rfloor$, there exist $n', m' \in \mathbb{N}$, where $n \leq n' \leq 2n$ and $\lceil m/9 \rceil \leq m' \leq 6m$, and a set $\mathcal{D}$ of n' unit disks in the plane with m' intersecting pairs that have the following property. There exists a centerpoint p of the centers of unit disks in $\mathcal{D}$ such that any line ℓ that passes through p intersects $\Omega(\sqrt{m} \log(n/\sqrt{m}))$ disks of $\mathcal{D}$.*

Proof (sketch). Given n and m, we choose $k \in \mathbb{N}$ as the smallest natural number k' such that $k' \geq \sqrt{\frac{6m}{1+\ln(n/k')}}$. First observe that such k exists and satisfies $k \leq n$. Indeed, this follows from the assumption $m \leq n^2/6$.

Let $\ell = \lceil n/k \rceil$, and consider a sufficiently small positive real number $\varepsilon \leq \frac{1}{2\pi}$. Consider now the ℓ concentric circles C_i centered at the origin with radius $2i(1 + \varepsilon)$ for $i = 1, \ldots, \ell$. On each such circle, we place k unit disks uniformly (i.e., the arc spacing between the centers of two consecutive disks is $4\pi i(1+\varepsilon)/k$). Let $\mathcal{D}$ be the collection of these disks, see Fig. 4 (left).

Let $n' = k\ell$. By construction, $\mathcal{D}$ has n' disks, and $n \leq n' \leq n + k \leq 2n$ as claimed. It turns out that the number m' of intersecting pairs in $\mathcal{D}$ satisfies $\lceil m/9 \rceil \leq m' \leq 6m$.

By symmetry, we can see that the origin is a centerpoint of the centers of disks in $\mathcal{D}$. Thus, it remains to show that any line that passes through the origin must intersect many disks of $\mathcal{D}$. In the following we show something stronger: any ray emanating from a point p inside C_1 will cross $\Omega(\sqrt{m} \log(n/\sqrt{m}))$ disks.

Partition the disks of $\mathcal{D}$ into ℓ layers $\mathcal{D}_1, \ldots, \mathcal{D}_\ell$ depending on which concentric circle their center lies on. Since we placed unit disks on circles that are $2(1+\varepsilon)$ units apart, only disks that belong to the same layer may have nonempty intersection.

Let γ_i be the maximum arc length of C_i such that two unit disks centered at two endpoints of the arc touch each other, see Fig. 4 (middle). Note that

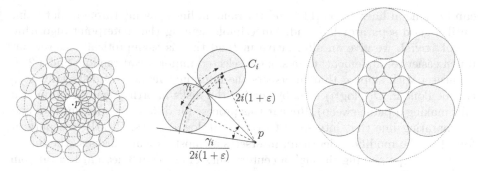

Fig. 4. Left: Almost tightness construction for $\ell = 3$ (and $k = 16$). Middle: upper and lower bounds for γ_i. Right: construction for disks of arbitrary radii.

two disks on the same layer overlap if and only if the arc distance between the centers is γ_i or less.

We count the number of intersections on each layer independently. Each unit disk in $\mathcal{D}_i$ covers $\frac{\gamma_i}{2i(1+\varepsilon)}$ radians of C_i. Since there are k evenly spread disks in $\mathcal{D}_i$ and $\varepsilon \leq \frac{1}{4\pi}$, each point of the circle C_i is contained in at least $\lfloor \frac{\gamma_i k}{2i(1+\varepsilon)} \cdot \frac{1}{2\pi} \rfloor > \frac{\gamma_i k}{i(4\pi+1)} - 1$ disks.

By substituting $k \geq \sqrt{\frac{6m}{1+\ln \ell}}$, $\gamma_i > 2$, $\ln(\ell+1) \geq (1+\ln \ell)/2$, $\ell = n'/k$ and $2m/9 \geq n'$, we obtain that the number of disks in $\mathcal{D}$ intersected by the ray is at least

$$\sum_{i=1}^{\ell} \left(\frac{\gamma_i k}{i(4\pi+1)} - 1 \right) > \frac{2}{4\pi+1} \left(\sqrt{\frac{6m}{1+\ln \ell}} \right) \left(\sum_{i=1}^{\ell} \frac{1}{i} \right) - \ell$$

$$= \Omega(\sqrt{m} \log(n/\sqrt{m})). \qquad \square$$

Theorem 4. *For any $n \in \mathbb{N}$ there exists a set $\mathcal{D}$ of $O(n)$ pairwise disjoint disks such that any line ℓ such that both halfplanes contain $\Omega(n)$ disks of $\mathcal{D}$ also intersects $\Omega(n)$ disks of $\mathcal{D}$.*

Proof (sketch). See Fig. 4 (right). $\qquad \square$

Note that in this construction the radii of the disks grow at an exponential rate. By using a bucketing technique we can show that our upper bounds work for disks of arbitrary radii at an expense of an additional $O(\log \Delta)$ multiplicative factor, where Δ is the ratio between the smallest and largest radius of the disks in $\mathcal{D}$. This shows that our upper bounds are optimal also in this sense.

4 Experiments

In our experiments, we evaluate the quality of separator algorithms by the separator size. Theorem 1 suggests a simple algorithm: find a centerpoint (which can be done in linear time [11]) and try random lines passing through that point until a good separator is found. Since implementing the centerpoint algorithm is not trivial, we give an alternative method that is asymptotically slower but much easier to implement: for a slope a selected uniformly at random find a 2/3-separator with slope a that intersects the minimum number of disks (this step can be done in $O(n \log n)$ time by sorting the disks in orthogonal direction of a and making a plane sweep). Repeat this process for different slopes until we find a separating line that intersects $O(\sqrt{(m + n) \log n})$ disks. Clearly, a separator found by the modified algorithm intersects at most as many disks as the line of the same slope passing through a centerpoint. Thus, as in Theorem 1, a random direction will be good with positive probability.

We compare our algorithm with the method by Fox and Pach [5] which guarantees the separator size of $O(\sqrt{m})$. For the implementation of our algorithm we use the simpler variation described above.

The Method by Fox and Pach. Fox and Pach [5] proved that the intersection graph of a set of Jordan curves in the plane has a 2/3-separator of size $O(\sqrt{m})$ if every pair of curves intersects in a constant number of points. Their proof is constructive, as outlined below.

First, we build the arrangement of curves, and obtain a plane graph whose vertex set are the vertices of the arrangements and and consecutive vertices on a curve are joined by an edge.[9] We triangulate the obtained plane graph to make it maximal planar. Then, we find a simple cycle 2/3-separator C (i.e., a 2/3-separator that forms a cycle in the graph) of size $O(\sqrt{m + n})$, which always exists [15]. We output all curves containing a vertex in C.

In our implementation, we construct the circle arrangement in a brute-force manner, and we use a simple cycle separator algorithm by Holzer *et al.* [10], called the fundamental cycle separator (FCS) algorithm. Although the FCS algorithm has no theoretical guarantee for the size of the obtained separator, the recent experimental study by Fox-Epstein *et al.* [6] showed that it has a comparable performance to the state-of-the-art cycle separator algorithm with theoretical guarantee for most of the cases.

Instance Generation and Experiment Setup. All instances for our experiments are randomly generated. We fix a $d \times d$ square S and generate n unit disks in S independently and uniformly at random. If the graph is disconnected, we discard it and generate again. All experiments have been performed on Intel ® Core™ i7-5600U CPU @2.60GHz × 4, with 7.7GB memory and 976.0GB hard disk, running Ubuntu 14.04.3 LTS 64bit.

Experiment 1: Quality of the Proposed Method. In the first experiment we empirically examine the size of a separator obtained by our proposed algorithm with the modification proposed at the beginning of this section. For instance

---

[9] The method of Fox and Pach needs to add a constant number of additional vertices, but the main feature is that the overall complexity of the graph is $O(m + n)$.

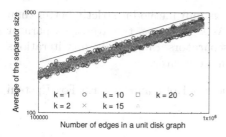 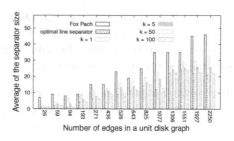

Fig. 5. Left: The average separator size of our algorithm for random instances. The horizontal axis shows the number of edges in logscale, and the vertical axis shows the average separator size in logscale when the algorithm is run k times, $k \subset \{1, 2, 10, 15, 20\}$. The solid line shows the square root of the number of edges. Right: Comparison of our method with the Fox-Pach method. The horizontal axis shows the number of edges, and the vertical axis shows the separator size; the red bar is obtained by the Fox-Pach method, the blue bar is an optimal line separator, and the remaining four bars are obtained by our method after trying k random directions ($k \in \{1, 5, 50, 100\}$) and returning the minimum size separators. For all methods we repeat this process 20 times and display the average minimum size.

generation, we fix $d = 100$, and vary n from $10,000$ to $30,000$ with an increment of 50. Since our algorithm is randomized, we run the algorithm k times, where $k \in \{1, 2, 5, 10, 15, 20\}$, and compute the average separator size.

Fig. 5 (left) shows the result. We can observe that as the number of edges increases, the size of the obtained line separators also increases, but the tendency is close to $\Theta(\sqrt{m})$ as the solid black line shows. We thus empirically conclude that the algorithm from Theorem 1 tends to output line separators of size $O(\sqrt{m})$ for random instances.

Experiment 2: Comparison with the Method by Fox and Pach. In the second experiment we compare our algorithm with the method by Fox and Pach in its separator size. For instance generation, we fix $d = 16$, and generated 14 random instances with various numbers of edges (the number of disks is not fixed). For each instance, we run our implementation of the Fox-Pach method as outlined above and our simplified algorithm described in Section 4. Our algorithm is iterated k times, where $k \in \{1, 5, 100, 500\}$, and we take the minimum separator size among k runs. In this way, we can observe how the minimum separator size converges to the real minimum. This process is repeated 20 times, and we compute the average of the minimum separator sizes over those 20 repetitions.

We also examine the size of an optimal line separator to investigate the limitation of the line-separator approach. Here, an *optimal line separator* means a line 2/3-separator of minimum size, which can be found in $\tilde{O}(n^2)$ time by looking at all possible bitangents to pairs of unit disks (but we implemented a simpler $O(n^3)$-time algorithm), where the $\tilde{O}(\cdot)$ notation suppresses logarithmic factors.

Fig. 5 (right) shows the result. On average, our method is strictly better than the Fox-Pach method even when $k = 1$. As k increases, the average separator size approaches to the size of optimal line separators. Empirically, fifty iterations are enough to obtain a reasonably good line separator.

Acknowledgments. The authors thank Michael Hoffmann and Eli Fox-Epstein for motivating discussion on the topic.

References

1. Alon, N., Katchalski, M., Pulleyblank, W.R.: Cutting disjoint disks by straight lines. Discrete & Computational Geometry 4, 239–243 (1989)
2. Alon, N., Seymour, P., Thomas, R.: A separator theorem for nonplanar graphs. J. Amer. Math. Soc. 3, 801–808 (1990)
3. Edelsbrunner, H., Guibas, L.J., Pach, J., Pollack, R., Seidel, R., Sharir, M.: Arrangements of curves in the plane—topology, combinatorics and algorithms. Theor. Comput. Sci. 92(2), 319–336 (1992)
4. Eppstein, D., Miller, G.L., Teng, S.: A deterministic linear time algorithm for geometric separators and its applications. Fundam. Inform. 22(4), 309–329 (1995)
5. Fox, J., Pach, J.: Separator theorems and Turán-type results for planar intersection graphs. Advances in Mathematics 219(3), 1070–1080 (2008)
6. Fox-Epstein, E., Mozes, S., Phothilimthana, P.M., Sommer, C.: Short and simple cycle separators in planar graphs. In: Proc. of the 15th Meeting on Algorithm Engineering and Experiments. pp. 26–40. SIAM (2013)
7. Gilbert, J.R., Hutchinson, J.P., Tarjan, R.E.: A separator theorem for graphs of bounded genus. J. Algorithms 5(3), 391–407 (1984)
8. Har-Peled, S., Quanrud, K.: Approximation algorithms for polynomial-expansion and low-density graphs. In: Proc. of the 23rd Annual European Symposium. pp. 717–728. Springer (2015)
9. Hoffmann, M., Kusters, V., Miltzow, T.: Halving balls in deterministic linear time. In: Proc. of the 22th Annual European Symposium. pp. 566–578. Springer (2014)
10. Holzer, M., Schulz, F., Wagner, D., Prasinos, G., Zaroliagis, C.D.: Engineering planar separator algorithms. ACM Journal of Experimental Algorithmics 14 (2009)
11. Jadhav, S., Mukhopadhyay, A.: Computing a centerpoint of a finite planar set of points in linear time. Discrete & Computational Geometry 12, 291–312 (1994)
12. Lipton, R.J., Tarjan, R.E.: A separator theorem for planar graphs. SIAM Journal on Applied Mathematics 36(2), 177–189 (1979)
13. Löffler, M., Mulzer, W.: Unions of onions: Preprocessing imprecise points for fast onion decomposition. JoCG 5(1), 1–13 (2014)
14. Matoušek, J.: Near-optimal separators in string graphs. Combinatorics, Probability & Computing 23(1), 135–139 (2014)
15. Miller, G.L.: Finding small simple cycle separators for 2-connected planar graphs. J. Comput. Syst. Sci. 32(3), 265–279 (1986)
16. Miller, G.L., Teng, S., Thurston, W.P., Vavasis, S.A.: Separators for sphere-packings and nearest neighbor graphs. J. ACM 44(1), 1–29 (1997)
17. Nešetřil, J., Ossona de Mendez, P.: Grad and classes with bounded expansion II. Algorithmic aspects. Eur. J. Comb. 29(3), 777–791 (2008)
18. Smith, W.D., Wormald, N.C.: Geometric separator theorems & applications. In: Proc. of the 39th Annual Symposium on Foundations of Computer Science. pp. 232–243 (1998)

All-Pairs Shortest Paths in Geometric Intersection Graphs

Timothy M. Chan[1] and Dimitrios Skrepetos[2]

[1] Department of Computer Science, University of Illinois at Urbana-Champaign,
tmc@illinois.edu
[2] Cheriton School of Computer Science, University of Waterloo,
dskrepet@uwaterloo.ca

Abstract. We address the All-Pairs Shortest Paths (APSP) problem for a number of unweighted, undirected geometric intersection graphs. We present a general reduction of the problem to static, offline intersection searching (specifically detection). As a consequence, we can solve APSP for intersection graphs of n arbitrary disks in $O\left(n^2 \log n\right)$ time, axis-aligned line segments in $O\left(n^2 \log \log n\right)$ time, arbitrary line segments in $O\left(n^{7/3} \log^{1/3} n\right)$ time, d-dimensional axis-aligned boxes in $O\left(n^2 \log^{d-1.5} n\right)$ time for $d \geq 2$, and d-dimensional axis-aligned unit hypercubes in $O\left(n^2 \log \log n\right)$ time for $d = 3$ and $O\left(n^2 \log^{d-3} n\right)$ time for $d \geq 4$.

In addition, we show how to solve the Single-Source Shortest Paths (SSSP) problem in unweighted intersection graphs of axis-aligned line segments in $O\left(n \log n\right)$ time, by a reduction to dynamic orthogonal point location.

Keywords: shortest paths, geometric intersection graphs, intersection searching data structures, disk graphs

1 Introduction

As a motivating example, consider the following toy problem: given a set S of n axis-aligned line segments in the plane representing a road network, and two points p_1 and p_2 lying on two segments of S, compute a path from p_1 to p_2 that stays on S while minimizing the number of turns. (See Figure 1.)

To solve the problem, we can create a vertex for each segment of S and an (unweighted, undirected) edge between two vertices if their corresponding segments intersect. This defines the *intersection graph* $G(S)$. Then given two points p_1 and p_2, lying on the segments s and t of S, a minimum-turn path from p_1 to p_2 corresponds precisely to an unweighted shortest path from s to t in $G(S)$. Naively constructing $G(S)$ and running breadth-first search (BFS) would require $O(n^2)$ worst-case time. In Section 4, however, we observe an $O(n \log n)$-time algorithm, which is new to the best of the authors's knowledge. In fact, the algorithm solves the more general, Single-Source Shortest Paths (SSSP) problem in $G(S)$, by an application of data structures for *dynamic orthogonal point location* [21,6].

© Springer International Publishing AG 2017
F. Ellen et al. (Eds.): WADS 2017, LNCS 10389, pp. 253–264, 2017.
DOI: 10.1007/978-3-319-62127-2_22

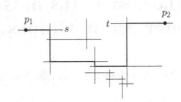

Fig. 1. A set S of axis-aligned line segments is shown. The path from p_1, lying on segment s, to p_2, lying on segment t, staying on S and using the minimum number of turns is marked in bold.

Our main focus in this paper will be a similar problem, namely the All-Pairs Shortest Path (APSP) problem in geometric intersection graphs. More generally, given a set S of n geometric objects, its intersection graph $G(S)$ is defined by creating one vertex for every object of S and an (undirected, unweighted) edge between two vertices if their corresponding objects intersect. We want to compute a representation of an unweighted shortest path between s and t for every pair of objects $s, t \in S$. For general unweighted, undirected graphs the problem can be solved in $O(n^{\omega})$ time (e.g., see [5,26]), where $\omega < 2.373$ is the matrix multiplication exponent [28], but better results are possible for geometric intersection graphs.

Our main results are as follows:

- For arbitrary disks, we solve APSP in $O(n^2 \log n)$ time. The disk case is naturally motivated by applications in ad hoc communication networks. Following work by Cabello and Jejčič [8] on SSSP for unit-disk graphs, a previous paper by the authors [15] studied APSP for unit-disk graphs and gave an $O\left(n^2 \sqrt{\frac{\log \log n}{\log n}}\right)$-time algorithm, but the approach cannot be extended to arbitrary disks. A paper by Kaplan et al. [23] contains an algorithm for SSSP for disks (which can be used for APSP), but this is for a directed variant of intersection graphs (called "transmission graphs"), and the running time has multiple logarithmic factors unless we assume that the maximum-to-minimum radius ratio is bounded.

- For axis-aligned line segments, we solve APSP in $O(n^2 \log \log n)$ time, which is better than running n times the $O(n \log n)$-time algorithm for SSSP that we have mentioned for the toy problem at the beginning. No previous results have been reported, to the best of the authors's knowledge.

- When the line segments are not axis-aligned but have arbitrary orientations instead, we solve APSP in $O\left(n^{7/3} \log^{1/3} n\right)$ time, which is a little better than the general $O(n^{\omega})$ result, at least with the current upper bound on ω. (Regardless, our algorithm has the advantage of being combinatorial.)

- See Table 1 for further results on axis-aligned boxes, unit hypercubes, and fat triangles of roughly equal size.

All these results stem from one single, general technique, which reduces APSP to the design of data structures for static, offline intersection *detection*, i.e., given a query object, decide whether there is an input object intersecting it (and report

Geometric Objects	Running Time
arbitrary disks	$O(n^2 \log n)$
axis-aligned line segments	$O\left(n^2 \log \log n\right)$
arbitrary line segments	$O\left(n^{7/3} \log^{1/3} n\right)$
d-dimensional axis-aligned boxes	$O\left(n^2 \log^{d-1.5} n\right)$ for $d \geq 2$
d-dimensional axis-aligned unit hypercubes	$O\left(n^2 \log \log n\right)$ for $d = 3$ and $O\left(n^2 \log^{d-3} n\right)$ for $d \geq 4$
fat triangles of roughly equal size	$O\left(n^2 \log^4 n\right)$

Table 1. The results for APSP

one if the answer is yes). Our technique, described in Section 2, works by visiting vertices in an order prescribed by a spanning tree; given the BFS tree from a source vertex s as a guide, we can generate the BFS tree from an adjacent source vertex s' quickly, by exploiting the fact that distances to s' are approximately known up to ± 1, and by using the right geometric data structures. Some form of this simple idea has appeared before for general graphs (e.g., see [4,10]), but it is somehow overlooked by previous researchers in the context of geometric APSP.

To appreciate the advantages of the new technique, we should compare it with other known general approaches:

- First, a naive approach is to solve SSSP n times from every source independently, i.e., generate the BFS trees from each source from scratch. Geometric SSSP problems can often be reduced to *dynamic* data structuring problems, for example, as observed in Chan and Efrat's paper [12] (the reduction is much simplified in the unweighted, undirected setting). In fact, our solution to the toy problem at the beginning is done via this approach. However, dynamic data structures for geometric intersection or range searching usually are more complicated and have slower query times than their static counterparts, sometimes by multiple logarithmic factors. For example, the arbitrary disk case requires dynamic data structures for additively weighted nearest neighbor search, and a BFS therein takes nearly $O\left(n \log^{10} n\right)$ time [24]. Our reduction to *static* data structuring problems yields better results.
- Another general approach is to employ *biclique covers* [19,2] to sparsify the intersection graph first and then solve the problem on the sparsified graph. Biclique covers are related to static, offline intersection searching data structures (e.g., as noted in [9]). However, the complexity of biclique covers also tends to generate extra logarithmic factors. For example, for d-dimensional boxes, the sparsified graph has $O\left(n \log^d n\right)$ edges, leading to an $O\left(n^2 \log^d n\right)$-time algorithm, but our solution requires $O\left(n^2 \log^{d-1.5} n\right)$ time. For arbitrary disks, the complexity of the biclique covers is even worse $\left(O\left(n^{3/2+\varepsilon}\right)\ [3]\right)$, leading to an $O\left(n^{5/2+\varepsilon}\right)$-time algorithm, which is much

slower than our $O\left(n^2 \log n\right)$ result. The underlying issue is that intersection searching (as implicitly needed in biclique covers) may in general be harder than intersection detection.

In deriving our result for axis-aligned boxes, we also obtain a new $O\left(n\sqrt{\log n}\right)$-time algorithm for offline rectangle stabbing in two dimensions (pre-process n axis-aligned rectangles so that we can find a rectangle stabbing each query point). This result (see the full paper) may be of independent interest.

For the rest of the paper, for $s, t \in S$, where S is a set of geometric objects, let $dist[s,t]$ denote the distance of the shortest path from s to t in the intersection graph of S and $pred[s,t]$ denote the predecessor of t in that path. In SSSP we want to compute $dist[s,t]$ and $pred[s,t]$ for a given $s \in S$ and $\forall t \in S$, while in APSP we want to compute $dist[s,t]$ and $pred[s,t]$ $\forall s, t \in S$. All algorithms assume the standard unit-cost RAM model of computation where the word size is at least $\log n$ in bits.

2   Reducing APSP to static, offline intersection detection

In this section, we reduce the problem of solving APSP in unweighted, undirected geometric intersection graphs of objects of constant-description complexity to static, offline intersection detection. We assume that the graph is connected; if not, then we can simply work with every connected component independently. We first compute an arbitrary spanning tree T_0 of $G(S)$, root it at an arbitrary object $s_0 \in S$, and then compute the shortest path tree of s_0. Then, we visit each object s of T_0 in a pre-order manner, and compute the shortest path tree of s by using the shortest path tree of s' as a guide, where s' is the parent of s in T_0. The pseudocode of the algorithm is given in Algorithm 1. The initial call is APSP(S, s_0).

Algorithm 1: APSP(S, s_0)

---

1 build $G(S)$
2 compute any spanning tree T_0 of $G(s)$ and root it at any $s_0 \in S$
3 compute the shortest path tree of s_0
4 **for** *each* $s \in S - \{s_0\}$ *following a pre-order traversal of* T_0 **do**
5   compute the shortest path tree $T(s)$ of s, using the shortest path tree
         $T(s')$ of its parent s' in T_0, by calling $SSSP(S, s, T(s'))$

It remains to describe how to compute the shortest path tree of a vertex $s \in S$, given the shortest path tree of a vertex t at unit distance from it, i.e., how to implement Line 5 in Algorithm 1. From the triangle inequality and from $dist[s,s'] = 1$, we know that if $dist[s',z] = \ell$ for an object $z \in S$, then $\ell - 1 \leq dist[s,z] \leq \ell + 1$. Thus we already have an 1-additive approximation of the distances $dist[s', z]$ for any $z \in S$. To compute the exact distances from s to any object $z \in S$, we follow the procedure of the next paragraph.

As in classical BFS, we proceed in $n - 1$ steps, where in step ℓ we assume that we have found all the objects at distance at most $\ell - 1$ from s and want to produce the objects at distance exactly ℓ. The objects at distance exactly $\ell - 1$ from s are called the *frontier* objects, while the ones whose distance has not yet been found are called the *undiscovered* objects. Then we need to procure quickly all the undiscovered objects that intersect the frontier objects. Because of the 1-additive approximation, an object z can be at distance ℓ from s only if it is at distance $\ell - 1$, ℓ, or $\ell + 1$ from s'. These points are called the *candidate* objects. Hence we need to determine, for each candidate object, whether it intersects any frontier object. This is an instance of intersection searching, or more specifically, *intersection detection*:

> Preprocess a set of input objects into a data structure so that we can quickly decide if a given query object intersects any input object, and report one such input object if it exists.

In our application, the input objects are *static*, and the query objects are *offline*, i.e., are all given in advance.

To summarize, the pseudocode is presented in Algorithm 2. Thus we obtain the following theorems:

Algorithm 2: $\mathrm{SSSP}(S, s, T(s'))$

1   $dist[s, s] \leftarrow 0$
2   $dist[s, z] = \infty \ \forall z \in S - \{s\}$
3   $pred[s, z] = NULL \ \forall z \in S$
4   **for** $\ell = 0$ *to* $n - 1$ **do**
5      $A_\ell = \{z \mid dist[s', z] = \ell\}$ // objects at distance ℓ from s'
6   **for** $\ell = 1$ *to* $n - 1$ **do**
7      $F = \{z \in S \mid dist[s, z] = \ell - 1\}$ // frontier objects
8      $C = A_{\ell-1} \cup A_\ell \cup A_{\ell+1}$ // candidate objects
9   build a static, offline intersection detection data structure for F and C
10   **for** $z \in C$ **do**
11   **if** $dist[s, z] = \infty$ **then**
12   query the data structure for z
13   let w be the answer
14   **if** w *not* $NULL$ **then**
15               $dist[s, z] = \ell$
16               $pred[s, z] = w$

Theorem 1. *Given a set S of n objects of constant-description complexity and the shortest path tree of an object $s' \in S$ in the unweighted, undirected intersection graph of S, we can compute the shortest path tree of an object $s \in S$, where*

$dist[s, s'] = 1$, *in the same graph in* $O(SI(n, n))$ *time, where* $SI(n, m)$ *is the time to construct a static, offline intersection detection data structure for* n *objects and query it* m *times, assuming the property that* $SI(n_1, m_1) + SI(n_2, m_2) \leq SI(n_1 + n_2, m_1 + m_2)$.

Proof. Let n_ℓ (resp. m_ℓ) be the number of frontier (resp. candidate) objects in step ℓ of the BFS. During the algorithm, an object is in the frontier exactly once and in the candidate at most thrice; in other words, $\sum_{\ell=1}^{n-1} n_\ell \leq n$ and $\sum_{\ell=1}^{n-1} m_\ell \leq 3n$. Then the time to compute the shortest path tree of s is $O\left(\sum_{\ell=1}^{n-1} SI(n_\ell, m_\ell)\right) = O(SI(n, n))$.   □

Theorem 2. *We can solve APSP in an unweighted geometric intersection graph of* n *objects of constant-description complexity in* $O\left(n^2 + nSI(n, n)\right)$ *time, where* $SI(\cdot, \cdot)$ *is defined as in Theorem 1.*

Proof. In Lines 1–3 of Algorithm 1, we can build $G(S)$ in $O\left(n^2\right)$ time, find a spanning tree T_0, and compute the shortest-path tree of s_0, in $O\left(n^2\right)$ time naively. In each of the $n - 1$ iterations, Line 3 of Algorithm 1 takes $O\left(SI(n, n)\right)$ time by Theorem 1.   □

3   Applications

In this section we apply Theorem 2 and known data structures for static, offline intersection detection to obtain efficient APSP algorithms in specific families of geometric intersection graphs. Some of the data structures we employ are in fact online.

Arbitrary disks in the plane. We first consider intersection graphs of disks of arbitrary radii, also known as disk graphs. The static intersection detection data structure for disks will be based on an additively weighted Voronoi diagram, where the distance between a site w corresponding to a disk of radius r_w and a point x is defined as $d(w, x) = ||w - x|| - r_w$. This Voronoi diagram allows us to determine the disk whose boundary is closest to a query point. We construct the Voronoi diagram for the centers of the frontier disks and a point location data structure for the diagram's cells. Then we query the Voronoi diagram with the center of each query disk. We can check if the query disk and the disk returned by the query intersect in constant time.

    The time for building the additively weighted Voronoi diagram of n disks is $O(n \log n)$ [20]. We build a point location data structure in $O(n \log n)$ time, so that (online) queries take $O(\log n)$ time [27]. Therefore, $SI(n, n) = O(n \log n)$.

Theorem 3. *We can solve APSP in an unweighted intersection graph of* n *disks in* $O\left(n^2 \log n\right)$ *time.*

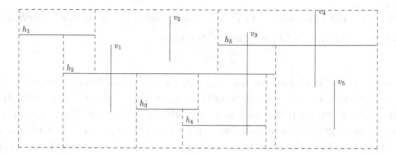

Fig. 2. This figure depicts a set of horizontal input segments, its vertical decomposition (shown by the dashed lines), and a set of vertical query segments.

Axis-aligned line segments in the plane. We now turn our attention to intersection graphs of axis-aligned line segments. We describe a static intersection detection data structure for horizontal input segments and vertical query segments. (Vertical input segments and horizontal query segments can be handled by a symmetric structure.) The data structure is composed of the vertical decomposition of the horizontal input segments, stored in a point location data structure. Given a vertical query segment, we perform a point location query for its bottom endpoint. If the top endpoint lies in the same cell, there is no intersection; otherwise, we can report the segment bounding the top side of the cell. (See Figure 2 for an example.)

We can apply the static orthogonal point location data structure of Chan [11] (Theorem 2.1), with $O(n \log \log U)$ preprocessing time and $O(\log \log U)$ query time, under the assumption that all coordinates are integers bounded by U. Thus, $SI(n, n) = O(n \log \log U)$. This implies an APSP algorithm running in $O\left(n^2 \log \log U\right)$ time. At the beginning, we can presort all coordinates in $O(n \log n)$ time and replace each coordinate value with its rank; this ensures that $U = n$. Thus, we obtain:

Theorem 4. *We can solve APSP in an unweighted intersection graph of n axis-aligned line segments in $O\left(n^2 \log \log n\right)$ time.*

The result can be easily be extended to any set of line segments with a constant number of different orientations.

Arbitrary line segments. Next we consider the case of arbitrary line segments. Chazelle [16] (Theorem 4.4) has given an $O\left(n^{4/3} \log^{1/3} n\right)$-time algorithm to count the number of intersections among n line segments. The algorithm can be modified to count the number of intersections between n red (input) line segments and n blue (offline query) line segments. In fact, it is straightforward to adapt the algorithm to decide, for each blue segment, whether it intersects any red segment and, if yes, report one such red segment. Thus, $SI(n, n) = O\left(n^{4/3} \log^{1/3} n\right)$.

Theorem 5. *We can solve APSP in an unweighted intersection graph of n arbitrary line segments in $O\left(n^{7/3}\log^{1/3} n\right)$ time.*

Axis-aligned boxes in d dimensions. For the case of axis-aligned rectangles in $d = 2$ dimensions, offline rectangle intersection counting is known to be reducible [18] to offline orthogonal range counting, for which Chan and Pătraşcu [14] have given an $O\left(n\sqrt{\log n}\right)$-time algorithm, under the assumption that all coordinates have been presorted. Consequently, we can decide for each query box whether it intersects any input box. With more effort, we can adapt their technique to report a witness input box for each query box with a yes answer, and thus solve offline intersection detection in $SI(n, n) = O\left(n\sqrt{\log n}\right)$ time; see full paper for details. At the beginning, we can presort all coordinates in $O(n\log n)$ time.

For axis-aligned boxes in $d \geq 3$ dimensions, we can use standard range trees [17] with the above $d = 2$ base case to obtain $SI(n, n) = O\left(n\log^{d-1.5} n\right)$.

Theorem 6. *We can solve APSP in an unweighted intersection graph of n d-dimensional axis-aligned boxes in $O\left(n^2\log^{d-1.5} n\right)$ time for $d \geq 3$.*

Axis-aligned unit hypercubes in d dimensions. When the axis-aligned boxes are unit hypercubes, the time bound for offline intersection detection can be improved. We build a uniform grid with unit side length and solve the problem inside each grid cell separately. Each input or query unit hypercube participates in at most a constant (2^d) number of grid cells. Inside a grid cell, each unit hypercube is effectively unbounded along d sides. Without loss of generality, we may assume that each input box is of the form $(-\infty, a_1] \times \cdots \times (-\infty, a_d]$ and each query box is of the form $[b_1, \infty) \times \cdots \times [b_d, \infty)$. Thus, the problem reduces to offline *dominance* detection: decide for each query point $(b_1, \ldots, b_d)$ whether it is dominated by some input point $(a_1, \ldots, a_d)$ and, if yes, report one such input point.

For $d = 3$, Gupta et al. [22] gave an algorithm to answer n offline dominance reporting queries in $O((n + K)\log\log U)$ time where K is the total output size, under the assumption that all coordinates are integers bounded by U. Their algorithm can be easily adapted to answer n offline dominance detection queries in $O(n\log\log U)$ time. This implies an APSP algorithm running in $O\left(n^2\log\log U\right)$ time. At the beginning, we can presort all coordinates in $O(n\log n)$ time and replace each coordinate value with its rank; this ensures that $U = n$.

For $d \geq 4$, Afshani et al. [1] (following Chan et al. [13]) gave a deterministic algorithm to answer n offline dominance reporting queries in $O\left(n\log^{d-3} n + K\right)$ time where K is the total output size. It can be checked that their algorithm can answer n offline dominance detection queries in $O\left(n\log^{d-3} n\right)$ time. (One step in their algorithm which involves reversing the role between input and query points becomes unnecessary for the detection problem.)

Theorem 7. *We can solve APSP in an unweighted intersection graph of n d-dimensional axis-aligned unit hypercubes in $O\left(n^2 \log \log n\right)$ time for $d = 3$ and in $O\left(n^2 \log^{d-3} n\right)$ time for $d \geq 4$.*

Fat triangles in the plane. Finally we consider the intersection graph of *fat triangles* (i.e., triangles that have bounded inradius-to-circumradius ratios) with roughly equal size. Katz [25] (Theorem 4.1 (i) and (iii)) has given an (online) data structure achieving $SI(n, n) = O\left(n \log^4 n\right)$. Thus:

Theorem 8. *We can solve APSP in an unweighted intersection graph of n fat triangles with roughly equal size in $O\left(n^2 \log^4 n\right)$ time.*

4   Reducing SSSP to decremental intersection detection

We give in this section a reduction of SSSP in intersection graphs to dynamic intersection detection.

We will emulate the classic BFS algorithm in the following way. Let $s \in S$ be the given source vertex. We proceed iteratively in $n - 1$ steps and follow the same process in each one. In step ℓ we assume that we have found all the distances and predecessors for all the objects that are at distance no more than $\ell - 1$ from s. We employ the definitions of the frontier and undiscovered objects as given in Section 2. The goal is to compute the distances and predecessors for all the undiscovered objects that are at distance ℓ from s. Those objects are the ones that have at least one intersection with a frontier object. To find those intersections we maintain an intersection detection data structure for the undiscovered objects that supports deletion—a deletion-only dynamic data structure is often referred to as a *decremental* data structure. We query the structure with the frontier objects; each time we detect an intersection of a frontier object with an undiscovered one, we properly update the latter's distance and predecessor and delete it from the data structure. The pseudocode of the algorithm is given in Algorithm 3.

We conclude this section with the following theorem.

Theorem 9. *We can solve SSSP in an unweighted, undirected geometric intersection graph in $O(DI(n, n))$ time, where $DI(n, m)$ is the time to construct a decremental intersection detection data structure of n objects and perform n deletions and m queries.*

Proof. The correctness of the algorithm can be easily proved by induction.

For the running time, we notice that an object can be in the frontier in only one step of the algorithm. In the beginning all the objects except the source are undiscovered (thus in the decremental intersection detection data structure as well), and once an object is deleted from that set, it is never inserted again. When querying the intersection detection data structure with a frontier object t there are two possible outcomes. If the query returns an undiscovered object z that intersects t, then z is deleted from the data structure, and since it is

---

Algorithm 3: SSSP(S, s)

---

1 $dist[s, s] = 0$
2 $dist[s, t] = \infty \; \forall t \in S - \{s\}$
3 $pred[s, t] = NULL \; \forall t \in S$
4 build a decremental intersection detection data structure for $S - \{s\}$
        // i.e., for undiscovered objects
5 **for** $\ell = 1$ *to* $n - 1$ **do**
6 $F = \{t \in S \mid dist[s, t] = \ell - 1\}$ // frontier
7   **for** *each* $t \in F$ **do**
8   **while** *true* **do**
9   query the data structure with t
10   let z be the answer
11   **if** z *not NULL* **then**
12 $dist[s, z] = \ell$
13 $pred[s, z] = t$
14   delete z from the data structure
15   **else**
16   break

---

never reinserted, this type of query happens only once $\forall z \in S$. If the query returns nothing, then this is the last query that t performs in that step, and since t can be in the frontier at most once, this type of query happens only once $\forall t \in S$. Consequently the total number of queries in the data structure is $O(n)$. Furthermore, the number of deletions in the decremental data structure is obviously $O(n)$. Thus the total running time is $O(DI(n, n))$.   □

Application to axis-aligned line segments. We need a decremental intersection detection data structure for horizontal input segments and vertical query segments. (Vertical input segments and horizontal query segments can be handled by a symmetric structure.) Giyora and Kaplan [21] (Theorem 5.3) and Blelloch [6] (Theorem 6.1) provided a data structure for supporting vertical ray shooting queries in $O(\log n)$ time and insertions and deletions of horizontal segments in $O(\log n)$ time—the problem is sometimes referred to as *dynamic orthogonal point location*. This immediately implies $DI(n, n) = O(n \log n)$. Thus:

Theorem 10. *We can solve SSSP in an unweighted intersection graph of n axis-aligned line segments in $O(n \log n)$ time.*

The result can be easily be extended to any set of line segments with a constant number of different orientations.

5   Conclusion

Interesting open problems in unweighted, undirected geometric intersection graphs include constructing efficient distance oracles and computing the diameter in truly subquadratic $O\left(n^{2-\varepsilon}\right)$ time for some $\varepsilon > 0$, in view of Cabello's recent breakthrough for the diameter problem in planar graphs [7]. For certain geometric objects such as arbitrary line segments, even a quadratic-time APSP algorithm is already open. Finally, solving APSP in the weighted case seems to be more difficult, as we can no longer exploit the general reduction from APSP to static, offline intersection detection.

References

1. Peyman Afshani, Timothy M. Chan, and Konstantinos Tsakalidis. Deterministic rectangle enclosure and offline dominance reporting on the RAM. In *Proceedings of the Forty-First Annual International Colloquium on Automata, Languages, and Programming*, pages 77–88, 2014.
2. Pankaj K. Agarwal, Noga Alon, Boris Aronov, and Subhash Suri. Can visibility graphs be represented compactly? *Discrete & Computational Geometry*, 12(3):347–365, 1994.
3. Pankaj K. Agarwal, Marco Pellegrini, and Micha Sharir. Counting circular arc intersections. *SIAM Journal on Computing*, 22(4):778–793, 1993.
4. Donald Aingworth, Chandra Chekuri, Piotr Indyk, and Rajeev Motwani. Fast estimation of diameter and shortest paths (without matrix multiplication). *SIAM Journal on Computing*, 28(4):1167–1181, 1999.
5. Noga Alon, Zvi Galil, Oded Margalit, and Moni Naor. Witnesses for boolean matrix multiplication and for shortest paths. In *Proceedings of the Thirty-Third Annual IEEE Symposium on Foundations of Computer Science*, pages 417–426, 1992.
6. Guy E. Blelloch. Space-efficient dynamic orthogonal point location, segment intersection, and range reporting. In *Proceedings of the Nineteenth Annual ACM-SIAM Symposium on Discrete Algorithms*, pages 894–903, 2008.
7. Sergio Cabello. Subquadratic algorithms for the diameter and the sum of pairwise distances in planar graphs. In *Proceedings of the Twenty-Eighth Annual ACM-SIAM Symposium on Discrete Algorithms*, pages 2143–2152. SIAM, 2017.
8. Sergio Cabello and Miha Jejčič. Shortest paths in intersection graphs of unit disks. *Computational Geometry*, 48(4):360–367, 2015.
9. Timothy M. Chan. Dynamic subgraph connectivity with geometric applications. *SIAM Journal on Computing*, 36(3):681–694, 2006.
10. Timothy M. Chan. All-pairs shortest paths for unweighted undirected graphs in $o(mn)$ time. *ACM Transactions on Algorithms*, 8:1–17, 2012.
11. Timothy M. Chan. Persistent predecessor search and orthogonal point location on the word RAM. *ACM Transactions on Algorithms*, 9(3):22, 2013.
12. Timothy M. Chan and Alon Efrat. Fly cheaply: On the minimum fuel consumption problem. *Journal of Algorithms*, 41(2):330–337, 2001.
13. Timothy M. Chan, Kasper Green Larsen, and Mihai Pătraşcu. Orthogonal range searching on the RAM, revisited. In *Proceedings of the Twenty-Seventh Annual ACM Symposium on Computational Geometry*, pages 1–10, 2011.
14. Timothy M. Chan and Mihai Pătraşcu. Counting inversions, offline orthogonal range counting, and related problems. In *Proceedings of the Twenty-First Annual ACM-SIAM Aymposium on Discrete Algorithms*, pages 161–173, 2010.
15. Timothy M. Chan and Dimitrios Skrepetos. All-pairs shortest paths in unit-disk graphs in slightly subquadratic time. In *Proccedings of the Twenty-Seventh Annual International Symposium on Algorithms and Computation*, pages 24:1–24:13, 2016.

16. Bernard Chazelle. Cutting hyperplanes for divide-and-conquer. *Discrete & Computational Geometry*, 9(2):145–158, 1993.
17. Herbert Edelsbrunner and Hermann A. Maurer. On the intersection of orthogonal objects. *Information Processing Letters*, 13(4/5):177–181, 1981.
18. Herbert Edelsbrunner and Mark H. Overmars. On the equivalence of some rectangle problems. *Information Processing Letters*, 14(3):124–127, 1982.
19. Tomás Feder and Rajeev Motwani. Clique partitions, graph compression and speeding-up algorithms. *Journal of Computer and System Sciences*, 51(2):261–272, 1995.
20. Steven Fortune. A sweepline algorithm for Voronoi diagrams. *Algorithmica*, 2:153–174, 1987.
21. Yoav Giora and Haim Kaplan. Optimal dynamic vertical ray shooting in rectilinear planar subdivisions. *ACM Transactions on Algorithms*, 5(3):28, 2009.
22. Prosenjit Gupta, Ravi Janardan, Michiel H. M. Smid, and Bhaskar DasGupta. The rectangle enclosure and point-dominance problems revisited. *International Journal of Computational Geometry and Applications*, 7(5):437–455, 1997.
23. Haim Kaplan, Wolfgang Mulzer, Liam Roditty, and Paul Seiferth. Spanners and reachability oracles for directed transmission graphs. In *LIPIcs-Leibniz International Proceedings in Informatics*, volume 34, 2015.
24. Haim Kaplan, Wolfgang Mulzer, Liam Roditty, Paul Seiferth, and Micha Sharir. Dynamic planar Voronoi diagrams for general distance functions and their algorithmic applications. In *Proceedings of the Twenty-Eighth Annual ACM-SIAM Symposium on Discrete Algorithms*, pages 2495–2504, 2017.
25. Matthew J. Katz. 3-D vertical ray shooting and 2-D point enclosure, range searching, and arc shooting amidst convex fat objects. *Computational Geometry*, 8(6):299–316, 1997.
26. Raimund Seidel. On the all-pairs-shortest-path problem in unweighted undirected graphs. *Journal of Computer and System Sciences*, 51(3):400–403, 1995.
27. Jack Snoeyink. Point location. In *Handbook of Discrete and Computational Geometry*, pages 767–785. CRC Press, second edition, 2004.
28. Virginia Vassilevska Williams. Multiplying matrices faster than Coppersmith-Winograd. In *Proceedings of the Forty-Fourth Annual ACM Symposium on Theory of Computing*, pages 887–898, 2012.

The Complexity of Drawing Graphs on Few Lines and Few Planes

Steven Chaplick[1], Krzysztof Fleszar[2*], Fabian Lipp[1**], Alexander Ravsky[3], Oleg Verbitsky[4***], and Alexander Wolff[1†]

[1] Lehrstuhl für Informatik I, Universität Würzburg, Germany,
www1.informatik.uni-wuerzburg.de/en/staff, first.last@uni-wuerzburg.de
[2] Department of Mathematical Engineering, Universidad de Chile, Chile,
kfleszar@dim.uchile.cl.
[3] Pidstryhach Institute for Applied Problems of Mechanics and Mathematics,
National Academy of Sciences of Ukraine, Lviv, Ukraine, oravsky@mail.ru
[4] Institut für Informatik, Humboldt-Universität zu Berlin, Germany,
verbitsk@informatik.hu-berlin.de

Abstract. It is well known that any graph admits a crossing-free straight-line drawing in $\mathbb{R}^3$ and that any planar graph admits the same even in $\mathbb{R}^2$. For a graph G and $d \in \{2, 3\}$, let $\rho_d^1(G)$ denote the minimum number of lines in $\mathbb{R}^d$ that together can cover all edges of a drawing of G. For $d = 2$, G must be planar. We investigate the complexity of computing these parameters and obtain the following hardness and algorithmic results.

- For $d \in \{2, 3\}$, we prove that deciding whether $\rho_d^1(G) \leq k$ for a given graph G and integer k is $\exists\mathbb{R}$-complete.
- Since NP $\subseteq \exists\mathbb{R}$, deciding $\rho_d^1(G) \leq k$ is NP-hard for $d \in \{2, 3\}$. On the positive side, we show that the problem is fixed-parameter tractable with respect to k.
- Since $\exists\mathbb{R} \subseteq$ PSPACE, both $\rho_2^1(G)$ and $\rho_3^1(G)$ are computable in polynomial space. On the negative side, we show that drawings that are optimal with respect to ρ_2^1 or ρ_3^1 sometimes require irrational coordinates.
- Let $\rho_3^2(G)$ be the minimum number of planes in $\mathbb{R}^3$ needed to cover a straight-line drawing of a graph G. We prove that deciding whether $\rho_3^2(G) \leq k$ is NP-hard for any fixed $k \geq 2$. Hence, the problem is not fixed-parameter tractable with respect to k unless P = NP.

* K. Fleszar was supported by Conicyt PCI PII 20150140 and Millennium Nucleus Information and Coordination in Networks RC130003.
** F. Lipp was supported by Cusanuswerk. ORCID: orcid.org/0000-0001-7833-0454
*** O. Verbitsky was supported by DFG grant VE 652/1-2.
† ORCID: orcid.org/0000-0001-5872-718X

© Springer International Publishing AG 2017
F. Ellen et al. (Eds.): WADS 2017, LNCS 10389, pp. 265–276, 2017.
DOI: 10.1007/978-3-319-62127-2_23

1   Introduction

As is well known, any graph can be drawn in $\mathbb{R}^3$ without crossings so that all
edges are segments of straight lines. Suppose that we have a supply $\mathcal{L}$ of lines in
$\mathbb{R}^3$, and the edges are allowed to be drawn only on lines in $\mathcal{L}$. How large does
$\mathcal{L}$ need to be for a given graph G? For planar graphs, a similar question makes
sense also in $\mathbb{R}^2$, since planar graphs admit straight-line drawings in $\mathbb{R}^2$ by the
Wagner–Fáry–Stein theorem. Let $\rho_3^1(G)$ denote the minimum size of $\mathcal{L}$ which is
sufficient to cover a drawing of G in $\mathbb{R}^3$. For a planar graph G, we denote the corresponding parameter in $\mathbb{R}^2$ by $\rho_2^1(G)$. The study of these parameters was posed
as an open problem by Durocher et al. [10]. The two parameters are related to
several challenging graph-drawing problems such as small-area or small-volume
drawings [9], layered or track drawings [8], and drawing graphs with low visual
complexity. Recently, we studied the extremal values of $\rho_3^1(G)$ and $\rho_2^1(G)$ for
various classes of graphs and examined their relations to other characteristics
of graphs [6]. In particular, we showed that there are planar graphs where the
parameter $\rho_3^1(G)$ is much smaller than $\rho_2^1(G)$. Determining the exact values of
$\rho_3^1(G)$ and $\rho_2^1(G)$ for particular graphs seems to be tricky even for trees.

In fact, the setting that we suggested is more general [6]. Let $1 \leq l < d$. We
define the *affine cover number* $\rho_d^l(G)$ as the minimum number of l-dimensional
planes in $\mathbb{R}^d$ such that G has a straight-line drawing that is contained in the
union of these planes. We suppose that $l \leq 2$ as otherwise $\rho_d^l(G) = 1$.

Moreover, we can focus on $d \leq 3$ as every graph can be drawn in 3-space
as efficiently as in higher dimensions, that is, $\rho_d^l(G) = \rho_3^l(G)$ if $d \geq 3$ [6]. This
implies that, besides the *line cover numbers* in 2D and 3D, $\rho_2^1(G)$ and $\rho_3^1(G)$,
the only interesting affine cover number is the *plane cover number* $\rho_3^2(G)$. Note
that $\rho_3^2(G) = 1$ if and only if G is planar. Let K_n denote the complete graph
on n vertices. For the smallest non-planar graph K_5, we have $\rho_3^2(K_5) = 3$. The
parameters $\rho_3^2(K_n)$ are not so easy to determine even for small values of n. We
have shown that $\rho_3^2(K_6) = 4$, $\rho_3^2(K_7) = 6$, and $6 \leq \rho_3^2(K_8) \leq 7$ [6]. It is not hard
to show that $\rho_3^2(K_n) = \Theta(n^2)$, and we determined the asymptotics of $\rho_3^2(K_n)$ up
to a factor of 2 using the relations of these numbers to Steiner systems.

The present paper is focused on the computational complexity of the affine
cover numbers. A good starting point is to observe that, for given G and k, the
statement $\rho_d^l(G) \leq k$ can be expressed by a first-order formula about the reals
of the form $\exists x_1 \ldots \exists x_m \Phi(x_1, \ldots, x_m)$, where the quantifier-free subformula Φ is
written using the constants 0 and 1, the basic arithmetic operations, and the
order and equality relations. If, for example, $l = 1$, then we just have to write
that there are k pairs of points, determining a set $\mathcal{L}$ of k lines, and there are n
points representing the vertices of G such that the segments corresponding to
the edges of G lie on the lines in $\mathcal{L}$ and do not cross each other. This observation
shows that deciding whether or not $\rho_d^l(G) \leq k$ reduces in polynomial time to
the decision problem (Hilbert's *Entscheidungsproblem*) for *the existential theory
of the reals*. The problems admitting such a reduction form the complexity class
$\exists \mathbb{R}$ introduced by Schaefer [23], whose importance in computational geometry
has been recognized recently [4,16,24]. In the complexity-theoretic hierarchy,

this class occupies a position between NP and PSPACE. It possesses natural complete problems like the decision version of the rectilinear crossing number [1], the recognition of segment intersection graphs [15] or unit disk graphs [13].

Below, we summarize our results on the computational complexity of the affine cover numbers.

The complexity of the line cover numbers in 2D and 3D. We begin by showing that it is ∃ℝ-hard to compute, for a given graph G, its line cover numbers $\rho_2^1(G)$ and $\rho_3^1(G)$; see Section 2.

Our proof uses some ingredients from a paper of Durocher et al. [10] who showed that it is NP-hard to compute the *segment number* segm(G) of a graph G. This parameter was introduced by Dujmović et al. [7] as a measure of the visual complexity of a planar graph. A *segment* in a straight-line drawing of a graph G is an inclusion-maximal connected path of edges of G lying on a line, and the *segment number* segm(G) of a planar graph G is the minimum number of segments in a straight-line drawing of G in the plane. Note that while $\rho_2^1(G) \leq$ segm(G), the parameters can be far apart, e.g., as shown by a graph with m isolated edges. For connected graphs, we have shown earlier [6] that segm$(G) \in O(\rho_2^1(G)^2)$ and that this bound is optimal as there exist planar triangulations with $\rho_2^1(G) \in O(\sqrt{n})$ and segm$(G) \in \Omega(n)$. Still, we follow Durocher et al. [10] to some extent in that we also reduce from ARRANGEMENT GRAPH RECOGNITION (see Theorem 1).

Parameterized complexity of computing the line cover numbers in 2D and 3D. It follows from the inclusion NP ⊆ ∃ℝ that the decision problems $\rho_2^1(G) \leq k$ and $\rho_3^1(G) \leq k$ are NP-hard if k is given as a part of the input. On the positive side, in Section 3, we show that both problems are fixed-parameter tractable. To this end, we first describe a linear-time kernelization procedure that reduces the given graph to one of size $O(k^4)$. Then, in $k^{O(k^2)}$ time, we carefully solve the problem on this reduced instance by using the exponential-time decision procedure for the existential theory of the reals by Renegar [20,21,22] as a subroutine. To the best of our knowledge, this is the first application of Renegar's algorithm for obtaining an FPT result, in particular, in the area of graph drawing where FPT algorithms are widely known.

The space complexity of ρ_d^1-optimal drawings. Since ∃ℝ belongs to PSPACE (as shown by Canny [3]), the parameters $\rho_d^1(G)$ for both $d = 2$ and 3 are computable in polynomial space. On the negative side, we construct a graph G with a ρ_2^1-optimal drawing requiring irrational coordinates; we provide a more complex argument to show that *any* ρ_2^1-optimal drawing of G requires irrational coordinates; for details see the full version [5].

The complexity of the plane cover number. Though the decision problem $\rho_3^2(G) \leq k$ also belongs to ∃ℝ, its complexity status is different from that of the line cover numbers. In Section 4, we establish the NP-hardness of deciding whether $\rho_3^2(G) \leq k$ for any fixed $k \geq 2$, which excludes an FPT algorithm for this problem unless P = NP. To show this, we first prove NP-hardness of POSITIVE PLANAR CYCLE 1-IN-3-SAT (a new problem of planar 3-SAT type), which we think is of independent interest.

Weak affine cover numbers. We previously defined the *weak affine cover number* $\pi_d^l(G)$ of a graph G similarly to $\rho_d^l(G)$ but under the weaker requirement that the l-dimensional planes in $\mathbb{R}^d$ whose number has to be minimized contain the *vertices* (and not necessarily the edges) of G [6]. Based on our combinatorial characterization of π_3^1 and π_3^2 [6], we show in Section 5 that the decision problem $\pi_3^1(G) \leq 2$ is NP-complete, and that it is NP-hard to approximate $\pi_3^1(G)$ within a factor of $O(n^{1-\epsilon})$, for any $\epsilon > 0$. Asymmetrically to the affine cover numbers ρ_2^1, ρ_3^1, and ρ_3^2, here it is the parameter π_2^1 (for planar graphs) whose complexity remains open. For more open problems, see Section 6.

2   Computational Hardness of the Line Cover Numbers

In this section, we show that deciding, for a given graph G and integer k, whether $\rho_2^1(G) \leq k$ or $\rho_3^1(G) \leq k$ is an $\exists\mathbb{R}$-complete problem. The $\exists\mathbb{R}$-hardness results are often established by a reduction from the PSEUDOLINE STRETCHABILITY problem: Given an arrangement of pseudolines in the projective plane, decide whether it is *stretchable*, that is, equivalent to an arrangement of lines [17,18]. Our reduction is based on an argument of Durocher et al. [10] who designed a reduction of the ARRANGEMENT GRAPH RECOGNITION problem, defined below, to the problem of computing the segment number of a graph.

A *simple line arrangement* is a set $\mathcal{L}$ of k lines in $\mathbb{R}^2$ such that each pair of lines has one intersection point and no three lines share a common point. In the following, we assume that every line arrangement is simple. We define the *arrangement graph* for a set of lines as follows [2]: The vertices correspond to the intersection points of lines and two vertices are adjacent in the graph if and only if they are adjacent along some line. The ARRANGEMENT GRAPH RECOGNITION problem is to decide whether a given graph is the arrangement graph of some set of lines.

Bose et al. [2] showed that this problem is NP-hard by reduction from a version of PSEUDOLINE STRETCHABILITY for the Euclidean plane, whose NP-hardness was proved by Shor [25]. It turns out that ARRANGEMENT GRAPH RECOGNITION is actually an $\exists\mathbb{R}$-complete problem [11, page 212]. This stronger statement follows from the fact that the Euclidean PSEUDOLINE STRETCHABILITY is $\exists\mathbb{R}$-hard as well as the original projective version [16,23].

Theorem 1. *Given a planar graph G and an integer k, it is $\exists\mathbb{R}$-hard to decide whether $\rho_2^1(G) \leq k$ and whether $\rho_3^1(G) \leq k$.*

Proof. We first treat the 2D case. We show hardness by a reduction from ARRANGEMENT GRAPH RECOGNITION. Let G be an instance of this problem. If G is an arrangement graph, there must be an integer ℓ such that G consists of $\ell(\ell-1)/2$ vertices and $\ell(\ell-2)$ edges, and each of its vertices has degree d where

$d \in [2, 4]$. So, we first check these easy conditions to determine ℓ and reject G if one of them fails. Let G' be the graph obtained from G by adding one tail (i.e., a degree-1 vertex) to each degree-3 vertex and two tails to each degree-2 vertex. So every vertex of G' has degree 1 or 4. Note that, if G is an arrangement graph, then there are exactly 2ℓ tails in G' (2 for each line) – if this is not true we can already safely reject G. We now pick $k = \ell$, and show that G is an arrangement graph if and only if $\rho_2^1(G') \leq k$.

For the first direction, let G be an arrangement graph. By our choice of k, it is clear that G corresponds to a line arrangement of k lines. Clearly, all edges of G lie on these k lines and the tails of G' can be added without increasing the number of lines. Hence, $\rho_2^1(G') \leq k$.

For the other direction, assume $\rho_2^1(G') \leq k$ and let Γ' be a straight-line drawing of G' on $\rho_2^1(G')$ lines. The graph G' contains $\binom{k}{2}$ degree-4 vertices. As each of these vertices lies on the intersection of two lines in Γ', we need k lines to get enough intersections, that is, $\rho_2^1(G') = k$. Additionally, there are no intersections of more than two lines. The most extreme points on any line have degree 1, that is, they are tails, because degree 4 would imply a more extreme vertex. We can assume that there are exactly $2k$ tails, otherwise G would have been rejected before as it could not be an arrangement graph. Each line contains exactly two of them. Let n_2 (resp. n_3) be the number of degree-2 (resp. degree-3) vertices. As we added 2 (resp. 1) tails to each of these vertices, we have $2k = 2n_2 + n_3$. By contradiction, we show that the edges on each line form a single segment. Otherwise, there would be a line with two segments. Note that the vertices at the ends of each segment have degree less than 4 (that is, degree 1). This would imply more than two degree-1 vertices on one line, a contradiction. So Γ' is indeed a drawing of G' using k segments. By removing the tails, we obtain a straight-line drawing of G using $k = n_2 + n_3/2$ segments. The result by Durocher et al. [10, Lemma 2] implies that G is an arrangement graph.

Now we turn to 3D. Let G be a graph and let G' be the augmented graph as above. We show that $\rho_3^1(G') = \rho_2^1(G')$, which yields that deciding $\rho_3^1(G')$ is also NP-hard. Clearly, $\rho_3^1(G') \leq \rho_2^1(G')$. Conversely, assume that G' can be drawn on k lines in 3-space. Since G' has $\binom{k}{2}$ vertices of degree 4, each of them must be a crossing point of two lines. It follows that each of the k lines crosses all the others. Fix any two of the lines and consider the plane that they determine. Then all k lines must lie in this plane, which shows that $\rho_2^1(G') \leq \rho_3^1(G')$. □

It remains to notice that the decision problems under consideration lie in the complexity class $\exists \mathbb{R}$. To this end, we transform the inequalities $\rho_d^l(G) \leq k$ into first-order existential expressions about the reals. For details, see the full version [5].

Lemma 2. *Each of the following decision problems belongs to the complexity class* $\exists \mathbb{R}$

(a) *deciding, for a planar graph G and an integer k, whether $\rho_2^1(G) \leq k$;*

(b) *deciding, for a graph G and an integer k, whether $\rho_3^1(G) \leq k$;*

(c) deciding, for a graph G and an integer k, whether $\rho_3^2(G) \leq k$.

3   Fixed-Parameter Tractability of the Line Cover Numbers

In this section we show that, for an input graph G and integer k, both testing whether $\rho_2^1(G) \leq k$, and testing whether $\rho_3^1(G) \leq k$ are decidable in FPT time (in k). Moreover, for both the 2D and 3D cases, for positive instances (G, k), we can compute the *combinatorial description* of a solution also in time FPT in k. One subtle point here is that there are graphs where *each ρ_2^1-optimal drawing* requires irrational coordinates; see the full version for details [5]. Thus, in some sense, a combinatorial description of a solution can be seen as a best possible output from an algorithm for these problems. Note that, by a k-line cover in $\mathbb{R}^d$ of a graph G, we mean a drawing D of G together with a set $\mathcal{L}$ of k lines such that $(D, \mathcal{L})$ certifies $\rho_d^1(G) \leq k$.

Our FPT algorithm follows from a simple *kernelization/pre-processing* procedure in which we reduce a given instance (G, k) to a reduced instance (H, k) where H has $O(k^4)$ vertices and edges, and G has a k-line cover if and only if H does as well. After this reduction, we can then apply any decision procedure for the existential theory of the reals since we have shown in Lemma 2 that both k-line cover problems are indeed both members of this complexity class. Our kernelization approach is given as Theorem 3 and our FPT result follows as described in Corollary 4. We denote the number of vertices and the number of edges in the input graph by n and m respectively.

Theorem 3. *For each $d \in \{2, 3\}$, graph G, and integer k, the problem of deciding whether $\rho_d^1(G) \leq k$ admits a kernel of size $O(k^4)$, i.e., we can produce a graph H such that H has $O(k^4)$ vertices and edges and $\rho_d^1(G) \leq k$ if and only if $\rho_d^1(H) \leq k$. Moreover, H can be computed in $O(n + m)$ time.*

Proof. For a graph G, if G is going to have a k-line cover $(D, \mathcal{L})$, then there are several necessary conditions about G which we can exploit to shrink G. First, notice that any connected components of G which are paths can easily be placed on any line in $\mathcal{L}$ without interfering with the other components, i.e., these can be disregarded. This provides a new instance G'. Second, there are at most $\binom{k}{2}$ intersection points among the lines in $\mathcal{L}$. Thus, G has at most $\binom{k}{2}$ vertices with degree larger than two. Moreover, each line $\ell \in \mathcal{L}$ will contain at most $k - 1$ of these vertices. Thus, the total number of edges which are incident to vertices with degree larger than two, is at most $2 \cdot (k - 1)$ per line, or $2 \cdot (k^2 - k)$ in total. Thus, G' contains at most $2 \cdot (k^2 - k)$ vertices of degree one (since each one occurs at the end of a path originating from a vertex of degree larger than two where all the internal vertices have degree 2). Similarly, G' contains at most $2 \cdot (k^2 - k)$ paths where every internal vertex has degree two and the end vertices either have degree one or degree larger than two. Finally, for each such path, at most $\binom{k}{2}$ vertices are mapped to intersection points in $\mathcal{L}$. Thus, any path with more than $\binom{k}{2}$ vertices can be safely contracted to a path with at most $\binom{k}{2}$ vertices. This results in our final graph G'' which can easily be seen to have $O(k^4)$ vertices and $O(k^4)$ edges (when G has a k-line cover). Now, if G'' does not satisfy one of the necessary conditions described above, we use the graph $K_{1,2k+1}$ as H, i.e., this way H has no k-line cover.

We conclude by remarking that this transformation of G to G'' can be performed in $O(n+m)$ time. The transformation from G to G' is trivial. The transformation from G' to G'' can be performed by two traversals of the graph (e.g., breadth first searches) where we first measure the lengths of the paths of degree-2 vertices, then we shrink them as needed.   □

In the notation of the above proof, note that the statement $\rho_d^1(G'') \leq k$ can be expressed as a prenex formula Φ in the existential first-order theory of the reals. The proof of Lemma 2 shows that such a formula can be written using $O(k^4)$ first-order variables and involving $O(k^4)$ polynomial inequalities, each of total degree at most 4 and with coefficients ± 1. We could now directly apply the decision procedure of Renegar [20,21,22] to Φ and obtain an FPT algorithm for deciding whether $\rho_d^1(G) \leq k$, but that would only provide a running time of $(k^{O(k^4)} + O(n+m))$. We can be a little more clever and reduce the exponent from $O(k^4)$ to $O(k^2)$. This is described in the proof of the following corollary.

Corollary 4. *For each $d \in \{2,3\}$, graph G, and integer k, we can decide whether $\rho_d^1(G) < k$ in $k^{O(k^2)} + O(n+m)$ time, i.e., FPT time in k.*

Proof. First, we apply to the given graph G the kernelization procedure from the proof of Theorem 3 to obtain a reduced graph G''. Now, notice that G'' has at most $O(k^4)$ vertices of degree two, but only $\binom{k}{2}$ of these can be bend points and are actually important in a solution, i.e., at most $\binom{k}{2}$ of these vertices are mapped to intersection points of the lines. Thus, we can simply enumerate all possible $O\left(\binom{k^4}{\binom{k}{2}}\right)$ subsets which will occur as intersection points, and, for each of these, test whether this further reduced instance has a k-line cover using Renegar's decision algorithm. This leads to a total running time of $k^{O(k^2)} + O(n+m)$ as needed.   □

We have now seen how to decide if a given graph G has a k-line cover in both 2D and 3D. Moreover, when G is a positive instance, our approach provides a reduced graph G''' where G''' also has a k-line cover, G''' has $O(k^2)$ vertices and edges, and any k-line cover of G''' naturally induces a k-line cover of G. In the following theorem, whose proof can be found in the full version [5], we show that we can further determine the combinatorial structure of some k-line cover of G''' in $k^{O(k^2)}$ time and use this to recover a corresponding combinatorial structure for G. Here, the combinatorial structure is a set of k *linear forests* since each line in a k-line cover naturally induces a linear forest in G. Recall that a linear forest is a forest whose connected components are paths.

Theorem 5. *For each $d \in \{2,3\}$, graph G, and integer k, in $2^{O(k^3)} + O(n+m)$ time we can not only decide whether $\rho_d^1(G) \leq k$ but, if so, also partition the edge set of G into linear forests accordingly to a k-line cover of G in $\mathbb{R}^d$.*

4 Computational Complexity of the Plane Cover Number

While graphs with ρ_3^2-value 1 are exactly the planar graphs, recognizing graphs with ρ_3^2-value k, for any $k > 1$, immediately becomes NP-hard. This requires a detour via the NP-hardness of a new problem of planar 3-SAT type, which we think is of independent interest. Full proofs for this section are given in the full version [5].

Definition 6 ([19]). *Let Φ be a Boolean formula in 3-CNF. The associated graph of Φ, $G(\Phi)$, has a vertex v_x for each variable x in Φ and a vertex v_c for each clause c in Φ. There is an edge between a variable-vertex v_x and a clause-vertex v_c if and only if x or $\neg x$ appears in c. The Boolean formula Φ is called planar if $G(\Phi)$ is planar.*

Kratochvíl et al. [14] proved NP-hardness of PLANAR CYCLE 3-SAT, which is a variant of PLANAR 3-SAT where the clauses are connected by a simple cycle in the associated graph without introducing crossings. Their reduction even shows hardness of a special case, where all clauses consist of at least two variables. We consider only this special case. Mulzer and Rote [19] proved NP-hardness of POSITIVE PLANAR 1-IN-3-SAT, another variant of PLANAR 3-SAT where all literals are positive and the assignment must be such that, in each clause, exactly one of the three variables is true. We combine proof ideas from the two to show NP-hardness of the following new problem.

Definition 7. *In the POSITIVE PLANAR CYCLE 1-IN-3-SAT problem, we are given a collection Φ of clauses each of which contains exactly three variables, together with a planar embedding of $G(\Phi) + C$ where C is a cycle through all clause-vertices. Again, all literals are positive. The problem is to decide whether there exists an assignment of truth values to the variables of Φ such that exactly one variable in each clause is true.*

Lemma 8. POSITIVE PLANAR CYCLE 1-IN-3-SAT *is NP-complete.*

Proof (sketch). We reduce from PLANAR CYCLE 3-SAT. We iteratively replace the clauses by positive 1-IN-3-SAT clauses while maintaining the cycle through these clauses. Our reduction uses some of the gadgets from the proof of Mulzer and Rote [19]. We show how to maintain the cycle when inserting these gadgets.

We consider the interaction between the cycle and the clauses. Every clause consists of two or three literals and thus there are two or three faces around a clause in the drawing. There are two options for the cycle: (O1) it can "touch" the clause, that is, the incoming and the outgoing edge are drawn in the same face; (O2) it can "pass through" the clause, that is, incoming and outgoing edge are drawn in different faces. As an example, Fig. 1a shows how we weave the cycle through the inequality gadget by Mulzer and Rote. As a replacement for the clauses with 2 variables we cannot use the gadget described by Mulzer and Rote as it does not allow us to add a cycle through the clauses. Therefore, we use a new gadget that is depicted in Fig. 1b.   □

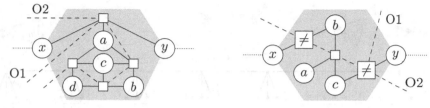

(a) Mulzer and Rote's gadget for $x \neq y$.   (b) Our gadget for the clause $(x \vee y)$.

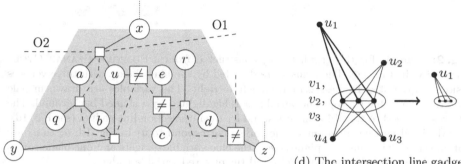

(c) Mulzer and Rote's gadget for the clause: $x \vee y \vee z$.

(d) The intersection line gadget and how it is depicted in Fig. 2.

Fig. 1: Gadgets for our NP-hardness proof. Variables are drawn in circles, clauses are represented by squares. The boxes with the inequality sign represent the inequality gadget. The dashed line shows how we weave the cycle through the clauses. There are two variants of the cycle, which differ only in one edge: (O1) The cycle touches the gadget; (O2) the cycle passes through the gadget.

We now introduce what we call the *intersection line gadget*; see Fig. 1d. It consists of a $K_{3,4}$ in which the vertices in the smaller set of the bipartition—denoted by v_1, v_2, and v_3—are connected by a path. We denote the vertices in the other set by u_1, u_2, u_3, and u_4.

Lemma 9. *If a graph containing the intersection line gadget can be embedded on two non-parallel planes, the vertices v_1, v_2, and v_3 must be drawn on the intersection line of the two planes while the vertices u_1, u_2, u_3, and u_4 cannot lie on the intersection line.*

Theorem 10. *Let G be a graph. Deciding whether $\rho_3^2(G) = 2$ is NP-hard.*

Proof (sketch). We show NP-hardness by reduction from POSITIVE PLANAR CYCLE 1-IN-3-SAT. We build the graph $G^*(\Phi) = (V, E)$ for formula Φ that consists of n clauses as follows: Each clause c is represented by a clause gadget that consists of three vertices v_c^1, v_c^2, and v_c^3 that are connected by a path. Let x be a variable that occurs in the clauses $c_{i_1}, c_{i_2}, \ldots, c_{i_l}$ with $i_1 < i_2 < \cdots < i_l$. Each variable x is represented by a tree with the vertices $w_1^x, w_2^x, \ldots, w_l^x$ that are connected to the relevant clauses, and the vertices $v_1^x, v_2^x, \ldots, v_l^x$ that lie on a path and are connected to these vertices. To each of the vertices $v_1^x, v_2^x, \ldots, v_l^x$ one

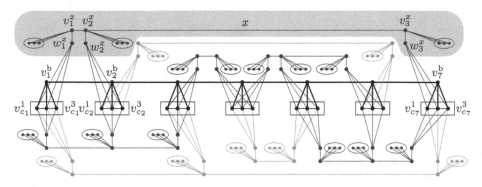

Fig. 2: Example for the graph $G^*(\Phi)$ constructed from a POSITIVE PLANAR CYCLE 1-IN-3-SAT instance Φ. The clauses are depicted by the black boxes with three vertices inside and denoted by $c_1, \ldots, c_7$ from left to right. The variables are drawn in pale red (true) and blue (false). The variable x is highlighted by a shaded background. The ellipses attached to variable-vertices stand for the intersection line gadget (see Fig. 1d). The depicted vertices incident to the gadget correspond to u_1 in Fig. 1d; u_2 to u_4 are not shown. If Φ is true, one plane covers the blue variable gadgets and one plane covers the blocking caterpillar (bold black) and the pale red variable gadgets.

instance of the intersection line gadget is connected. Finally, we add a blocking caterpillar consisting of the vertices $v_1^b, \ldots, v_n^b$ that are connected to the clauses in their cyclic order, which exists for the POSITIVE PLANAR CYCLE 1-IN-3-SAT instance by definition. See Fig. 2 for an example of this construction.

We show that the formula Φ has a truth assignment with exactly one true variable in each clause if and only if the graph $G^*(\Phi)$ can be drawn onto two planes. The idea of our construction is that only two variables can be connected to a clause gadget on each of the planes. One plane contains the blocking caterpillar and one variable per clause (corresponding to true variables). The other plane contains two variables per clause (corresponding to false variables). Our construction ensures that the vertices of a variable cannot be partitioned onto both planes in any drawing. $\qquad\square$

Corollary 11. *Deciding whether $\rho_3^2(G) = k$ is NP-hard for any $k \geq 2$.*

Proof (sketch). We add a blocking gadget for each additional plane. $\qquad\square$

5   Complexity of the Weak Affine Cover Numbers π_3^1 / π_3^2

Recall that a *linear forest* is a forest whose connected components are paths. The *linear vertex arboricity* lva(G) of a graph G equals the smallest size r of a partition $V(G) = V_1 \cup \cdots \cup V_r$ such that every V_i induces a linear forest. The *vertex thickness* vt(G) of a graph G is the smallest size r of a partition $V(G) = V_1 \cup \cdots \cup V_r$ such that $G[V_1], \ldots, G[V_r]$ are all planar. Obviously, vt(G) $\leq$ lva(G). We recently used these notions to characterize the 3D weak affine cover numbers in purely combinatorial terms [6]: $\pi_3^1(G) = $ lva(G) and $\pi_3^2(G) = $ vt(G).

Theorem 12. *For $l \in \{1, 2\}$,*

(a) deciding whether or not $\pi_3^l(G) \leq 2$ is NP-complete, and

(b) approximating $\pi_3^l(G)$ within a factor of $O(n^{1-\epsilon})$, for any $\epsilon > 0$, is NP-hard.

Proof. (a) The membership in NP follows directly from the above combinatorial characterization [6], which also allows us to deduce NP-hardness from a much more general hardness result by Farrugia [12]: For any two graph classes $\mathcal{P}$ and $\mathcal{Q}$ that are closed under vertex-disjoint unions and taking induced subgraphs, deciding whether the vertex set of a given graph G can be partitioned into two parts X and Y such that $G[X] \in \mathcal{P}$ and $G[Y] \in \mathcal{Q}$ is NP-hard unless both $\mathcal{P}$ and $\mathcal{Q}$ consist of all graphs or all empty graphs. To see the hardness of our two problems, we set $\mathcal{P} = \mathcal{Q}$ to the class of linear forests (for $l = 1$) and to the class of planar graphs (for $l = 2$).

(b) The combinatorial characterization [6] given above implies that $\chi(G) \leq 4\,\mathrm{vt}(G) = 4\pi_3^2(G)$ (by the four-color theorem). Note that each color class can be placed on its own line, so $\pi_3^1(G) \leq \chi(G)$. As $\pi_3^2(G) \leq \pi_3^1(G)$, both parameters are linearly related to the chromatic number of G. Now, the approximation hardness of our problems follows from that of the chromatic number [26]. $\qquad\square$

6 Conclusion and Open Problems

1. We have determined the computational complexity of the affine cover numbers ρ_2^1 and ρ_3^1. The corresponding decision problems $\rho_2^1(G) \leq k$ and $\rho_3^1(G) \leq k$ turn out to be $\exists\mathbb{R}$-complete. On the positive side, these problems admit an FPT algorithm (Corollary 4). This is impossible for the plane cover number ρ_3^2, unless $P = NP$, because the decision problem $\rho_3^2(G) \leq k$ is NP-hard even for $k = 2$ (Theorem 10 in Section 4). If k is arbitrary and given as a part of the input, then this problem is in $\exists\mathbb{R}$ (Lemma 2)—but is it $\exists\mathbb{R}$-hard?
2. Is the segment number $\mathrm{segm}(G)$ introduced in [7] fixed-parameter tractable?
3. Our proof of Theorem 1 implies that computing $\rho_2^1(G)$ and $\rho_3^1(G)$ is hard even for planar graphs of maximum degree 4. Can $\rho_2^1(G)$ and $\rho_3^1(G)$ be computed efficiently for trees? This is true for the segment number $\mathrm{segm}(G)$ [7].
4. How hard is it to approximate ρ_2^1, ρ_3^1, and ρ_3^2?

References

1. Bienstock, D.: Some provably hard crossing number problems. Discrete Comput. Geom. 6, 443–459 (1991)
2. Bose, P., Everett, H., Wismath, S.K.: Properties of arrangement graphs. Int. J. Comput. Geom. Appl. 13(6), 447–462 (2003)
3. Canny, J.F.: Some algebraic and geometric computations in PSPACE. In: Proc. 20th Annu. ACM Symp. Theory Comput. (STOC'88). pp. 460–467 (1988)
4. Cardinal, J.: Computational geometry column 62. SIGACT News 46(4), 69–78 (2015)
5. Chaplick, S., Fleszar, K., Lipp, F., Ravsky, A., Verbitsky, O., Wolff, A.: The complexity of drawing graphs on few lines and few planes (2016), arxiv.org/1607.06444

6. Chaplick, S., Fleszar, K., Lipp, F., Ravsky, A., Verbitsky, O., Wolff, A.: Drawing graphs on few lines and few planes. In: Hu, Y., Nöllenburg, M. (eds.) GD 2016. LNCS, vol. 9801, pp. 166–180. Springer-Verlag (2016), arxiv.org/1607.01196
7. Dujmović, V., Eppstein, D., Suderman, M., Wood, D.R.: Drawings of planar graphs with few slopes and segments. Comput. Geom. Theory Appl. 38(3), 194–212 (2007)
8. Dujmović, V., Pór, A., Wood, D.R.: Track layouts of graphs. Discrete Math. & Theor. Comput. Sci. 6(2), 497–522 (2004)
9. Dujmović, V., Whitesides, S.: Three-dimensional drawings. In: Tamassia, R. (ed.) Handbook of Graph Drawing and Visualization, chap. 14, pp. 455–488. CRC Press (2013)
10. Durocher, S., Mondal, D., Nishat, R.I., Whitesides, S.: A note on minimum-segment drawings of planar graphs. J. Graph Alg. Appl. 17(3), 301–328 (2013)
11. Eppstein, D.: Drawing arrangement graphs in small grids, or how to play planarity. J. Graph Alg. Appl. 18(2), 211–231 (2014)
12. Farrugia, A.: Vertex-partitioning into fixed additive induced-hereditary properties is NP-hard. Electr. J. Comb. 11(1) (2004)
13. Kang, R.J., Müller, T.: Sphere and dot product representations of graphs. Discrete Comput. Geom. 47(3), 548–568 (2012)
14. Kratochvíl, J., Lubiw, A., Nešetřil, J.: Noncrossing subgraphs in topological layouts. SIAM J. Discrete Math. 4(2), 223–244 (1991)
15. Kratochvíl, J., Matoušek, J.: Intersection graphs of segments. J. Comb. Theory, Ser. B 62(2), 289–315 (1994)
16. Matoušek, J.: Intersection graphs of segments and $\exists\mathbb{R}$. arxiv.org/1406.2636 (2014)
17. Mněv, N.E.: On manifolds of combinatorial types of projective configurations and convex polyhedra. Sov. Math., Dokl. 32, 335–337 (1985)
18. Mněv, N.E.: The universality theorems on the classification problem of configuration varieties and convex polytopes varieties. In: Viro, O. (ed.) Topology and Geometry, Rohlin Seminar 1984–1986, Lect. Notes Math., vol. 1346, pp. 527–543. Springer-Verlag, Berlin (1988)
19. Mulzer, W., Rote, G.: Minimum-weight triangulation is NP-hard. J. ACM 55(2), 11 (2008)
20. Renegar, J.: On the computational complexity and geometry of the first-order theory of the reals, part I: Introduction. preliminaries. the geometry of semi-algebraic sets. the decision problem for the existential theory of the reals. J. Symb. Comput. 13(3), 255–300 (1992)
21. Renegar, J.: On the computational complexity and geometry of the first-order theory of the reals, part II: The general decision problem. preliminaries for quantifier elimination. J. Symb. Comput. 13(3), 301–328 (1992)
22. Renegar, J.: On the computational complexity and geometry of the first-order theory of the reals, part III: Quantifier elimination. J. Symb. Comput. 13(3), 329–352 (1992)
23. Schaefer, M.: Complexity of some geometric and topological problems. In: Eppstein, D., Gansner, E.R. (eds.) GD 2009. LNCS, vol. 5849, pp. 334–344. Springer-Verlag (2010)
24. Schaefer, M., Štefankovič, D.: Fixed points, Nash equilibria, and the existential theory of the reals. Theory Comput. Syst. pp. 1–22 (2015)
25. Shor, P.W.: Stretchability of pseudolines is NP-hard. In: Gritzmann, P., Sturmfels, B. (eds.) Applied Geometry and Discrete Mathematics—The Victor Klee Festschrift, DIMACS Series, vol. 4, pp. 531–554. Amer. Math. Soc. (1991)
26. Zuckerman, D.: Linear degree extractors and the inapproximability of max clique and chromatic number. Theory Comput. 3(1), 103–128 (2007)

Modular Circulation and Applications to Traffic Management*

Philip Dasler and David M. Mount

Department of Computer Science and
Institute for Advanced Computer Studies
University of Maryland, College Park, Maryland 20742
{daslerpc,mount}@cs.umd.edu

Abstract. We introduce a variant of the well-known minimum-cost circulation problem in directed networks, where vertex demand values are taken from the integers modulo λ, for some integer $\lambda \geq 2$. More formally, given a directed network $G = (V, E)$, each of whose edges is associated with a weight and each of whose vertices is associated with a demand taken over the integers modulo λ, the objective is to compute a flow of minimum weight that satisfies all the vertex demands modulo λ. This problem is motivated by a problem of computing a periodic schedule for traffic lights in an urban transportation network that minimizes the total delay time of vehicles. We show that this modular circulation problem is solvable in polynomial time when $\lambda = 2$ and that the problem is NP-hard when $\lambda = 3$. We also present a polynomial time algorithm that achieves a $4(\lambda - 1)$-approximation.

Keywords: Network flows and circulations, Traffic management, Approximation algorithms, NP-hard problems

1 Introduction

Minimum (and maximum) cost network flows and the related concept of circulations are fundamental computational problems in discrete optimization. In this paper, we introduce a variant of the circulation problem, where vertex demand values are taken from the integers modulo λ, for some integer $\lambda \geq 2$. For example, if $\lambda = 10$ a vertex with demand 6 can be satisfied by any net incoming flow of 6, 16, 26 and so on or a net outgoing flow of 4, 14, 24, and so on. Our motivation in studying this problem stems from an application in synchronizing the traffic lights of an urban transportation system.

Throughout, let $G = (V, E)$ denote a directed graph, and let $\lambda \geq 2$ be an integer. Each edge $(u, v) \in E$ is associated with a nonnegative integer *weight*, $wt(u, v)$, and each vertex $u \in V$ is associated with a *demand*, $d(u)$, which is an integer drawn from $\mathbb{Z}_\lambda$, the integers modulo λ. Let f be an assignment of values

* Research supported by NSF grant CCF-1618866.

© Springer International Publishing AG 2017
F. Ellen et al. (Eds.): WADS 2017, LNCS 10389, pp. 277–288, 2017.
DOI: 10.1007/978-3-319-62127-2_24

from $\mathbb{Z}_\lambda$ to the edges of G. For each vertex $v \in V$, define

$$f_{in}(v) = \sum_{(u,v)\in E} f(u,v) \quad \text{and} \quad f_{out}(v) = \sum_{(v,w)\in E} f(v,w),$$

and define the *net flow* into a vertex v to be $f_{in}(v) - f_{out}(v)$. We say that f is a *circulation with λ-modular demands*, or λ-*CMD* for short, if it satisfies the *modular flow-balance constraints*, which state that for each $v \in V$,

$$f_{in}(v) - f_{out}(v) \equiv d(v) \pmod{\lambda}.$$

Observe that a demand of $d(v)$ is equivalent to the modular "supply" requirement that the net flow out of this vertex modulo λ is $\lambda - d(v)$.

Define the *cost* of a circulation f to be the weighted sum of the flow values on all the edges, that is,

$$\text{cost}(f) = \sum_{(u,v)\in E} wt(u,v) \cdot f(u,v).$$

Given a directed graph G and the vertex demands d, the λ-*CMD problem* is that of computing a λ-CMD of minimum cost. (Observe that there is no loss in generality in restricting the flow value on each edge to $\mathbb{Z}_\lambda$, since the cost could be reduced by subtracting λ from this value without affecting the flow's validity.)

The standard minimum-cost circulation problem (without the modular aspect) is well studied. We refer the reader to any of a number of standard sources on this topic, for example, [1, 2, 6, 8]. In contrast, λ-CMD is complicated by the "wrap-around" effect due to the modular nature of the demand constraints. A vertex's demand of $d(u)$ units can be satisfied in the traditional manner by having a net incoming flow of $d(u)$, but it could also be met by generating a net outgoing flow of $\lambda - d(u)$ (not to mention all variants thereof that involve adding multiples of λ). Our main results are:

- 2-CMD can be solved exactly in polynomial time (see Section 4).
- 3-CMD is NP-hard (see Section 5).
- There is a polynomial time $4(\lambda-1)$-approximation to λ-CMD (see Section 6).

In Section 2 we discuss the relevance of the λ-CMD problem to a traffic-management problem. In Section 3 we present some preliminary observations regarding this problem. In Sections 4–6 we present each of our three main results.

2   Application to Traffic Management

Our motivation in studying the λ-CMD problem arises from an application in traffic management. In urban settings, intersections are the shared common resource between vehicles traveling in different directions, and their control is essential to maximizing the utilization of a transportation network [9]. There are numerous approaches to modeling traffic flow and diverse computational approaches to solve and analyze the associated traffic management problems [4,11]. Despite the popular interest in automated traffic systems, there has been relatively little work on this problem from the perspective of algorithm design.

In an earlier paper [3], we considered the problem of scheduling the movements of a collections of vehicles through a system of unregulated crossing. Our approach was based on the idealized assumption that the motion of individual vehicles in the system is controlled by a central server. A more practical approach is based on aggregating vehicles into groups, or *platoons*, and planning motion at the motion of these groups [7,10].

We consider the problem in this aggregated form, but from a periodic perspective. Consider an urban transportation network consisting of a grid of horizontal and vertical roads as laid out on a map. Each pair of horizontal and vertical roads meets at a unique intersection controlled by a traffic light that alternates between horizontal and vertical traffic, such that the pattern repeats over a time interval λ. We assume throughout that λ has been discretized to a reasonably small integer value, say in terms of seconds or tens of seconds.

More formally, we say that a traffic-light schedule is λ-*periodic* if repeats every λ time units. We consider a traffic management system of the foreseeable future where the traffic light schedule is transmitted to the vehicles, which in turn may adjust their speeds to avoid excessive waiting at intersections. While vehicles may turn at intersections, the schedule is designed to minimize the delay of straight-moving traffic.

To motivate the connection with modular circulations, consider a four-sided city block (see Fig. 1). Let a, b, c, and d denote the intersections, and let t_{ab}, t_{bc}, t_{cd}, t_{ad} denote the travel times between successive intersections along each road segment. If the road segment is oriented counterclockwise around the block (as shown in our example), these travel times are positive, and otherwise they are negative. Suppose that the traffic-light schedule is λ-periodic, and that at time $t = t_a$ the light at intersection a transitions so that the eastbound traffic can move horizontally through the intersection (see Fig. 1(a)). In order for these vehicles to proceed without delay through intersection b, this light must transition from vertical to horizontal at time $t_b = t_a + t_{ab}$ (see Fig. 1(b)).

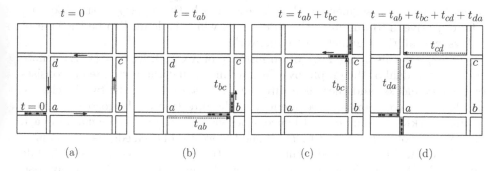

Fig. 1. Delay-free traffic-light schedule.

Reasoning analogously, for the other intersections, it follows that the vertical-to-horizontal transition times for intersections c and d are $t_c = t_b + t_{bc}$ and $t_d = t_c + t_{cd}$, respectively (see Fig. 1(c)). On returning to a (see Fig. 1(d)), we find that

$$t_a \equiv t_a + (t_{ab} + t_{bc} + t_{cd} + t_{da}) \pmod{\lambda}.$$

Thus, in order to achieve delay-free flow around the intersection in a λ-periodic context, we must satisfy the constraint

$$t_{ab} + t_{bc} + t_{cd} + t_{da} \equiv 0 \pmod{\lambda}.$$

Since the transportation times along the road segments are not under our control, in order to satisfy this constraint, we introduce an (ideally small) delay $\delta_{ij} \geq 0$ along each road segment ij. This yields the new constraint

$$(t_{ab} + \delta_{ab}) + (t_{bc} + \delta_{bc}) + (t_{cd} + \delta_{cd}) + (t_{da} + \delta_{da}) \equiv 0 \pmod{\lambda},$$

or equivalently, if we define $T = t_{ab} + t_{bc} + t_{cd} + t_{da}$ to be the sum of (signed) travel times of the road segments around this block, we have

$$\delta_{ab} + \delta_{bc} + \delta_{cd} + \delta_{da} \equiv -T \pmod{\lambda}. \tag{1}$$

The upshot is that if vehicles travel at a reduced speed so that the transit time along each of the road segments includes the associated delay, then the straight-line vehicular traffic along each road need never pause or wait at any traffic signal. The objective is to minimize the sum of delay values over all the road segments in the network, which we refer to as the *total delay*.

More formally, the transportation network is modeled as a set of horizontal and vertical roads. This defines a directed grid graph whose vertices are the intersections, whose edges are the *road segments*, and whose (bounded) faces are the *blocks* of the city. For each pair of adjacent intersections i and j, let t_{ij} denote the delay-free travel time along this road segment. For each block u, define the *total signed travel time* about u to be the sum of the travel times for each of the road segments bounding u, where the travel time is counted positively if the segment is oriented counterclockwise about u and negatively otherwise. Let $T(u)$ denote this value modulo λ. A λ-*periodic traffic-light schedule* assigns a *delay* to each road segment so that for each block, these delays satisfy Eq. (1). The objective is to minimize the *total delay*, which is defined to be the sum of delays over all the segments in the network.

To express this in the form of an instance of λ-CMD, let $G = (V, E)$ denote the directed dual of the graph, by which we mean that the vertex set V consists of the city blocks, and there is a directed edge $(u, v) \in E$ if the two blocks are incident to a common road segment, and the direction of the road segment is counterclockwise about u (and hence, clockwise about v). The demand of each vertex u, denoted $d(u)$, is set to $T(u)$, and the weight of each edge is set to unity.

There remains one impediment to linking the λ-periodic traffic-light schedule and the λ-CMD problems. The issue is that the delay associated with any road segment (which may be as large as $\lambda - 1$) can be significantly larger than the time to traverse the road segment. If so, the capacity of the road segment to hold the vehicles that are waiting for the next signal may spill backwards and block the preceding intersection. In order to deal with this issue without complicating our model, we introduce the assumption that λ is smaller than the time to traverse any road segment. The link between the two problems is presented in the following lemma. Due to space limitations, the proofs of this and many other lemmas have been omitted and will appear in the full version of the paper.

Lemma 1. *Given a transportation network and integer $\lambda \geq 2$, let G be the associated directed graph with vertex demands and edge weights as described above.*

(i) *If there exists a λ-periodic traffic-light schedule with total delay Δ, then there exists a λ-CMD for G of cost Δ.*

(ii) *If there exists a λ-CMD for G of cost Δ and for all road segments ij, $t_{ij} \geq \lambda$, then there exists a λ-periodic traffic-light schedule with total delay Δ.*

3 Preliminaries

In this section we present a few definitions and observations that will be used throughout the paper. Given an instance $G = (V, E)$ of the λ-CMD problem, consider any subset $V' \subseteq V$. Let $G' = (V', E')$ be the associated induced subgraph of G, and let $d(V')$ denote the sum of demands of all the nodes in V'. We refer to E' as the *internal edges* of this subgraph, and we refer to the edges of G that cross the cut $(V', V \setminus V')$ as the *interface*. Given such a subgraph and any flow f on G, define its *internal flow* to be only the flow on the internal edges, and define the *internal cost* to be the cost of the flow restricted to these edges. Define the *interface flow* and *interface cost* analogously for the interface edges. Define $f_{\text{in}}(V')$ to be the sum of flow values on the interface edges that are directed into V', and define $f_{\text{out}}(V')$ analogously for outward directed edges. The following lemma provides necessary and sufficient conditions for the existence of a λ-CMD.

Lemma 2. *Given an instance $G = (V, E)$ of the λ-CMD problem:*

(i) *For any induced subgraph $G' = (V', E')$ and any λ-CMD f, we have*

$$f_{\text{in}}(V') - f_{\text{out}}(V') \equiv d(V') \pmod{\lambda}.$$

(ii) *If G is weakly connected, then a λ-CMD exists for G if and only if $d(V) \equiv 0 \pmod{\lambda}$.*

It follows from this lemma the λ-CMD instance associated with any traffic-light scheduling problem has a solution. The reason is that each edge (u, v) contributes its travel time t_{uv} positively to $d(u)$ and negatively to $d(v)$, and therefore the sum of demands over all the vertices of the network is zero, irrespective of the travel times.

4 Polynomial Time Solution to 2-CMD

In this section we show that 2-CMD, which we also call *binary CMD*, can be solved in polynomial time by a reduction to minimum-cost matching in general graphs. Intuitively, the binary case is simpler because the edge directions are not significant. If a vertex is incident to an even number of flow-carrying edges (whether directed into or out of this vertex), then the net flow into this vertex modulo λ is zero, and otherwise it is one. Thus, solving the problem reduces to computing a minimum-cost set of paths that connect each pair of vertices of nonzero demand, which is essentially a minimum-cost perfect matching in a complete graph whose vertex set consists of the subset vertices of nonzero demand and whose edge weights are distances between vertices ignoring edge directions. The remainder of this section is devoted to providing a formal justification of this intuition.

Recall $G = (V, E)$ is a directed graph, and $d(v)$ denotes the demand of vertex v. Since $\lambda = 2$, for each $v \in V$, we have $d(v) \in \{0, 1\}$. Let $G' = (V, E')$ denote the graph on the same vertices as G but with directions removed from all the edges. We may assume that G' is connected, for otherwise it suffices to solve the problem separately on each connected component of G'. We set the weight of each edge of G' to the weight of the corresponding edge of G. If there are two oppositely directed edges joining the same pair of vertices, the weight is set to the minimum of the two.

Let $U = U(G)$ denote the subset of vertices of V whose demand values are equal to 1. By Lemma 2(ii), we may assume that $d(V) \equiv 0 \pmod{\lambda}$. Therefore, $d(V)$ is even, which implies that $|U|$ is also even. For each $u, v \in U$, let $\pi(u, v)$ denote the shortest weight path between them in G', and let $wt(\pi(u, v))$ denote this weight. Define $\widehat{G} = (U, \widehat{E})$ to be a complete, undirected graph on the vertex set U, where for each $u, v \in U$, the weight of this edge $wt(\pi(u, v))$. (This is well defined by our assumption that G' is connected.) Since $\widehat{G}$ is complete and has an even number of vertices, it has a perfect matching. The reduction of 2-CMD to the minimum-cost perfect matching problem is implied by the following lemma.

Lemma 3. *lem:2-cmd Given an instance $G = (V, E)$ of the 2-CMD problem, the minimum cost of any 2-CMD for G is equal to the minimum cost of a perfect matching in $\widehat{G}$.*

Since $\widehat{G}$ is dense, a minimum-cost perfect matching can be constructed in $O(|U|^3)$ time. The graph can be computed in $O(n^3)$ time, where $n = |V|$, by applying the Floyd-Warshall algorithm for computing shortest paths [2]. Thus, the overall running time is $O(n^3)$.

Theorem 1. *It is possible to solve the 2-CMD problem in $O(n^3)$ time on any instance $G = (V, E)$, where $n = |V|$.*

5 Hardness of 3-CMD

In this section, we present the following hardness result for λ-CMD.

Theorem 2. *For $\lambda \geq 3$, the λ-CMD problem is NP-hard.*

The reduction is from positive 1-in-3-SAT [5]. For the sake of brevity and simplicity, we show the proof for the case $\lambda = 3$ here, but the method easily generalizes (as will be explained in the full version of the paper). Let F denote a boolean formula in 3-CNF, where each literal is in positive form. Throughout, for $\alpha \in \{0, 1, \ldots, \lambda - 1\}$, we use the term α-vertex to denote a vertex whose demand is α. The reduction involves two principal components, a *variable gadget* which associates truth values with the variables of F and a *clause gadget* which enforces the condition that exactly one variable in each clause is assigned the value **True**.

5.1 Variable Gadget

Before discussing the general gadget, we describe a fundamental building block from which all variables will be constructed. The block consists of six vertices, three 1-vertices and three $(\lambda - 1)$-vertices (i.e., 2-vertices), connected together with edges as shown in Figure 2(a). Edges connecting 1-vertices have weight $wt(u, v) = 1.5$, while all other edges are of weight $wt(u, v) = 1$.

If a flow of 1 is sent from each 2-vertex to its connected 1-vertex, the 2-vertices overflow and all demands are satisfied with $cost(f) = 3$ (see Figure 2(b)). This flow, in which there is no flow across the interface edges, represents a logical value of **False**.

If instead a flow of 1 is sent across the interface edges, then the demands of the 2-vertices are satisfied. A flow of 1 across each edge originating at the central 1-vertex will cause it to overflow and will satisfy each of the connected 1-vertices. This flow, in which each interface edge carries a flow of 1, represents a logical value of **True** and again has $cost(f) = 3$ (see Figure 2(c)).

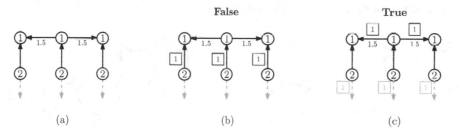

Fig. 2. (a) The fundamental building block used to build variable gadgets. Interface edges are dashed gray segments. (b,c) CMDs representing the assignment of **False** and **True** values, respectively, with the flow values in boxes.

If every interface edge of a variable gadget carries the same flow and that flow is either 0 or 1, that variable is said to be *interface-consistent*. If all variables are consistent for a given flow, then that flow is said to be *variable-consistent*. Notice that both logical values above are realized via interface-consistent flows.

Lemma 4. *Given a fundamental block, a satisfying flow has cost ≤ 3 if and only if that flow is interface-consistent.*

Proof. Each 2-vertex can only be satisfied by: (1) sending a flow of 1 across one of its edges or (2) sending a flow of 2 across both of its edges (in both cases the vertex's demand overflows).

In the second case, the 2-vertex sends a flow of 2 to its neighboring 1-vertex. As per Lemma 2(i), that vertex now requires a flow of 2 across its other edge in order to have its demand satisfied. Together these flows come at a cost of 5 (one of these edges has a weight of 1.5), therefor no 2-vertex may be satisfied by a flow greater than 1 without a cost greater than 3.

There exists a satisfying flow for a fundamental block if and only if the total flow across its interface edges is equivalent to 0 (mod λ) (see Lemma 2). Given this and the fact that no single interface flow may equal 2, the flows across the interface edges must either all be 0 or all be 1, i.e., the over all flow must be interface consistent for it to be a satisfying flow.

As variables may appear in multiple clauses, we need a mechanism by which existing variables can be expanded. For this, we create an expansion module, any number of which can be added to a variable so that there are three interface edges for each clause in which that variable appears.

To understand how this module functions, let us first look at the case when two fundamental blocks are connected together. This connection occurs through a shared 2-vertex, so that what was an interface edge for one block becomes the connection between a 2-vertex and 1-vertex in the other block (see Figure 3). Recall that the value assignment of a variable is determined by the direction of flow from the 2-vertices, with flow along the interface representing **True** and internal flow (i.e., flow to the connected 1-vertices) representing **False**. Because the outgoing edges of the shared 2-vertex are simultaneously an interface edge of one block and an internal edge of the other, pushing a flow across either edge will assign opposing values to the blocks.

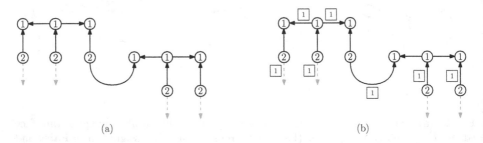

(a)   (b)

Fig. 3. Two fundamental blocks connected via a shared 2-vertex, with figure (b) showing a flow that is satisfying but not interface-consistent.

Knowing this, the module is constructed as a double-negative, ensuring that it is assigned the same value as the variable it extends. A fundamental block is used as a hub and to this hub we attach two more fundamental blocks (see Figure 4). When attached to a variable, this module creates four new interface edges and consumes one, thus extending the variable by three interface edges.

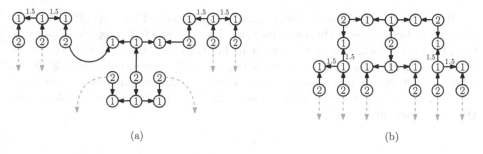

(a)   (b)

Fig. 4. (a) A fundamental block with a single extension module attached. (b) The same structure rearranged to emphasize the two clause outputs, each with three interface edges.

While the structure of the fundamental blocks is as described above, the weighting must be adjusted to maintain equal costs between the **True** and **False** states. Rather than each fundamental block having two edges of weight 1.5, only the rightmost block in the expansion module has such edges; all others are of weight 1. In this way, the minimal cost of a consistent satisfying flow across the module is 8, regardless of the value assigned to the variable. This is easily verified by assigning a truth value to the gadget (fixing an interface-consistent flow on the interface edges of 0 or 1) and then traversing the structure, satisfying the demand in each vertex by assigning flow to its unused edge.

Given this, Lemma 4, and the fact that the expansion module is constructed from fundamental blocks, we have the following.

Lemma 5. *Given an expansion module, there exists an interface-consistent flow of internal cost 8 (in either the **True** or **False** cases), and any other satisfying flow has a strictly larger internal cost.*

To construct a gadget for a variable v_i, appearing in $c(v_i)$ clauses, begin with a fundamental block and connect $c(v_i) - 1$ expansion modules to it. Doing so provides $c(v_i)\lambda$ interface edges and yields the following result:

Lemma 6. *Given a variable gadget, there exists an interface-consistent flow of internal cost $3 + 8[c(v_i) - 1]$ (in either the **True** or **False** cases), and any other satisfying flow has a strictly larger internal cost.*

5.2 Clause Gadget

The basis for the clause gadget is a single 1-vertex with three incoming edges, one for each literal. These edges have a weight $wt(u, v) = \gamma$ and are connected to the appropriate variables as their outgoing interface edges. If a single literal is **True**, one of these edges will carry a flow of 1, satisfying the demand of the clause vertex. If more literals are **True**, the demand underflows and the vertex is left unsatisfied. It is possible to satisfy the vertex by creating flows on these edges greater than 1, but such flows can be made cost-prohibitive by setting the edge weights γ sufficiently high.

Recall that each variable gadget produces λ copies of its respective variable (three interface edges in this example) per clause in which it appears. Because of this, the clause gadget must also be created in triplicate. Every clause consists of three 1-vertices, each with an incoming edge from its three literals (see Figure 5). Their weighting and behavior are as described above. Since there are no internal edges in the clause gadgets, they do not contribute to the cost of the flow (but their interface edges will).

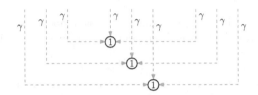

Fig. 5. A full clause gadget, with three inputs from each of three literals.

5.3   Final Construction

Each variable in F is represented by a fundamental block connected to $c(v_i) - 1$ expansion modules, creating $c(v_i)\lambda$ outputs. Thus, λ outputs are linked to each of the appropriate clause gadgets. The size of the variable gadget is a linear function of the number of clauses in which that variable appears and can thus be constructed in polynomial time.

If F is satisfiable, then a 3-CMD exists that is variable-consistent. In this case, each fundamental block incurs a cost of 3, and each expansion module incurs an additional cost of 8 for a flow representing a consistent truth value across its interface and the interfaces of the modules/fundamental block to which it is attached, as per Lemma 5.

For each clause, create λ 1-vertices, each connected to the clause's three literals by incoming edges. As there are no edges between these vertices, there is no flow possible within the clause gadget, resulting in an internal cost of 0. The size of the clause gadget is constant.

Finally, the flow on the edges between the variable gadgets and clause gadgets has yet to be counted as they are interface edges for both gadgets. Each clause gadget contains λ 1-vertices, with each receiving a flow of 1 across edges of weight γ. Thus, these add a cost of $3|C|\gamma$, where $|C|$ is the number of clauses in F.

If F is not satisfiable, then some set of variables must have inconsistent outputs in order to create a valid CMD. As shown in Lemma 4, these inconsistencies will always lead to a strictly greater cost. Thus:

Lemma 7. *Given a positive boolean formula F in 3-CNF, in polynomial time it is possible to construct an instance of 3-CMD that has a satisfying flow with* $cost(f) \leq \sum_{v_i \in V}(3 + 8[c(v_i) - 1] + 3|C|\gamma)$ *if and only if F is 1-in-3 satisfiable.*

6   Approximation Algorithm

In this section, we present an $4(\lambda-1)$-factor approximation to the λ-CMD problem for $\lambda \geq 2$. Before presenting the algorithm, we introduce some terminology. Consider an instance $G = (V, E)$ of the λ-CMD problem, with vertex demands d. Let $G' = (V, E')$ be (as defined in Section 4) the undirected version of G. Let us assume that G' is connected. Let $U = U(G)$ denote the subset of vertices of V whose demand values are nonzero. Define $SMT(U)$ to be a *Steiner minimal tree* in G' whose terminal set is U (that is, a connected subgraph of G' of minimum weight that contains all the vertices of U).

As in Section 4, define $\widehat{G} = (U, \widehat{E})$ to be the complete, undirected graph over the vertex set U, where for each $u, v \in U$, the weight of this edge is the weight of a minimum weight path between u and v in G'. Given any $U \subseteq V$, let $MST(U)$ denote any minimum spanning tree on the subgraph of G' induced on U. From standard results on Steiner and minimum spanning trees we have the following.

Lemma 8. *For any $U \subseteq V$, $wt(MST(U)) \leq 2 \cdot wt(SMT(U))$.*

Define a *balanced partition* to be a partition $\{U_1, \ldots, U_k\}$ of U such that for $1 \leq i \leq k$, the total demand within U_i (that is, $d(U_i)$) is equivalent to zero modulo λ. By Lemma 2(ii), we may assume that $d(V) \equiv 0 \pmod{\lambda}$, and so there is always a trivial partition, namely $\{V\}$ itself. Define $cost(U_i)$ to be $cost(SMT(U_i))$, and define the *cost* of a balanced partition to be the sum of costs over its components. A *minimum balanced partition* for G is a balanced partition of minimum cost. The following lemma establishes the connection between balanced partitions and minimum modular circulations.

Lemma 9. *Consider an instance $G = (V, E)$ of λ-CMD. Let $\Psi = (U_1, \ldots, U_k)$ denote a minimum balanced partition of G, as defined above, and let f denote any minimum cost λ-CMD for G. Then $cost(\Psi) \leq |f| \leq (\lambda - 1) \cdot cost(\Psi)$.*

By the above lemma, it suffices to compute a balanced partition for G of low cost. We will present a simple approximation algorithm that outputs a balanced partition whose cost is within a factor of 4 of the optimum.

The construction begins with the metric closure $\widehat{G}$ defined above. In a manner similar to Kruskal's algorithm, we sort the edges of $\widehat{G}$ in increasing order, and start with each vertex of $\widehat{G}$ in a separate component. All these components are labeled as *active*. We process the edges one by one. Letting (u, v) denote the next edge being processed, if u and v are in distinct components, and both components are active, we merge these components into a single component. If the sum of the demands of the vertices within this component is equivalent to zero modulo λ, we label the resulting component as *finished*, and output its set of vertices. Because the total sum of demands of all the nodes is equivalent to zero modulo λ, it follows that every vertex is placed within a finished component, and therefore the algorithm produces a balanced partition of $\widehat{G}$ (and by extension, a balanced partition of G).

This algorithm has the same running time as Kruskal's algorithm. (Observe that we can associate each component with its sum of demands, thus enabling us to determine the sum of merged components in constant time.) The following lemma establishes the approximation factor for this construction.

Lemma 10. *Let Ψ' denote the balanced partition generated by the above algorithm, and let Ψ denote the optimum balanced partition. Then $\mathrm{cost}(\Psi') \leq 4 \cdot \mathrm{cost}(\Psi)$.*

Combining Lemmas 9 and 10(ii), it follows that our algorithm achieves an approximation factor of $4(\lambda - 1)$. While obtaining the best running time has not been a focus of this work, it is easy to see that this procedure runs in polynomial time. Let $n = |V|$. The graph $\widehat{G}$ can be computed in $O(n^3)$ time by the Floyd-Warshall algorithm [2]. The Kruskal-like algorithm for computing the balanced partition can be performed in $O(n^2 \log n)$ time, as can the algorithm of Lemma 9. Thus, the overall running time is $O(n^3)$.

Theorem 3. *Given an instance $G = (V, E)$ of the λ-CMD problem for $\lambda \geq 2$, it is possible to compute a $4(\lambda - 1)$-approximation in time $O(n^3)$, where $n = |V|$.*

References

1. R. K. Ahuja, T. L. Magnanti, and J. B. Orlin. *Network Flows: Theory, Algorithms, and Applications.* Prentice-Hall, 1993.
2. T. H. Cormen, C. Stein, R. L. Rivest, and C. E. Leiserson. *Introduction to Algorithms.* McGraw-Hill Higher Education, 2nd edition, 2001.
3. P. Dasler and D. M. Mount. On the complexity of an unregulated traffic crossing. In *Proc. 14th Internat. Sympos. Algorithms Data Struct.*, volume 9214 of *Lecture Notes Comput. Sci.*, pages 224–235. Springer-Verlag, 2015.
4. K. M. Dresner and P. Stone. A multiagent approach to autonomous intersection management. *J. Artif. Int. Res.*, 31:591–656, 2008.
5. M. R. Garey and D. S. Johnson. *Computers and Intractability: A Guide to the Theory of NP-Completeness.* W. H. Freeman & Co., 1979.
6. A. V. Goldberg and R. E. Tarjan. Finding minimum-cost circulations by canceling negative cycles. *Journal of the ACM*, 36:873–886, 1989.
7. S. I. Guler, M. Menendez, and L. Meier. Using connected vehicle technology to improve the efficiency of intersections. *Transportation Research Part C: Emerging Technologies*, 46:121–131, 2014.
8. J. B. Orlin. A polynomial time primal network simplex algorithm for minimum cost flows. *Mathematical Programming*, 78:109–129, 1997.
9. R. Tachet, P. Santi, S. Sobolevsky, L. I. Reyes-Castro, E. Frazzoli, D. Helbing, and C. Ratti. Revisiting street intersections using slot-based systems. *PLOS ONE*, 11(3):e0149607, 2016.
10. J. J. B. Vial, W. E. Devanny, D. Eppstein, and M. T. Goodrich. Scheduling autonomous vehicle platoons through an unregulated intersection. In M. Goerigk and R. Werneck, editors, *16th Wkshp. Alg. Approaches Transport. Model., Opt., and Syst. (ATMOS 2016)*, volume 54, pages 1–14, 2016.
11. J. Yu and S. M. LaValle. Multi-agent path planning and network flow. In E. Frazzoli, T. Lozano-Perez, N. Roy, and D. Rus, editors, *Algorithmic Foundations of Robotics X*, pages 157–173, Berlin, Heidelberg, 2013. Springer Berlin Heidelberg.

The Homogeneous Broadcast Problem in Narrow and Wide Strips*

Mark de Berg[1], Hans L. Bodlaender[12], and Sándor Kisfaludi-Bak[1]

[1] TU Eindhoven, The Netherlands
m.t.d.berg@tue.nl; h.l.bodlaender@tue.nl; s.kisfaludi.bak@tue.nl
[2] Utrecht University, The Netherlands

Abstract. Let P be a set of nodes in a wireless network, where each node is modeled as a point in the plane, and let $s \in P$ be a given source node. Each node p can transmit information to all other nodes within unit distance, provided p is activated. The (homogeneous) broadcast problem is to activate a minimum number of nodes such that in the resulting directed communication graph, the source s can reach any other node. We study the complexity of the regular and the hop-bounded version of the problem (in the latter, s must be able to reach every node within a specified number of hops), with the restriction that all points lie inside a strip of width w. We almost completely characterize the complexity of both the regular and the hop-bounded versions as a function of the strip width w.

1 Introduction

Wireless networks give rise to a host of interesting algorithmic problems. In the traditional model of a wireless network each node is modeled as a point $p \in \mathbb{R}^2$, which is the center of a disk $\delta(p)$ whose radius equals the transmission range of p. Thus p can send a message to another node q if and only if $q \in \delta(p)$. Using a larger transmission radius may allow a node to transmit to more nodes, but it requires more power and is more expensive. This leads to so-called range-assignment problems, where the goal is to assign a transmission range to each node such that the resulting communication graph has desirable properties, while minimizing the cost of the assignment. We are interested in broadcast problems, where the desired property is that a given source node can reach any other node in the communication graph. Next, we define the problem more formally.

Let P be a set of n points in $\mathbb{R}^d$ and let $s \in P$ be a source node. A *range assignment* is a function $\rho : P \to \mathbb{R}_{\geq 0}$ that assigns a transmission range $\rho(p)$ to each point $p \in P$. Let $\mathcal{G}_\rho = (P, E_\rho)$ be the directed graph where $(p, q) \in E_\rho$ iff $|pq| \leq \rho(p)$. The function ρ is a *broadcast assignment* if every point $p \in P$ is reachable from s in $\mathcal{G}_\rho$. If every $p \in P$ is reachable within h hops, for a given parameter h, then ρ is an *h-hop broadcast assignment*. The (h-hop) broadcast

---

* This research was supported by the Netherlands Organization for Scientific Research (NWO) under project no. 024.002.003.

F. Ellen et al. (Eds.): WADS 2017, LNCS 10389, pp. 289–300, 2017.
DOI: 10.1007/978-3-319-62127-2_25

problem is to find an (h-hop) broadcast assignment whose cost $\sum_{p \in P} \text{cost}(\rho(p))$ is minimized. Often the cost of assigning transmission radius x is defined as $\text{cost}(x) = x^\alpha$ for some constant α. In $\mathbb{R}^1$, both the basic broadcast problem and the h-hop version are solvable in $O(n^2)$ time [7]. In $\mathbb{R}^2$ the problem is NP-hard for any $\alpha > 1$ [6,10], and in $\mathbb{R}^3$ it is even APX-hard [10]. There are also several approximation algorithms [1,6]. For the 2-hop broadcast problem in $\mathbb{R}^2$ an $O(n^7)$ algorithm is known [2] and for any constant h there is a PTAS [2]. Interestingly, the complexity of the 3-hop broadcast problem is unknown.

An important special case of the broadcast problem is where we allow only two possible transmission ranges for the points, $\rho(p) = 1$ or $\rho(p) = 0$. In this case the exact cost function is irrelevant and the problem becomes to minimize the number of active points. This is called the *homogeneous broadcast problem* and it is the version we focus on. From now on, all mentions of broadcast and h-hop broadcast refer to the homogeneous setting. Observe that if $\rho(p) = 1$ then (p, q) is an edge in $\mathcal{G}_\rho$ if and only if the disks of radius $1/2$ centered at p and q intersect. Hence, if all points are active then $\mathcal{G}_\rho$ in the intersection graph of a set of congruent disks or, in other words, a *unit-disk graph (UDG)*. Because of their relation to wireless networks, UDGs have been studied extensively.

Let $\mathcal{D}$ be a set of congruent disks in the plane, and let $\mathcal{G}_\mathcal{D}$ be the UDG induced by $\mathcal{D}$. A *broadcast tree* on $\mathcal{G}_\mathcal{D}$ is a rooted spanning tree of $\mathcal{G}_\mathcal{D}$. To send a message from the root to all other nodes, each internal node of the tree has to send the message to its children. Hence, the cost of broadcasting is related to the internal nodes in the broadcast tree. A cheapest broadcast tree corresponds to a minimum-size *connected dominating set* on $\mathcal{G}_\mathcal{D}$, that is, a minimum-size subset $\Delta \subset \mathcal{D}$ such that the subgraph induced by Δ is connected and each node in $\mathcal{G}_\mathcal{D}$ is either in Δ or a neighbor of a node in Δ. The broadcast problem is thus equivalent to the following: given a UDG $\mathcal{G}_\mathcal{D}$ with a designated source node s, compute a minimum-size connected dominated set $\Delta \subset \mathcal{D}$ such that $s \in \Delta$.

In the following we denote the dominating set problem by DS, the connected dominating set problem by CDS, and we denote these problems on UDGs by DS-UDG and CDS-UDG, respectively. Given an algorithm for the broadcast problem, one can solve CDS-UDG by running the algorithm n times, once for each possible source point. Consequently, hardness results for CDS-UDG can be transferred to the broadcast problem, and algorithms for the broadcast problem can be transferred to CDS-UDG at the cost of an extra linear factor in the running time. It is well known that DS and CDS are NP-hard, even for planar graphs [11]. DS-UDG and CDS-UDG are also NP-hard [13,15]. The parameterized complexity of DS-UDG has also been investigated: Marx [14] proved that DS-UDG is W[1]-hard when parameterized by the size of the dominating set. (The definition of W[1] and other parameterized complexity classes can be found in the book by Flum and Grohe [9].)

Our contributions. Knowing the existing hardness results for the broadcast problem, we set out to investigate the following questions. Is there a natural special case or parameterization admitting an efficient algorithm? Since the broadcast problem is polynomially solvable in $\mathbb{R}^1$, we study how the complexity of the problem changes as we go from the 1-dimensional problem to the 2-dimensional problem. To do this, we assume the points (that is, the disk centers) lie in a strip of width w, and we study how the problem complexity changes as we increase w. We give an almost complete characterization of the complexity, both for the general and for the hop-bounded version of the problem. More precisely, our results are as follows.

We first study strips of width at most $\sqrt{3}/2$. Unit disk graphs restricted to such *narrow strips* are a subclass of co-comparability graphs [16], for which an $O(nm)$ time CDS algorithm is known [12,3]. (Here m denotes the number of edges in the graph.) The broadcast problem is slightly different because it requires s to be in the dominating set; still, one would expect better running times in this restricted graph class. Indeed, we show that for narrow strips the broadcast problem can be solved in $O(n \log n)$ time. The hop condition in the h-hop broadcast problem has not been studied yet for co-comparability graphs to our knowledge. This condition complicates the problem considerably. Nevertheless, we show that the h-hop broadcast problem in narrow strips is solvable in polynomial time. Our algorithm runs in $O(n^6)$ and uses a subroutine for 2-hop broadcast, which may be of independent interest: we show that the 2-hop broadcast problem is solvable in $O(n^4)$ time. Our subroutine is based on an algorithm by Ambühl et al. [2] for the non-homogeneous case, which runs in $O(n^7)$ time. This result is can be found in the full version of this paper.

Second, we investigate what happens for wider strips. We show that the broadcast problem has an $n^{O(w)}$ dynamic-programming algorithm for strips of width w. We prove a matching lower bound of $n^{\Omega(w)}$, conditional on the Exponential Time Hypothesis (ETH). Interestingly, the h-hop broadcast problem has no such algorithm (unless P = NP): we show this problem is already NP-hard on a strip of width 40. One of the gadgets in this intricate construction can also be used to prove that a CDS-UDG and the broadcast problem are W[1]-hard parameterized by the solution size k. The W[1]-hardness proof is discussed only in the full version. It is a reduction from GRID TILING based on ideas by Marx [14], and it implies that there is no $f(k)n^{o(\sqrt{k})}$ algorithm for CDS-UDG unless ETH fails.

2   Algorithms for broadcasting inside a narrow strip

In this section we present polynomial algorithms (both for broadcast and for h-hop broadcast) for inputs that lie inside a strip $\mathcal{S} := \mathbb{R} \times [0, w]$, where $0 < w \leqslant \sqrt{3}/2$ is the width of the strip. Without loss of generality, we assume that the source lies on the y-axis. Define $\mathcal{S}_{\geqslant 0} := [0, \infty) \times [0, w]$ and $\mathcal{S}_{\leqslant 0} := (-\infty, 0] \times [0, w]$.

Let P be the set of input points. We define $x(p)$ and $y(p)$ to be the x- and y-coordinate of a point $p \in P$, respectively, and $\delta(p)$ to be the unit-radius disk centered at p. Let $\mathcal{G} = (P, E)$ be the graph with $(p, q) \in E$ iff $q \in \delta(p)$, and let $P' := P \setminus \delta(s)$ be the set of input points outside the source disk. We say that a point $p \in P$ is *left-covering* if $pp' \in E$ for all $p' \in P'$ with $x(p') < x(p)$; p is *right-covering* if $p'p \in E$ for all $p' \in P'$ with $x(p') > x(p)$. We denote the set of left-covering and right-covering points by Q^- and Q^+ respectively. Finally, the *core area* of a point p, denoted by core(p), is $[x(p) - \frac{1}{2}, x(p) + \frac{1}{2}] \times [0, w]$. Note that core($p$) $\subset \delta(p)$ because $w \leqslant \sqrt{3}/2$, i.e., the disk of p covers a part of the strip that has horizontal length at least one. This is a key property of strips of width at most $\sqrt{3}/2$, and will be used repeatedly.

We partition P into levels $L_0, L_1, \ldots L_t$, based on hop distance from s in $\mathcal{G}$. Thus $L_i := \{p \in P : d_{\mathcal{G}}(s, p) = i\}$, where $d_{\mathcal{G}}(s, p)$ denotes the hop-distance. Let L_i^- and L_i^+ denote the points of L_i with negative and nonnegative coordinates, respectively. We will use the following observation multiple times.

Observation 1. *Let $\mathcal{G} = (P, E)$ be a unit disk graph on a narrow strip $\mathcal{S}$.*

(i) Let π be a path in $\mathcal{G}$ from a point $p \in P$ to a point $q \in P$. Then the region $[x(p) - \frac{1}{2}, x(q) + \frac{1}{2}] \times [0, w]$ is fully covered by the disks of the points in π.

(ii) The overlap of neighboring levels is at most $\frac{1}{2}$ in x-coordinates: $\max\{x(p)|p \in L_{i-1}^+\} \leqslant \min\{x(q)|q \in L_i^+\} + \frac{1}{2}$ for any $i > 0$ with $L_i^+ \neq \emptyset$; similarly, $\min\{x(p)|p \in L_{i-1}^-\} \geqslant \max\{x(q)|q \in L_i^-\} - \frac{1}{2}$ for any $i > 0$ with $L_i^- \neq \emptyset$.

(iii) Let p be an arbitrary point in L_i^+ for some $i > 0$. Then the disks of any path $\pi(s, p)$ cover all points in all levels $L_0 \cup L_1 \cup L_2^+ \cup \cdots \cup L_{i-1}^+$. A similar statement holds for points in L_i^-.

2.1 Minimum broadcast set in a narrow strip

A *broadcast set* is a point set $D \subseteq P$ that gives a feasible broadcast, i.e., a connected dominating set of $\mathcal{G}$ that contains s. Our task is to find a minimum broadcast set inside a narrow strip. Let $p, p' \in P$ be points with maximum and minimum x-coordinate, respectively. Obviously there must be paths from s to p and p' in $\mathcal{G}$ such that all points on these paths are active, except possibly p and p'. If p and p' are also active, then these paths alone give us a feasible broadcast set: by Observation 1(i), these paths cover all our input points. Instead of activating p and p', it is also enough to activate the points of a path that reaches Q^- and a path that reaches Q^+. In most cases it is sufficient to look for broadcast sets with this structure.

Lemma 1. *If there is a minimum broadcast set with an active point on L_2, then there is a minimum broadcast set consisting of the disks of a shortest path π^- from s to Q^- and a shortest path π^+ from s to Q^+. These two paths share s and they may or may not share their first point after s.*

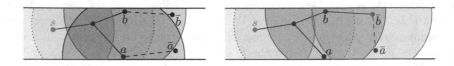

Fig. 1. A swap operation. The edges of the broadcast tree are solid lines.

Proof sketch. If a minimum broadcast set does not contain a point from Q^+, then we can find two active points a and b whose disks uniquely cover two non-active points $\bar{a}$ and $\bar{b}$, respectively; see Fig. 1. By deactivating a and activating $\bar{b}$ we get a new feasible solution, since $\delta(\bar{b})$ covers all points previously only covered by $\delta(a)$. By using such operations repeatedly, we can find a solution containing a point from Q^+. Using similar arguments, we can find a solution also containing a point from Q^-. Finally, using Observation 1(iii), we can show that a shortest path π^+ from s to Q^+ and a shortest path π^- from s to Q^- together form a feasible and minimum-size solution. □

Lemma 2 below fully characterizes optimal broadcast sets. To deal with the case where Lemma 1 does not apply, we need some more terminology. We say that the disk $\delta(q)$ of an active point q in a feasible broadcast set is *bidirectional* if there are two input points $p^- \in L_2^-$ and $p^+ \in L_2^+$ that are covered only by $\delta(q)$. See points p and p' in Fig. 2 for an example. Note that $q \in \text{core}(s)$, because $\text{core}(s) = [-\frac{1}{2}, \frac{1}{2}] \times [0, w]$ is covered by $\delta(s)$, and our bidirectional disk has to cover points both in $(-\infty, -\frac{1}{2}] \times [0, w]$ and $[\frac{1}{2}, \infty) \times [0, w]$. Active disks that are not the source disk and not bidirectional are called *monodirectional.*

Lemma 2. *For any input P that has a feasible broadcast set, there is a minimum broadcast set D that has one of the following structures.*

(i) Small: $|D| \leqslant 2$.

(ii) Path-like: $|D| \geqslant 3$, and D consists of a shortest path π^- from s to Q^- and a shortest path π^+ from s to Q^+; π^+ and π^- share s and may or may not share their first point after s.

(iii) Bidirectional: $|D| = 3$, and D contains two bidirectional disk centers and s.

As it turns out, the bidirectional case is the most difficult one to compute efficiently. (It is similar to CDS-UDG in co-comparability graphs, where the case of a connected dominating set of size at most 3 dominates the running time.)

Lemma 3. *In $O(n \log n)$ time we can find a bidirectional broadcast if it exists.*

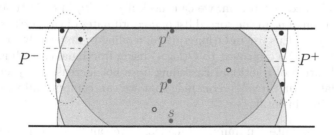

Fig. 2. A bidirectional broadcast.

Proof. Let $P^- := \{u_1, u_2, \ldots, u_k\}$ be the set of points to the left of the source disk $\delta(s)$, where the points are sorted in increasing y-order with ties broken arbitrarily. Similarly, let $P^+ := \{v_1, v_2, \ldots, v_l\}$ be the set of points to the right of $\delta(s)$, again sorted in order of increasing y-coordinate. Define $P^-_{\leqslant i} := \{u_1, \ldots, u_i\}$, and define $P^-_{>i}$, and $P^+_{\leqslant i}$ and $P^+_{>i}$ analogously. Our algorithm is based on the following observation: There is a bidirectional solution if and only if there are indices i, j and points $p, p' \in \text{core}(s)$ such that $\delta(p)$ covers $P^-_{\leqslant i} \cup P^+_{\leqslant j}$ and $\delta(p')$ covers $P^-_{>i} \cup P^+_{>j}$; see Fig. 2.

Now for a point $p \in \text{core}(s)$, define $Z^-_{\leqslant}(p) := \max\{i : P^-_{\leqslant i} \subset \delta(p)\}$ and $Z^-_{>}(p) := \min\{i : P^-_{>i} \subset \delta(p)\}$, and $Z^+_{\leqslant}(p) := \max\{i : P^+_{\leqslant i} \subset \delta(p)\}$, and $Z^+_{>}(p) := \min\{i : P^+_{>i} \subset \delta(p)\}$. Then the observation above can be restated as:

There is a bidirectional solution if and only if there are points $p, p' \in \text{core}(s)$ such that $Z^-_{\leqslant}(p) \geqslant Z^-_{>}(p')$ and $Z^+_{\leqslant}(p) \geqslant Z^+_{>}(p')$.

It is easy to find such a pair—if it exists—in $O(n \log n)$ time once we have computed the values $Z^-_{\leqslant}(p)$, $Z^-_{>}(p)$, $Z^+_{\leqslant}(p)$, and $Z^+_{>}(p)$ for all points $p \in \delta(s)$. It remains to show that these values can be computed in $O(n \log n)$ time.

Consider the computation of $Z^-_{\leqslant}(p)$; the other values can be computed similarly. Let $\mathcal{T}$ be a balanced binary tree whose leaves store the points from P^- in order of their y-coordinate. For a node ν in $\mathcal{T}$, let $F(\nu) := \{\delta(u_i) : u_i \text{ is stored in the subtree rooted at } \nu\}$. We start by computing at each node ν the intersection of the disks in $F(\nu)$. More precisely, for each ν we compute the region $I(\nu) := \text{core}(s) \cap \bigcap F(\nu)$. Notice that $I(\nu)$ is y-monotone and convex, and each disk $\delta(u_i)$ contributes at most one arc to $\partial I(\nu)$. (Here $\partial I(\nu)$ refers to the boundary of $I(\nu)$ that falls inside $\mathcal{S}$.) Moreover, $I(\nu) = I(\text{left-child}(\nu)) \cap I(\text{right-child}(\nu))$. Hence, we can compute the regions $I(\nu)$ of all nodes ν in $\mathcal{T}$ in $O(n \log n)$ time in total, in a bottom-up manner. Using the tree $\mathcal{T}$ we can now compute $Z^-_{\leqslant}(p)$ for any given $p \in \text{core}(s)$ by searching in $\mathcal{T}$, as follows. Suppose we arrive at a node ν. If $p \in I(\text{left-child}(\nu))$, then descend to right-child(ν), otherwise descend to left-child(ν). The search stops when we reach a leaf, storing a point u_i. One easily verifies that if $p \in \delta(u_i)$ then $Z^-_{\leqslant}(p) = i$, otherwise $Z^-_{\leqslant}(p) = i - 1$.

Since $I(\nu)$ is a convex region, we can check if $p \in I(\nu)$ in $O(1)$ time if we can locate the position of p_y in the sorted list of y-coordinates of the vertices of $\partial I(\nu)$. We can locate p_y in this list in $O(\log n)$ time, leading to an overall query time of $O(\log^2 n)$. This can be improved to $O(\log n)$ using fractional cascading [5]. Note that the application of fractional cascading does not increase the preprocessing time of the data structure. We conclude that we can compute all values $Z^-_{\leqslant}(p)$ in $O(n \log n)$ time in total. $\square$

In order to compute a minimum broadcast, we can first check for small and bidirectional solutions. To find path-like solutions, we first compute the sets Q^- and Q^+, and compute shortest paths starting from these sets back to the source disk. The path computation is very similar to the shortest path algorithm in UDGs by Cabello and Jejčič [4].

Theorem 2. *The broadcast problem inside a strip of width at most $\sqrt{3}/2$ can be solved in $O(n \log n)$ time.*

Remark 1. If we apply this algorithm to every disk as source, we get an $O(n^2 \log n)$ algorithm for CDS in narrow strip UDGs. We can compare this to $O(mn)$, the running time that we get by applying the algorithm for co-comparability graphs [3]. Note that in the most difficult case, when the size of the minimum connected dominating set is at most 3, the unit disk graph has constant diameter, which implies that the graph is dense, i.e., the number of edges is $m = \Omega(n^2)$. Hence, we get an (almost) linear speedup for the worst-case running time.

2.2 Minimum-size h-hop broadcast in a narrow strip

In the hop-bounded version of the problem we are given P and a parameter h, and we want to compute a broadcast set D such that every point $p \in P$ can be reached in at most h hops from s. In other words, for any $p \in P$, there must be a path in $\mathcal{G}$ from s to p of length at most h, all of whose vertices, except possibly p itself, are in D. We start by investigating the structure of optimal solutions in this setting, which can be very different from the non-hop-bounded setting.

As before, we partition P into levels L_i according to the hop distance from s in the graph $\mathcal{G}$, and we define L_i^+ and L_i^- to be the subsets of points at level i with positive and nonnegative x-coordinates, respectively. Let L_t be the highest non-empty level. If $t > h$ then clearly there is no feasible solution.

If $t < h$ then we can safely use our solution for the non-hop-bounded case, because the non-hop-bounded algorithm gives a solution which contains a path with at most $t+1$ hops to any point in P. This follows from the structure of the solution; see Lemma 2. (Note that it is possible that the solution given by this algorithm requires $t+1$ hops to some point, namely, if $Q^+ \cup Q^- \subseteq L_t$.) With the $t < h$ case handled by the non-hop-bounded algorithm, we are only concerned with the case $t = h$.

We deal with *one-sided* inputs first, where the source is the leftmost input point. Let $\mathcal{G}^*$ be the directed graph obtained by deleting edges connecting points inside the same level of $\mathcal{G}$, and orienting all remaining edges from lower to higher levels. A *Steiner arborescence* of $\mathcal{G}^*$ for the terminal set L_h is a directed tree rooted at s that contains a (directed) path π_p from s to p for each $p \in L_h$. From now on, whenever we speak of *arborescence* we refer to a Steiner arborescence in $\mathcal{G}^*$ for terminal set L_h. We define the *size* of an arborescence to be the number of internal nodes of the arborescence. Note that the leaves in a minimum-size arborescence are exactly the points in L_h: these points must be in the arborescence by definition, they must be leaves since they have out-degree zero in $\mathcal{G}^*$, and leaves that are not in L_h can be removed.

Remark 2. In the minimum Steiner Set problem, we are given a graph G and a vertex subset T of terminals, and the goal is to find a minimum-size vertex subset S such that $T \cup S$ induces a connected subgraph. This problem has a polynomial algorithm in co-comparability graphs [3], and therefore in narrow

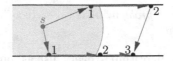

Fig. 3. Two different arborescences, with vertices labeled with their level. The arborescence made of the bottom path does not define a feasible broadcast for $h = 3$, since it would take four hops to reach the top right node.

strip unit disk graphs. However, the broadcast set given by a solution does not fit our hop bound requirements. Hence, we have to work with a different graph (e.g. the edges within each level L_i have been removed), and this modified graph is not necessarily a co-comparability graph.

Lemma 4 below states that either we have a path-like solution—for the one-sided case a path-like solution is a shortest $s \to Q^+$ path— or any minimum-size arborescence defines a minimum-size broadcast set. The latter solution is obtained by activating all non-leaf nodes of the arborescence. We denote the broadcast set obtained from an arborescence A by D_A.

Lemma 4. *Any minimum-size Steiner arborescence for the terminal set L_h defines a minimum broadcast set, or there is a path-like minimum broadcast set.*

Notice that a path-like solution also corresponds to an arborescence. However, it can happen that there are minimum-size arborescences that do not define a feasible broadcast; see Fig. 3. Lemma 4 implies that if this happens, then there must be an optimal path-like solution. The lemma also implies that for non-path-like solutions we can use the Dreyfus-Wagner dynamic-programming algorithm to compute a minimum Steiner tree [8], and obtain an optimal solution from this tree.[3] Unfortunately the running time is exponential in the number of terminals, which is $|L_h|$ in our case. However, our setup has some special properties that we can use to get a polynomial algorithm.

We define an arborescence A to be *nice* if the following holds. For any two arcs uu' and vv' of A that go between the same two levels, with $u \neq v$, we have: $y(u') < y(v') \Rightarrow y(u) < y(v)$. Intuitively, a nice arborescence is one consisting of paths that can be ordered vertically in a consistent manner, see the left of Fig. 4. We define an arborescence A to be *compatible* with a broadcast set D if $D = D_A$. Note that there can be multiple arborescences—that is, arborescences with the same node set but different edge sets—compatible with a given broadcast set D.

Lemma 5. *Every optimal broadcast set D has a nice compatible arborescence.*

Proof sketch. To find a nice compatible arborescence we will associate a unique arborescence with D. To this end, we define for each $p \in (D \cup L_h) \setminus \{s\}$ a unique

---

[3] The Dreyfus-Wagner algorithm minimizes the number of edges in the arborescence. In our setting the number of edges equals the number of internal nodes plus $|L_h| - 1$, so this also minimizes the number of internal nodes.

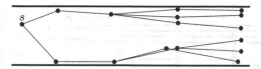
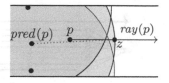

Fig. 4. Left: A nice Steiner arborescence. Note that arc crossings are possible. Right: Defining the *pred* function.

predecessor $pred(p)$, as follows. Let ∂_i^* be the boundary of $\bigcup\{\delta(p)|p \in L_i \cap D\}$. The two lines bounding the strip $\mathcal{S}$ cut ∂_i^* into four parts: a top and a bottom part that lie outside the strip, and a left and a right part that lie inside the strip. Let ∂_i be the part on the right inside the strip. We then define the function $pred : (D \cup L_h) \setminus \{s\} \to D$ the following way. Consider a point $p \in (D \cup L_h) \setminus \{s\}$ and let i be its level. Let $ray(q)$ be the horizontal ray emanating from q to the right; see the right of Fig. 4. It follows from Observation 1(iii) that $ray(q)$ cannot enter any disk from level $i - 1$. We can prove that any point $p \in D \cap L_h$ is contained in a disk from p's previous level, so $pred(p)$ is well defined for these points. The edges $pred(p)p$ for $p \in D \cap L_h$ thus define an arborescence. We can prove that it is nice by showing that the y-order of the points in a level L_i corresponds to the vertical order in which the boundaries of their disks appear on $\bigcup\{\delta(p) : p \in L_i \cap D\}$. □

Let $q_1, q_2, \ldots, q_m$ be the points of L_h in increasing y-order. The crucial property of a nice arborescence is that the descendant leaves of a point p in the arborescence form an interval of $q_1, q_2, \ldots, q_m$. Using the above lemmas, we can adapt the Dreyfus-Wagner algorithm and get the following theorem.

Theorem 3. *The one-sided h-hop broadcast problem inside a strip of width at most $\sqrt{3}/2$ can be solved in $O(n^4)$ time.*

In the general (two-sided) case, we can have path-like solutions and arborescence-based solutions on both sides, and the two side solutions may or may not share points in L_1. We also need to handle "small" solutions—now these are 2-hop solutions—separately.

Theorem 4. *The h-hop broadcast problem inside a strip of width at most $\sqrt{3}/2$ can be solved in $O(n^6)$ time.*

3 Broadcasting in a wide strip

Theorem 5. *The broadcast problem and CDS-UDG can be solved in $n^{O(w)}$ time on a strip of width w. Moreover, there is no algorithm for CDS-UDG or the broadcast problem with runtime $f(w)n^{o(w)}$ unless ETH fails.*

Surprisingly, the h-hop version has no $n^{O(w)}$ algorithm (unless P =NP).

Fig. 5. The gadget representing the variables. The dotted paths form the x_2-string.

Theorem 6. *The h-hop broadcast problem is* NP-*complete in strips of width* 40.

(The theorem of course refers to the decision version of the problem: given a point set P, a hop bound h, and a value K, does P admit an h-hop broadcast set of size at most K?) Our reduction is from 3-SAT. Let $x_1, x_2, \ldots x_n$ be the variables and $C_1, \ldots, C_m$ be the clauses of a 3-CNF.

Fig. 5 shows the structural idea for representing the variables, which we call the *base bundle*. It consists of $(2h-1)n+1$ points arranged as shown in the figure, where h is an appropriate value. The distances between the points are chosen such that the graph $\mathcal{G}$, which connects two points if they are within distance 1, consists of the edges in the figure plus all edges between points in the same level. Thus (except for the intra-level edges, which we can ignore) $\mathcal{G}$ consists of n pairs of paths, one path pair for each variable x_i. The i-th pair of paths represents the variable x_i, and we call it the x_i-*string*. By setting the target size, K, of the problem appropriately, we can ensure the following for each x_i: any feasible solution must use either the top path of the x_i-string or the bottom path, but it cannot use points from both paths. Thus we can use the top path of the x_i-path to represent a TRUE setting of the variable x_i, and the bottom path to represent a FALSE setting. A group of consecutive strings is called a *bundle*. We denote the bundle containing all x_t-strings with $t = i, i+1, \ldots, j$ by $bundle(i, j)$.

The clause gadgets all start and end in the base bundle, as shown in Fig. 6. The gadget to check a clause involving variables x_i, x_j, x_k, with $i < j < k$, roughly works as follows; see also the lower part of Fig. 6, where the strings for x_i, x_j, and x_k are drawn with dotted lines.

First we split off $bundle(1, i-1)$ from the base bundle, by letting the top $i-1$ strings of the base bundle turn left. (In Fig. 6 this bundle consists of two strings.) We then separate the x_i-string from the base bundle, and route the x_i-string into a *branching gadget*. The branching gadget creates a branch consisting of two *tapes*—this branch will eventually be routed to the *clause-checking gadget*—and a branch that returns to the base bundle. Before the tapes can be routed to the clause-checking gadget, they have to cross each of the strings in $bundle(1, i-1)$. For each string that must be crossed we introduce a *crossing gadget*. A crossing gadget lets the tapes continue to the right, while the string being crossed can return to the base bundle. The final crossing gadget turns the tapes into a *side string* that can now be routed to the clause-checking gadget. The construction guarantees that the side string for x_i still carries the truth value that was selected for the x_i-string in the base bundle. Moreover, if the TRUE path (resp. FALSE

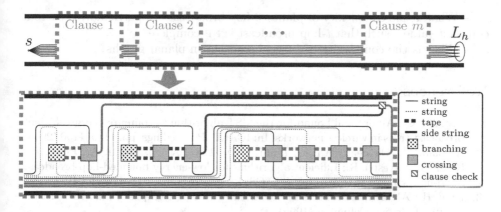

Fig. 6. The overall construction, and the way a single clause is checked. Note that in this figure each string (which actually consists of two paths) is shown as a single curve.

path) of the x_i-string was selected to be part of the broadcast set initially, then the TRUE path (resp. FALSE path) of the rest of the x_i-string that return to the base bundle must be in the minimum broadcast set as well.

After we have created a side string for x_i, we create side strings for x_j and x_k in a similar way. The three side strings are then fed into the clause-checking gadget. The clause checking gadget is a simple construction of four points. Intuitively, if at least one side string carries the correct truth value—TRUE if the clause contains the positive variable, FALSE if it contains the negated variable—, then we activate a single disk in the clause check gadget that corresponds to a true literal. Otherwise we need to change truth value in at least one of the side strings, which requires an extra disk.

The final construction contains $\Theta(n^4 m)$ points that all fit into a strip of width 40. The details are given in the full version.

4 Conclusion

We studied the complexity of the broadcast problem in narrow and wider strips. For narrow strips we obtained efficient polynomial algorithms, both for the non-hop-bounded and for the h-hop version, thanks to the special structure of the problem inside such strips. On wider strips, the broadcast problem has an $n^{O(w)}$ algorithm, while the h-hop broadcast becomes NP-complete on strips of width 40. With the exception of a constant width range (between $\sqrt{3}/2$ and 40) we characterized the complexity when parameterized by strip width. We have also proved that the planar problem (and, similarly, CDS-UDG) is W[1]-hard when parameterized by the solution size. The problem of finding a planar h-hop broadcast set seems even harder: we can solve it in polynomial time for $h = 2$ (see full version) but already for $h = 3$ we know no better algorithm than brute force. Interesting open problems include:

- What is the complexity of planar 3-hop broadcast? In particular, is there a constant value t such that t-hop broadcast is NP-complete?
- What is the complexity of h-hop broadcast in planar graphs?

References

1. Ambühl, C.: An optimal bound for the MST algorithm to compute energy efficient broadcast trees in wireless networks. In: ICALP, Proceedings. pp. 1139–1150 (2005)
2. Ambühl, C., Clementi, A.E.F., Ianni, M.D., Lev-Tov, N., Monti, A., Peleg, D., Rossi, G., Silvestri, R.: Efficient algorithms for low-energy bounded-hop broadcast in ad-hoc wireless networks. In: STACS, Proceedings. pp. 418–427 (2004)
3. Breu, H.: Algorithmic aspects of constrained unit disk graphs. Ph.D. thesis, University of British Columbia (1996)
4. Cabello, S., Jejčič, M.: Shortest paths in intersection graphs of unit disks. Comput. Geom. 48(4), 360–367 (2015)
5. Chazelle, B., Guibas, L.J.: Fractional cascading: I. A data structuring technique. Algorithmica 1(1), 133–162 (1986)
6. Clementi, A.E., Crescenzi, P., Penna, P., Rossi, G., Vocca, P.: A worst-case analysis of an mst-based heuristic to construct energy-efficient broadcast trees in wireless networks. In: STACS, Proceedings. pp. 121–131 (2001)
7. Das, G.K., Das, S., Nandy, S.C.: Range assignment for energy efficient broadcasting in linear radio networks. Theor. Comput. Sci. 352(1-3), 332–341 (2006)
8. Dreyfus, S.E., Wagner, R.A.: The Steiner problem in graphs. Networks 1(3), 195–207 (1971)
9. Flum, J., Grohe, M.: Parameterized Complexity Theory. Texts in Theoretical Computer Science. An EATCS Series, Springer (2006)
10. Fuchs, B.: On the hardness of range assignment problems. Networks 52(4), 183–195 (2008)
11. Garey, M.R., Johnson, D.S.: Computers and Intractability: A Guide to the Theory of NP-Completeness. W. H. Freeman (1979)
12. Kratsch, D., Stewart, L.: Domination on cocomparability graphs. SIAM J. Discrete Math. 6(3), 400–417 (1993)
13. Lichtenstein, D.: Planar formulae and their uses. SIAM J. Comput. 11(2), 329–343 (1982)
14. Marx, D.: Parameterized complexity of independence and domination on geometric graphs. In: Parameterized and Exact Computation, Second International Workshop, IWPEC, Proceedings. pp. 154–165 (2006)
15. Masuyama, S., Ibaraki, T., Hasegawa, T.: The computational complexity of the m-center problems on the plane. IEICE Transactions 64(2), 57–64 (1981)
16. Matsui, T.: Approximation algorithms for maximum independent set problems and fractional coloring problems on unit disk graphs. In: JCDCG, Proceedings. pp. 194–200. Springer (1998)

Minimizing the Continuous Diameter when Augmenting a Tree with a Shortcut[*]

Jean-Lou De Carufel[1], Carsten Grimm[2,3], Stefan Schirra[3], and Michiel Smid[2]

[1] School of Electrical Engineering and Computer Science, University of Ottawa
800 King Edward Avenue, Ottawa, Ontario, K1N 6N5, Canada
[2] School of Computer Science, Carleton University
1125 Colonel By Drive, Ottawa, Ontario, K1S 5B6, Canada
[3] Institut für Simulation und Graphik, Otto-von-Guericke-Universität Magdeburg
Universitätsplatz 2, D-39106 Magdeburg, Germany

Abstract. We augment a tree T with a shortcut pq to minimize the largest distance between any two points along the resulting augmented tree $T + pq$. We study this problem in a continuous and geometric setting where T is a geometric tree in the Euclidean plane, a shortcut is a line segment connecting any two points along the edges of T, and we consider all points on $T + pq$ (i.e., vertices and points along edges) when determining the largest distance along $T + pq$. The *continuous diameter* is the largest distance between any two points along edges. We establish that a single shortcut is sufficient to reduce the continuous diameter of a geometric tree T if and only if the intersection of all diametral paths of T is neither a line segment nor a point. We determine an optimal shortcut for a geometric tree with n straight-line edges in $O(n \log n)$ time.

Keywords: Network Augmentation · Continuous Diameter Minimization

1 Introduction

A *network* is a connected, undirected graph with positive edge weights. A curve is rectifiable if it has a finite length. A *geometric network* is a network that is embedded in the Euclidean plane whose edges are rectifiable curves weighted with their length. We describe our algorithmic results for straight-line edges, even though they extend to more general edges. We say that a point p lies on a geometric network G and write $p \in G$ when there is an edge e of G such that p is a point along the embedding of e. A point p on an edge e of length l subdivides e into two sub-edges of lengths $(1 - \lambda) \cdot l$ and $\lambda \cdot l$ for some value $\lambda \in [0, 1]$. We represent the points on G in terms of their relative positions (expressed by λ) along their containing edges, thereby avoiding ambiguity in case of crossings.

The *network distance* between any two points p and q on a geometric network G is the length of a shortest weighted path from p to q in G and it is denoted by $d_G(p, q)$. The *continuous diameter* of G is the largest network

---

[*] This work was partially supported by NSERC and FQRNT.

F. Ellen et al. (Eds.): WADS 2017, LNCS 10389, pp. 301–312, 2017.
DOI: 10.1007/978-3-319-62127-2_26

distance between any two points on G, and it is denoted by $\mathrm{diam}(G)$, i.e., $\mathrm{diam}(G) = \max_{p,q \in G} d_G(p,q)$. In contrast, for a network with vertex set V, the *discrete diameter* is the largest distance between any two vertices, i.e., $\max_{u,v \in V} d_G(u,v)$. A pair $p,q \in G$ is *diametral* when their distance is the continuous diameter, i.e., $\mathrm{diam}(G) = d_G(p,q)$. A *diametral path* in G is a shortest weighted path in G that connects a diametral pair of G. We denote the Euclidean distance between two points p and q in the plane by $|pq|$. A line segment pq with endpoints $p,q \in G$ is a *shortcut* for G. We *augment* a geometric network G with a shortcut pq as follows. If they do not exist already, we introduce vertices at p and at q, thereby subdividing the edges containing p and q. We add the line segment pq as an edge of length $|pq|$ to G without introducing vertices at crossings between pq and other edges. We denote the resulting network by $G + pq$.

Our goal is to locate a shortcut pq for a geometric tree T that minimizes the continuous diameter of the augmented tree $T + pq$, as illustrated in Fig. 1. This means we seek two points $p,q \in T$ with $\mathrm{diam}(T+pq) = \min_{r,s \in T} \mathrm{diam}(T+rs)$. We call a shortcut that minimizes the continuous diameter an *optimal shortcut*.

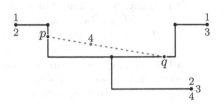

Fig. 1. An optimal shortcut pq for a geometric tree T with the diametral pairs of $T+pq$ marked by matching numbers.

The *backbone* of a tree T is the intersection of all diametral paths of T; we denote the backbone of T by $\mathcal{B}$. The *absolute center* of a geometric tree T is the unique point $c \in T$ that minimizes the largest network distance from c, i.e., $\max_{q \in T} d_T(c,q) = \min_{p \in T} \max_{q \in T} d_T(p,q)$. Note that we always have $c \in \mathcal{B}$. It takes $O(n)$ time to determine the absolute center—and, thereby, the backbone—of a geometric tree with n vertices [9].

1.1 Related Work

We summarize related work on minimum-diameter network augmentation.

In the *abstract and discrete setting*, the goal is to minimize the discrete diameter of an abstract graph with positive weights for the edges and non-edges by inserting non-edges as shortcuts. If the edges and non-edges have unit weight, then it is NP-hard to decide whether the diameter can be reduced below $D \geq 2$ by adding at most k shortcuts [2, 10, 13]. This problem remains NP-hard, even for restricted graph classes such as trees [2]. The weighted version of this problem falls into the parameterized complexity class W[2]-hard [5, 6]. Minimum-diameter augmentation has also been studied as a bicriteria optimization in which both the diameter and the number (or cost) of the additional edges are minimized. For an overview on bicriteria approximation algorithms refer to Frati *et al.* [5].

Große *et al.* [7] introduce the *geometric and discrete setting* in which the problem is to minimize the discrete diameter of a geometric network by connecting vertices with line segments. Große *et al.* [7] determine an optimal shortcut for a polygonal path with n vertices in $O(n \log^3 n)$ time. The stretch factor, i.e., the largest ratio of the network distance between any two vertices and their Euclidean distance, has also been considered as a target function [4, 11].

In the *geometric and continuous setting* [3], the task is to minimize the continuous diameter of a geometric network by inserting line segments that may connect any two points along the edges. For a polygonal path of length n, one can determine an optimal shortcut in $O(n)$ time. For a cycle, one shortcut can never decrease the continuous diameter while two always suffice. For convex cycles with n vertices, one can determine an optimal pair of shortcuts in $O(n)$.

In the model studied in this work, a crossing of a shortcut with an edge or another shortcut is not a vertex: a path may only enter edges at their endpoints. In the *planar model* [1, 14], every crossing is a vertex of the resulting network, which leads to a different graph structure and, thus, continuous diameter. In the planar model, Yang [14] characterizes optimal shortcuts for a polygonal path. Cáceres *et al.* [1] determine in polynomial time whether the continuous diameter of a plane geometric network can be reduced with a single shortcut.

Recently, Oh and Ahn [12] develop an $O(n^2 \log^3 n)$-time algorithm to determine an optimal shortcut for a tree with n vertices in the discrete setting.

1.2 Structure and Results

We present the following structural results about optimal shortcuts for geometric trees. A shortcut pq is *useful* for a geometric tree T when augmenting T with pq decreases the continuous diameter. A geometric tree T admits a useful shortcut if and only if its backbone B is not a line segment. We consider a point to be a degenerate line segment. Every geometric tree T has an optimal shortcut pq with $p, q \in B$ such that the absolute center c lies on the path from p to q in T.

Based on these insights, we determine an optimal shortcut for a geometric tree T with n vertices in $O(n \log n)$ time. Conceptually, we slide a candidate shortcut pq along the backbone of T while balancing the diametral paths in $T + pq$. The diametral pairs of the augmented tree guide our search: each diametral pair in $T + pq$ rules out a better shortcut in one of four possible directions of our search. To implement this approach, we discretize this movement such that the shortcut jumps from one change in the diametral paths to the next.

2 Structural Results

We say a shortcut pq is *useful* for T when $\text{diam}(T + pq) < \text{diam}(T)$, we say pq is *indifferent* for T when $\text{diam}(T + pq) = \text{diam}(T)$, and we say pq is *useless* for T when $\text{diam}(T + pq) > \text{diam}(T)$. In the discrete setting, every shortcut is useful or indifferent, as the discrete diameter only considers vertices of T. In the continuous setting, a shortcut may be useless for T, since the points on the shortcut pq matter as well, as exemplified in Fig. 2. For some trees, we cannot reduce the continuous diameter with a single shortcut, as demonstrated in Fig. 3. In this example, the diametral paths of T intersect in a single vertex—the absolute center—and, thus, the backbone is a degenerate line segment.

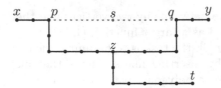

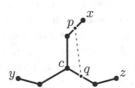

Fig. 2. A geometric tree T with a short-cut pq that is useless for T. Each edge has weight one and $|pq| = 8$. The continuous diameter increases when we augment T with pq, since $\text{diam}(T) = d_T(x, y) = 16$ and $\text{diam}(T + pq) = d_{T+pq}(s, t) = 17$.

Fig. 3. A geometric tree T where no shortcut decreases the continuous diameter. Every shortcut pq is useless for at least one pair among the three leaves x, y, and z.

Theorem 1. *A geometric tree has a useful shortcut if and only if its backbone is not a line segment.*

Proof (Sketch). If the backbone $\mathcal{B}$ of a geometric tree T consists of a single point (i.e., $\mathcal{B}$ is a degenerate line segment), then T does not possess a useful shortcut. Thus, if T does possess a useful shortcut pq, then $\mathcal{B}$ is a path with distinct endpoints a and b such that pq is useful for $\{a, b\}$, i.e., $d_{T+pq}(a, b) < d_T(a, b)$. This means that the path from a to b in T, i.e., $\mathcal{B}$, cannot be a line segment.

Conversely, if $\mathcal{B}$ is not a line segment, then $\mathcal{B}$ is a path from a to b with $|ab| < d_T(a, b)$. If ab is not a useful shortcut for T, then $T + ab$ has a diametral pair with one partner on ab and one partner in a sub-tree attached to $\mathcal{B}$. We find $p, q \in \mathcal{B}$ where the path from p to q is not a line segment and the cycle in $T + pq$ is sufficiently small to rule out diametral pairs in $T + pq$ with a partner on pq. Therefore, if $\mathcal{B}$ is not a line segment, then T has a useful shortcut. □

Let T be a geometric tree where the backbone $\mathcal{B}$ is a path from a to b with $a \neq b$. This path contains the absolute center c of T. We prove that there is an optimal shortcut pq for T such that p lies on the path from a to c along $\mathcal{B}$ and q lies on the path from c to b along $\mathcal{B}$. Suppose we pick the absolute center $c \in \mathcal{B}$ as the root of T. Let S_a and S_b be the sub-trees of T attached to the backbone at a and at b, respectively. We refer to S_a and S_b as the *primary $\mathcal{B}$-sub-trees* of T; we refer to any other sub-trees of T attached to $\mathcal{B}$ as *secondary $\mathcal{B}$-sub-trees*. The *root* of a $\mathcal{B}$-sub-tree S is the vertex r connecting S and the backbone $\mathcal{B}$. As illustrated in Fig. 4, every diametral pair in T consists of a leaf x in S_a and a leaf y in S_b.

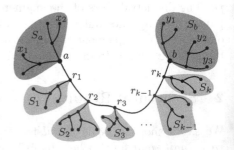

Fig. 4. A geometric tree T with its backbone $\mathcal{B}$ and the primary $\mathcal{B}$-sub-trees S_a and S_b, as well as the secondary $\mathcal{B}$-sub-trees $S_1, S_2, \ldots, S_k$.

Theorem 2. *For every geometric tree T, there exists an optimal shortcut pq where both endpoints lie on the backbone $\mathcal{B}$ of T, i.e., $p, q \in \mathcal{B}$, and where the path from p to q along $\mathcal{B}$ contains the absolute center c of T.*

Proof (Sketch). If T has no useful shortcuts, then the degenerate shortcut cc satisfies the claim, since it is indifferent and therefore optimal for T with $c \in \mathcal{B}$.

Suppose T does possess a useful shortcut, i.e., the backbone $\mathcal{B}$ of T is a path from a to b with $a \neq b$ and $|ab| < d_T(a, b)$. Let S be a $\mathcal{B}$-sub-tree with root r. First, we show that if pq is useful for T, then p and q cannot lie in the same $\mathcal{B}$-sub-tree of T, i.e., we have $p \notin S$ or $q \notin S$. Then, we show that if $p \in S$ then the diameter never increases if we move p to the root r of S, i.e., $\mathrm{diam}(T + rq) \leq \mathrm{diam}(T + pq)$. Since our argument does not depend on the position of q, we conclude that there is an optimal shortcut for T with both endpoints on the backbone $\mathcal{B}$.

To prove $\mathrm{diam}(T + rq) \leq \mathrm{diam}(T + pq)$ we consider a diametral pair s, t of $T + rq$. We distinguish the three cases: (1) $s, t \in T$ with $d_T(s, t) = d_{T+pq}(s, t)$, (2) $s, t \in T$ with $d_T(s, t) < d_{T+pq}(s, t)$, and (3) $s \notin T$ or $t \notin T$. Große et al. [8] show the claim for Cases (1) and (2). The argument for Case (3) is as follows.

(3) Suppose $s \notin T$ or $t \notin T$. We assume, without loss of generality, that $s \notin T$; otherwise, we swap s and t. Then $s \in rq$ with $r \neq s \neq q$, as in Fig. 5.

Let $C(r, q)$ be the simple cycle in $T + rq$. The path from s to t leaves $C(r, q)$ at the point $\bar{s}$ that is the farthest point from s on $C(r, q)$. Note that $r \neq \bar{s} \neq q$, because $r \neq s \neq q$ and $|rq| \leq d_T(r, q)$. Since $p \in S$, we have $q \notin S$. This means the path connecting r and q in T lies outside of S and, therefore, $\bar{s} \notin S$ and $t \notin S$. The cycle $C(p, q)$ is formed by pq and the path from p to q in T. Since r lies on the path from p to q, we know that $s \in C(p, q)$. Let s' be the farthest point from $\bar{s}$ on $C(p, q)$. We obtain

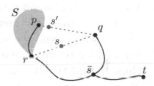

Fig. 5. A diametral pair s, t of $T + rq$ with $s \in rq$.

$$
\begin{aligned}
\mathrm{diam}(T + rq) &= d_{T+rq}(s, t) &&\text{(s and t are diametral in $T + rq$)} \\
&= d_{T+rq}(s, \bar{s}) + d_T(\bar{s}, t) &&\text{($\bar{s}$ is on any path from t to s in $T + rq$)} \\
&= \frac{d_T(r, q) + |rq|}{2} + d_T(\bar{s}, t) &&\text{(s and $\bar{s}$ are antipodals on $C(r, q)$)} \\
&\leq \frac{d_T(r, q) + d_T(r, p) + |pq|}{2} + d_T(\bar{s}, t) &&\text{(triangle inequality)} \\
&= \frac{d_T(p, q) + |pq|}{2} + d_T(\bar{s}, t) &&\text{(r is on the path from p to q in T)} \\
&= d_{T+pq}(s', \bar{s}) + d_T(\bar{s}, t) &&\text{(s' and $\bar{s}$ are antipodals on $C(p, q)$)} \\
&= d_{T+pq}(s', t) &&\text{($\bar{s}$ is on any path from t to s' in $T + pq$)} \\
&\leq \mathrm{diam}(T + pq) \ .
\end{aligned}
$$

Suppose pq is an optimal shortcut for T with $p, q \in \mathcal{B}$ where c does not lie on the path from p to q. We argue that if q lies on the path from p to c, then the diameter never increases as q moves to c, i.e., $\mathrm{diam}(T + pc) \leq \mathrm{diam}(T + pq)$. □

3   Preparations for the Algorithm

Initially, we place the endpoints of the shortcut, p and q, at the absolute center c of T. Then, we move p and q along the backbone $\mathcal{B}$ balancing the diametral paths in $T + pq$. During this movement p remains along the path from a to c and q remains on the path from c to b. The diametral pairs in $T + pq$ guide our search: each diametral pair in $T + pq$ rules out some direction in which we could search for a better shortcut. We describe our algorithm along the following steps.

1. We simplify the geometric tree T by compressing the $\mathcal{B}$-sub-trees, thereby simplifying the discussion about diametral pairs and paths in $T + pq$.
2. We define algorithm states in terms of the diametral paths and diametral pairs that are present in the augmented tree, and we distinguish four types of movements for the shortcut as the operations of our algorithm.
3. Each type of diametral pair rules out a better shortcut in some direction; some combinations of pair types imply that the current shortcut is optimal.
4. We describe the continuous, conceptual movement of the shortcut that is guided by the set of types of diametral pairs that are present in $T + pq$. We identify the invariants that guarantee that we find an optimal shortcut.
5. We determine the speeds at which the endpoints of the shortcut would move in the continuous algorithm, depending on the diametral paths in $T + pq$.
6. We bound the number of events of the discrete algorithm by $O(n)$ by ruling out some transitions between the algorithm states and by identifying when we can safely ignore events without compromising optimality.
7. Finally, we process each of the $O(n)$ events in $O(\log n)$ amortized time and, thus, bound the running time of our algorithm by $O(n \log n)$.

Simplifying the Tree Let S_a and S_b be the primary $\mathcal{B}$-sub-trees of a geometric tree T and let $S_1, S_2, \ldots, S_k$ be the secondary $\mathcal{B}$-sub-trees of T that are attached to the backbone $\mathcal{B}$ at their roots $r_1, r_2, \ldots, r_k$, respectively, as illustrated in Fig. 6. Let x be a farthest leaf from a in S_a, let y be a farthest leaf from b in S_b and, for every $i = 1, 2, \ldots, k$, let s_i be a farthest leaf from r_i in S_i. We replace each $\mathcal{B}$-sub-tree S_i with an edge from r_i to a vertex representing s_i of length $d_T(r_i, s_i)$.

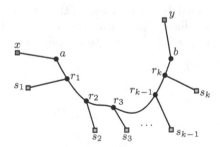

Fig. 6. The perspective from the backbone for the geometric tree from Fig. 4.

Likewise, we replace S_a and S_b with edges of appropriate lengths.

There is no need to distinguish diametral pairs from two different $\mathcal{B}$-sub-trees: for any $i, j = 1, 2, \ldots, k$ with $i \neq j$, the pair s_i, s_j is diametral in $T + pq$ if and only if every farthest leaf from r_i in S_i forms a diametral pair with every farthest leaf from r_j in S_j. There is no need to consider diametral pairs from the same $\mathcal{B}$-sub-tree S: if the augmented tree $T + pq$ has a diametral pair u, v such that u and v lie in the same $\mathcal{B}$-sub-tree S of T, then the shortcut pq is optimal.

Every diametral pair of T consists of $x \in S_a$ and $y \in S_b$. In the following, the symbols x and y indicate diametral partners in S_a and in S_b, respectively.

Pair Types Every point v of a diametral pair u, v in $T + pq$ is either: (x, y) a point in the tree that occurs in a diametral pair, (■) some other point on the tree, or (○) a point on the shortcut. More specifically, v is either (x, y) a leaf of a primary $\mathcal{B}$-sub-tree, (▲) a leaf of a secondary $\mathcal{B}$-sub-tree, or v is a point on the simple cycle $C(p, q)$ in the augmented tree that (•) lies in the original tree or (○) on the shortcut. This leads to the following distinction of the diametral pairs.

1. $(x\text{-}y)$: Diametral pairs x, y of $T + pq$ with $x \in S_a$ and $y \in S_b$.
2. $(x\text{-}■)$: Diametral pairs x, v of $T + pq$ with $x \in S_a$ and $v \in T \setminus (S_a \cup S_b)$.
   (a) $(x\text{-}▲)$: Pairs x, s_j with $x \in S_a$ and $s_j \in S_j$ for some $j = 1, 2, \ldots, k$.
   (b) $(x\text{-}•)$: Pairs $x, \bar{x}$ with $x \in S_a$ and where $\bar{x}$ is farthest from x on $C(p, q)$.
3. $(■\text{-}y)$: Diametral pairs u, y of $T + pq$ with $u \in T \setminus (S_a \cup S_b)$ and $y \in S_b$.
   (a) $(▲\text{-}y)$: Pairs s_i, y with $s_i \in S_i$ and $y \in S_b$ for some $i = 1, 2, \ldots, k$.
   (b) $(•\text{-}y)$: Pairs $y, \bar{y}$ with $y \in S_b$ and where $\bar{y}$ is farthest from y on $C(p, q)$.
4. $(■\text{-}■)$: Diametral pairs u, v of $T + pq$ with $u, v \in T \setminus (S_a \cup S_b)$.
   (a) $(▲\text{-}▲)$: Pairs s_i, s_j with $s_i \in S_i$ and $s_j \in S_j$ for $i, j = 1, 2, \ldots, k$.
   (b) $(▲\text{-}•)$: Pairs $s_i, \bar{s}_i$ with $s_i, \bar{s}_i \in T$ where $s_i \in S_i$ for some $i = 1, 2, \ldots, k$ and $\bar{s}_i \in T$ is farthest from s_i on $C(p, q)$.
5. $(■\text{-}○)$: Diametral pairs u, v of $T + pq$ with $v \notin T$, i.e., $v \in pq$ with $p \neq v \neq q$.
   (a) $(▲\text{-}○)$: Pairs $s_i, \bar{s}_i$ with $s_i \in S_i$, $\bar{s}_i \in pq$, and $p \neq \bar{s}_i \neq q$, for $i = 1, 2, \ldots, k$, and where $\bar{s}_i$ is farthest from s_i on $C(p, q)$.

There is no need to consider diametral pairs of type •-• or • ○: the distance from x or y to their respective farthest points on the cycle $C(p, q)$ is always larger than the distance between any two points on $C(p, q)$—unless $T + pq = C(p, q)$. There are no diametral pairs of type x-○ or ○-y, since $\bar{x}, \bar{y} \in T$: If x, v is of type x-■ in $T + pq$, then $v \in T$, and if u, y is of type ■-y in $T + pq$, then $u \in T$.

The *pair state* is the set of types of diametral pairs in $T + pq$. If $T + pq$ has the diametral pairs x, y; x, s_3; x, s_5; and $x, \bar{x}$, then $T + pq$ is in pair state $\{x\text{-}y, x\text{-}■\}$.

Path Types Every diametral path that connects $u, v \in T$ contains the shortcut or not. Every diametral path that connects $u \in T$ with $v \notin T$ contains p or q. There are no diametral paths connecting $u, v \notin T$. We distinguish the following.

1. $(*\text{-}pq\text{-}*)$: A diametral path that *does* contain pq and connects $u, v \in T$.
2. $(*\text{-}T\text{-}*)$: A diametral path that *does not* contain pq and connects $u, v \in T$.
3. $(*\text{-}p\text{-}*)$: A diametral path that contains p and connects $u \in T$ with $v \notin T$.
4. $(*\text{-}q\text{-}*)$: A diametral path that contains q and connects $u \in T$ with $v \notin T$.

Any type of endpoint (e.g., x, y, ■, ▲, •, ○) may appear in place of $*$. For instance, we denote a diametral path from x to $\bar{x}$ via the shortcut by $x\text{-}pq\text{-}•$.

The *path state* is the set of types of diametral paths that are present in $T + pq$. For instance, if $T + pq$ has the diametral pairs x, y; x, s_3; x, s_5; and $x, \bar{x}$ such that pq is useful for x, y and x, s_3 but useless for x, s_5 then $T + pq$ is in path state $\{x\text{-}pq\text{-}y, x\text{-}pq\text{-}▲, x\text{-}T\text{-}▲, x\text{-}pq\text{-}•, x\text{-}T\text{-}•\}$. Figure 7 illustrates the distinction between diametral paths of type ■-pq-■ and ■-T-■ when $x\text{-}pq\text{-}y$ is diametral.

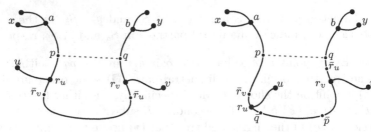

(a) The diametral path of type ■-pq-■ connects u with v via pq.

(b) The diametral path of type ■-T-■ connects u with v via T.

Fig. 7. An illustration of an augmented tree $T + pq$ with (a) diametral paths of type x-pq-y and ■-pq-■ and (b) diametral paths of type x-pq-y and ■-T-■. The points $\bar{r}_u$ and $\bar{r}_v$ mark the farthest points from r_u and r_v, respectively along the simple cycle in $T + pq$. Their position indicates wether pq is useful or useless for the pair u, v.

Operations We distinguish the four types of movements for the shortcut. Suppose we move a shortcut pq for T with $p, q \in \mathcal{B}$ such that $d(a, p) \leq d(a, q)$ to $p'q'$ with $p', q' \in \mathcal{B}$ such that $d(a, p') \leq d(a, q')$. The movement from pq to $p'q'$ is

- an *out-shift* if p moves to a and q moves to b, i.e., if $d(a, p') \leq d(a, p)$ and $d(b, q') \leq d(b, q)$ and, thus $d(a, p') \leq d(a, p) \leq d(a, q) \leq d(a, q')$,
- an *in-shift* if p moves away from a and q moves away from b, i.e., if $d(a, p) \leq d(a, p')$ and $d(b, q) \leq d(b, q')$ and, thus, $d(a, p) \leq d(a, p') \leq d(a, q') \leq d(a, q)$,
- an *x-shift* if p moves to a and q moves away from b, i.e., if $d(a, p') \leq d(a, p)$ and $d(b, q) \leq d(b, q')$ and, thus, $d(a, p') \leq d(a, p) \leq d(a, q') \leq d(a, q)$, or
- a *y-shift* if p moves away from a and q moves to b, i.e., if $d(a, p) \leq d(a, p')$ and $d(b, q') \leq d(b, q)$ and, thus, $d(a, p) \leq d(a, p') \leq d(a, q) \leq d(a, q')$.

The types of movements intentionally overlap when one of the endpoints remains stationary, e.g., when $p = p'$ every shift towards y is also an outwards shift.

Lemma 3 (Blocking Lemma). *Let pq be a shortcut for a tree T with $p, q \in \mathcal{B}$.*

1. *If $T + pq$ has a diametral pair of type x-y, then $\mathrm{diam}(T + pq) \leq \mathrm{diam}(T + p'q')$ for every shortcut $p'q'$ reached with an in-shift from pq.*
2. *If $T + pq$ has a diametral pair of type x-■, then $\mathrm{diam}(T + pq) \leq \mathrm{diam}(T + p'q')$ for every shortcut $p'q'$ reached with a y-shift from pq.*
3. *If $T + pq$ has a diametral pair of type ■-y, then $\mathrm{diam}(T + pq) \leq \mathrm{diam}(T + p'q')$ for every shortcut $p'q'$ reached with an x-shift from pq.*
4. *If $T + pq$ has a diametral pair of type ■-■ or ■-○, then $\mathrm{diam}(T + pq) \leq \mathrm{diam}(T + p'q')$ for every shortcut $p'q'$ reached with an out-shift from pq.*

As an immediate consequence of Lemma 3, the shortcut pq must be optimal for T when each of the four types of movements is blocked by some diametral pair $T + pq$. If the augmented tree $T + pq$ has a diametral pair of type x-y, then there is no need to consider diametral pairs of type ■-■, due to the following.

Lemma 4. *If $T + pq$ has diametral pairs x-y and ■-■ then $T + pq$ also has diametral pairs of type x-■ and ■-y and, thus, the shortcut pq is optimal for T.*

4   Continuous Algorithm

Inspired by the plane-sweep paradigm, we—conceptually—move the shortcut continuously while changing its speed and direction at certain events, i.e., when the pair state or path state changes. Figure 8 describes the continuous algorithm in terms of the pair states. Initially, we place the shortcut with both endpoints on the absolute center c and, thus, we start in pair state $\{x\text{-}y\}$. The algorithm consists of at most three phases: an outwards shift, a shift towards x or a shift towards y, and another outwards shift. For the sake of simplicity, we omit pair states containing $x\text{-}y$ and ■-■ in Fig. 8, due to Lemma 4, we omit transitions implied by transitivity, and we omit pair states that are supersets of final states.

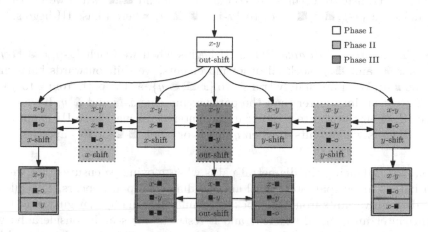

Fig. 8. The pair states during our search for an optimal shortcut for a tree. There are three types of states: First, regular states (solid) indicate the pair state and the operation applied (out-shift, x-shift, or y-shift). Second, transition states (dotted) are visited only momemtarily while transitioning from one regular state to another. Third, final states (double) where we terminate our search and report the best shortcut encountered. Under certain conditions, the search may terminate in non-final states. We start in state $\{x\text{-}y\}$ with an outward shift. When we reach the pair state $\{x\text{-}y, ■\text{-}\circ\}$ from $\{x\text{-}y\}$ then we perform both a shift towards x *and* separately a shift towards y.

Phase I: Shifting Outwards In Phase I, we shorten all diametral paths of type $x\text{-}y$ with an outwards shift, i.e., we move p with unit speed from c towards a and we move q with unit speed from c towards b. If p reaches a before q reaches b, then p stays at a while q continues towards b. Likewise, p continues towards a if q reaches b. In Phase I, the current shortcut is the best shortcut encountered so far. Phase I ends when the shortcut reaches the end of the backbone, i.e., $pq = ab$, or when a second type of diametral pair appears and Phase II begins.

Phase II: Shifting Sideways Phase II begins when we leave pair state $\{x\text{-}y\}$. If we transit from $\{x\text{-}y\}$ to $\{x\text{-}y, x\text{-}■\}$, then we shift towards x. If we transit from $\{x\text{-}y\}$ to $\{x\text{-}y, \blacksquare\text{-}y\}$, then we shift towards y. If we transit from $\{x\text{-}y\}$ to $\{x\text{-}y, \blacksquare\text{-}\circ\}$, then we branch into a shift towards x and a shift towards y. Suppose we reach the pair state $\{x\text{-}y, x\text{-}\blacksquare\}$ from $\{x\text{-}y\}$. All diametral paths in $T + pq$ contain the path from a to p. We move p closer to a and we move q with a speed towards a that keeps all diametral paths in balance. This ensures that we remain in the current pair state until another diametral pair appears. In this state, the current shortcut is the best shortcut encountered so far. When we reach the pair state $\{x\text{-}y, \blacksquare\text{-}\circ\}$, we move p towards a and adjust the position of q to balance the diametral paths of type $x\text{-}pq\text{-}y$ with those of type $\blacksquare\text{-}p\text{-}\circ$ and $\blacksquare\text{-}q\text{-}\circ$. In this state, the diameter shrinks and grows with $|pq|$ and the shortest shortcut since we entered this state is the best shortcut so far. Phase II ends when p reaches a, when we transit to a pair state containing $x\text{-}y$ and $\blacksquare\text{-}\blacksquare$, when we transit to the final state $\{x\text{-}y, \blacksquare\text{-}\circ, \blacksquare\text{-}y\}$ or to $\{x\text{-}y, x\text{-}\blacksquare, \blacksquare\text{-}y\}$ where Phase III begins.

Phase III: Shifting Outwards Phase III begins when we reach $\{x\text{-}y, x\text{-}\blacksquare, \blacksquare\text{-}y\}$. As $x\text{-}y$, $x\text{-}\blacksquare$, and $\blacksquare\text{-}y$ block all other movements, we shift outwards balancing $x\text{-}\blacksquare$ and $\blacksquare\text{-}y$. We immediately transit to $\{x\text{-}\blacksquare, \blacksquare\text{-}y\}$, as the path from x to y via the shortcut shrinks faster than the paths connecting $x\text{-}\blacksquare$ and $\blacksquare\text{-}y$. If we reach Phase III, then the shortest shortcut encountered during Phase III is optimal. Phase III ends when p hits a, when q hits b, or when $\blacksquare\text{-}\blacksquare$ or $\blacksquare\text{-}\circ$ appears.

Optimality Balancing the diametral paths when moving pq ensures that we remain in the current pair state until another diametral pair appears. This allows us to derive invariants from the blocking lemma (Lemma 3) that guarantee that we encounter an optimal shortcut and it restricts our search considerably: we are essentially conducting a linear search, since the speed of q is determined by the speed of p, the path state, and the change in the length of the shortcut.

We elucidate our invariants and their use through an example: In pair state $\{x\text{-}y, \blacksquare\text{-}\circ\}$ of Phase II, the blocking lemma implies that there is an optimal shortcut p^*q^* that we reach from the current shortcut pq with an x-shift, with a y-shift, or without moving ($pq = p^*q^*$). Suppose we miss p^*q^* during an x-shift from pq to the next change in the pair state at $p'q'$. This means we reach p^*q^* with an x-shift from pq and with an y-shift from $p'q'$. Let $p''q''$ be the last position during the shift from pq to $p'q'$ where the shift from $p''q''$ to p^*q^* leads towards x. Then $T + p''q''$ is in pair state $\{x\text{-}y, \blacksquare\text{-}\circ\}$ and we have $p'' = p^*$ or $q'' = q^*$. If $p'' = p^*$, then the movement from $p''q''$ to p^*q^* is an inward shift towards x. Since $x\text{-}y$ blocks any inward shift, we have $\operatorname{diam}(T + p''q'') \leq \operatorname{diam}(T + p^*q^*)$. If $q'' = q^*$, then the movement from $p''q''$ to p^*q^* is an outward shift towards x. Since $\blacksquare\text{-}\circ$ blocks any outward shift, we have $\operatorname{diam}(T + p''q'') \leq \operatorname{diam}(T + p^*q^*)$. Since p^*q^* is optimal, $p''q''$ is an optimal shortcut that we encounter.

If we could implement and run it, the continuous algorithm would produce an optimal shortcut pq for T with p on the path from a to c and q on the path from c to b: In Phase I and III p and q move away from c. In Phase II a diametral pair of type $\blacksquare\text{-}y$ appears before q reaches c during a shift towards x and a diametral pair of type $x\text{-}\blacksquare$ appears before p reaches c during a shift towards y.

5 Discretization

To discretize the continuous algorithm, we subdivide the continuous motion of the shortcut with events such that we can calculate the next event and the change in the continuous diameter of $T + pq$ between subsequent events. We introduce events, for instance, when the shortcut hits a vertex, when the path state changes, and when the shortcut changes between shrinking and growing.

During Phase I, the continuous diameter decreases or remains constant. Therefore, it is sufficient for the discrete algorithm to determine where Phase I of the continuous algorithm ends. At the end of Phase I, we have either reached the end of the backbone, i.e., $pq = ab$, or a diametral pair of type x-■, ■-y, or ■-○ has appeared alongside the diametral pairs of type x-y. We may ignore diametral pairs of type ■-■ as they appear together with x-■ and ■-y when x-y is diametral, as shown in Lemma 4. We detect changes in the path state by monitoring the candidates for each type of diametral path—except for those connecting diametral pairs of type ■-■. We argue that there are $O(n)$ events where we spend $O(\log n)$ time to detect whether Phase I is about to end.

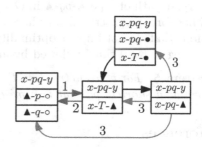

Fig. 9. The transitions between the path states during Phase II. We omit transitory states like $\{x\text{-}pq\text{-}y, x\text{-}pq\text{-}▲, x\text{-}pq\text{-}●, x\text{-}T\text{-}●\}$. Transition 1 is possible when the shortcut grows. Transition 2 is possible when the shortcut shrinks. If we take any of the transitions with label 3 from $\{x\ pq\ y, x\text{-}pq\text{-}s_j\}$ for $j = 1, 2, \ldots, k$, then the pair x, s_j ceases to be diametral and cannot become diametral in Phase II again.

To discretize Phase II, we calculate the speeds at which p and q move to balance the diametral paths and how this impacts the diameter. For instance, if we balance the paths x-pq-y, x-pq-●, and x-T-● during an x-shift from pq to $p'q'$, then $d_T(q, q') = \frac{1}{3}(d_T(p, p') + |pq| - |p'q'|)$ and the diameter changes by $\frac{2}{3}(d_T(p, p') + |pq| - |p'q'|)$. With the assumption that p moves with unit speed, this allows us to predict the position of q. The respective speeds for the path states and their dependence on the length of the shortcut allow us to bound the number of path state changes in Phase II by $O(n)$, as hinted in Fig. 9.

During Phase III, the path state has two components: the diametral path of type x-■ and the diametral path of type ■-y. We encounter two challenges when discretizing Phase III: First, the path state might change $\Omega(n^2)$ times. This occurs, for instance, when the shortcut alternates $\Omega(n)$ times between growing and shrinking such that there are $\Omega(n)$ candidates for diametral pairs, for each of which the shortcut becomes useful whenever it shrinks and useless whenever it grows. We circumvent this issue by ignoring certain superfluous path state events. Second, since x-y is no longer diametral, we need to detect diametral pairs of type ■-■ of which there are $\Omega(k^2)$ candidates, where k is the number of $\mathcal{B}$-sub-trees of T. The difficulty lies in detecting diametral paths of type ▲-pq-▲. It suffices to find the first position $\hat{p}\hat{q}$ where a path s_i-pq-s_j, for $i, j \in \{1, 2, \ldots, k\}$, becomes a diametral path of type ▲-pq-▲: If we shift outwards from $\hat{p}\hat{q}$, then s_i-pq-s_j will remain diametral while increasing in length. We proceed as follows. First, we simulate the modified Phase III without attempting to detect if a diametral path of type ▲-pq-▲ appears. We record the sequence of edge pairs that we visit during this simulation. This sequence contains $O(n)$ edge pairs. After the simulation, we perform a binary search for $\hat{p}\hat{q}$ in the sequence of visited edge pairs. The binary search for $\hat{p}\hat{q}$ takes $O(n \log n)$ time, since we can determine the largest path of type ▲-pq-▲ in $O(n)$ time for a fixed position of the shortcut.

Therefore, we discretize all phases of the continuous algorithm—with modifications that do not impact optimality—with $O(n)$ events that we can process in $O(n \log n)$ total time, followed by an $O(n \log n)$-time post-processing step.

Theorem 5. *For a geometric tree T with n vertices, it takes $O(n \log n)$ time to determine a shortcut pq that minimizes the continuous diameter of $T + pq$.* □

References

1. Cáceres, J., Garijo, D., González, A., Márquez, A., Puertas, M.L., Ribeiro, P.: Shortcut Sets for Plane Euclidean Networks. Electron Notes Discrete Math 54, 163–168 (2016)
2. Chepoi, V., Vaxès, Y.: Augmenting Trees to Meet Biconnectivity and Diameter Constraints. Algorithmica 33(2), 243–262 (2002)
3. De Carufel, J.L., Grimm, C., Maheshwari, A., Smid, M.: Minimizing the Continuous Diameter when Augmenting Paths and Cycles with Shortcuts. In: SWAT 2016. pp. 27:1–27:14 (2016)
4. Farshi, M., Giannopoulos, P., Gudmundsson, J.: Improving the Stretch Factor of a Geometric Network by Edge Augmentation. SIAM J. Comp. 38(1), 226–240 (2008)
5. Frati, F., Gaspers, S., Gudmundsson, J., Mathieson, L.: Augmenting Graphs to Minimize the Diameter. Algorithmica 72(4), 995–1010 (2015)
6. Gao, Y., Hare, D.R., Nastos, J.: The Parametric Complexity of Graph Diameter Augmentation. Discrete Appl Math 161(10-11), 1626–1631 (2013)
7. Große, U., Gudmundsson, J., Knauer, C., Smid, M., Stehn, F.: Fast Algorithms for Diameter-Optimally Augmenting Paths. In: ICALP 2015. pp. 678–688 (2015)
8. Große, U., Gudmundsson, J., Knauer, C., Smid, M., Stehn, F.: Fast Algorithms for Diameter-Optimally Augmenting Paths and Trees (2016), arXiv:1607.05547
9. Hakimi, S.L.: Optimum locations of switching centers and the absolute centers and medians of a graph. Operations Research 12(3), 450–459 (1964)
10. Li, C.L., McCormick, S., Simchi-Levi, D.: On the Minimum-Cardinality-Bounded-Diameter and the Bounded-Cardinality-Minimum-Diameter Edge Addition Problems. Operations Research Letters 11(5), 303–308 (1992)
11. Luo, J., Wulff-Nilsen, C.: Computing Best and Worst Shortcuts of Graphs Embedded in Metric Spaces. In: ISAAC 2008. pp. 764–775 (2008)
12. Oh, E., Ahn, H.K.: A Near-Optimal Algorithm for Finding an Optimal Shortcut of a Tree. In: ISAAC 2016. pp. 59:1–59:12 (2016)
13. Schoone, A.A., Bodlaender, H.L., van Leeuwen, J.: Diameter Increase Caused by Edge Deletion. Journal of Graph Theory 11(3), 409–427 (1987)
14. Yang, B.: Euclidean Chains and their Shortcuts. Theor Comput Sci 497, 55–67 (2013)

Inapproximability of the Standard Pebble Game and Hard to Pebble Graphs

Erik D. Demaine, Quanquan C. Liu

MIT CSAIL, Cambridge, Massachusetts

Abstract. Pebble games are single-player games on DAGs involving placing and moving pebbles on nodes of the graph according to a certain set of rules. The goal is to pebble a set of target nodes using a minimum number of pebbles. In this paper, we present a possibly simpler proof of the result in [4] and strengthen the result to show that it is PSPACE-hard to determine the minimum number of pebbles to an additive $n^{1/3-\varepsilon}$ term for all $\varepsilon > 0$, which improves upon the currently known additive *constant* hardness of approximation [4] in the standard pebble game. We also introduce a family of explicit, constant indegree graphs with n nodes where there exists a graph in the family such that using $0 < k < \sqrt{n}$ pebbles requires $\Omega((n/k)^k)$ moves to pebble in both the standard and black-white pebble games. This independently answers an open question summarized in [14] of whether a family of DAGs exists that meets the upper bound of $O(n^k)$ moves using constant k pebbles with a different construction than that presented in [1].

1 Introduction

Pebble games were originally introduced to study compiler operations and programming languages. For such applications, a DAG represents the computational dependency of each operation on a set of previous operations and pebbles represent register allocation. Minimizing the amount of resources allocated to perform a computation is accomplished by minimizing the number of pebbles placed on the graph [16]. The *standard pebble game* (also known as the *black pebble game*) is traditionally used to model such behavior. In the standard pebble game, one is given a DAG, $G = (V, E)$, with n nodes and constant indegree and told to perform a set of *pebbling moves* that places, removes, or slides pebbles around the nodes of G.

The premise of such games is given some input modeled by *source* nodes $S \subseteq V$ one should compute some set of outputs modeled as target nodes $T \subseteq V$. In terms of G, S is typically the set of nodes without incoming edges and T is typically the set of nodes without outgoing edges. The rules of the standard pebble game are as follows:

© Springer International Publishing AG 2017
F. Ellen et al. (Eds.): WADS 2017, LNCS 10389, pp. 313–324, 2017.
DOI: 10.1007/978-3-319-62127-2_27

STANDARD PEBBLE GAME
Input: Given a DAG, $G_{n,\delta} = (V, E)$. Let $\mathsf{pred}(v) = \{u \in V : (u, v) \in E\}$. Let $S \subseteq V$ be the set of sources of G and $T \subseteq V$ be the set of targets of G. Let $\mathcal{P} = \{P_0, \ldots, P_\tau\}$ be a valid pebbling strategy that obeys the following rules where P_i is a set of nodes containing pebbles at timestep i and $P_0 = \emptyset$ and $P_\tau = \{T\}$. Let $\mathsf{Peb}(G, \mathcal{P}) = \max_{i \in [\tau]} \{|P_i|\}$.

Rules:

1. At most one pebble can be placed or removed from a node at a time.
2. A pebble can be placed on any source, $s \in S$.
3. A pebble can be removed from any vertex.
4. A pebble can be placed on a non-source vertex, v, at time i if and only if its direct predecessors are pebbled, $\mathsf{pred}(v) \in P_{i-1}$.
5. A pebble can slide from vertex v to vertex w at time i if and only if $(v, w) \in E$ and $\mathsf{pred}(w) \in P_{i-1}$.

Goal: Determine $\min_{\mathcal{P}} \{\mathsf{Peb}(G, \mathcal{P})\}$ using a valid strategy $\mathcal{P}$.

In addition to the standard pebble game, other pebble games are useful for studying computation. The *red-blue pebble game* is used to study I/O complexity [12], the *reversible pebble game* is used to model reversible computation [3], and the *black-white pebble game* is used to model non-deterministic straight-line programs [5]. Although we will be proving a result about the black-white pebble game in Section 4, we will defer introducing the rules of the game to our full paper [6] since the black-white pebble game is not central to the main results of this paper.

Much previous research has focused on proving lower and upper bounds on the *pebbling space cost* (i.e. the maximum number of pebbles over time) of pebbling a given DAG under the rules of each of these games. For all of the aforementioned pebble games (except the red-blue pebble game since it relies on a different set of parameters), any DAG can be pebbled using $O(n/\log n)$ pebbles [9,11,15]. Furthermore, there exist DAGs for each of the games that require $\Omega(n/\log n)$ pebbles [9,11,15].

It turns out that finding a strategy to optimally pebble a graph in the standard pebble game is computationally difficult even when each vertex is allowed to be pebbled only once. Specifically, finding the minimum number of black pebbles needed to pebble a DAG in the standard pebble game is PSPACE-complete [8] and finding the minimum number of black pebbles needed in the one-shot case is NP-complete [16]. In addition, finding the minimum number of pebbles in both the black-white and reversible pebble games have been recently shown to be both PSPACE-complete [4,10]. But the result for the black-white pebble game is proven for unbounded indegree [10]. A key open question in the field is whether hardness results can be obtained for constant indegree graphs for the black-white pebble game. However, whether it is possible to find good approximate solutions to the minimization problem has barely been studied. In fact, it was not known until this paper whether it is hard to find the minimum number of pebbles within even a non-constant *additive* factor [4]. The best known multiplicative approximation factor is the very loose $\Theta(n/\log n)$ which is the pebbling space upper bound [11], leaving much room for improvement.

Our results deal primarily with the standard pebble game, but we believe that the techniques could be extended to show hardness of approximation for other pebble games. We prove the following:

Theorem 1. *The minimum number of pebbles needed in the standard pebble game on DAGs with maximum indegree 2 is PSPACE-hard to approximate to within an additive $n^{1/3-\varepsilon}$ for any $\varepsilon > 0$.*

In addition to determining the pebbling space cost, we sometimes also care about pebbling time which refers to the number of operations (placements, removals, or slides) that a strategy uses. For example, such a situation arises if we care not only about the memory used in computation but also the time of computation. It is previously known that there exists a family of graphs such that, given $\Theta(\frac{n}{\log n})$ pebbles, one is required to use $\Omega(2^{\Theta(\frac{n}{\log n})})$ moves to pebble any graphs with n nodes in the family [13].

Less is known about the trade-offs when a small number (e.g. constant k) of pebbles is used until the very recent, independent result presented in [1]. It can be easily shown through a combinatorial argument that the maximum number of moves necessary using $k = O(1)$ pebbles to pebble n nodes is $O(n^k)$ [14]. It is an open question whether it is possible to prove a time-space trade-off such that using $k = O(1)$ pebbles requires $\Omega(n^k)$ time. In this paper, we resolve this open question for both the standard pebble and the black-white pebble games using an independent construction from that presented in [1].

Theorem 2. *There exists a family of graphs with n vertices and maximum indegree 2 such that $\Omega((\frac{n-k^2}{k})^k)$ moves are necessary to pebble any graph with n vertices in the family using $k < \sqrt{n}$ pebbles in both the standard and black-white pebble games.*

In particular, when $k = O(1)$, the number of moves necessary to pebble a graph in the family is $\Theta(n^k)$.

The organization of the paper is as follows. First, in Section 2, we provide the definitions and terminology we use in the remaining parts of the paper. Then, in Section 3, we provide a proof for the inapproximability of the standard pebble game to an $n^{1/3-\varepsilon}$ additive factor.

In Section 4, we present our hard to pebble graph families using $k < \sqrt{n}$ pebbles and prove that the family takes $\Omega(n^k)$ moves to pebble in both the standard and black-white pebble games when $k = O(1)$.

Finally, in Section 5, we discuss some open problems resulting from this paper.

2   Definitions and Terminology

In this section, we define the terminology we use throughout the rest of the paper. All of the pebble games we consider in this paper are played on directed acyclic graphs (DAGs). In this paper, we only consider DAGs with maximum indegree 2. We define such a DAG as $G = (V, E)$ where $|V| = n$ and $|E| = m$.

The purpose of any pebble game is to pebble a set of targets $T \subseteq V$ using minimum number of pebbles. In all pebble games we consider, a player can always place a pebble on any source node, $S \subseteq V$. Usually, S consists of all nodes with indegree 0 and T consists of all nodes with outdegree 0.

A *sequential pebbling strategy*, $\mathcal{P} = [P_0, \ldots, P_\tau]$ is a series of configurations of pebbles on G where each P_i is a set of pebbled vertices $P_i \subseteq V$. P_i follows from P_{i-1} by the rules of the game and $P_0 = \emptyset$ and $P_\tau = T$. Then, by definition, $|P_i|$ is the number of pebbles used in configuration P_i. For a *sequential* strategy, $|P_{i-1}| - 1 \leq |P_i| \leq |P_{i-1}| + 1$ for all $i \in [\tau] = [1, \ldots, \tau]$ (i.e. at most one pebble can be placed, removed, or slid on the graph at any time). In this paper, we only consider sequential strategies.

Given any strategy $\mathcal{P}$ for pebbling G, the *pebbling space cost*, $\mathsf{Peb}(G, \mathcal{P})$, of $\mathcal{P}$ is defined as the maximum number of pebbles used by the strategy at any time: $\mathsf{Peb}(G, \mathcal{P}) = \max_{i \in [\tau]} \{|P_i|\}$.

The *minimum pebbling space cost* of G, $\mathsf{Peb}(G)$, is defined as the smallest space cost over the set of all valid strategies, $\mathbb{P}$, for G:

Definition 1 (Minimum Pebbling Space Cost).

$$\mathsf{Peb}(G) = \min_{\mathcal{P} \in \mathbb{P}} \{\mathsf{Peb}(G, \mathcal{P})\}.$$

The *pebbling time cost*, $\mathsf{Time}(G, \mathcal{P}, s) = |\mathcal{P}| - 1$, of a strategy $\mathcal{P}$ using s pebbles is the number of moves used by the strategy. The *minimum pebbling time cost* of any strategy that has pebbling space cost s is the minimum number of moves used by any such strategy.

Definition 2 (Minimum Pebbling Time Cost).

$$\mathsf{Time}(G, s) = \min_{\mathcal{P}' \in \{\mathcal{P} \in \mathbb{P} : |\mathcal{P}| \leq s\}} \{\mathsf{Time}(G, \mathcal{P}', s)\} \geq n.$$

3   Inapproximability of the Standard Pebble Game

In this section, we provide an alternative proof of the result presented in [4] that the standard pebble game is inapproximable to any constant additive term. Then, we show that our proof technique can be used to show our main result stated in Theorem 1. We make modifications to the proof presented by [8] to obtain our main result. A quick explanation of the relevant results presented in [8] can be found in our full paper [6].

3.1   Inapproximability to $n^{1/3-\varepsilon}$ additive term for any $\varepsilon > 0$

We now prove our main result. For our reduction we modify the variable, clause, and quantifier gadgets in [8] to produce a gap reduction from the PSPACE-hard problem, QBF.

Important Graph Components Before we dive into the details of our construction, we first mention two subgraphs and the properties they exhibit.

The first graph is the *pyramid graph* (please refer to [6] for a figure), Π_h with height h, which requires a number of pebbles that is equal to the height, h, of the pyramid to pebble [8]. Therefore, in order to pebble the apex of such a graph, at least h pebbles must be available. As in [8], we depict such pyramid graphs by a triangle with a number indicating the height (hence number of pebbles) needed to pebble the pyramid (see Figure 1 for an example of the triangle symbolism).

We make use of the following definition and lemma (restated and adapted) from [8] in our proofs:

Definition 3 (Frugal Strategy [8]). *A pebbling strategy, $\mathcal{P}$, is frugal if the following are true:*

1. *Suppose vertex $v \in G$ is pebbled for the first time at time t'. Then, for all times, $t > t'$, some path from v to q_1 (the only target node) contains a pebble.*
2. *At all times after v is pebbled for the last time, all paths from v to q_1 contain a pebble.*
3. *The number of pebble placements on any vertex $v \in G$ where $v \neq q_1$, is bounded by the number of pebble placements on $\mathsf{pred}(v)$.*

Lemma 1 (Normal Pebbling Strategy [8]). *If the target vertex is not inside a pyramid, Π_h, and each of the vertices in the bottom level of the pyramid has at most one predecessor each, then any pebbling strategy can be transformed into a normal pebbling strategy without increasing the number of pebbles used. A normal pebbling strategy is one that is frugal and after the first pebble is placed on any pyramid, Π_h, no placements of pebbles occurs outside Π_h until the apex of Π_h is pebbled and all other pebbles are removed from Π_h.*

The other important subgraph is the *road graph* (see [6] for figure), R_w with width w, which requires a number of pebbles that is the width of the graph to pebble *any* of the outputs [7,14]. Therefore, we state as an immediately corollary of their proof:

Corollary 1 (Road Graph Pebbling). *To pebble $O \subseteq \{o_1, \ldots, o_w\}$ of the outputs of R_w, with a valid strategy, $\mathcal{P} = [P_0, \ldots, P_\tau]$ where $P_\tau = O$, requires $w + |O| - 1$ pebbles.*

We define a *regular* pebbling strategy for road graphs similarly to the *normal* pebbling strategy for pyramids.

Lemma 2 (Regular Pebbling Strategy). *If each input, $i_j \in \{i_1, \ldots, i_w\}$, to the road graph has at most 1 predecessor, any pebbling strategy can be transformed into a* regular *pebbling strategy without increasing the number of pebbles used. A* regular *pebbling strategy is one that is frugal and after the first pebble is placed on any road graph, R_w, no placements of pebbles occurs outside R_w until a set of desired outputs, $O \subseteq \{o_1, \ldots, o_w\}$, of R_w all contain pebbles and all other pebbles are removed from R_w.*

We immediately obtain the following corollary from Lemmas 1 and 2:

Corollary 2. *Any pebbling strategy, $\mathcal{P}$, can be transformed into a pebbling strategy, $\mathcal{P}'$, that is normal and regular if no target vertices lie inside a pyramid or road graph and each input node to either the pyramid or road graph has at most one predecessor.*

Modified Graph Constructions We first describe the changes we made to each of the gadgets used in the PSPACE-completeness proof presented by [8] and then prove our inapproximability result using these gadgets in Section 3.1. Given a QBF instance, $B = Q_1x_1 \cdots Q_ux_uF$, with c clauses, we create the following gadgets:

Variable Nodes: We replace all variable nodes in the proof provided in [8] with road graphs each of width K. The modified variable nodes are shown in Figure 1. Each variable node as in the original proof by [8] has 3 possible configurations which are also shown in Figure 1.

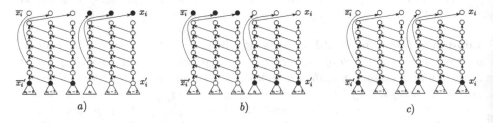

Fig. 1: Modified variable gadget with 3 possible configurations using the road graph subgraph previously described. Here $K = 3$. a) x_i is True. b) $\overline{x_i}$ is True. c) Double false.

Quantifier Blocks: Each universal and existential quantifier blocks are also modified to account for the new variable nodes. See Figure 2 and Figure 3 which depict the new quantifier gadgets that use the new variable nodes. Note that instead of each quantifier gadget requiring 3 total pebbles, each gadget requires $3K$ pebbles to remain on each block before the clauses are pebbled. The basic idea is to expand all nodes $a_i, b_i, c_i...etc.$ into a path of length K to account for each of the K copies of x_i and each of the K copies of $\overline{x_i}$. Each $s_i = s_{i-1} - 3K$ and $s_1 = 3Ku + 3K + 1$.

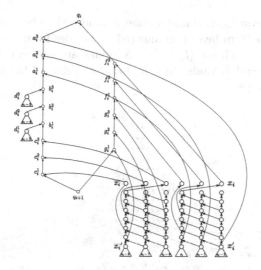

Fig. 2: Modified universal quantifier block. Here $K = 3$.

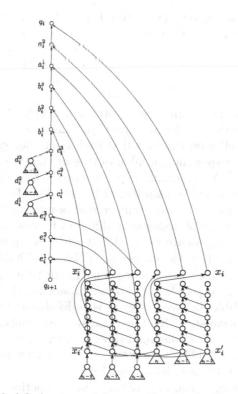

Fig. 3: Modified existential quantifier block. Here $K = 3$.

Clause Gadgets: Each clause gadget is modified to be a pyramid of height $3K + 1$ where the bottom layer is connected to 2 nodes from two different literals. Therefore, for a given clause (l_i, l_j, l_k), K nodes are connected to l_i and l_j, K nodes to l_j and l_k, and K nodes to l_i and l_k. See Figure 4 for an example of the modified clause gadget.

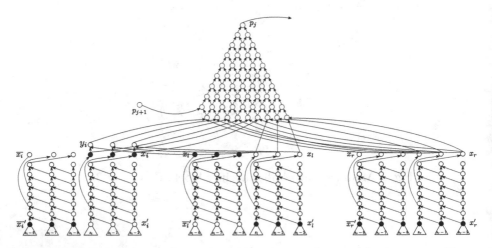

Fig. 4: Modified clause gadget. The clause here is (x_i, x_l, x_r) where $x_i = True$, $x_l = False$, and $x_r = False$. Here $K = 3$.

Proofs of the Construction We construct a graph G using the gadgets described above in Section 3.1 for any given QBF instance, $B = Q_1 x_1 \cdots Q_k x_u F$. In short, the proof relies on the fact that each quantifier gadget requires $2K$ pebbles to set the corresponding variable to true or false (i.e. the corresponding literals to true or false). An additional K pebbles need to remain on each quantifier in order to be able to repebble quantifiers when checking for universal variables' satisfaction. Furthermore, a clause would consist of modified pyramids of height $3K + 1$ connected to pairs of nodes from different literals. Following the proof in [8], the quantifier gadgets are pebbled first with $3Ku$ pebbles remaining on the quantifier gadgets. Then, the clauses are pebbled with $3K + 1$ pebbles.

If B is satisfiable, then clauses can be pebbled with $3K + 1$ pebbles. Otherwise, $4K$ pebbles are needed to pebble one or more unsatisfied clauses in G, resulting in a gap of $K - 1$ pebbles between when B is satisfiable and unsatisfiable. Thus, if given an approximation algorithm that estimates the number of pebbles needed within additive $K - 1$, we can distinguish between the case when B is satisfiable (at most $3Km + 3K + 1$ pebbles are needed) and the case when B is unsatisfiable (when at least $3Km + 4K$ pebbles are needed).

In this construction, K can be any polynomial function of u where u is the number of variables in B and c is the number of clauses (in other words, $K = u^a c^b$

for any constants a and b). The total minimum number of pebbles necessary is $O(Ku)$ and the total number of nodes in the graph is $O(K^3(u+c))$.

We first prove that the number of pebbles needed to pebble each quantifier gadget is $3K$ and $3K$ pebbles remain on the quantifier blocks throughout the pebbling of the clauses.

Lemma 3. *Every regular and normal strategy, $\mathcal{P}'$, must be one where each quantifier gadget must be pebbled with $3K$ pebbles before the clauses are pebbled. Furthermore, each quantifier gadget must be pebbled with $3K$ pebbles when q_u is pebbled.*

Next we prove that provided $3Ku$ pebbles stay on the quantifier blocks, each unsatisfied clause requires $4K$ pebbles.

Lemma 4. *Given a clause gadget, C_i, its corresponding variable, c_i is true if and only if C_i can be pebbled with $3K+1$ pebbles. Furthermore, if c_i is false and all literals in C_i are set in the false configuration, then at least $4K$ pebbles are necessary to pebble the clause.*

Given the previous proofs, we now prove the following key lemmas:

Lemma 5. *Given G which is constructed from the provided QBF instance, $B = Q_1 x_1 \cdots Q_u x_u F$, using our modified reduction in Section 3.1, B is satisfiable if and only if $\mathsf{Peb}(G) \leq 3Ku + 3K + 1$.*

Before, we prove the next crucial lemma (Lemma 7), we first prove the following lemma which will help us prove Lemma 7:

Lemma 6. *Let N_i be the configuration such that some number of pebbles are on the first $i-1$ quantifier blocks and the i-th is being pebbled. Therefore, N_{u+1} is the configuration when some number of pebbles are on all u quantifier blocks and the first clause gadget is being pebbled. There does not exist a frugal strategy, $\mathcal{P}$, that can pebble our reduction construction, G, such that N_i contains less than $s - s_i$ pebbles on the first $i-1$ quantifier blocks when the i-th quantifier block or when the first clause is being pebbled.*

Lemma 7. *Given G which is constructed from the provided QBF instance, $B = Q_1 x_1 \cdots Q_u x_u F$, using our modified reduction in Section 3.1, B is unsatisfiable if and only if $\mathsf{Peb}(G) \geq 3Ku + 4K$.*

Proof of Inapproximability Using Lemmas 5 and 7, we prove that it is PSPACE-hard to approximate the minimum number of black pebbles needed given a DAG, G, to an additive $n^{1-\varepsilon}$ factor.

Theorem 3 (Restatement of Theorem 1). *The minimum number of pebbles needed in the standard pebble game on DAGs with maximum indegree 2 is PSPACE-hard to approximate to additive $n^{1/3-\varepsilon}$ for $\varepsilon > 0$.*

Proof. From Lemmas 5 and 7, the cost of pebbling a graph constructed from a satisfiable B is at most $3Ku + 3K + 1$ whereas the cost of pebbling a graph constructed from an unsatisfiable B is at least $3Ku + 4K$. As we can see, the aforementioned reduction is a gap-producing reduction with a gap of $K - 1$ pebbles. Then, all that remains to be shown is that for any $\varepsilon > 0$, it is the case that $K \geq (K^3(u+c))^{(1/3-\varepsilon)}$. (Note that for $\varepsilon > 1/3$, setting K to any positive integer achieves this bound.) Suppose we set $K = \max(u, c)^a$ where $a > 0$. Given an $0 < \varepsilon \leq 1/3$, $a = \frac{1/3-\varepsilon}{1+3\varepsilon} \geq 0$ precisely when ε is in the stated range.

For values of $a \geq 0$, we can duplicate the clauses and variables gadgets so that u and c are large enough such that $K = \max(u, c)^a \geq 2$. Let $d = \max(u, c)$. Then, we need d to be large enough so that $d^a \geq 2$ (i.e. we want d^a to be some integer). Then, we can set $d \geq 2^{1/a}$. Thus, we can duplicate the number of variables and clauses so that $d \geq 2^{\frac{1+3\varepsilon}{1/3-\varepsilon}}$.

Therefore, for every $\varepsilon > 0$, we can construct a graph with a specific K calculated from ε such that it is PSPACE-hard to find an approximation within an additive $n^{1/3-\varepsilon}$ factor where n is the number of nodes in the graph.

4   Hard to Pebble Graphs for Constant k Pebbles

It is long known that the maximum number of moves necessary to pebble any graph with constant k pebbles is $O(n^k)$. (Note that the maximum number of moves necessary to pebble any graph is either $O(n^{k-1})$ or $O(n^k)$ depending on whether or not sliding is allowed. Here, we allow sliding in all of our games. The bound of $O(n^{k-1})$ proven in [14] is one for the case when sliding is not allowed.) The upper bound of $O(n^k)$ for any constant k number of pebbles submits to a simple combinatorial proof adapted from [14]) to account for sliding. However, to the best of the author's knowledge, examples of such families of graphs that require $O(n^k)$ moves to pebble using k pebbles did not exist until very recently in an independent work [1]. In this section, we present an independent, simple to construct family of graphs that require $\Theta(n^k)$ time for constant k number of pebbles in both the standard and black-white pebble games. We further reduce the indegree of nodes in this family of graphs to 2 and show that our results still hold. Furthermore, we show this family of graphs to exhibit a steep time-space trade-off (from exponential in k to linear) even when k is not constant. Such families of graphs could potentially have useful applications in cryptography in the domain of proofs of space and memory-hard functions [2].

We construct the following family of graphs, $\mathbb{H}_{n,\delta}$, below with n nodes and indegree δ and show that for constant k pebbles, the number of steps it takes to pebble the graph $H_{n,k} \in \mathbb{H}_{n,\delta}$ with k pebbles is $\Omega(n^k)$. We also show a family of graphs, $\mathbb{H}_{n,2}$ with indegree 2 that shows the same asymptotic tradeoff.

We construct the family of graphs in the following way.

Definition 4. *Given a set of n nodes and maximum number of pebbles k where $k < \sqrt{n}$, we lexicographically order the nodes (from 1 to n) and create the following set of edges between the nodes where directed edges are directed from v_i to v_j where $i < j$:*

1. v_i and v_{i+1} for all $i \in [k, n]$
2. v_i and v_j for all $i \in [l - 1]$ for all $2 \leq l \leq k$ and $j \in \{f(l) + 2r - 2\}$ for all $r \in [\frac{n-k}{2k} + 1]$ where $f(l) = k + (l - 1)(\frac{n-k}{k}) + 1$.
3. v_i and v_j for all $i = f(l) - 1$ and $j \in \{f(l) + 2r - 1\}$ for all $i \in [\frac{n-k}{2k} + 1]$ where $l \in [1, k - 1]$.

The target node (the only sink) is v_n. Note that the sources in our construction are v_j for all $j \in [1, k]$.

Due to the space constraints, we leave all proofs of the properties of the graph family as well as an example figure of a member of the family in our full paper [6].

5   Open Problems

There are a number of open questions that naturally follow the content of this paper.

The first obvious open question is whether the techniques introduced in this paper can be tweaked to allow for a PSPACE-hardness of approximation to an $n^{1-\varepsilon}$ additive factor for any $\varepsilon > 0$. We note that the trivial method of attempting to reduce the size of the subgraph gadgets used in the variables (i.e. use a different construction than the road graph such that less than K^3 nodes are used) is not sufficient since the number of nodes in the graph is still $O(K^3(u+c))$. This is not to say that such an approach is not possible; simply that more changes need to be made to all of the other gadgets. The next logical step is to determine whether $\text{Peb}(G)$ can be approximated to a constant 2 factor multiplicative approximation.

Another open question is whether the techniques introduced in this paper can be applied to show hardness of approximation results for other pebble games such as the black-white or reversible pebble games. The main open question in the topic of hardness of approximation of pebble games is whether the standard pebble game can be approximated to any factor smaller than $n/\log n$ or whether the games are PSPACE-hard to approximate to any constant factor, perhaps even logarithmic factors.

With regard to hard to pebble graphs, we wonder if our graph family could be improved to show $\Omega(n^k)$ for any $0 < k \leq n/\log n$. This would be interesting because to the best of the authors' knowledge we do not yet know of any graph families that exhibit sharp (asymptotically tight) time-space trade-offs for this entire range of pebble number.

References

1. Joël Alwen, Susanna F. de Rezende, Jakob Nordström, and Marc Vinyals. Cumulative space in black-white pebbling and resolution. In *Innovations in Theoretical Computer Science, ITCS 2017, Berkeley, CA, USA, 9-11 January, 2017*, 2017.
2. Joël Alwen and Vladimir Serbinenko. High parallel complexity graphs and memory-hard functions. In *Proceedings of the Forty-Seventh Annual ACM on Symposium on Theory of Computing, STOC 2015, Portland, OR, USA, June 14-17, 2015*, pages 595–603, 2015. Available from: http://doi.acm.org/10.1145/2746539.2746622.

3. Charles H. Bennett. Time/space trade-offs for reversible computation. *SIAM J. Comput.*, 18(4):766–776, August 1989. Available from: http://dx.doi.org/10.1137/0218053.

4. Siu Man Chan, Massimo Lauria, Jakob Nordström, and Marc Vinyals. Hardness of approximation in PSPACE and separation results for pebble games. In *IEEE 56th Annual Symposium on Foundations of Computer Science, FOCS 2015, Berkeley, CA, USA, 17-20 October, 2015*, pages 466–485, 2015. Available from: http://dx.doi.org/10.1109/FOCS.2015.36.

5. Stephen Cook and Ravi Sethi. Storage requirements for deterministic / polynomial time recognizable languages. In *Proceedings of the Sixth Annual ACM Symposium on Theory of Computing*, STOC '74, pages 33–39, New York, NY, USA, 1974. ACM. Available from: http://doi.acm.org/10.1145/800119.803882.

6. Erik D. Demaine and Quanquan C. Liu. Inapproximability of the standard pebble game and hard to pebble graphs. *CoRR*, 2017.

7. Peter Emde Boas and Jan Leeuwen. *Theoretical Computer Science 4th GI Conference: Aachen, March 26-28, 1979*, chapter Move rules and trade-offs in the pebble game, pages 101–112. Springer Berlin Heidelberg, Berlin, Heidelberg, 1979. Available from: http://dx.doi.org/10.1007/3-540-09118-1_12.

8. John R. Gilbert, Thomas Lengauer, and Robert Endre Tarjan. The pebbling problem is complete in polynomial space. In *Proceedings of the Eleventh Annual ACM Symposium on Theory of Computing*, STOC '79, pages 237–248, New York, NY, USA, 1979. ACM. Available from: http://doi.acm.org/10.1145/800135.804418.

9. John R. Gilbert and Robert E Tarjan. Variations of a pebble game on graphs. Technical report, Stanford, CA, USA, 1978.

10. Philipp Hertel and Toniann Pitassi. The PSPACE-completeness of black-white pebbling. *SIAM J. Comput.*, 39(6):2622–2682, April 2010. Available from: http://dx.doi.org/10.1137/080713513.

11. John Hopcroft, Wolfgang Paul, and Leslie Valiant. On time versus space. *J. ACM*, 24(2):332–337, April 1977. Available from: http://doi.acm.org/10.1145/322003.322015.

12. Hong Jia-Wei and H. T. Kung. I/O complexity: The red-blue pebble game. In *Proceedings of the Thirteenth Annual ACM Symposium on Theory of Computing*, STOC '81, pages 326–333, New York, NY, USA, 1981. ACM. Available from: http://doi.acm.org/10.1145/800076.802486.

13. Thomas Lengauer and Robert Endre Tarjan. Upper and lower bounds on time-space tradeoffs. In *Proceedings of the Eleventh Annual ACM Symposium on Theory of Computing*, STOC '79, pages 262–277, New York, NY, USA, 1979. ACM. Available from: http://doi.acm.org/10.1145/800135.804420.

14. Jakob Nordstrom. New wine into old wineskins: A survey of some pebbling classics with supplemental results. 2015.

15. Wolfgang J. Paul, Robert Endre Tarjan, and James R. Celoni. Space bounds for a game on graphs. In *Proceedings of the Eighth Annual ACM Symposium on Theory of Computing*, STOC '76, pages 149–160, New York, NY, USA, 1976. ACM. Available from: http://doi.acm.org/10.1145/800113.803643.

16. Ravi Sethi. Complete register allocation problems. *SIAM J. Comput.*, 4(3):226–248, 1975. Available from: http://dx.doi.org/10.1137/0204020.

Capacitated Center Problems with Two-Sided Bounds and Outliers

Hu Ding[1], Lunjia Hu[2], Lingxiao Huang[2] and Jian Li[2]

[1] Computer Science and Engineering, Michigan State University, East Lansing, MI, USA *

[2] Institute for Interdisciplinary Information Sciences, Tsinghua University, Beijing, China**

Abstract. In recent years, the capacitated center problems have attracted a lot of research interest. Given a set of vertices V, we want to find a subset of vertices S, called centers, such that the maximum cluster radius is minimized. Moreover, each center in S should satisfy some capacity constraint, which could be an upper or lower bound on the number of vertices it can serve. Capacitated k-center problems with one-sided bounds (upper or lower) have been well studied in previous work, and a constant factor approximation was obtained.

We are the first to study the capacitated center problem with both capacity lower and upper bounds (with or without outliers). We assume each vertex has a uniform lower bound and a non-uniform upper bound. For the case of opening exactly k centers, we note that a generalization of a recent LP approach can achieve constant factor approximation algorithms for our problems. Our main contribution is a simple combinatorial algorithm for the case where there is no cardinality constraint on the number of open centers. Our combinatorial algorithm is simpler and achieves better constant approximation factor compared to the LP approach.

1 Introduction

The k-center clustering is a fundamental problem in theoretical computer science and has numerous applications in a variety of fields. Roughly speaking, given a metric space containing a set of vertices, the k-center problem asks for a subset of k vertices, called centers, such that the maximum radius of the induced k clusters is minimized. Actually k-center clustering falls in the umbrella of the general *facility location* problems which have been extensively studied in the past decades. Many operation and management problems can be modeled as facility location problems, and usually the input vertices and selected centers are also called "clients" and "facilities" respectively. In this paper, we consider a

* HD is supported by the start-up fund from Michigan State University.
** LJH, LXH and JL are supported in part by the National Basic Research Program of China grants 2015CB358700, 2011CBA00300, 2011CBA00301, and the National NSFC grants 61033001, 61632016, 61361136003.

F. Ellen et al. (Eds.): WADS 2017, LNCS 10389, pp. 325–336, 2017.
DOI: 10.1007/978-3-319-62127-2_28

significant generalization of the k-center problem, where each vertex is associated with a capacity interval; that is, the cardinality of the resulting cluster centered at the vertex should satisfy the given lower and upper capacity bounds (the formal definition is shown in Section 1.2). In addition, we also consider the case where a given number of vertices may be excluded as outliers.

Besides being a natural combinatorial problem on its own, the k-center problem with both capacity upper and lower bounds is also strongly motivated by several realistic issues raised in a variety of application contexts.

1. In the context of facility location, each open facility may be constrained by the maximum number of clients it can serve. The capacity lower bounds also come naturally, since an open facility needs to serve at least a certain number of clients in order to generate profit.
2. Several variants of the k-center clustering have been used in the context of preserving privacy in publication of sensitive data (see e.g., [1, 23, 26]). In such applications, it is important to have an appropriate lower bound for the cluster sizes, in order to protect the privacy to certain extent (roughly speaking, it would be relatively easier for an adversary to identify the clients inside a too small cluster).
3. Consider the scenario where the data is distributed over the nodes in a large network. We would like to choose k nodes as central servers, and aggregate the information of the entire network. We need to minimize the delay (i.e., minimize the cluster radius), and at the same time consider the balancedness, for the obvious reason that the machines receiving too much data could be the bottleneck of the system and the ones receiving too little data is not sufficiently energy-efficient [11].

Our problem generalizes the classic k-center problem as well as many important variants studied by previous authors. The optimal approximation results for the classic k-center problem appeared in the 80's: Gonzalez [15] and Hochbaum and Shmoys [17] provided a 2-approximation in a metric graph; moreover, they proved that any approximation ratio $c < 2$ would imply $P = NP$. The first study on capacitated (with only upper bounds) k-center clustering is due to Bar-Ilan et al. [5] who provided a 10-approximation algorithm for uniform capacities (i.e., all the upper bounds are identical). Further, Khuller and Sussmann [20] improved the approximation ratio to be 6 and 5 for hard and soft uniform capacities, respectively. [3] The recent breakthrough for non-uniform (upper) capacities is due to Cygan et al. [10]. They developed the first constant approximation algorithm based on LP rounding, though their approximation ratio is about hundreds. Following this work, An et al. [3] provided an approximation algorithm with the much lower approximation ratio 9. On the imapproximability side, it is impossible to achieve an approximation ratio lower than 3 for non-uniform capacities unless $P = NP$ [10].

---

[3] We can open more than one copies of a facility in the same node in the soft capacity version. But in the hard capacity version, we can only open at most one copy.

For the ordinary k-center with outliers, a 3-approximation algorithm was obtained by Charikar et al. [8]. Kociumaka and Cygan [21] studied k-center with non-uniform upper capacities and outliers, and provided a 25-approximation algorithm.

k-center clustering with lower bounds on cluster sizes was first studied in the context of privacy-preserving data management [26]. Aggarwal et al. [1] provided a 2-approximation and a 4-approximation for the cases without and with outliers, respectively. Further, Ene et al. [13] presented a near linear time $(4 + \varepsilon)$-approximation algorithm in constant dimensional Euclidean space. Note that both [1, 13] are only for uniform lower bounds. Recently, Ahmadian and Swamy [2] provided a 3-approximation and a 5-approximation for the non-uniform lower bound case without and with outliers.

Our main results. To the best of our knowledge, we are the first to study the capacitated center with both capacity lower and upper bounds (with or without outliers). Recently, Ding [12] also studies k-center clustering with two-sided bounds in high dimension or any metric space when k is a constant, and provides a nearly linear time 4-approximation. Given a set V of n vertices, we focus on the case where the capacity of each vertex $u \in V$ has a uniform lower bound $L_u = L$ and a non-uniform upper bound U_u. Sometimes, we consider a generalized supplier version where we are only allowed to open centers among a facility set $\mathcal{F}$, see Definition 1 for details. We mainly provide first constant factor approximation algorithms for the following variants, see Table 1 for other results. Due to the lack of space, we defer many details and proofs to a full version.

1. $(L,U,\text{soft-}\emptyset,p)$-CENTER (Section 2.2): In this problem, both the lower bounds and the upper bounds are uniform, i.e., $L_u = L, U_u = U$ for all $u \in V$. The number of open centers can be arbitrary, i.e., there is no requirement to choose exactly k open centers. Moreover, we allow multiple open centers at a single vertex $u \in V$ (i.e., soft capacity). We may exclude $n - p$ outliers. We provide the first polynomial time combinatorial algorithm which can achieve an approximate factor of 5.

2. $(L,\{U_u\},\emptyset,p)$-CENTER(Section 2.3): In this problem, the lower bounds are uniform, i.e., $L_u = L$ for all $u \in V$, but the upper bound can be nonuniform. The number of open centers can be arbitrary. We may exclude $n - p$ outliers. We provide the first polynomial time combinatorial 11-approximation for this problem.

3. $(L,\{U_u\},k)$-CENTER (Section 3.3): In this problem, we would like to open exactly k centers, such that the maximum cluster radius is minimized. All vertices have the same capacity lower bounds, i.e., $L_u = L$ for all $u \in V$. But the capacity upper bounds may be nonuniform, i.e., each vertex u has an individual capacity upper bound U_u. Moreover, we do not exclude any outlier. We provide the first polynomial time 9-approximation algorithm for this problem, based on LP rounding.

4. $(L,\{U_u\},k,p)$-CENTER (Section 3.3): This problem is the outlier version of the $(L,\{U_u\},k)$-CENTER problem. The problem setting is exactly the same

except that we can exclude $n-p$ vertices as outliers. We provide a polynomial time 25-approximation algorithm for this problem.

Problem Setting		Approximation Ratio	
		Center Version	Supplier Version
Without k Constraint	$(L,U,\text{soft-}\emptyset,p)$	5	5
	$(L,U,\emptyset,p)$	10	23
	$(L,\{U_u\},\text{soft-}\emptyset,p)$	11	11
	$(L,\{U_u\},\emptyset,p)$	11	25
With k Constraint	(L,U,k)	6	9
	$(L,\{U_u\},k)$	9	13
	$(L,U,\text{soft-}k,p)$	13	13
	(L,U,k,p)	23	23
	$(L,\{U_u\},\text{soft-}k,p)$	25	25
	$(L,\{U_u\},k,p)$	25	25

Table 1. A summarization table for our results in this paper.

Our main techniques. In Section 2, we consider the first two variants which allow to open arbitrarily many centers. We design simple and faster combinatorial algorithms which can achieve better constant approximation ratios compared to the LP approach. For the simpler case $(L,U,\text{soft-}\emptyset,p)$-CENTER, we construct a data structure for all possible open centers. We call it a *core-center tree* (CCT). Our greedy algorithm mainly contains two procedures. The first procedure *pass-up* greedily assigns vertices to open centers from the leaves of CCT to the root. After this procedure, there may exist some unassigned vertices around the root. We then introduce the second procedure called *pass-down*, which assigns these vertices in order by finding an *exchange route* each time. For the more general case $(L,\{U_u\},\emptyset,p)$-CENTER, our greedy algorithm is similar but somewhat more subtle. We still construct a CCT and run the pass-up procedure. Then we obtain an open center set F, which may contain redundant centers. However, since we deal with hard capacities and outliers, we need to find a non-redundant open center set which is not 'too far' from F (see Section 2.3 for details) and have enough total capacities. Then by a pass-down procedure, we can assign enough vertices to their nearby open centers.

In Section 3 and 3.3, we consider the last two variants which require to open exactly k centers. We generalized the LP approach developed for k-center with only capacity upper bounds [3, 21] and obtain constant approximation schemes for two-sided capacitated bounds. The omitted proofs can be found in the full version in this paper.

1.1 Other Related Work

The classic k-center problem is quite fundamental and has been generalized in many ways, to incorporate various constraints motivated by different application scenarios. Recently, Fernandes et al. [14] also provided constant approximations for the fault-tolerant capacitated k-center clustering. Chen et al. [9] studied the matroid center problem where the selected centers must form an independent set of a given matroid, and provided constant factor approximation algorithms (with or without outliers).

There is a large body of work on approximation algorithms for the facility location and k-median problems (see e.g., [4, 6, 7, 16, 18, 19, 22, 24, 25]). Moreover, Dick et al. [11] studied multiple balanced clustering problems with uniform capacity intervals, that is, all the lower (upper) bounds are identical; they also consider the problems under the stability assumption.

1.2 Preliminaries

In this paper, we usually work with the following more general problem, called the capacitated k-supplier problem. It is easy to see it generalizes the capacitated k-center problem since we can not open centers at any vertex. The formal definition is as follows.

Definition 1. *(Capacitated k-supplier with two-sided bounds and outliers) Suppose that we have*

1. *Two integers $k, p \in \mathbb{Z}_{\geq 0}$;*
2. *A finite set C of clients, and a finite set $\mathcal{F}$ of facilities;*
3. *A symmetric distance function $d : (C \cup \mathcal{F}) \times (C \cup \mathcal{F}) \to \mathbb{R}_{\geq 0}$ satisfying the triangle inequality;*
4. *A capacity interval $[L_u, U_u]$ for each facility $u \in \mathcal{F}$, where $L_u, U_u \in \mathbb{Z}_{\geq 0}$ and $L_u \leq U_u$.*

Our goal is to find a client set $C' \subseteq C$ of size at least p, an open facility set $F \subseteq \mathcal{F}$ of size exactly k, and a function $\phi : C' \to F$ satisfying that $L_u \leq |\phi^{-1}(u)| \leq U_u$ for each $u \in F$, which minimize the maximum cluster radius $\max_{v \in C'} d(v, \phi(v))$. If the maximum cluster radius is at most r, we call the tuple (C', F, ϕ) a distance-r solution.

By the similar approach of Cygan et al. [21], we can reduce the $(\{L_u\}, \{U_u\}, k, p)$-SUPPLIER problem to a simpler case. We first introduce some definitions.

Definition 2. *(Induced distance function) We say the distance function $d_G : (C \cup \mathcal{F}) \times (C \cup \mathcal{F}) \to \mathbb{R}_{\geq 0}$ is induced by an undirected unweighted connected graph $G = (C \cup \mathcal{F}, E)$ if*

1. *$\forall (u, v) \in E$, we have $u \in \mathcal{F}$ and $v \in C$.*
2. *$\forall a_1, a_2 \in C \cup \mathcal{F}$, the distance $d_G(a_1, a_2)$ between a_1 and a_2 equals to the length of the shortest path from a_1 to a_2.*

Definition 3. *(Induced $(\{L_u\}, \{U_u\}, k, p)$-SUPPLIER instance) An $(\{L_u\}, \{U_u\}, k, p)$-SUPPLIER instance is called an induced $(\{L_u\}, \{U_u\}, k, p)$-SUPPLIER instance if the following properties are satisfied:*

1. *The distance function d_G is induced by an undirected connected graph $G = (C \cup \mathcal{F}, E)$.*
2. *The optimal capacitated k-supplier value is at most 1.*

Moreover, we say this instance is induced by G.

When the graph of interest G is clear from the context, we will use d instead of d_G for convenience. We then show a reduction from solving the generalized $(\{L_u\},\{U_u\},k,p)$-SUPPLIER problem to solving induced $(\{L_u\},\{U_u\},k,p)$-SUPPLIER instances by Lemma 1.

Lemma 1. *Suppose we have a polynomial time algorithm A that takes as input any induced $(\{L_u\},\{U_u\},k,p)$-SUPPLIER instance, and outputs a distance-ρ solution. Then, there exists a ρ-approximation algorithm for the $(\{L_u\},\{U_u\},k,p)$-SUPPLIER problem with polynomial running time.*

2 Capacitated Center with Two-Sided Bounds and Outliers

In this section, we consider the version that the number of open centers can be arbitrary. By the LP approach in Section 3.3 and enumerating the number of open centers, we can achieve approximation algorithms for different variants in this case. However, the approximation factor is not small enough. In this section, we introduce a new greedy approach in order to achieve better approximation factors. Since our algorithm is combinatorial, it is easier to be implemented and saves the running time compared to the LP approach.

2.1 Core-center tree (CCT)

Consider the $(L,\{U_u\},\emptyset,p)$-SUPPLIER problem. By Lemma 1, we only need to consider induced $(L,\{U_u\},\emptyset,p)$-SUPPLIER instances induced by an undirected unweighted connected graph $G = (\mathcal{C} \cup \mathcal{F}, E)$. We first propose a new data structure called *core-center tree (CCT)* as follows.

Definition 4. *(Core-center tree (CCT)) Given an induced $(L,\{U_u\},\emptyset,p)$-SUPPLIER instance induced by an undirected unweighted connected graph $G = (\mathcal{C} \cup \mathcal{F}, E)$, we call a tree $T = (\mathcal{F}, E_T)$ a core-center tree(CCT) if the following properties hold.*

1. *For each edge $(u, u') \in E_T$, we have $d_G(u, u') \leq 2$;*
2. *Suppose the root of T is at layer 0. Denote I to be the set of vertices in the even layers of T. We call I the core-center set of T. For any two distinct vertices $u, u' \in I$, we have $d_G(u, u') \geq 3$.*

Lemma 2. *Given an induced $(L,\{U_u\},\emptyset,p)$-SUPPLIER instance induced by an undirected unweighted connected graph $G = (\mathcal{C} \cup \mathcal{F}, E)$, we can construct a CCT in polynomial time.*

For any $u \in \mathcal{F}$, denote $N_G[u] = \{v \in \mathcal{C} : (u, v) \in E\}$ to be the collection of all neighbors of $u \in \mathcal{F}$. [4] W.l.o.g., we assume that $U_u \leq |N_G(u)|$ for every facility $u \in \mathcal{F}$ in this section. In fact, we can directly delete all $u \in \mathcal{F}$ satisfying that $|N_G[u]| < L$ from the facility set $\mathcal{F}$, since u can not be open in any optimal feasible solution. [5] Otherwise if $L \leq |N_G[u]| < U_u$, we set

---

[4] If $u \in \mathcal{C}$ is also a client, then $u \in N_G[u]$.

[5] If this deletion causes the induced graph unconnected, similar to Lemma 6 in [21], we divide the graph into different connected components, and consider each smaller induced instance based on different connected components.

$U_u \leftarrow \min\{U_u, |N_G[u]|\}$, which has no influence on any optimal feasible solution of the induced $(L, \{U_u\}, \emptyset, p)$-SUPPLIER instance. The following lemma gives a useful property of CCT.

Lemma 3. *Given an induced $(L, \{U_u\}, \emptyset, p)$-SUPPLIER instance induced by an undirected unweighted connected graph $G = (\mathcal{C} \cup \mathcal{F}, E)$, and a core-center tree $T = (\mathcal{F}, E_T)$, suppose I is the core-center set of T. Then, we can construct a function $\xi : \mathcal{C} \to \mathcal{F}$ satisfying the following properties in polynomial time.*

1. *For all $v \in \mathcal{C}$, we have $(\xi(v), v) \in E$;*
2. *For all $u \in I$, we have $|\xi^{-1}(u)| \geq L$.*

2.2 A Simple Case: $(L,U,\text{soft-}\emptyset,p)$-Supplier

We first consider a simple case where the capacity bounds (upper and lower) are uniform and soft. In this setting, we want to find an open facility set $F = \{u_i \mid u_i \in \mathcal{F}\}_i$. Note that we allow multiple open centers in F. We also need to find an assignment function $\phi : \mathcal{C} \to F$, representing that we assign every client $v \in \mathcal{C}$ to facility $\phi(v)$. The main theorem is as follows.

Theorem 1. *(main theorem) There exists a 5-approximation polynomial time algorithm for the $(L,U,\text{soft-}\emptyset,p)$-SUPPLIER problem.*

By Lemma 1, we only consider induced $(L,U,\text{soft-}\emptyset,p)$-SUPPLIER instances. Given an induced $(L,U,\text{soft-}\emptyset,p)$-SUPPLIER instance induced by an undirected unweighted connected graph $G = (\mathcal{C} \cup \mathcal{F}, E)$, recall that we can assume $|N_G[u]| \geq U_u \geq L$ for each $u \in \mathcal{F}$. We first construct a CCT $T - (\mathcal{F}, E_T)$ rooted at node u^*, and a function $\xi : \mathcal{C} \to \mathcal{F}$ satisfying Lemma 3. For a facility set $P \subseteq \mathcal{F}$, we denote $\xi^{-1}(P) = \bigcup_{u \in P} \xi^{-1}(u)$ to be the collection of clients assigned to some facility in P by ξ.

Our algorithm mainly includes two procedures. The first procedure is called *pass-up*, which is a greedy algorithm to map clients to facilities from the leaves of T to the root. After the 'pass-up' procedure, we still leave some unassigned clients nearby the root. Then we use a procedure called *pass-down* to allocate those unassigned clients by iteratively finding an *exchange route*. In the following, we give the details of both procedures.

Procedure Pass-Up. Assume that $|\mathcal{C}| = aL + b$ for some $a \in \mathbb{N}$ and $0 \leq b \leq L - 1$. In this procedure, we will find an open facility set F of size a. We also find an assignment function ϕ which assigns aL clients to some nearby facility in F except a client set $S \subseteq \mathcal{C}$. Here, S is a collection of b clients in $\xi^{-1}(u^*)$ nearby the root u^*. Our main idea is to open facility centers from the leaves of CCT T to the root iteratively. During opening centers, we assign exactly L 'close' clients to each center. Thus, there are b unassigned clients after the whole procedure.

We then describe an iteration of pass-up. Assume that I is the core-center set of T. At the beginning, we find a non-leaf vertex $u \in I$ satisfying that all of its grandchildren (if exists) are leaves. [6] We denote $P \subseteq \mathcal{F}$ to be the collection of all

---

[6] If multiple non-leaf nodes satisfy this property, we choose an arbitrary one.

children and all grandchildren of u. In the next step, we consider all unscanned clients in $\xi^{-1}(P)$, [7] and assign them to the facility u by ϕ. Note that we may open multiple centers at u. We want that each center at u serves exactly L centers. However, there may exist one center at u serving less than L unscanned clients in $\xi^{-1}(P)$. We assign some clients in $\xi^{-1}(u)$ to this center such that it also serves exactly L clients. After this iteration, we delete the subtree rooted at u from T except u itself.

Finally, the root u^* will become the only remaining node in T. We open multiple centers at u^*, each serving exactly L clients in $\xi^{-1}(u^*)$, until there are less than L unassigned clients. We denote S to be the collection of those unassigned clients. At the end of pass-up, we output an open facility set F, an unassigned client set S and an assigned function $\phi : C \setminus S \to F$. We have the following lemma by the algorithm.

Lemma 4. *Given an induced $(L,U,soft\text{-}\emptyset,p)$-SUPPLIER instance induced by an undirected unweighted connected graph $G = (C \cup F, E)$, assume that $|C| = aL + b$ for some $a \in \mathbb{N}$ and $0 \leq b \leq L - 1$. The output of pass-up satisfies the following properties:*

1. *Each open facility $u_j \in F$ satisfies that $u_j \in I$, and $|F| = a$;*
2. *The unassigned client set $S \subseteq \xi^{-1}(u^*)$, and $|S| = b$;*
3. *For each facility $u_i \in F$, we have $|\phi^{-1}(u_i)| = L$.*
4. *For each client $v \in C \setminus S$, $\phi(v)$ is either $\xi(v)$, or the parent of $\xi(v)$ in T, or the grandparent of $\xi(v)$ in T. Moreover, we have $d_G(v, \phi(v)) \leq 5$.*

Procedure Pass-Down. After the procedure *pass-up*, we still leave an unassigned client set S of size b. However, our goal is to serve at least p clients. Therefore, we need to modify the assignment function ϕ and serve more clients.

The procedure pass-down handles the remaining b clients in S one by one. At the beginning of pass-down, we initialize an 'unscanned' client set $B \leftarrow C \setminus S$, i.e., B is the collection of those clients allowing to be reassigned by pass-down. In each iteration, we arbitrarily pick a client $v \in S$ and assign it to the root node u^*. However, if each open facility at u^* has already served U_{u^*} clients by ϕ, assigning v to u^* will violate the capacity upper bound. In this case, we actually find an open center $u_j \in F$ such that $|\phi^{-1}(u_j)| < U_j$, i.e., there are less than U_j clients assigned to u_j by ϕ. We then construct an *exchange route* consisting of open facilities in F. We first find a sequence of nodes $w_0 = u^*, w_1, \cdots, w_m = u_j$ in T satisfying that w_i is the grandparent of w_{i+1} in the core-center tree T for all $0 \leq i \leq m - 1$. Then for each node w_i $(1 \leq i \leq m - 1)$, we pick a client $v_i \in \xi^{-1}(w_i)$ which has not been reassigned so far. We call such a sequence of clients $v, v_1, \ldots, v_{m-1}$ an exchange route. Our algorithm is as follows: 1) we assign v to $\phi(v_1)$; 2) we iteratively reassign v_i to $\phi(v_{i+1})$ in order $(1 \leq i \leq m-2)$; 3) finally we reassign v_{m-1} to u_j. We then mark all clients v_i $(1 \leq i \leq m - 1)$ in the exchange route by removing them from the 'unscanned' client set B, and

---

[7] Here, unscanned clients are those clients that have not been assigned by ϕ before this iteration.

remove the client $\{v\}$ from the unassigned client set S. Note that our exchange route only increases the number of clients assigned to u_j by one. In fact, such an exchange route always exists in each iteration. Thus in each iteration, the procedure pass-down assigns one more client $v \in S$ to some open facility in F. At the end of pass-down, we output a client set $C \leftarrow C \setminus S$ of size at least p, an open facility set F and an assigned function $\phi : C \to F$.

Now we prove the following lemma. Note that Theorem 1 can be directly obtained by Lemma 1 and Lemma 5.

Lemma 5. *The procedure pass-down outputs a distance-5 solution (C, F, ϕ) of the given induced $(L, U, soft-\emptyset, p)$-SUPPLIER instance induced by $G = (C \cup F, E)$ in polynomial time.*

2.3 $(L, \{U_u\}, \emptyset, p)$-Center

In this subsection, we consider a more complicated case where the capacity upper bounds are non-uniform, and each vertex has a hard capacity.

Theorem 2. *(main theorem) There exists an 11-approximation polynomial time algorithm for the $(L, \{U_u\}, \emptyset, p)$-CENTER problem.*

By Lemma 1, we only need to consider induced $(L, \{U_u\}, \emptyset, p)$-SUPPLIER instances. For an induced $(L, \{U_u\}, \emptyset, p)$-SUPPLIER instance induced by an undirected unweighted connected graph $G = (V = C \cup F, E)$, recall that we can assume $U_u \leq |N_G(u)|$ for every vertex $u \subset \mathcal{F}$. [8] Since we consider the center version, every vertex $v \in C$ has an individual capacity interval $[L, U_v]$ and can be opened as a center as well.

Similar to $(L, U, soft-\emptyset, p)$-CENTER, our algorithm first computes a core-center tree $T = (\mathcal{F}, E)$ rooted at u^*, a core-center set I and a function ξ described as in Lemma 3. Assume that $|C| = aL + b$ for some $a \in \mathbb{N}$ and $0 \leq b \leq L - 1$. We still use the procedure *pass-up* to compute an open set $F = \{u_1, u_2, \cdots, u_a\}$, an unassigned set $S \subseteq \xi^{-1}(u^*)$ of size $b < L$, and a function $\phi : (C \setminus S) \to F$.

However, we can not apply *pass-down* directly since we consider non-uniform hard capacity upper bounds. Thus, we need the following lemma to modify the open center set F. We prove this lemma by Hall's theorem in the full version.

Lemma 6. *Given an induced $(L, \{U_u\}, \emptyset, p)$-CENTER instance induced by $G = (V = C \cup F, E)$ where $|N_G(u)| \geq U_u$ for each $u \in F$ and an open set $F = \{u_1, u_2, \cdots, u_a\}$ computed by pass-up, there exists a polynomial time algorithm that finds another open set $F' = \{u'_1, u'_2, \ldots, u'_a\}$ such that:*

1. *F' is a single set.*
2. *For all $1 \leq i \leq a$, we have $d_G(u_i, u'_i) \leq 6$.*
3. *$\sum_{i=1}^{a} U_{u'_i} \geq p$.*

---

[8] Recall that we may remove some facilities from $\mathcal{F}$ such that this assumption is satisfied. Thus, the set $\mathcal{F}$ may be a subset of V.

Proof of Theorem 2. By Lemma 6, we obtain another open set $F' = \{u'_1, u'_2, \ldots, u'_a\}$. We first modify U_{u_i} to be $U_{u'_i}$ for all $1 \leq i \leq a$. Then we apply the procedure *pass-down* according to the modified capacities. By Lemma 5, we obtain a distance-5 solution (C, F, ϕ). Since $\sum_{i=1}^{a} U_{u'_i} \geq p$, at least p vertices are served by ϕ. Finally, for each vertex $v \in C$ and $u_i \in F$ such that $\phi(v) = u_i$, we reassign v to $u'_i \in F'$, i.e., let $\phi(v) = u'_i$. By Lemma 6, we obtain a feasible solution for the given induced $(L, \{U_u\}, \emptyset, p)$-CENTER instance. Since $d(u_i, u'_i) \leq 6$ $(1 \leq i \leq a)$, the capacitated center value of our solution is at most $5 + 6 = 11$. Combining with Lemma 1, we finish the proof.

3   Capacitated k-Center with Two-Sided Bounds and Outliers

Now we study the capacitated k-center problems with two-sided bounds. We consider that all vertices have a uniform capacity lower bound $L_v = L$, while the capacity upper bounds can be either uniform or non-uniform. Similar to [3, 21], we use the LP relaxation and the rounding procedure distance-r transfer.

3.1   LP Formulation

We first give a natural LP relaxation for $(\{L_u\}, \{U_u\}, k, p)$-SUPPLIER.

Definition 5. *(LP$_r(G)$) Given an $(\{L_u\}, \{U_u\}, k, p)$-SUPPLIER instance, the following feasibility LP$_r(G)$ that fractionally verifies whether there exists a solution that assigns at least p clients to an open center of distance at most r:*

$$
\begin{array}{ll}
0 \leq x_{uv}, y_u \leq 1, & \forall u \in \mathcal{F}, v \in \mathcal{C}; \\
x_{uv} = 0, & \text{if } d(u, v) > r; \\
x_{uv} \leq y_u, & \forall u \in \mathcal{F}, v \in \mathcal{C}; \\
\sum_{u \in \mathcal{F}} y_u = k; & \\
\sum_{u \in \mathcal{F}, v \in \mathcal{C}} x_{uv} \geq p; & \\
\sum_{u \in \mathcal{F}} x_{uv} \leq 1, & \forall v \in \mathcal{C}; \\
L_u y_u \leq \sum_{v \in \mathcal{C}} x_{uv} \leq U_u y_u, & \forall u \in \mathcal{F}.
\end{array}
$$

Here we call x_{uv} an assignment variable representing the fractional amount of assignment from client v to center u, and y_u the opening variable of $u \in \mathcal{F}$. For convenience, we use x, y to represent $\{x_{uv}\}_{u \in \mathcal{F}, v \in \mathcal{C}}$ and $\{y_u\}_{u \in \mathcal{F}}$, respectively.

By Definition 3, LP$_1(G)$ must have a feasible solution for any induced $(\{L_u\}, \{U_u\}, k, p)$-SUPPLIER instance. We recall a rounding procedure called distance-r transfer.

3.2   Distance-r Transfer

We first extend the definition of distance-r transfer proposed in [3, 21] by adding the third condition. For a vertex $a \in \mathcal{C} \cup \mathcal{F}$ and a set $B \subseteq \mathcal{C} \cup \mathcal{F}$, we define $d(a, B) = \min_{b \in B} d(a, b)$.

Definition 6. *Given an $(\{L_u\}, \{U_u\}, k, p)$-SUPPLIER instance and $y \in \mathbb{R}_{\geq 0}^{\mathcal{F}}$, a vector $y' \in \mathbb{R}_{\geq 0}^{\mathcal{F}}$ is a distance-r transfer of y if*

1. $\sum_{u \in \mathcal{F}} y'_u = \sum_{u \in \mathcal{F}} y_u$;

2. $\sum_{w \in \mathcal{F}: d(w,W) \leq r} U_w y'_w \geq \sum_{u \in W} U_u y_u$ for all $W \subseteq \mathcal{F}$;
3. $\sum_{w \in \mathcal{F}: d(w,W) \leq r} L_w y_w \geq \sum_{u \in W} L_u y'_u$ for all $W \subseteq \mathcal{F}$.

If y' is a characteristic vector of $F \subseteq \mathcal{F}$, we say that F is an integral distance-r transfer of y.

In this paper, we add the third condition to satisfy the capacity lower bounds. Like in [3, 21], we still have the following lemma.

Lemma 7. *Given an $(\{L_u\}, \{U_u\}, k, p)$-SUPPLIER problem, assume (x, y) is a feasible solution of $\mathsf{LP}_1(G)$ and $F \subseteq \mathcal{F}$ is an integral distance-r transfer of y. Then one can find a distance-$(r + 1)$ solution (C, F, ϕ) in polynomial time.*

3.3 Capacitated k-Center with Two-Sided Bounds and Outliers

Now we are ready to solve the $(L, \{U_u\}, k, p)$-SUPPLIER problem. By Lemma 7, we only need to find an integral distance-r transfer satisfying Definition 6 given a feasible fractional solution (x, y) of $\mathsf{LP}_1(G)$. Fortunately, the rounding schemes in [3, 21] have this property. Thus, we have the following theorem by [3].

Theorem 3. *There is a polynomial time 9-approximation algorithm for the $(L, \{U_u\}, k)$-CENTER problem. For the uniform capacity upper bound version, the (L, U, k)-CENTER problem admits a 6-approximation.*

Theorem 4. *There is a polynomial time 13-approximation algorithm for the $(L, \{U_u\}, k)$-SUPPLIER problem. For the uniform capacity upper bound version, the (L, U, k)-SUPPLIER problem admits a 9-approximation.*

By [21], we have the following theorem.

Theorem 5. *There is a polynomial time 25-approximation algorithm for the $(L, \{U_u\}, k, p)$-SUPPLIER problem and the $(L, \{U_u\}, soft-k, p)$-SUPPLIER problem. For the uniform capacity upper bound version, the (L, U, k, p)-SUPPLIER problem admits a 23-approximation, and the $(L, U, soft-k, p)$-SUPPLIER problem admits a 13-approximation.*

References

1. Aggarwal, G., Panigrahy, R., Feder, T., Thomas, D., Kenthapadi, K., Khuller, S., Zhu, A.: Achieving anonymity via clustering. ACM Trans. Algorithms 6(3), 49:1–49:19 (2010)
2. Ahmadian, S., Swamy, C.: Approximation algorithms for clustering problems with lower bounds and outliers. arXiv preprint arXiv:1608.01700 (2016)
3. An, H.C., Bhaskara, A., Chekuri, C., Gupta, S., Madan, V., Svensson, O.: Centrality of trees for capacitated k-center. Mathematical Programming 154(1-2), 29–53 (2015)
4. Arya, V., Garg, N., Khandekar, R., Meyerson, A., Munagala, K., Pandit, V.: Local search heuristics for k-median and facility location problems. SIAM Journal on computing 33(3), 544–562 (2004)

5. Barilan, J., Kortsarz, G., Peleg, D.: How to allocate network centers. Journal of Algorithms 15(3), 385–415 (1993)
6. Charikar, M., Guha, S.: Improved combinatorial algorithms for facility location problems. SIAM Journal on Computing 34(4), 803–824 (2005)
7. Charikar, M., Guha, S., Tardos, É., Shmoys, D.B.: A constant-factor approximation algorithm for the k-median problem. In: Proceedings of the thirty-first annual ACM symposium on Theory of computing. pp. 1–10. ACM (1999)
8. Charikar, M., Khuller, S., Mount, D.M., Narasimhan, G.: Algorithms for facility location problems with outliers. In: Proceedings of the twelfth annual ACM-SIAM symposium on Discrete algorithms. pp. 642–651. Society for Industrial and Applied Mathematics (2001)
9. Chen, D.Z., Li, J., Liang, H., Wang, H.: Matroid and knapsack center problems. Algorithmica 75(1), 27–52 (2016)
10. Cygan, M., Hajiaghayi, M., Khuller, S.: Lp rounding for k-centers with non-uniform hard capacities. In: Proceedings of the 2012 IEEE 53rd Annual Symposium on Foundations of Computer Science. pp. 273–282. IEEE Computer Society (2012)
11. Dick, T., Li, M., Pillutla, V.K., White, C., Balcan, M.F., Smola, A.: Data driven resource allocation for distributed learning. arXiv preprint arXiv:1512.04848 (2015)
12. Ding, H.: Balanced k-center clustering when k is a constant. arXiv preprint arXiv:1704.02515 (2017)
13. Ene, A., Har-Peled, S., Raichel, B.: Fast clustering with lower bounds: No customer too far, no shop too small. arXiv preprint arXiv:1304.7318 (2013)
14. Fernandes, C.G., de Paula, S.P., Pedrosa, L.L.: Improved approximation algorithms for capacitated fault-tolerant k-center. In: Latin American Symposium on Theoretical Informatics. pp. 441–453. Springer (2016)
15. Gonzalez, T.F.: Clustering to minimize the maximum intercluster distance. Theoretical Computer Science 38, 293–306 (1985)
16. Guha, S., Khuller, S.: Greedy strikes back: Improved facility location algorithms. Journal of algorithms 31(1), 228–248 (1999)
17. Hochbaum, D.S., Shmoys, D.B.: A best possible heuristic for the k-center problem. Mathematics of operations research 10(2), 180–184 (1985)
18. Jain, K., Mahdian, M., Saberi, A.: A new greedy approach for facility location problems. In: Proceedings of the thiry-fourth annual ACM symposium on Theory of computing. pp. 731–740. ACM (2002)
19. Jain, K., Vazirani, V.V.: Approximation algorithms for metric facility location and k-median problems using the primal-dual schema and lagrangian relaxation. Journal of the ACM (JACM) 48(2), 274–296 (2001)
20. Khuller, S., Sussmann, Y.J.: The capacitated k-center problem. SIAM Journal on Discrete Mathematics 13(3), 403–418 (2000)
21. Kociumaka, T., Cygan, M.: Constant factor approximation for capacitated k-center with outliers. arXiv preprint arXiv:1401.2874 (2014)
22. Korupolu, M.R., Plaxton, C.G., Rajaraman, R.: Analysis of a local search heuristic for facility location problems. Journal of algorithms 37(1), 146–188 (2000)
23. Li, J., Yi, K., Zhang, Q.: Clustering with diversity. In: International Colloquium on Automata, Languages, and Programming. pp. 188–200. Springer (2010)
24. Li, S.: A 1.488 approximation algorithm for the uncapacitated facility location problem. Information and Computation 222, 45–58 (2013)
25. Li, S., Svensson, O.: Approximating k-median via pseudo-approximation. SIAM Journal on Computing 45(2), 530–547 (2016)
26. Sweeney, L.: k-anonymity: A model for protecting privacy. International Journal of Uncertainty, Fuzziness and Knowledge-Based Systems 10(05), 557–570 (2002)

Faster Randomized Worst-Case Update Time for Dynamic Subgraph Connectivity

Ran Duan[1,2] and Le Zhang[1,3]

[1] Institute for Interdisciplinary Information Sciences, Tsinghua University, China
[2] duanran@mail.tsinghua.edu.cn
[3] le-zhang12@mails.tsinghua.edu.cn

Abstract. Real-world networks are prone to breakdowns. Typically in the underlying graph G, besides the insertion or deletion of edges, the set of active vertices changes overtime. A vertex might work actively, or it might fail, and gets isolated temporarily. The active vertices are grouped as a set S. The set S is subjected to updates, i.e., a failed vertex restarts, or an active vertex fails, and gets deleted from S. Dynamic subgraph connectivity answers the queries on connectivity between any two active vertices in the subgraph of G induced by S. The problem is solved by a dynamic data structure, which supports the updates and answers the connectivity queries. In the general undirected graph, we propose a randomized data structure, which has $\widetilde{O}(m^{3/4})$ worst-case update time. The former best results for it include $\widetilde{O}(m^{2/3})$ deterministic *amortized* update time by Chan, Pătraşcu and Roditty [4], $\widetilde{O}(m^{4/5})$ by Duan [8] and $\widetilde{O}(\sqrt{mn})$ by Baswana, Chaudhury, Choudhary and Khan [2] deterministic *worst-case* update time.

1 Introduction

Dynamic subgraph connectivity is defined as follows: Given an undirected graph $G = (V, E)$ having m edges, n vertices with $m = \Omega(n)$, there is a subset $S \subseteq V$. The set E is subjected to edge updates of the forms $insert(e, E)$ or $delete(e, E)$, where e is an edge. There are vertex updates of the forms $insert(v, S)$ or $remove(v, S)$. Through vertex updates, S changes overtime. The query is on whether any two vertices s and t are connected in the subgraph of G induced by S.

The problem was first proposed by Frigioni and Italiano [12], and poly-logarithmic algorithms on connectivity were described for the special case of planar graphs. As to the general graphs, Chan [3] first described an algorithm of deterministic amortized update time $\widetilde{O}(m^{4\omega/(3\omega+3)})^4$, where ω is the matrix multiplication exponent. Adopting FMM (*Fast Matrix Multiplication*) algorithm of [6], the update time is $O(m^{0.94})$. Its query time and space complexity are $\widetilde{O}(m^{1/3})$ and linear, respectively. Later Chan, Pătraşcu, and Roditty [4] proposed a simpler algorithm with the improved update time of $\widetilde{O}(m^{2/3})$. The

---

[4] $\widetilde{O}(\cdot)$ hides poly-logarithmic factors.

© Springer International Publishing AG 2017
F. Ellen et al. (Eds.): WADS 2017, LNCS 10389, pp. 337–348, 2017.
DOI: 10.1007/978-3-319-62127-2_29

space complexity of the new algorithm increases to $\widetilde{O}(m^{4/3})$. The new algorithm is of compact description, getting rid of the use of FMM. With the same update time, Duan [8] presented new data structures occupying linear space. Also a worst-case deterministic $\widetilde{O}(m^{4/5})$ algorithm was proposed by Duan [8]. Via an application of dynamic DFS tree [2], Baswana et al. discussed a new algorithm with $\widetilde{O}(\sqrt{mn})$ deterministic worst-case update time. Its query time is $O(1)$. An improvement of it is discussed in [5]. These results are summarized in Table 1.

A close related problem is dynamic graph connectivity, which cares only about the edge updates. Poly-logarithmic amortized update time was first achieved by Henzinger and King [15]. The algorithm is randomized Las Vegas. Inspired by it, Holm et al. [16] proposed a deterministic algorithm with $O(\lg^2 n)^5$ amortized update time, which is now one of the classic results in the field. A cell-probe lower bound of $\Omega(\lg n)$ was proved by Pătrașcu and Demaine [21]. The lower bound is amortized randomized. Near-optimal results were considered by Thorup [22], where a randomized Las Vegas algorithm was described with $O(\lg n(\lg \lg n)^3)$ amortized update time. The upper bound is recently improved to $O(\lg n(\lg \lg n)^2)$ by Huang et al. [17]. Besides the classic deterministic $O(\lg^2 n)$ result, a faster deterministic algorithm was proposed by Wulff-Nilsen [24], of which the update time is $O(\lg^2 n/\lg \lg n)$. Turning to the worst-case dynamic connectivity, a deterministic $O(\sqrt{n})$ update-time algorithm is Frederickson's $O(\sqrt{m})$ worst-case algorithm [11] sped up via *sparsification* technique proposed by Eppstein et al. [10]. The result holds for online updating of minimum spanning trees. With roughly the same structure, but different and simpler techniques, Kejlberg-Rasmussen et al. [19] provided the so far best deterministic worst-case bound of $O(\sqrt{n(\lg \lg n)^2/\lg n})$ for dynamic connectivity. After the discovery of $O(\sqrt{n})$ update-time algorithm, people were wondering whether any poly-logarithmic worst-case update time algorithm is possible, even randomized. The open problem stands firmly for many years. A breakthrough should be attributed to Kapron et al. [18]. Their algorithm is Monte-Carlo, with poly-logarithmic worst-case update time. It has several improvements until now, as done in [13, 23]. For subgraph connectivity, the trivial update time of $\widetilde{O}(n)$ follows from Kapron et al.'s algorithm. The query time of it for subgraph connectivity can also be improved to $O(1)$, as the explicit maintenance of connected components can be done without blowing up the $\widetilde{O}(n)$ update time. Very recently, Wulff-Nilsen [25] gave a Las Vegas data structure maintaining a minimum spanning forest in expected worst-case time polynomially faster than $\Theta(n^{1/2})$ w.h.p. per edge update. An independent work of Nanongkai and Saranurak [20] showed an algorithm with $O(n^{0.49306})$ worst-case update time w.h.p..

1.1   Our Results

The former $\widetilde{O}(m^{4/5})$ deterministic worst-case *subgraph* connectivity structure adopted as a sub-routine the $O(\sqrt{n})$ deterministic worst-case algorithm for dynamic *graph* connectivity. Now the randomized poly-logarithmic worst-case con-

---

[5] We use $\lg x$ to denote $\log_2 x$.

Table 1. Results on Dynamic Subgraph Connectivity

Update time	Query time	Notes
$\widetilde{O}(m^{4\omega/(3\omega+3)})$	$\widetilde{O}(m^{1/3})$	Amortized, deterministic, linear space [3]
$\widetilde{O}(\sqrt{mn})$	$O(1)$	Worst case, deterministic, space $\widetilde{O}(m)$ [2, 5]
$\widetilde{O}(m^{2/3})$	$\widetilde{O}(m^{1/3})$	Amortized, deterministic, space $\widetilde{O}(m^{4/3})$ [4]
$\widetilde{O}(m^{2/3})$	$\widetilde{O}(m^{1/3})$	Amortized, deterministic, linear space [8]
$\widetilde{O}(m^{4/5})$	$\widetilde{O}(m^{1/5})$	Worst case, deterministic, space $\widetilde{O}(m)$ [8]
$\widetilde{O}(m^{3/4})$	$\widetilde{O}(m^{1/4})$	Worst case, randomized, linear space, this paper

nectivity structures for dynamic *graph* connectivity are discovered. We consider the question of whether it brings progress in *subgraph* connectivity. The answer is affirmative. But it does not come by simple replacement. More precisely, we tried in vain to get an improvement by carefully tuning the former setting of the $\widetilde{O}(m^{4/5})$ algorithm. Intuitively, the amortized $\widetilde{O}(m^{2/3})$ update time was achieved partially because it uses the connectivity structure of poly-logarithmic amortized update time. Now poly-logarithmic worst-case algorithms are discovered, it seems that the $\widetilde{O}(m^{2/3})$ worst-case update time is in sight. Nonetheless, we found that it is still hard to get the $\widetilde{O}(m^{2/3})$ update time. Until now we obtain the update time of $\widetilde{O}(m^{3/4})$. The main contribution is a new organization of the auxiliary data structures.

The $\widetilde{O}(\sqrt{mn})$ result comes from dynamic DFS tree [2, 5], which is a periodic rebuilding technique with fault tolerant DFS trees. Our result is always no worse than $\widetilde{O}(\sqrt{mn})$ as $n = \Omega(m^{1/2})$. Faster query time can be traded with slower update time for the bottom four results in Table 1. As to our result, $\widetilde{O}(m^{3/4+\epsilon})$ update time and $\widetilde{O}(m^{1/4-\epsilon})$ query time can be implemented. Note that the trade-offs are in *one direction*, i.e. better query time with worse update time, but not vice-versa. Consequently, the former $\widetilde{O}(m^{4/5})$ algorithm never gives update time of $\widetilde{O}(m^{3/4})$. The trade-off phenomenon is definitely hard to break, as indicated by the OMv (*Online Boolean Matrix-Vector Multiplication*) conjecture proposed by Henzinger et al. [14], which rules out *polynomial* pre-processing time algorithms with the product of amortized update and query time being $o(m)$. Based on the conjecture of no truly subcubic combinatorial boolean matrix multiplication, Abboud and Williams [1] showed that any combinatorial dynamic algorithm with truly sublinear in m query time and truly subcubic in n preprocessing time must have $\Omega(m^{1/2-\delta})$ update time for all $\delta > 0$ unless the conjecture is false. Our result is grouped as the following theorem.

Theorem 1.1 (Main Theorem) *Given a graph $G = (V, E)$, there is a data structure for the dynamic subgraph connectivity, which has the worst-case vertex (edge) update time $\widetilde{O}(m^{3/4})$, query time $\widetilde{O}(m^{1/4})$, where m is the number of edges in G, rather than in the subgraph of G induced by S. The answer to each query is correct if the answer is "yes", and is correct w.h.p. if the answer is "no." The pre-processing time is $\widetilde{O}(m^{5/4})$, and the space usage is linear.*

2 Preliminaries

Theorem 2.1 ([19]) *A spanning forest F of G can be maintained by a deterministic data structure of linear space, with $O(\sqrt{m}(\lg \lg n)^2 / \lg n)$ worst-case update time for an edge update in G, and constant query time to determine whether two vertices are connected in G.*

Theorem 2.2 ([18, 23]) *There is a randomized data structure on dynamic graph connectivity, which supports the worst-cast time $O(\lg^4 n)$ per edge insertion, $O(\lg^5 n)$ per edge deletion, and $O(\lg n / \lg \lg n)$ per query. For any constant c the answer to each query is correct if the answer is "yes" and is correct with probability $\geq 1 - 1/n^c$ if the answer is "no." The pre-processing time of it is $O(m \lg^3 n + n \lg^4 n)$.*

Moving to subgraph connectivity, here we consider only the case of vertex updates, with the extension to edge updates deferred to the full paper [9]. Hence temporarily G is assumed to be static, as E does not change if there are no edge updates. The vertex updates change S. Initially, G is slightly modified to keep $m = \Omega(n)$ during its lifetime, i.e., for every $v \in V$, insert a new vertex v' and a new edge (v, v'). The variant graph has $m = \Omega(n)$, which facilitates the presentation of time and space complexity as functions of m in the case of degenerate graphs.

3 The Data Structure

We give some high-level ideas, which originate from [4]. Main difficulties are the update of S (recall that S is the set of active vertices) incurred by the high-degree vertices, as their degrees are too high to explicitly delete their incident edges one by one. Nonetheless, if the low-degree vertices had been removed, the graph became smaller, and consequently former high-degree vertices were not high-degree anymore. Hence our aim is to remove the low-degree vertices. After that, some artificial edges are added to restore the loss of connectivity due to the removal of the low-degree vertices. Next a dynamic connectivity data structure is maintained on the modified graph, i.e., the graph with the low-degree vertices removed, and the artificial edges added. Besides, as S evolves dynamically, we need to update the artificial edges accordingly. Hence the *point* is how to maintain these artificial edges consistently and efficiently. We now move to the details. We partition V according to their degrees in G. Use $\deg_G(v)$ to denote the degree of v in G.

- C: Vertices with $\deg_G(v) > m^{1/2}$
- B: Vertices with $m^{1/4} < \deg_G(v) \le m^{1/2}$
- A: Vertices with $\deg_G(v) \le m^{1/4}$

Denote $C \cap S$, $B \cap S$, and $A \cap S$ as V_C, V_B, and V_A respectively. Consider the subgraph G_A of G induced by V_A. Define the *degree* of a component as the sum of $\deg_G(v)$'s for v's in it. According to the degrees of the components, partition the components of G_A into two types: *high* component, with its degree $> m^{1/4}$; *low* component, with its degree $\le m^{1/4}$. A spanning forest F_A of G_A is maintained by the deterministic connectivity structure of Theorem 2.1.

3.1 Path Graph

A path graph inserts some artificial edges to reflect the "are connected" relation of the vertices within V_B via directly linking with a component of G_A. We give a more elaborate analysis based on [8]. W.l.o.g. assume $V = \{0, \ldots, n-1\}$. Consider a spanning tree T of F_A.

- *subpath tree*: For $v \in T$, identify the set of vertices in V_B that are adjacent to v. Store the set of vertices in a balanced search tree, which has the worst-case $O(\lg n)$ update time for the well-known search-tree operations [7]. Name the search tree as the subpath tree of v. Given the subpath tree of v, a sequence of artificial edges is added to link the vertices stored in the subpath tree of v. The sequence of artificial edges constitutes a subpath.
- *path tree*: Given $T \in F_A$, group all $v \in T$ with the *non-empty* subpath tree as a balanced search tree, ordered by the Euler-tour order of T. Name it as the path tree of T. As each vertex stored in the path tree of T has an associated subpath, these subgraphs are also concatenated one by one via the artificial edges, generating a path. To emphasize its difference from an ordinary path, it is referred to as the *path graph* of V_B w.r.t. T. An example is shown in Fig. 1.

Lemma 3.1 *The path graphs can be updated in $\widetilde{O}(m^{1/2})$ time for a vertex update in V_B, and in $\widetilde{O}(1)$ time for a link or cut on F_A.*

Proof. We categorize the analysis into two cases.

- Reflect a vertex update in V_B: Suppose $v \in V_B$ is removed from S. The case of insertion is similar. v has $\le m^{1/2}$ edges adjacent to F_A. Consider (v, w) with $w \in T$. We locate w in the path tree of T. Now the subpath associated with w is known. Update the subpath of w by removing v from the subpath. If v happens to be the first or the last vertex on the subpath, the path graph of T is also updated. As the subpaths and the path graph are concerned with the nodes stored in the subpath trees and the path tree respectively, which are all balanced search trees, the removal of (v, w) needs $\widetilde{O}(1)$ time. The removal of all such (v, w)'s requires $\widetilde{O}(m^{1/2})$ time.

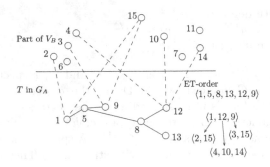

Fig. 1. The path graph of V_B w.r.t. a spanning tree T in F_A. The *dashed* edges represent edges between V_B and V_A. The path tree is on sequence $\langle 1, 12, 9 \rangle$, and three subpath trees are on sequences $\langle 2, 15 \rangle$, $\langle 4, 10, 14 \rangle$, and $\langle 3, 15 \rangle$ respectively. The resulted path graph is a path $\langle 2, 15, 4, 10, 14, 3, 15 \rangle$.

- Reflect a link or cut on F_A: We only discuss the edge cut on F_A. The edge link is similar. Assume the edge cut is $(v, w) \in T$, and the Euler tour of T is $\langle L_1, (v, w), L_2, (w, v), L_3 \rangle$ (The details can be found in the full paper [9].). After the cut of (v, w), the Euler tours for the two resulted trees are $\langle L_1, L_3 \rangle$ and $\langle L_2 \rangle$. We can determine the first vertex a and the last vertex b of $\langle L_2 \rangle$. With the order tree of T (discussed in the full paper [9]), the predecessor of a and the successor of b in the path tree of T can be found in $O(\lg^2 n)$ time. With the predecessor and the successor, the path tree of T is split. After the split, $O(1)$ edges in the path graph are removed to reflect the split of the path tree of T. As a conclusion, the path graph can be updated in $\widetilde{O}(1)$ time to reflect a link or cut on F_A.

$\square$

3.2 Adjacency Structure

Given $T \in F_A$ and $v \in C$, we want a data structure that provides the fast query of whether T and v are adjacent, i.e., whether an edge (u, v) exists with $u \in T$. We give a more elaborate analysis based on [8]. Assuming $v \in C$, the adjacency structure of v contains the following search trees.

- *sub-adjacency tree*: Given $T \in F_A$, identify the set of vertices in T that are adjacent to v. Store the set of vertices as a balanced search tree, ordered by the Euler-tour order of T. Name the balanced search tree as the sub-adjacency tree of v w.r.t. T.
- *adjacency tree*: Identify $T \in F_A$ by the smallest vertex in T. Group all $T \in F_A$, w.r.t. which v has non-empty sub-adjacency trees, as a balanced search tree. Name the balanced search tree as the adjacency tree of v.

The sub-adjacency trees and the adjacency tree of v constitute the adjacency structure of v w.r.t. F_A. The query aforementioned is answered by checking whether T is in the adjacency tree of v. Note $v \in C$, rather than $\in V_C$. The adjacency structure of $v \in C$ w.r.t. F_A is maintained even if $v \notin S$.

Lemma 3.2 *The adjacency structures of C w.r.t. F_A can be renewed in $\widetilde{O}(m^{1/2})$ time for a link or cut on F_A. Given a query of whether $v \in C$ is adjacent to $T \in F_A$, it can be answered in $\widetilde{O}(1)$ time.*

Proof. We only discuss the edge cut on F_A. The edge link is similar. The adjacency structures of the vertices in C are renewed one by one. Consider $v \in C$. Suppose the edge cut occurs on T, splitting T into T_1 and T_2. We check whether T is in the adjacency tree of v. If "no", the update is done; if "yes", remove T from it, and update the sub-adjacent tree of v w.r.t. T to reflect the edge cut on T. For T_j ($j = 1, 2$), add T_j into the adjacent tree of v if it is adjacent to v (determined by whether a sub-adjacent tree of v exists w.r.t. T_j). For every vertex in C, we need to check and update when necessary. Hence the total update time is $\widetilde{O}(m^{1/2})$, since $|C|$ is $O(m^{1/2})$. The query is answered by checking whether T is in the adjacency tree of v. $\qquad\square$

3.3 The Whole Structure

Now we turn to the discussion of the whole structure of our result. First, V_A is removed. After that some artificial vertices and edges are added to the subgraph of G induced by $V_B \cup C$, resulting in a graph H. (Note that we include the vertices in $C \setminus S$, rather than just V_C, which is $C \cap S$.) The artificial vertices and edges are used to restore the loss of connectivity due to the removal of V_A. Recall that the components of G_A are either low or high. We describe how the artificial edges or vertices are added as follows.

- Added by the path graphs: For $T \in F_A$, construct the path graph of V_B w.r.t. T.
- Added by the high components: For a high component $P \in G_A$, add a *meta-vertex*. For $v \in C$ adjacent to P, add an artificial edge between v and the meta-vertex. Identify the first vertex of the path graph of V_B w.r.t. T, where T is the spanning tree of P. Add an artificial edge between *the* first vertex and the meta-vertex.
- Added by the low components: For a low component $Q \in G_A$, construct a complete graph within the vertices in C that are adjacent to Q. Similarly as above, identify the first vertex of the path graph of V_B w.r.t. T, where T is the spanning tree of Q. Add the artificial edges between *the* first vertex and the vertices in C that are adjacent to Q.

After these, H can be defined as follows.

- The vertex set $V(H)$ of H: $V_B \cup C \cup M$, where M is the set of meta-vertices. Since the degree of a high component is $> m^{1/4}$, and the vertices in $V_B \cup C$ are of degree $> m^{1/4}$, H has $O(m^{3/4})$ vertices.

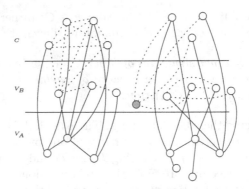

Fig. 2. An example of the whole structure. The irrelevant edges within V_A, V_B, and C are omitted for clarity. The *solid* edges are the edges in G, while the *dotted* edges denote the artificial edges. The *grey* vertex in the V_B layer indicates a meta-vertex. The left component of V_A is low; whereas the right one is high. We construct a complete graph within the vertices in C w.r.t. the low component.

- The edge set $E(H)$ of H: The original edges of G within $V_B \cup C$, and the artificial edges.

Figure 2 gives an example for the construction. H is a multigraph. Use $D[u, v] > 0$ of edge multiplicity to represent the edge $(u, v) \in E(H)$. The maintenance of $D[u, v]$'s is discussed later. Now we construct a graph G^*, based on H.

- The vertex set $V(G^*)$ of G^*: $V_B \cup V_C \cup M$.
- The edge set $E(G^*)$ of G^*: The edges (u, v)'s with $D[u, v] > 0$, where $u, v \in V(G^*), u \neq v$.

G^* is a variant of the subgraph of H induced by $V_B \cup V_C \cup M$. It excludes the vertices in $C \setminus S$, i.e., only the vertices in V_C of C are contained. Besides, the multiple edges are substituted by the single ones. G^* is a simple graph. The randomized connectivity structure of Theorem 2.2 is maintained on G^*.

About the $D[u, v]$'s aforementioned, a balanced search tree is used to store them, with $D[u, v]$ indexed by $u + nv$ (assuming $u \leq v$). Only $D[u, v] > 0$ is stored in the search tree. Along the process of the updates, we might increment or decrement $D[u, v]$'s. When $D[u, v]$ decrements to 0, we remove it from the search tree. If both u and v are the vertices in G^* and $u \neq v$, the edge (u, v) is deleted from G^*. Similar updates works for incrementing. G^* captures the property of connectivity, which is stated in the following lemma.

Lemma 3.3 *For any two vertices $u, v \in V_B \cup V_C$, they are connected in the subgraph of G induced by S if and only if they are connected in G^*.*

Proof. G^* is a variant of the subgraph of G induced by S. G^* removes V_A from the subgraph. Connectivity within V_B via V_A is restored by the path graphs. Connectivity within V_C via V_A is restored *either* by linking with the same meta-vertex, *or* by the complete graph constructed. Lastly, for the connectivity between V_C and V_B via V_A, it is restored by the first vertex of the path graph linking with the meta-vertex, or with all the relevant vertices in V_C. Consider a path between u and v in the subgraph induced by S, the segments of the path consisting only of the vertices in V_A can be eliminated, as the "via V_A" connectivity is restored as discussed. The lemma follows.   □

3.4   Update and Query

The difficulty of the vertex updates is to keep $D[u,v]$'s being consistent with S. As $E(G^*)$ is a subset of the (u,v)'s with $D[u,v] > 0$, it might also need to be updated.

Lemma 3.4 *The whole structure constructed has the worst-case vertex update time $\widetilde{O}(m^{3/4})$.*

Proof. We discuss the various cases of vertex updates, categorized according to whether $v \in A$, or $\in B$, or $\in C$.

– $v \in A$: Consider the case of inserting v into S. v is first inserted as a singleton component containing only v in G_A. Next the edges incident on v are restored in the following order: First, the edges between v and C; second, the edges between v and V_B; third, the edges between v and V_A
   Restore the edges between v and C: For every u adjacent to v where $u \subset C$, construct a sub-adjacency tree (containing only v) of u, and insert v into the adjacency tree of u. Next the complete graph within these u's in C is constructed. Because $\deg_G(v) \le m^{1/4}$, i.e. a low component, the update time is $\widetilde{O}(m^{1/2})$, dominated by constructing the complete graph.
   Restore the edges between v and V_B: Construct the subpath tree and the path tree of v. Add the path-graph edges associated with v (*Add* means incrementing the corresponding entry $D[u,v]$), and the edges *between* the first vertex of the path graph *and* the vertices in C that are adjacent to v. The update time is $\widetilde{O}(m^{1/4})$.
   Restore the edges between v and V_A: $\widetilde{O}(\sqrt{m})$ deterministic data structure maintaining F_A is updated in $\widetilde{O}(m^{3/4})$ time. As $\deg_G(v) \le m^{1/4}$, the link or cut on F_A happens $O(m^{1/4})$ times. Consequently, according to Lemma 3.1, the path graphs are updated in $\widetilde{O}(m^{1/4})$ time. According to Lemma 3.2, the adjacency structures are updated in $\widetilde{O}(m^{3/4})$ time.
   $O(m^{1/4})$ components of G_A are affected. For every high component, using the adjacency structures, the edges between C and the meta-vertex (corresponding to the high component) can be determined in $\widetilde{O}(m^{1/2})$ time according to Lemma 3.2, since $|C| = O(m^{1/2})$; for every low component, as the degree of a low component is $\le m^{1/4}$, $\widetilde{O}(m^{1/2})$ time suffices to construct the complete graph within the vertices in C that are adjacent to the low component, and $\widetilde{O}(m^{1/4})$ time suffices to construct the edges *between* the first vertex of the

path graph w.r.t. the low component *and* the vertices in C that are adjacent to the low component. Hence no matter whether the component is low or high, the update time is $\widetilde{O}(m^{1/2})$. The time needed to update all these components is $\widetilde{O}(m^{3/4})$. Deleting of $v \in S$ from S is a reverse process. In summary, a vertex update of $v \in A$ requires $\widetilde{O}(m^{3/4})$ time.

- $v \in B$: Consider the case when $v \in S$ is removed. The case of insertion is the reverse. First destroy the edges between v and V_A. According to Lemma 3.1, the path graphs can be updated in $\widetilde{O}(m^{1/2})$ time. Besides, v might be the first vertex of some path graphs. We see how it is updated. v can be adjacent to $\leq m^{1/2}$ components of G_A, as $\deg_G(v) \leq m^{1/2}$. For a high component, as only one edge linking v with the meta-vertex, the update is easy; for a low component, since only $\leq m^{1/4}$ edges can be outward for a low component, $\widetilde{O}(m^{1/4})$ time suffices for updating the edges between v and the vertices in C that are adjacent to the low component. Hence the update time for v being the first vertex of some path graphs is $\widetilde{O}(m^{3/4})$. Until now the artificial edges concerning v are removed. Other edges concerning v are the original edges in G. Hence we can remove these original edges one by one in $\widetilde{O}(m^{1/2})$ time as $\deg_G(v) \leq m^{1/2}$. In summary, the total update time of $v \in B$ is $\widetilde{O}(m^{3/4})$.
- $v \in C$: As there are only $O(m^{3/4})$ vertices in G^*, the update time is $\widetilde{O}(m^{3/4})$. The relevant $D[u, v]$'s are left intact, and the adjacency structure of v is not destroyed (if v is removed from S). The total update time is $\widetilde{O}(m^{3/4})$.

□

The query algorithm is as follows: Given $u, v \in S$, the goal is to substitute them with the *equivalent* vertices in G^*, where an *equivalent* vertex of u (or v) is a vertex in G^* that is connected with u (or v). As $V(G^*) = V_B \cup V_C \cup M$, if $u, v \in V_B \cup V_C$, the search for the equivalent vertices is done. Otherwise, if u (or v) is in a high component, replace u (or v) with the meta-vertex corresponding to the high component; if u (or v) is in a low component, exhaustively search the outward edges of the low component for a vertex of G^*. When the equivalent vertex of u (or v) cannot be found, it indicates that u (or v) is in a low component of G_A, and the low component is not connected with any vertex in $V_B \cup V_C$. Intuitively u (or v) is on an "island" of G_A.

Lemma 3.5 *The time complexity of the query algorithm is $\widetilde{O}(m^{1/4})$. The answer to every query is correct if the answer is "yes", and is correct w.h.p. if the answer is "no".*

Proof. Connectivity within G^* is answered by the randomized connectivity structure on G^*; whereas for the other cases, u and v are connected if and only if they are in the same component of G_A, of which the queries can be answered by the deterministic connectivity structure on G_A. The time complexity is dominated by the exhaustive search if u (or v) is in a low component, and thus is $\widetilde{O}(m^{1/4})$.

The correctness can be analyzed as follows. If $u, v \in V_B \cup V_C$, it follows from Lemma 3.3; otherwise, for any one not in, we only replace it with an equivalent vertex of G^*. If such an equivalent vertex cannot be found, the queried vertex is on an island aforementioned of G_A. Then u and v are connected if and only if they are on the same island. We analyze the error probability. A deterministic connectivity structure is adopted for G_A. F_A is always a spanning forest of G_A. The queries are answered *either* by the deterministic connectivity structure if at least one queried vertex is on an island aforementioned of G_A, *or* by the randomized connectivity structure if both queried vertices are (replaced with) the vertices in G^*. The deterministic connectivity structure always gives the right answer; whereas the randomized one might answer erroneously. The randomized algorithm of [18] maintains a private *witness* of a spanning forest of G^*. The algorithm has the property that after every update, the witness is a spanning forest of G^* with probability $\geq 1 - 1/n^c$. It is the property which ensures the answers are correct w.h.p.. Here, after every vertex update (which is transformed into a sequence of edge updates in G^*), the witness for G^* is also a spanning forest of G^* w.h.p. after the vertex update. We can just focus on the correctness of the witness at the point after the last transformed edge update. Consequently, the error probability is negligible, i.e., $\leq 1/n^c$ for any constant c. $\square$

The proofs of the pre-processing time being $\widetilde{O}(m^{5/4})$, and the space usage being linear can be found in the full paper [9]. Hence Theorem 1.1 follows.

Acknowledgments. This work was supported in part by the National Basic Research Program of China Grant 2011CBA00300, 2011CBA00301, the National Natural Science Foundation of China Grant 61033001, 61361136003. R. Duan is supported by a China Youth 1000-Talent grant.

References

[1] Amir Abboud and Virginia Vassilevska Williams. Popular conjectures imply strong lower bounds for dynamic problems. In *IEEE 55th Annual Symposium on Foundations of Computer Science*, pages 434–443. IEEE, 2014.

[2] Surender Baswana, Shreejit Ray Chaudhury, Keerti Choudhary, and Shahbaz Khan. Dynamic DFS in undirected graphs: breaking the $O(m)$ barrier. In *Proceedings of the twenty-seventh Annual ACM-SIAM Symposium on Discrete Algorithms*, pages 730–739. SIAM, 2016.

[3] Timothy M. Chan. Dynamic subgraph connectivity with geometric applications. In *Proceedings of the thiry-fourth annual ACM Symposium on Theory of Computing*, pages 7–13. ACM, 2002.

[4] Timothy M. Chan, Mihai Pătraşcu, and Liam Roditty. Dynamic connectivity: Connecting to networks and geometry. *SIAM Journal on Computing*, 40(2):333–349, 2011.

[5] Lijie Chen, Ran Duan, Ruosong Wang, and Hanrui Zhang. Improved algorithms for maintaining DFS tree in undirected graphs. *CoRR*, abs/1607.04913, 2016.

[6] Don Coppersmith and Shmuel Winograd. Matrix multiplication via arithmetic progressions. *Journal of Symbolic Computation*, 9(3):251 – 280, 1990. Computational algebraic complexity editorial.

[7] Thomas H. Cormen, Charles E. Leiserson, Ronald L. Rivest, and Clifford Stein. *Introduction to Algorithms, Third Edition*. The MIT Press, 3rd edition, 2009.

[8] Ran Duan. New data structures for subgraph connectivity. In *Automata, Languages and Programming*, pages 201–212. Springer, 2010.

[9] Ran Duan and Le Zhang. Faster worst-case update time for dynamic subgraph connectivity. *CoRR*, abs/1611.09072, 2016.

[10] David Eppstein, Zvi Galil, Giuseppe F. Italiano, and Amnon Nissenzweig. Sparsification–a technique for speeding up dynamic graph algorithms. *Journal of the ACM*, 44(5):669–696, 1997.

[11] Greg N. Frederickson. Data structures for on-line updating of minimum spanning trees, with applications. *SIAM Journal on Computing*, 14(4):781–798, 1985.

[12] Daniele Frigioni and Giuseppe F. Italiano. Dynamically switching vertices in planar graphs. *Algorithmica*, 28(1):76–103, 2000.

[13] David Gibb, Bruce M. Kapron, Valerie King, and Nolan Thorn. Dynamic graph connectivity with improved worst case update time and sublinear space. *CoRR*, abs/1509.06464, 2015.

[14] Monika Henzinger, Sebastian Krinninger, Danupon Nanongkai, and Thatchaphol Saranurak. Unifying and strengthening hardness for dynamic problems via the online matrix-vector multiplication conjecture. In *Proceedings of the forty-seventh Annual ACM on Symposium on Theory of Computing*, pages 21–30. ACM, 2015.

[15] Monika R. Henzinger and Valerie King. Randomized fully dynamic graph algorithms with polylogarithmic time per operation. *Journal of the ACM*, 46(4):502–516, 1999.

[16] Jacob Holm, Kristian De Lichtenberg, and Mikkel Thorup. Poly-logarithmic deterministic fully-dynamic algorithms for connectivity, minimum spanning tree, 2-edge, and biconnectivity. *Journal of the ACM*, 48(4):723–760, 2001.

[17] Shang-En Huang, Dawei Huang, Tsvi Kopelowitz, and Seth Pettie. Fully dynamic connectivity in $O(\log n(\log \log n)^2)$ amortized expected time. In *Proceedings of the twenty-eighth Annual ACM-SIAM Symposium on Discrete Algorithms*. SIAM, 2017.

[18] Bruce M. Kapron, Valerie King, and Ben Mountjoy. Dynamic graph connectivity in polylogarithmic worst case time. In *Proceedings of the twenty-fourth Annual ACM-SIAM Symposium on Discrete Algorithms*, pages 1131–1142. SIAM, 2013.

[19] Casper Kejlberg-Rasmussen, Tsvi Kopelowitz, Seth Pettie, and Mikkel Thorup. Faster worst case deterministic dynamic connectivity. In *Proceedings of the twenty-fourth Annual European Symposium on Algorithms*, 2016.

[20] Danupon Nanongkai and Thatchaphol Saranurak. Dynamic spanning forest with worst-case update time: Adaptive, Las Vegas, and $O(n^{1/2-\epsilon})$-time. In *Proceedings of the forty-ninth Annual ACM on Symposium on Theory of Computing*, 2017.

[21] Mihai Pătraşcu and Erik D. Demaine. Logarithmic lower bounds in the cell-probe model. *SIAM Journal on Computing*, 35(4):932–963, 2006.

[22] Mikkel Thorup. Near-optimal fully-dynamic graph connectivity. In *Proceedings of the thirty-second annual ACM Symposium on Theory of Computing*, pages 343–350, 2000.

[23] Zhengyu Wang. An improved randomized data structure for dynamic graph connectivity. *CoRR*, abs/1510.04590, 2015.

[24] Christian Wulff-Nilsen. Faster deterministic fully-dynamic graph connectivity. In *Proceedings of the twenty-fourth Annual ACM-SIAM Symposium on Discrete Algorithms*, pages 1757–1769, 2013.

[25] Christian Wulff-Nilsen. Fully-dynamic minimum spanning forest with improved worst-case update time. In *Proceedings of the forty-ninth Annual ACM on Symposium on Theory of Computing*, 2017.

Improved distance sensitivity oracles via tree partitioning

Ran Duan and Tianyi Zhang

Tsinghua university

Abstract. We introduce an improved structure of *distance sensitivity oracle* (DSO). The task is to pre-process a non-negatively weighted graph so that a data structure can quickly answer replacement path length for every triple of source, terminal and failed vertex. The previous best algorithm [Bernstein and Karger, 2009] constructs in time [1]$\tilde{O}(mn)$ a distance sensitivity oracle of size $O(n^2 \log n)$ that processes queries in $O(1)$ time. As an improvement, our oracle takes up $O(n^2)$ space, while preserving $O(1)$ query efficiency and $\tilde{O}(mn)$ preprocessing time. One should notice that space complexity and query time of our novel data structure are asymptotically optimal.

1 Introduction

The objective of *distance sensitivity oracles* (DSO) is to pre-process the graph so that pairwise shortest distances can be answered by a static data structure in constant time even when one vertex or edge crashes. More precisely, every query for the DSO is composed of three fields: a source vertex, a terminal vertex, and a vertex or an edge that is presumed failed, and then the DSO is supposed to compute the length of shortest source-terminal replacement path that circumvents the failed vertex or edge. In this paper, we are only concerned with vertex failures, as all vertex-failure DSOs can be easily extended to handle edge-failure queries without any asymptotic loss in space / time efficiency [10].

Motivations for distance sensitivity oracles mainly come from practical scenarios like network routing where some nodes occasionally undergo crash failures. As recomputing all-pair shortest paths from scratch every time a node or link crashes is expensive, a static data structure like DSOs that plans for emergency is highly desirable. Vickery pricing [19] is an example that motivates the DSOs from a theoretical perspective, where one wishes to measure, for every pair of source and target as well as a failed edge, by how much the shortest distance would rise if this designated edge were to shut down.

1.1 Related work

The naive approach is that we pre-compute and store the length of all $O(n^3)$ possible replacement path lengths, which incurs intolerable space complexity.

---

[1] $\tilde{O}(\cdot)$ suppresses poly-logarithmic factors.

© Springer International Publishing AG 2017
F. Ellen et al. (Eds.): WADS 2017, LNCS 10389, pp. 349–360, 2017.
DOI: 10.1007/978-3-319-62127-2_30

Authors of [12] proposes the first DSO that occupies only near-quadratic space. More specifically, The DSO in [12] has space complexity $O(n^2 \log n)$ and $O(1)$ query time. Space complexity of constant query time DSO has not been improved ever since.

DSO in [12] demands a somewhat high preprocessing time complexity of $O(mn^2)$, which was improved to $\tilde{O}(mn^{\frac{3}{2}})$ in the journal version [10] while the space complexity was blown up to $O(n^{2.5})$. Cubic time preprocessing algorithm was first obtained by [6] and shortly improved from $\tilde{O}(n^2\sqrt{m})$ to $\tilde{O}(mn)$ in [7], while maintaining $O(n^2 \log n)$ space and $O(1)$ query time. Note that $O(n^2 \log n)$ and $\tilde{O}(mn)$ are basically optimal up to poly-logarithmic factors, as discussed in [7]. Therefore, surpassing [7]'s construction time has been deemed hard from then.

Since the publication of [7], the community's interest has diverged to seeking truly sub-cubic preprocessing time algorithms. F. Grandoni and V. Williams [15] obtained truly sub-cubic preprocessing time bound $O(Mn^{2.88})$, if one should tolerate a sub-linear query time of $O(n^{0.7})$; here all edge weights are assumed to be integers within interval $[-M, M]$.

There are several generalizations of the distance sensitivity problem. In [13], the authors considered the scenario where two instead of one vertices could fail. The paper presented a distance oracle with $O(n^2 \log^3 n)$ space complexity and $O(\log n)$ query time. As it turned out, things got far more complicated than single vertex-failures, and sadly no non-trivial polynomial preprocessing algorithms were known.

There are papers (e.g., [22, 9, 11, 3, 2]) mainly concerned with dynamically maintaining all pairs shortest paths (APSP). Such data structures can solve distance problems if subsequent failures are cumulative, but update time would be as large as $O(n^{\frac{8}{3}})$.

If one should sacrifice preciseness for space efficiency, one may consider approximate distance oracles for vertex failures. Authors of [20] considered approximating the replacement path lengths where a single vertex could crash. In [8], the authors focused on data structures that approximately answer minimal length of paths that do not pass through a designated set of failed edges. For fully dynamic approximate APSP, one can refer to [16]; especially, for planar graphs, [1] may provide useful results.

There are remotely related problems such as (partially) dynamic single-source shortest paths and reachability. Papers [18] and [17] discussed these topics in depth. Some other loosely related work concerns the construction of spanners and distance preservers resilient to one edge / node failure.

1.2 Our result

In this paper we present a DSO construction that improves upon [7].

Theorem 1. *For any directed non-negatively weighted graph $G = (V, E)$ with weight function $\omega : E \to R^+ \cup \{0\}$, with m, n referring to total number of edges* *and vertices, a DSO with $O(n^2)$ space complexity and $O(1)$ query time exists. Also, such DSO can be preprocessed in time $\tilde{O}(mn)$.*

In Bernstein & Karger's work [7], the space / query time was $O(n^2 \log n)$ and $O(1)$. So compared with [7]'s result, our construction shaves off the last $\log n$ factor in the space complexity, leading to a quadratic space consumption, while preserving constant time query efficiency. Plus, our DSO can also be constructed in $\tilde{O}(mn)$ time as in [7], which is nearly optimal. We emphasize that the space complexity of our DSO is asymptotically **optimal** even for sparse graphs, assuming hardness of set intersection [21].

The observation comes from Demetrescu and Thorup's original work [12]. Its construction basically applies the idea of sparse table, where each pair of source and target are associated with $O(\log n)$ sparse table entries, thus resulting in a total storage of $O(n^2 \log n)$. Our preliminary idea is that we do not store sparse table entries for every source-terminal pair. More specifically, for each single source shortest path (SSSP) tree, only a proportion of all tree vertices will be associated with sparse table entries, hence making our data structure even sparser. The set of all designated vertices should be carefully chosen with respect to the topological structure of the SSSP tree.

With the sparser data structure, we can answer queries (s, t, f) when f keeps distance from both of s and t. So the bottleneck lies in degenerated cases where f is very close to one of the endpoints s or t. For degenerated cases, we can use much smaller sparse tables to cover short paths, resulting in a DSO that only occupies $O(n^2 \log \log n)$ space. To obtain an optimum of $O(n^2)$ space complexity, we would need to apply a tabulation technique (otherwise known as the "Four Russians" [4]).

Our preprocessing algorithm heavily relies on the notion of *admissible functions* from [6]. The idea is that we substitute "bottleneck" vertices for intervals in the original construction, without harming the correctness of query algorithm. This would largely facilitate the preprocessing algorithm of DSOs.

2 Preliminaries

Suppose we are given a directed graph $G = (V, E)$ with non-negative edge weights $\omega : E \to R^+ \cup \{0\}$. In this section, we summarize the notations or assumptions that are used throughout this paper. More or less, we inherit the conventions from [13].

- Our data structures & algorithms are implemented on $\Omega(\log n)$-word RAM machines. We will leverage its strength in computing the most significant set bit [14]. Although in previous works on DSO ([9, 10, 6, 7]) $\Omega(\log n)$-word RAM model was not explicitly assumed, this assumption is not dispensable in these papers since their algorithms required memory indexing and computing logarithms in constant time.
- We call a data structure $\langle f(n), g(n) \rangle$, if its space complexity is at most $f(n)$ and its query time is at most $g(n)$.

- For each pair of $s, t \in V$, the weighted shortest path from s to t is unique. This assumption is without loss of generality since we can add small perturbations to break ties (e.g., [10, 7]).
- For each pair of $s, t \in V$, let st denote the weighted shortest path from s to t.
- Let p be a simple path. Denote by $\|p\|$ and $|p|$ the weighted and un-weighted length of path p.
- For each $s \in V$, let T_s be the single-source shortest path tree rooted at s; let $\widehat{T}_s$ be the single-source shortest path tree rooted at s in the reverse graph $\widehat{G}$ where every directed edge in G is reversed.
- For each query (s, t, f), we only consider the case when f lies on the path st, because verification can be done by checking whether $\|st\| = \|sf\| + \|ft\|$.
- For each vertex set A, let $st \diamond A$ denote the weighted shortest path from s to t that avoids the entire set A. For instance, $st \diamond \{f\}$ (abbreviated as $st \diamond f$) denotes the replacement path, and $st \diamond [u, v]$ refers to the weighted shortest path that skips over an entire interval $[u, v] \subseteq st$. Here for $[u, v]$ to be a properly defined interval on st, it is required that both of u and v are on st, and either $u = v$, or u lying between s and v.
- Let $s \oplus i$ and $s \ominus i$ be the ith vertex after and before s on some path that can be learnt from context.

By uniqueness of shortest paths, it is easy to verify that any path of the form $st \diamond f$ ($st \diamond [u, v]$) must diverge from and converge with st for **at most** once [10], with divergence on path sf (su) and convergence on ft (vt). We denote Δ and ∇ to be the vertices at which divergence and convergence take place, respectively, when the path can be learnt from context.

3   Admissible functions and the triple path lemma

Our query algorithms will be frequently using the *triple path lemma* from [6]. This lemma is based on the notion of *admissible functions*.

Definition 1 ([6]). A function $F_{s,t}^{[u,v]}$ is admissible if $\forall f \in [u, v]$, $[u, v]$ an interval on st, we have $\|st \diamond [u, v]\| \geq F_{s,t}^{[u,v]} \geq \|st \diamond f\|$.

Definition 2 ([6]). Two important admissible functions are $\max_{f \in [u,v]} \{\|st \diamond f\|\}$ and $\|st \diamond [u, v]\|$. We call them **bottleneck** and **interval** admissible functions, respectively.

Lemma 2 (The triple path lemma [6]). *Let $[u, v]$ be an interval on st, and $f \in [u, v]$ be a vertex. For any admissible function $F_{s,t}^{[u,v]}$, $\|st \diamond f\| = \min\{\|su\| + \|ut \diamond f\|, \|sv \diamond f\| + \|vt\|, F_{s,t}^{[u,v]}\}$.*

Proof. On the one hand, $\|st \diamond f\|$ is always smaller or equal to $\min\{\|su\| + \|ut \diamond f\|, \|sv \diamond f\| + \|vt\|, F_{s,t}^{[u,v]}\}$. This is because, by definition 1 $F_{s,t}^{[u,v]} \geq \|st \diamond f\|$; also it is easy to see that both of $\|su\| + \|ut \diamond f\|$ and $\|sv \diamond f\| + \|vt\|$ are $\geq \|st \diamond f\|$.

On the other hand, we argue $\|st \diamond f\| \geq \min\{\|su\| + \|ut \diamond f\|, \|sv \diamond f\| + \|vt\|, F_{s,t}^{[u,v]}\}$. If $st \diamond f$ passes through either u or v, then $\|st \diamond f\|$ would be equal to either $\|su\| + \|ut \diamond f\|$ or $\|sv \diamond f\| + \|vt\|$, which is $\geq \min\{\|su\| + \|ut \diamond f\|, \|sv \diamond f\| + \|vt\|, F_{s,t}^{[u,v]}\}$. Otherwise $st \diamond f$ skips over the entire interval $[u, v]$ and thus $\|st \diamond f\| = \|st \diamond [u, v]\| \geq F_{s,t}^{[u,v]} \geq \min\{\|su\| + \|ut \diamond f\|, \|sv \diamond f\| + \|vt\|, F_{s,t}^{[u,v]}\}$.

4 The tree partition lemma

Our novel DSO begins with the following lemma. Paper [15] has a similar lemma, but it is actually different from ours.

Lemma 3 (Tree partition). *Given a rooted tree $\mathcal{T}$, and any integer $2 \leq k \leq n = |V(\mathcal{T})|$, there exists a subset of vertices $M \subseteq V(\mathcal{T})$, $|M| \leq 3k - 5$, such that after removing all vertices in M, the tree $\mathcal{T}$ is partitioned into sub-trees of size $\leq n/k$. We call every $u \in M$ an M-marked vertex, and M a marked set. Plus, such M can be computed in $O(n \log k)$ time.*

A detailed proof can be found in the full version of this paper.

The high-level idea of our data structure is that we reduce the computation of an arbitrary $\|st \diamond f\|$ to a "shorter" $\|uv \diamond f\|$; here we say "short" in the sense that either $|uf|$ or $|fv|$ is small. The tree partition lemma helps us with the reduction. Basically, we apply the lemma **twice** with different parameters so that either uf or fv becomes "short" enough, and then we can directly retrieve the length of replacement path from storage.

4.1 Shortness of replacement paths

Our new data structure will heavily rely on the notion of "shortness" of replacement paths, which we describe below. On a high-level, for any replacement path $st \diamond f$, shortness measures how close f is to either s or t on the weighted shortest path st. For those replacement paths where f lies near to the middle of st, we think of them as long ones, and for those where f is very close to one of the endpoints, we view them as short paths. The rough idea is that, when we compute the length of a general replacement path, we reduce it to a constant number of shorter paths and conquer them separately. Now we propose a formal definition of shortness.

Definition 3. Given a vertex subset M, whose removal breaks T (either T_s or $\hat{T}_s$) into subtrees of size $< L$, L being a fixed parameter. Then for any t, as well as a failed vertex f on path st, we say $st \diamond f$ is L-short with respect to M, if t and f lie in the same subtree of T induced by removal of M. We often do not explicitly refer to M when it can be learnt from context.

5 Reducing to $\log^2 n$-short paths

We devise an $O(n^2)$-space data structure that computes any non-$\log^2 n$-short path in constant time.

5.1 Data structure

Our DSO first pre-computes all values of $\|st\|$ and $|st|$, which accounts for $O(n^2)$ space. Then, for each $s \in V$, apply lemma **3** in T_s ($\widehat{T}_s$) to obtain a marked set M_s ($\widehat{M}_s$) with parameter $\lceil n/(L-1) \rceil$, where $2 \le L \le n$ is an integer to be set later. So M_s ($\widehat{M}_s$) is of size $O(n/L)$, and the size of each sub-tree is $< L$. Consider the following structures.

For any pair of s, t such that $t \in M_s$, suppose we are met with M_s-marked vertices $u_1 \to u_2 \to \cdots \to u_k = t$ along the path st in T_s. Our data structure consists of several parts. Note that we also build the symmetric structures of (i) to (iv) for every pair of s, t where $t \in \widehat{M}_s$.

  (i) For each $k - 2^i \in [1, k-1]$, the value of $\|st \diamond u_{k-2^i}\|$.
 (ii) For each $k - 2^i \in [1, k-2]$, the value of $\|st \diamond [u_{k-2^i}, u_{k-2^i+1}]\|$.
(iii) Let $v_l \to \cdots \to v_1$ be the sequence of all $\widehat{M}_t$-marked vertices along the path st. Then for each properly defined interval $[v_{l-2^i}, u_{k-2^j}]$ on the path st, store the value of $\|st \diamond [v_{l-2^i}, u_{k-2^j}]\|$.
 (iv) For each f such that $|ft| \le 2L$ or $|sf| \le 2L$, the value of $\|st \diamond f\|$.
      From now on we drop the assumption that t is M_s-marked.
  (v) For every pair (s, t) of different vertices, let x be t's nearest M_s-marked T_s-ancestor, and y be s's nearest $\widehat{M}_t$-marked $\widehat{T}_t$-ancestor. If intervals $(s, x]$ and $[y, t)$ intersect, then we pre-compute and store $\|st \diamond [y, x]\|$. Also, store addresses of x, y, if such ancestors exist.
 (vi) Build a tree upon all $M_s \cup \{s\}$'s vertices as follows. In this tree, u is v's parent if and only if in T_s u is v's nearest ancestor that belongs to $M_s \cup \{s\}$. Then pre-compute and store the level-ancestor [5] data structure of this tree. Note again that we also build similar structures for $\widehat{M}_s \cup \{s\}$ in the reverse graph.

Note 1. The un-weighted distances between two adjacent marked vertices in T_s ($\widehat{T}_s$) are $\le L$.

Conduct a simple space complexity analysis for each part of the data structure.

  (i) takes up space $O(\frac{n^2 \log n}{L})$ since we have $O(\log n)$ choices for the index i, and every $|M_s| = O(n/L)$.
 (ii) uses $O(\frac{n^2 \log n}{L})$ for a similar reason as in the previous part.
(iii) demands $O(\frac{n^2 \log^2 n}{L})$ space since we have $O(\log^2 n)$ choices for the pair of (i, j).
 (iv) entails an $O(n^2)$ space consumption since each M_s-marked t is associated with $O(L)$ entries, and there are $O(n/L)$ M_s-marked vertices t.
  (v) induces $O(n^2)$ space complexity.
 (vi) takes $O(n^2/L)$ total space, each tree of size $O(n/L)$.

Therefore, the overall space complexity from (i) through (vi) is equal to $O(\frac{n^2 \log^2 n}{L} + n^2)$. Taking $L = \log^2 n$, it becomes $O(n^2)$.

5.2 Query algorithm

We prove the following reduction lemma in this sub-section.

Lemma 4. *The data structure specified in section* **5.1** *can compute* $\|st \diamond f\|$ *in* $O(1)$ *time if* $st \diamond f$ *is not L-short with respect to* M_s *or* $\widehat{M}_t$.

Proof. A constant time verification for L-shortness is easy: we check if f lies below x, which is the nearest M_s-marked T_s-ancestor of t, and similarly if f lies below y, which is the nearest $\widehat{M}_t$-marked $\widehat{T}_t$-ancestor of s.

Firstly we argue that it is without loss of generality to assume that t is M_s-marked. The reduction proceeds as follows.

Let x and y be vertices defined as in (v) from section **5.1**. Since $st \diamond f$ is not L-short, $f \in (s, x] \cap [y, t)$, and thus $[y, x]$ is a properly defined interval on path st. By the *triple path lemma*, one has:

$$\|st \diamond f\| = \min\{\|sx \diamond f\| + \|xt\|, \|sy\| + \|yt \diamond f\|, \|st \diamond [y, x]\|\}$$

Here we use the interval admissible function $\|st \diamond [y, x]\|$. Noticing that the third term $\|st \diamond [y, x]\|$ is already covered in (v) from section **5.1**, we are left with $\|sx \diamond f\|$ and $\|yt \diamond f\|$. By definition, x is M_s-marked and y is $\widehat{M}_t$-marked, and thus we complete our reduction.

Let $u_1 \to u_2 \to \cdots \cdots \to u_p = t$ be the sequence of all M_s-marked vertices along st. We can assume that $p > 1$; otherwise $\|st \diamond f\|$ has already been computed in structure (iv).

It is not hard to find the interval $[u_a, u_{a+1})$ that contains f. On the one hand, u_a is easily retrieved: if f not M_s-marked, then u_a is its nearest marked ancestor stored in (v); otherwise, $u_a = f$. On the other hand, u_{a+1} can be found by querying the level-ancestor data structure (vi) at node t in tree rooted at s.

Let $v_q \to v_{q-1} \to \cdots \cdots \to v_1$ be the sequence of all $\widehat{M}_t$-marked vertices on st. It is also safe to assume $q > 1$; otherwise, we have $|st| \leq 2L$ and then using (iv) we can directly compute $\|st \diamond f\|$. Similar to the previous paragraph, locate the interval $(v_{b+1}, v_b]$ that includes f. We only need to consider the case when $a + 1 < p$ and $b + 1 < q$, since otherwise $\|st \diamond f\|$ can be directly retrieved from structure (iv).

Find maximum indices $i, j \geq 0$ such that $q - 2^i \geq b + 1$, $p - 2^j \geq a + 1$. Applying the *triple path lemma* with respect to $[v_{q-2^i}, u_{p-2^j}]$ in terms of interval admissible function, $\|st \diamond f\|$ must be the minimum among the following three distances.

(1) $\|st \diamond [v_{q-2^i}, u_{p-2^j}]\|$.
   This value is directly retrievable from (iii) in section **5.1**.
(2) $\|su_{p-2^j} \diamond f\| + \|u_{p-2^j}t\|$.
   Note that since we are interested in the minimum among (1)(2)(3) which gives us $\|st \diamond f\|$, we can substitute any value for (2) that lies in range $[\|st \diamond f\|, \|su_{p-2^j} \diamond f\| + \|u_{p-2^j}t\|]$.

By definition of j, u_{a+2^j} lies between u_{p-2^j} and t, and hence we know that the concatenation of paths $su_{p-2^j} \diamond f$ and $u_{p-2^j}t$ passes through u_{a+2^j}. Thus, it must be

$$\|su_{p-2^j} \diamond f\| + \|u_{p-2^j}t\| \geq \|su_{a+2^j} \diamond f\| + \|u_{a+2^j}t\| \geq \|st \diamond f\|$$

So instead of computing the original (2), we are actually calculating $\|su_{a+2^j} \diamond f\| + \|u_{a+2^j}t\|$.

We focus on the case when $f \neq u_a$; the case where $f = u_a$ is easy in that we can directly query $\|su_{a+2^j} \diamond u_a\|$ using (i).

Applying the *triple path lemma* for a third time to $su_{a+2^j} \diamond f$ and interval $[u_a, u_{a+1}]$, we further divide it into three cases.

(a) $\|su_{a+2^j} \diamond [u_a, u_{a+1}]\| + \|u_{a+2^j}t\|$.
    This can be computed by a single table lookup in (ii).

(b) $\|su_{a+1} \diamond f\| + \|u_{a+1}t\|$.
    Since u_{a+1} is M_s-marked, $\|su_{a+1} \diamond f\|$ is stored in structure (iv), and thereby $\|su_{a+1} \diamond f\| + \|u_{a+1}t\|$ is computed effortlessly.

(c) $\|su_a\| + \|u_at \diamond f\|$.
    If u_a itself is $\widehat{M_t}$ marked, then (iv) directly help us out since $\|u_at \diamond f\|$ is already pre-computed as $|u_af| \leq L$.
    Otherwise, suppose v is u_a's nearest $\widehat{M_t}$-marked ancestor in T_s (if any). To locate such v, we can try to find the interval $(v_{c+1}, v_c]$ that contains u_a, in a similar fashion of finding intervals $[u_a, u_{a+1})$ and $(v_{b+1}, v_b]$; after that we assign $v \leftarrow v_{c+1}$.
    If such v does not exist, then s and u_a lie in the same sub-tree of $\widehat{T_t}$ after removing $\widehat{M_t}$. Noticing that $|sf| < |su_{a+1}| = |su_a| + |u_au_{a+1}| \leq 2L$, (iv) can finish up $\|st \diamond f\|$ by a single table look-up. If v exists, then by $\|su_a\| + \|u_at \diamond f\| \geq \|sv\| + \|vt \diamond f\| \geq \|st \diamond f\|$, it suffices to compute $\|vt \diamond f\|$, which also has already been pre-computed in (iv) due to $|vf| = |vu_a| + |u_af| \leq 2L$.

(3) $\|sv_{q-2^i}\| + \|v_{q-2^i}t \diamond f\|$.
    The only difficult part is $\|v_{q-2^i}t \diamond f\|$. Similar arguments as in the previous case (2) would still work.

6   An $\langle O(n^2 \log \log n), O(1) \rangle$ construction

In this section, we present an ordinary way of handling L-short paths, resulting in an $\langle O(n^2 \log \log n), O(1) \rangle$ DSO. On a high level, we directly apply the sparse table construction as in [10]. But since the sparse table only needs to cover L-short paths, the space requirement shrinks to $O(\log L) = O(\log \log n)$ for every pair of $s, t \in V$. Hence the total space complexity would be $O(n^2 \log \log n)$.

6.1   Data structure

For any pair of $s, t \in V$, besides $\|st\|, |st|$, build the following structures.

(i) For every $2^i \leq \min\{4L, |st| - 1\}$, store $\|st \diamond (s \oplus 2^i)\|$ and $\|st \diamond (t \ominus 2^i)\|$.

(ii) For every $2^{i+1} \leq \min\{4L, |st| - 1\}$, store $\|st \diamond [s \oplus 2^i, s \oplus 2^{i+1}]\|$ and $\|st \diamond [t \ominus 2^{i+1}, t \ominus 2^i]\|$.

(iii) Level ancestor data structures of T_s and $\widehat{T}_s$.

Since $L = \log^2 n$, the total space of this structure is equal to $O(n^2 \log L) = O(n^2 \log \log n)$. Note that this structure is basically identical to [10], except for the additional bound $4L$ on the power-of-two's 2^i. Therefore, when $|st| \leq 4L$, $\|st \diamond f\|$ can be retrieved in $O(1)$ time according to the correctness guaranteed by [10].

6.2 Query algorithm

We prove the following lemma, showing how our data structure covers all L-short paths.

Lemma 5. *For any* $st \diamond f$ *such that* $|sf| \leq L$ *or* $|ft| \leq L$, $\|st \diamond f\|$ *can be computed in* $O(1)$ *time by the data structure presented in section* **6.1**.

For details of the proof, please refer to the full version of this paper.

From definition **3**, we can see every L-short path $st \diamond f$ satisfies $|sf| \leq L$ or $|ft| \leq L$. So by the above lemma, data structure introduced in section **6.1** answers every L-short path query. Together with the structures in section **5.1**, it makes an $\langle O(n^2 \log \log n), O(1) \rangle$ DSO.

7 Two-level partition

In this section, we obtain an $\langle O(n^2), O(1) \rangle$ construction of DSO. The high-level idea is that we further partition every sub-tree into even smaller ones, and then we apply a tabulation ("Four Russians") technique to store all answers. More specifically, we apply the tree-partitioning lemma for the second time and break each SSSP tree into sub-trees of size $\leq \log \log^2 n$. Then we devise data structures to reduce $\log^2 n$-short paths to $\log \log^2 n$-short paths. Finally, the tabulation technique kicks in when it comes to $\log \log^2 n$-short paths.

7.1 Data structure

Let $L' \leq L$ be a parameter to be set later. For each SSSP tree T_s ($\widehat{T}_s$), compute its tree partition with parameter $\lceil n/(L' - 1) \rceil$ by Lemma 3, and let M'_s ($\widehat{M'_s}$) be the corresponding marked set. For any t, let r be the root of the sub-tree, induced by removal of M_s, that contains vertex t. We build the following data structures. (Note that similar structures are also built for the reverse graph where t is the source and s is the terminal.)

(i) If t is not M'_s-marked.
Let u be t's nearest M'_s-marked ancestor below r (if such u exists), and store the value of $\|st \diamond [r, u]\|$.

(ii) If t is M_s'-marked.

Let $u_1 \rightarrow u_2 \rightarrow \cdots \cdots \rightarrow u_k = t$ be the sequence of all M_s'-marked ancestor along the directed path rt. Note that $k < L = O(\log^2 n)$. Then for each $k - 2^i \in [1, k-1]$, store the value of $\|st \diamond [r, u_{k-2^i}]\|$. After that, for each $f \in [u_{k-1}, u_k)$ (define $u_0 = r$), store the value of $\|st \diamond f\|$.

(iii) Build upon the marked set M_s' all structures from (i) to (vi) in section **5.1**. The only difference is that we impose an additional constraint that $|st| \leq L$ on structures (i) through (iii). It is not hard to verify that the space complexity of this part becomes $O(\frac{n^2 \log^2 L}{L'} + n^2) = O(\frac{n^2 \log \log^2 n}{L'} + n^2)$. So if $st \diamond f$ is non-L'-short, with $|st| \leq L$, then applying lemma **5.1**, $\|st \diamond f\|$ can be answered in constant time.

Note that the space complexity of (i) and (ii) in section **7.1** is equal to $O(\frac{n^2 \log \log n}{L'} + n^2)$. Together with (iii), the overall space complexity of the data structure is $O(n^2 + \frac{n^2 \log \log n}{L'} + \frac{n^2 \log \log^2 n}{L'})$. Taking $L' = \log \log^2 n$, the space becomes $O(n^2)$.

7.2 Reduction algorithm

We claim the following lemma; due to page limits, its full proof is presented in the full version.

Lemma 6. *Given an L-short replacement path $st \diamond f$, the data structure in sections **7.1** and **5.1** can reduce $\|st \diamond f\|$ to a constant number of $\|uv \diamond f\|$'s, where $uv \diamond f$'s are L'-short with respect to M_u' or $\widehat{M_v'}$.*

7.3 Tabulation

In this sub-section, we handle all L'-short paths. Recall that the notation Δ, ∇ refers to divergence and convergence of replacement path $st \diamond f$. when s, t, f can be learnt from context.

Let $st \diamond f$ be an L'-short path; without loss of generality assume that t, f lie in the same sub-tree, the corresponding marked set being M_s'. One observation is that we only need to focus on cases where $|s\Delta| \leq L$: if the divergence comes after $s \oplus L$, then it admits the decomposition $\|st \diamond f\| = \|su\| + \|ut \diamond f\|$, u being s's nearest $\widehat{M_t}$-marked ancestor in tree $\widehat{T_t}$. Since $|ft| \leq L' < L < 2L$, $\|ut \diamond f\|$ can be found in (iv) from section **5.1**.

For each sub-tree T partitioned by marked set M_s', we in-order sort all its vertices. The aggregate divergence / convergence information within this sub-tree can be summarized as an $L' \times L'$ matrix, each element being a pair $(|s\Delta|, |\nabla t|)$ corresponding to a replacement path $st \diamond f, \forall t, f \in V(T)$. Since we only consider the case when $|s\Delta| \leq L$, the total number of choices for this matrix is no more than $(L \cdot L')^{(L')^2} < L^{2(L')^2} = 2^{4 \log \log^5 n} = o(n)$. Recall we are running on $\Omega(\log n)$-word RAM machines, so this matrix admits random accesses.

Construct an indexable table of all possible configurations of such matrices. The space of this table is $\leq o(n \cdot (L')^2) = o(n \log \log^2 n) = o(n^{1.1})$. Then associate each sub-tree with an index of its corresponding matrix in the table, which demands a storage of $O(n/L')$ indices, totalling $o(n)$ space for every s. Thus the overall space complexity associated with tabulation is $o(n^2)$.

Now the L'-short $\|st \diamond f\|$ can be computed effortlessly. After indexing the corresponding matrix in the table, we can extract $(|s\Delta|, |\nabla t|)$ directly from this matrix, and then recover Δ, ∇ from level-ancestor data structures. Finally, decompose the replacement path as $\|st \diamond f\| = \|s\Delta\| + \|\Delta\nabla \diamond (\Delta, \nabla)\| + \|\nabla t\|$. Noticing that $\Delta\nabla \diamond f = \Delta\nabla \diamond (\Delta, \nabla)$, thereby the value of $\|\Delta\nabla \diamond f\|$ is equal to any admissible function value $F_{\Delta,\nabla}^{[\Delta\oplus 1, \nabla\ominus 1]}$. Hence, storing a $\|uv \diamond [u\oplus 1, v\ominus 1]\|$ for every pair of u, v will suffice for querying $\|st \diamond f\|$ once divergence and convergence vertices are known.

8 Concluding remarks

From the *triple path lemma* (Lemma 1), we can see:

Remark 1. We can obtain a DSO with the same space and query time if all interval admission functions of the form $\|st \diamond [u, v]\|$ are replaced by corresponding bottleneck admission functions $\max_{f \in [u,v]} \{\|st \diamond f\|\}$.

So far we have devised $\langle O(n^2), O(1)\rangle$ DSOs. Clearly both of the space complexity and query efficiency have reached asymptotic optima; also its preprocessing time (for the bottleneck admission functions form) is $\tilde{O}(mn)$ (see full version of this paper), which is nearly optimal.

References

1. Abraham, I., Chechik, S., Gavoille, C.: Fully dynamic approximate distance oracles for planar graphs via forbidden-set distance labels. In: Proceedings of the Forty-fourth Annual ACM Symposium on Theory of Computing. pp. 1199–1218. STOC '12, ACM, New York, NY, USA (2012), http://doi.acm.org/10.1145/2213977.2214084
2. Abraham, I., Chechik, S., Krinninger, S.: Fully dynamic all-pairs shortest paths with worst-case update-time revisited. In: Proceedings of the Twenty-Eighth Annual ACM-SIAM Symposium on Discrete Algorithms. pp. 440–452. SIAM (2017)
3. Abraham, I., Chechik, S., Talwar, K.: Fully dynamic all-pairs shortest paths: Breaking the $o(n)$ barrier. In: APPROX-RANDOM. pp. 1–16 (2014)
4. Arlazarov, V.L., Dinic, E.A., Kronrod, M.A., Faradžev, I.A.: On economical construction of the transitive closure of a directed graph. Soviet Mathematics—Doklady 11(5), 1209–1210 (1970)
5. Bender, M.A., Farach-Colton, M.: The level ancestor problem simplified. Theoretical Computer Science 321(1), 5–12 (2004)
6. Bernstein, A., Karger, D.: Improved distance sensitivity oracles via random sampling. In: Proceedings 19th ACM-SIAM Symposium on Discrete Algorithms (SODA). pp. 34–43 (2008)
7. Bernstein, A., Karger, D.: A nearly optimal oracle for avoiding failed vertices and edges. In: Proceedings 41st Annual ACM Symposium on Theory of Computing (STOC). pp. 101–110 (2009)

8. Chechik, S., Langberg, M., Peleg, D., Roditty, L.: f-sensitivity distance oracles and routing schemes. Algorithmica 63(4), 861–882 (Aug 2012), http://dx.doi.org/10.1007/s00453-011-9543-0

9. Demetrescu, C., Italiano, G.F.: A new approach to dynamic all pairs shortest paths. J. ACM 51(6), 968–992 (2004)

10. Demetrescu, C., Thorup, M., Chowdhury, R.A., Ramachandran, V.: Oracles for distances avoiding a failed node or link. SIAM J. Comput. 37(5), 1299–1318 (2008)

11. Demetrescu, C., Italiano, G.F.: Fully dynamic all pairs shortest paths with real edge weights. J. Comput. Syst. Sci. 72(5), 813–837 (Aug 2006), http://dx.doi.org/10.1016/j.jcss.2005.05.005

12. Demetrescu, C., Thorup, M.: Oracles for distances avoiding a link-failure. In: Proceedings of the Thirteenth Annual ACM-SIAM Symposium on Discrete Algorithms. pp. 838–843. SODA '02, Society for Industrial and Applied Mathematics, Philadelphia, PA, USA (2002), http://dl.acm.org/citation.cfm?id=545381.545490

13. Duan, R., Pettie, S.: Dual-failure distance and connectivity oracles. In: Proceedings 20th ACM-SIAM Symposium on Discrete Algorithms (SODA). pp. 506–515 (2009)

14. Fredman, M.L., Willard, D.E.: Surpassing the information-theoretic bound with fusion trees. J. Comput. Syst. Sci. 47(3), 424–436 (1993)

15. Grandoni, F., Williams, V.V.: Improved distance sensitivity oracles via fast single-source replacement paths. In: FOCS. pp. 748–757. IEEE Computer Society (2012), http://dblp.uni-trier.de/db/conf/focs/focs2012.htmlGrandoniW12

16. Henzinger, M., Krinninger, S., Nanongkai, D.: Dynamic approximate all-pairs shortest paths: Breaking the o(mn) barrier and derandomization. In: FOCS 2013 54th Annual IEEE Symposium on Foundations of Computer Science. pp. 538–547. Proceedings 2013 IEEE 54th Annual Symposium on Foundations of Computer Science FOCS 2013, IEEE, Los Alamitos, CA (October 2013), http://eprints.cs.univie.ac.at/3747/

17. Henzinger, M., Krinninger, S., Nanongkai, D.: Sublinear-time decremental algorithms for single-source reachability and shortest paths on directed graphs. In: 46th ACM Symposium on Theory of Computing (STOC 2014) (June 2014), http://eprints.cs.univie.ac.at/4042/

18. Henzinger, M., Krinninger, S., Nanongkai, D.: A subquadratic-time algorithm for decremental single-source shortest paths. In: SODA 2014. SIAM, Philadelphia (January 2014), http://eprints.cs.univie.ac.at/3785/

19. Hershberger, J., Suri, S.: Vickrey prices and shortest paths: what is an edge worth? In: Proceedings 42nd IEEE Symposium on Foundations of Computer Science (FOCS). pp. 252–259 (2001), erratum, Proc. 43rd FOCS, p. 809, 2002

20. Khanna, N., Baswana, S.: Approximate shortest paths avoiding a failed vertex: Optimal size data structures for unweighted graphs. In: 27th International Symposium on Theoretical Aspects of Computer Science, STACS 2010, March 4-6, 2010, Nancy, France. pp. 513–524 (2010), http://dx.doi.org/10.4230/LIPIcs.STACS.2010.2481

21. Patrascu, M., Roditty, L.: Distance oracles beyond the thorup-zwick bound. In: Proceedings of the 2010 IEEE 51st Annual Symposium on Foundations of Computer Science. pp. 815–823. FOCS '10, IEEE Computer Society, Washington, DC, USA (2010), http://dx.doi.org/10.1109/FOCS.2010.83

22. Thorup, M.: Worst-case update times for fully-dynamic all-pairs shortest paths. In: Proceedings 37th ACM Symposium on Theory of Computing (STOC). pp. 112–119 (2005)

Delta-Fast Tries: Local Searches in Bounded Universes with Linear Space[*]

Marcel Ehrhardt and Wolfgang Mulzer

Institut für Informatik, Freie Universität Berlin, Germany
[marehr,mulzer]@inf.fu-berlin.de

Abstract. Let $w \in \mathbb{N}$ and $U = \{0, 1, \dots, 2^w - 1\}$ be a bounded universe of w-bit integers. We present a dynamic data structure for predecessor searching in U. Our structure needs $O(\log \log \Delta)$ time for queries and $O(\log \log \Delta)$ expected time for updates, where Δ is the difference between the query element and its nearest neighbor in the structure. Our data structure requires linear space. This improves a result by Bose *et al.* [CGTA, 46(2), pp. 181–189].
The structure can be applied for answering approximate nearest neighbor queries in low dimensions and for dominance queries on a grid.

1   Introduction

Predecessor searching is one of the oldest problems in theoretical computer science [5, 12]. Let U be a totally ordered universe. The task is to maintain a set $S \subseteq U$, while supporting *predecessor* and *successor* queries: given $q \in U$, find the largest element in S smaller than q (q's predecessor) or the smallest element in S larger than q (q's successor). In the *dynamic* version of the problem, we also want to be able to modify S by inserting and/or deleting elements.

In the *word-RAM* model of computation, all input elements are w-bit words, where $w \in \mathbb{N}$ is a parameter. Without loss of generality, we may assume that w is a power of 2. We are allowed to manipulate the input elements at the bit level, in constant time per operation. In this case, we may assume that the universe is $U = \{0, \dots, 2^w - 1\}$. A classic solution for predecessor searching on the word-RAM is due to van Emde Boas, who described a dynamic data structure that requires space $O(n)$ and supports insertions, deletions, and predecessor queries in $O(\log \log |U|)$ time [9, 10].

In 2013, Bose *et al.* [3] described a word-RAM data structure for the predecessor problem that is *local* in the following sense. Suppose our data structure currently contains the set $S \subseteq U$, and let $q \in U$ be a query element. Let $q^+ := \min\{s \in S \mid s \geq q\}$ and $q^- := \max\{s \in S \mid s \leq q\}$ be the successor and the predecessor of q in S, and let $\Delta = \min\{|q - q^-|, |q - q^+|\}$ be the distance between q and its nearest neighbor in S. Then, the structure by Bose *et al.* can answer predecessor and successor queries in $O(\log \log \Delta)$ time. Their solution requires $\mathcal{O}(n \log \log \log |U|)$ words of space, where $n = |S|$ is the size of

---

[*] Supported in part by DFG project MU/3501-1.

© Springer International Publishing AG 2017
F. Ellen et al. (Eds.): WADS 2017, LNCS 10389, pp. 361–372, 2017.
DOI: 10.1007/978-3-319-62127-2_31

the current set. Bose *et al.* apply their structure to obtain a fast data structure for approximate nearest neighbor queries in low dimensions and for answering dominance and range searching queries on a grid.

Here, we show how to obtain a data structure with similar guarantees for the query and update times that reduces the space requirement to $O(n)$. This solves an open problem from [3]. Furthermore, this also improves the space requirement for data structures for nearest neighbor searching and dominance reporting. Full details and pseudocode for all the algorithms and data structures described here can be found in the Master's thesis of the first author [8]. Belazzougui *et al.* give a linear space bound for distance-sensitive queries in the static setting, using almost the same techniques as in the present paper [2]. Our result was obtained independently from the work of Belazzougui *et al.*

2   Preliminaries

We begin by listing some known structures and background information required for our data structure.

Compressed Tries. Our data structure is based on *compressed tries* [5]. These are defined as follows: we interpret the elements from S as bitstrings of length w (the most significant bit being in the leftmost position). The *trie* T' for S is a binary tree of height w. Each node $v \in T'$ corresponds to a bitstring $p_v \in \{0,1\}^*$. The root r has $p_r = \varepsilon$. For each inner node v, the left child u of v has $p_u = p_v 0$, and the right child w of v has $p_w = p_v 1$ (one of the two children may not exist). The bitstrings of the leaves correspond to the elements of S, and the bitstrings of the inner nodes are prefixes for the elements in S, see Figure 1.

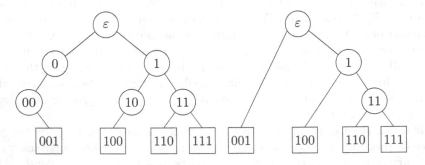

Fig. 1. A trie (left) and a compressed trie (right) for the set 000, 100, 110, 111. The longest common prefix of 101 is 10. The lca of 101 in the compressed trie is the node labeled 1.

The *compressed trie* T for S is obtained from T' by contracting each maximal path of nodes with only one child into a single edge. Each inner node in T has exactly two children, and consequently T has $O(n)$ nodes. Maybe somewhat unusually, in the following, the *height* and *depth* of a node v in T will refer to the corresponding height and depth in the (uncompressed) trie T'. This convention will make the description of the operations more convenient.

Let $q \in \{0,1\}^*$ be a bitstring of length at most w. The *longest common prefix* of q with S, $\mathrm{lcp}_S(q)$, is the longest prefix that q shares with an element in S. We say that q *lies on an edge* $e = (u,v)$ of T if p_u is a prefix of q and q is a proper prefix of p_v. If $\mathrm{lcp}_S(q)$ lies on the edge (u,v), we call u the *lowest common ancestor* of q in T, denoted by $\mathrm{lca}_T(q)$. One can show that $\mathrm{lca}_T(q)$ is uniquely defined.

Associated Keys. Our algorithm uses the notion of *associated keys*. This notion was introduced in the context of *z-fast tries* [1,16], and it is also useful in our data structure.

Associated keys provide a quick way to compute $\mathrm{lca}_T(q)$, for any element $q \in U$. A natural way to find $\mathrm{lca}_T(q)$ is to do binary search on the depth of $\mathrm{lca}_T(q)$: we initialize $(l, r) = (0, w)$ and let $m = (l + r)/2$. We denote by $q' = q_0 \ldots q_{m-1}$ the leftmost m bits of q, and we check whether T has an edge $e = (u,v)$ such that q' lies on e. If not, we set $r = m$, and we continue. Otherwise, we determine if u is $\mathrm{lca}_T(q)$, by testing whether p_v is not a prefix of q. If u is not $\mathrm{lca}_T(q)$, we set $l = m$ and continue. In order to perform this search quickly, we need to find the edge e that contains a given prefix q', if it exists. For this, we precompute for each edge e of T the first time that the binary search encounters a prefix that lies on e. This prefix is uniquely determined and depends only on e, not on the specific string q that we are looking for. We let α_e be this prefix, and we call α_e the *associated key* for $c = (u,v)$, see Figure 2.

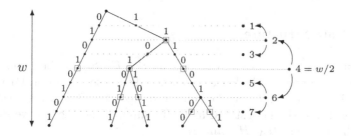

Fig. 2. The associated key α_e of an edge e: we perform a binary search on the height of $\mathrm{lcp}_S(q)$ in T. The *associated key* of an edge e is the prefix of $\mathrm{lcp}_S(q)$ in which the search first encounters the edge e.

The binary search needs $\log w$ steps, and since we assumed that w is a power of two, each step determines the next bit in the binary expansion of the *length* of $\mathrm{lcp}_S(q)$. Thus, the associated key of an edge e can be computed in $O(1)$ time on a word RAM as follows: consider the $\log w$-bit binary expansions $\ell_u = |p_u|_2$ and $\ell_v = |p_v|_2$ of the *lengths* of the prefixes p_u and p_v, and let ℓ' be the longest common prefix of ℓ_u and ℓ_v. We need to determine the first step when the binary search can distinguish between ℓ_u and ℓ_v. Since $\ell_u < \ell_v$, and since the two binary expansions differ in the first bit after ℓ', it follows that ℓ_u begins with $\ell'0$ and ℓ_v begins with $\ell'1$. Thus, let ℓ be obtained by taking ℓ', followed by 1 and enough 0's to make a $\log w$-bit word. Let l be the number encoded by ℓ. Then, the associated key α_e consists of the first l bits of p_v; see [1, 8, 16] for more details.

Hash Maps. Our data structure also makes extensive use of hashing. In particular, we will maintain several succinct hashtables that store additional information for supporting fast queries. For this, we will use a hashtable described by Demaine *et al.* [7]. The following theorem summarizes the properties of their data structure.

Theorem 2.1. *For any $r \geq 1$, there exists a dynamic dictionary that stores entries with keys from U and with associated values of r bits each. The dictionary supports updates and queries in $O(1)$ time, using $O(n \log \log(|U|/n) + nr)$ bits of space. The bounds for the space and the queries are worst-case, the bounds for the updates hold with high probability.* □

3   Static Δ-fast Tries

We are now ready to describe our data structure for the static case. In the next section, we will discuss how to add support for insertions and deletions.

3.1   The Data Structure

Our data structure is organized as follows: let $S \subseteq U$, $|S| = n$, be given. We store S in a compressed trie T. The leaves of T are linked in sorted order. Furthermore, for each node v of T, let T_v be the subtree rooted at v. Then, v stores pointers to the smallest and the largest leaf in T_v. To support the queries, we store three additional hash maps: H_Δ, H_z, and H_b.

First, we describe the hash map H_Δ. Set $m = \log \log w$. For $i = 0, \ldots, m$, we let $h_i = 2^{2^i}$ and $d_i = w - h_i$. The hash map H_Δ stores the following information: for each $s \in S$ and each d_i, $i = 1, \ldots, m$, let $s_i = s_0 \ldots s_{d_i-1}$ be the leftmost d_i-bits of s and let $e = (u, v)$ be the edge of T such that s_i lies on e. Then, H_Δ stores the entry $s_i \mapsto u$.

Next, we describe the hash map H_z. It is defined similarly as the hash map used for z-fast tries [1, 16]. For each edge e of T, let α_e be the associated key of e, as explained in Section 2. Then, H_z stores the entry $\alpha_e \mapsto e$.

Finally, the hash map H_b is used to implement a second layer of indirection that lets us achieve linear space. It will be described below.

3.2   The Predecessor Query

Let $q \in U$ be the query, and let q^- and q^+ be the predecessor and the successor of q in S, as described above. We first show how to get a running time of $O(\log \log \Delta)$ for the queries, with $\Delta = |q - q^+|$. In Theorem 3.2, we will see that this can easily be improved to $\Delta = \min\{|q - q^-|, |q - q^+|\}$.

The predecessor search works in several *iterations*. In iteration i, we consider the prefix q_i that consists of the first d_i bits of q.

First, we check whether H_Δ contains an entry for q_i. If so, we know that T contains an edge e such that q_i lies on e. Hence, q_i must be a prefix of $\mathrm{lcp}_S(q)$. If one of the endpoints of e happens to be $\mathrm{lca}_T(q)$, we are done. Otherwise, we consider the two edges emanating from the lower endpoint of e, finding the edge e' that lies on the path to q. We take the associated key $\alpha_{e'}$ of e', and we use it to continue the binary search for $\mathrm{lca}_T(q)$, as described in Section 2. Since $|q_i| = d_i$, this binary search takes $O(\log(w - d_i)) = O(\log h_i)$ steps to complete. Once the lowest common ancestor $v = \mathrm{lca}_T(q)$ is at hand, we can find the predecessor of q in $O(1)$ additional time: it is either the rightmost element in T_v, the predecessor of the leftmost element in T_v, or the rightmost element in the left subtree of v. Given the pointers stored with v and the leaves of T, all these nodes can be found in $O(1)$ time.

If H_Δ contains no entry for q_i and if q_i does not consist of all 1's, we check if H_Δ contains an entry for $q_i + 1$. Notice that $q_i + 1$ is the successor of q_i. If such an entry exists, we first obtain $u = H_\Delta[q_i + 1]$, and the child v of u such that $q_i + 1$ lies on the edge $e = (u, v)$. Then, we follow the pointer to the leftmost element of T_v. This is the successor q^+ of q. The predecessor q^- can then be found by following the leaf pointers. This takes $O(1)$ time overall.

Finally, if there is neither an entry for q_i nor for $q_i + 1$, we continue with iteration $i + 1$, see Figure 3.

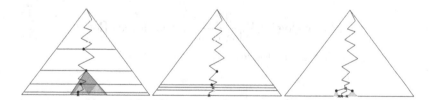

Fig. 3. The query algorithm: first we perform an exponential search from the lowest level, to find a prefix of q_k or $q_k + 1$ (left). If a prefix q_k is found, we perform a binary search for $\mathrm{lca}_T(q)$ (middle), which can then be used to find the predecessor and successor of q (right). If a prefix $q_k + 1$ is found, the successor and predecessor can be found immediately (not shown).

From the above discussion, it follows that the total time for the predecessor query is $O(k + \log h_k)$, where k is the number of iterations and $\log h_k$ is the worst-case time for the predecessor search once one of the lookups in an iteration succeeds. By our predecessor algorithm, we know that S contains no element with prefix q_{k-1} or $q_{k-1} + 1$, but an element with prefix q_k or $q_k + 1$. Thus, there must be at least $2^{w-d_k} = 2^{h_k}$ consecutive elements in $U \setminus S$ following q. By our definition of h_k, it follows that $\Delta \geq 2^{h_{k-1}} = 2^{2^{2^{k-1}}}$, so $k \leq 1 + \log \log \log \Delta$. Furthermore, since $h_k = 2^{2^k} = \left(2^{2^{k-1}}\right)^2 = (h_{k-1})^2$, it follows that $h_k = O(\log^2 \Delta)$.

3.3 Obtaining Linear Space

We now analyze the space requirement for our data structure. Clearly, the trie T and the hash map H_z require $O(n)$ words of space. Furthermore, as described so far, the number of words needed for H_Δ is $O(n \log \log w)$, since we store at most n entries for each height h_i, $i = 0, \ldots, m = \log \log w$.

Using a trick due to Pătrașcu [15], we can introduce another level of indirection to reduce the space requirement to $O(n)$. The idea is to store in H_Δ the *depth* d_u of each branch node u in T_Δ, instead of storing u itself (here, we mean the depth in the original trie, i.e., the length of the prefix p_u). We then use an additional hash map H_b to obtain u. This is done as follows: when trying to find the branch node u for a given prefix q_i, we first get the depth $d_u = |p_u|$ of u from H_Δ. After that, we look up the branch node $u = H_b[q_0 \ldots q_{d_u-1}]$ from the hash map H_b. Finally, we check whether u is actually the lowest branch node of q_i. If any of those steps fails, we return $\bot$.

Let us analyze the needed space: clearly, H_b needs $O(n)$ words, since it stores $O(n)$ entries. Furthermore, we have to store $O(n \log \log w)$ entries in H_Δ, each mapping a prefix q_i to the depth of its lowest branch node. This depth requires $\lceil \log w \rceil$ bits. By Theorem 2.1, a retrieval only hash map for n' items and r bits of data needs $O(n' \log \log \frac{|U|}{n'} + n'r)$ bits. Therefore, the space *in bits* for H_Δ is proportional to

$$n \log \log w \cdot \log \log \frac{|U|}{n \log \log w} + n \log \log w \cdot \lceil \log w \rceil$$
$$= O(n \log \log w \cdot \log w)$$
$$= o(n \cdot w),$$

using $n' = n \log \log w$, $r = \lceil \log w \rceil$ and $w = \log |U|$. Thus, we can store H_Δ in $O(n)$ words of w bits each. The following lemma summarizes the discussion

Lemma 3.1. *The Δ-fast trie needs $O(n)$ words space.*

3.4 Putting it Together

We can now obtain our result for the static predecessor problem.

Theorem 3.2. *Let $U = \{0, \ldots, 2^w - 1\}$ and let $S \subseteq U$, $|S| = n$. The static Δ-fast trie for S requires $O(n)$ words of space, and it can answer a static predecessor*

query for an element $q \in U$ on S in time $O(\log \log \min\{|q - q^-|, |q - q^+|\})$, where q^- and q^+ denote the predecessor and successor of q in S. The preprocessing time is $O(n \log \log \log |U|)$, assuming that S is sorted.

Proof. The regular search for $q \in S$ can be done in $O(1)$ time by a lookup in H_z. We have seen that the predecessor of q can be found in $O(\log \log |q - q^+|)$ time. A symmetric result also holds for successor queries. In particular, we can achieve query time $O(\log \log |q - q^-|)$ by checking for $H_\Delta[q_i - 1]$ instead of $H_\Delta[q_i + 1]$ in the query algorithm.

By interleaving the two searches, we obtain the desired running time of $O(\log \log \min\{|q - q^-|, |q - q^+|\})$. Of course, in a practical implementation, it would be more efficient to check directly for $H_\Delta[q_i - 1]$ and $H_\Delta[q_i + 1]$ in the query algorithm.

The trie T and the hash maps H_z and H_b can be computed in $O(n)$ time, given that S is sorted. Thus, the preprocessing time is dominated by the time to fill the hash map H_Δ. Hence, the preprocessing needs $O(n \log \log \log |U|)$ steps, because $\mathcal{O}(n \log \log w)$ nodes have to be inserted into H_Δ. By Lemma 3.1, the space requirement is linear.   □

4   Dynamic Δ-fast tries

We will now explain how to extend our data structure to the dynamic case. The basic data structure remains the same, but we need to update the hashtables and the trie T after each insertion and deletion. In particular, our data structure requires that for each v in T_v, we can access the leftmost and the rightmost node in the subtree T_v. In the static case, this could be done simply by maintaining explicit pointers from each node $v \in T$ to these nodes in T_v, letting us find the nodes in $O(1)$ time. In the dynamic case, we will maintain a data structure which allows finding and updating these nodes in in $O(\log \log \Delta)$ time.

4.1   Computing Lowest Common Ancestor

To perform the update operation, we need a procedure to compute the lowest common ancestor $\mathrm{lca}_T(q)$ for any given element $q \in U$. For this, we proceed as in the query algorithm from Section 3.2, but skipping the lookups for $H_\Delta[q_i - 1]$ and $H_\Delta[q_i + 1]$. By the analysis in Section 3.2, this will find $\mathrm{lca}_T(q)$ in time $O(\log \log l)$, where l is height of $\mathrm{lca}_T(q)$ in T.

Unfortunately, it may happen that this height l is as large as w, even if q is close to an element in the current set S. To get around this, we use a trick of Bose *et al.* [3]. Namely, their idea is to perform a random shift of the universe. More precisely, we pick a random number $r \in U$, and we add r to all query and update elements that appear in the data structure (modulo $|U|$).

Lemma 4.1 (Lemma 4 in [3]). *Let $x, y \in U$ be two fixed elements in U. Let $r \in U$ be picked uniformly at random. After a random shift of U by r, the expected height of the lowest common ancestor of x and y in a compressed trie is $O(\log |x - y|)$.*   □

Corollary 4.1. *Let $S \subseteq U$ and let T be a randomly shifted Δ-fast trie storing S. Let $q \in U$. We can find $\mathrm{lca}_T(q)$ in expected time $O(\log \log \Delta)$, where $\Delta = \min\{|q - q^+|, |q - q^-|\}$, the elements q^+ and q^- being the predecessor and successor of q in S. The expectation is over the random choice of the shift r.*

Proof. Suppose without loss of generality that $\Delta = |q - q^+|$. By Lemma 4.1, the expected height h_k of the lowest common ancestor of q and q^+ is $O(\log \Delta)$. We perform the doubly exponential search on the prefixes of q, as in Section 3.2 (without checking $q_i + 1$) to find the height h_k. After that, we resume the search for $\mathrm{lca}_T(q)$ on the remaining h_k bits. Since $h_k = O(\log \Delta)$ in expectation, it follows by Jensen's inequality that the number k of loop iterations to find h_k is $O(\log \log \log \Delta)$ in expectation. Thus, the expected running time is proportional to $k + \log h_k = O(\log \log \Delta)$.   □

4.2   Managing the Left- and Rightmost Elements of the Subtrees

We also need to maintain for each node $v \in T$ the leftmost and the rightmost element in the subtree T_v. In the static case, it suffices to have direct pointers from v to the respective leaves, but in the dynamic case, we need an additional data structure.

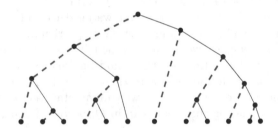

Fig. 4. For each leaf v' of T, the nodes $v \in T$ for which v is the leftmost leaf in T_v if a subpath of a root-to-leaf path in T. Considering these subpaths for all leaves in T, we obtain a *path decomposition* of T (shown in bold).

To do this, we observe the following: let $v' \in T$ be a leaf in T. Then, v' is the leftmost (or rightmost) leaf in the subtrees of at most w ancestors v of v'. Furthermore, all these nodes form a subpath (more precisely, a prefix) of the path from v to the root, see Figure 4. Hence, if we maintain the nodes of this subpath in a concatenable queue data structure (realized by, e.g., a balanced binary tree) [14], we can obtain $O(\log w)$ update and query time to find the leftmost (or rightmost) element in T_v for each $v \in T$. However, we need that the update and query time for this data structure depend on the height h_i (i.e, the remaining bits) of the query node v. Thus, we partition the possible heights $\{0, 1, \ldots, w\}$ of the nodes on a subpath into the sets $T_{-1} = \{0\}$, $T_i = [2^i, 2^{i+1})$, for $i = 0, \ldots, \log w - 1$, and $T_{\log w} = \{w\}$. Each set is managed by a balanced binary tree, and the roots of the trees are linked together. The height of the i-th binary search tree is $\log |T_i| = O(i)$. Furthermore, if a query node of height h is given, the set $T_{\lfloor \log h \rfloor}$ is responsible for it, see Figure 5.

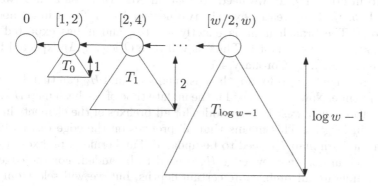

Fig. 5. The data structure for a subpath. We group the nodes in the subpath according to their heights, where the groups grow exponentially in size. Each group is represented by a balanced tree. The roots are joined in a linked list. With this data structure, a node v of height h can find the leftmost leaf in the subtree T_v in time $O(\log h)$.

Moreover, T_{-1} is a leaf (the depth of that node is w) in the trie and therefore the minimum of the whole subpath. Thus, the minimum of a subpath can be found from a given node $v \in T_i$ in $O(i)$ time by following the pointers to the root of T_i and the pointers down to T_{-1}.

If a node v has $h_k = O(\log \Delta)$ height (remaining bits), the node is within the tree $T_{\lfloor \log h_k \rfloor}$. Thus, it takes $O(\log h_k) = O(\log \log \Delta)$ time to find the leftmost or rightmost leaf in T_v.

Furthermore, we can support the following update operations: (i) **split**: given a subpath π and a node v on π, split the representation of π into two representations, one for the *lower* subpath from the leaf up to the child of v, and one for the *upper* subpath starting from v; and (ii) **join**: given a representation of an upper subpath starting at a node v obtained from an operation of type (i), and a representation for a lower subpath up to a child of v, join the two representations into the representation for a joint subpath. Given the data structure, we can support both **split** and **join** in $O(\log h)$ time, where h is the height of the node v where the operation occurs. This decomposition of T into dynamically changing suppaths is similar to the *preferred paths decomposition* of Tango trees [6].

4.3 Performing an Update

We know from the Lemma 4.1, that the lowest common ancestor of a query element q has expected height $h_k = O(\log \Delta)$.

Lemma 4.2. *Let $S \subseteq U$, and let T be a randomly shifted Δ-fast tree for S. Let $q \in U$ be fixed. We can insert or delete q into T in $\mathcal{O}(\log \log \Delta)$ expected time, where the expectation is over the random choice of the shift r.*

Proof. To insert q into T, we need to split an edge (u, v) of T into two edges (u, b) and (b, v). This creates exactly two new nodes in T, an inner node and a leaf node. The branch node is exactly $\mathrm{lca}_T(q)$, and it has expected height $h_k = O(\log \Delta)$, by Lemma 4.1. Thus, it will take $O(\log \log \Delta)$ expected time to find the edge (u, v), by Corollary 4.1.

Once the edge (u, v) is found, the hash maps H_z and H_u can then be updated in constant time. Now let us consider the update time of the hash map H_Δ. Recall that H_Δ stores the lowest branch nodes for all prefixes of the elements in S that have certain lengths. This means that all prefixes on the edge (b, v) which are stored in the hash map T_Δ need to be updated. Furthermore, prefixes at certain depths which are on the new edge (b, q) need to be added. For the edge (b, v), we will enumerate all prefixes at certain depths, but we will select only those that lie on the edge (b, v). This needs $O(\log \log \log \Delta)$ insertions and updates in total: we have to insert the prefixes $q_0 \ldots q_{d_i}$ for all $i \geq 1$ with $d_i < |b|$. Since we defined $d_i = w - h_i = w - 2^{2^i}$, and since $|b| = w - O(\log \Delta)$, we have that $d_i \leq |b|$ as soon as $c \log \Delta < 2^{2^i}$. This holds for $i > \log \log(c \log \Delta)$, and hence $i = \Theta(\log \log \log \Delta)$.

After that, the leftmost and rightmost elements for the subtrees of T have to be updated. For this, we need to add one subpath for the new leaf q, and we may need to split a subpath at a node of height $h_k = O(\log \Delta)$ and join the resulting upper path with the newly created subpath. As we have seen, this takes $O(\log h_k) = O(\log \log \Delta)$ time.

The operations for deleting an element q from S are symmetric. $\square$

The following theorem summarizes our result.

Theorem 4.3. *Let $r \in U$ be picked uniformly at random. After performing a shift of U by r, the Δ-fast trie provides a data structure for the dynamic predecessor problem such that the query operations take $O(\log \log \Delta)$ worst-case time and the update operations need $O(\log \log \Delta)$ expected time, for $\Delta = \min\{|q - q^+|, |q - q^-|\}$, where q is the requested element and q^+ and q^- are the predecessor and successor of q in the current set S. At any point in time, the data structure needs $O(n)$ words of space, where $n = |S|$.*

5   Applications

Bose *et al.* [3] describe how to combine their structure with a technique of Chan [4] and random shifting [11, Chapter 11] for obtaining a data structure for distance-sensitive approximate nearest neighbor queries on a grid. More precisely, let $d \in \mathbb{N}$ be the fixed dimension, $U = \{0, \ldots, 2^w - 1\}$ be the universe, and let $\varepsilon > 0$ be given. The goal is to maintain a dynamic set $S \subseteq U^d$ under insertions, deletions, and ε-*approximate nearest neighbor queries*: given a query point $q \in U^d$, find a $p \in S$ with $d_2(p,q) \le (1+\varepsilon)d_2(p,S)$. Plugging our Δ-fast tries into the structure of Bose *et al.* [3, Theorem 9], we can immediately improve the space requirement of their structure to linear:

Theorem 5.1. *Let* $U = \{0,\ldots,2^w-1\}$ *and let* d *be a constant. Furthermore, let* $\varepsilon > 0$ *be given. There exists a data structure that supports* $(1+\varepsilon)$-*approximate nearest neighbor queries over a subset* $S \subseteq U^d$ *in* $(1/\varepsilon^d)\log\log\Delta)$ *expected time and insertions and deletions of elements of* U^d *in* $O(\log\log\Delta)$ *expected time. Here,* Δ *denotes the Euclidean distance between the query element and* S. *At any point in time, the data structure requires* $O(n)$ *words of space, where* $n = |S|$.

As a second application, Bose *et al.* [3] present a data structure for dominance queries on a grid, based on a technique of Overmars [13]. Again, let $U = \{0,\ldots,2^w-1\}$, and let $S \subseteq U^2$, $|S| = n$ be given. The goal is to construct a data structure for *dominance queries* in S. That is, given a query point $q \in U^2$, find all points p in S that *dominate* q, i.e., for which we have $p_x \ge q_x$ and $p_y \ge q_y$, there p_x, p_y and q_x, q_y are the x- and y-coordinates of p and q.

Again, using Δ-fast tries, we can immediately improve the space requirement for the result of Bose *et al.* [3, Theorem 10, Corollary 13].

Theorem 5.2. *Let* $U = \{0,\ldots,2^w-1\}$, *and let* $S \subseteq U^2$, $|S| = n$ *be given. There exists a data structure that reports the points in* S *that dominate a given query point* $q = (a,b) \subset U^2$ *in expected time* $O(\log\log(h+v)+k)$, *where* $h = 2^w - a$, $v = 2^2 - b$, *and* k *is the number of points in* S *dominated by* q. *The data structure uses* $O(n\log n)$ *space.*

6 Conclusion

We present a new data structure for local searches in bounded universes. This structure now interpolates seamlessly between hashtables and van-Emde-Boas trees, while requiring only a linear number of words. This provides an improved, and in our opinion also slightly simpler, version of a data structure by Bose *et al.* [3]. All the operations of our structure can be presented explicitly in pseudocode. This can be found in the Master's thesis of the first author [8].

Acknowledgments. We thank the anonymous reviewers for numerous insightful comments that improved the quality of the paper. In particular, we would like to thank the anonymous reviewers for pointing us to [2].

References

1. Belazzougui, D., Boldi, P., Vigna, S.: Dynamic z-fast tries. In: Proc. 17th Int. Symp. String Processing and Information Retrieval (SPIRE). pp. 159–172 (2010)
2. Belazzougui, D., Boldi, P., Vigna, S.: Predecessor search with distance-sensitive query time (2012), arXiv:1209.5441
3. Bose, P., Douïeb, K., Dujmovic, V., Howat, J., Morin, P.: Fast local searches and updates in bounded universes. Comput. Geom. Theory Appl. 46(2), 181–189 (2013)

4. Chan, T.M.: Closest-point problems simplified on the RAM. In: Proc. 13th Annu. ACM-SIAM Sympos. Discrete Algorithms (SODA). pp. 472–473 (2002)
5. Cormen, T.H., Leiserson, C.E., Rivest, R.L., Stein, C.: Introduction to algorithms. MIT Press, third edn. (2009)
6. Demaine, E.D., Harmon, D., Iacono, J., Pătrașcu, M.: Dynamic optimality – almost. SIAM J. Comput. 37(1), 240–251 (2007)
7. Demaine, E.D., Meyer auf der Heide, F., Pagh, R., Pătrașcu, M.: De dictionariis dynamicis pauco spatio utentibus. In: Proc. 7th Latin American Symp. Theoretical Inf. (LATIN). pp. 349–361 (2006)
8. Ehrhardt, M.: An In-Depth Analysis of Data Structures Derived from van-Emde-Boas-Trees. Master's thesis, Freie Universität Berlin (2015), http://www.mi.fu-berlin.de/inf/groups/ag-ti/theses/download/Ehrhardt15.pdf
9. van Emde Boas, P., Kaas, R., Zijlstra, E.: Design and implementation of an efficient priority queue. Math. Systems Theory 10(2), 99–127 (1976)
10. van Emde Boas, P.: Preserving order in a forest in less than logarithmic time and linear space. Inform. Process. Lett. 6(3), 80–82 (1977)
11. Har-Peled, S.: Geometric approximation algorithms, Mathematical Surveys and Monographs, vol. 173. American Mathematical Society (2011)
12. Knuth, D.E.: The art of computer programming. Vol. 3. Sorting and searching. Addison-Wesley, second edn. (1998)
13. Overmars, M.H.: Efficient data structures for range searching on a grid. J. Algorithms 9(2), 254–275 (1988)
14. Preparata, F.P., Shamos, M.I.: Computational geometry. An introduction. Springer Verlag (1985)
15. Pătrașcu, M.: vEB space: Method 4 (2010), http://infoweekly.blogspot.de/2010/09/veb-space-method-4.html
16. Ružić, M.: Making deterministic signatures quickly. ACM Transactions on Algorithms 5(3), 26:1–26:26 (2009)

Split Packing: Packing Circles into Triangles with Optimal Worst-Case Density

Sándor P. Fekete, Sebastian Morr, and Christian Scheffer

Department of Computer Science, TU Braunschweig, Germany
s.fekete@tu-bs.de, sebastian@morr.cc, scheffer@ibr.cs.tu-bs.de

Abstract. In the *circle packing problem for triangular containers*, one asks whether a given set of circles can be packed into a given triangle. Packing problems like this have been shown to be NP-hard. In this paper, we present a new sufficient condition for packing circles into any right or obtuse triangle using only the circles' combined area: It is possible to pack any circle instance whose combined area does not exceed the triangle's incircle. This area condition is tight, in the sense that for any larger area, there are instances which cannot be packed.

A similar result for square containers has been established earlier this year, using the versatile, divide-and-conquer based Split Packing algorithm. In this paper, we present a generalized, weighted version of this approach, allowing us to construct packings of circles into asymmetric triangles. It seems crucial to the success of these results that Split Packing does not depend on an orthogonal subdivision structure. Beside realizing all packings below the critical density bound, our algorithm can also be used as a constant-factor approximation algorithm when looking for the smallest non-acute triangle of a given side ratio in which a given set of circles can be packed.

An interactive visualization of the Split Packing approach and other related material can be found at https://morr.cc/split-packing/.

1 Introduction

Given a set of circles, can you decide whether it is possible to pack these circles into a given container without overlapping one another or the container's boundary? This naturally occurring *circle packing problem* has numerous applications in engineering, science, operational research and everyday life. Examples include packaging cylinders [2], bundling tubes or cables [16, 18], the cutting industry [17], the layout of control panels [2], the design of digital modulation schemes [14], or radio tower placement [17]. Further applications stem from chemistry [19], foresting [17], and origami design [9].

Despite its simple formulation, packing problems like these were shown to be NP-hard in 2010 by Demaine, Fekete, and Lang [3], using a reduction from 3-PARTITION. Additionally, due to the irrational coordinates which arise when packing circular objects, it is also surprisingly hard to solve circle packing problems in practice. Even when the input consists of equally-sized circles, exact

© Springer International Publishing AG 2017
F. Ellen et al. (Eds.): WADS 2017, LNCS 10389, pp. 373–384, 2017.
DOI: 10.1007/978-3-319-62127-2_32

boundaries for the smallest square container are currently only known for up to 35 circles, see [10]. For right isosceles triangular containers, optimal results have been published for up to 7 equal circles, see [20].

The related problem of packing square objects has long been studied. Already in 1967, Moon and Moser [12] found a sufficient condition: They proved that it is possible to pack a set of squares into the unit square in a shelf-like manner if their combined area does not exceed 1/2, see Figure 2. At the same time, 1/2 is the *largest upper area bound* you could hope for, because two squares larger than the quarter-squares depicted in Figure 1 cannot be packed anymore. We call the ratio between the largest combined object area that can always be packed and the area of the container the problem's *critical density*, or *worst-case density*.

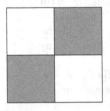

Fig. 1. Worst-case instance for packing squares into a square.

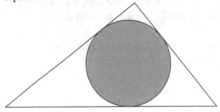

Fig. 2. Example of Moon and Moser's shelf-packing.

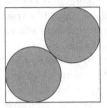

Fig. 3. Worst-case instance for packing circles into a square.

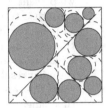

Fig. 4. Example packing produced by Split Packing.

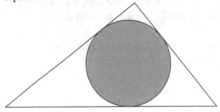

Fig. 5. Suspected worst-case instance for packing circles into a non-acute triangle.

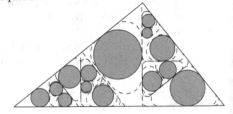

Fig. 6. Example packing produced by Split Packing.

We recently showed a similar result for circular objects: Each circle instance not exceeding the area of the instance shown in Figure 3 can be packed, and this area condition is tight [13]. Proving this required a fundamentally different approach than Moon and Moser's orthogonal shelf-packing, compare Figure 4.

In this paper, we consider the problem of packing circles into non-acute triangular containers. It is obvious that circles larger than a triangle's incircle cannot be packed (compare Figure 5), but is it also possible to pack all circle instances of up to that combined area? We will answer this question affirmatively and introduce a weighted modification of the Split Packing algorithm, allowing us to pack circles into asymmetric non-acute triangles with critical density. See Figure 6 for an example packing.

Many authors have considered heuristics for circle packing problems, see [7, 17] for overviews of numerous heuristics and optimization methods. The best known solutions for packing equal circles into squares, triangles and other shapes are continuously published on Specht's website http://packomania.com [15].

That being said, the literature on exact approximation algorithms which actually give performance guarantees is small. Miyazawa et al. [11] devised asymptotic polynomial-time approximation schemes for packing circles into the smallest number of unit square bins. And recently, Hokama, Miyazawa, and Schouery [8] developed an asymptotic approximation algorithm for the online version of that problem. To the best of our knowledge, this paper presents the first approximation algorithm for packing circles into triangular containers.

1.1 Results

We show that, for any right or obtuse triangle, any circle instance with a combined area of up to the triangle's incircle can be packed into that triangle. At the same time, for any larger area, there are instances which cannot be packed, making the ratio between the incircle's and the triangle's area the triangle's *critical density*. For a right isosceles triangle, this density is approximately 53.91%. In the general case, the critical density of a non-acute triangle with side lengths a, b, and c is

$$\sqrt{\frac{-(a-b-c)(a+b-c)(a-b+c)}{(a+b+c)^3}}\pi.$$

Our proof is constructive: The *Split Packing algorithm* can be used to construct the packings in polynomial time. Split Packing can also be used as a constant-factor approximation algorithm of the smallest-area non-acute triangle of a given side ratio which can pack a given set of circles. The approximation factor is the reciprocal of the critical density.

While we focus on triangular containers in this paper, we see more opportunities to generalize the Split Packing approach for other container and object types. We discuss some of these extensions in the conclusion on page 11.

1.2 Key ideas

The *Split Packing* approach, which we successfully used for packing circles into square containers earlier this year [13], is built on two basic ideas:

First, it applies a recursive subdivision strategy, which cuts the container into smaller triangles, while keeping the combined area of the triangles' incircles constant. And second, it performs the splitting of the circle instance into subgroups using an algorithm which resembles greedy scheduling. This makes sure the resulting subgroups are *close* to equal in terms of their combined area. If the groups' areas deviate from the targeted 1:1 ratio, we can gain information

about the minimum circle size in the larger group, allowing us to round off the
subcontainer triangles.

In this paper, we introduce a weighted generalization of the Split Packing
approach: When packing into asymmetric triangles, we do not want the resulting
groups to have equal area, as it is not possible to cut the container into two
subtriangles of equal size. Instead, we target a different area ratio, defined by the
incircles of the two triangles created by cutting the container orthogonally its
the base through its tip, see Figure 8 on page 8. We call this desired area ratio
the *split key*.

The rest of the paper will detail this process.

2   Greedy splitting

The following definitions makes it easier to talk about the properties of circle
instances:

Definition 1. *A circle instance is a multiset of nonnegative real numbers, which
define the circles' areas. For any circle instance C, $\mathrm{sum}(C)$ is the combined area
of the instance's circles and $\min(C)$ is the area of the smallest circle contained
in the instance.*

Definition 2. $\mathbb{C}$ *is the set of all circle instances.* $\mathbb{C}(a)$ *consists of exactly those
circle instances C with $\mathrm{sum}(C) \leq a$. Finally, $\mathbb{C}(a,b)$ consists of exactly those
circle instances $C \in \mathbb{C}(a)$ with $\min(C) \geq b$.*

Algorithm 1 takes a circle instance C, and splits it into two groups according
to the *split key* F, which determines the targeted ratio of the resulting groups'
combined areas. Because the method resembles a greedy scheduling algorithm, we
call the process *greedy splitting*. The algorithm first creates two empty "buckets",
and in each step adds the largest remaining circle of the input instance to the
"relatively more empty" bucket:

Algorithm 1 SPLIT(C, F)

Input: A circle instance C, sorted by size in descending order, and a split key
$\quad F = (f_1, f_2)$
Output: Circle instances C_1, C_2
$\quad C_1 \leftarrow \emptyset$
$\quad C_2 \leftarrow \emptyset$
$\quad$**for all** $c \in C$ **do**
$\quad\quad j = \arg\min_i \frac{\mathrm{sum}(C_i)}{f_i}$ $\qquad\qquad$ ▷ Find the index of the more empty bucket.
$\quad\quad C_j \leftarrow C_j \cup \{c\}$
$\quad$**end for**

If the resulting groups' area ratio deviates from the area ratio targeted by
the split key, we gain additional information about the "relatively larger" group:

The more this group exceeds its targeted ratio, the larger the minimum size of its elements, allowing a "more rounded" subcontainer in the packing. See Figure 9 on page 9 for an illustration.

Lemma 1. *For any C_1 and C_2 produced by* SPLIT$(C, (f_1, f_2))$:

$$\min(C_i) \geq \text{sum}(C_i) - f_i \frac{\text{sum}(C_j)}{f_j}$$

Proof. If $\frac{\text{sum}(C_i)}{f_i} < \frac{\text{sum}(C_j)}{f_j}$, then the lemma says that $\min(C_i)$ is larger than a negative number, which is certainly true.

Otherwise, set $r := \frac{\text{sum}(C_j)}{f_j}$. This value describes the smaller "relative filling level" by the time the algorithm ends. Now assume for contradiction C_i contained an element smaller than $\text{sum}(C_i) - f_i r$. As the elements were inserted by descending size, all elements which were put into C_i after that element would have to be at least as small. So the final element put into C_i (let us call it c) would be smaller than $\text{sum}(C_i) - f_i r$, as well.

But this means that

$$\frac{\text{sum}(C_i) - c}{f_i} > \frac{\text{sum}(C_i) - (\text{sum}(C_i) - f_i r)}{f_i} = r,$$

meaning that at the moment before c was inserted, the relative filling level of C_i would already have been larger than r. Recall that r is the smallest filling level of any group by the time the algorithm ends, meaning that at the time when c is inserted, C_i's filling level is already larger than the filling level of the other group. This is a contradiction, as the greedy algorithm would choose to put c not into C_i, but into the other group with the smaller filling level in this case. □

We are now going to define a term which encapsulates all properties of the circle instances output by SPLIT. These properties depend on the used split key F, and also on the combined area a and the minimum circle size b of the circle instance, which is why it the term has three parameters.

Definition 3. *For any $0 \leq b \leq a$ and any split key $F = (f_1, f_2)$, we say that the tuples $(a_1, b_1), (a_2, b_2)$ are (a, b, F)-conjugated if*

- $a_1 + a_2 = a$,
- $b_i \geq b$, *and*
- $b_i \geq a_i - f_i \frac{a_j}{f_j}$.

Two circle instances C_1 and C_2 are (a, b, F)-conjugated if there are any (a, b, F)-conjugated tuples (a_1, b_1) and (a_2, b_2) so that $C_1 \in \mathbb{C}(a_1, b_1)$ and $C_2 \in \mathbb{C}(a_2, b_2)$.

We can now associate this property with SPLIT in the following theorem:

Theorem 1. *For any $C \in \mathbb{C}(a, b)$ and any split key $F = (f_1, f_2)$, SPLIT(C, F) always produces two (a, b, F)-conjugated subinstances.*

Proof. That the subinstances' combined areas add up to a follows directly from the algorithm. As the minimum size of all circles in C is b, this must also be true for the subinstances, so $\min(C_i) \geq b$. The other minimum-size property follows from Lemma 1.   □

3   Split Packing

The SPLIT algorithm presented in the previous section, in addition to the properties of the instances it produces, are the foundations on which we now build the central theorem of this paper. Split Packing by itself is a general framework to pack circles and other shapes into containers. We will apply the Split Packing theorem to triangular containers in the next section.

We will often want to state that a shape can pack all circle instances which belong to a certain class. For this, we define the term C-*shape*:

Definition 4. *For any* $C \subseteq \mathbb{C}$, *a* C-shape *is a shape in which each* $C \in C$ *can be packed.*

For example, if a shape is a $\mathbb{C}(a)$-shape, it means that it can pack all circle instances with a combined area of a. And a $\mathbb{C}(a, b)$-shape can pack all circle instances with a combined area of a, whose circles each have an area of at most b.

We can now state our central theorem: If it is possible to find two subcontainers which fit in a given shape, and which can pack all possible subinstances produced by SPLIT, it is possible to pack the original class of circle instances into that shape.

Theorem 2 (Split Packing). *A shape* s *is a* $\mathbb{C}(a, b)$-*shape if there is a split key* F, *so that for all* (a, b, F)-*conjugated tuples* (a_1, b_1) *and* (a_2, b_2) *one can find a* $\mathbb{C}(a_1, b_1)$-*shape and a* $\mathbb{C}(a_2, b_2)$-*shape which can be packed into* s.

Proof. Consider an arbitrary $C \in \mathbb{C}(a, b)$. We use SPLIT(C, F) to produce two subinstances C_1 and C_2. We know from Theorem 1 that those subinstances will always be (a, b, F)-conjugated. So if we can indeed find two shapes which can pack these subinstances, and if we can pack these two shapes into s, then we also can pack the original circle instance C into s.

Note that in the special case that C consists of a single circle, SPLIT(C, F) will yield two circle instances $C_1 = \{C\}$ and $C_2 = \emptyset$. For this case, Theorem 1 guarantees a minimum size of a for the first group, and the associated $\mathbb{C}(a_1, b_1)$-shape is just an a-circle. This means that we can simply place the input circle in the container, and stop the recursion at this point.   □

Written as an algorithm, Split Packing looks like this:

Algorithm 2 SPLITPACK(s, C)

Input: A $\mathbb{C}(a,b)$-shape s and a circle instance $C \in \mathbb{C}(a,b)$, sorted by size in descending order

Output: A packing of C into s

    Determine split key F for shape s

    $(C_1, C_2) \leftarrow$ SPLIT(C, F)   ▷ See Algorithm 1.

    for all $i \in \{1, 2\}$ **do**

        $a_i \leftarrow$ sum(C_i)

        $b_i \leftarrow$ minimum guarantee for C_i ▷ See Lemma 1.

        Determine a $\mathbb{C}(a_i, b_i)$-shape s_i

        SPLITPACK(s_i, C_i)

    end for

    Pack s_1, s_2, and their contents into s

Note that the Split Packing algorithm can easily be extended to allow splitting into more than two subgroups. For simplicity, we only describe the case of two subgroups here, as this suffices for the shapes we discuss in this paper.

3.1 Analysis

The analysis of the Split Packing approach follows exactly the same lines as in our previous paper [13]. We will repeat the result here without proof.

Theorem 3. *Split Packing requires $\mathcal{O}(n)$ basic geometric constructions and $\mathcal{O}(n^2)$ numerical operations.*

Theorem 4. *Split Packing, when used to pack circles into a $\mathbb{C}(a,b)$-shape of area A, is an approximation algorithm with an approximation factor of $\frac{A}{a}$, compared to the container of minimum area.*

4 Packing into hats

After this general description of Split Packing, we will now apply it to concrete containers. We start with an observation:

    If all circles which we want to pack have a certain minimum size, sharp corners of the container cannot be utilized anyway. This observation motivates a family of shapes which resemble rounded triangles. We call these shapes *hats*:

Definition 5. *For each $0 \leq b \leq a$, an (a,b)-hat is a non-acute triangle with an incircle of area a, whose corners are rounded to the radius of a b-circle, see Figure 7. Call the two smaller angles of the original triangle left-angle and right-angle. If we say right hat, the hat is based on a right triangle.*

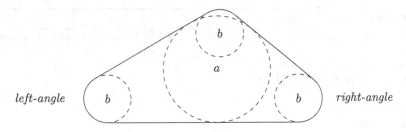

Fig. 7. An (a, b)-hat.

We will now proceed to show that all circle instances with a combined area of up to a with a minimum circle size of b can be packed into an (a, b)-hat.

First, it is important to choose the correct split key when packing into asymmetric hats. We are aiming for a group ratio which will lead to a cut through the hat's tip if it is reached exactly:

Definition 6. *To get a hat's associated split key, split the underlying triangle orthogonally to its base through its tip, and inscribe two circles in the two sides, see Figure 8. The areas of these circles are the two components of the hat's split key.*

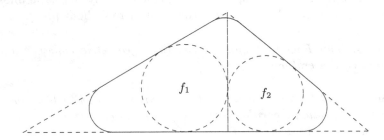

Fig. 8. A hat's *associated split key* equals (f_1, f_2)

Lemma 2. *Consider an $(a, 0)$-hat with the associated split key $F = (f_1, f_2)$, and call its left- and right-angles α and β. For all $(a, 0, F)$-conjugated tuples (a_1, b_1) and (a_2, b_2), the following two shapes can be packed into the hat:*

– *a right (a_1, b_1)-hat with a right-angle of α and*
– *a right (a_2, b_2)-hat with a left-angle of β.*

The proof of this theorem is rather technical in nature. We omit it here due to space constraints, refer to the full version [6]. See Figure 9 for an intuition of what the resulting hats look like. Note that, as the hats' incircles are getting larger than the targeted area ratio, their corners become more rounded so that they don't overlap the container's boundary.

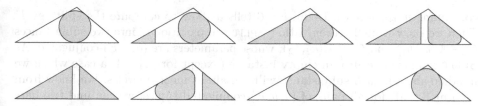

Fig. 9. Hat-in-hat packings for different ratios of a_1 and a_2.

In the previous lemma, the container is always an $(a, 0)$-hat, which is essentially a non-rounded triangle with an incircle of a. The next lemma extends this idea to hats which are actually rounded. It is identical to Lemma 2, except that the rounding of the container hat is no longer 0, but b.

Lemma 3. *Consider an (a, b)-hat with the associated split key $F = (f_1, f_2)$, and call its left- and right-angles α and β. For all (a, b, F)-conjugated tuples (a_1, b_1) and (a_2, b_2) with $a_1 + a_2 \leq a$, the following two shapes can be packed into the hat:*

 - a right (a_1, b_1)-hat with a right-angle of α and
 - a right (a_2, b_2)-hat with a left-angle of β.

Proof. Lemma 2 tells us that this theorem is true for $b = 0$. Now, the container's corners can be rounded to the radius of a b-circle, and we need to show that the two hats from the previous construction still fit inside. But all of the two hat's corners are also rounded to (at least) the same radius (see Theorem 1), so they will never overlap the container, see Figure 10. □

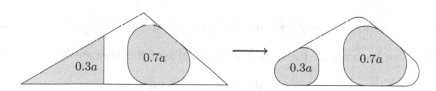

Fig. 10. Rounding all hats' corners by the same radius does not affect the packing.

With these preparations, we can apply Split Packing to hats:

Theorem 5. *Given an (a, b)-hat, all circle instances with a combined area of at most a and a minimum circle size of at least b can be packed into that hat.*

Proof. We proof by induction that we can pack each $C \in \mathbb{C}(a, b)$ into the hat:

If C only consists of a single circle, it can be packed into the hat, as it is at most as big as the hat's incircle.

Now assume that for any $0 \leq b \leq a$, any (a, b)-hat could pack all circle instances into $\mathbb{C}(a, b)$ with at most n circles. Consider a circle instance $C \in \mathbb{C}(a, b)$

containing $n + 1$ circles. Definition 6 tells us how to compute the split key F. Then we know from Theorem 1 that SPLIT will partition C into two subinstances $C_1 \in \mathbb{C}(a_1, b_1)$ and $C_2 \in \mathbb{C}(a_2, b_2)$, whose parameters are (a, b, F)-conjugated. As SPLIT can never return an empty instance (except for $|C| = 1$, a case which we handled above), each subinstance will contain at most n circles. We know from Lemma 3 that, for all pairs of (a, b, F)-conjugated tuples, we can find two hats with matching parameters which fit into the container hat. By assumption, these hats can now pack all instances from $\mathbb{C}(a_1, b_1)$ and $\mathbb{C}(a_2, b_2)$, respectively, which means that they can especially also pack C_1 and C_2. If we then pack the two hats into the container, we have constructed a packing of C into the container hat.

By induction, we can pack each $C \in \mathbb{C}(a, b)$ into the (a, b)-hat. $\qquad\square$

Finally, we can state this paper's central result:

Theorem 6. *Given a non-acute triangle with an incircle of area a, all circle instances with a combined area of up to a can be packed into the triangle, and this bound is tight. Expressed algebraically, for a triangle with side lengths a, b, and c, the critical density is*

$$\sqrt{\frac{-(a - b - c)(a + b - c)(a - b + c)}{(a + b + c)^3}}\, \pi.$$

Proof. The triangle is an $(a, 0)$-hat, which by Theorem 5 is a $\mathbb{C}(a)$-shape.

On the other hand, a single circle of area $a + \varepsilon$ cannot be packed, as the incircle is by definition the largest circle which fits into the triangle.

As for the algebraic formulation of the critical density, the area of the triangle can be calculated using Heron's formula:

$$\Delta(a, b, c) := \sqrt{s(s - a)(s - b)(s - c)} \text{ with } s = \frac{a + b + c}{2}$$

It is also known that the radius of the incircle of this triangle is

$$R(a, b, c) := \frac{\Delta(a, b, c)}{s} \text{ with } s = \frac{a + b + c}{2},$$

so the incircle has an area of

$$I(a, b, c) = \pi R(a, b, c)^2 = \frac{(a + b - c)(c + a - b)(b + c - a)}{4(a + b + c)}.$$

Finally, the ratio between the areas of the circle and the triangle can be calculated to be

$$\frac{I(a, b, c)}{\Delta(a, b, c)} = \sqrt{\frac{-(a - b - c)(a + b - c)(a - b + c)}{(a + b + c)^3}}\, \pi,$$

For a right isosceles triangle, this density is approximately 53.91%. $\qquad\square$

5   Conclusion

In this paper, we presented a constructive proof of the critical densities when packing circles into right or obtuse triangles, using a weighted Split Packing technique. We see more opportunities to apply this approach in the context of other packing and covering problems.

It is possible to use Split Packing to pack into **more container types**. At this point, we can establish the critical densities for packing circles into equilateral triangles and rectangles exceeding a certain aspect ratio. The case of acute triangles is still open, we discuss why the approach presented in this paper does not work there in the full version [6].

Split Packing can also be extended to pack **objects other than circles**. We can establish the critical densities for packing octagons into squares, and think we can describe the maximum shape which can be packed into squares using Split Packing.

Another natural extension is the **online version** of the problem. The current best algorithm that packs squares into a square in an online fashion by Brubach [1], based on the work by Fekete and Hoffmann [4, 5], gives a density guarantee of $\frac{2}{5}$. It is possible to directly use this algorithm to pack circles into a square in an online situation with a density of $\frac{\pi}{10} \approx 0.3142$. It would be interesting to see whether some form of online Split Packing would give better results.

A related problem asks for the smallest area so that we can always **cover the container** with circles of that combined area. For example, we conjecture that for an isosceles right triangle, any circle instance with a total area of at least its excircle's area is sufficient to cover it.

Acknowledgements
We thank the three anonymous reviewers for their helpful comments.

References

[1]   Brian Brubach. "Improved Bound for Online Square-into-Square Packing". In: *Approximation and Online Algorithms*. Springer, 2014, pp. 47–58.

[2]   Ignacio Castillo, Frank J. Kampas, and János D. Pintér. "Solving circle packing problems by global optimization: numerical results and industrial applications". In: *European Journal of Operational Research* 191(3) (2008), pp. 786–802.

[3]   Erik D. Demaine, Sándor P. Fekete, and Robert J. Lang. "Circle Packing for Origami Design is Hard". In: *5th International Conference on Origami in Science, Mathematics and Education*. AK Peters/CRC Press. 2011, pp. 609–626.

[4]   Sándor P. Fekete and Hella-Franziska Hoffmann. "Online Square-into-Square Packing". In: *APPROX-RANDOM*. 2013, pp. 126–141.

[5]   Sándor P. Fekete and Hella-Franziska Hoffmann. "Online Square-into-Square Packing". In: *Algorithmica* 77(3) (2017), pp. 867–901.

[6] Sándor P. Fekete, Sebastian Morr, and Christian Scheffer. "Split Packing: Algorithms for Packing Circles with Optimal Worst-Case Density". In: *CoRR* abs/1705.00924 (2017). http://arxiv.org/abs/1705.00924.

[7] Mhand Hifi and Rym M'hallah. "A literature review on circle and sphere packing problems: models and methodologies". In: *Advances in Operations Research* Article ID 150624 (2009).

[8] Pedro Hokama, Flávio K. Miyazawa, and Rafael C. S. Schouery. "A bounded space algorithm for online circle packing". In: *Information Processing Letters* 116(5) (May 2016), pp. 337–342. ISSN: 0020-0190.

[9] Robert J. Lang. "A computational algorithm for origami design". In: *Proceedings of the Twelfth Annual Symposium on Computational Geometry (SoCG)* (1996), pp. 98–105.

[10] Marco Locatelli and Ulrich Raber. "Packing equal circles in a square: a deterministic global optimization approach". In: *Discrete Applied Mathematics* 122(1) (2002), pp. 139–166.

[11] Flávio K. Miyazawa, Lehilton L.C. Pedrosa, Rafael C.S. Schouery, Maxim Sviridenko, and Yoshiko Wakabayashi. "Polynomial-time approximation schemes for circle packing problems". In: *European Symposium on Algorithms (ESA)*. Springer. 2014, pp. 713–724.

[12] John W. Moon and Leo Moser. "Some packing and covering theorems". In: *Colloquium Mathematicae*. Vol. 17. 1. Institute of Mathematics, Polish Academy of Sciences. 1967, pp. 103–110.

[13] Sebastian Morr. "Split Packing: An Algorithm for Packing Circles with Optimal Worst-Case Density". In: *Proceedings of the Twenty-Eighth Annual ACM-SIAM Symposium on Discrete Algorithms (SODA)*. 2017, pp. 99–109.

[14] R. Peikert, D. Würtz, M. Monagan, and C. de Groot. "Packing circles in a square: A review and new results". In: *Proceedings of the 15th IFIP Conference*. 1992, pp. 45–54.

[15] Eckard Specht. *Packomania*. 2015. URL: http://www.packomania.com/.

[16] Kokichi Sugihara, Masayoshi Sawai, Hiroaki Sano, Deok-Soo Kim, and Donguk Kim. "Disk packing for the estimation of the size of a wire bundle". In: *Japan Journal of Industrial and Applied Mathematics* 21(3) (2004), pp. 259–278.

[17] Péter Gábor Szabó, Mihaly Csaba Markót, Tibor Csendes, Eckard Specht, Leocadio G. Casado, and Inmaculada García. *New Approaches to Circle Packing in a Square*. Springer US, 2007.

[18] Huaiqing Wang, Wenqi Huang, Quan Zhang, and Dongming Xu. "An improved algorithm for the packing of unequal circles within a larger containing circle". In: *European Journal of Operational Research* 141(2) (Sept. 2002), pp. 440–453. ISSN: 0377-2217.

[19] D. Würtz, M. Monagan, and R. Peikert. "The history of packing circles in a square". In: *Maple Technical Newsletter* (1994), pp. 35–42.

[20] Yinfeng Xu. "On the minimum distance determined by $n(\leq 7)$ points in an isoscele right triangle". In: *Acta Mathematicae Applicatae Sinica* 12(2) (1996), pp. 169–175.

Fast and Compact Planar Embeddings[*]

Leo Ferres[1], José Fuentes[2], Travis Gagie[3], Meng He[4], and Gonzalo Navarro[2]

[1] Faculty of Engineering, Universidad del Desarrollo
[2] CeBiB; Department of Computer Science, University of Chile
[3] CeBiB; EIT, Diego Portales University
[4] Faculty of Computer Science, Dalhousie University

Abstract. There are many representations of planar graphs but few are as elegant as Turán's (1984): it is simple and practical, uses only four bits per edge, can handle multi-edges and can store any specified embedding. Its main disadvantage has been that "it does not allow efficient searching" (Jacobson, 1989). In this paper we show how to add a sublinear number of bits to Turán's representation such that it supports fast navigation, thus overcoming this disadvantage. Other data structures for planar embeddings may be asymptotically faster or smaller but ours is simpler, and that can be a theoretical as well as a practical advantage: e.g., we show how our structure can be built efficiently in parallel.

1 Introduction

The rate at which we store data is increasing even faster than the speed and capacity of computing hardware. Thus, if we want to use what we store efficiently, we need to represent it in better ways. The surge in the number and complexity of the maps we want to have available on mobile devices is particularly pronounced and has resulted in a bewildering number of ways to store planar graphs. Each of these representations has its disadvantages, however: e.g., some do not support fast navigation, some are large, some cannot represent multi-edges or certain embeddings, and some are complicated to build in practice, especially in parallel, which is a concern when dealing with massive datasets.

Tutte [26] showed that representing a specified embedding of a connected planar multi-graph with n vertices and m edges takes $m \lg 12 \approx 3.58m$ bits in the worst case. Turán [25] gave a very simple representation that uses $4m$ bits, but Jacobson [15] noted that it "does not allow fast searching" and proposed

---

[*] The second and fifth authors received travel funding from EU grant H2020-MSCA-RISE-2015 BIRDS GA No. 690941. The second, third and fifth authors received funding from Basal Funds FB0001, Conicyt, Chile. The third author received funding from Academy of Finland grant 268324. Early parts of this work were done while the third author was at the University of Helsinki and while the third and fifth authors were visiting the University of A Coruña. Many thanks to Jérémy Barbay, Luca Castelli Aleardi, Arash Farzan, Ian Munro, Pat Nicholson and Julian Shun. The third author is grateful to the late David Gregory for his course on graph theory.

© Springer International Publishing AG 2017
F. Ellen et al. (Eds.): WADS 2017, LNCS 10389, pp. 385–396, 2017.
DOI: 10.1007/978-3-319-62127-2_33

one that instead uses $\mathcal{O}(m)$ bits and supports fast navigation. Keeler and West-brook [17] noted in turn that "the constant factor in [Jacobson's] space bound is relatively large" and gave a representation that uses $m \lg 12 + \mathcal{O}(1)$ bits when the graph contains either no self-loops or no vertices with degree 1, but gave up fast navigation again. Chiang, Lin and Lu [8] gave a representation that uses $2m + 3n + o(m)$ bits with fast navigation, but it is based on orderly spanning trees; although all planar graphs can be represented with orderly spanning trees, some planar embeddings cannot. Blelloch and Farzan [6] extended work by Blandford et al. [5] and gave a representation that uses $m \lg 12 + o(m)$ bits with fast navigation of any specified embedding, but it is complicated and has not been implemented. Barbay et al. [3] gave a data structure that uses $\mathcal{O}(n)$ bits to represent a simple planar graph on n nodes with fast navigation, but the hidden coefficient is about 18. Other authors (see, e.g., [7, 13, 14]) have considered special kinds of planar graphs, notably as tri-connected planar graphs and triangulations. We refer the reader to Munro and Nicholson's [20] and Navarro's [21, Chapter 9] recent surveys for further discussion of compact data structures for graphs.

In this paper we show how to add $o(m)$ bits to Turán's representation such that it supports fast navigation: we can list the edges incident to any vertex in counter-clockwise order using constant time per edge, and determine whether two vertices are neighbours or find a vertex's degree in $\mathcal{O}(f(m))$-time for any given function $f(m) \in \omega(1)$. Our data structure is faster, smaller or more expressive than any of the structures listed above except Blelloch and Farzan's, and it is much simpler than theirs. Our structure's simplicity is a theoretical as well as a practical advantage, in that we can build it in parallel with linear work and logarithmic span (albeit without support for fast neighbour and degree queries). We summarize our construction algorithm in this paper and will provide details in a subsequent paper. In contrast, we do not have such efficient parallel algorithms for finding the book embeddings [27], orderly spanning trees and triangulations of planar subdivisions required by, respectively, Jacobson's, Chiang et al.'s and Barbay et al.'s constructions. Blandford et al.'s and Blelloch and Farzan's constructions are based on finding small vertex separators [19] and, although Kao et al. [16] designed a linear-work and logarithmic-span algorithm for computing a cycle separator of a planar graph, both Blandford et al.'s and Blelloch and Farzan's constructions decompose the input graph by repeatedly computing separators until each piece is sufficiently small, which increases the total work to $\mathcal{O}(n \log n)$ even when this optimal parallel algorithm is used.

Turán chooses an arbitrary spanning tree of the graph, roots it at a vertex on the outer face and traverses it, writing its balanced-parentheses representation as he goes and interleaving that with a sequence over a different binary alphabet, consisting of an occurrence of one character for the first time he sees each edge not in the tree and an occurrence of the other character for the second time he sees that edge. These two sequences can be written as three sequences over $\{0, 1\}$: one of length $2n - 2$ encoding the balanced-parentheses representation of the tree; one of length $2m - 2n + 2$ encoding the interleaved sequence;

and one of length $2m$ indicating how they are interleaved. Our extension of his representation is based on the observation that the interleaved sequence encodes the balanced-parentheses representation of the complementary spanning tree of the dual of the graph. By adding a sublinear number of bits to each balanced-parentheses representation, we can support fast navigation in the trees, and by storing the sequence indicating the interleaving as a bitvector, we can support fast navigation in the graph.

In Section 2 we briefly describe bitvectors and the balanced-parentheses representation of trees, which are the building blocks of our extension of Turán's representation. For further discussion of these data structures, we again direct the reader to Navarro's text [21]. In Section 3 we prove the observation mentioned above. In Section 4 we describe our data structure and how we implement queries. We summarize our parallel construction algorithm in Section 5 and report the results of our preliminary experiments.

2 Preliminaries

A bitvector is a binary string that supports the queries rank and select in addition to random access, where $rank_b(i)$ returns the number of bits set to b in the prefix of length ℓ of the string and $select_b(j)$ returns the position of the jth bit set to b. For convenience, we define $select_b(0) = 0$. There are many different implementations that represent a bitvector of length n in $n + o(n)$ bits and support random access, rank and select in constant time.

With bitvectors we can represent an ordered tree or forest on n vertices using $2n + o(n)$ bits and support natural navigation queries quickly. One of the most popular such representations is as a string of balanced parentheses: we traverse each tree from left to right, writing an opening parenthesis when we first visit a vertex (starting at the root) and a closing parenthesis when we leave it for the last time (or, in the case of the root, when we finish the traversal). We can encode the string of parentheses as a bitvector, with 0s encoding opening parentheses and 1s encoding closing parentheses, and achieve the space bound stated above while supporting each of the follow queries used by our solution in constant time:

- match(i), locates the parenthesis matching the ith parenthesis,
- parent(v), returns the parent of v, given as its pre-order rank in the traversal, or 0 if v is the root of its tree.

3 Spanning trees of planar graphs

It is well known that for any spanning tree T of a connected planar graph G, the edges dual to T are a spanning tree T^* of the dual of G, with T and T^* interdigitating; see Figure 1 for an illustration (including multi-edges and a self-loop) and, e.g., [4, 11, 22] for discussions. If we choose T as the spanning tree of G for Turán's representation, then we store a 0 and a 1, in that order, for each edge in T^*. We now show that these bits encode a traversal of T^*.

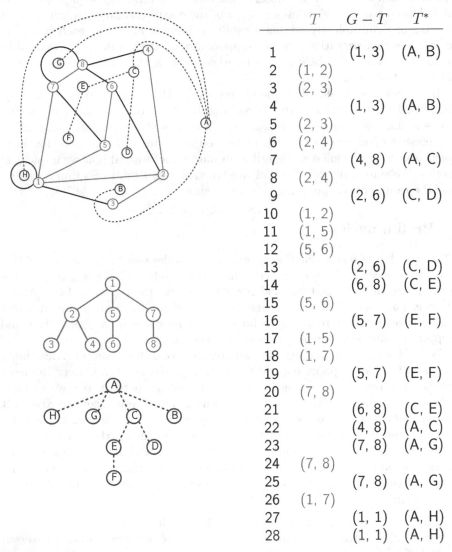

	T	$G-T$	T^*
1		(1, 3)	(A, B)
2	(1, 2)		
3	(2, 3)		
4		(1, 3)	(A, B)
5	(2, 3)		
6	(2, 4)		
7		(4, 8)	(A, C)
8	(2, 4)		
9		(2, 6)	(C, D)
10	(1, 2)		
11	(1, 5)		
12	(5, 6)		
13		(2, 6)	(C, D)
14		(6, 8)	(C, E)
15	(5, 6)		
16		(5, 7)	(E, F)
17	(1, 5)		
18	(1, 7)		
19		(5, 7)	(E, F)
20	(7, 8)		
21		(6, 8)	(C, E)
22		(4, 8)	(A, C)
23		(7, 8)	(A, G)
24	(7, 8)		
25		(7, 8)	(A, G)
26	(1, 7)		
27		(1, 1)	(A, H)
28		(1, 1)	(A, H)

Fig. 1. Top left: A planar embedding of a planar graph G, with a spanning tree T of G shown in red and the complementary spanning tree T^* of the dual of G shown in blue with dashed lines. **Bottom left:** The two spanning trees, with T rooted at the vertex 1 on the outer face and T^* rooted at the vertex A corresponding to the outer face. **Right:** The list of edges we process while traversing T starting at 1 and processing edges in counter-clockwise order, with the edges in T shown in red and the ones in $G-T$ shown in black; the edges of T^* corresponding to the edges in $G-T$ are shown in blue.

Lemma 1. *Consider any planar embedding of a planar graph G, any spanning tree T of G and the complementary spanning tree T^* of the dual of G. If we perform a depth-first traversal of T starting from any vertex on the outer face of G and always process the edges (of the graph) incident to the vertex v we are visiting in counter-clockwise order (starting from the edge immediately after the one to v's parent or, if v is the root of T, from immediately after any incidence of the outer face), then each edge not in T corresponds to the next edge we cross in a depth-first traversal of T^*.*

Proof. Suppose the traversal of T^* starts at the vertex of the dual of G corresponding to the outer face of G. We now prove by induction that the vertex we are visiting in T^* always corresponds to the face of G incident to the vertex we are visiting in T and to the previous and next edges in counter-clockwise order.

Our claim is true before we process any edges, since we order the edges starting from an incidence of the outer face to the root of T. Assume it is still true after we have processed $i < m$ edges, and that at this time we are visiting v in T and v^* in T^*. First suppose that the $(i+1)$st edge (v, w) we process is in T. We note that $w \neq v$, since otherwise (v, w) could not be in T. We cross from v to w in T, which is also incident to the face corresponding to v^*. Now (v, w) is the previous edge — considering their counter-clockwise order at w, starting from (v, w) — and the next edge (which is (v, w) again if w has degree 1) is also incident to v^*. This is illustrated on the left side of Figure 2. In fact, the next edge is the one after (v, w) in a clockwise traversal of the edges incident to the face corresponding to v^*.

Now suppose (v, w) is not in T and let w^* be the vertex in T^* corresponding to the face on the opposite side of (v, w), which is also incident to v. We note that $w^* \neq v^*$, since otherwise (v, w) would have to be in T. We cross from v^* to w^* in T^*. Now (v, w) is the previous edge — this time still considering their counter-clockwise order at v — and the next edge (which may be (v, w) again if it is a self-loop) is also incident to w^*. This is illustrated on the right side of Figure 2. In fact, the next edge is the one that follows (v, w) in a clockwise traversal of the edges incident to the face corresponding to w^*.

Since our claim remains true in both cases after we have processed $i + 1$ edges, by induction it is always true. In other words, whenever we should process next an edge e in G that is not in T, we are visiting in T^* one of the vertices corresponding to the faces incident to e (i.e., one of the endpoints of the edge in the dual of G that corresponds to e). Since we process each edge in G twice, once at each of its endpoints or twice at its unique endpoint if it is a self-loop, it follows that the list of edges we process that are not in T, corresponds to the list of edges we cross in a traversal of T^*.   □

We process the edges in counter-clockwise order so that the traversals of T and T^* are from left to right and from right to left, respectively; processing them in clockwise order would reverse those directions. For example, for the embedding in Figure 1, if we start the traversal of the red tree T at vertex 1 and start processing the edges at $(1, 3)$, then we process them in the order shown at the right of the figure.

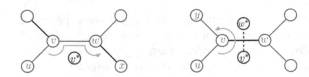

Fig. 2. Left: If we process an edge (v, w) in T, then we move to w in our traversal of T and the next edge, (w, x) in this case, is also incident to the vertex v^* we are visiting in our traversal of T^*. **Right:** If (v, w) is not in T, then in T^* we move from v^* to the vertex w^* corresponding to the face on the opposite side of (v, w) in G. The next edge, (v, y) in this case, is also incident to w^*.

4   Data Structure

Our extension of Turán's representation of a planar embedding of a connected planar graph G with n vertices and m edges consists of the following components, which take $4m + o(m)$ bits:

- a bitvector $A[1..2m]$ in which $A[i]$ indicates whether the ith edge we process in the traversal of T described in Lemma 1, is in T;
- a bitvector $B[1..2(n-1)]$ in which $B[i]$ indicates whether the ith time we process an edge in T during the traversal, is the second time we process that edge;
- a bitvector $B^*[1..2(m-n+1)]$ in which $B^*[i]$ indicates whether the ith time we process an edge not in T during the traversal, is the second time we process that edge.

Notice B encodes the balanced-parentheses representation of T except that it lacks the leading 0 and trailing 1 encoding the parentheses for the root. By Lemma 1, B^* encodes the balanced-parentheses representation of a traversal of the spanning tree T^* of the dual of G complementary to T (the right-to-left traversal of T^*, in fact) except that it also lacks the leading 0 and trailing 1 encoding the parentheses for the root. Therefore, since B and B^* encode forests, we can support match and parent with them.

   To build A, B and B^* given the embedding of G and T, we traverse T as in Lemma 1. Whenever we process an edge, if it is in T then we append a 1 to A and append the edge to a list L; otherwise, we append a 0 to A and append the edge to another list L^*. When we have finished the traversal, we replace each edge in L or L^* by a 0 if it is the first occurrence of that edge in that list, and by a 1 if it is the second occurrence; this turns L and L^* into B and B^*, respectively. For the example shown in Figure 1, L and L^* eventually contain the edges shown in the columns labelled T and $G - T$, respectively, in the table on the on the right side of the figure, and

$$A[1..28] = 0110110101110010110100010100$$
$$B[1..14] = 00101100110011$$
$$B^*[1..14] = 01001001110101 \, .$$

We identify each vertex v in G by its pre-order rank in our traversal of T. Consider the following queries:

first(v): return i such that the first edge we process while visiting v is the ith we process during our traversal;

next(i): return j such that if we are visiting v when we process the ith edge during our traversal, then the next edge incident to v in counter-clockwise order is the one we process jth;

mate(i): return j such that we process the same edge ith and jth during our traversal;

vertex(i): return the vertex v such that we are visiting v when we process the ith edge during our traversal.

With these it is straightforward to reenact our traversal of T and recover the embedding of G. For example, with the following queries we can list the edges incident to the root of T in Figure 1 and determine whether they are in T:

$$\begin{array}{llll}
\text{first}(1) = 1 & \text{mate}(1) = 4 & \text{vertex}(4) = 3 & A[1] = 0 \\
\text{next}(1) = 2 & \text{mate}(2) = 10 & \text{vertex}(10) = 2 & A[2] = 1 \\
\text{next}(2) = 11 & \text{mate}(11) = 17 & \text{vertex}(17) = 5 & A[11] = 1 \\
\text{next}(11) = 18 & \text{mate}(18) = 26 & \text{vertex}(26) = 7 & A[18] = 1 .
\end{array}$$

To see why we can recover the embedding from the traversal, consider that if we have already correctly embedded the first i edges processed in the traversal, then we can embed the $(i+1)$st correctly given its endpoints and its rank in the counter-clockwise order at those vertices.

We now explain our constant-time implementations of first, next, mate and vertex. If $m = 0$ then first(v) is undefined, which we indicate by returning 0. Otherwise, we first process an edge at v immediately after first arriving at v. Since we identify v with its pre-order rank in our traversal of T and B lacks the opening parenthesis for the root, while first arriving at any vertex v other than the root we write the $(v-1)$st 0 in B and, thus, the $B.\text{select}_0(v-1)$st 1 in A. If v is the root then first(v) $= 1$ and so, since $\text{select}_x(0) = 0$, this case is also handled by the formula below:

$$\text{first}(v) = \begin{cases} A.\text{select}_1(B.\text{select}_0(v-1)) + 1 & \text{if } m \geq 1 \\ 0 & \text{otherwise.} \end{cases}$$

In our example,

$$\text{first}(5) = A.\text{select}_1(B.\text{select}_0(4)) + 1 = A.\text{select}_1(7) + 1 = 12$$

and indeed the twelfth edge we process, $(5, 6)$, is the first one we process at vertex 5.

If the ith edge we process is the last edge we process at a vertex v then next(i) is undefined, which we again indicate by returning 0. This is the case when $i = 2m$, or $A[i] = 1$ and $B[A.\text{rank}_1(i)] = 1$. Otherwise, if the ith edge we process is not in T, then $A[i] = 0$, and we process the next edge at v one time step later. Finally, if the ith edge e we process is in T and not the last one we process at v, then we next process an edge at v immediately after returning to v by processing e again at time $\mathsf{mate}(i)$. This is the case when $A[i] = 1$ and $B[A.\mathsf{rank}_1(i)] = 0$. In other words,

$$\mathsf{next}(i) = \begin{cases} i+1 & \text{if } A[i] = 0 \text{ and } i < 2m \\ \mathsf{mate}(i) + 1 & \text{if } A[i] = 1 \text{ and } B[A.\mathsf{rank}_1(i)] = 0 \\ 0 & \text{otherwise.} \end{cases}$$

In our example, since $A[12] = 1$, $B[A.\mathsf{rank}_1(12)] = B[8] = 0$, the twelfth edge we process is $(5, 6)$ and it is also the fifteenth edge we process,

$$\mathsf{next}(12) = \mathsf{mate}(12) + 1 = 16\,,$$

and indeed the second edge we process at vertex 5 is $(5, 7)$.

To implement $\mathsf{mate}(i)$, we check $A[i]$ and use rank to determine whether we wrote a bit in B or in B^* while processing the ith edge, and to find that bit. We use match to find the bit encoding the matching parenthesis, and then use select on A to find the bit we wrote in A when we wrote that matching bit. Therefore,

$$\mathsf{mate}(i) = \begin{cases} A.\mathsf{select}_0(B^*.\mathsf{match}(A.\mathsf{rank}_0(i))) & \text{if } A[i] = 0 \\ A.\mathsf{select}_1(B.\mathsf{match}(A.\mathsf{rank}_1(i))) & \text{otherwise.} \end{cases}$$

To compute $\mathsf{mate}(12)$ for our example, since $A[12] = 1$,

$$\begin{aligned} &\mathsf{mate}(12) \\ &= A.\mathsf{select}_1(B.\mathsf{match}(A.\mathsf{rank}_1(12))) \\ &= A.\mathsf{select}_1(B.\mathsf{match}(8)) \\ &= A.\mathsf{select}_1(9) \\ &= 15\,. \end{aligned}$$

Suppose the ith edge e we process is not in T and we process it at vertex v. If the preceding time we processed an edge in T was the first time we processed that edge, we then wrote a 0 in B, encoding the opening parenthesis for v; otherwise, we then wrote a 1 in B, encoding the closing parenthesis for one of v's children. Now suppose e is in T. If that is the first time we process e, we move to the other endpoint w of e — which is a child of v — and write a 0 in B, encoding the opening parenthesis for w. If it is the second time we process e, then we write a 1 in B, encoding the closing parenthesis for v itself. Therefore,

$$\mathsf{vertex}(i) = \begin{cases} B.\mathsf{rank}_0(A.\mathsf{rank}_1(i)) + 1 \\ \quad \text{if } A[i] = 0 \text{ and } B[A.\mathsf{rank}_1(i)] = 0 \\ B.\mathsf{parent}(B.\mathsf{rank}_0(B.\mathsf{match}(A.\mathsf{rank}_1(i)))) + 1 \\ \quad \text{if } A[i] = 0 \text{ and } B[A.\mathsf{rank}_1(i)] = 1 \\ B.\mathsf{parent}(B.\mathsf{rank}_0(A.\mathsf{rank}_1(i))) + 1 \\ \quad \text{if } A[i] = 1 \text{ and } B[A.\mathsf{rank}_1(i)] = 0 \\ B.\mathsf{rank}_0(B.\mathsf{match}(A.\mathsf{rank}_1(i))) + 1 \\ \quad \text{otherwise.} \end{cases}$$

In our example, since $A[16] = 0$ and $B[A.\mathrm{rank}_1(16)] = B[9] = 1$,

$$\mathsf{vertex}(16)$$
$$= B.\mathsf{parent}(B.\mathrm{rank}_0(B.\mathsf{match}(A.\mathrm{rank}_1(16)))) + 1$$
$$= B.\mathsf{parent}(B.\mathrm{rank}_0(B.\mathsf{match}(9))) + 1$$
$$= B.\mathsf{parent}(B.\mathrm{rank}_0(8)) + 1$$
$$= B.\mathsf{parent}(5) + 1$$
$$= 5 \,,$$

and indeed we process the sixteenth edge $(5,7)$ while visiting 5.

We remind the reader that since B lacks parentheses for the root of T, $B.\mathsf{parent}(5)$ refers to the parent of the fifth vertex in an in-order traversal of T not including the root, i.e., the parent vertex 5 of vertex 6. Adding 1 includes the root in the traversal, so the final answer correctly refers to vertex 5. The lack of parentheses for the root also means that, e.g., $B.\mathsf{parent}(4)$ refers to the parent of vertex 5 and returns 0 because vertex 5 is the root of its own tree in the forest encoded by B, without vertex 1. Adding 1 to that 0 also correctly turns the final value into 1, the in-order rank of the root. Of course, we have the option of prepending and appending bits to A, B and B^* to represent the roots of T and T^*, but that slightly confuses the relationship between the positions of the bits and the time steps at which we process edges.

Clearly we can determine whether two vertices u and v are neighbours by listing the neighbours of each in parallel in $\mathcal{O}(\min(\mathsf{degree}(u), \mathsf{degree}(v)))$ time, and we can find $\mathsf{degree}(v)$ in $\mathcal{O}(\mathsf{degree}(v))$ time. Moreover, given a function $f(m) \in \omega(1)$, we can make both kinds of queries take $\mathcal{O}(f(m))$ time. To do this, we store a bitvector marking the $\mathcal{O}(m/f(m)) = o(m)$ vertices with degree at least $f(m)$, which takes $o(m)$ bits. To be able to answer neighbour queries quickly, we consider the graph induced by those high-degree vertices and eliminate multi-edges and self-loops. The resulting simple graph G' is still planar — so it has average degree less than 6 and thus $o(m)$ edges — and preserves the neighbour relation between those vertices. We can store G' using Blelloch and Farzan's representation or, since the neighbour relation does not depend on the embedding, using one of the other compact representations of planar graphs that supports constant-time neighbour queries, which also takes $o(m)$ bits. To answer $\mathsf{neighbour}(u, v)$ now, we check whether either u or v is low-degree and, if so, list its neighbours in $\mathcal{O}(f(m))$ time; if not, we query our auxiliary representation in $\mathcal{O}(1)$ time. To be able to answer degree queries quickly, we simply store the degrees of the high-degree vertices in unary using a bitvector, which takes $o(m)$ bits. To find $\mathsf{degree}(v)$ now, we check whether v is low-degree and, if so, list and count its incident edges; if not, we look up $\mathsf{degree}(v)$.

Summarizing our results so far, we have the following theorem:

Theorem 1. *We can store a given planar embedding of a connected planar graph G with m edges in $4m + o(m)$ bits such that later, given a vertex v, we can list the edges incident to v in counter-clockwise order (optionally, starting at a* *given edge e incident to v) using constant time per edge, and determine whether two vertices are neighbours or find a vertex's degree in $\mathcal{O}(f(m))$-time for any given function $f(m) \in \omega(1)$.*

5   Parallel Construction and Experiments

Due to space constraints, in this section we can only summarize our parallel algorithm and then briefly report the result of our experiments on construction and query times. We will provide the full details of the algorithm in the subsequent paper mentioned in Section 1, of which a preprint is available at http://arxiv.org/abs/1705.00415 .

We construct our extension of Turán's representation in parallel as follows: given a planar graph with a planar embedding G, we first compute a spanning tree T of G in parallel. In our experiments we used Bader and Cong's algorithm [2] because it works well in practice, but its theoretical bounds are for random graphs. To obtain good worst-case bounds, we could use Shiloach and Vishkin's [23] or Awerbuch and Shiloach's [1] algorithms, which use linear work with logarithmic span in the CRCW PRAM model. We recently learned that Shun, Dhulipala and Blelloch's [24] practical connectivity algorithm can be made to return a spanning tree with linear work and polylogarithmic span. As a by-product of the computation of T, we obtain an array C of length $2n - 2$ that stores the number of edges of $G \setminus T$ between two consecutive edges in T, in counter-clockwise order. Notice the starting vertex for the spanning tree must be in the outer face of G.

We construct bitvectors A, B and B^* by performing a parallel Euler Tour over T [9]. During the tour, we obtain B by writing a 0 for each forward (parent-to-child) edge and a 1 for each backward (child-to-parent) edge. We obtain A by counting the number of edges of $G \setminus T$ between two consecutive edges of T (stored in C). We represent the former with 0's and the edges of T with 1's. The visiting order of edges of $G \setminus T$ encoded in B^* is implicit in the previous Euler Tour. Therefore, with the Euler Tour and the array C, we have enough information to compute the position of each bit in the bitvector B^*. We can decide if an edge of $G \setminus T$ is a forward or backward edge by checking its relative position and the position of its complement edge on the Euler tour. Finally, in order to support operations on A, B and B^*, we used Labeit et al.'s algorithm for succinct bitvectors [18], and Ferres et al.'s algorithm for succinct trees [12].

We analyze our algorithm (after the computation of the spanning tree) in the *Dynamic Multithreading (DyM)* model of parallel computation [10]. The DyM model relies on two parameters: the work T_1, i.e., the running time on a single core; and the span T_∞, i.e., the complexity of the intrinsically sequential part of the parallel computation. The time T_p needed to execute the computation on p cores is bounded by $T_p = \Theta(T_1/p + T_\infty)$. The Euler tour and the array C can be computed in parallel in $T_1 = O(n)$ and $T_\infty = O(\lg n)$ time. Assigning values to A and B can be done independently for each entry of the bitvectors, which gives us $T_1 = O(n)$ and $T_\infty = O(1)$ time, while B^* takes $T_1 = O(m - n)$ and $T_\infty = O(1)$

time. Rank/select structures can be constructed in parallel with $T_1 = O(m)$ and $T_\infty = O(\lg m)$ time. Succinct trees can be constructed in $T_1 = O(m)$ and $T_\infty = O(\lg m)$ time.

Theorem 2. *Given a spanning tree of a planar embedding, the compact representation from Theorem 1 (without the auxiliary data structures for fast* neighbour *and* degree *queries) can be constructed in parallel with linear work and logarithmic span.*

To provide some grounds for comparison, we also implemented a sequential algorithm based on our parallel one, but instead of using Bader and Cong's algorithm, we used depth-first search to compute T. This sequential implementation of Turán's representation is simpler than our parallel algorithm running on a single core, avoiding the additional steps needed for the parallel computation. To test both the sequential and parallel implementations, we synthetically generated a planar graph (represented as an adjacency list) by computing the Delaunay Triangulation of 25,000,000 random coordinates, yielding 25,000,000 vertices and 74,999,979 edges, with a minimum degree of 3 and a maximum degree of 15. The experiments were carried out on a 28-core machine (two processors with 14 physical cores each) with hyperthreading turned on (for a total of 56 cores), per-core L1 and L2 caches of sizes 64KB and 256KB, respectively and a per-processor shared L3 cache of 35MB and a total of 768GB DDR3 RAM memory.

Results show that the sequential algorithm took 71.8 seconds to construct the representation of Theorem 1, while the parallel implementation took 5.4 seconds with 28 threads and 3.3 seconds with 56 threads. The space used by the adjacency list representation was 1.02 GB, 117.3 bits per edge. The space used by our compact representation was 44.7 MB, 5 bits per edge, which matches Theorem 1. Memory consumption of our parallel algorithm peaked at 1.4 GB.

With respect to queries, we tested `counting` (number of neighbors) and `listing` (list of neighbors in counter-clockwise order) queries. For the former, the adjacency-list representation took 0.047 microseconds per node and the compact representation took 4.6 microseconds per node. For `listing`, the adjacency-list representation took 0.046 microseconds per node listed and the compact representation took 3.69 microseconds per node listed.

In summary, our parallel algorithm achieves a reasonable speed up in terms of the number of threads and is an already order of magnitude faster than the sequential algorithm when using 28 threads; our compact representation is between one and two orders of magnitude smaller than the adjacency-list representation, but takes two orders of magnitude more time to answer queries. Our code and datasets are available at `https://users.dcc.uchile.cl/~jfuentess/pemb/` .

References

1. B. Awerbuch and Y. Shiloach. New connectivity and MSF algorithms for shuffle-exchange network and PRAM. *IEEE Trans. Computers*, 36(10):1258–1263, 1987.

2. D. A. Bader and G. Cong. A fast, parallel spanning tree algorithm for symmetric multiprocessors (SMPs). *J. Parallel and Distributed Computing*, 65:994–1006, 2005.

3. J. Barbay, L. C. Aleardi, M. He, and J. I. Munro. Succinct representation of labeled graphs. *Algorithmica*, 62:224–257, 2012.

4. N. Biggs. Spanning trees of dual graphs. *J. Comb. Theory, Series B*, 11:127–131, 1971.

5. D. K. Blandford, G. E. Blelloch, and I. A. Kash. Compact representations of separable graphs. In *SODA*, pages 679–688, 2003.

6. G. E. Blelloch and A. Farzan. Succinct representations of separable graphs. In *CPM*, pages 138–150, 2010.

7. Castelli Aleardi, L, O. Devillers, and G. Schaeffer. Succinct representations of planar maps. *TCS*, 408:174–187, 2008.

8. Y.-T. Chiang, C.-C. Lin, and H.-I. Lu. Orderly spanning trees with applications. *SIAM J. Comp.*, 34:924–945, 2005.

9. G. Cong and D. A. Bader. The Euler tour technique and parallel rooted spanning tree. In *ICPP*, pages 448–457, 2004.

10. T. H. Cormen, C. E. Leiserson, R. L. Rivest, and C. Stein. Multithreaded algorithms. In *Introduction to Algorithms*, pages 772–812. MIT Press, 2009.

11. D. Eppstein. Dynamic generators of topologically embedded graphs. In *SODA*, pages 599–608, 2003.

12. L. Ferres, J. Fuentes-Sepúlveda, M. He, and N. Zeh. Parallel construction of succinct trees. In *SEA*, pages 3–14, 2015.

13. É. Fusy, G. Schaeffer, and D. Poulalhon. Dissections, orientations, and trees with applications to optimal mesh encoding and random sampling. *TALG*, 4:19, 2008.

14. X. He, M. Kao, and H. Lu. Linear-time succinct encodings of planar graphs via canonical orderings. *SIAM J. Discrete Math.*, 12:317–325, 1999.

15. G. Jacobson. Space-efficient static trees and graphs. In *FOCS*, pages 549–554, 1989.

16. M. Kao, S. Teng, and K. Toyama. An optimal parallel algorithm for planar cycle separators. *Algorithmica*, 14:398–408, 1995.

17. K. Keeler and J. Westbrook. Short encodings of planar graphs and maps. *DAM*, 58:239–252, 1995.

18. J. Labeit, J. Shun, and G. E. Blelloch. Parallel lightweight wavelet tree, suffix array and FM-index construction. In *DCC*, pages 33–42, 2016.

19. R. J. Lipton and R. E. Tarjan. A separator theorem for planar graphs. *SIAM J. Applied Math.*, 36:177–189, 1979.

20. J. I. Munro and P. K. Nicholson. Compressed representations of graphs. In *Encyclopedia of Algorithms*, pages 382–386. Springer, 2016.

21. G. Navarro. *Compact Data Structures: A Practical Approach*. Cambridge University Press, 2016.

22. T. R. Riley and W. P. Thurston. The absence of efficient dual pairs of spanning trees in planar graphs. *Electronic J. Comb.*, 13, 2006.

23. Y. Shiloach and U. Vishkin. An o(log n) parallel connectivity algorithm. *J. Algorithms*, 3(1):57–67, 1982.

24. J. Shun, L. Dhulipala, and G. E. Blelloch. A simple and practical linear-work parallel algorithm for connectivity. In *SPAA*, pages 143–153, 2014.

25. G. Turán. On the succinct representation of graphs. *DAM*, 8:289–294, 1984.

26. W. T. Tutte. A census of planar maps. *Canadian J. Math.*, 15:249–271, 1963.

27. M. Yannakakis. Embedding planar graphs in four pages. *JCSS*, 38:36–67, 1989.

When Can Graph Hyperbolicity be Computed in Linear Time?*

Till Fluschnik[1,**], Christian Komusiewicz[2,***], George B. Mertzios[3],
André Nichterlein[1,3,†], Rolf Niedermeier[1], and Nimrod Talmon[4,‡]

[1] Institut für Softwaretechnik und Theoretische Informatik, TU Berlin, Germany,
{till.fluschnik, andre.nichterlein, rolf.niedermeier}@tu-berlin.de
[2] Institut für Informatik, Friedrich-Schiller-Universität Jena, Germany,
christian.komusiewicz@uni-jena.de
[3] School of Engineering and Computing Sciences, Durham University, UK,
george.mertzios@durham.ac.uk
[4] Weizmann Institute of Science, Rehovot, Israel,
nimrodtalmon77@gmail.com

Abstract. Hyperbolicity measures, in terms of (distance) metrics, how close a given graph is to being a tree. Due to its relevance in modeling real-world networks, hyperbolicity has seen intensive research over the last years. Unfortunately, the best known practical algorithms for computing the hyperbolicity number of a n-vertex graph have running time $O(n^4)$. Exploiting the framework of parameterized complexity analysis, we explore possibilities for "linear-time FPT" algorithms to compute hyperbolicity. For instance, we show that hyperbolicity can be computed in time $2^{O(k)} + O(n + m)$ (m being the number of graph edges, k being the size of a vertex cover) while at the same time, unless the SETH fails, there is no $2^{o(k)}n^2$-time algorithm.

1 Introduction

(Gromov) hyperbolicity [16] of a graph is a popular attempt to capture and measure how *metrically* close a graph is to being a tree. The study of hyperbolicity is motivated by the fact that many real-world graphs are tree-like from a distance metric point of view [2, 3]. This is due to the fact that many of these graphs (including Internet application networks or social networks) possess certain geometric and topological characteristics. Hence, for many applications (cf., e.g. [3]), including the design of (more) efficient algorithms, it is useful to know

---

* This work was initiated at the yearly research retreat of the Algorithmics and Computational Complexity (AKT) group of TU Berlin, held in in Krölpa, Thuringia, Germany, from April 3rd till April 9th, 2016.
** Supported by the DFG, project DAMM (NI 369/13-2).
*** Supported by the DFG, project MAGZ (KO 3669/4-1).
† Supported by a postdoctoral fellowship of the DAAD while at Durham University.
‡ Supported by a postdoctoral fellowship from I-CORE ALGO.

© Springer International Publishing AG 2017
F. Ellen et al. (Eds.): WADS 2017, LNCS 10389, pp. 397–408, 2017.
DOI: 10.1007/978-3-319-62127-2_34

the hyperbolicity of a graph. The hyperbolicity of a graph is a nonnegative number δ; the smaller δ is, the more tree-like the graph is; in particular, $\delta = 0$ means that the graph metric indeed is a tree metric. Typical hyperbolicity values for real-world graphs are below 5 [2].

Hyperbolicity can be defined via a four-point condition: Considering a size-four subset $\{a, b, c, d\}$ of the vertex set of a graph, one takes the (nonnegative) difference between the two largest of the three sums $\overline{ab} + \overline{cd}$, $\overline{ac} + \overline{bd}$, and $\overline{ad} + \overline{bc}$, where, e.g., $\overline{ab}$ denotes the length of the shortest path between vertices a and b in the given graph. The hyperbolicity is the maximum of these differences over all size-four subsets of the vertex set of the graph. For an n-vertex graph, this characterization of hyperbolicity directly implies a simple (brute-force) $O(n^4)$-time algorithm to compute its hyperbolicity. It has been observed that this running time is too slow for computing the hyperbolicity of large graphs as occurring in applications [2, 3, 4, 13]. On the theoretical side, it was shown that relying on some (rather impractical) matrix multiplication results, one can improve the upper bound to $O(n^{3.69})$ [13]. Moreover, roughly quadratic lower bounds are known [4, 13]. In practice, however, the best known algorithm still has an $O(n^4)$-time worst-case bound but uses several clever tricks when compared to the straightforward brute-force algorithm [3]. Indeed, based on empirical studies an $O(mn)$ running time is claimed, where m is the number of edges in the graph. Furthermore, there are heuristics for computing the hyperbolicity of a given graph [7].

To explore the possibility of faster algorithms for hyperbolicity in relevant special cases is the guiding principle of this work. More specifically, introducing some graph parameters, we investigate whether one can compute hyperbolicity in linear time when these parameters take small values. In other words, we employ the framework of parameterized complexity analysis (so far mainly used for studying NP-hard problems) applied to the polynomial-time solvable hyperbolicity problem. In this sense, we follow the recent trend of studying "FPT in P" [15]. Indeed, other than for NP-hard problems (where parameterized complexity is typically applied), for some parameters we achieve not only exponential dependence on the parameter but also polynomial ones. Note that such algorithms are unlikely for metric parameters like diameter or hyperbolicity.

Our contributions. Table 1 summarizes our main results. On the positive side, for a number of natural graph parameters we can attain "linear FPT" running times. Our "positive" graph parameters here are the following:

- the *covering path* number, that is, the minimum number of paths where only the endpoints have degree greater than two and which cover all vertices;
- the *feedback edge* number, that is, the minimum number of edges to delete to obtain a forest;
- the number of graph vertices of *degree at least three*;
- the *vertex cover* number, that is the minimum number of vertices needed to cover all edges in the graph;

Table 1. Summary of our algorithmic results. Herein, k denotes the parameter and n and m denote the number of vertices and edges, respectively.

Parameter	Running time	
covering path number	$O(k^4(n+m))$	[Theorem 5]
feedback edge number	$O(k^4(n+m))$	[Theorem 6]
number of $\geq$ 3-degree vertices	$O(k^8(n+m))$	[Theorem 8]
vertex cover number	$2^{O(k)} + O(n+m)$	[Theorem 10]
distance to cographs	$O(4^{4k} \cdot k^7 \cdot (n+m))$	[Theorem 15]

- the *distance to cographs*, that is, the minimum number of vertices to delete to obtain a cograph.[5]

On the negative side we prove that, with respect to the parameter vertex cover number k, we cannot hope for any $2^{o(k)}n^{2-\epsilon}$ algorithm unless the Strong Exponential Time Hypothesis (SETH) fails. We also obtain a "quadratic-time FPT" lower bound with respect to the parameter *maximum vertex degree*, again assuming SETH. Finally, we show that computing the hyperbolicity is at least as hard as computing a *size-four independent set* in a graph. It is conjectured that computing size-four independent sets needs $\Omega(n^3)$ time [20]. Due to lack of space, many details and proofs (marked with $(\star)$) had to be deferred.[6]

2 Preliminaries and Basic Observations

We write $[n] := \{1,\dots,n\}$ for every $n \in \mathbb{N}$. For a function $f : X \to Y$ and $X' \subseteq X$ we set $f(X') := \{y \in Y \mid \exists x \in X' : f(x) = y\}$.

Graph theory. Let $G = (V, E)$ be a graph. We define $|G| = |V| + |E|$. For $W \subseteq V$, we denote by $G[W]$ the graph *induced* by W. We use $G - W := G[V \setminus W]$ to denote the graph obtained from G by deleting the vertices of $W \subseteq V$. A *path* $P = (v_1, \dots, v_k)$ in G is a tuple of distinct vertices in V such that $\{v_i, v_{i+1}\} \in E$ for all $i \in [k-1]$; we say that such a path P has endpoints v_1 and v_k, we call the other vertices of P inner vertices, and we say that P is a v_1-v_k path. We denote by $\overline{ab}$ the length of a shortest a-b path if such a path exists; otherwise, that is, if a and b are in different connected components, we define $\overline{ab} := \infty$. Let $P = (v_1, \dots, v_k)$ be a path and v_i, v_j two vertices on P. We denote by $\overline{v_i v_j}|_P$ the distance of v_i to v_j on the path P, that is, $\overline{v_i v_j}|_P = |j - i|$. For a graph G we denote by $V_G^{\geq 3}$ the set of vertices of G that have degree at least three.

Hyperbolicity. Let $G = (V, E)$ be graph and $a, b, c, d \in V$. We define $D_1 := \overline{ab} + \overline{cd}$, $D_2 := \overline{ac} + \overline{bd}$, and $D_3 := \overline{ad} + \overline{bc}$ (referred to as *distance sums*). Moreover, we

---

[5] Cographs are the graphs without induced P_4s. Distance to cographs is upper-bounded by the parameter distance to cluster graph [10] and thus also by the parameter vertex cover number.

[6] A full version is available at https://arxiv.org/abs/1702.06503.

define $\delta(a, b, c, d) := |D_i - D_j|$ if $D_k \leq \min\{D_i, D_j\}$, for pairwise distinct $i, j, k \in \{1, 2, 3\}$. If any two vertices of the quadruple $\{a, b, c, d\}$ are not connected, we set $\delta(a, b, c, d) = 0$.[7] The *hyperbolicity* of $G = (V, E)$ is defined as $\delta(G) := \max_{a,b,c,d \in V}\{\delta(a, b, c, d)\}$. Note that by our definition, if G is not connected, $\delta(G)$ computes the maximal hyperbolicity over all connected components of G. We say that the graph is δ-*hyperbolic* for some $\delta \in \mathbb{N}$ if it has hyperbolicity at most δ. That is, a graph is δ-hyperbolic[8]. if for each 4-tuple $a, b, c, d \in V$ we have

$$\overline{ab} + \overline{cd} \leq \max\{\overline{ac} + \overline{bd}, \overline{ad} + \overline{bc}\} + \delta.$$

Formally, the HYPERBOLICITY problem is defined as follows.

HYPERBOLICITY
Input: An undirected graph $G = (V, E)$ and a positive integer δ.
Question: Is G δ-hyperbolic?

The following lemma will be useful later. For any quadruple $\{a, b, c, d\}$, Lemma 1 upper bounds $\delta(a, b, c, d)$ by twice the distance between any pair of vertices of the quadruple.

Lemma 1 ([7, Lemma 3.1]). $\delta(a, b, c, d) \leq 2 \cdot \min_{u \neq v \in \{a,b,c,d\}}\{\overline{uv}\}$

Reduction Rule 1. As long as there are more than four vertices, remove vertices of degree one.

Lemma 2 ($\star$). *Reduction Rule 1 is correct and can be exhaustively applied in linear time.*

3   Polynomial Linear-Time Parameterized Algorithms

In this section, we provide *polynomial linear-time parameterized* algorithms with respect to the parameters feedback edge number and number of vertices with degree at least three; that is, we present algorithms with running time having a linear-time dependence on the input size times a polynomial-time dependence on the parameter value (to which we refer to as PL-FPT running time).

To this end, we first introduce an auxiliary parameter, the *minimum maximal path cover number*, which we formally define below and also describe a polynomial linear-time parameterized algorithm for it.

Building upon this result, for the parameter feedback edge number we then show that, after applying Reduction Rule 1, the number of maximal paths can be upper-bounded by a polynomial of the feedback edge number. This implies a polynomial linear-time parameterized algorithm for the feedback edge number as well. For the parameter number of vertices with degree at least three, we

---

[7] This case is often left undefined in the literature. Our definition however enables to consider also disconnected graphs.

[8] Note that there is also a slightly different definition where graphs we call δ-hyperbolic are called 2δ-hyperbolic [7, 17]; we follow the definition of Brinkmann et al. [6].

introduce an additional reduction rule to achieve that the number of maximal paths is upper-bounded in a polynomial of this parameter. Again, this implies an algorithm with PL-FPT running time.

Minimum maximal path cover number.

Definition 3 (Maximal path). *Let G be a graph and P be a path in G. Then, P is a maximal path if the following holds: (1) P contains at least two vertices; (2) all its inner vertices have degree two in G; (3) either both its endpoints have degree at least three in G, or one of its endpoints has degree at least three in G while the other endpoint is of degree two in G; and (4) P is size-wise maximal with respect to these properties.*

We will be interested in the minimum number of maximal paths needed to cover the vertices of a given graph; we call this number the *minimum maximal path cover number*. While not all graphs can be covered by maximal paths (e.g., edgeless graphs), graphs which have minimum degree two and contain no isolated cycles, i.e. components that form induced cycles, can be covered by maximal paths (this follows by, e.g., a greedy algorithm which iteratively starts a path with an arbitrary uncovered vertex and exhaustively extends it arbitrarily; since there are no isolated cycles and the minimum degree is two, we are bound to eventually hit at least one vertex of degree three). Based on the approximation algorithm given in the next lemma, we assume in the following that we are given a maximal path cover.

Lemma 4 (⋆). *There is a linear-time 2-approximation algorithm for the minimum maximal path cover number for graphs which have minimum degree two and contain no isolated cycles.*

Now we are ready to design a polynomial linear-time parameterized algorithm for HYPERBOLICITY with respect to the minimum maximal path cover number.

Theorem 5 (⋆). *Let $G = (V, E)$ be a graph and k be its minimum maximal path cover number. Then, HYPERBOLICITY can be solved in $O(k^4(n + m))$ time.*

Feedback edge number. We next present a polynomial linear-time parameterized algorithm with respect to the parameter feedback edge number k. The idea is to show that a graph that is reduced with respect to Reduction Rule 1 contains $O(k)$ maximal paths.

Theorem 6 (⋆). *HYPERBOLICITY can be computed in $O(k^4(n + m))$ time, where k is the feedback edge number.*

Number of vertices with degree at least three. We finally show a polynomial linear-time parameterized algorithm with respect to the number k of vertices with degree three or more. To this end, we use the following data reduction rule additionally to Reduction Rule 1 to bound the number of maximal paths in the graph by $O(k^2)$ (in order to make use of Theorem 5).

Reduction Rule 2. Let $G = (V, E)$ be a graph, $u, v \in V_G^{\geq 3}$ be two vertices of degree at least three, and $\mathcal{P}_{uv}$ be the set of maximal paths in G with endpoints u and v. Let $\mathcal{P}_{uv}^9 \subseteq \mathcal{P}_{uv}$ be the set containing the shortest path, the four longest even-length paths, and the four longest odd-length paths in $\mathcal{P}_{uv}$. If $\mathcal{P}_{uv} \setminus \mathcal{P}_{uv}^9 \neq \emptyset$, then delete in G all inner vertices of the paths in $\mathcal{P}_{uv} \setminus \mathcal{P}_{uv}^9$.

Lemma 7 ($\star$). *Reduction Rule 2 is correct and can be exhaustively applied in linear time.*

Observe that if the graph G is reduced with respect to Reduction Rule 2 after Reduction Rule 1 was applied, then for each pair $u, v \in V_G^{\geq 3}$ there exist at most nine maximal paths with endpoints u and v. Thus, G contains at most $O(k^2)$ maximal paths and using Theorem 5 we arrive at the following.

Theorem 8. HYPERBOLICITY *can be solved in $O(k^8(n + m))$ time, where k is the number of vertices with degree at least three.*

4 Parameter Vertex Cover

A *vertex cover* of a graph $G = (V, E)$ is a subset $W \subseteq V$ of vertices of G such that each edge in G is incident to at least one vertex in W. Deciding whether a graph G has a vertex cover of size at most k is NP-complete in general [14]. There is, however, a simple linear-time factor-2 approximation (see, e.g., [18]). In this section, we consider the size k of a vertex cover as the parameter. We show that we can solve HYPERBOLICITY in time linear in $|G|$, but exponential in k; further, we show that, unless SETH fails, we cannot do asymptotically better.

A Linear-Time Algorithm Parameterized by the Vertex Cover Number. We prove that HYPERBOLICITY can be solved in time linear in the size of the graph and exponential in the size k of a vertex cover. This result is based on a linear-time computable problem kernel of size $O(2^k)$ that can be obtained by exhaustively applying the following reduction rule.

Reduction Rule 3. If there are at least five vertices $v_1, v_2, \ldots, v_\ell \in V$, $\ell > 4$, with the same (open) neighborhood $N(v_1) = N(v_2) = \ldots = N(v_\ell)$, then delete $v_5, \ldots, v_\ell$.

We next show that the above rule is correct, can be applied in linear time, and leads to a problem kernel for the parameter vertex cover number.

Lemma 9 ($\star$). *Reduction Rule 3 is correct and can be applied exhaustively in linear time. Furthermore, if Reduction Rule 3 is not applicable, then the graph contains at most $k + 4 \cdot 2^k$ vertices and $O(k \cdot 2^k)$ edges, where k is the vertex cover number.*

With Reduction Rule 1 we can compute in linear time an equivalent instance having a bounded number of vertices. Applying to this instance the trivial $O(n^4)$-time algorithm yields the following.

Theorem 10. HYPERBOLICITY *can be computed in $O(2^{4k} + n + m)$ time, where k denotes the size of a vertex cover of the input graph.*

SETH-based Lower Bounds. We show that, unless SETH breaks, the $2^{O(k)} + O(n + m)$-time algorithm obtained in the previous subsection cannot be improved to an algorithm even with running time $2^{o(k)} \cdot n^{2-\epsilon}$. This also implies, that, assuming SETH, there is no problem kernel with $2^{o(k)}$ vertices computable in $O(n^{2-\epsilon})$ time, i. e., the kernel obtained by applying Reduction Rule 3 cannot be improved significantly. The proof follows by a many-one reduction from the problem ORTHOGONAL VECTORS: herein, given two sets $\vec{A}$ and $\vec{B}$ each containing n binary vectors of length $\ell = O(\log n)$, the question is whether there are two vectors $\vec{a} \in \vec{A}$ and $\vec{b} \in \vec{B}$ such that $\vec{a}$ and $\vec{b}$ are orthogonal, that is, such that there is no position i for which $\vec{a}[i] = \vec{b}[i] = 1$.

Williams and Yu [19] proved that, if ORTHOGONAL VECTORS can be solved in $O(n^{2-\epsilon})$ time, then SETH breaks. We provide a linear-time reduction from ORTHOGONAL VECTORS to HYPERBOLICITY where the graph G constructed in the reduction contains $O(n)$ vertices and admits a vertex cover of size $O(\log n)$ (and thus contains $O(n \cdot \log n)$ edges). The reduction then implies that, unless SETH breaks, there is no algorithm solving HYPERBOLICITY in time polynomial in the size of the vertex cover and linear in the size of the graph. We mention that Borassi et al. [4] showed that under the SETH HYPERBOLICITY cannot be solved in $O(n^{2-\epsilon})$. However, the instances constructed in their reduction have a minimum vertex cover of size $\Omega(n)$. Note that our reduction is based on ideas from the reduction of Abboud et al. [1] for the DIAMETER problem.

Theorem 11. *Assuming SETH, HYPERBOLICITY cannot be solved in $2^{o(k)} \cdot (n^{2-\epsilon})$ time, even on graphs with $O(n \log n)$ edges, diameter four, and domination number three. Here, k denotes the vertex cover number of the input graph.*

Proof. We reduce any instance $(\vec{A}, \vec{B})$ of ORTHOGONAL VECTORS to an instance (G, δ) of HYPERBOLICITY, where we construct the graph G as follows (we refer to Figure 1 for a sketch of the construction).

Make each $\vec{a} \in \vec{A}$ a vertex a and each $\vec{b} \in \vec{B}$ a vertex b of G, and denote these vertex sets by A and B, respectively. Add two vertices for each of the ℓ dimensions, that is, add the vertex set $C := \{c_1, \ldots, c_\ell\}$ and the vertex set $D = \{d_1, \ldots, d_\ell\}$ to G and make each of C and D a clique. Next, connect each $a \in A$ to the vertices of C in the natural way, that is, add an edge between a and c_i if and only if $\vec{a}[i] = 1$. Similarly, add an edge between $b \in B$ and $d_i \in D$ if and only if $\vec{b}[i] = 1$. Moreover, add the edge set $\{\{c_i, d_i\} \mid i \in [\ell]\}$. This part will constitute the central gadget of our construction.

Our aim is to ensure that the maximum hyperbolicity is reached for 4-tuples (a, b, c, d) such that $a \in A$, $b \in B$, and $\vec{a}$ and $\vec{b}$ are orthogonal vectors. The construction of G is completed by adding two paths (u_A, u, u_B) and (v_A, v, v_B), and making u_A and v_A adjacent to all vertices in $A \cup C$ and u_B and v_B adjacent to all vertices in $B \cup D$.

Observe that G contains $O(n)$ vertices, $O(n \cdot \log n)$ edges, and that the set $V \setminus (A \cup B)$ forms a vertex cover in G of size $O(\log n)$. Moreover, observe that G has diameter four. Note that each vertex in $A \cup B \cup C \cup D$ is at distance two to each

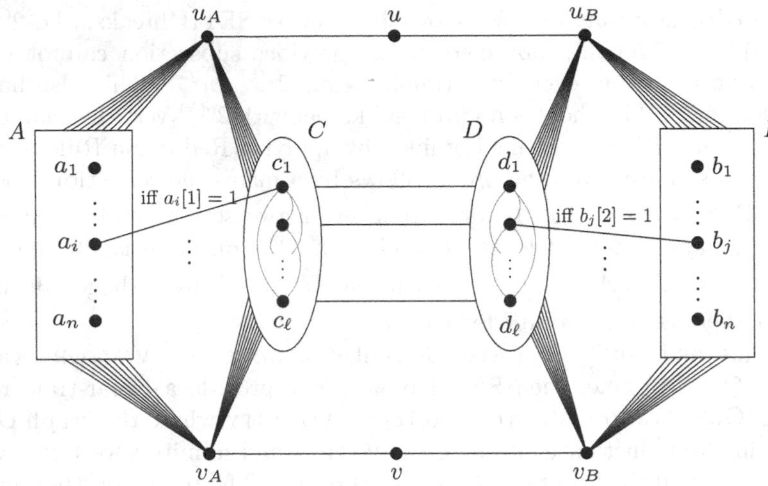

Fig. 1. Sketch of the construction described in the proof of Theorem 11. Ellipses indicate cliques, rectangles indicate independent sets. Multiple edges to an object indicate that the corresponding vertex is incident to each vertex enclosed within that object.

of u and v. Moreover, v_A and v_B are at distance three to u. Analogously, u_A, u_B are at distance three to v. Furthermore u and v are at distance four. Finally, observe that $\{u_A, u_B, v\}$ forms a dominating set in G.

We complete the proof by showing that $(\vec{A}, \vec{B})$ is a yes-instance of ORTHOGONAL VECTORS if and only if G has hyperbolicity at least $\delta = 4$.

($\Rightarrow$) Let $(\vec{A}, \vec{B})$ be a yes-instance, and let $\vec{a} \in \vec{A}$ and $\vec{b} \in \vec{B}$ be a pair of orthogonal vectors. We claim that $\delta(a, b, u, v) = 4$. Since $\vec{a}$ and $\vec{b}$ are orthogonal, there is no $i \in [\ell]$ with $\vec{a}[i] = \vec{b}[i] = 1$ and, hence, there is no path connecting a and b only containing two vertices in $C \cup D$, and it holds that $\overline{ab} = 4$. Moreover, we know that $\overline{uv} = 4$ as that $\overline{au} = \overline{bu} = \overline{av} = \overline{av} = 2$. Thus, $\delta(a, b, u, v) = 8 - 4 = 4$, and G is 4-hyperbolic.

($\Leftarrow$) Let $S = \{a, b, c, d\}$ be a set of vertices such that $\delta(a, b, c, d) \geq 4$. By Lemma 1, it follows that no two vertices of S are adjacent. Hence, we assume without loss of generality that $\overline{ab} = \overline{cd} = 4$. Observe that all vertices of C and D have distance at most three to all other vertices. Similarly, each vertex of $\{u_A, v_A, u_B, v_B\}$ has distance at most three to all other vertices. (Consider for example u_A. By construction, u_A is a neighbor of all vertices in $A \cup C \cup \{u\}$ and, hence, u_A has distance at most two to v_A and to all vertices in D. Thus, u_A has distance at most three to v, B, u_B and v_B and therefore to all vertices of G. The arguments for v_A, u_B, and v_B are symmetric.)

It follows that $S \subseteq A \cup B \cup \{u, v\}$, and therefore at least two vertices in S are from $A \cup B$. Thus, assume without loss of generality that a is contained in A. By the previous assumption, we have that $\overline{ab} = 4$. This implies that $b \in B$ and $\vec{a}$ and $\vec{b}$ are orthogonal vectors, as every other vertex in $V \setminus B$ is at distance three to a and each $b' \in B$ with $\overrightarrow{b'}$ being non-orthogonal to $\overrightarrow{d}$ is at distance three to a. Hence, $(\overrightarrow{A}, \overrightarrow{B})$ is a yes-instance. □

We remark that, with the above reduction, the hardness also holds for the variants in which we fix one vertex (u) or two vertices (u and v). The reduction also shows that approximating the hyperbolicity of a graph within a factor of $4/3 - \epsilon$ cannot be done in strongly subquadratic time or with a PL-FPT running time with respect to the vertex cover number.

Next, we adapt the above reduction to obtain the following hardness result on graphs of bounded maximum degree.

Theorem 12 ($\star$). *Assuming SETH,* HYPERBOLICITY *cannot be solved in* $f(\Delta) \cdot (n^{2-\epsilon})$ *time, where* Δ *denotes the maximum degree of the input graph.*

5 Parameter Distance to Cographs

In this section we describe a *linear-time parameterized* algorithm for HYPERBOLICITY parameterized by the vertex deletion distance k to cographs; that is, we present an algorithm with linear dependence on the input size but arbitrary dependence on the parameter (to which we refer to as L-FPT). A graph is a cograph if and only if it is P_4-free. Given a graph G we can determine in linear time whether it is a cograph and return an induced P_4 if this is not the case. This implies that in $O(k \cdot (m+n))$ time we can compute a set $X \subseteq V$ of size at most $4k$ such that $G - X$ is a cograph.

A further characterization is that a cograph can be obtained from graphs consisting of one single vertex via unions and joins [5].

- A *union* of two graphs $G_1 = (V_1, E_1)$ and $G_2 = (V_2, E_2)$ is the graph $(V_1 \cup V_2, E_1 \cup E_2)$.
- A *join* of two graphs $G_1 = (V_1, E_1)$ and $G_2 = (V_2, E_2)$ is the graph $(V_1 \cup V_2, E_1 \cup E_2 \cup \{\{v_1, v_2\}|v_1 \in V_1, v_2 \in V_2\})$.

The union of t graphs and the join of t graphs are defined by taking successive unions or joins, respectively, of the t graphs in an arbitrary order. Each cograph G can be associated with a rooted cotree T_G. The leaves of T_G are the vertices of V. Each internal node of T_G is labeled either as a union or join node. For node v in T_G, let $L(v)$ denote the leaves of the subtree rooted at v. For a union node v with children $u_1, \ldots, u_t$, the graph $G[L(v)]$ is the union of the graphs $G[L(u_i)]$, $1 \le i \le t$. For a join node v with children $u_1, \ldots, u_t$, the graph $G[L(v)]$ is the join of the graphs $G[L(u_i)]$, $1 \le i \le t$.

The cotree of a cograph can be computed in linear time [8]. In a subroutine in our algorithm for HYPERBOLICITY we need to solve the following variant of SUBGRAPH ISOMORPHISM.

COLORED INDUCED SUBGRAPH ISOMORPHISM

Input: An undirected graph $G = (V, E)$ with a vertex-coloring $\gamma : V \to \mathbb{N}$ and an undirected graph $H = (W, F)$, where $|W| = k$, with a vertex-coloring $\chi : W \to \mathbb{N}$.

Question: Is there a vertex set $S \subseteq V$ such that there is an isomorphism f from $G[S]$ to H such that $\gamma(v) = \chi(f(v))$ for all $v \in S$?

Informally, the condition that $\gamma(v) = \chi(f(v))$ means that every vertex is mapped to a vertex of the same color. We say that such an isomorphism *respects the colorings*. As shown by Damaschke [9], INDUCED SUBGRAPH ISOMORPHISM on cographs is NP-complete. Since this is the special case of COLORED INDUCED SUBGRAPH ISOMORPHISM where all vertices in G and H have the same color, COLORED INDUCED SUBGRAPH ISOMORPHISM is also NP-complete (containment in NP is obvious). In the following, we show that on cographs COLORED INDUCED SUBGRAPH ISOMORPHISM can be solved by an L-FPT algorithm when the parameter k is the order of H.

Lemma 13 ($\star$). COLORED INDUCED SUBGRAPH ISOMORPHISM *can be solved in* $O(3^k(n+m))$ *time in cographs.*

We now turn to the algorithm for HYPERBOLICITY on graphs that can be made into cographs by at most k vertex deletions. The final step is to reduce HYPERBOLICITY to the problem DISTANCE-CONSTRAINED 4-TUPLE: herein, given an undirected graph $G = (V, E)$ and six integers $d_{\{a,b\}}$, $d_{\{a,c\}}$, $d_{\{a,d\}}$, $d_{\{b,c\}}$, $d_{\{b,d\}}$, and $d_{\{c,d\}}$, the question is whether there is a set $S \subseteq V$ of four vertices and a bijection $f \colon S \to \{a, b, c, d\}$ such that for each $x, y \in S$ we have $\overline{xy} = d_{\{f(x),f(y)\}}$.

Lemma 14 ($\star$). DISTANCE-CONSTRAINED 4-TUPLE *can be solved in* $O(4^{4k} \cdot k \cdot (n+m))$ *time if* $G - X$ *is a cograph for some* $X \subseteq V$ *of size* k.

We solve HYPERBOLICITY by creating $O(k^6)$ instances of DISTANCE-CONSTRAINED 4-TUPLE as shown below.

Theorem 15. HYPERBOLICITY *can be solved in* $O(4^{4k} \cdot k^7 \cdot (n+m))$ *time, where* k *is the vertex deletion distance of* G *to cographs.*

Proof. Let $G = (V, E)$ be the input graph and $X \subseteq V$, $|X| \le k$, such that $G - X$ is a cograph and observe that X can be computed in $O(4^k \cdot (n+m))$ time. Since every connected component of $G - X$ has diameter at most two, the maximum distance between any pair of vertices in the same component of G is at most $4k + 2$: any shortest path between two vertices u and v visits at most k vertices in X, at most three vertices between every pair of vertices x and x' from X and at most three vertices before encountering the first vertex of X and at most three vertices before encountering the last vertex of X.

Consequently, for the 4-tuple (a, b, c, d) that maximizes $\delta(a, b, c, d)$, there are $O(k^6)$ possibilities for the pairwise distances between the four vertices. Thus, we may compute whether there is a 4-tuple such that $\delta(a, b, c, d) = \delta$ by checking for each of the $O(k^6)$ many 6-tuples of possible pairwise distances of four vertices in G whether there are 4 vertices in G with these six pairwise distances and whether this implies $\delta(a, b, c, d) \ge \delta$. The latter check can be performed in $O(1)$ time, and the first is equivalent to solving DISTANCE-CONSTRAINED 4-TUPLE which can be done in $O(4^{4k} \cdot k \cdot (n+m))$ time by Lemma 14. The overall running time follows. □

6 Reduction from 4-Independent Set

In this section, we provide a further relative lower bound for HYPERBOLIC-ITY. Specifically, we prove that, if the running time is measured in terms of n, then HYPERBOLICITY is at least as hard as the problem of finding an independent set of size four in a graph. The currently best running time for this problem is $O(n^{3.257})$ [11, 20]. Hence, any improvement on the running time of HYPERBOL-ICITY which breaks this bound (e.g., an algorithm running in $o(n^3)$ time), would also yield a substantial improvement for the 4-INDEPENDENT SET problem.

To this end, we reduce from a 4-partite (or 4-colored) variant of the INDE-PENDENT SET problem. The standard reduction [12] from INDEPENDENT SET to MULTICOLORED INDEPENDENT SET shows that this 4-colored variant has the same asymptotic running time lower bound as 4-INDEPENDENT SET.

Theorem 16 (⋆). *Any algorithm solving* HYPERBOLICITY *in* $O(n^c)$ *time for some constant c yields an* $O(n^c)$*-time algorithm solving* 4-INDEPENDENT SET.

7 Conclusion

To efficiently compute the hyperbolicity number, parameterization sometimes may help. In this respect, perhaps our practically most promising results relate to the $O(k^4(n + m))$ running times (for the parameters covering path number and feedback edge number, see Table 1). Note that they clearly improve on the standard algorithm when $k = O(n^{1/4})$. Moreover, the linear-time data reduc-tion rules we presented may be of independent practical interest. On the lower bound side, together with the work of Abboud et al. [1] our SETH-based lower bound with respect to the parameter vertex cover number is among few known "exponential lower bounds" for a polynomial-time solvable problem.

As to future work, we particularly point to the following open questions. First, we left open whether there is an L-FPT algorithm exploiting the parameter feedback vertex number for computing the hyperbolicity number. Second, for parameter vertex cover number we have an SETH-based exponential lower bound for the parameter function in any L-FPT algorithm. This does not imply that it is impossible to achieve a polynomial parameter dependence when asking for algorithms with running time factors such as $O(n^2)$ or $O(n^3)$.

Bibliography

[1] A. Abboud, V. Vassilevska Williams, and J. R. Wang. Approximation and fixed parameter subquadratic algorithms for radius and diameter in sparse graphs. In *Proc. 27th SODA*, pages 377–391. SIAM, 2016.

[2] M. Abu-Ata and F. F. Dragan. Metric tree-like structures in real-world networks: an empirical study. *Networks*, 67(1):49–68, 2016.

[3] M. Borassi, D. Coudert, P. Crescenzi, and A. Marino. On computing the hyperbolicity of real-world graphs. In *Proc. 23rd ESA*, volume 9294 of *LNCS*, pages 215–226, 2015.

[4] M. Borassi, P. Crescenzi, and M. Habib. Into the square: On the complexity of some quadratic-time solvable problems. *Electronic Notes in Theoretical Computer Science*, 322:51–67, 2016.

[5] A. Brandstädt, V. B. Le, and J. P. Spinrad. *Graph Classes: a Survey*, volume 3 of *SIAM Monographs on Discrete Mathematics and Applications*. SIAM, 1999.

[6] G. Brinkmann, J. H. Koolen, and V. Moulton. On the hyperbolicity of chordal graphs. *Annals of Combinatorics*, 5(1):61–69, 2001.

[7] N. Cohen, D. Coudert, and A. Lancin. On computing the Gromov hyperbolicity. *ACM Journal of Experimental Algorithmics*, 20:1.6:1–1.6:18, 2015.

[8] D. G. Corneil, Y. Perl, and L. K. Stewart. A linear recognition algorithm for cographs. *SIAM Journal on Computing*, 14(4):926–934, 1985.

[9] P. Damaschke. Induced subgraph isomorphism for cographs is NP-complete. In *Proc. 16th WG*, volume 484 of *LNCS*, pages 72–78. Springer, 1991.

[10] M. Doucha and J. Kratochvíl. Cluster vertex deletion: A parameterization between vertex cover and clique-width. In *Proc. 37th MFCS*, volume 7464 of *LNCS*, pages 348–359. Springer, 2012.

[11] F. Eisenbrand and F. Grandoni. On the complexity of fixed parameter clique and dominating set. *Theoretical Computer Science*, 326(1-3):57–67, 2004.

[12] M. Fellows, D. Hermelin, F. Rosamond, and S. Vialette. On the parameterized complexity of multiple-interval graph problems. *Theoretical Computer Science*, 410(1):53–61, 2009.

[13] H. Fournier, A. Ismail, and A. Vigneron. Computing the Gromov hyperbolicity of a discrete metric space. *Information Processing Letters*, 115(6-8): 576–579, 2015.

[14] M. R. Garey and D. S. Johnson. *Computers and Intractability: A Guide to the Theory of NP-Completeness*. Freeman, 1979.

[15] A. C. Giannopoulou, G. B. Mertzios, and R. Niedermeier. Polynomial fixed-parameter algorithms: A case study for longest path on interval graphs. In *Proc. 10th IPEC*, volume 43 of *LIPIcs*, pages 102–113. Schloss Dagstuhl - Leibniz-Zentrum fuer Informatik, 2015.

[16] M. Gromov. Hyperbolic groups. In *Essays in Group Theory, MSRI Publ., vol. 8*, pages 75–263. Springer New York, 1987.

[17] D. Mitsche and P. Pralat. On the hyperbolicity of random graphs. *The Electronic Journal of Combinatorics*, 21(2):P2.39, 2014.

[18] C. H. Papadimitriou and K. Steiglitz. *Combinatorial Optimization: Algorithms and Complexity*. Prentice-Hall, 1982.

[19] R. Williams and H. Yu. Finding orthogonal vectors in discrete structures. In *Proc. 25th SODA*, pages 1867–1877. SIAM, 2014.

[20] V. V. Williams, J. R. Wang, R. Williams, and H. Yu. Finding four-node subgraphs in triangle time. In *Proc. 26th SODA*, pages 1671–1680. SIAM, 2015.

Optimal Query Time for Encoding Range Majority

Paweł Gawrychowski and Patrick K. Nicholson

University of Haifa, Israel and Nokia Bell Labs, Ireland

Abstract. We revisit the range τ-majority problem, which asks us to preprocess an array A$[1..n]$ for a fixed value of $\tau \in (0, \frac{1}{2}]$, such that for any query range $[i, j]$ we can return a position in A of each distinct τ-majority element. A τ-majority element is one that has relative frequency at least τ in the range $[i, j]$: i.e., frequency at least $\tau(j - i + 1)$. Belazzougui et al. [WADS 2013] presented a data structure that can answer such queries in $\mathcal{O}(1/\tau)$ time, which is optimal, but the space can be as much as $\Theta(n \lg n)$ bits. Recently, Navarro and Thankachan [Algorithmica 2016] showed that this problem could be solved using an $\mathcal{O}(n \lg(1/\tau))$ bit encoding, which is optimal in terms of space, but has suboptimal query time. In this paper, we close this gap and present a data structure that occupies $\mathcal{O}(n \lg(1/\tau))$ bits of space, and has $\mathcal{O}(1/\tau)$ query time. We also show that this space bound is optimal, even for the much weaker query in which we must decide whether the query range contains at least one τ-majority element.

1 Introduction

Misra and Gries [14] generalized a classic 2-pass algorithm by Boyer and Moore [3] for finding majorities in lists of elements. Formally, a τ-majority of a list of length n (or τ-heavy-hitter) is an element that appears with frequency at least $\tau \cdot n$. More recent variants and improvements [5,12] to the Misra-Gries algorithm have become standard tools in a wide variety of applications involving streaming analytics, such as IP traffic monitoring, data mining, etc.

In this paper we consider the data structure variant of the problem. Suppose we are given an array A of n elements. The goal is to preprocess the array into a data structure that supports *range τ-majority queries*: given an arbitrary subarray A$[i..j]$, return all distinct elements that are τ-majorities in A$[i..j]$. As an example, we may wish to construct such a structure on network traffic logs, to perform an analysis of how the set of frequent users changes over time.

In the last few years, this problem has received a lot of attention [2,6,8,13], finally leading to a recent result of Belazzougui et al. [1,2]: these queries can be supported in $\mathcal{O}(1/\tau)$ time, using $(1 + \varepsilon)nH_0 + o(n)$ bits of space, where H_0 is the zero-th order empirical entropy of the array A, and ε is an arbitrary positive constant.[1] Since, for an arbitrary τ-majority query, there can be $\lfloor 1/\tau \rfloor$

---

[1] Note that, for this and all forthcoming results discussed, we assume the word-RAM model of computation with word-size $w = \Omega(\lg n)$ bits; we use $\lg x$ to denote $\log_2 x$.

© Springer International Publishing AG 2017
F. Ellen et al. (Eds.): WADS 2017, LNCS 10389, pp. 409–420, 2017.
DOI: 10.1007/978-3-319-62127-2_35

answers, there is not much hope for significantly improving the query time of $\mathcal{O}(1/\tau)$, except perhaps to make the time bound output-sensitive on the number of results returned [1, Sec. 7].

On the other hand, much more can be said about the space bound. Note that, in general, if A contains elements drawn from the alphabet $[1, \sigma]$, then we can represent it using $n\lceil \lg \sigma \rceil$ bits. If f_i is the frequency of element $i \in [1, \sigma]$, then we have $nH_0 = n \sum_i ((f_i/n) \lg(n/f_i)) \leq n\lceil \lg \sigma \rceil$.[2] Since the bound of Belazzougui et al. [1] depends on the entropy of the elements in A, it can therefore can be $\Theta(n \lg n)$ bits, if $\sigma = \Omega(n^c)$ for any constant $c \leq 1$, and the distribution is close to uniform. However, quite recently, Navarro and Thankachan [15] showed that this space bound can be improved significantly in the *encoding model.*

In the encoding model, given array A as input, we are allowed to construct an encoding that supports a specific query operation on A. After constructing the encoding, the array A is deleted, and queries must be supported by accessing only the encoding. For many query operations, we can achieve space bounds that are much smaller than the space required to store A. One issue is that for range τ-majority queries, if we return the actual element which is a τ-majority, then we must store at least as many bits as are required to represent A. This follows since an encoding supporting such queries can be used to return the contents of the array A by querying the range $A[i..i]$ for each $1 \in [1, n]$.

Navarro and Thankachan [15] therefore considered a different query, in which, for each τ-majority a in the query range $A[i..j]$, we instead return an arbitrary position ℓ in A such that $A[\ell] = a$ and $i \leq \ell \leq j$. In the remainder of the paper, we use *range τ-majority position query* to refer to this positional variant of the query operation. Navarro and Thankachan [15] showed two main results:

Theorem 1 ([15], Theorems 1 and 2).

1. *For any $\tau \in (0, 1)$, there is an encoding that occupies $\mathcal{O}(n\lceil \lg(1/\tau) \rceil)$ bits of space that supports range τ-majority position queries in:*
   (a) $\mathcal{O}((1/\tau) \lg n)$ time if $1/\tau = o(\mathrm{polylog}(n))$.
   (b) $\mathcal{O}(1/\tau)$ time if $1/\tau = \Theta(\mathrm{polylog}(n))$.
   (c) $\mathcal{O}(1/\tau \lg \lg_w(1/\tau))$ time if $1/\tau = \omega(\mathrm{polylog}(n))$.
2. *Any encoding that can support range τ-majority counting queries (i.e., return the total the number of τ-majorities) in an arbitrary query range $A[i..j]$ occupies space (in bits) at least $\frac{n}{4} \left(\lg \left(\frac{1}{2\tau} - 1 \right) - \lg e \right) = \Omega(n \lg(1/\tau))$.*

Thus, their lower bound implies that their space bound, which depends only on n and τ rather than elements in the input array A, is optimal. However, there is gap between the query time of their encoding and the data structure of Belazzougui et al. [1] for the case where $1/\tau$ is not $\Theta(\mathrm{polylog}(n))$. Crucially, this does not yield optimal time in the important case where $1/\tau$ is a constant. In this paper, we close this time gap, and prove the following theorem:

We also note that Belazzougui et al. [1] also considered a slightly more difficult problem in which τ can be specified at query time, rather than fixed once-and-for-all before constructing the data structure.

[2] We follow the convention that $(f_i/n) \lg(n/f_i) = 0$ if $f_i = 0$.

Theorem 2. *For any $\tau \in (0, 1/2]$, there is an encoding that occupies $\mathcal{O}(n \lg(1/\tau))$ bits of space that can answer range τ-majority position queries in $\mathcal{O}(1/\tau)$ time.*[3]

Of course one could ask if $\mathcal{O}(1/\tau)$ is the right bound for the query time at all. In the output-sensitive variant of the problem the query time should depend on the number of results returned, which might be up to $\mathcal{O}(1/\tau)$ but possibly smaller. However, we note that a straightforward reduction from the set intersection conjecture indicates that a significantly smaller query time cannot be guaranteed even if the size of the output is 0 or 1 [9].

In terms of techniques, our approach uses the level-based decomposition of Durocher et al. [6], but with three significant improvements. We define two new methods for pruning their data structure to reduce space, and one method to speed up queries. The first pruning method is a top-down approach that avoids replicating data structures at more than one level and is analysed using a charging argument. The second pruning method is bottom-up, operating on small ranges of the input array, that we call micro-arrays, and applies one of two strategies, depending on the parameter τ. One of these strategies involves bootstrapping an optimal space (but suboptimal query time) encoding by combining it with pre-computed lookup tables in order to speed up queries on the micro-arrays. The other strategy stores (a rank reduced) copy of the micro-array and solves queries in a brute-force manner. Finally, the last improvement uses wavelet trees [11] in a non-trivial way in order to build fast ranking data structures to improve query time for the case when $1/\tau = \omega(\text{polylog}(n))$.

Implications. Since the encoding yields the positions of each distinct τ-majority element in the query range, we can use our optimal encoding as an alternative to the non-encoding data structure of Belazzougui et al. This is done by first compressing the original array A using any compressor that supports access in $\mathcal{O}(1)$ to the underlying elements.

Theorem 3. *Let $\mathcal{S}(n)$ be the space required to store the input array in a compressed form such that each position can be accessed in $\mathcal{O}(1)$ time. Then there is a data structure that occupies $\mathcal{S}(n) + \mathcal{O}(n \lg(1/\tau))$ bits of space, and can return the range τ-majorities for an arbitrary range $[i, j]$ in $\mathcal{O}(1/\tau)$ time.*

For example, using results for higher order entropy compression with $\mathcal{O}(1)$ access time [7, 10] yields the following:

Corollary 1. *Let $A[1..n]$ be an array with elements drawn from $[1, \sigma]$. There is a data structure that occupies $nH_k + o(n \lg \sigma) + \mathcal{O}(n \lg(1/\tau))$ bits of space[4], and can support arbitrary range τ-majority queries in time $\mathcal{O}(1/\tau)$, for any $k = o(\log_\sigma n)$.*

---

[3] For $\tau \in (1/2, 1)$ the structure for $1/2$-majorities can answer queries in $\mathcal{O}(1)$ time.

[4] H_k denotes the k-th order empirical entropy of the sequence of elements in A: a lower bound on the space achievable by any compressor that considers the previous k elements in the sequence. For all $k \geq 1$ we have $nH_k \leq nH_{k-1}$.

Lower Bound. Recall the lower bound of $\Omega(n \lg(1/\tau))$ bits holds for any encoding supporting range τ-majority counting queries. We consider an easier problem that we call *range τ-majority decision queries*. The query asks "Is there at least one element in the query range $A[i..j]$ which is a τ-majority?". Since the previous lower bound does not rule out a better encoding for these decision queries, it is natural to ask whether a better encoding exists. We prove the following:

Theorem 4. *Any data structure that can be used to represent an array $A[1..n]$ and support $1/k$-majority decision queries, for any integer $k \geq 2$, on any arbitrary query range $[i, j]$, requires $n \lg \frac{k}{e} - \Theta(k^4 \lg k)$ bits of space.*

Thus, we answer this question in the negative by showing a lower bound of $\Omega(n \lg(1/\tau))$ bits for any encoding that supports these queries, which proves our structure is space-optimal for even these restricted types of queries. Moreover, we note that our lower bound has an improved constant factor compared to the previous lower bound.

Related Work. Finally, we remark that the area of range queries on arrays is quite vast, and there are many interesting related types of queries that have been studied in the both the non-encoding and encoding models; we refer the reader to surveys on the topics [17,19]. The most closely related problem to the range τ-majority problem is the *range mode problem* [4]: given a query range $[i, j]$ return the most frequently appearing element in the range. In contrast with range τ-majority, this type of query is significantly less efficient, with the best $\Theta(n \lg n)$ bit data structures having $\mathcal{O}(\sqrt{n/\lg n})$ query time.

2 Preliminaries

Lemma 1 ([16]). *Let V be a bit vector of length n bits in which m of the bits are set to one. There is a data structure for representing V that uses $m \lg(n/m) + \mathcal{O}(n/\lg^c(n))$ bits for any constant $c \geq 1$ such that the following queries can be answered in $\mathcal{O}(1)$ time:*

- access(V, i): *returns bit $V[i]$.*
- rank(V, i): *returns the number of ones in the prefix $V[1..i]$.*
- select(V, j): *returns the index of the j-th one in V, if it exists, and -1 otherwise. In other words, the inverse of the rank operation: if select$(V, j) = i$, then rank$(V, i) = j$.*

Since our proof makes heavy use of this lemma, we distinguish the $m \lg(n/m)$ term in the space bound by calling it the *leading term*, and the other term the *redundancy*. If we do not need the full power of rank, then we can use the following lemma to reduce the redundancy:

Lemma 2 ([18]). *If only the constant time select and access operations are required, then we can represent V using $m \lg(n/m) + o(m) + \mathcal{O}(\lg \lg n)$ bits.*

A useful fact about applying these previous Lemmas to bit vectors is that concatenation is often helpful: if we apply either Lemma to two bit vectors separately, both of length n containing at least m bits, then the sum of the leading terms is no more than $2m \lg(n/m)$. If we concatenate the bit vectors before applying the lemma, the upper bound on the leading term is the same.

3 Upper Bound

3.1 Quadruple Decomposition

The upper bound makes use of the *quadruple decomposition* of Durocher et al. [6]. For ease of description, we assume that n is a power of 2, but note that decomposition works in general. First, at a conceptual level we build a balanced binary tree over the array $A[1..n]$. Each leaf represents an element $A[i]$. On the k-th level of the tree $T(k)$, counting from the leaves at level 0, the nodes represent a partition of $A[1..n]$ into $n_k = n/2^k$ contiguous *blocks* of length 2^k. Second, consider all levels containing at least four blocks. At each such level, consider the blocks $B_1, \ldots, B_{n_k}$. We create a list of *quadruples* (i.e., groups of four consecutive blocks) at each such level:

$$\mathcal{D}(k) = [(B_1, B_2, B_3, B_4), (B_3, B_4, B_5, B_6), \ldots, (B_{n_k-1}, B_{n_k}, B_1, B_2)].$$

Thus, each index in A is contained in exactly two quadruples at each level, and there is one quadruple that wraps-around to handle corner cases. The quadruples are staggered at an offset of two blocks from each other. Moreover, given a quadruple $D = (B_{2\ell+1}, B_{2\ell+2}, B_{2\ell+3}, B_{2\ell+4})$, the two middle blocks $B_{2\ell+2}$ and $B_{2\ell+3}$ are not siblings in the binary tree T. We call the range spanned by these two middle blocks the *middle part* of D.

As observed by Durocher et al. [6], for every query range $[i, j]$ there exists a unique level k in the tree such that $[i, j]$ contains at least one and at most two consecutive blocks in $T(k)$, and, if $[i, j]$ contains two blocks, then the nodes representing these blocks are not siblings in the tree T. Thus, based on our quadruple decomposition, for every query range $[i, j]$ we can *associate* it with exactly one quadruple $D = (B_{2\ell+1}, B_{2\ell+2}, B_{2\ell+3}, B_{2\ell+4})$ such that

$$((B_{2\ell+2} \subseteq [i, j]) \vee (B_{2\ell+3} \subseteq [i, j])) \wedge ([i, j] \subset B_{2\ell+1} \cup B_{2\ell+2} \cup B_{2\ell+3} \cup B_{2\ell+4}.)$$

Moreover, Durocher et al. [6] proved the following lemma:

Lemma 3 ([6]). *For each query range $[i, j]$, in $\mathcal{O}(1)$ time we can compute the level k, as well as the offset of the quadruple associated with $[i, j]$ in the list $\mathcal{D}(k)$, using $o(n)$ bits of space.*

Furthermore, if we consider any arbitrary query range $[i, j]$ that is associated with a quadruple D, there are at most $4/\tau$ elements in the range represented by D that could be τ-majorities for the query range. Following Durocher et al., we refer to these elements as *candidates* for the quadruple D.

For each quadruple, we compute and store all of its candidates, so that, by Lemma 3, in $\mathcal{O}(1)$ time we can obtain $\mathcal{O}(1/\tau)$ candidates. It remains to show how to verify that a candidate is in fact a τ-majority in $A[i..j]$. At this point, our approach deviates from Durocher et al. [6], who make use of a wavelet tree for verification, and end up with a space bound of $\mathcal{O}(n \lg n \lg(1/\tau))$ bits.

Consider such a candidate y for quadruple $D = (B_{2\ell+1}, \ldots, B_{2\ell+4})$. Our goal is to count the number of occurrences of y in the query range $[i, j]$. To do this we store a bit vector $V(D, y)$, that represents the (slightly extended) range $B_{2\ell} \cup \ldots \cup B_{2\ell+5}$ and marks all occurrences of y in this range with a one bit. By counting the number of ones in the range corresponding to $[i, j]$ in $V(D, y)$, we can determine if the number of occurrences exceeds the threshold $\tau(j - i + 1)$. If the threshold is exceeded, then we can return the first one bit in the range, as that position in A contains element y. Note that we have extended the range of the bit vector beyond the range covered by D by one extra block to the left and right. We call this extended range the *extent* of D, and observe the following:

Observation 5. *Let $E(D)$ be the extent of quadruple D at level k. Then for all quadruples D' at level $k' < k$ such that the range of D' has non-empty intersection with the range of D, we have that $D' \subset E(D)$.*

We now briefly analyze the total space of this method, under the assumption that we can store a bit vector of length n with m one bits using $\mathcal{O}(m \lg(n/m))$ bits. This crude analysis is merely to illustrate that additional tricks are needed to achieve optimal space. The quadruple decomposition consists of $\lg n$ levels. On each level, we store a number of bit vectors. For each quadruple we have up to $\mathcal{O}(1/\tau)$ candidates $\mathcal{Y}$. Thus, if f_y represents the frequency of candidate y in extent of quadruple D, then the space bound, for each quadruple at level k, is $\sum_{y \in \mathcal{Y}} \mathcal{O}(f_c \lg(2^k/f_c))$, which, by the concave version of Jensen's inequality, is bounded by $\mathcal{O}((2^k) \lg(1/\tau))$. So each level uses $\mathcal{O}(n \lg(1/\tau))$ bits, for a total of $\mathcal{O}(n \lg n \lg(1/\tau))$ bits over all levels.

3.2 Optimal Space with Suboptimal Query Time

To achieve space $\mathcal{O}(n \lg(1/\tau))$ bits, the intuition is that we should avoid duplicating the same bit vectors between levels. It is easy to imagine a case where element y is a candidate at every level and in every quadruple of the decomposition, which results in many duplicated bit vectors. To avoid this duplication problem, we propose a top-down algorithm for pruning the bit vectors. Initially, all indices in A are *active* at the beginning. Our goal is to charge at most $\mathcal{O}(\lg(1/\tau))$ bits to each active index in A, which achieves the desired space bound.

Let k be the current level of the quadruple decomposition, as we proceed top-down. We maintain the invariant that for any element y in a block B_i, either all indices storing occurrences of y are active in B_i (in which case we say y is active in B_i), or none are (in which case we say y is inactive in B_i). Consider a candidate y associated with quadruple $D = (B_{2\ell+1}, B_{2\ell+2}, B_{2\ell+3}, B_{2\ell+4})$. Then:

1. If y is active in blocks $B_{2\ell+1}, \ldots, B_{2\ell+4}$, then we store the bit vector $V(D, y)$, and (conceptually) mark all occurrences of y inactive in these blocks *after* we finish processing level k. This makes y inactive in all blocks contained in D at lower levels. Since a block B_i is contained in two quadruples at level k, a position storing y in B_i may be made inactive for two reasons: this is why we mark positions inactive after processing all quadruples at level k.

2. If y is inactive in some block $B_i \subset D$, then it is the case that we have computed and stored the bit vector $V(D', y)$ for some quadruple D' at level $k' > k$, such that $D \cap D' \neq 0$. Therefore, Observation 5 implies that D is contained in the extent of D', and thus the bit vector associated with D' can be used to answer queries for D. For D we need not to store $V(D, y)$, though for now we do not address how to efficiently answer these queries.

Next we analyse the total cost of the bit vectors that we stored during the top-down construction. The high level idea is that we can charge the cost of bit vector $V(D, y)$ to the indices in D that store occurrences of y. Call these the indices the *sponsors* of $V(D, y)$. Since y is a τ-majority, it occurs at least $\mathcal{O}(\tau \cdot 2^k)$ times in D, which has length $\mathcal{O}(2^k)$. Thus, we can expect to charge $\mathcal{O}(\lg(1/\tau))$ bits to each sponsor: the expected gap between one bits is $\mathcal{O}(1/\tau)$ and therefore can be recorded using $\mathcal{O}(\lg(1/\tau))$ bits. There are some minor technicalities that must be addressed, but this basic idea leads to the following intermediate result, in which we don't concern ourselves with the query time:

Lemma 4. *There is an encoding of size $\mathcal{O}(n \lg(1/\tau))$ bits such that the answer to all range τ-majority position queries can be recovered.*

Proof. Consider candidate y and its occurrences in extent $E(D)$ of quadruple D at level k, for which we stored the bit vector $V(D, y)$. Suppose there are f_y occurrences of y in $E(D)$. If at least one third of the occurrences of y are contained in D, then we charge the cost of the bit vector to the (at least) $f_y/3$ sponsor indices in D. Otherwise, this implies one of the two blocks, call it B_i such that $B_i \subset E(D)$ but $B_i \not\subset D$ contains at least $f_y/3$ occurrences of y. Therefore, y must also be an active candidate for the unique quadruple D' that has non-empty intersection with both B_i and D: this follows since y occurs more times in D' than in D, and y is a candidate for D. In this case we charge the cost of the bit vector to the sponsor indices in *neighbouring quadruple D'*.

Suppose we store the bit vectors using Lemma 2: for now ignore the $\mathcal{O}(\lg \lg n)$ term in the space bound as we deal with it in the next paragraph. Using Lemma 2, the cost of the bit vector $V(D, y)$ associated with D is at most $\mathcal{O}(f_y \lg(1/\tau))$, since y is a $(\tau/4)$-majority in D. Thus, $\mathcal{O}(f_y)$ sponsors in D pay for at most three bit vectors: $V(D, y)$ and possibly the two other bit vectors that cost $\mathcal{O}(f_y \lg(1/\tau))$ bits, charged by neighbouring quadruples. Since this charge can only occur at one level in the decomposition (the index becomes inactive at lower levels after the first charge occurs), each sponsor is charged $\mathcal{O}(\lg(1/\tau))$, making the total amount charged $\mathcal{O}(n \lg(1/\tau))$ bits overall.

To make answering queries actually possible, we make use of the same technique used by Durocher et al. [6], which is to concatenate the bit vectors at level

k. The candidates have some implicit ordering in each quadruple, $[1, ..., \mathcal{O}(1/\tau)]$: the ordering can in fact be arbitrary. For each level k, we concatenate the bit vectors associated with quadruple according to this implicit ordering of the candidates. Thus, since there are $\mathcal{O}(\lg n)$ bit vectors (one per level), the $\mathcal{O}(\lg \lg n)$ term for Lemma 2 contributes $\mathcal{O}(\lg n \lg \lg n)$ to the overall space bound.

Given a query $[i, j]$, Lemma 3 allows us to compute the level, k, and offset, ℓ, of the quadruple associated with $[i, j]$. Our goal is to remap $[i, j]$ to the relevant query range in the concatenated bit vector at level k. Since all bit vectors $V(D, y)$ at level k have the same length, we only need to know how many bit vectors are stored for quadruples $1, ..., \ell - 1$: call this quantity X. Thus, at level k we construct and store a bit vector L_k of length $\mathcal{O}(n_k/\tau)$ in which we store the number of bit vectors associated with the quadruples in unary. So, if the first three quadruples have $2, 6, 4$ candidates (respectively), we store $L_k = 1001000000100001 \ldots$. Overall, the space for L_k is $\mathcal{O}(n_k \lg(1/\tau))$, or $\mathcal{O}(n \lg(1/\tau))$ overall, if we represent each L_k using Lemma 2.

Given an offset ℓ, we can perform $\mathsf{select}(L_k, \ell) - \ell$ to get X. Once we have X, we use the fact that all extents have fixed length at a level in order to remap the query $[i, j]$ to the appropriate range $[i', j']$ in the concatenated bit vector for each candidate. We can then use binary search and the select operation to count the number of 1 bits corresponding to each candidate in the remapped range $[i', j']$ in $\mathcal{O}(\lg n)$ time per candidate. Since some of the candidates for the D associated with $[i, j]$ may have been inactive, we also must compute the frequency of each candidate in quadruples at higher levels that contain D. Since there are $\mathcal{O}(\lg n)$ levels, $\mathcal{O}(1)$ quadruples that overlap D per level, and $\mathcal{O}(1/\tau)$ candidates per quadruple, we can answer range τ-majority position queries in $\mathcal{O}(\lg^2 n/\tau)$ time. Note that we have to be careful to remove possible duplicate candidates (at each level the quadruples that overlap D may share candidates).

3.3   Optimal Space with Optimal Query Time

In Lemma 4 there are two issues that make querying inefficient: 1) we have to search for inactive candidates in $\mathcal{O}(\lg n)$ levels; and 2) we used Lemma 2 which does not support $\mathcal{O}(1)$ time rank queries. The solutions to both of these issues are straightforward. For the first issue, we store pointers to the appropriate bit vector at higher levels, allowing us to access them in $\mathcal{O}(1)$ time. For the second issue we can use Lemma 1 to support rank in $\mathcal{O}(1)$ time. However, both of these solutions raise their own technical issues that we must resolve in this section.

Pointers to higher levels. Consider a quadruple D at level k for which candidate y is inactive in some block contained in D. Recall that this implies the existence of some bit vector $V(D', y)$ for some D' at level $k' > k$ that can be used to count occurrences of y in D. In order to access this bit vector in $\mathcal{O}(1)$ time, the only information that we need to store is the number k' and also the offset of y in the list of candidates for D': D' might have a different ordering on its candidates than D. Thus, in this case we store $\mathcal{O}(\lg \lg n + \lg(1/\tau))$ bits per quadruple as we have $\mathcal{O}(\lg n)$ levels and $\mathcal{O}(1/\tau)$ candidates per quadruple. This is a problem, because there are $\mathcal{O}(n)$ quadruples, which means these pointers can occupy $\mathcal{O}(n/\tau(\lg\lg n + \lg(1/\tau)))$ bits overall.

To deal with this problem, we simply reduce the number of quadruples using a bottom-up pruning technique: all data associated with quadruples spanning a range of size Z or smaller is deleted. This is good as it limits the space for the pointers to at most $\mathcal{O}(n(\lg\lg n + \lg(1/\tau))/(\tau \cdot Z))$ bits, as there are $\mathcal{O}(n/Z)$ quadruples of length greater than Z. However, we need to come up with an alternative approach for queries associated with these small quadruples.

The value we select for Z, as well as the strategy to handle queries associated with quadruples of size Z or smaller, depends on the value of $1/\tau$:

1. If $1/\tau \geq \sqrt{\lg n}$: then we set $Z = 1/\tau$. Thus, the pointers occupy $\mathcal{O}(n(\lg\lg n + \lg(1/\tau)) = \mathcal{O}(n\lg(1/\tau))$ bits (since $\lg(1/\tau) = \Omega(\lg\lg n)$). Consider the maximum level k such that the quadruples are of size Z or smaller. For each quadruple D in level k, we construct a new *micro-array* of length 2^k by copying the range spanned by D from A. Thus, any query $[i,j]$ associated with a quadruple at levels k or lower can be reduced to a query on one of these micro-arrays. Since the micro-arrays have length $1/\tau$, we preprocess the elements in the array by replacing them by their ranks (i.e., we reduce the elements to rank space). Storing the micro-array therefore requires only $\mathcal{O}(n_k 2^k \lg(1/\tau)) = \mathcal{O}(n\lg(1/\tau))$ bits. Moreover, since we have access to the ranks of the elements directly, we can answer any query on the micro-array directly by scanning it in $\mathcal{O}(1/\tau)$ time. Thus, in this case, the space for the micro-arrays and pointers is $\mathcal{O}(n\lg(1/\tau))$.

2. If $1/\tau < \sqrt{\lg n}$: in this branch we use the encoding of Lemma 4 that occupies $c \cdot n \lg(1/\tau)$ bits of space for an array of length n, for some constant $c \geq 1$. We set $Z = \lg n/(2c\lg(1/\tau))$, so that the space for the pointers becomes:

$$\mathcal{O}(n(\lg\lg n + \lg(1/\tau))\lg(1/\tau)/(\tau \cdot \lg n)) = \mathcal{O}(n(\lg\lg n)^2/\sqrt{\lg n}) = o(n).$$

As in the previous case, we construct the micro-arrays for the appropriate quadruples based on the size Z. However, this time we encode each micro-array using Lemma 4. This gives us a set of n_k encodings, taking a total $\mathcal{O}(n_k 2^k \lg(1/\tau)) = \mathcal{O}(n\lg(1/\tau))$ bits. Moreover, the answer to a query is fully determined by the encoding and the endpoints i, j. Since i and j are fully contained in the micro-array, their description takes $\lg Z$ bits. Thus, using an auxiliary lookup table of size $\mathcal{O}(2^{c \cdot Z \lg(1/\tau)} \times 2^{\lg^2 Z})$ we can preprocess the answer for every possible encoding and positions i, j so that a query takes $\mathcal{O}(1)$ time. Because $1/\tau < \sqrt{\lg n}$ the space for this lookup table is:

$$\mathcal{O}(2^{c(\lg n/(2c\lg(1/\tau)))\lg(1/\tau)+\lg^2(\lg n/(2c\lg(1/\tau)))}) = \mathcal{O}(2^{\lg n/2+(\lg\lg n)^2}) = o(n).$$

In summary, we can apply level-based pruning to reduce the space required by the pointers to at most $\mathcal{O}(n\lg(1/\tau))$. Note that we must be able to quickly access the pointers associated with each quadruple D. To do this, we concatenate the pointers at level k, and construct yet another bit vector L'_k having a similar format as L_k. The bit vector L'_k allows us to easily determine how many pointers are stored for the quadruples to the left of D at the current level, as well as how many are stored for D. Thus, these additional bit vectors occupy $\mathcal{O}(n \lg(1/\tau))$ bits of space, and allow accessing an arbitrary pointer in $\mathcal{O}(1)$ time.

Using the faster ranking structure. When we use the faster rank structure of Lemma 1, we immediately get that we can verify the frequency of each candidate in $\mathcal{O}(1)$ time, rather than $\mathcal{O}(\lg n)$ time. Recall that the bit vectors are concatenated at each level. In the structure of Lemma 2, the redundancy at each level was merely $\mathcal{O}(\lg \lg n)$ bits. However, with Lemma 1 we end up with a redundancy of $\mathcal{O}(n/(\tau \lg^c(n)))$ bits per level, for a total of $\mathcal{O}(n \lg n/(\tau \lg^c(n)))$ bits. So, if $1/\tau = \mathcal{O}(\mathrm{polylog}(n))$, then we can choose the constant c to be sufficiently large so that this term is sublinear. Immediately, this yields:

Lemma 5. *If $1/\tau = \mathcal{O}(\mathrm{polylog}(n))$, there is an encoding that supports range τ-majority position queries in $\mathcal{O}(1/\tau)$ time, and occupies $\mathcal{O}(n \lg(1/\tau))$ bits.*

When $1/\tau$ is $\omega(\mathrm{polylog(n)})$, we require a more sophisticated data structure to achieve $\mathcal{O}(n \lg(1/\tau))$ bits of space. Basically, we have to replace the data structure of Lemma 1 representing the bit vectors with a more space-efficient batch structure that groups all candidates together. The details required to finish the proof of Theorem 2 can be found in the full version of this paper [9].

4   Lower Bound

In this section we prove Theorem 4. The idea is to show that a sequence of permutations, each of length roughly $1/\tau$, can be recovered using queries.

Formally, we will describe a *bad string*, defined using concatenation, in which array $A[i]$ will store the i-th symbol in the string. Conceptually, this bad string is constructed by concatenating some *padding*, denoted L, before a sequence of m permutations over the alphabet $[\alpha_1, \ldots, \alpha_k]$, denoted $R = \pi_1 \cdot \ldots \cdot \pi_m$. Notationally, we use α_i^c to denote a concatenation of the symbol α_i c times, and $a \cdot b$ to denote the concatenation of the strings a and b. In the construction we make use of *dummy symbols*, β, which are defined to be symbols that occur exactly one time in the bad string. A sequence of ℓ dummy symbols, written β^ℓ, should be taken to mean: a sequence of ℓ characters, each of which are distinct from any other symbol in A.

Padding definition. Key to defining L is a gadget $G(k, i)$, that is defined for any integer $k \geq 2$ using concatenation as follows: $G(k, i) = \alpha_1^{k'} \cdot \alpha_2^{k'} \cdot \ldots \cdot \alpha_{i-1}^{k'} \cdot \alpha_{i+1}^{k'} \cdot \ldots \cdot \alpha_k^{k'} \cdot (\alpha_i \cdot \beta^{k-2})^{k-1} \cdot \alpha_i \beta^k$, where $k' = k^2 - k + 2$. Suppose we define A such that $A[x..y]$ contains gadget $G(k, i)$. Let $f(x, y, \alpha)$ denote the number of occurrences of symbol α in range $[x, y]$. We define the *density* of symbol α in the query range $[x, y]$ to be $\delta(x, y, \alpha) = f(x, y, \alpha)/(y - x + 1)$. We observe the following:

1. The length of the gadget $G(k, i)$ is $k(k^2 - k + 2)$ for all $i \in [1, k]$. This fact will be useful later when we bound the total size of the padding L.

2. $\delta(x, y, \alpha_j) = 1/k$ for all $j \neq i$. This follows from the previous observation and that, for all $j \neq i$, the number of occurrences of α_j in $G(k, i)$ is $k^2 - k + 2$.

Next, we finish defining our array A by defining L to be the concatenation $G(k, k) \cdot G(k, k - 1) \cdot \ldots \cdot G(k, 1)$. Thus, our array is obtained by embedding the string $L \cdot R$ into an array A. Note that the total length of the array is $k^2(k^2 - k + 2) + mk$. Thus, the padding is of length $\Theta(k^4)$.

Query Procedure. The following procedure can recover the position of symbol α_i in π_j, for any $i \in [1, k]$ and $j \in [1, m]$. This procedure uses $\Theta(k)$ $(1/k)$-majority decision queries: overall, recovering the contents of R uses $\Theta(k^2m)$ queries.

Let $r_{j,1}, \ldots, r_{j,k}$ denote the indices of A containing the symbols in π_j from left-to-right. Moreover, consider the indices of the k occurrences of symbol α_i in $G(k, i)$, from left-to-right, and denote these as $\ell_{i,k}, \ldots, \ell_{i,1}$, respectively (note that the rightmost occurrence is marked with subscript 1). Formally, the query procedure will perform a sequence of queries, stopping if the answer is YES, and continuing if the answer is NO. The ordered sequence of queries we execute is $[\ell_{i,1}, r_{j,1}], [\ell_{i,2}, r_{j,2}], \ldots, [\ell_{i,k}, r_{j,k}]$.

We now claim that if the answer to a query $[\ell_{i,x}, r_{j,x}]$ is NO, then $A[r_{j,x}] \neq \alpha_i$. This follows since the density of symbol α_i in the query range is:

$$\frac{x + (i - 1)(k^2 - k + 2) + (j - 1)}{k(x + (i - 1)(k^2 - k + 2) + (j - 1)) + 2} < \frac{1}{k}$$

On the other hand, if the answer is YES, we have that the symbol α_i must be a $(1/k)$-majority for the following reasons:

1. No other symbol α_j where $j \neq i$ can be a $(1/k)$-majority. To see this, divide the query range into a middle-part, consisting of $G(k, i-1) \cdot \ldots \cdot G(k, 1) \cdot \pi_1 \cdot \ldots \cdot \pi_{k-1}$, as well as a prefix (which is a suffix of $G(k, i)$), and a suffix (which is a prefix of π_j). The prefix of the query range contains no occurrence of α_j and is at least of length $k + 1$. The suffix contains at most one occurrence of α_j. Thus, the density of α_j is strictly less than $1/k$ in the union of the prefix and suffix, exactly $1/k$ in the middle part, and strictly less than $1/k$ overall.
2. No dummy symbol β can be an $(1/k)$-majority, since these symbols appear one time only, and all query ranges have length strictly larger than k.
3. Finally, if $A[r_{j,x}] = \alpha_i$, then the density $\delta(\ell_{i,x}, r_{j,x}, \alpha_i)$ is:

$$\frac{x + (i - 1)(k^2 - k + 2) + (j - 1) + 1}{k(x + (i - 1)(k^2 - k + 2) + (j - 1)) + 2} \geq 1/k,$$

since $k \geq 2$. Since we stop immediately after the first YES, the procedure therefore is guaranteed to identify the correct position of α_i.

As we stated, the length of the array is $k^2(k^2 - k + 2) + mk = n$, and for n large enough the queries allow us to recover $\frac{n - \Theta(k^4)}{k} \lg(k!)$ bits of information using $(1/k)$-majority queries for any integer $k \geq 2$, which is at least $(n/k - \Theta(k^3))k \lg(k/e) = n \lg(k/e) - \Theta(k^4 \lg k)$ bits. Since there exists a unit fraction $\tau' = 1/\lfloor 1/\tau \rfloor$ (if $\tau \in (0, 1/2]$), there also exists a bad input of length n in which $k = 1/\tau'$. Therefore, we have proved Theorem 4.

References

1. Belazzougui, D., Gagie, T., Munro, J.I., Navarro, G., Nekrich, Y.: Range majorities and minorities in arrays. CoRR abs/1606.04495 (2016)
2. Belazzougui, D., Gagie, T., Navarro, G.: Better space bounds for parameterized range majority and minority. In: Proc. WADS 2013. LNCS, vol. 8037, pp. 121–132. Springer (2013)
3. Boyer, R.S., Moore, J.S.: MJRTY: A fast majority vote algorithm. In: Automated Reasoning: Essays in Honor of Woody Bledsoe. pp. 105–118. Automated Reasoning Series, Kluwer Academic Publishers (1991)
4. Chan, T.M., Durocher, S., Larsen, K.G., Morrison, J., Wilkinson, B.T.: Linear-space data structures for range mode query in arrays. Theory Comput. Syst. 55(4), 719–741 (2014)
5. Demaine, E.D., López-Ortiz, A., Munro, J.I.: Frequency estimation of internet packet streams with limited space. In: Proc. ESA 2002. LNCS, vol. 2461, pp. 348–360. Springer (2002)
6. Durocher, S., He, M., Munro, J.I., Nicholson, P.K., Skala, M.: Range majority in constant time and linear space. Inf. Comput. 222, 169–179 (2013)
7. Ferragina, P., Venturini, R.: A simple storage scheme for strings achieving entropy bounds. Theor. Comput. Sci. 372(1), 115–121 (2007)
8. Gagie, T., He, M., Munro, J.I., Nicholson, P.K.: Finding frequent elements in compressed 2d arrays and strings. In: Proc. SPIRE 2011. LNCS, vol. 7024, pp. 295–300. Springer (2011)
9. Gawrychowski, P., Nicholson, P.K.: Optimal query time for encoding range majority. CoRR arXiv:1704.06149 (2017), http://arxiv.org/abs/1704.06149
10. González, R., Navarro, G.: Statistical encoding of succinct data structures. In: Proc. CPM 2006. LNCS, vol. 4009, pp. 294–305. Springer (2006)
11. Grossi, R., Gupta, A., Vitter, J.S.: High-order entropy-compressed text indexes. In: Proc. SODA 2003. pp. 841–850. ACM/SIAM (2003)
12. Karp, R.M., Shenker, S., Papadimitriou, C.H.: A simple algorithm for finding frequent elements in streams and bags. ACM Trans. Database Syst. 28, 51–55 (2003)
13. Karpinski, M., Nekrich, Y.: Searching for frequent colors in rectangles. In: Proc. CCCG 2008 (2008)
14. Misra, J., Gries, D.: Finding repeated elements. Sci. Comput. Program. 2(2), 143–152 (1982)
15. Navarro, G., Thankachan, S.V.: Optimal encodings for range majority queries. Algorithmica 74(3), 1082–1098 (2016)
16. Patrascu, M.: Succincter. In: Proc. FOCS 2008. pp. 305–313. IEEE (2008)
17. Raman, R.: Encoding data structures. In: Proc. WALCOM 2015. LNCS, vol. 8973, pp. 1–7. Springer (2015)
18. Raman, R., Raman, V., Satti, S.R.: Succinct indexable dictionaries with applications to encoding k-ary trees, prefix sums and multisets. ACM Trans. Algorithms 3(4), 43 (2007)
19. Skala, M.: Array range queries. In: Space-Efficient Data Structures, Streams, and Algorithms - Papers in Honor of J. Ian Munro on the Occasion of His 66th Birthday. LNCS, vol. 8066, pp. 333–350. Springer (2013)

Conditional Lower Bounds for Space/Time Tradeoffs

Isaac Goldstein[*1], Tsvi Kopelowitz[**2], Moshe Lewenstein[***1], and Ely Porat [***1]

[1]Bar-Ilan University , {goldshi,moshe,porately}@cs.biu.ac.il
[2]University of Waterloo , kopelot@gmail.com

Abstract. In recent years much effort has been concentrated towards achieving polynomial time lower bounds on algorithms for solving various well-known problems. A useful technique for showing such lower bounds is to prove them conditionally based on well-studied hardness assumptions such as 3SUM, APSP, SETH, etc. This line of research helps to obtain a better understanding of the complexity inside Γ.

A related question asks to prove conditional *space* lower bounds on data structures that are constructed to solve certain algorithmic tasks after an initial preprocessing stage. This question received little attention in previous research even though it has potential strong impact.

In this paper we address this question and show that surprisingly many of the well-studied hard problems that are known to have conditional polynomial *time* lower bounds are also hard when concerning *space*. This hardness is shown as a tradeoff between the space consumed by the data structure and the time needed to answer queries. The tradeoff may be either smooth or admit one or more singularity points.

We reveal interesting connections between different space hardness conjectures and present matching upper bounds. We also apply these hardness conjectures to both static and dynamic problems and prove their conditional space hardness.

We believe that this novel framework of polynomial space conjectures can play an important role in expressing polynomial space lower bounds of many important algorithmic problems. Moreover, it seems that it can also help in achieving a better understanding of the hardness of their corresponding problems in terms of time.

---

[*]This research is supported by the Adams Foundation of the Israel Academy of Sciences and Humanities

[**]Part of this work took place while the second author was at University of Michigan. This work is supported in part by the Canada Research Chair for Algorithm Design, NSF grants CCF-1217338, CNS-1318294, and CCF-1514383

[***]This work was partially supported by an ISF grant #1278/16

F. Ellen et al. (Eds.): WADS 2017, LNCS 10389, pp. 421–436, 2017.
DOI: 10.1007/978-3-319-62127-2_36

1   Introduction

1.1   Background

Lately there has been a concentrated effort to understand the time complexity within P, the class of decision problems solvable by polynomial time algorithms. The main goal is to explain why certain problems have time complexity that seems to be non-optimal. For example, all known efficient algorithmic solutions for the 3SUM problem, where we seek to determine whether there are three elements x, y, z in input set S of size n such that $x + y + z = 0$, take $\tilde{O}(n^2)$ time[1]. However, the only real lower bound that we know is the trivial $\Omega(n)$. Likewise, we know how to solve the *all pairs shortest path*, APSP, problem in $\tilde{O}(n^3)$ time but we cannot even determine whether it is impossible to obtain an $\tilde{O}(n^2)$ time algorithm. One may note that it follows from the time-hierarchy theorem that there exist problems in P with complexity $\Omega(n^k)$ for every fixed k. Nevertheless, such a separation for natural practical problems seems to be hard to achieve.

The collaborated effort to understand the internals of P has been concentrated on identifying some basic problems that are conjectured to be hard to solve more efficiently (by polynomial factors) than their current known complexity. These problems serve as a basis to prove conditional hardness of other problems by using reductions. The reductions are reminiscent of NP-complete reductions but differ in that they are restricted to be of time complexity strictly smaller (by a polynomial factor) than the problem that we are reducing to. Examples of such hard problems include the well-known 3SUM problem, the fundamental APSP problem, (combinatorial) Boolean matrix multiplication, etc. Recently, conditional time lower bounds have been proven based on the conjectured hardness of these problems for graph algorithms [4, 30], edit distance [12], longest common subsequence (LCS) [3, 14], dynamic algorithms [5, 25], jumbled indexing [11, 19], and many other problems [1, 2, 6, 7, 13, 20, 23, 24, 29].

1.2   Motivation

In stark contrast to polynomial *time* lower bounds, little effort has been devoted to finding polynomial *space* conditional lower bounds. An example of a space lower bound appears in the work of Cohen and Porat [17] and Pătraşcu and Roditty [27] where lower bounds are shown on the size of a distance oracle for sparse graphs based on a conjecture about the best possible data structure for a set intersection problem (which we call set disjointness in order to differ it from its reporting variant).

A more general question is, for algorithmic problems, what conditional lower bounds of a space/time tradeoff can be shown based on the set disjointness (intersection) conjecture? Even more general is to discover what space/time tradeoffs can be achieved based on the other algorithmic problems that we assumed are

---

[1] The $\tilde{O}$ and $\tilde{\Omega}$ notations suppress polylogarithmic factors

hard (in the time sense)? Also, what are the relations between these identified "hard" problems in the space/time tradeoff sense? These are the questions which form the basis and framework of this paper.

Throughout this paper we show connections between different hardness assumptions, show some matching upper bounds and propose several conjectures based on this accumulated knowledge. Moreover, we conjecture that there is a strong correlation between polynomial hardness in time and space. We note that in order to discuss space it is often more natural to consider data structure variants of problems and this is the approach we follow in this paper.

1.3 Our Results

Set Disjointness. In the SetDisjointness problem mentioned before, it is required to preprocess a collection of m sets $S_1, \cdots, S_m \subset U$, where U is the universe of elements and the total number of elements in all sets is N. For a query, a pair of integers (i, j) $(1 \leq i, j \leq m)$ is given and we are asked whether $S_i \cap S_j$ is empty or not. A folklore conjecture, which appears in [16, 27], suggests that to achieve a constant query time the space of the data structure constructed in the preprocessing stage needs to be $\tilde{\Omega}(N^2)$. We call this conjecture the SetDisjointness conjecture. This conjecture does not say anything about the case where we allow *higher* query time. Therefore, we suggest a stronger conjecture which admits a *full tradeoff* between the space consumed by the data structure (denoted by S) and the query time (denoted by T). This is what we call the Strong SetDisjointness conjecture. This conjecture states that for solving SetDisjointness with a query time T our data structure needs $\tilde{\Omega}(N^2/T^2)$ space. A matching upper bound exists for this problem by generalizing ideas from [16] (see also [22]). Our new SetDisjointness conjecture can be used to admit more expressive space lower bounds for a full tradeoff between space and query time.

3SUM Indexing. One of the basic and frequently used hardness conjectures is the celebrated 3SUM conjecture. This conjecture was used for about 20 years to show many conditional *time* lower bounds on various problems. However, we focus on what can be said about its *space* behavior. To do this, it is natural to consider a data structure version of 3SUM which allows one to preprocess the input set S. Then, the query is an external number z for which we need to answer whether there are $x, y \in S$ such that $x + y = z$. It was pointed out by Chan and Lewenstein [15] that all known algorithms for 3SUM actually work within this model as well. We call this problem *3SUM Indexing*. On one hand, this problem can easily be solved using $O(n^2)$ space by sorting $x + y$ for all $x, y \in S$ and then searching for z in $\tilde{O}(1)$ time. On the other hand, by just sorting S we can answer queries by a well-known linear time algorithm. The big question is whether we can obtain better than $\tilde{\Omega}(n^2)$ space while using just $\tilde{O}(1)$ time query? Can it be done even if we allow $\tilde{O}(n^{1-\Omega(1)})$ query time? This leads us to our two new hardness conjectures. The 3SUM-Indexing conjecture states that when using $\tilde{O}(1)$ query time we need $\tilde{\Omega}(n^2)$ space to solve 3SUM-Indexing. In the Strong 3SUM-Indexing conjecture we say that even when using $\tilde{O}(n^{1-\Omega(1)})$ query time we need $\tilde{\Omega}(n^2)$ space to solve 3SUM-Indexing.

3SUM Indexing and Set Disjointness. We prove connections between the SetDisjointness conjectures and the 3SUM-Indexing conjectures. Specifically, we show that the Strong 3SUM-Indexing conjecture implies the Strong SetDisjointness conjecture, while the SetDisjointness conjecture implies the 3SUM-Indexing conjecture. This gives some evidence towards establishing the difficulty within the 3SUM-Indexing conjectures. The usefulness of these conjectures should not be underestimated. As many problems are known to be 3SUM-hard these new conjectures can play an important role in achieving *space* lower bounds on their corresponding data structure variants. Moreover, it is interesting to point on the difference between SetDisjointness which admits smooth tradeoff between space and query time and 3SUM-Indexing which admits a big gap between the two trivial extremes. This may explain why we are unable to show full equivalence between the hardness conjectures of the two problems. Moreover, it can suggest a separation between problems with smooth space-time behavior and others which have no such tradeoff but rather two "far" extremes.

Generalizations. Following the discussion on the SetDisjointness and the 3SUM-Indexing conjectures we investigate their generalizations.

I. k-Set Disjointness and (k+1)-SUM Indexing. The first generalization is a natural parametrization of both problems. In the SetDisjointness problem we query about the emptiness of the intersection between *two* sets, while in the 3SUM-Indexing problem we ask, given a query number z, whether *two* numbers of the input S sum up to z. In the parameterized versions of these problems we are interested in the emptiness of the intersection between k sets and ask if k numbers sum up to a number given as a query. These generalized variants are called k-SetDisjointness and (k+1)-SUM-Indexing respectively. For each problem we give corresponding space lower bounds conjectures which generalize those of SetDisjointness and 3SUM-Indexing. These conjectures also have corresponding strong variants which are accompanied by matching upper bounds. We prove that the k-SetDisjointness conjecture implies (k+1)-SUM-Indexing conjecture via a novel method using linear equations.

II. k-Reachability. A second generalization is the problem we call k-Reachability. In this problem we are given as an input a directed sparse graph $G = (V, E)$ for preprocessing. Afterwards, for a query, given as a pair of vertices u, v, we wish to return if there is a path from u to v consisting of at most k edges. We provide an upper bound on this problem for every fixed $k \geq 1$. The upper bound admits a tradeoff between the space of the data structure (denoted by S) and the query time (denoted by T), which is $ST^{2/(k-1)} = O(n^2)$. We argue that this upper bound is tight. That is, we conjecture that if query takes T time, the space must be $\tilde{\Omega}(\frac{n^2}{T^{2/(k-1)}})$. We call this conjecture the k-Reachability conjecture.

We give three indications towards the correctness of this conjecture. First, we prove that the base case, where $k = 2$, is equivalent to the SetDisjointness problem. This is why this problem can be thought of as a generalization of SetDisjointness.

Second, if we consider non-constant k then the smooth tradeoff surprisingly disappears and we get "extreme behavior" as $\tilde{\Omega}(\frac{n^2}{T^{2/(k-1)}})$ eventually becomes $\tilde{\Omega}(n^2)$. This means that to answer reachability queries for non-constant path length, we can either store all answers in advance using n^2 space or simply answer queries from scratch using a standard graph traversal algorithm. The general problem where the length of the path from u to v is unlimited in length is sometimes referred to as the problem of constructing efficient reachability oracles. Pătraşcu in [26] leaves it as an open question if a data structure with less than $\tilde{\Omega}(n^2)$ space can answer reachability queries efficiently. Moreover, Pătraşcu proved that for constant time query, truly superlinear space is needed. Our k-Reachability conjecture points to this direction, while admitting full space-time tradeoff for constant k.

The third indication for the correctness of the k-Reachability conjecture comes from a connection to distance oracles. A *distance oracle* is a data structure that can be used to quickly answer queries about the shortest path between two given nodes in a preprocessed undirected graph. As mentioned above, the Set-Disjointness conjecture was used to exclude some possible tradeoffs for sparse graphs. Specifically, Cohen and Porat [17] showed that obtaining an approximation ratio smaller than 2 with constant query time requires $\tilde{\Omega}(n^2)$ space. Using a somewhat stronger conjecture Pătraşcu and Roditty [27] showed that a $(2,1)$-distance oracle for unweighted graphs with $m = O(n)$ edges requires $\tilde{\Omega}(n^{1.5})$ space. Later, this result was strengthened by Pătraşcu et al. [28]. However, these results do not exclude the possibility of compact distance oracles if we allow higher query time. For stretch-2 and stretch-3 in sparse graphs, Agarwal et. al. [9, 10] achieved a space-time tradeoff of $S \times T = O(n^2)$ and $S \times T^2 = O(n^2)$, respectively. Agarwal [8] also showed many other results for stretch-2 and below. We use our k-Reachability conjecture to prove that for stretch-less-than-$(1+2/k)$ distance oracles $S \times T^{2/(k-1)}$ is bounded by $\tilde{\Omega}(n^2)$. This result is interesting in light of Agarwal [8] where a stretch-$(5/3)$ oracle was presented which achieves a space-time tradeoff of $S \times T = O(n^2)$. This matches our lower bound, where $k = 3$, if our lower bound would hold not only for stretch-less-than-$(5/3)$ but also for stretch-$(5/3)$ oracles. Consequently, we see that there is strong evidence for the correctness of the k-Reachability conjecture.

Moreover, these observations show that on one hand k-Reachability is a generalization of SetDisjointness which is closely related to 3SUM-Indexing. On the other hand, k-Reachability is related to distance oracles which solve the famous APSP problem using smaller space by sacrificing the accuracy of the distance between the vertices. Therefore, the k-Reachability conjecture seems as a conjecture corresponding to the APSP hardness conjecture, while also admitting some connection with the celebrated 3SUM hardness conjecture.

SETH and Orthogonal Vectors. After considering space variants of the 3SUM and APSP conjectures it is natural to consider space variants for the Strong Exponential Time Hypothesis (SETH) and the closely related conjecture of orthogonal vectors. SETH asserts that for any $\epsilon > 0$ there is an integer $k > 3$ such that k-SAT cannot be solved in $2^{(1-\epsilon)n}$ time. The orthogonal vectors time conjecture states that there is no algorithm that for every $c \geq 1$, finds if there are at least two orthogonal vectors in a set of n Boolean vectors of length $c \log n$ in $\tilde{O}(n^{2-\Omega(1)})$ time. A discussion about the space variants of these conjectures will appear in the full version of this paper. However, we note that we are unable to connect these conjectures and the previous ones. This is perhaps not surprising as the connection between SETH and the other conjectures even in the time perspective is very loose (see, for example, discussions in [5, 20]).

Boolean Matrix Multiplication. Another problem which receives a lot of attention in the context of conditional time lower bounds is calculating Boolean Matrix Multiplication (BMM). We give a data structure variant of this well-known problem. We then demonstrate the connection between this problem and the problems of SetDisjointness and k-Reachability. The discussion about BMM and its data structure variant will appear in the full version of this paper.

Applications. Finally, armed with the *space* variants of many well-known conditional *time* lower bounds, we apply this conditional space lower bounds to some static and dynamic problems. This gives interesting space lower bound results on these important problems which sometimes also admits clear space-time tradeoff. The list of problems that we prove their conditional space-time hardness includes: edge triangles, histogram indexing, distance oracles for colors, two patterns document retrieval, forbidden pattern document retrieval, (s,t)-reachability, bipartite perfect matching and strong connectivity. All the results regarding the applications of our framework will appear in the full version of this paper. We believe that this is just a glimpse of space lower bounds that can be achieved based on our new framework and that many other interesting results are expected to follow this promising route.

2   Set Intersection Hardness Conjectures

We first give formal definitions of the SetDisjointness problem and its enumeration variant:

Problem 1 (SetDisjointness Problem). Preprocess a family F of m sets, all from universe U, with total size $N = \sum_{S \in F} |S|$ so that given two query sets $S, S' \in F$ one can determine if $S \cap S' = \emptyset$.

Problem 2 (SetIntersection Problem). Preprocess a family F of m sets, all from universe U, with total size $N = \sum_{S \in F} |S|$ so that given two query sets $S, S' \in F$ one can enumerate the set $S \cap S'$.

Conjectures. The SetDisjointness problem was regarded as a problem that admits space hardness. The hardness conjecture of the SetDisjointness problem has received several closely related formulations. One such formulation, given by Pătraşcu and Roditty [27], is as follows:

Conjecture 1. SetDisjointness Conjecture [Formulation 1]. Any data structure for the SetDisjointness problem where $|U| = \log^c m$ for a large enough constant c and with a constant query time must use $\tilde{\Omega}(m^2)$ space.

Another formulation is implicitly suggested in Cohen and Porat [16]:

Conjecture 2. SetDisjointness Conjecture [Formulation 2]. Any data structure for the SetDisjointness problem with constant query time must use $\tilde{\Omega}(N^2)$ space.

There is an important distinction between the two formulations, which is related to the sparsity of SetDisjointness instances. This distinction follows from the following upper bound: store an $m \times m$ matrix of the answers to all possible queries, and then queries will cost constant time. The first formulation of the SetDisjointness conjecture states that if we want constant (or poly-logaritmic) query time, then this is the best we can do. At a first glance this makes the second formulation, whose bounds are in terms of N and not m, look rather weak. In particular, why would we ever be interested in a data structure that uses $O(N^2)$ space when we can use one with $O(m^2)$ space? The answer is that the two conjectures are the same if the sets are very sparse, and so at least in terms of N, if one were to require a constant query time then by the second formulation the space must be at least $\Omega(N^2)$ (which happens in the very sparse case).

Nevertheless, we present a more general conjecture, which in particular captures a tradeoff curve between the space usage and query time. This formulation captures the difficulty that is commonly believed to arise from the SetDisjointness problem, and matches the upper bounds of Cohen and Porat [16] (see also [22]).

Conjecture 3. Strong SetDisjointness Conjecture. Any data structure for the SetDisjointness problem that answers queries in T time must use $S = \tilde{\Omega}(\frac{N^2}{T^2})$ space.

For example, a natural question to ask is "what is the smallest query time possible with linear space?". This question is addressed, at least from a lower bound perspective, by the Strong SetDisjointness conjecture.

Conjecture 4. Strong SetIntersection Conjecture. Any data structure for the SetIntersection problem that answers queries in $O(T + op)$ time, where op is the size of the output of the query, must use $S = \tilde{\Omega}(\frac{N^2}{T})$ space.

3   3SUM-Indexing Hardness Conjectures

In the classic 3SUM problem we are given an integer array A of size n and
we wish to decide whether there are 3 distinct integers in A which sum up to
zero. Gajentaan and Overmars [18] showed that an equivalent formulation of
this problem receives 3 integer arrays A_1, A_2, and A_3, each of size n, and the
goal is to decide if there is a triplet $x_1 \in A_1, x_2 \in A_2$, and $x_3 \in A_3$ that sum up
to zero.

We consider the data structure variant of this problem which is formally
defined as follows:

Problem 3 (3SUM-Indexing Problem). Preprocess two integer arrays A_1 and A_2,
each of length n, so that given a query integer z we can decide whether there
are $x \in A_1$ and $y \in A_2$ such that $z = x + y$.

It is straightforward to maintain all possible $O(n^2)$ sums of pairs in quadratic
space, and then answer a query in $\tilde{O}(1)$ time. On the other extreme, if one does
not wish to utilize more than linear space then one can sort the arrays separately
during preprocssing time, and then a query can be answered in $\tilde{O}(n)$ time by
scanning both of the sorted arrays in parallel and in opposite directions.

We introduce two conjectures with regards to the 3SUM-Indexing problem,
which serve as natural candidates for proving polynomial space lower bounds.

Conjecture 5. **3SUM-Indexing Conjecture**: There is no solution for the 3SUM-
Indexing problem with truly subquadratic space and $\tilde{O}(1)$ query time.

Conjecture 6. **Strong 3SUM-Indexing Conjecture**: There is no solution for
the 3SUM-Indexing problem with truly subquadratic space and truly sublinear
query time.

Notice that one can solve the classic 3SUM problem using a data structure
for 3SUM-Indexing by preprocessing A_1 and A_2, and answering n 3SUM-Indexing
queries on all of the values in A_3.

Next, we prove theorems that show tight connections between the 3SUM-
Indexing conjectures and the SetDisjointness conjectures. We note that the proofs
of the first two theorems are similar to the proofs of [23], but with space inter-
pretation. These proofs will appear in the full version of this paper.

Theorem 1. *The Strong 3SUM-Indexing Conjecture implies the Strong SetDisjointness Conjecture.*

Theorem 2. *The Strong 3SUM-Indexing Conjecture implies the Strong SetIntersection Conjecture.*

Theorem 3. *The SetDisjointness Conjecture implies the 3SUM-Indexing Conjecture.*

Proof. Given an instance of SetDisjointness, we construct an instance of 3SUM-Indexing as follows. Denote with M the value of the largest element in the Set-Disjointness instance. Notice that we may assume that $M \leq N$ (otherwise we can use a straightforward renaming). For every element $x \in U$ that is contained in at least one of the sets we create two integers x_A and x_B, which are represented by $2\lceil \log m \rceil + \lceil \log N \rceil + 3$ bits each (recall that m is the number of sets).

The $\lceil \log N \rceil$ least significant bits in x_A represent the value of x. The following bit is a zero. The following $\lceil \log m \rceil$ bits in x_A represent the index of the set containing x, and the rest of the $2 + \lceil \log m \rceil$ are all set to zero. The $\lceil \log N \rceil$ least significant bits in x_B represent the value of $M - x$. The following $2 + \lceil \log m \rceil$ are all set to zero. The following $\lceil \log m \rceil$ bits in x_B represent the index of the set containing x, and the last bit is set to zero. Finally, the integer x_A is added to A_1 of the 3SUM-Indexing instance, while the integer x_B is added to A_2.

We have created two sets of $n \leq M$ integers. We then preprocess them to answer 3SUM-Indexing queries. Now, to answer a SetDisjointness query on sets S_i and S_j, we query the 3SUM-Indexing data structure with an integer z which is determined as follows. The $\lceil \log N \rceil$ least significant bits in z represent the value of M. The following bit is a zero. The following $\lceil \log m \rceil$ bits represent the index i and are followed by a zero. The next $\lceil \log m \rceil$ bits represent the index j and the last bit is set to zero.

It is straightforward to verify that there exists a solution to the 3SUM-Indexing problem on z if and only if the sets S_i and S_j are not disjoint. Therefore, if there is a solution to the 3SUM-Indexing problem with less than $\tilde{\Omega}(n^2)$ space and constant query time then there is a solution for the SetDisjointness problem which refutes the SetDisjointness Conjecture. □

4 Parameterized Generalization: k-Set Intersection and (k+1)-SUM

Two parameterized generalizations of the SetDisjointness and 3SUM-Indexing problems are formally defined as follows:

Problem 4 (k-SetDisjointness Problem). Preprocess a family F of m sets, all from universe U, with total size $N = \sum_{S \in F} |S|$ so that given k query sets $S_1, S_2, \ldots, S_k \in F$ one can quickly determine if $\cap_{i=1}^{k} S_i = \emptyset$.

Problem 5 ((k+1)-SUM-Indexing Problem). Preprocess k integer arrays $A_1, A_2, \ldots, A_k$, each of length n, so that given a query integer z we can decide if there is $x_1 \in A_1, x_2 \in A_2, \ldots, x_k \in A_k$ such that $z = \sum_{i=1}^{k} x_i$.

It turn out that a natural generalization of the data structure of Cohen and Porat [16] leads to a data structure for k-SetDisjointness as shown in the following lemma.

Lemma 1. *There exists a data structure for the k-SetDisjointness problem where the query time is T and the space usage is $S = O((N/T)^k)$.*

Proof. We call the f largest sets in F *large sets*. The rest of the sets are called *small sets*. In the preprocessing stage we explicitly maintain a k-dimensional table with the answers for all k-SetDisjointness queries where all k sets are large sets. The space needed for such a table is $S = f^k$. Moreover, for each set (large or small) we maintain a look-up table that supports disjointness queries (with this set) in constant time. Since there are f large sets and the total number of elements is N, the size of each of the small sets is at most N/f.

Given a k-SetDisjointness query, if all of the query sets are large then we look up the answer in the k-dimensional table. If at least one of the sets is small then using a brute-force search we look-up each of the at most $O(N/f)$ elements in each of the other $k-1$ sets. Thus, the total query time is bounded by $O(kN/f)$, and the space usage is $S = O(f^k)$. The rest follows.

Notice that for the case of $k = 2$ in Lemma 1 we obtain the same tradeoff of Cohen and Porat [16] for SetDisjointness. The following conjecture suggests that the upper bound of Lemma 1 is the best possible.

Conjecture 7. **Strong k-SetDisjointness Conjecture.** Any data structure for the k-SetDisjointness problem that answers queries in T time must use $S = \tilde{\Omega}(\frac{N^k}{T^k})$ space.

Similarly, a natural generalization of the Strong 3SUM-Indexing conjecture is the following.

Conjecture 8. **Strong (k+1)-SUM-Indexing Conjecture.** There is no solution for the (k+1)-SUM-Indexing problem with $\tilde{O}(n^{k-\Omega(1)})$ space and truly sublinear query time.

We also consider some weaker conjectures, similar to the SetDisjointness and 3SUM-Indexing conjectures.

Conjecture 9. **k-SetDisjointness Conjecture.** Any data structure for the k-SetDisjointness problem that answers queries in constant time must use $\tilde{\Omega}(N^k)$ space.

Conjecture 10. **(k+1)-SUM-Indexing Conjecture.** There is no solution for the (k+1)-SUM-Indexing problem with $\tilde{O}(n^{k-\Omega(1)})$ space and constant query time.

Similar to Theorem 3, we prove the following relationship between the k-SetDisjointness conjecture and the (k+1)-SUM-Indexing conjecture.

Theorem 4. *The k-SetDisjointness conjecture implies the (k+1)-SUM-Indexing conjecture*

Proof. Given an instance of k-SetDisjointness, we construct k instances of (k+1)-SUM-Indexing as follows. Denote by M the value of the largest element in the SetDisjointness instance. Notice that we may assume that $M \leq N$ (otherwise we use a straightforward renaming). For every element $x \in U$ that is contained in at least one of the sets we create k^2 integers in a matrix $X = \{x_{i,j}\}$ of size $k \times k$, where each integer is represented by $(k-1)\lceil \log m \rceil + \lceil \log N \rceil + k$ bits.

For integer $x_{i,j}$, the $\lceil \log N \rceil + 1$ least significant bits represent the value of $(k-1)x$ if $i = j$, and the value of $M - x$ otherwise. The $(k-1)\lceil \log m \rceil + k - 1$ following bits are all set to zero, except for the bits in indices $(j-1)(\lceil \log m \rceil + 1) + 1, ..., j(\lceil \log m \rceil + 1)$ which represent the index of the set containing x.

We now create k instances of (k+1)-SUM-Indexing where the jth input array A_j for the ith instance is the set of integers $x_{i,j}$ for all $x \in U$ that are contained in at least one set of our family. Thus, the size of each array is at most N. Now, given a k-SetDisjointness query $(i_1, i_2, ..., i_k)$ we must decide if $S_{i_1} \cap S_{i_2} \cap ... \cap S_{i_k} = \emptyset$. To answer this query we will query each of the k instances of (k+1)-SUM-Indexing with an integer z whose binary representation is as follows: the $\lceil \log N \rceil + 1$ least significant bits represent the value of $(k-1)M$, and the bits at locations $(j-1)(\lceil \log m \rceil + 1) + 1, ..., j(\lceil \log m \rceil + 1)$ representing i_j (for $1 \le j \le k$). The rest of the bits are padding zero bit (in between representations of various i_j).

If $S_{i_1} \cap S_{i_2} \cap ... \cap S_{i_k} \ne \emptyset$ then by our construction it is straightforward to verify that all of the k (k+1)-SUM-Indexing queries on z will return that there is a solution. If $S_{i_1} \cap S_{i_2} \cap ... \cap S_{i_k} = \emptyset$ then at least one (k+1)-SUM-Indexing query will not be able to find a solution. This is because we can view each instance and query as solving a linear equation. As we construct k instances which represent k independent linear equations, we are guaranteed that only one solution exists. This solution is exactly the one that corresponds to finding a specific x which is contained in all of the k sets. Therefore, we get a correct answer to a k-SetDisjointness query by answering k (k+1)-SUM-Indexing queries.

Consequently, if for some specific constant k there is a solution to the (k+1)-SUM-Indexing problem with less than $\tilde{\Omega}(n^k)$ space and constant query time, then with this reduction we refute the k-SetDisjointness conjecture.   □

5   Directed Reachability Oracles as a Generalization of Set Disjointness Conjecture

An open question which was stated by Pătrașcu in [26] asks if it is possible to preprocess a sparse directed graph in less than $\Omega(n^2)$ space so that Reachability queries (given two query vertices u and v decide whether there is a path from u to v or not) can be answered efficiently. A partial answer, given in [26], states that for constant query time truly superlinear space is necessary. In the undirected case the question is trivial and one can answer queries in constant time using linear space. This is also possible for planar directed graphs (see Holm et al. [21]).

We now show that Reachability oracles for sparse graphs can serve as a generalization of the SetDisjointness conjecture. We define the following parameterized version of Reachability. In the k-Reachability problem the goal is to preprocess a directed sparse graph $G = (V, E)$ so that given a pair of distinct vertices $u, v \in V$ one can quickly answer whether there is a path from u to v consisting of at most k edges. We prove that 2-Reachability and SetDisjointness are tightly connected.

Lemma 2. *There is a linear time reduction from SetDisjointness to 2-Reachability and vice versa which preserves the size of the instance.*

Proof. Given a graph $G = (V, E)$ as an instance for 2-Reachability, we construct a corresponding instance of SetDisjointness as follows. For each vertex v we create the sets $V_{in} = \{u | (u, v) \in E\}$ and $V_{out} = \{u | (v, u) \in E\} \cup \{v\}$. We have $2n$ sets and $2m + n$ elements in all of them ($|V| = n$ and $|E| = m$). Now, a query u, v is reduced to determining if the sets U_{out} and V_{in} are disjoint or not. Notice, that the construction is done in linear time and preserves the size of the instance. In the opposite direction, we are given m sets $S_1, S_2, ..., S_m$ having N elements in total $e_1, e_2, ..., e_N$. We can create an instance of 2-Reachability in the following way. For each set S_i we create a vertex v_i. Moreover, for each element e_j we create a vertex u_j. Then, for each element e_j in a set s_i we create two directed edges (v_i, u_j) and (u_j, v_i). These vertices and edges define a directed graph, which is preprocessed for 2-Reachability queries. It is straightforward to verify that the disjointness of S_i and S_j is equivalent to determining if there is a path of length at most 2 edges from v_i to v_j. Moreover, the construction is done in linear time and preserves the size of the instance. $\qquad\square$

Furthermore, we consider k-Reachability for $k \geq 3$. First we show an upper bound on the tradeoff between space and query time for solving k-Reachability.

Lemma 3. *There exists a data structure for k-Reachability with S space and T query time such that $ST^{2/(k-1)} = O(n^2)$.*

Proof. Let $\alpha > 0$ be an integer parameter to be set later. Given a directed graph $G = (V, E)$, we call vertex $v \in V$ a *heavy vertex* if $deg(v) > \alpha$ and a vertex $u \in V$ a *light vertex* if $deg(u) \leq \alpha$. Notice that the number of heavy vertices is at most n/α. For all heavy vertices in V we maintain a matrix containing the answers to any k-Reachability query between two heavy vertices. This uses $O(n^2/\alpha^2)$ space.

Next, we recursively construct a data structure for (k-1)-Reachability. Given a query u, v, if both vertices are heavy then the answer is obtained from the matrix. Otherwise, either u or v is light vertex. Without loss of generality, say u is a light vertex. We consider each vertex $w \in N_{out}(u)$ ($N_{out}(u) = \{v | (u, v) \in E\}$) and query the (k-1)-Reachability data structure with the pair w, v. Since u is a light node, there are no more than α queries. One of the queries returns a positive answer if and only if there exists a path of length at most k from u to v.

Denote by $S(k, n)$ the space used by our k-Reachability oracle on a graph with n vertices and denote by $Q(k, n)$ the corresponding query time. In our construction we have $S(k, n) = n^2/\alpha^2 + S(k - 1, n)$ and $Q(k, n) = \alpha Q(k - 1, n) + O(1)$. For $k = 1$ it is easy to construct a linear space data structure using hashing so that queries can be answered in constant time. Thus, $S = S(k, n) = O((k - 1)n^2/\alpha^2)$ and $T = Q(k, n) = O(\alpha^{k-1})$. $\qquad\square$

Notice that for the case of $k = 2$ the upper bounds from Lemma 3 exactly match the tradeoff of the Strong SetDisjointness Conjecture ($ST^2 = \tilde{O}(n^2)$). We expand this conjecture by considering the tightness of our upper bound for k-Reachability, which then leads to some interesting consequences with regard to distance oracles.

Conjecture 11. **Directed k-Reachability Conjecture.** Any data structure for the k-Reachability problem with query time T must use $S = \tilde{\Omega}(\frac{n^2}{T^{2/(k-1)}})$ space.

Notice that when k is non-constant then by our upper bound $\tilde{O}(n^2)$ space is necessary independent of the query time. This fits nicely with what is currently known about the general question of Reachability oracles: either we spend n^2 space and answer queries in constant time or we do no preprocessing and then answer queries in linear time. This leads to the following conjecture.

Conjecture 12. **Directed Reachability Hypothesis.** Any data structure for the Reachability problem must either use $\tilde{\Omega}(n^2)$ space, or linear query time.

The conjecture states that in the general case of Reachability there is no full tradeoff between space and query time. We believe the conjecture is true even if the path is limited to lengths of some non-constant number of edges.

6 Distance Oracles and Directed Reachability

There are known lower bounds for constant query time distance oracles based on the SetDisjointness hypothesis. Specifically, Cohen and Porat [16] showed that stretch-less-than-2 oracles need $\Omega(n^2)$ space for constant queries. Patrascu et al. [28] showed a conditional space lower bound of $\Omega(m^{5/3})$ for constant-time stretch-2 oracles. Applying the Strong SetDisjointness conjecture to the same argument as in [16] we can prove that for stretch-less-than-2 oracles the tradeoff between S (the space for the oracle) and T (the query time) is by $S \times T^2 = \Omega(n^2)$.

Recent effort was taken toward constructing compact distance oracles where we allow non-constant query time. For stretch-2 and stretch-3 Agarwal et al. [10] [9] achieves a space-time tradeoff of $S \times T = O(n^2)$ and $S \times T^2 = O(n^2)$, respectively, for sparse graphs. Agarwal [8] also showed many other results for stretch-2 and below. Specifically, Agarwal showed that for any integer k a stretch-$(1+1/k)$ oracle exhibits the following space-time tradeoff: $S \times T^{1/k} = O(n^2)$. Agarwal also showed a stretch-$(1+1/(k+0.5))$ oracle that exhibits the following tradeoff: $S \times T^{1/(k+1)} = O(n^2)$. Finally, Agarwal gave a stretch-$(5/3)$ oracle that achieves a space-time tradeoff of $S \times T = O(n^2)$. Unfortunately, no lower bounds are known for non-constant query time.

Conditioned on the directed k-Reachability conjecture we prove the following lower bound.

Lemma 4. *Assume the directed k-Reachability conjecture holds. Then stretch-less-than-$(1 + 2/k)$ distance oracles with query time T must use $S \times T^{2/(k-1)} = \tilde{\Omega}(n^2)$ space.*

Proof. Given a graph $G = (V, E)$ for which we want to preprocess for k-Reachability, we create a layered graph with k layers where each layer consists of a copy of all vertices of V. Each pair of neighboring layers is connected by a copy of all edges in E. We omit all directions from the edges. For every fixed integer k, the layered graph has $O(|V|)$ vertices and $O(|E|)$ edges. Next, notice that if we construct a distance oracle that can distinguish between pairs of vertices of distance at most k and pairs of vertices of distance at least $k + 2$, then we can answer k-Reachability queries. Consequently, assuming the k-Reachability conjecture we have that $S \times T^{2/(k-1)} = \Omega(n^2)$ for stretch-less-than-$(1 + 2/k)$ distance oracles (For $k = 2$ this is exactly the result we get by the SetDisjointness hypothesis). □

Notice, that the stretch-$(5/3)$ oracle shown by Agarwal [8] achieves a space-time tradeoff of $S \times T = O(n^2)$. Our lower bound is very close to this upper bound since it applies for any distance oracle with stretch-less-than-$(5/3)$, by setting $k = 3$.

References

1. Amir Abboud, Arturs Backurs, Thomas Deuholm Hansen, Virginia Vassilevska Williams, and Or Zamir. Subtree isomorphism revisited. In *Proc. of 27th ACM-SIAM Symposium on Discrete Algorithms, SODA*, pages 1256–1271, 2016.
2. Amir Abboud, Arturs Backurs, and Virginia Vassilevska Williams. If the current clique algorithms are optimal, so is Valiant's parser. *55th IEEE Annual Symposium on Foundations of Computer Science, FOCS*, pages 98–117, 2015.
3. Amir Abboud, Arturs Backurs, and Virginia Vassilevska Williams. Quadratic-time hardness of LCS and other sequence similarity measures. *55th IEEE Annual Symposium on Foundations of Computer Science, FOCS*, pages 59–78, 2015.
4. Amir Abboud, Fabrizio Grandoni, and Virginia Vassilevska Williams. Subcubic equivalences between graph centrality problems, APSP and diameter. In *Proceedings of the Twenty-Sixth Annual ACM-SIAM Symposium on Discrete Algorithms, SODA 2015, San Diego, CA, USA, January 4-6, 2015*, pages 1681–1697, 2015.
5. Amir Abboud and Virginia Vassilevska Williams. Popular conjectures imply strong lower bounds for dynamic problems. In *55th IEEE Annual Symposium on Foundations of Computer Science, FOCS 2014, Philadelphia, PA, USA, October 18-21, 2014*, pages 434–443, 2014.
6. Amir Abboud, Virginia Vassilevska Williams, and Oren Weimann. Consequences of faster alignment of sequences. In *Automata, Languages, and Programming - 41st International Colloquium, ICALP 2014, Copenhagen, Denmark, July 8-11, 2014, Proceedings, Part I*, pages 39–51, 2014.
7. Amir Abboud, Virginia Vassilevska Williams, and Huacheng Yu. Matching triangles and basing hardness on an extremely popular conjecture. In *Proceedings of the Forty-Seventh Annual ACM on Symposium on Theory of Computing, STOC 2015, Portland, OR, USA, June 14-17, 2015*, pages 41–50, 2015.
8. Rachit Agarwal. The space-stretch-time tradeoff in distance oracles. In *Algorithms - ESA 2014 - 22th Annual European Symposium on Algorithms, Wroclaw, Poland, September 8-10, 2014. Proceedings*, pages 49–60, 2014.
9. Rachit Agarwal, Brighten Godfrey, and Sariel Har-Peled. Faster approximate distance queries and compact routing in sparse graphs. *CoRR*, abs/1201.2703, 2012.

10. Rachit Agarwal, Philip Brighten Godfrey, and Sariel Har-Peled. Approximate distance queries and compact routing in sparse graphs. In *INFOCOM 2011. 30th IEEE International Conference on Computer Communications*, pages 1754–1762, 2011.

11. Amihood Amir, Timothy M. Chan, Moshe Lewenstein, and Noa Lewenstein. On hardness of jumbled indexing. In *Automata, Languages, and Programming - 41st International Colloquium, ICALP 2014, Copenhagen, Denmark, July 8-11, 2014, Proceedings, Part I*, pages 114–125, 2014.

12. Arturs Backurs and Piotr Indyk. Edit distance cannot be computed in strongly subquadratic time (unless SETH is false). In *Proceedings of the Forty-Seventh Annual ACM on Symposium on Theory of Computing, STOC 2015, Portland, OR, USA, June 14-17, 2015*, pages 51–58, 2015.

13. Karl Bringmann. Why walking the dog takes time: Frechet distance has no strongly subquadratic algorithms unless SETH fails. In *55th IEEE Annual Symposium on Foundations of Computer Science, FOCS 2014, Philadelphia, PA, USA, October 18-21, 2014*, pages 661–670, 2014.

14. Karl Bringmann and Marvin Künnemann. Quadratic conditional lower bounds for string problems and dynamic time warping. *55th IEEE Annual Symposium on Foundations of Computer Science, FOCS*, 2015.

15. Timothy M. Chan and Moshe Lewenstein. Clustered integer 3SUM via additive combinatorics. In *Proceedings of the Forty-Seventh Annual ACM on Symposium on Theory of Computing, STOC 2015, Portland, OR, USA, June 14-17, 2015*, pages 31–40, 2015.

16. Hagai Cohen and Ely Porat. Fast set intersection and two-patterns matching. *Theor. Comput. Sci.*, 411(40-42):3795–3800, 2010.

17. Hagai Cohen and Ely Porat. On the hardness of distance oracle for sparse graph. *CoRR*, abs/1006.1117, 2010.

18. A. Gajentaan and M. H. Overmars. On a class of $O(n^2)$ problems in computational geometry. *Comput. Geom.*, 5:165–185, 1995.

19. Isaac Goldstein, Tsvi Kopelowitz, Moshe Lewenstein, and Ely Porat. How hard is it to find (honest) witnesses? In *European Symposium on Algorithms, ESA 2016*, pages 45:1–45:16, 2016.

20. Monika Henzinger, Sebastian Krinninger, Danupon Nanongkai, and Thatchaphol Saranurak. Unifying and strengthening hardness for dynamic problems via the online matrix-vector multiplication conjecture. In *Proceedings of the Forty-Seventh Annual ACM on Symposium on Theory of Computing, STOC 2015, Portland, OR, USA, June 14-17, 2015*, pages 21–30, 2015.

21. Jacob Holm, Eva Rotenberg, and Mikkel Thorup. Planar reachability in linear space and constant time. In *56th Annual Symposium on Foundations of Computer Science, FOCS 2015, Berkeley, CA, USA, 17-20 October, 2015*, pages 370–389, 2015.

22. Tsvi Kopelowitz, Seth Pettie, and Ely Porat. Dynamic set intersection. In *Proceedings 14th Int'l Symposium on Algorithms and Data Structures (WADS)*, pages 470–481, 2015.

23. Tsvi Kopelowitz, Seth Pettie, and Ely Porat. Higher lower bounds from the 3SUM conjecture. In *Proceedings of the Twenty-Seventh Annual ACM-SIAM Symposium on Discrete Algorithms, SODA 2016, Arlington, VA, USA, January 10-12, 2016*, pages 1272–1287, 2016.

24. Kasper Green Larsen, J. Ian Munro, Jesper Sindahl Nielsen, and Sharma V. Thankachan. On hardness of several string indexing problems. *Theor. Comput. Sci.*, 582:74–82, 2015.

25. Mihai Patrascu. Towards polynomial lower bounds for dynamic problems. In *Proceedings of the 42nd ACM Symposium on Theory of Computing, STOC 2010, Cambridge, Massachusetts, USA, 5-8 June 2010*, pages 603–610, 2010.
26. Mihai Patrascu. Unifying the landscape of cell-probe lower bounds. *SIAM J. Comput.*, 40(3):827–847, 2011.
27. Mihai Patrascu and Liam Roditty. Distance oracles beyond the Thorup-Zwick bound. *SIAM J. Comput.*, 43(1):300–311, 2014.
28. Mihai Patrascu, Liam Roditty, and Mikkel Thorup. A new infinity of distance oracles for sparse graphs. In *53rd Annual IEEE Symposium on Foundations of Computer Science, FOCS 2012, New Brunswick, NJ, USA, October 20-23, 2012*, pages 738–747, 2012.
29. Mihai Patrascu and Ryan Williams. On the possibility of faster SAT algorithms. In *Proceedings of the Twenty-First Annual ACM-SIAM Symposium on Discrete Algorithms, SODA 2010, Austin, Texas, USA, January 17-19, 2010*, pages 1065–1075, 2010.
30. Virginia Vassilevska Williams and Ryan Williams. Subcubic equivalences between path, matrix and triangle problems. In *51th Annual IEEE Symposium on Foundations of Computer Science, FOCS 2010, October 23-26, 2010, Las Vegas, Nevada, USA*, pages 645–654, 2010.

Posimodular Function Optimization [*]

Magnús M. Halldórsson[1], Toshimasa Ishii[2], Kazuhisa Makino[3], and
Kenjiro Takazawa[4]

[1] ICE-TCS, School of Computer Science, Reykjavik University, Iceland
mmh@ru.is
[2] Graduate School of Economics, Hokkaido University, Sapporo, Japan
ishii@econ.hokudai.ac.jp
[3] Research Institute for Mathematical Sciences, Kyoto University, Japan
makino@kurims.kyoto-u.ac.jp
[4] Faculty of Science and Engineering, Hosei University, Japan
takazawa@hosei.ac.jp

Abstract. A function $f : 2^V \to \mathbb{R}$ on a finite set V is *posimodular* if
$f(X) + f(Y) \geq f(X \setminus Y) + f(Y \setminus X)$, for all $X, Y \subseteq V$. Posimodular
functions often arise in combinatorial optimization such as undirected
cut functions. We consider the problem of finding a nonempty subset X
minimizing $f(X)$, when the posimodular function f is given by oracle
access.

We show that posimodular function minimization requires exponential
time, contrasting with the polynomial solvability of submodular function
minimization that forms another generalization of cut functions. On the
other hand, the problem is fixed-parameter tractable in terms of the size
of the image (or range) of f.

In more detail, we show that $\Omega(2^{0.3219n}T_f)$ time is necessary and $O(2^{0.92n}T_f)$
sufficient, where T_f denotes the time for one function evaluation. When
the image of f is $D = \{0, 1, \ldots, d\}$, $O(2^{1.271d}nT_f)$ time is sufficient and
$\Omega(2^{0.1609d}T_f)$ necessary. We can also generate all sets minimizing f in
time $2^{O(d)}n^2T_f$.

Finally, we also consider the problem of maximizing a given posimodular
function, showing that it requires at least $2^{n-1}T_f$ time in general, while
it has time complexity $\Theta(n^{d-1}T_f)$ when $D = \{0, 1, \ldots, d\}$ is the image
of f, for integer d.

1 Introduction

Let V denote a finite set with $n = |V|$. A set function $f : 2^V \to \mathbb{R}$ is called
posimodular if

$$f(X) + f(Y) \geq f(X \setminus Y) + f(Y \setminus X) \tag{1.1}$$

[*] This research was partially supported by Icelandic Research Fund grants 152679-05
and 174484-05, MEXT KAKENHI Grant Numbers JP24106002, JSPS KAKENHI
Grant Numbers JP25280004, JP26280001 and JP16K00001, and JST CREST Grant
Number JPMJCR1402, Japan.

© Springer International Publishing AG 2017
F. Ellen et al. (Eds.): WADS 2017, LNCS 10389, pp. 437–448, 2017.
DOI: 10.1007/978-3-319-62127-2_37

for all $X, Y \subseteq V$, where $\mathbb{R}$ denotes the set of all reals. Posimodularity is a fundamental property in combinatorial optimization [6, 7, 12, 14, 15, 17] and is typically the key for efficiently solving undirected network optimization and related problems, since cut functions for undirected networks are posimodular. In comparison, cut functions for directed networks are not posimodular.

There are numerous network optimization problems where posimodularity leads to gaps between the complexity for undirected and directed variants. One example is the local edge-connectivity augmentation problem, which is polynomially solvable in undirected networks but NP-hard in directed networks [5]. Similarly, undirected versions of the source location problem with uniform demands or with uniform costs can be solved in polynomial time, [1, 19], while the directed versions are NP-hard [8]. More generally, the current fastest algorithm for minimizing a submodular and posimodular function runs in $O(n^3 T_f)$ time [13], while the one for minimizing a submodular function requires $O(n^5 T_f + n^6)$ time [16], where a set function $f : 2^V \to \mathbb{R}$ is called *submodular* if

$$f(X) + f(Y) \geq f(X \cap Y) + f(X \cup Y) \tag{1.2}$$

for all $X, Y \subseteq V$, and T_f denotes the time needed to evaluate the function value $f(X)$ for a given $X \subseteq V$. The submodular multiway partition problem, which is a generalization of the graph multiway cut problems, is 2-approximable in polynomial time, while the symmetric submodular multiway partition problem is 1.5-approximable [2], where a set function $f : 2^V \to \mathbb{R}$ is called *symmetric* if $f(X) = f(V \setminus X)$ holds for any $X \subseteq V$. We note that a function is symmetric posimodular if and only if it is symmetric submodular, since the symmetricity of f implies that $f(X) + f(Y) = f(V \setminus X) + f(Y)$ and $f(X \setminus Y) + f(Y \setminus X) = f((V \setminus X) \cup Y) + f((V \setminus X) \cap Y)$.

These phenomena can be partially explained by the following three structural properties on posimodular functions. The first structural property is used under the name of uncrossing techniques. There are many variants of partition problems that ask for a partition $\{V_1, V_2, \ldots, V_k\}$ of V minimizing $\sum_{i=1}^{k} f(V_i)$, for a given set function f. This includes the graph multiway cut problem, the graph k-way cut problem, and the submodular multiway partition problem. If f is posimodular, then after obtaining a family $\{V_1', V_2', \ldots, V_k'\}$ of subsets that covers V but may not be disjoint, we can apply uncrossing techniques to obtain a partition $\{V_1, V_2, \ldots, V_k\}$ of V without increasing the cost (i.e., $\sum_{i=1}^{k} f(V_i) \leq \sum_{i=1}^{k} f(V_i')$). This is because the posimodularity of f implies that $f(X) + f(Y) \geq \min\{f(X) + f(Y \setminus X), f(Y) + f(X \setminus Y)\}$ for any two sets X and Y. Indeed, this uncrossing technique results in a better approximation ratio for the symmetric submodular multiway partition problem than the (non-symmetric) submodular multiway partition problem [2] (recall that a symmetric submodular function is posimodular). Similar uncrossing techniques have been utilized in other partition problems [10, 18].

The second structural property holds for extreme sets. A subset X of V is called *extreme* if every nonempty proper subset Y of X satisfies $f(Y) > f(X)$. It is known that when f is posimodular, the family $\mathcal{X}(f)$ of extreme sets is *laminar* (i.e., every two members X and Y of $\mathcal{X}(f)$ satisfy $X \cap Y = \emptyset$, $X \subseteq Y$, or $X \supseteq Y$). Note that if $X, Y \in \mathcal{X}(f)$ would satisfy $X \cap Y, X \setminus Y, Y \setminus X \neq \emptyset$, then we have $f(X) + f(Y) \geq f(X \setminus Y) + f(Y \setminus X) > f(X) + f(Y)$, a contradiction. The family $\mathcal{X}(f)$ of extreme sets for an undirected cut function f represents the connectivity structure of a given network and helps to design many efficient network algorithms [9, 21]. For example, the undirected source location problem with uniform demands can be solved in $O(n)$ time, if the family $\mathcal{X}(f)$ is known in advance, where n corresponds to the number of vertices in the network [11]. In fact, $\mathcal{X}(f)$ can be computed in $O(n(m + n \log n))$ time for any undirected cut function [11], where m denotes the number of edges in the network. We note that $\mathcal{X}(f)$ can be found in $O(n^3 T_f)$ time if f is posimodular and submodular [12].

The third structural property holds for solid sets, where a subset X of V is said to be v-*solid* for an element $v \in V$, if $v \in X$ and every nonempty proper subset Y of X that contains v satisfies $f(Y) > f(X)$. Let $\mathcal{S}(f)$ denote the family of all solid sets, i.e., $\mathcal{S}(f) = \bigcup_{v \in V} \{X : X \text{ is } v\text{-solid}\}$. It is known [17] that the family $\mathcal{S}(f)$ forms a tree hypergraph if f is posimodular. Similarly to the previous case for $\mathcal{X}(f)$, if a host tree T of $\mathcal{S}(f)$ is known in advance, this structure enables us to construct a polynomial time algorithm for the minimum transversal problem for posimodular functions f, which is an extension of the undirected source location problem with uniform costs [19] and the undirected external network problem [20]. If f is in addition submodular, a host tree T can be computed in polynomial time.

We remark that the above structural properties of $\mathcal{X}(f)$ and $\mathcal{S}(f)$ follow from the posimodularity of f, while the submodularity is used to derive such structures efficiently.

On the other hand, to our best knowledge, all previous results for posimodular optimization also make use of submodularity or symmetry, since undirected cut functions, the most representative posimodular functions, are also submodular and symmetric.

In this paper, we focus on the posimodular function minimization defined as follows.

POSIMODULAR FUNCTION MINIMIZATION

Input: A posimodular function $f : 2^V \to \mathbb{R}$,   (1.3)

Output: A nonempty subset X^* of V such that $f(X^*) = \min_{X \subseteq V : X \neq \emptyset} f(X)$.

Here an input function f is given by an oracle that answers $f(X)$ for a given subset X of V, and we assume that the optimal value $f(X^*)$ is also output. The problem was posed as an open problem on the Egres open problem list [3] in 2010, as negamodular function maximization, where a set function f is *negamodular*, if $-f$ is posimodular. We also consider the posimodular function maximization, as submodular function maximization has been intensively studied in recent years.

Our Contributions

The main results obtained in this paper can be summarized as follows.

1. **Intractability:** We show that any algorithm for posimodular function minimization requires $\Omega(2^{0.3219n}T_f)$ time. On the other hand, we show that it
   is possible to beat the trivial $2^n T_f$ upper bound, giving an $O(2^{0.92n}T_f)$-time
   algorithm.
2. **Tractability on small images:** We consider functions $f : 2^V \to D$ with
   a restricted image. Our main positive result is that the problem is fixed-
   parameter tractable in terms of the image size $|D|$. Specifically, we give
   an algorithm with complexity $O(2^{3|D|}nT_f)$. For the case of the image $D =
   \{0, 1, \ldots, d\}$, we obtain an improved bound of $O(2^{1.271d}nT_f)$. This is matched
   with an exponential lower bound of $\Omega(2^{0.1609d}T_f)$ time.
   The most technical part of the paper is the extension of the parameterized al-
   gorithm to generate all minimizers with linear delay, after initial $2^{O(|D|)}n^2T_f$-
   time.
3. **Hardness of Maximization:** We show that posimodular function maximization requires at least $2^{n-1}T_f$ time, and thus only trivial solutions are
   possible. For image restricted to (a subset of) $D = \{0, 1, \ldots, d\}$, for a constant d, we obtain a tight bound of $\Theta(n^{d-1}T_f)$ on the time complexity.

We also obtain implications for related problems. For instance, we can com-
pute all extreme sets in $O(|D|2^{3|D|}nT_f)$ time, which implies that the source
location problem for posimodular functions can be solved in $O(|D|2^{3|D|}nT_f)$
time.

We note that no complexity-theoretic assumptions are needed for the lower
bounds. For related results, Feige et al. [4] showed that at least $e^{\epsilon^2 n/8}$ oracle
calls are necessary for obtaining a solution of at least $(1/2 + \epsilon)$ times optimal for
symmetric submodular function maximization, which is equivalent to symmetric
posimodular function maximization.

The rest of this paper is organized as follows. In Section 2, we give the
hardness results and a $o(2^n T_f)$-time algorithm for posimodular function minimization. In Section 3, we consider the case where the image of f is bounded
or given by $D = \{0, 1, \ldots, d\}$ and show hardness results and a fixed parameter
algorithm in terms of the image size. Section 4 treats the posimodular function
maximization. Due to space limitations, some proofs are omitted.

2   General Posimodular Function Minimization

2.1   Hardness Results

Let V be a finite set with $n = |V|$ and $f : 2^V \to \mathbb{R}$ be a posimodular function.
Notice that f satisfies

$$f(X) \geq f(\emptyset) \text{ for all } X \subseteq V, \tag{2.1}$$

since $f(X) + f(X) \geq f(\emptyset) + f(\emptyset)$. Throughout the paper, we assume that $f(\emptyset) = 0$, since otherwise, we can replace $f(X)$ by $f(X) - f(\emptyset)$ for all $X \subseteq V$.

In this section, we analyze the number of oracle calls necessary for posimodular function minimization. An optimal solution to the posimodular function minimization (1.3) is referred to as *a minimizer of f* (among nonempty subsets).

Let $g : 2^V \to \mathbb{R}_+$ be the cardinality function defined by $g(X) = |X|$. Clearly, g is posimodular since g is monotone, i.e., $g(X) \geq g(Y)$ holds for all two subsets X and Y of V with $X \supseteq Y$.

For a given positive integer k, we construct the family $\mathcal{G}_k = \{g\} \cup \{g_S \mid S \subseteq V, |S| = 2k\}$ of functions, where $g_S : 2^V \to \mathbb{R}_+$ is defined by

$$g_S(X) = \begin{cases} 2k - |X| & \text{if } X \subseteq S \text{ and } |X| \geq k+1, \\ g(X) = |X| & \text{otherwise.} \end{cases}$$

We can see that each g_S is a posimodular function close to g. We show below that exponential number of oracles queries are necessary to distinguish between the posimodular functions in $\mathcal{G}_k$.

Let $\mathcal{S}_k = \{S \subseteq V \mid |S| = 2k\}$ and $\mathcal{T}_k = \{T \subseteq V \mid k+1 \leq |T| \leq 2k\}$. Consider the following integer program that formulates the hitting set problem, which asks for a minimum cardinality subset of $\mathcal{T}_k$ that hits each set in $\mathcal{S}_k$.

$$\begin{aligned} \text{minimize} \quad & \textstyle\sum_{T \in \mathcal{T}_k} z_T \\ \text{subject to} \quad & \textstyle\sum_{T \subset \mathcal{T}_k, T \subseteq S} z_T \geq 1 \text{ for each } S \subset \mathcal{S}_k, \\ & z_T \in \{0, 1\} \qquad \text{for each } T \in \mathcal{T}_k. \end{aligned} \qquad (2.2)$$

Note that every posimodular function f in $\mathcal{G}_k$ satisfies $f(X) = g(X)$ if $|X| \leq k$ or $|X| \geq 2k + 1$. Oracle calls for such sets X do not help to distinguish among posimodular functions in $\mathcal{G}_k$. Therefore, we can restrict our attention to subsets T in $\mathcal{T}_k$ for oracle calls.

Then, we have the following lemma, whose proof is omitted.

Lemma 2. *Let q_k denote the optimal value for (2.2).*
(i) At least q_k oracle calls are necessary to distinguish among posimodular functions in $\mathcal{G}_k$.
(ii) For $2 \leq k \leq n/2$, we have $c_1 \frac{n^n}{(4k)^k (n-k)^{n-k}} \leq q_k \leq c_2 2^{o(n)} \frac{n^n}{(4k)^k (n-k)^{n-k}}$ for some positive constants c_1 and c_2.

Observe that when $n = 5k$, we have $\frac{n^n}{(4k)^k (n-k)^{n-k}} = 1.25^n = 2^{0.3219n}$. By Lemma 2 and this, we have $q_k \in [c_1 2^{0.3219n}, c_2 2^{0.3219n + o(n)}]$ for some positive constants c_1 and c_2. Thus, we have the following theorem.

Theorem 1. *Every algorithm for posimodular function minimization makes $\Omega(2^{0.3219n})$ oracle calls in the worst case.*

Finally, it follows that there is no fixed parameter algorithm in terms of the solution size $|S|$. Indeed, Lemma 2 implies that for the above instance, any algorithm requires at least $\frac{c_1}{(2|S|)^{|S|/2}} \cdot n^{|S|/2}$ oracle calls (note that $|S| = 2k$).

2.2 $o(2^n T_f)$-time algorithm

Posimodular function minimization can trivially be solved in $2^n T_f$ time, since the number of subsets of V is 2^n. In this subsection, we give a $c^n T_f$ time algorithm for the problem, for $c < 2$.

Theorem 2. *Posimodular function minimization can be solved in* $O\left(\binom{n}{n/3} n \log n \cdot T_f\right) = O(2^{0.92n} T_f)$ *time.*

We say that a set $X \subseteq S$ is a *splitter with respect to* S if $f(X \cup \{v\}) > f(X \cup \{u\})$ for all $v \in V \setminus S$ and all $u \in S \setminus X$. Let $v_1, v_2, \ldots, v_{|V \setminus X|}$ be an ordering of $V \setminus X$ such that $f(X \cup \{v_1\}) \leq f(X \cup \{v_2\}) \leq \cdots \leq f(X \cup \{v_{|V \setminus X|}\})$. Note that for each i with $f(X \cup \{v_i\}) < f(X \cup \{v_{i+1}\})$, X is a splitter with respect to $X \cup \{v_1, v_2, \ldots, v_i\}$ and that X is not a splitter with respect to any other subset of V.

It turns out that a small splitter exists, as long as there is no singleton minimizer. The search then reduces to finding either a small splitter or a very large minimizer.

Lemma 5. *Suppose no singleton is a minimizer of f. Then, there exists a splitter of cardinality at most* $\lceil (|S| - 1)/2 \rceil$ *with respect to a minimizer S.*

Based on this lemma, we can prove Theorem 2.

Proof of Theorem 2. Lemma 5 shows that there exists either: (i) a singleton minimizer, (ii) a minimizer of size at least $2n/3$, or (iii) a splitter of size at most $n/3$. The number of sets to be examined are n, $O\left(\binom{n}{2n/3}\right) = O\left(\binom{n}{n/3}\right)$, and $O\left(\binom{n}{n/3}\right)$, respectively. Recall that in the case a splitter is found, the minimizer can be found in at most n oracle calls. $\square$

3   Minimization on Small Images

In light of the hardness of minimizing general posimodular functions, we turn our attention to parameterized algorithms. The intractability results still hold in terms of the solution size (cardinality of the minimizer) or the value of the minimum solution. Instead, we treat in this section the parameter $|D|$, the cardinality of the image D of the function f, with a particular focus on the case $D = \{0, 1, \ldots, d\}$, for a number d.

3.1   Fixed Parameter Algorithm

We propose a bounded-depth tree search algorithm. Each tree node corresponds to an invocation of a recursive procedure given three set parameters, A, B, and C. In each invocation, the algorithm either produces a solution and terminates, or it selects an element v with which it makes two recursive call: adding v to A, and adding v to either B or C. The crucial property maintained is that whenever

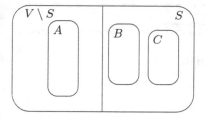

an element is added to a set, the value of the set increases. It follows immediately that the recursion depth is at most $3|D|$, and since the tree is binary, the number of recursive calls, which dominates the time complexity, is at most $2^{3|D|}$.

The challenge is in showing that a solution can be obtained once no options for recursing remain. The parameter triplet (A, B, C) forms a *valid tuple* with respect to a minimizer S, if A is disjoint from S while B and C are contained in S ($A \subseteq (V \setminus S) \wedge B \cup C \subseteq S$). One can show by induction that exactly one leaf in the search tree corresponds to a valid tuple (for a given minimizer S). The crucial characterization that we obtain is that once we reach a leaf with a valid tuple, we can easily identify the rest of S: namely, $S \setminus (B \cup C)$ is given by those elements in $V \setminus (A \cup B \cup C)$ whose addition to neither B nor C increases the values of those sets.

The algorithm MINPOSIMODULAR-D(f) operates in more detail as follows. It uses a global variable SS, initially set as the singleton set of smallest function value, which progressively improves to a minimizer. The algorithm calls a recursive subroutine SOLVE with parameters A, B, and C, initially empty sets. Note that the algorithm need not know anything about the image D.

Algorithm 1 MINPOSIMODULAR-D(f)

1: $SS \leftarrow \{v^*\}$, where $v^* = \arg\min_{v \in V} f(v)$ is a singleton of smallest value
2: Call SOLVE$(\emptyset, \emptyset, \emptyset)$
3: Output SS

We say that an element v *increases* a set X if $f(X \cup \{v\}) > f(X)$, and *increases by two* if $f(X \cup \{v\}) > f(X) + 1$. Let $incA(v)$ denote the predicate that element v increases A, i.e., $f(A \cup \{v\}) > f(A)$. Similarly, let $incBC(v)$ denote the predicate that v increases B or v increases C (or both).

Let TRY(v) denote the following inline macro: Recursively call SOLVE$(A \cup \{v\}, B, C)$; then, call SOLVE$(A, B \cup \{v\}, C)$ if v increases B, and otherwise call SOLVE$(A, B, C \cup \{v\})$.

The core tools for our results are the following two lemmas. They show that if given a set that is either properly inside or outside a minimizer, then adding elements from the other side must increase the value of the set. This helps us find sets with successively larger values; once we obtain sets of maximum value,

---

Algorithm 2 Procedure SOLVE(A, B, C)

Require: Disjoint subsets A, B, C of V; SS is a global variable representing the current champion
1: **if** $\exists v \in V \setminus (A \cup B \cup C)$, incA($v$) $\wedge$ incBC(v) **then**
2:     TRY(v);
3: **else**
4:     $\hat{S} \leftarrow B \cup C \cup \{v \in V \setminus A : \neg\text{incBC}(v)\}$
5:     **if** $f(\hat{S}) < f(SS)$ **then**
6:        $SS \leftarrow \hat{S}$

---

the remaining elements can be quickly assigned, based on the characterization of the lemmas.

A subset X is called *locally minimal* if $f(X) < f(X \setminus \{v\})$ holds for every $v \in X$.

Lemma 7. *Let S be a locally minimal set (for f) with $|S| \geq 2$. If $X \subseteq V \setminus S$ and v is an element in S, then v increases X.*

Lemma 8. *Let S be a set with smaller value than any singleton set. Let B and C be disjoint sets within S and let v be an element in $V \setminus S$. Then, v increases B or it increases C by two.*

We first argue correctness.

Lemma 9. *The algorithm* MINPOSIMODULAR-D(f) *finds a (minimal) minimizer of f.*

Proof. If there is a singleton minimizer, then it is found in the first step of MINPOSIMODULAR-D(f). Suppose then that there is a minimal minimizer S with $|S| \geq 2$. Recall that there is leaf in the recursion tree with a valid tuple (A, B, C) for S. Since the recurrence ended at this leaf, each element v outside of $A \cup B \cup C$ fails either incA(v) or incBC(v). Each element v failing incA(v) must be in $V \setminus S$, by Lemma 7, while an element w failing incBC(w) must be in S, by Lemma 8. Then, S is given by $S = B \cup C \cup \{v \in V \setminus A : \neg\text{incBC}(v)\}$, which is found in line 4 of Solve. $\qquad\square$

The bound $O(2^{3|D|} \cdot nT_f)$ on the time complexity is immediate. We argue a stronger bound for when the image is a small range.

Theorem 3. *Minimization of a posimodular function $f : 2^V \to D$. can be solved in time $O(2^{3|D|}nT_f)$. For $D = \{0, 1, \ldots, d\}$, the complexity improves to $O(2^{1.271d}nT_f)$.*

Finding all minimal minimizers. We can generalize the algorithm to find all *minimal* minimizers, i.e., all minimizers S such that no proper subset of S is also a minimizer.

Corollary 1. *All minimal minimizers can be found in time $O(2^{\min(3|D|, 1.271d)}nT_f)$.*

3.2 Related Problems: Extreme Sets, Transversals, Approximation

Extreme sets. We first show that the family $\mathcal{X}(f)$ of all extreme sets can be obtained as an application of Theorem 3. Recall that a subset X of V is called *extreme* if every nonempty proper subset Y of X satisfies $f(Y) > f(X)$. By definition, $\mathcal{X}(f)$ contains all singletons $\{v\}$, $v \in V$, and any extreme set X with $|X| \geq 2$ is locally minimal.

Consider the subfamily $\mathcal{X}_p(f) = \{X : X \text{ is extreme with } f(X) = p\}$ of extreme sets with value p, for $p \in D$. The singleton sets in $\mathcal{X}_p(f)$ are given precisely by $V_p = \{v \in V : f(v) = p\}$, while the non-singleton sets in $\mathcal{X}_p(f)$ can only contain elements v with $f(v) > p$. Thus, to find non-singleton sets in $\mathcal{X}_p(f)$, we can restrict attention to the universe $V_{>p} = \{v \in V : f(v) > p\}$.

We observe that MINPOSIMODULAR-D(f) restricted to $V_{>p}$ identifies all non-singleton sets in $\mathcal{X}_p(f)$, since Lemmas 7 and 8 hold for all non-singleton extreme sets. Thus, by iterating over the $|D|$ possible values of p, we can produce all extreme sets in $O(|D|2^{3|D|}nT_f)$ time.

In summary, we have the following result.

Corollary 2. *For a posimodular function $f : 2^V \to D$, we can compute the family $\mathcal{X}(f)$ of all extreme sets of f in $O(|D|2^{3|D|}nT_f)$ time.*

Minimum traversal. Consider the following problem:

MINIMUM TRANSVERSAL(f, c, r)

Input: A posimodular function $f : 2^V \to D$, a cost function
$c : V \to \mathbb{R}$, and a demand function $r : 2^V \to \mathbb{R}$, $\qquad$ (3.1)

Output: A nonempty subset S of V minimizing $\sum_{v \in S} c(v)$ such that
$f(X) \geq r(X)$ for every nonempty subset $X \subseteq V \setminus S$.

We note that undirected source location problem with uniform demands is a special case of this problem where f is a cut function in an undirected graph and r is uniform [1, 11]. As is the case with the source location [11], we can find an optimal solution in linear time if the family $\mathcal{X}(f)$ of extreme sets is known in advance. Thus, we have the following result.

Corollary 3. *The minimum transversal problem (3.1) for a posimodular function $f : 2^V \to D$ can be solved in $O(|D|2^{3|D|}nT_f)$ time if r is uniform.*

We note that the minimum transversal problem (3.1) for a uniform cost function c and a modulotone demand function r is studied in [17] as a generalization of source location problem with uniform costs [19] or external network problem [20], where a set function $r : 2^V \to \mathbb{R}$ is called *modulotone* if for every nonempty subset X of V, there exists an element $v \in X$ such that all proper subsets Y of X with $v \in Y$ satisfy $r(Y) > r(X)$. As observed in [17], this problem can be solved if solid sets can be computed efficiently. More precisely, we need to compute solid sets including u and v but w for all three distinct elements u, v, and w in V. Therefore, Corollary 2 does not imply the tractability of this problem.

Approximation. We can speed up the search significantly when seeking (additive) approximations.

Theorem 4. *There is an algorithm for finding an additive ρ-approximate solution, that uses $\rho^{O(d/\rho)}n$ oracle queries, for any given $\rho \geq 1$.*

3.3 Generating All Minimizers

We can extend the parameterized algorithm to generate all minimizers of a posimodular function.

Theorem 5. *For a posimodular function $f : 2^V \rightarrow D$, we can generate all minimizers of f with $O(nT_f)$ delay, after $O(5^{3|D|}nT_f)$ time to compute the first minimizer.*

This part is the most technical; we briefly summarize the approach.

We first observe that it suffices to find locally minimal minimizers, as other minimizers can be quickly generated from those. The main challenge is dealing with the set Z of singleton minimizers – without those, the previous algorithm suffices. We can treat the elements of $V \setminus Z$ as before, and also those elements of Z that don't increase A. We partition the remaining elements of Z into a collection $\mathcal{Z}$ of maximal minimizers, and observe that these sets don't cross (or overlap with) other minimizers. Those sets in $\mathcal{Z}$ that increase both A and one of B, C can be treated the same way as before (with TRY). The rest is split into two: $\mathcal{Z}_0$, those that increase only B or C, and $\mathcal{Z}_1$, those that increase only A. We find that a minimizer contains at most two sets from $\mathcal{Z}_0$. The key idea is to examine pairs of sets from $\mathcal{Z}_1$: if there are three sets in $\mathcal{Z}_1$ such that any pair increases B or C, then some pair has to be on the same side of a minimizer S, which allows us to make progress in one of the recursive calls. Otherwise, we can show that $\mathcal{Z}_1$ contains at most two sets that are not inside S, and can try all such possibilities.

3.4 Hardness Results

We complement the parameterized algorithm with the following lower bound. It shows that the time complexity must both be exponential in d and involve a factor linear in n.

Theorem 6. (*i*) *Posimodular function minimization requires $\Omega(2^{0.1609d})$ oracle calls.*
(*ii*) *Posimodular function minimization requires $\Omega(n)$ oracle calls, even when restricted to functions with image $D = \{0,1\}$.*

First, we show an exponential lower bound in a similar way to the proof of Theorem 1. Let T be a subset of V with $|T| = \lfloor d/2 \rfloor$. Define $g : 2^V \rightarrow D$ by

$$g(X) = \begin{cases} 0 & \text{if } X = \emptyset, \\ |X| & \text{if } \emptyset \neq X \subseteq T, \\ |T| + |T \cap X| & \text{otherwise.} \end{cases}$$

For a positive integer k with $2k \leq |T|$, let S be a subset of T with $|S| = 2k$. Define a function $g_S : 2^V \to D$ by

$$g_S(X) = \begin{cases} 2k - |X| & \text{if } X \subseteq S \text{ and } |X| \geq k+1, \\ g(X) & \text{otherwise.} \end{cases}$$

We can see that g and g_S are both posimodular.

If $k \geq 2$ and $|T| = 5k$ (and hence $d = 10k$), then by applying an argument similar to Lemma 2, we can observe that $\Omega(2^{0.1609d})$ oracle calls are necessary for solving the problem, which proves Theorem 6 (i).

Finally, we argue Theorem 6 (ii). Let $g : 2^V \to \{0,1\}$ be a function defined by $g(X) = 1$ if $X \neq \emptyset$, and $g(\emptyset) = 0$. For each element $v \in V$, define $g_v : 2^V \to \{0,1\}$ by

$$g_v(X) = \begin{cases} 0 & \text{if } X = \emptyset \text{ or } X = \{v\} \\ 1 & \text{otherwise.} \end{cases}$$

Note that both g and g_v are monotone and thus posimodular. Also note that the minimum g-value is 1, and each function g_v has exactly one minimizer $\{v\}$ with $g_v(v) = 0$ for $v \in V$. Observe that n oracle calls are necessary to distinguish functions in $\{g\} \cup \{g_v \mid v \in V\}$.

Remark. Modifying the construction by multiplying the value of function by a factor of $\rho + 1$ shows that obtaining an additive ρ-approximation also requires $\exp(d/(\rho + 1))$ oracle calls, matching Thm. 4.

4 Posimodular Function Maximization

In this section, we consider posimodular function maximization defined as follows.

> POSIMODULAR FUNCTION MAXIMIZATION
>
> Input: A posimodular function $f : 2^V \to \mathbb{R}_+$, (4.1)
>
> Output: A nonempty subset X of V maximizing f.

Here we assume that the optimal value $f(X^*)$ is also output. Similar to posimodular function minimization, the problem (4.1) is in general intractable.

Theorem 7. *Every algorithm for posimodular function maximization requires at least 2^{n-1} oracle calls in the worst case.*

For the case $f : 2^V \to \{0, 1, \ldots, d\}$ we have the following tight bound.

Theorem 8. *Posimodular function maximization for $f : 2^V \to \{0, 1, \ldots, d\}$ can be solved in $\Theta(n^{d-1} T_f)$ time.*

Acknowledgments: We would like to express our thanks to S. Fujishige, M. Grötschel, and S. Tanigawa for their helpful comments.

References

[1] K. Arata, S. Iwata, K. Makino, and S. Fujishige. Locating sources to meet flow demands in undirected networks. *Journal of Algorithms*, 42:54–68, 2002.

[2] C. Chekuri and A. Ene. Approximation algorithms for submodular multiway partition. In *IEEE 52nd Annual Symposium on Foundations of Computer Science*, pages 807–816, 2011.

[3] Egres open problem list. http://lemon.cs.elte.hu/egres/open/Maximizing_a_skew-supermodular_function.

[4] U. Feige, V. S. Mirrokni, and J. Vondrák. Maximizing non-monotone submodular functions. *SIAM J. Comput.*, 40:1133–1153, 2011.

[5] A. Frank. Augmenting graphs to meet edge-connectivity requirements. *SIAM Journal on Discrete Mathematics*, 5(1):25–53, 1992.

[6] S. Fujishige. A laminarity property of the polyhedron described by a weakly posi-modular set function. *Discrete Applied Mathematics*, 100(1-2):123–126, 2000.

[7] T. Ishii and K. Makino. Posi-modular systems with modulotone requirements under permutation constraints. *Discrete Mathematics, Algorithms and Applications*, 2(1):61–76, 2010.

[8] H. Ito, K. Makino, K. Arata, S. Honami, Y. Itatsu, and S. Fujishige. Source location problem with flow requirements in directed networks. *Optimization Methods and Software*, 18:427–435, 2003.

[9] E. L. Lawler. Cutsets and partitions of hypergraphs. *Networks*, 3(3):275–285, 1973.

[10] D. Lokshtanov and D. Marx. Clustering with local restrictions. *Information and Computation*, 222:278–292, 2013.

[11] H. Nagamochi. Graph algorithms for network connectivity problems. *Journal of the Operations Research Society of Japan*, 47(4):199–223, Dec 2004.

[12] H. Nagamochi. Minimum degree orderings. *Algorithmica*, 56(1):17–34, 2010.

[13] H. Nagamochi and T. Ibaraki. A note on minimizing submodular functions. *Inf. Process. Lett.*, 67(5):239–244, 1998.

[14] H. Nagamochi and T. Ibaraki. Polyhedral structure of submodular and posi-modular systems. *Discrete Applied Mathematics*, 107(1-3):165–189, 2000.

[15] H. Nagamochi, T. Shiraki, and T. Ibaraki. Augmenting a submodular and posi-modular set function by a multigraph. *Journal of Combinatorial Optimization*, 5(2):175–212, 2001.

[16] J. Orlin. A faster strongly polynomial time algorithm for submodular function minimization. *Mathematical Programming*, 118(2):237–251, 2009.

[17] M. Sakashita, K. Makino, H. Nagamochi, and S. Fujishige. Minimum transversals in posi-modular systems. *SIAM Journal on Discrete Mathematics*, 23:858–871, 2009.

[18] Z. Svitkina and É. Tardos. Min-max multiway cut. In *Approximation, Randomization, and Combinatorial Optimization*, pages 207–218, 2004.

[19] H. Tamura, H. Sugawara, M. Sengoku, and S. Shinoda. Plural cover problem on undirected flow networks. *IEICE Transactions*, J81-A:863–869, 1998. (in Japanese).

[20] J. van den Heuvel and M. Johnson. The external network problem with edge- or arc-connectivity requirements. In *Combinatorial and Algorithmic Aspects of Networking*, volume 3405 of *Lecture Notes in Computer Science*, pages 114–126. Springer, 2004.

[21] T. Watanabe and A. Nakamura. Edge-connectivity augmentation problems. *Journal of Computer System Sciences*, 35:96–144, 1987.

How to play hot and cold on a line

Herman Haverkort[1], David Kübel[2], Elmar Langetepe[2], and Barbara Schwarzwald[2]

[1] TU Eindhoven, Department of Mathematics and Computer Science
[2] University of Bonn, Department of Computer Science

Abstract. Suppose we are searching for a target point t in the unit interval. To pinpoint the location of t, we issue query points $q_1, \ldots, q_n \in [0, 1]$. As a response, we obtain an ordering of the query points by distance to t. This restricts possible locations of t to a subinterval. We define the *accuracy* of a query strategy as the reciprocal of the size of the subinterval to which we can pinpoint t in the worst case. We describe a strategy with accuracy $\Theta(n^2)$, which is at most a factor two from optimal if all query points are generated at once. With query points generated one by one depending on the response received on previous query points, we achieve accuracy $\Omega(2.29^n)$, and prove that no strategy can achieve $\Omega(3.66^n)$.

Keywords: search games, combinatorial optimization, target localization, online/offline strategies

1 Introduction

Imagine we want to set up receivers to locate an animal that carries a tracking device that sends signals. The strength of the signals may vary, but we know one thing: the further the distance from the animal, the weaker the signal. Or imagine a researcher conducting a survey, who wants to summarize respondents' political preferences by scoring them on several scales (for example, from conservative to progressive, or from favouring a small state to a large state). Respondents may not be able to score themselves but, given a number of hypothetical party programmes, they can rank them and say which one they like best.

In such settings, we are essentially searching for a target that is a point in a one- or higher-dimensional space. To pinpoint the location of the target, we issue queries (receivers, party programmes) that are points in the same configuration space. As a response, we obtain an ordering of the query points according to their distance to the target. From this we try to derive the location of the target with the highest possible accuracy—or conversely, we try to reach high accuracy with as few (expensive) queries as possible.

Searching for a stationary target is a common problem in computer science and applied mathematics. Strategies such as evenly distributed query points or binary search spring to mind immediately, but, as we will see in this paper, at least in certain abstract settings of the problem we can do much better. The problem of efficiently obtaining an order on a set of objects has been studied

© Springer International Publishing AG 2017
F. Ellen et al. (Eds.): WADS 2017, LNCS 10389, pp. 449–460, 2017.
DOI: 10.1007/978-3-319-62127-2_38

before in very general settings [3], but note that in our case, the cost measure is the number of query points that are used, not the number of comparisons that are made between them. The reconstruction of geometric objects based on a sequence of geometric probes (points, lines, hyperplanes, wedges, etc.) has also been investigated: the problem was introduced by Cole and Yap [2] and the main focus is also on the number of queries [5].

Specifically, we focus on the following setting. The target point t is a point located at an unknown position in the unit interval $[0, 1]$. To pinpoint the location of t, we may query the interval at points $q_1, \ldots, q_n \in [0, 1]$. As a response, we obtain an ordering of the points by ascending distance to t. This restricts possible positions of t to a subinterval bounded by bisectors of query points or an endpoint of the initial interval. We measure the efficiency or quality of a query strategy in terms of the reciprocal of the size of the subinterval in which the target t is found to lie. The worst-case of this reciprocal, that is, the minimum over all possible locations of t, is called the *accuracy* of the query strategy.

With respect to the frequency of the responses, we distinguish two variants. In the *one-shot* variant (Sec. 2), the response is only given after all points $q_1, \ldots q_n$ have been placed. In this case one needs to maximize the number of different, well-spaced bisectors. Such combinatorial questions are classical problems in discrete geometry; see for example [4]. In the *incremental* variant (Sec. 3), a response is given after each point placement, and may affect the choice of the next point. This enables a binary search strategy, but we can do much better than that. The problem can be interpreted as a game where an adversary tries to hide the target in the largest possible area. Geometric games about area optimization have a tradition in Computational Geometry; see for example [1].

For both variants we present an efficient strategy and an upper bound on the accuracy that can be achieved. In the second part of Section 2, we also address lower and upper bounds for the one-shot variant in two dimensions. In Section 4, we briefly discuss room for improvement and how the strategies can be extended to higher-dimensional settings.

2   One-shot strategies

First we consider the one-shot variant of the problem: only after generating n query points, we get to hear their ordering by distance to the target t. This pinpoints t to a subinterval bounded by bisectors of query points or an end point of the interval. As the target may lie in any subinterval, our problem is equivalent to minimizing the maximum size of such an interval.

2.1   The one-shot strategies on the unit interval

As n query points can produce at most $n(n-1)/2$ distinct bisectors, there are at most $n(n-1)/2 + 1$ intervals. Thus, we get the following (trivial) upper bound for the accuracy of one-shot strategies:

Theorem 1. *The accuracy of any one-shot strategy with n points is at most $\frac{1}{2} \cdot (n^2 - n + 2) \in O(n^2)$.*

We will now develop a strategy to get close to this upper bound. As a starting point, consider the following simple strategy, which we call EQUIDIST(n): place n evenly spaced query points $(q_1, q_2, \ldots, q_i, \ldots, q_n) := (0, 1/(n-1), \ldots, (i-1)/(n-1), \ldots, 1)$. The accuracy of EQUIDIST(n) is only $2(n-1)$ because many bisectors conincide. However, for $n \geq 7$ it is possible to forgo some of these query points, while the number of distinct bisectors, and so the accuracy, stays the same; see Figure 1a. Hence, for some function $\varphi(n)$ with $\varphi(n) > n$, we can achieve the same accuracy as EQUIDIST($\varphi(n)$) with n query points. Specifically, we now introduce a strategy GAP$_x(n)$ that will form a subset of the query points defined by EQUIDIST($\varphi_x(n)$) for $\varphi_x(n) := n(x+1) - 2x^2 - x - 2$. It uses only the following n points from EQUIDIST($\varphi_x(n)$): $q_1, q_2, \ldots, q_{x+1}$, as well as $q_{2x+1+k \cdot (x+1)}$ for $k \in \{0, \ldots, n - (2x+3)\}$, and $q_{\varphi_x(n)-x}, \ldots, q_{\varphi_x(n)-1}, q_{\varphi_x(n)}$. This results in widely spaced query points throughout the search range, and tightly spaced points near both ends, omitting at most x consecutive query points from EQUIDIST($\varphi_x(n)$). Most bisectors will then be formed by one of the tightly spaced points and one of the widely spaced points.

Lemma 1. *For $x, n \in \mathbb{N}$ with $n \geq (2x+3)$, the one-shot query strategy GAP$_x$ has accuracy $2(\varphi_x(n) - 1)$.*

Proof. It suffices to show that every distinct bisector from EQUIDIST($\varphi_x(n)$) is also created by GAP$_x(n)$, as the points chosen by GAP$_x(n)$ are a subset of those chosen by EQUIDIST($\varphi_x(n)$). Hence, for each $1 \leq i < \varphi_x(n)$, GAP$_x(n)$ must choose two points that form a bisector in the middle between q_i and q_{i+1}. For each $1 < j < \varphi_x(n)$, GAP$_x(n)$ must choose two points that form a bisector directly on q_j.

For $i \leq x$ and $j \leq x$ this is given by $q_1, q_2, \ldots, q_{x+1}$. For $x < i < \varphi_x(n)/2$, EQUIDIST($\varphi_x(n)$) forms a bisector between q_i and q_{i+1} with all of the pairs $(q_1, q_{2i}), (q_2, q_{2i-1}), \ldots, (q_{x+1}, q_{2i-x})$. One of these pairs has to be in the chosen subset of GAP$_x(n)$, since the choosing process omits at most x consecutive points. Likewise, one of the pairs $(q_1, q_{2j-1}), (q_2, q_{2j-2}), \ldots, (q_{x+1}, q_{2j-(x+1)})$ must have been chosen by GAP$_x(n)$. Those pairs form a bisector on q_j for $x < j \leq \varphi_x(n)/2$. The existence of the remaining bisectors, that is, those in the right half of the unit interval, follows by the symmetry of the strategy.

Hence GAP$_x(n)$ produces the same set of $2(\varphi_x(n) - 2)$ distinct equidistant bisectors as EQUIDIST($\varphi_x(n)$), resulting in an accuracy of $2(\varphi_x(n) - 1)$. $\square$

It remains to choose the optimal x, given n, maximizing $\varphi_x(n)$. As the slope of the linear function $\varphi_x(n)$ increases with x, GAP$_x$ is eventually surpassed by GAP$_{x+1}$. The break-even point between GAP$_x$ and GAP$_{x+1}$ is at $n = 4x + 3$. Hence, the optimal $x = \lceil (n-3)/4 \rceil$ and we get:

Theorem 2. *For any number of points $n \geq 3$ and $x = \lceil (n-3)/4 \rceil$, the one-shot strategy GAP$_x(n)$ has an accuracy of $2(\varphi_x(n) - 1) \geq (n^2 + 6n - 27)/4 \in \Omega(n^2)$.*

This leaves only a gap of a constant factor two between the lower and upper bound. Note that it is (already) impossible to place four query points in such a way that their bisectors are all distinct and divide the unit interval into equal parts. This shows that the upper bound for the one-shot strategy is not tight.

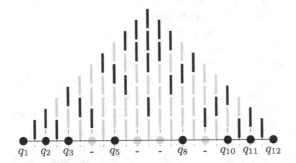
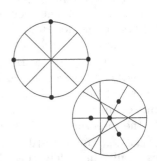

(a) Removing q_4, q_6, q_7 and q_9 from EQUIDIST(12) on the unit interval does not affect the accuracy.

(b) GAP(4) 2d-version above, an optimal placement below?

Fig. 1: Examples for the one-shot variant of the problem: 1a considers the strategies EQUIDIST(12) and GAP$_2$(8) on the unit interval. Possible placements of 4 query points on the unit disk are given in 1b.

2.2   One-shot strategies in two dimensions

Now let us turn our attention to the two dimensional case, where we use the diameter of the resulting region as the measure for accuracy[3]. So assume the target hides somewhere in the unit square or the unit disk. Similar to the one-dimensional case, we obtain quadratic lower and upper bounds for the accuracy of one-shot strategies for both types.

Theorem 3. *The accuracy of any two-dimensional one-shot strategy on the unit square (or unit disk) that places n query points is $O(n^2)$.*

Proof. Using n query points, one can construct at most $n(n-1)/2$ bisectors. Consequently, the number of cells in the arrangement of bisectors is at most $(n^4 - 2n^3 + 3n^2 - 2n + 8)/8$. A cell of diameter δ covers an area of at most $\pi\delta^2/4$. This gives us an upper bound on the area that all cells can cover together if their maximum diameter is δ. In turn, this gives us a lower bound $\delta \geq 4\sqrt{2} \cdot \left(\pi(n^4 - 2n^3 + 3n^2 - 2n + 8)\right)^{-1/2} \in \Omega(n^{-2})$ on the smallest maximum diameter that can be achieved. Hence the accuracy is $O(n^2)$.

The same argument can be applied for the unit circle.   □

---

[3] One can argue to use the area of the region instead. This however can result in very thin and long regions contradicting the aim to precisely locate the target, as the diameter limits the area but not vice versa.

Theorem 4. *Any target on the unit square (or unit disk) can be located with accuracy* $\Omega(n^2)$ *in the one-shot variant with n query points.*

Proof. Define $k := \lfloor n/2 \rfloor$. As we want to prove a lower bound for the accuracy, we may choose to use $2k$ query points and ignore the remaining ones. We may choose our coordinate system such that the center of the unit square (or unit disk) lies at $(0,0)$. Then, we apply $\text{GAP}_x(k)$ along each axis of the coordinate system. Now consider only the bisectors perpendicular to the coordinate axis. These bisectors subdivide the unit square (unit disk) into small rectangles (or parts of small rectangles), each of height and width at most $1/2 \cdot (\varphi_x(k) - 1)^{-1}$, by Lemma 1. Consequently, the diameter of each square is at most $4\sqrt{2}\left(k^2 + 6k - 27\right)^{-1}$, which gives the quadratic lower bound for the accuracy in terms of n. $\qquad\square$

It is not clear whether GAP_x is optimal in the one-dimensional case. It is certain however, that the GAP-strategy extended to the unit disk is not, not even for even n. Instead of placing the queries along the coordinate axis, it might be better to place one of the query points in the center of the disk; see Figure 1b.

3 Incremental and online strategies

In this section we consider incremental strategies, that is, before placing any query point q_i we may learn the ordering of $q_1, ..., q_{i-1}$ by distance to the target, and we may choose the location of q_i depending on that information. We will show upper and lower bounds on the accuracy that can be achieved.

3.1 A strategy with high accuracy

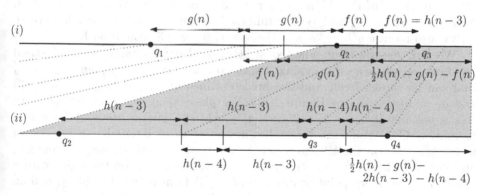

Fig. 2: Location of the first query points for our incremental strategy to locate a target with accuracy $h(n)$.

A simple incremental strategy could choose the query points such that the interval containing the target is halved in each step, except the first. Thus, with

n query points, we achieve accuracy 2^{n-1}. But we can do much better, with a recursive strategy that takes advantage of more than one new bisector in many steps. Let $h(n)$ be the accuracy we aim for, as a function of the number n of query points we get to place. For ease of description, we use the interval $[0, h(n)]$ instead of the unit interval and pinpoint the target to an interval of length 1. The recursion in our strategy depends on a number of conditions, labelled A to E and presented below.

Our overall strategy to locate a target in an interval $[0, h(n)]$ with n query points starts with placing the first two query points symmetrically, otherwise the overall strategy can easily be improved.

Let $g(n)$ be the distance of the first two query points to their common bisector, so we place q_1 at $\frac{1}{2}h(n) - g(n)$ and q_2 at $\frac{1}{2}h(n) + g(n)$. Now suppose the target lies to the right of the bisector (the other case is symmetric). We now place q_3 at some distance $2f(n)$ to the right of q_2, and find that the target lies in one of three intervals (see Figure 2(i)):

(i) $[\mathrm{bs}(q_1, q_2), \mathrm{bs}(q_1, q_3)] = [\frac{1}{2}h(n), \frac{1}{2}h(n) + f(n)]$, of size $f(n)$.
(ii) $[\mathrm{bs}(q_1, q_3), \mathrm{bs}(q_2, q_3)] = [\frac{1}{2}h(n) + f(n), \frac{1}{2}h(n) + g(n) + f(n)]$, of size $g(n)$, or
(iii) $[\mathrm{bs}(q_2, q_3), h(n)] = [\frac{1}{2}h(n) + g(n) + f(n), h(n)]$, of size $\frac{1}{2}h(n) - g(n) - f(n)$.

To be able to apply our strategy recursively in the first interval, using the remaining $n - 3$ query points, we choose $f(n) = h(n - 3)$. For the third interval, of width $\frac{1}{2}h(n) - g(n) - h(n - 3)$, we use an adapted strategy, explained below, that places the remaining $n - 3$ query points in a way that exploits the previously placed query points q_2 and q_3 at distance $f(n) = h(n - 3)$ left and right of its left boundary. The same strategy can be applied symmetrically to the second interval, provided the second interval is not larger than the third, that is, $g(n) \le \frac{1}{4}h(n) - \frac{1}{2}h(n - 3)$ for $n \ge 3$ (condition **A**). Note that condition A also ensures that q_1 and q_2 lie within the interval $[0, h(n)]$. To be able to place q_3, we also require $g(n) + 2h(n - 3) \le \frac{1}{2}h(n)$ for $n \ge 3$ (condition **B**).

We now describe our adapted strategy to locate a target in an interval $[0, \frac{1}{2}h(n) - g(n) - h(n-3)]$ (modulo translation) with $n-3$ query points $q_4, \ldots, q_n$ that can be chosen freely and two predetermined query points $q_2 = -h(n-3)$ and $q_3 = h(n-3)$ (see Figure 2(ii)). We place q_4 at $h(n-3) + 2h(n-4)$; this is possible if $h(n-3) + 2h(n-4) \le \frac{1}{2}h(n) - g(n) - h(n-3)$ for $n \ge 4$ (condition **C**). Again, the target lies in one of three intervals: an interval of width $h(n-4)$ on the left, to which we apply the overall strategy recursively, an interval of width $\frac{1}{2}h(n) - g(n) - 2h(n-3) - h(n-4)$ on the right, with a predetermined query point at distance $h(n-4)$ from its left boundary, and an interval of width $h(n-3)$ in the middle, with a predetermined query point at distance $h(n-4)$ from its right boundary. We can apply the adapted strategy recursively to the rightmost interval, using the remaining $n-4$ points, provided $\frac{1}{2}h(n) - g(n) - 2h(n-3) - h(n-4) = \frac{1}{2}h(n-1) - g(n-1) - h(n-4)$ for $n \ge 4$ (condition **D**), and we can apply the adapted strategy recursively to the middle interval if $h(n-3) \le \frac{1}{2}h(n-1) - g(n-1) - h(n-4)$ for $n \ge 4$ (condition **E**).

It remains to choose h and g as functions of n:

- such that the conditions for recursive application are satisfied, so for $n \geq 4$ we must have:
  - (A) $g(n-1) \leq 1/4 \cdot h(n-1) - 1/2h(n-4)$;
  - (B) $g(n-1) \leq 1/2 \cdot h(n-1) - 2h(n-4)$;
  - (C) $g(n) \leq 1/2 \cdot h(n) - 2h(n-3) - 2h(n-4)$;
  - (D) $1/2 \cdot h(n) - g(n) - 2h(n-3) = 1/2 \cdot h(n-1) - g(n-1)$;
  - (E) $g(n-1) \leq 1/2 \cdot h(n-1) - h(n-3) - h(n-4)$;
- such that, when no further recursion is possible because we have run out of query points, the remaining interval is small enough: $h(0) \leq 1$, $h(1) \leq 1$, $h(2) \leq 2$, and $\frac{1}{2}h(3) - g(3) - h(0) \leq 1$.

Theorem 5. *There is an incremental strategy to locate a target in a unit interval with accuracy $\Omega(b^n)$, where $b > 2.2993$ is the largest root of $b^4 - b^3 - 6b - 2$. The accuracy $h(n)$ is given by the recursion formula $h(0) = 1$, $h(1) = 1$, $h(2) = 2$, $h(3) = 6$ and $h(n) = h(n-1) + 6h(n-3) + 2h(n-4)$ for $n \geq 4$.*

Proof. We will prove that for the above recursion formula we can find suitable values $g(3), g(4), \ldots$ such that condition A to E and $\frac{1}{2}h(3) - g(3) - h(0) \leq 1$ hold. Thus by construction, we obtain a strategy of accuracy $h(n)$.

Since h is a non-decreasing function of n, condition B is redundant, as it is already implied by condition E. Rewriting condition D to $g(n) = g(n-1) + \frac{1}{2}(h(n) - h(n-1)) - 2h(n-3)$ and substituting it into condition C, learns that condition C is equivalent to condition B, and equally redundant. We now satisfy condition A by simply choosing equality $g(n-1) = \frac{1}{4}h(n-1) - \frac{1}{2}h(n-4)$ for $n \geq 4$. Substituting this in the remaining conditions we get the following. Condition D becomes $h(n) = h(n-1) + 6h(n-3) + 2h(n-4)$ for $n \geq 4$ (which is among the conditions of the recursion formula in the theorem). Condition E becomes $h(n-1) \geq 4h(n-3) + 2h(n-4)$, by using $g(n-1)$ this is equivalent to $h(n-3) \leq g(n-1)$.

Altogether, the remaining task is that for $h(0) = 1$, $h(1) = 1$, $h(2) = 2$, and $g(n-1) = \frac{1}{4}h(n-1) - \frac{1}{2}h(n-4)$ and $h(n) = h(n-1) + 6h(n-3) + 2h(n-4)$ for all $n \geq 4$ we have to prove that $\frac{1}{2}h(3) - g(3) - h(0) \leq 1$ and $h(n-3) \leq g(n-1)$ for $n \geq 4$ or equivalently $g(n) \geq h(n-2)$ for $n \geq 3$.

First, we require suitable values for $g(3)$ and $h(3)$ which are derived from $g(3) = \frac{1}{4}h(3) - \frac{1}{2}h(0)$ and $\frac{1}{2}h(3) - g(3) - h(0) \leq 1$. The statement holds for $g(3) = 1$ and $h(3) = 6$. The final task is to prove condition E or $h(n-3) \leq g(n-1)$ for $n \geq 4$. From the above formulas we have $h(2) = 2$, $h(3) = 6$, $h(4) = 14$, $g(4) = 3$ and $g(5) = 6$. This means $g(3) \geq h(1)$, $g(4) \geq h(2)$ and $g(5) \geq h(3)$. (Due to space limitations, the proof of this condition is omitted.)

Thus, the strategy attains accuracy $h(n) = h(n-1) + 6h(n-3) + 2h(n-4)$. A closer analysis of the recursion formula reveals that the growth is dominated by $b = 2.29936\ldots$, the greatest root of the polynomial $b^4 - b^3 - 6b - 2$. $\square$

Note that for sake of simplicity the above strategy is not quite optimal. If we recurse with only few points left to place, we can sometimes do slightly better by taking some already placed points into account. For example, $h(2) = 2$ because, if we get to place only two points, we can only create one bisector and therefore we cannot achieve accuracy better than 2. However, if we start with more than 3 points and eventually we get to a final interval in which we get to place the last two points, then each of these two points can form a bisector together with a previously placed query point outside the interval. This allows us to assume $h(2) = 3$ in the recursive definition of $h(n)$ for $n \geq 4$, which slightly improves the constant factor in the asymptotic bound.

In the on-line setting, the strategy is to take $h(n) = 2b^{n-2}$ and $g(n) = 1/4 \cdot h(n) - 1/2 \cdot h(n-3) = (1/2 - b^{-3})b^{n-2}$. Thus, the locations of $q_1, \ldots, q_4$, relative to the size of the interval, $h(n)$, are independent of n, and can therefore be decided without advance knowledge of n.

Corollary 1. *Theorem 5 also holds in the on-line setting, where n is unknown until the last query point has been placed.*

3.2 Upper bound

By placing a query point q_i, we obtain $i - 1$ new bisectors of q_i with previously placed query points. We learn the rank of q_i among $q_1, \ldots, q_i$ with respect to the distance to the target, which tells us where the target is with respect to the new bisectors. This reduces the interval R where the target must be to one of at most i subintervals of R, at least one of which must have size at least $1/i$ times the size of R. Therefore, for any incremental strategy, there must be targets for which we can increase the accuracy with a factor at most i with every query point q_i, leading to an accuracy of at most $n!$ after placing n query points. However, this bound is far from tight. We can show a much stronger upper bound on the accuracy that can be achieved:

Theorem 6. *No deterministic, incremental strategy can locate an arbitrary target with an accuracy $\Omega(3.66^n)$, where n is the number of query points used.*

The proof of this theorem is based on the following insight. Suppose, for example, that we have placed $n - 1$ query points, and placing another query point q_n generates two new bisectors, which divide the target interval into three equal parts. Thus, the current accuracy improves by a factor three.

However, this large increase in accuracy requires two previously placed query points q_i and q_j that satisfy the following conditions. First, the points q_i and q_j are consecutive in the left-to-right order of all previously placed points. Second, the distance between the bisectors $\mathrm{bs}(q_n, q_i)$ and $\mathrm{bs}(q_n, q_j)$ is $1/3$ of the current size of the target interval. Therefore, the distance between q_i and q_j must be $2/3$ of the current size of the target interval.

The previous example illustrates that, for a large increase in accuracy, one needs to have *close pairs*: pairs of previously placed query points with a (comparatively) small distance in between. The example also illustrates that, as the accuracy increases, close pairs can stop being close as their distance relative to the current accuracy increases. Moreover, with the placement of each query point, one can create at most two new close pairs: one with the left neighbour and one with the right neighbour among all query points placed so far.

Definition 1 (accuracy gain). *Let R_n denote the interval in which the target is known to lie after q_n has been placed. Further, let r_n be the size of R_n and a_n the corresponding accuracy $1/r_n$. With $g_n := a_n/a_{n-1} = r_{n-1}/r_n$, we denote the increase of accuracy achieved by placing q_n.*

To create a close pair, one must place a query point q_n close to a previously placed query point, but then the newly created bisectors are also close to the previously created bisectors. In particular, if a close pair is created, most new bisectors inside R_{n-1} will be relatively close to the boundary of R_{n-1} and away from the centre of R_{n-1}. Thus, creating close pairs cannot go together with substantial accuracy gains if the target is near the centre of R_{n-1}. In effect, close pairs are a resource whose scarcity limits the accuracy that can be obtained. We will now make this insight more precise.

Definition 2 (d-close pair). *A d-close pair is a pair of previously placed query points that are consecutive in their left-to-right order and have a distance strictly less than d divided by the current accuracy. Further, let $p_n(d)$ denote the number of d-close pairs after q_n has been placed.*

The following three lemmas capture the trade-off between on the one hand, creating close pairs, and on the other hand, using and losing them while pin-pointing the location of the target. Each lemma assumes that query points $q_1, ..., q_n$ have been placed in response to learning $R_0, ..., R_{n-1}$, which we now consider to be fixed. Within this context, R_n is still a variable: the bisectors of q_n with the previously placed query points subdivide R_{n-1} into a set $\mathcal{R}$ of subintervals, and $R_n \in \mathcal{R}$ is the subinterval among them that contains t. In turn, g_n and $p_n(d)$ are functions of R_n. Each lemma assumes a certain accuracy gain g_n and derives a bound for the net change in the number of d-close pairs $p_n(d)$, for all or for some possible values of R_n. The lemmas address different levels of accuracy gains: *small, medium* and *large*. (Some of the proofs are omitted due to space limitations.)

Lemma 2 (small accuracy gain). *If $g_n \geq 1 + d/2$ for all possible $R_n \in \mathcal{R}$, then $p_n(d) \leq p_{n-1}(d) + 1$ for all possible $R_n \in \mathcal{R}$.*

Now, assume that the accuracy gain is somewhat higher than what we assumed in the previous lemma. In this case no additional d-close pairs can be created at all, as the following lemma shows. Note that it might even be the case that several d-close pairs are lost.

Lemma 3 (medium accuracy gain). *If $g_n \geq 2 + d/2$ for all possible $R_n \in \mathcal{R}$, then $p_n(d) \leq p_{n-1}(d)$ for all possible $R_n \in \mathcal{R}$.*

Lemma 4 (large accuracy gain). *For any $c > 1$ and h, d such that $h \geq 2 + d/2$ and $2/h \leq d \leq 2$, the following implication holds:*
If $g_n > h$ for all possible $R_n \in \mathcal{R}$, then there is a possible interval $R_n \in \mathcal{R}$ and a $k > 0$, such that $g_n < c^k/(c-1)(1-2/h)$ and $p_n(d) \leq p_{n-1}(d) - k$.

Proof. Suppose that the placement of q_n subdivides R_{n-1} into $m+2$ regions. Note that $m \geq 1$, since otherwise it would not be possible to achieve an increase in accuracy that is strictly greater than $h > 2$. Among these $m+2$ regions, each of size less than r_{n-1}/h, there is a leftmost region and a rightmost region. We call the other m regions *interior* regions and label them $I_1, \ldots, I_m$, in order of decreasing size. Subtracting the size of the leftmost and rightmost region from r_{n-1}, we conclude that the total size of the interior regions is more than $(1 - 2/h)r_{n-1}$. Since $c > 1$ we may also conclude

$$\sum_{k=1}^{m} |I_k| > \left(1 - \frac{2}{h}\right) r_{n-1} > \left(1 - \frac{2}{h}\right) r_{n-1} \cdot \left(1 - \frac{1}{c^m}\right)$$

$$= \left(1 - \frac{2}{h}\right) r_{n-1} \cdot (c - 1) \cdot \left(\frac{c^{-(m+1)} - 1}{c^{-1} - 1} - 1\right)$$

$$= \sum_{k=1}^{m} \frac{(c - 1)(1 - 2/h)r_{n-1}}{c^k},$$

by using the geometric series. Therefore, there must be a k such that the k-th largest interior region has size more than $c^{-k}(c-1)(1-2/h)r_{n-1}$. If R_n is that k-th largest region, the accuracy gain is upper-bounded by $c^k(c-1)^{-1}(1-2/h)^{-1}$.

It remains to show that in this case, the number of d-close pairs decreases by at least k. Since $h \geq 2 + 2/d$, by Lemma 3, no d-close pairs are created. Each interior region I_i with $i \in \{1, \ldots, k\}$ is bounded by the bisectors of q_n with two previously placed query points q_- and q_+, which were neighbours in the left-to-right order. The distance between q_- and q_+ is

- less than $2r_{n-1}/h \leq d \cdot r_{n-1}$, since, by the conditions of the lemma, I_i has size less than r_{n-1}/h;
- at least $2r_n \geq d \cdot r_n$, since I_i is at least as big as $R_n = I_k$.

Thus, q_- and q_+ constitute a d-close pair before, but not after placing q_n.   □

We can now derive the following formula that expresses the cumulative effects of Lemmas 2, 3 and 4 for a complete deterministic query strategy:

Lemma 5. *For any h, d, c, and w such that $h \geq 2 + d/2$ and $2/h \leq d \leq 2$ and $1 < c \leq w$, for each non-negative integer n, it holds that*

$$a_n w^{p_n(d)} \leq b^n$$

for certain locations of the target, where

$$b = \max\left\{ \left(1 + \frac{d}{2}\right) w^2, \quad \left(2 + \frac{d}{2}\right) w, \quad h, \quad \frac{c}{w(c-1)(1 - 2/h)} \right\}.$$

Proof (by induction). If $n = 0$, we have $a_0 = 1, p_0(d) = 0$ and the claim holds.

Now suppose we add another query point q_n. Distinguishing by the minimum accuracy gain that is achieved for any target in R_{n-1}, one of the following cases must hold:

- Lemma 4 applies. Then, there is a possible $R_n \in \mathcal{R}$ and a k such that $g_n < c^k(c-1)^{-1}(1-2/h)^{-1}$ and $p_n(d) \leq p_{n-1}(d) - k$. Using the induction hypothesis, we conclude

$$a_n w^{p_n(d)} < \frac{c^k \cdot w^{-k}}{(c-1)(1-2/h)} \cdot a_{n-1} w^{p_{n-1}(d)} \leq \frac{c}{w(c-1)(1-2/h)} \cdot b^{n-1} \leq b^n.$$

Note that here we use the condition $c \leq w$; without that condition we would not be able to bound $c^k w^{-k}$ to c/w, or to any other constant.
- Lemma 3 applies. So $p_n(d) \leq p_{n-1}(d)$ and, for some $R_n \in \mathcal{R}$, we have $g_n \leq h$, which implies $a_n \leq h \cdot a_{n-1}$. Using the induction hypothesis, we may conclude $a_n w^{p_n(d)} \leq h a_{n-1} w^{p_{n-1}(d)} \leq b \cdot b^{n-1} = b^n$.
- Lemma 2 applies. So $p_n(d) \leq p_{n-1}(d) + 1$ and, for some $R_n \in \mathcal{R}$, we have $g_n < (2 + d/2)$, which implies $a_n < (2 + d/2)a_{n-1}$. With the induction hypothesis, we get $a_n w^{p_n(d)} < (2 + d/2)w a_{n-1} w^{p_{n-1}(d)} \leq b \cdot b^{n-1} = b^n$.
- None of the Lemmas 2, 3, 4 apply. We have $p_n(d) \leq p_{n-1}(d) + 2$, because each new query point creates at most two d-close pairs, and $g_n < (1 + d/2)$, so $a_n w^{p_n(d)} < (1 + d/2)w^2 a_{n-1} w^{p_{n-1}(d)} \leq b \cdot b^{n-1} = b^n$.

This establishes the induction step, and proves the lemma. $\qquad\square$

It remains to pick h, d, c and w such that we get the most out of Lemma 5. Our goal is to prove a strong upper bound, so we want to minimize b. As can be seen directly in the definition of b, it cannot hurt to choose d as small as possible and c as large as possible, so we choose $c = w$ and $d = 2/h$ (this also makes the condition $h \geq 2 + d/2$ as permissive as possible). It remains to minimize:

$$\max \left\{ \left(1 + \frac{1}{h}\right) w^2, \quad \left(2 + \frac{1}{h}\right) w, \quad h, \quad \frac{1}{(w-1)(1-2/h)} \right\},$$

subject to $h \geq 2 + 1/h$ and $w > 1$. The maximum of the last three terms is minimized when all are equal—if either one or two of them would be higher than the other, it would be possible to adjust h and w slightly such that the higher terms are lowered. Solving for w and h gives us that the last three terms are equal when h is the largest root of $h^3 - 4h^2 + h + 1$, so h is slightly smaller than 3.6511 and $w = h/(2 + 1/h)$ is approximately 1.6057. The first term is of no concern since for these values of h and w, which minimize the other three terms, the first term is even smaller. Thus we get $b = h \approx 3.6511$.

Since $p_n(d)$ cannot be negative, this implies that there are targets for which $a_n \leq a_n w^{p_n(d)} \leq b^n$ for all n, and thus, such targets are not located with accuracy better than $\Omega(b^n)$. This concludes the proof of Theorem 6.

4   Conclusions and outlook

For the one-shot variant of our location problem, we presented an upper bound and a constructive lower bound on the accuracy that can be achieved. The remaining constant-factor gap between these bounds may be due to the fact that most bisectors formed by pairs of both tightly or both widely spaced points do not contribute anything to the accuracy. However, the upper bound might also be improved.

For the incremental variant, we obtained non-trivial upper and lower bounds but a gap that is exponential in n still remains. Our search strategy does not seem to leave any space for substantial improvements, so further progress must come from an entirely new search strategy, or from tightening the upper bound. Observe that in each step, our search strategy only uses the bisectors that are formed with the two previously placed query points that are closest to the target. If we could prove that there is an optimal search strategy with this property, then this would immediately improve the upper bound to $O(3^n)$.

To search a d-dimensional unit cube, we could place roughly n/d query points on each coordinate axis in a round-robin fashion; on each axis, we place the points according to the one-shot, incremental, or on-line incremental strategy. Thus, where we obtain accuracy $h(n)$ in one dimension, we can pinpoint a target to a cube of width $1/h(\lfloor n/d \rfloor)$ in d dimensions. This approach, however, fails to take advantage of bisectors between query points placed on different axes, not to mention query points placed more freely. Indeed, for $d = 2$, there are better solutions for $n \in \{3, 4\}$.

Acknowledgements

We thank all participants of the 2016 Lorentz Center Workshop on Search Games, and, in particular, Bengt Nilsson and Endre Csóka, for their invaluable contributions in defining the problem and initial solutions.

References

1. H.-K. Ahn, S.-W. Cheng, O. Cheong, M. Golin, and R. van Oostrum. Competitive facility location: the Voronoi game. *Th. Comp. Sc.*, 310(1):457–467, 2004.
2. R. Cole R. and Ch. K. Yap. Shape from probing. *J. Alg.*, 8(1):19–38,1987.
3. J. Kahn and M. Saks Balancing poset extensions *Order*, 1: 113–126, 1984.
4. J. Matoušek. *Lectures on discrete geometry*, Springer, New York, 2002.
5. S. Skiena. Interactive reconstruction via geometric probing. *Proc. of the IEEE*, 80(9):1364–1383, 1992.

Faster Algorithm for Truth Discovery via Range Cover*

Ziyun Huang[1]   Hu Ding[2]   Jinhui Xu[1]

[1] Department of Computer Science and Engineering
State University of New York at Buffalo
{ziyunhua, jinhui}@buffalo.edu
[2] Department of Computer Science and Engineering
Michigan State University
huding@msu.edu

Abstract. Truth discovery is a key problem in data analytics which
has received a great deal of attention in recent years. In this problem,
we seek to obtain trustworthy information from data aggregated from
multiple (possibly) unreliable sources. Most of the existing approaches
for this problem are of heuristic nature and do not provide any quality
guarantee. Very recently, the first quality-guaranteed algorithm has been
discovered. However, the running time of the algorithm depends on the
spread ratio of the input points and is fully polynomial only when the
spread ratio is relatively small. This could severely restrict the applicability of the algorithm. To resolve this issue, we propose in this paper
a new algorithm which yields a $(1 + \epsilon)$-approximation in near quadratic
time for any dataset with constant probability. Our algorithm relies on
a data structure called range cover, which is interesting in its own right.
The data structure provides a general approach for solving some high
dimensional optimization problems by breaking them down into a small
number of parametrized cases.

1 Introduction

Truth discovery is an important problem arising in data analytics, and has received a great deal of attentions in recent years in the fields of data mining,
database, and big data [3, 6–8, 4, 9–11]. Truth discovery seeks to find trustworthy information from a dataset acquired from a number of sources which may
contain false or inaccurate information. There are numerous applications for this
problem. For example, the latest search engines are able to answer user queries
directly, instead of simply listing webpages that might be relevant to the query.
This process involves retrieving answers from potentially a large number of related webpages. It is quite common that these webpages may provide inaccurate
or inconsistent information. Thus a direct answer to the query needs the search

---

* The research of the first and third authors was supported in part by NSF through
grants CCF-1422324, IIS-1422591, and CNS-1547167. The research of the second
author was supported by a start-up fund from Michigan State University.

F. Ellen et al. (Eds.): WADS 2017, LNCS 10389, pp. 461–472, 2017.
DOI: 10.1007/978-3-319-62127-2_39

engine to be able to extract the most trustworthy information from all these webpages, which is exactly the problem of truth discovery.

Truth discovery is an unsupervised learning problem. Besides the input data, no prior knowledge about the reliability of each data source is provided. In such settings, an intuitive approach is to view all data sources equally reliable and obtain the solution by averaging or majority rule. A major issue of this approach is that the yielded answer may be quite far away from the truth. This is because a small number of unreliable data sources could significantly deviate the final solution. To deal with this issue, truth discovery treats data sources differently by estimating the reliability for each of them. This greatly increases the level of challenge for the problem. Moreover, since the truth discovery problem often occurs in big data scenarios, the number of data sources could be quite large and the dimensionality of the data could be rather high, which brings another dimension of challenge to the problem.

A widely accepted geometric modeling of the truth discovery problem is the follows. Data from each source is formulated as a set of real number attributes, and thus can be viewed as a vector in $\mathbb{R}^d$, where d is the number of attributes. Each data source is associated with a positive variable (or weight) representing its reliability. Formally, the truth discovery problem can be defined as follows.

Definition 1. *(Truth Discovery [4, 8]). Let $P = \{p_1, p_2, \ldots p_n\}$ be a set of points in $\mathbb{R}^d$ space, where each p_i represents the data acquired from the i-th source among a set of n sources. The truth discovery problem is to find the truth vector p^* and w_i (i.e., reliability) for each i-th source such that the following objective function is minimized.*

$$\min \Sigma_{i=1}^{n} w_i \|p_i - p^*\|^2, s.t. \Sigma_{i=1}^{n} e^{-w_i} = 1. \tag{1}$$

The meaning of the above truth discovery formulation was discussed in [1] from an information theory's point of view. It is shown that the constraint on w_i in Definition 1 ensures that the **entropy** is minimized when p^* approaches the truth vector. For this reason, the problem is also called *Entropy based Geometric Variance* problem [1].

Despite extensive studies on this problem, most of the existing techniques are of heuristic nature, and do not provide any guarantee on the quality of solution. It is not until very recently that the truth discovery problem has a theoretically guaranteed solution [1]. This result ensures that a $(1 + \epsilon)$-approximation of the problem can be achieved in $O(dn^2 + (n\Delta)^\sigma nd)$ time, where n is the number of input points (*i.e.*, data sources), d is the dimensionality of the space, Δ is the spread ratio of the input points (*i.e.* the ratio of the largest distance between any two input points to the smallest distance), and σ is any fixed small positive number. The result is based on an elegant sampling technique called Simplex Lemma [2] which is capable of handling high dimensional data. A main issue of this method is that its running time depends on the spread ratio of the input points, and is polynomial only when the spread ratio is relatively small (*i.e.*, $\Delta = O(\sqrt{n})$). This could severely restrict its applicability.

To overcome this main issue, we present in this paper a faster algorithm for the truth discovery problem. With constant probability, our algorithm achieves a $(1 + \epsilon)$-approximation in $O(dn^2(\log n + \log d))$ time, and is completely independent of the spread ratio. Our algorithm is also space efficient, using only near linear space, while the space complexity of [1] also depends on the spread ratio. Our algorithm relies on a new data structure called *range cover*, which is interesting in its own right. Roughly speaking, range cover is a data structure designed for a class of optimization problems (in high dimensional space) which are decomposable into a number of "easier" cases, where each case can be characterized by a parameterized assumption. For example, truth discovery can be formulated as a problem of finding a truth vector $p^* \in \mathbb{R}^d$ from a given set P of points in $\mathbb{R}^d$ so that a certain objective function (the exact formulation will be discussed later) is minimized. We are able to show that although directly optimizing the objective function is challenging, the problem is much easier to solve if some additional information (*e.g.*, the distance r between p^* and P) is known. Thus, by viewing the additional information as a parameterized assumption, we can solve the truth discovery problem by searching for the best assumption. The range cover data structure shows that even though the number of parameterized assumptions could be very large (or even infinite), it is sufficient to sample only a small number of assumptions to ensure an approximate solution. This leads to a small-size data structure (*i.e.*, $O(n \log n)$ space) and a faster algorithm for truth discovery. Since the idea of decomposing problem into cases is not restricted only to the truth discovery problem, we expect that this data structure will provide new approaches to other problems.

2 Range Cover Data Structure

In this section, we present the aforementioned range cover data structure.

Range cover is motivated by several high dimensional optimization problems (such as truth discovery). In these problems, an input point set P is given in $\mathbb{R}^d$ space, and the objective is to find a point q in $\mathbb{R}^d$ so that a certain objective function is optimized. A commonly used approach for such problems is to examine a number of candidate points selected by some algorithms. But directly applying such an approach could require too many (*e.g.*, exponential in d) points to be examined in high dimensional space. A possible way to overcome this difficulty is to characterize all possibilities of q into a small number of cases so that in each case q is associated with a certain parametrized assumption which could help solve the problem more efficiently. For instance, in some optimization problem, q could be much easier to obtain if we know in advance the nearest neighbor (say p) of q in P and its distance r to q (*i.e.*, $\|p - q\| = r$) for some parameter r. We expect that these parameterized assumptions form a space with much lower dimensionality than d, and thus the overall time complexity can be significantly reduced.

From the above discussion we know that for the range cover data structure to be efficient, the problem needs to be decomposable into a small number of "easier" cases. For this purpose, we will take advantage of the distribution of the points in P, such as their locality and point aggregation properties. To understand how point aggregation can be useful, consider the following parameterized assumption on q: Assume that p is the nearest neighbor of q in P and r is their distance. Denote this assumption by $\mathcal{NN}_q(p,r)$. If a subset of points, $v = \{p_1, p_2, \ldots, p_m\}$, are close to each other compared to r, i.e. their diameter $D(v)$ is no larger than λr for some predefined small constant $\lambda > 0$, then points in v can be viewed as a single 'heavy' point (simply denoted by v for convenience), and assumptions $\mathcal{NN}_q(p_1, r), \mathcal{NN}_q(p_2, r), \ldots, \mathcal{NN}_q(p_m, r)$ can be covered (or replaced) by a single assumption $\mathcal{NN}_q(v, r)$ without losing much quality. We formally define $\mathcal{NN}_q(v, r)$ for aggregated subset v as follows.

Assumption 1 $\mathcal{NN}_q(v, r)$: *For a subset v of P, $\mathcal{NN}_q(v, r)$ is an assumption made about q which says: $D(v) \leq \lambda r$ for some small constant $\lambda > 0$, where $D(v)$ is the diameter of v, and $r \leq \|p' - q\| \leq (1 + \lambda)r$ holds for p' which denotes the nearest neighbor of q in v.*

Another property of P which can be made use of is the *domination* relation. If q is very close to an aggregated subset of points $v \subseteq P$ compared to points in $P \setminus v$, it is often a degenerated case for the problem and relatively easy to solve. To cover such cases, we define the following assumption $\mathcal{DOM}_q(v)$ for predefined constants $\xi > 0$ and $\lambda > 0$.

Assumption 2 $\mathcal{DOM}_q(v)$: *For a subset v of P, $\mathcal{DOM}_q(v)$ is an assumption made about q, which says: there exists a point $p_v \in v$ such that $D(v) \leq \lambda \|q - p_v\|$ and $\|p_v - q\| \leq \xi \|p_{-v} - q\|$ for any point $p_{-v} \in P \setminus v$, where $D(v)$ is the diameter of v.*

With the above definitions of assumption, we know that the goal of the range cover data structure is to generate a small number of assumptions $\mathcal{DOM}_q(v_1)$, $\mathcal{DOM}_q(v_2), \ldots, \mathcal{DOM}_q(v_h)$ and $\mathcal{NN}_q(v'_1, r_1), \mathcal{NN}_q(v'_2, r_2), \ldots, \mathcal{NN}_q(v'_g, r_g)$, so that for any $q \in \mathbb{R}^d$, at least one of these assumptions holds. We call such a collection of assumptions an **assumption coverage**.

The main idea of range cover is to build a series of *views* of P formed by aggregated subsets from different scales of r, which is a controlling factor and can be interpreted as the distance of observation. Range cover identifies, for each r, a collection of disjoint aggregated subsets v of P with diameter no larger than λr for some predefined small constant $\lambda > 0$. The collection could be used as a sketch of P observed from distance r, which takes much less space than P. These views (from different distances r) jointly provide an easy way to access the "skeleton" information of P, which allow us to produce a smaller size assumption coverage. Particularly, for a given r, instead of generating assumptions $\mathcal{NN}_q(\{p\}, r)$ for each point $p \in P$, we produce coarse-grained assumptions $\mathcal{NN}_q(v, r)$ for every v in this view. Furthermore, by utilizing domination relation, we do not need to consider small values of r, and thus can further reduce the size of the assumption coverage. This is because the aggregation-based views of P from small enough r's correspond to situations where q is very close to some point and the domination relation holds. Note that when determining point aggregation, we need not to consider too large r as well, since for large enough r the whole point set P is an aggregated set.

To generate the assumption coverage, an obvious challenge is how to reduce the number of possible values for r for which we need to build a view of P. Even though there is no need to consider too large and too small values for r, the gap between the maximum and minimum values often depends on the spread ratio of P, which could lead to pseudo-polynomial running time for some algorithms using the range cover data structure. Below we will show how to overcome this challenge and obtain a small size range cover.

2.1 Range Cover and Assumption Coverage

The range cover data structure uses the aggregation tree as an ingredient. The aggregation tree is a version of Hierarchical Well-Separated Tree (HST)[5] which is defined conveniently for point aggregation in a well-behaved manner. The definition is as follows.

1. Every node v (called *aggregation node*) represents a subset $P(v)$ of P, and the root represents P.
2. Every aggregation node v is associated with a representative point $l(v) \in P(v)$ and a size $s(v)$ which is an upper bound on the diameter of $P(v)$.
3. Every leaf node corresponds to one point in P with size $s(v) = 0$, and each point appears in exactly one leaf node.
4. The two children v_1 and v_2 of any internal node v form a partition of v with $\max\{s(v_1), s(v_2)\} < s(v)$.
5. For every aggregation node v with parent v_p, $\frac{s(v_p)}{r_{out}}$ is bounded by a polynomial function $\mathcal{P}(n, d) \geq 1$ (called *distortion polynomial*), where r_{out} is the minimum distance between any point in $P(v)$ and any point in $P \setminus P(v)$.

The following theorem shows that an HST with polynomial distortion (therefore, the aggregation tree also) can be built within near linear time.

Theorem 1. *[5] An HST with distortion polynomial $O(\sqrt{d}n^5)$ can be built in $O(dn \log n)$ time with success probability $1 - 1/n$.*

Below we will show how to build a range cover data structure from a given aggregation tree T_p which ensures to form an assumption coverage.

Consider an aggregation node v from distance r. If the diameter of v is not larger than λr for a predefined constant $\lambda > 0$, all points in v can be viewed as an aggregated subset and thus is part of the view from r. If r is so large that even the parent v' of v in T_p is an aggregated subset, v can be replaced by v' in the view. This means that an aggregation node v should not appear in the view from a far enough distance r. Also if r is small, either v has a too large diameter and thus cannot be an aggregated subset or v dominates q (*i.e.* the solution point). In the former case, v should be replaced by one of its descendant in the view. In the latter case, we do not include v in the view from distance r, with the belief (which will be proved later) that the absence of v can be compensated by including the $\mathcal{DOM}_q(v)$ assumption in the assumption coverage.

The above observation implies that for any aggregation node v, there exists a range (r_L, r_H) of the value of r, such that v is only "visible" when r lies in the range. This immediately suggests the following scheme. Divide the set of all positive real numbers into intervals $((1+\lambda)^t, (1+\lambda)^{t+1}], t = \ldots, -2, -1, 0, 1, \ldots,$ and associate each of them with a bucket. If an interval $(a, b]$ lies within the interval (r_L, r_H) of a aggregation node v, then insert v into the bucket of $(a, b]$. The collection of these buckets is then the desired range cover data structure.

Algorithm 1 RangeCover(T_p, λ, ξ)

Input: A aggregation tree T_p built over a set P of points in $\mathbb{R}^d$; an approximation factor $0 < \lambda < \frac{1}{4}$, a controlling factor $0 < \xi < 1$.

Output: A number of sets of aggregation nodes, each of which is associated with an interval $((1+\lambda)^t, (1+\lambda)^{t+1}]$ for some integer t.

1: For every interval $((1+\lambda)^t, (1+\lambda)^{t+1}]$, create an empty bucket B_t. (Note that B_t will not be actually created until some aggregation node v is inserted into it.)

2: For every non-root node v of T_p, let v_p be its parent in T_p, r_H be $s(v_p)/\lambda$, and r_L be $\max\{s(v)/\lambda, \xi s(v_p)/(16\mathcal{P}(n, d))\}$. Do
   - For every integer t satisfying the condition of $r_L \leq (1+\lambda)^t < r_H$, insert v into bucket B_t.

Given input P, for any constant factors $0 < \lambda < 1/4$ and $\xi > 0$ in **Assumption 1** and **Assumption 2**, we build the aggregation tree T_p and the corresponding range cover data structure $\mathcal{R}$ by calling RangeCover(T_p, λ, ξ), and let the assumption coverage $\mathcal{A}_{\lambda,\xi}$ (or simply $\mathcal{A}$ for convenience) contain the following assumptions:

1. $\mathcal{DOM}_q(v)$, for every aggregation node v of T_p
2. $\mathcal{NN}_q(v, r)$, for every aggregation node v of T_p and r such that interval $(r, (1+\lambda)r]$ is one of the nonempty bucket in $\mathcal{R}$ and v is a aggregation node in this bucket.

Clearly obtaining $\mathcal{A}$ from $\mathcal{R}$ is quite straightforward, and $|\mathcal{A}|$ has a size no larger than that of $\mathcal{R}$.

The following theorem shows that $\mathcal{A}$ is indeed an assumption coverage.

Theorem 2. *For any q in $\mathbb{R}^d$, at least one of the assumptions in $\mathcal{A}$ holds.*

Proof. Let p' be the nearest neighbor of q in P. If $\|q - p'\| = 0$, $\mathcal{DOM}_q(\{p'\})$ holds. In the following we assume that $\|q - p'\| > 0$. Let t' be the integer such that $(1+\lambda)^{t'} < \|q - p'\| \leq (1+\lambda)^{t'+1}$. Let v' be a aggregation node of T_p which is the highest ancestor of $\{p'\}$ in T_p such that $s(v') \leq \lambda(1+\lambda)^{t'}$. Since $\{p'\}$ is a leaf of T_p and $s(\{p'\}) = 0 \leq \lambda(1+\lambda)^{t'}$, such a v' always exists.

Based on the relationship between v', t' and the range cover data structure, we have 4 cases to consider. (a) v' is the root of T_p, (b) $(1+\lambda)^{t'} <$

$\max\{s(v')/\lambda, \xi s(v_p')/(16\mathcal{P}(n, d))\}$, where v_p' is the parent of v' in T_p, (c) $(1 + \lambda)^{t'} \geq s(v_p')/\lambda$, and (d) $\max\{s(v')/\lambda, \xi s(v_p')/(16\mathcal{P}(n, d))\} \leq (1+\lambda)^{t'} < s(v_p')/\lambda$. Below we analyze each of them.

Case (a): Since $s(v') \leq \lambda(1 + \lambda)^{t'} \leq \lambda\|q - p'\|$ and v' represents the whole point set P (as it is the root of T_p), we have $P \setminus v'$ is empty. This means that the assumption $\mathcal{DOM}_q(v')$ holds for q.

Case (b): Note that by the definition of t', we know that $(1+\lambda)^{t'} \geq s(v')/\lambda$. Therefore if case (b) occurs, we have $(1+\lambda)^{t'} \leq \xi s(v_p')/(16\mathcal{P}(n, d))$. By $(1+\lambda)^{t'} < \|q - p'\| \leq (1 + \lambda)^{t'+1}$ and $\lambda < 1$, it follows that $\|q - p'\| \leq \xi s(v_p')/(8\mathcal{P}(n, d))$. Let p_o be any point in $P \setminus v'$. Then $\|p_o - p'\| \geq s(v_p')/\mathcal{P}(n, d)$ by the property of aggregation tree. Therefore, $\xi\|p_o - p'\| \geq 8\|q - p'\|$. Thus, $\|p_o - q\| \geq \|p_o - p'\| - \|q - p'\| \geq (8/\xi - 1)\|q - p'\|$. By the fact $\xi < 1$, we have $\|q - p'\| \leq \xi\|p_o - q\|$. Also since $(1 + \lambda)^{t'} \geq s(v')/\lambda$ and $(1 + \lambda)^{t'} < \|q - p'\| \leq (1 + \lambda)^{t'+1}$, we have $\|q - p'\| \geq s(v')/\lambda$. This indicates that $\mathcal{DOM}_q(v')$ holds for case (b).

Case (c): This case actually never occurs. This is because, by the definition of v', $s(v_p') > \lambda(1 + \lambda)^{t'}$, since otherwise v' cannot be the highest ancestor of $\{p'\}$ satisfying the inequality $s(v') \leq \lambda(1 + \lambda)^{t'}$.

Case (d): Note that this case means that v' is placed in bucket $((1+\lambda)^{t'}, (1+\lambda)^{t'+1}]$. Thus $\mathcal{NN}_q(v', (1 + \lambda)^{t'})$ is in $\mathcal{A}$. We show that $\mathcal{NN}_q(v', (1 + \lambda)^{t'})$ holds for q. Indeed, this follows immediately from previous discussion on v': $s(v') \leq \lambda(1 + \lambda)^{t'}$ and $(1 + \lambda)(1 + \lambda)^{t'} \geq \|p' - q\| > (1 + \lambda)^{t'}$.

Since in all cases at least one assumption in $\mathcal{A}$ holds for q, the theorem follows. □

The following theorem indicates that the size of the assumption coverage is small.

Theorem 3. *Given a aggregation tree T_p and factors $0 < \lambda < 1/4$ and $0 < \xi < 1$, the range cover data structure can be built in $O(1/\lambda \log(1/\xi)n(\log n + \log d))$ time and takes $O(1/\lambda \log(1/\xi)n(\log n + \log d))$ space. Consequently, $|\mathcal{A}| = O(1/\lambda \log(1/\xi)n(\log n + \log d))$.*

Proof. From **Algorithm 1**, we know that every aggregation node v is inserted into $O(\log_{1+\lambda} r_H/r_L)$ buckets (see Step 2 of the algorithm). Note that $\log_{1+\lambda} r_H/r_L$ is no larger than $\log_{1+\lambda}((s(v_p)/\lambda)/(\xi s(v_p)/16\mathcal{P}(n, d))) = O(1/\lambda \log(1/\xi)(\log n + \log d))$. Since the total number of aggregation node is $O(n)$, the theorem follows. □

3 Solving Truth Discovery with Assumption Coverage

In this section, we show how to use the assumption coverage to solve the truth discovery problem. Given any point set P in $\mathbb{R}^d$ and a small constant $0 < \epsilon < 1$, we first build an assumption coverage $\mathcal{A}$ with factors λ and ξ whose values depend on ϵ only and will be determined later. We then show how to obtain a $(1 + \epsilon)$-approximation of the problem in polynomial time. Let p^* be the truth vector (*i.e.*, optimal solution) of the problem.

We first borrow a useful lemma from [6]. It shows that once p^* is determined, the weights w_i can also be determined. Thus we only need to find an approximate truth vector p^*.

Lemma 1. *[6] If the truth vector p^* is fixed, the following value for each weight w_l minimizes the the objective function (1) (in Definition 1),*

$$w_l = \log\left(\frac{\sum_{i=1}^{n}\|p^* - p_i\|^2}{\|p^* - p_l\|^2}\right). \tag{2}$$

There are two types of assumptions about p^* in $\mathcal{A}$ which covers all possibilities of p^*: $\mathcal{NN}_{p*}(v,r)$ and $\mathcal{DOM}_{p*}(v)$. Below we discuss each of them.

The following lemma shows that $\mathcal{DOM}_{p*}(v)$ is easy to solve.

Lemma 2. *By setting $\lambda \leq 1/4$ and $\xi \leq \epsilon/4$, if $\mathcal{DOM}_{p*}(v)$ holds for the truth vector p^*, there exists a point $p' \in v \subseteq P$ such that p' is a $(1+\epsilon)$-approximation of the truth discovery problem (using the objective function (1) in **Definition 1**).*

From the above lemma, we know that if $\mathcal{DOM}_{p*}(v)$ holds for some v, then one of the input point in P will be a $(1+\epsilon)$-approximation. This means that we can handle all such cases by trying every input point as p^* by computing the objective function (1) in equation (2), and choosing the one with the minimum objective value as the solution. This takes $O(dn^2)$ time.

The following lemma shows that $\mathcal{NN}_{p*}(v,r)$ can also be handled efficiently. We leave the proof to the next subsection.

Lemma 3. *If $\mathcal{NN}_{p*}(v,r)$ holds for any factor $0 < \lambda < 1/4$, then a $(1+\epsilon)$-approximation can be computed in time $O(dn)$ with constant probability, where ϵ is a small constant in $(0,1)$.*

The above lemmas suggest that we can compute an approximate p^* by the following algorithm.

1. Compute an aggregation tree from P.
2. Set $\xi = \epsilon/4$, $\lambda = 1/5$, compute a range cover from the aggregation tree.
3. Compute $\mathcal{A}$ from the range cover.
4. Try every $p \in P$ as a candidate for the truth vector. Choose the one, say p_1, that minimizes the objective function.
5. For every $\mathcal{NN}_{p*}(v,r)$ in $\mathcal{A}$, compute a candidate for p^*. Choose the one, say p_2, that minimizes the objective function.
6. Choose from p_1 and p_2 the one that minimizes the objective function

In the above algorithm, Step 1 takes $O(dn\log n)$ time. Step 2 needs $O(n(\log n + \log d))$ time (where ϵ is hidden in the $O(\cdot)$ notion). Step 3 costs $O(n(\log n + \log d))$ time. Step 4 can be done in $O(dn^2)$ time. Step 5 takes $O(dn^2(\log n + \log d))$ time, since we test at most $O(n(\log n + \log d))$ assumptions in $\mathcal{A}$. Step 6 requires only $O(1)$ time. For the space usage, it can be computed $O(dn\log n) + O(n(\log n + \log d)) + O(n(\log n + \log d)) + O(dn) + O(dn) + O(1) = O(dn(\log n + \log d))$. Thus we have the following main theorem.

Theorem 4. *Given any set P of n points in $\mathbb{R}^d$, with constant probability, it is possible to compute a $(1+\epsilon)$-approximate solution for the truth discovery problem in $O(dn^2(\log n + \log d))$ time. The space usage can be made to $O(dn(\log n + \log d))$.*

3.1 Solving $\mathcal{NN}_{p*}(v, r)$

In this section we prove Lemma 3. We assume that $\mathcal{NN}_{p*}(v, r)$ holds for p^*, where $v \subseteq P$ and $r > 0$.

Lemma 1 reveals how the weight w_i of every $p_i \in P$ is related to p^*. It is clear from the objective function (1) and Lemma 1 that p^* is the weighted mean of P. Since we do not know p^* in advance, w_i is also unknown for every $p_i \in P$. The truth discovery problem can be viewed as a problem of finding the weighted mean of a point set with unknown weights. Our strategy for solving this problem consists of two main steps: (1) we partition P into a number of subsets (or sub-clusters), with each having some nice property. The weights of the points in some clusters are approximately known, while the weights of the points in other clusters are unknown, but have an upper and lower bound; (2) we apply a technique in [1] to find the approximate weighted mean point of each subset, and combine them to estimate p^*.

Partitioning P for Estimating Weights We first show how to estimate the weights of some points by $\mathcal{NN}_{p*}(v, r)$ without knowing p^*. This is crucial for our algorithm to be efficient for any point set P.

Let $p_1 \in v$ denotes the representative point $l(v)$ of v. We label the rest of points in P as $p_2, p_3, \ldots, p_n$. For each point $p_i \in P$, define $r'_i = \max(\|p_1 - p_i\|, r)$ and $r_i = \|p^* - p_i\|$. For $\mathcal{NN}_{p*}(v, r)$, let $p_{i_s} \in v$ be the nearest neighbor of p^* in P. Below we derive the relationship between r_i and r'_i.

First, we consider the case that $\max(\|p_1 - p_i\|, r) = r$. In this case, we have $r_i \geq \|p_{i_s} - p^*\| \geq r = r'_i$ by assumption $\mathcal{NN}_{p*}(v, r)$ and the fact that p_{i_s} is the nearest neighbor of p^*. Also we have $r_i \leq \|p_1 - p^*\| + \|p_1 - p_i\| \leq \|p_1 - p^*\| + r$, and

$$\|p_1 - p^*\| \leq \|p_1 - p_{i_s}\| + \|p^* - p_{i_s}\| \leq D(v) + (1 + \lambda)r \leq (1 + 2\lambda)r.$$

Thus, $r_i \leq (2 + 2\lambda)r = (2 + 2\lambda)r'_i$. Putting all together, we have $r'_i \leq r_i \leq (2 + 2\lambda)r'_i$.

Then, we consider the case that $\max(\|p_1 - p_i\|, r) = \|p_1 - p_i\|$. In this case, $r'_i = \|p_1 - p_i\| \geq r$. Again, we have $\|p_1 - p^*\| \leq \|p_1 - p_{i_s}\| + \|p^* - p_{i_s}\| \leq D(v) + (1 + \lambda)r \leq (1 + 2\lambda)r$. Therefore, $(1 + 2\lambda)r'_i \geq \|p_1 - p^*\|$. Thus,

$$r_i = \|p_i - p^*\| \leq \|p_1 - p_i\| + \|p_1 - p^*\| \leq \|p_1 - p_i\| + (1 + 2\lambda)r'_i = (2 + 2\lambda)r'_i.$$

Next, we consider 2 subcases, $r'_i \geq 2r$ and $r'_i < 2r$. If $r'_i < 2r$, since $r_i \geq r$, we have $r_i > r'_i/2$. If $r'_i \geq 2r$, since $\|p_1 - p^*\| \leq (1 + 2\lambda)r$, we have $\|p_1 - p^*\| \leq (1 + 2\lambda)r'_i/2$. This means that

$$r_i = \|p_i - p^*\| \geq \|p_1 - p_i\| - \|p_1 - p^*\| \geq r'_i - (1 + 2\lambda)r'_i/2 = (1 - 2\lambda)r'_i/2.$$

To conclude, we have $(1 - 2\lambda)r_i'/2 \leq r_i \leq (2 + 2\lambda)r_i'$.

From the above analysis and the fact that $\lambda < 1/4$, we can obtain the following.

$$r_i/4 \leq r_i' \leq 4r_i. \tag{3}$$

For each $p_i \in P$, let $w_i = \log((\sum_{p_j \in P} r_j^2)/(r_i^2))$, i.e., w_i is the optimal weight determined by Lemma 1. Let $w_i' = \log((\sum_{p_j \in P} r_j'^2)/(r_i'^2))$. From inequality (3), we obtain the following:

$$w_i - \log 256 \leq w_i' \leq w_i + \log 256. \tag{4}$$

This means that w_i' can be used as an approximation of w_i if w_i is large enough.

For any $p_i \in P$, if $w_i' \geq 8/\beta \geq \log 256/\beta$ for any $0 < \beta < 1$, we have the following (by (4))

$$(1 - \beta)w_i \leq w_i' \leq (1 + \beta)w_i.$$

This means that w_i can be well approximated by w_i' in this case. Let P_β denote the set $\{p_i \in P | w_i' \geq 8/\beta\}$.

Next, we further show that there is at most one point p_i in P with weight $w_i < \log 36/25$ which, if exists, can be identified by a simple procedure. By the definition of w_i, we know that $w_i < \log 36/25$ can happen only when $\|p^* - p_i\| > 5\|p^* - p_j\|$ for any $i \neq j$. This means that for any $j, l \neq i$,

$$\|p_j - p_l\| \leq \|p^* - p_j\| + \|p^* - p_l\| \leq 2\max(\|p^* - p_j\|, \|p^* - p_l\|). \text{ Thus, we have}$$

$$\|p_j - p_i\| \geq \|p^* - p_i\| - \|p^* - p_j\|$$
$$> 5\max(\|p^* - p_j\|, \|p^* - p_l\|) - \max(\|p^* - p_j\|, \|p^* - p_l\|)$$
$$= 4\max(\|p^* - p_j\|, \|p^* - p_l\|) \geq 2\|p_j - p_l\|.$$

Hence, for any $j, l \neq i$, the inequality $\|p_j - p_l\| < \|p_i - p_j\|/2$ holds. In other words, p_i is isolated from the rest of the points in P. It is easy to see that such a p_i is unique, if exists. The following procedure searches for such a p_i.

1. Choose an arbitrary point p from P.
2. Find a point p' in P farthest away from p.
3. Find a farthest point p'' from p' in P.
4. Compare the pairwise distances among the three points in $\{p, p', p''\}$. Throw away the pair of points with the smallest pairwise distance. Output the remaining point as $\hat{p}$.

From the above discussion, it is easy to see that if there is a point p_i with weight $w_i < \log 36/25$, it must be $\hat{p}$. Clearly, this procedure takes only $O(dn)$ time.

For a constant $0 < \beta < 1/2$ (whose value will be determined later), let $P_u = P \setminus (P_\beta \cup \{\hat{p}\})$ and $P_< = \{\hat{p}\} \setminus P_\beta$. Then, $P_u, P_<, P_\beta$ form a partition of P. P_β contains all points p_i in P whose weights w_i have already been roughly determined (i.e., approximated by w_i'); $P_<$ has at most one point, which will be

---

Algorithm 2 $(1 + O(1)\epsilon)$-approximate Truth Discovery from $\mathcal{NN}_{p*}(v, r)$

Input: A set P of n points in $\mathbb{R}^d$ space. Assumption $\mathcal{NN}_{p*}(v, r)$. $\beta = \epsilon^2$. Constants γ, k solved from $2\gamma\sqrt{k} \leq \epsilon^2$ and $k = \lceil \log_{1+\gamma} \frac{16}{\epsilon^2 \log 36/25} \rceil + 1$. $c_1 = \frac{4k}{\alpha\gamma^2} \log \frac{16k^2}{\gamma^2}$. $c_2 = \frac{4k}{\gamma^2}$. $\alpha = \epsilon^3\beta/48k$.

Output: An approximate truth vector.

1: Identify $P_<, P_\beta, P_u$ by computing w_i' for each $p_i \in P$.
2: Compute the weighted mean o_3' of P_β using weights w_i'.
3: Randomly sample c_1 points from P. Enumerate all subsets of c_2 points from the sample. Compute means of these subsets, and put all the means into a set M.
4: For every k-subset $\{o_1, \ldots o_k\}$ of M, apply SIMPLEX$(\epsilon^2, k, o_1, \ldots, o_k)$ to produce a grid. Put all grid points in into a point set G.
5: For every o_2' in G, if $P_<$ contains a point o_1', then build a grid by applying SIMPLEX$(\epsilon, 3, o_1', o_2', o_3')$; otherwise, build a grid using SIMPLEX$(\epsilon, 2, o_2', o_3')$.
6: Try all the grid points produced above. Output the one that minimized the objective function (1).

---

the one with weight smaller than $\log 36/25$, if exists; P_u contains all the remaining points whose weights are not known yet. $P_u, P_<, P_\beta$ together with w_i' can be obtained in $O(dn)$ time since it takes a total of $O(n)$ distance computations.

Following a similar idea in [1], we further decompose P_u by using the *log-partition* technique, where $\gamma > 0$ is a constant to be determined later. (Note that the log-partition cannot be explicitly obtained since we do not know the weights w_i. We assume that such a partition exists and will be used in our later analysis.)

Definition 2. *The log-partition of P_u divides points in P_u into k groups $\mathcal{G}_1, \ldots \mathcal{G}_k$ as follows, where $k = \lceil \log_{1+\gamma} \frac{16/\beta}{\log 36/25} \rceil + 1$: $\mathcal{G}_i = \{p_j \in P_u | (1+\gamma)^{i-1} \log 36/25 \leq w_j \leq (1+\gamma)^i \log 36/25\}$.*

Note that the above partition indeed involves all points in P_u. This is because by the definition of $P_<$ and P_β, and the fact that $(1 - \beta)w_i \leq w_i' \leq (1+\beta)w_i$ for all point $p_i \in P_\beta$, we know that $\log 36/25 \leq w_i \leq 16\beta$ for each point $p_i \in P_u$. This implies that $\mathcal{G}_1, \ldots \mathcal{G}_k, P_<, P_\beta$ form a partition of P. Also, we apply log-partition to P_u instead of P as in [1]. In this way the value of k is bounded, making our algorithm efficient for any data.

Applying the Simplex Lemma Roughly speaking, Simplex Lemma in [1] provides a procedure SIMPLEX$(\epsilon, k, o_1, \ldots, o_k)$ to approximate the weighted mean of a partitioned point set $Q = \bigcup Q_i$, where ϵ is an approximate factor, k is an integer and every o_i is a point in $\mathbb{R}^d$. The procedure outputs a grid of size $((8k/\epsilon)^k)$ within $O((8k/\epsilon)^k)$ time which ensures that at least one of the grid points is close to the weighted mean of Q, if o_i is a good approximation of the weighted mean of Q_i.

Algorithm 2 shows how to use SIMPLEX to produce an approximate truth vector, given P partitioned into $P_<, P_\beta, P_u$ as above. The running time and space usage match those appear in Lemma 3. To obtain a $(1 + \epsilon)$-approximation, we only need to do a scaling on the constants without affecting the asymptotic running time.

Below we briefly explain the main steps of Algorithm 2. In Step 1 we partition P into $P_<, P_\beta, P_u$ as mentioned before. In Step 2 an approximate weighted mean of P_β is computed. In Steps 3 and 4, we try to guess k weighted means $\{o_1, \dots o_k\}$ for the clusters $\mathcal{G}_1, \dots \mathcal{G}_k$ resulted from the log-partition of P_u by using random sampling. We apply SIMPLEX to these approximate means $\{o_1, \dots o_k\}$ to produce a small grid. The set G of grid points contains at least one point which is a good approximate weighted mean of P_u. In Steps 5 and 6, we already have approximate weighted means o'_1 and o'_3 of $P_<$ and P_β, respectively, and a set G which contains an approximate weighted mean o'_2 of P_u. We then try all possible o'_2 from G and use SIMPLEX on o'_1, o'_2, o'_3 to produce grids and one of such grids contains the desired approximation of the truth vector.

References

1. Ding, H., Gao, J., and Xu, J.: Finding Global Optimum for Truth Discovery: Entropy Based Geometric Variance. Leibniz International Proceedings in Informatics (LIPIcs), 32nd International Symposium on Computational Geometry (SoCG 2016), Vol. 51, 34:1-34:16(2016).
2. Ding, H. and Xu, J.: A Unified Framework for Clustering Constrained Data without Locality Property. Proceedings of ACM-SIAM Symposium on Discrete Algorithms (SODA 2015), pp. 1471-1490, January 4-6, 2015, San Diego, California, USA.
3. Dong, X.L., Berti-Equille, L., Srivastava, D.: Integrating conflicting data: The role of source dependence. PVLDB, 2(1): 550-561(2009).
4. Li, Y., Gao, J., Meng, C., Li, Q., Su, L., Zhao, B., Fan, W., Han, J.: A Survey on Truth Discovery, CoRR abs/1505.02463(2015).
5. Har-Peled, S.: Geometric approximation algorithms. Vol. 173. Boston: American mathematical society(2011).
6. Li, H., Zhao, B., Fuxman, A.: The Wisdom of Minority: Discovering And Targeting The Right Group of Workers for Crowdsourcing. Proc. of the International Conference on World Wide Web (WWW'14), pp. 165-176(2014).
7. Li, Q., Li, Y., Gao, J., Su, L., Zhao, B., Demirbas, M., Fan, W., Han, J.: A Confidence- Aware Approach for Truth Discovery on Long-Tail Data. PVLDB 8(4): 425-436(2014).
8. Li, Q., Li, Y., Gao, J., Zhao, B., Fan, W., Han, J.: Resolving Conflicts in Heterogeneous Data by Truth Discovery and Source Reliability Estimation. Proc. the 2014 ACM SIGMOD International Conference on Management of Data (SIGMOD'14), pp. 1187-1198(2014).
9. Pasternack, J., Roth, D.: Knowing what to believe (when you already know something). Proc. of the International Conference on Computational Linguistics (COLING'10), pp. 877-885(2010).
10. Whitehill, J., Ruvolo, P., Wu, T., Bergsma, J., Movellan, J.: Whose Vote Should Count More: Optimal Integration of Labelers of Unknown Expertise. Advances in Neural Information Processing Systems (NIPS'09), pp. 2035-2043(2009).
11. Yin, X., Han, J., and Yu, P.S.: Truth discovery with multiple conflicting information providers on the web: Proc. of the ACM SIGKDD International Conference on Knowledge Discovery and Data Mining (KDD'07), pp. 1048-1052(2007).

Searching edges in the overlap of two plane graphs

John Iacono[1], Elena Khramtcova[2], and Stefan Langerman[2]

[1] Department of Computer Science and Engineering, New York University
New York, USA, iacono@nyu.edu
[2] Computer Science Department, Université libre de Bruxelles (ULB)
Brussels, Belgium, {elena.khramtcova, stefan.langerman}@ulb.ac.be

Abstract. Consider a pair of plane straight-line graphs whose edges are colored red and blue, respectively, and let n be the total complexity of both graphs. We present a $O(n \log n)$-time $O(n)$-space technique to preprocess such a pair of graphs, that enables efficient searches among the red-blue intersections along edges of one of the graphs. Our technique has a number of applications to geometric problems. This includes: (1) a solution to the *batched red-blue search* problem [Dehne et al. 2006] in $O(n \log n)$ queries to the oracle; (2) an algorithm to compute the maximum vertical distance between a pair of 3D polyhedral terrains, one of which is convex, in $O(n \log n)$ time, where n is the total complexity of both terrains; (3) an algorithm to construct the Hausdorff Voronoi diagram of a family of point clusters in the plane in $O((n + m) \log^3 n)$ time and $O(n + m)$ space, where n is the total number of points in all clusters and m is the number of *crossings* between all clusters; (4) an algorithm to construct the farthest-color Voronoi diagram of the corners of n disjoint axis-aligned rectangles in $O(n \log^2 n)$ time; (5) an algorithm to solve the stabbing circle problem for n parallel line segments in the plane in optimal $O(n \log n)$ time. All these results are new or improve on the best known algorithms.

1 Introduction

Many geometric algorithms have subroutines that involve investigating intersections between two plane graphs, often assumed being colored red and blue respectively. Such subroutines differ in the questions that are asked about the red-blue intersections. The most well-studied questions are to report all red-blue intersections or to count them. It is shown how to report all the intersections in optimal $O(n \log n + k)$ time and $O(n)$ space [3, 4, 15, 16, 18], where n is the total complexity of both graphs, and k is the size of the output. Note that k may be $\Omega(n^2)$. Counting the red-blue intersections can be carried out in $O(n \log n)$ time and $O(n)$ space [4, 16].

In this paper, we consider the situation where one wants to *search* the red-blue intersections, though avoiding to compute all of them. Problems of this type appear as building blocks in diverse geometric algorithms. The latter include: distance measurement between polyhedral terrains [4], motion planning [12], construction of various generalized Voronoi diagrams (divide-and-conquer [9, 6] or randomized incremental [5] construction). Therefore solving such problems efficiently is of high importance.

Often it is guaranteed that each red edge contains at most one sought red-blue intersection, and an oracle is provided, that, given a red-blue intersection, is able to quickly

© Springer International Publishing AG 2017
F. Ellen et al. (Eds.): WADS 2017, LNCS 10389, pp. 473–484, 2017.
DOI: 10.1007/978-3-319-62127-2_40

determine to which side of that intersection the sought intersection lies along the same red edge (see Section 3.1 for more details on this setting). A particular case, when the red graph consists of a unique edge, appeared under the name of *segment query* in the randomized incremental construction algorithm for the Hausdorff Voronoi diagram [5], and under the name of *find-change* query in an algorithm to solve the stabbing circle problem for a set of line segments [8]. If the blue graph is a tree, it can be preprocessed in $O(n \log n)$ time using the *centroid decomposition* [2, 5]. Centroid decomposition supports segment (or find-change) queries for arbitrary line segments, requiring only $O(\log n)$ queries to the oracle [5, 7]. If the blue graph is not a tree, then in $O(n \log n)$ time it can be preprocessed for point location, and a nested point location along the red edge is performed, which requires $O(\log^2 n)$ queries to the oracle [6, 8]. For two general plane straight-line graphs (where the red graph is not necessarily one edge) the problem is called *batched red-blue intersection problem* (see Problem 2). It was formulated in Dehne et al. [9], and solved in $O(n \log^3 n)$ time and $O(n \log^2 n)$ space [9] using *hereditary segment trees* [4]. However, this is optimal in neither time nor space.

We present a data structure that provides a clear interface for efficient searches for red-blue intersections along a red edge. Our data structure can be used to improve the above result [9] (see Section 3.1), which includes an improvement on segment (or find-change) queries in plane straight-line graphs. Our data structure can also handle more general search problems, e.g., a setting when a red edge may have more than one sought red-blue intersection on it. Below we state our result and its applications.

1.1 Our result

Let R, B be a pair of plane straight-line[3] graphs. We address the following problem.

Problem 1 (RB-Preprocessing problem). Given graphs R, B, construct a data structure that for each edge e of R stores implicitly the intersections between e and the edges of B sorted according to the order, in which these intersections appear along e. Let T_e be a perfectly balanced binary search tree built on the sorted sequence of intersections along e. The data structure should answer efficiently the following *navigation queries* in T_e:

- Return the root of T_e;
- Given a non-root node of T_e, return the parent of this node;
- Given a non-leaf node of T_e, return the left (or the right) child of this node.

We provide a solution to the RB-Preprocessing problem, where each of the navigation queries can be answered in $O(1)$ time, and constructing the data structure requires $O(n \log n)$ time and $O(n)$ space, where n is the total number of vertices and edges in both R and B (see Section 2).

The resulting data structure allows for fast searches for *interesting* intersections between edges of R and the ones of B. We note that the notion of interesting is external to the data structure: It is not known at the time of preprocessing, but rather guides the searches on the data structure after it is built. In particular, for the input graphs R and

---

[3] Our technique can be trivially generalized to apply to x-monotone pseudoline arcs in place of straight-line edges of the graphs.

B, the data structure is always the same, while interesting intersections can be defined in several ways, which of course implies that the searches may have different outputs.

Our preprocessing technique can be applied to a number of geometric problems. We provide a list of applications, which is not exhaustive. For each application, we show how to reduce the initial problem to searching for interesting red-blue intersections, and how to navigate the searches, that is, how to decide, which subtree(s) of the current node of the (implicit) tree to search. Using our technique we are able to make the contributions listed below, and we expect it to be applicable to many more problems.

1. The *batched red-blue search problem* [9] for a pair of segment sets can be solved in $O(n)$ space and $O(n \log n)$ queries to the oracle, where n is the total number of segments in both sets (see Section 3.1). The problem is as follows. Given are: (1) two sets of line segments in the plane (colored red and blue, respectively), where no two segments in the same set intersect; and (2) an oracle that, given a point p of intersection between a red segment r and a blue segment b, determines to which side of segment r with respect to point p the interesting red-blue intersection lies. It is assumed that each segment contains at most one interesting intersection. The *batched red-blue search problem* is to find all interesting red-blue intersections. Our solution is an improvement on the one of Dehne et al. [9] which requires $O(n \log^2 n)$ space and $O(n \log^3 n)$ queries to the oracle.

2. The maximum vertical distance between a pair of polyhedral terrains, one of which is convex, can be computed in $O(n \log n)$ time and $O(n)$ space (see Section 3.2). Previously, a related notion of the minimum vertical distance between a pair of non-intersecting polyhedral terrains was considered, and it was shown how to find it in $O(n^{4/3+\epsilon})$ time and space for a pair of general polyhedral terrains [4], in $O(n \log n)$ time for one convex and one general terrain [22], and in $O(n)$ time for two convex terrains [22]. Our technique yields an alternative solution for the second case within the same time bound as in [22]. The maximum distance for non-intersecting polyhedra can be found by the above methods [4, 22], however it is different from the minimum distance for intersecting polyhedra: asking about the former is still interesting, while the latter is trivially zero.

3. The Hausdorff Voronoi diagram of a family of point clusters in the plane can be constructed in $O((n + m) \log^3 n)$ time, where m is the total number of pairwise *crossings* of the clusters (see Section 3.3). Parameter m can be $\Theta(n^2)$, but is small in practice [20, 19]. There is a deterministic algorithm to compute the diagram in $O(n^2)$ time [11]. All other known deterministic algorithms [19, 20] have a running time that depends on parameters of the input, that cannot be bounded by a function of m.[4] Each of them may take $\Omega(n^2)$ time even if $m = 0$. There is a recent randomized algorithm with expected time complexity $O((m + n \log n) \log n))$ [14]. For a simpler case of non-crossing clusters ($m = 0$), the diagram can be computed in deterministic $O(n \log^5 n)$ time[5] [9], or in expected $O(n \log^2 n)$ time [5, 14]. Thus our algorithm is the best deterministic algorithm for the case of small number of

---

[4] The algorithms have time complexity respectively $O(M + n \log^2 n + (m + K) \log n)$ and $O(M' + (n + m + K') \log n)$, where parameters M, M', K, K' reflect the number of pairs of clusters such that one is enclosed in a certain type of enclosing circle of the other.

[5] The time complexity claimed in [9] is $O(n \log^4 n)$. See the discussion in Section 3.3.

crossings. The time complexity of our algorithm is subquadratic in n and m and depends only on them, unlike any previous deterministic algorithm.

4. The farthest-color Voronoi diagram for a family of n point clusters, where each cluster is all the corners of an axis-aligned rectangle,[6] and these rectangles are pairwise disjoint, can be computed in $O(n \log^2 n)$ time and $O(n)$ space. Previous results on the topic are as follows. For arbitrary point clusters, the diagram may have complexity $\Theta(n^2)$ and can be computed in $O(n^2)$ time and space [1, 11], where n is the total number of points in all clusters. When clusters are pairs of endpoints of n parallel line segments, the diagram has $O(n)$ complexity and can be constructed in $O(n \log n)$ time and $O(n)$ space [8]. In this paper, we broaden the class of inputs, for which the diagram can be constructed in subquadratic time. We also show that the complexity of the diagram for such inputs is $O(n)$.

5. The stabbing circle problem for line segments in the plane can be solved in time $O(\mathcal{T}_{\mathsf{HVD}(S)} + \mathcal{T}_{\mathsf{FCVD}(S)} + (|\mathsf{HVD}(S)| + |\mathsf{FCVD}(S)| + m) \log n)$, where $|\mathsf{HVD}(S)|$ and $|\mathsf{FCVD}(S)|$ denote respectively the complexity of the Hausdorff and the farthest-color Voronoi diagram of the pairs of endpoints of segments in S, $\mathcal{T}_{\mathsf{HVD}(S)}$ and $\mathcal{T}_{\mathsf{FCVD}(S)}$ denote the time to compute these diagrams, and m is a parameter reflecting the number of "bad" pairs of segments in S. If all segments in S are parallel to each other, the stabbing circle problem can be solved in optimal $O(n \log n)$ time and $O(n)$ space. This is an improvement over the recent $O(\mathcal{T}_{\mathsf{HVD}(S)} + \mathcal{T}_{\mathsf{FCVD}(S)} + (|\mathsf{HVD}(S)| + |\mathsf{FCVD}(S)| + m) \log^2 n)$ time technique for general segments, which yielded an $O(n \log^2 n)$ time algorithm for parallel segments [8].

2   The technique to preprocess a pair of graphs

Suppose we are given two plane straight-line graphs R and B, and let n be the total number of vertices and edges in both R, B. We assume that no two vertices of the graphs have the same x coordinate. In this section, it is more convenient to treat R and B as two sets of line segments in the plane, where the segments in R are colored red, and the ones in B are colored blue. No two segments of the same color intersect, although they may share an endpoint.

Our preprocessing technique consists of three phases. In the first phase, we invoke an algorithm that finds the intersections between the edges of the two graphs (see Section 2.1). After that, in the second phase, we build a linearized *life table* for the red segments (see Section 2.2). Finally we sweep the life table with a line, which provides us the resulting data structure (see Section 2.3).

2.1   Finding red-blue intersections

For the sets R (red) and B (blue), we need to find all the intersections between segments of different color, i.e., all the red-blue intersections.

It is known how to count the red-blue intersections in optimal $O(n \log n)$ time [4, 16, 18], or report them in optimal $O(n \log n + k)$ time [3, 4, 15, 16, 18], where k is the

---

[6] A cluster is either the four corners of a non-degenerate axis-aligned rectangle, or the two endpoints of a horizontal/vertical segment, or a single point.

total number of the intersections. The space requirement of each of these algorithms is $O(n)$. The algorithm by Mantler and Snoeyink [16] processes the red-blue intersections in batches (called *bundle-bundle* intersections). In $O(n \log n)$ time and $O(n)$ space it can implicitly discover all the red-blue intersections, without reporting every one of them individually. The latter feature is useful for our technique, therefore we invoke the Mantler-Snoeyink algorithm in its first phase. We summarize the algorithm below.

To describe the algorithm, we need to define the following key notions: the *witness* of a (bichromatic) segment intersection, a *pseudoline at time i*, and a (monochromatic) *bundle* of segments at time i.

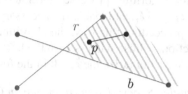

Fig. 1. Two line segments, r and b, the closed right wedge formed by them, and the witness p of their intersection

Given a red segment r that intersects a blue segment b, the *witness* of their intersection is the leftmost of the endpoints of segments in S that are contained in the closed right wedge formed by r and b. The closed right wedge formed by r and b is the intersection of two closed right halfplanes: the one bounded by the line through r, and the one bounded by the line through b, see the shaded area in Figure 1. Note that the witness always exists: it may be an endpoint of a segment different from r or b (as in Figure 1), or it may be an endpoint of either r or b.

Let n' be the total number of distinct endpoints of the segments in R and B. Let $p_1, p_2, \ldots, p_{n'}$ denote the sequence of these endpoints in the order of increasing x coordinate. The basis of the Mantler-Snoeyink algorithm can be formulated as follows.

Lemma 1. *For each $i, 1 \leq i \leq n'$ there is a y-monotone curve ℓ_i that passes through point p_i, and subdivides the plane into two open regions (the left and the right one), such that all the points $p_j, j < i$, and all the red-blue intersections witnessed by the points $p_j, j \leq i$ are contained in the left region, and all the points $p_k, k > i$ together with the intersections witnessed by them are contained in the right region, and ℓ_i intersects each segment in R or in B at most once.*

We call such curve ℓ_i a *pseudoline at time i*. Figure 2 shows a pseudoline at time 7, i.e., ℓ_7, in dashed black lines. Note that ℓ_7 cannot be replaced by a vertical straight line, because it must pass through the point 7, and to the left of the intersection point between the segments r_4 and b_4, and the latter point lies to the left of the former one.

A *blue bundle* at time i is a maximal contiguous sequence of blue segments that intersect the pseudoline ℓ_i.[7] See Figure 2, right. A *red bundle* is defined analogously.

---

[7] This definition can be seen as a generalization of the one of *single-edge bundles* in Mount [17].

The algorithm can be seen as a topological sweep with a pseudoline, where the only events are the endpoints $p_1, \ldots, p_{n'}$ of the segments in R and B. The sweepline at each moment i is a pseudoline ℓ_i such as defined in Lemma 1. The sweepline status structure maintains all the red and blue bundles that intersect the current sweepline. The sweepline status consists of (1) a balanced binary tree for each bundle, supporting insertion, deletion of segments, and a query for the topmost and the bottommost segment in the bundle, (2) a doubly-linked list for all the bundles intersecting the sweepline (bundles alternate colors), supporting insertion, deletion of the bundles, and sequential search, and (3) two balanced binary trees (one per color) storing all the red and blue bundles in order, and supporting splitting and merging of bundles.

At the event point p_i the algorithm processes the intersections witnessed by p_i, updates the sweepline from ℓ_{i-1} to ℓ_i, and makes the necessary changes to bundles (i.e., splits or merges them). By proceeding this way, the algorithm maintains the invariant that all the red-blue intersections whose witness is to the left of the current event point p_i are already encountered. We summarize the result in the following.

Theorem 1 ([16]). *The Mantler-Snoeyink algorithm runs in $O(n \log n)$ time, requires $O(n)$ space, and encounters $O(n)$ bundle-bundle intersections in total.*

2.2 Building the life table

In this section we describe our algorithm to build the life table for the sets R and B. Figure 2 illustrates the execution of the algorithm for a simple example.

Before we start our description, recall [10] that every pointer-based data structure with constant in-degree can be transformed into a partially persistent one. Such a persistent data structure allows accessing in constant time the data structure at any moment in the past, and performing pointer operations on it (but not modifying it); the total time and space required to make a data structure partially persistent is linear in the number of structural changes it underwent.

To build the life table for R and B, we first perform the Mantler-Snoeyink plane sweep algorithm (see Section 2.1), making the sweepline status structure partially persistent. This ensures that each blue bundle that has appeared during the algorithm, can afterwards be retrieved from the version of the sweepline status at the corresponding moment in the past. In particular, we are interested in the blue bundles that intersect red bundles. We assign each such blue bundle B_i a *timestamp* t_i reflecting the moment when the first bundle-bundle intersection involving B_i was witnessed. In order to distinguish between two different bundle-bundle intersections discovered at the same moment t_k (i.e., witnessed by the same point), we assign the moment $t_k + \epsilon$ to the intersection that has smaller y coordinate. Figure 2, right, lists all such blue bundles for the given example.

Observe that the plane sweep algorithm induces a partial order among the red segments: At any moment, the red segments crossed by the sweepline can be ordered from bottom to top. Since the red segments are pairwise non-intersecting, no two segments may swap their relative position. Let $r_1, \ldots, r_n$ be a total order consistent with the partial order along the sweepline at each moment. In Figure 2, the red segments are named according to such an order.

We now build the *life table* of red segments and blue bundles, see Figure 2, bottom. The life table is a graph defined as follows. On its y axis it has integers from 0 to n_R, where n_R is the number of red segments; the x axis of the life table coincides with the x axis of the original setting, i.e., of the plane $\mathbb{R}^2$. Each red segment r_i is represented by a horizontal line segment whose y coordinate equals i and whose endpoints' x coordinates coincide with the x coordinates of the endpoints of r_i. Each blue bundle B_j, that has participated in at least one bundle-bundle intersection, is retrieved from the version of the sweepline status at the moment t_j when the first such intersection has been witnessed; t_j is the timestamp of B_j. In the table, B_j is represented by a vertical line segment (that could possibly be a point), whose x coordinate is t_j. This vertical segment intersects exactly the segments representing all red segments intersected by bundle B_j (i.e., the segments of the red bundle(s) participating in the bundle-bundle intersection(s) with B_j). In particular, the bottom and the top endpoints of this segment lie respectively on the two red segments that represent the first and the last segment in R intersected by bundle B_j, according to the topological ordering of the red segments. If B_j intersects only one red segment, then in the life table B_j is represented by a point. In Figure 2 all the blue bundles except B_5 are represented by a point, but in a more complicated example many bundles might be represented by line segments. Note that instead of storing the segment list of each blue bundle explicitly, we just maintain a pointer to that bundle as it appears in (the corresponding version of) the sweepline status structure.

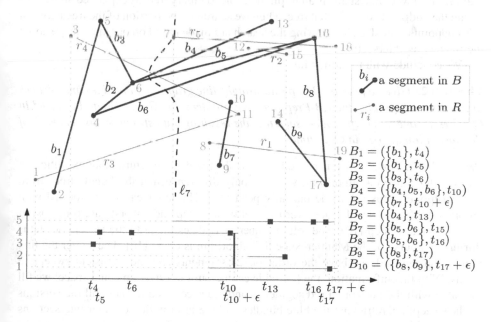

Fig. 2. Execution of the algorithm from Section 2.2 for $R = \{r_1, \ldots, r_5\}$ and $B = \{b_1, \ldots, b_{10}\}$. Above: the segments in R and B (solid lines, see the legend on the right); the events of the plane sweep in order (gray numbers), and the pseudoline ℓ_7 at time 7 (dashed line). Right: the blue bundles $\{B_1, \ldots, B_{10}\}$ encountered by the algorithm. Below: the life table for R and $\{B_1, \ldots, B_{10}\}$.

2.3   The resulting data structure

After the life table is built, we sweep it with a horizontal straight line from bottom to top, again making the sweepline status partially persistent. The events now correspond to red segments, and the version of the sweepline status at a time moment i contains all blue bundles crossing the horizontal line $y = i$, sorted by x coordinate and stored in a balanced binary tree.

Our ultimate data structure is the persistent sweepline status of the above (second) plane sweep. We are able, in $O(1)$ time, to retrieve the version of the sweepline status structure at any moment i. The sweepline status at the moment i is a tree storing the blue bundles whose beginning was witnessed before the moment i and whose end was witnessed after that moment (in other words, those blue bundles that intersect the horizontal line $y = i$ in the life table). See Figure 2. Since each single bundle is stored in a balanced binary tree, the tree of bundles is also a balanced binary tree. Therefore it has height $O(\log n)$. Moreover, it can be accessed in the same way as a standard binary tree: any navigation query in it (see Problem 1) can be performed in $O(1)$ time.

Now suppose we wish to perform a search on (a portion of) a red segment r_i among the blue segments that cross it. We are required to be able to quickly determine where the interesting intersection(s) lie with respect to p, for any point p of intersection between r_i and a blue segment. The search in our data structure is then done as follows. We retrieve the version of the sweepline status (of the second plane sweep) at the moment i. This sweepline status is an implicit balanced binary tree, as explained above. We locate the endpoints of r_i in that tree. Then we search in the portion of the tree between r_i's endpoints. The decisions during the search are made based on our knowledge about the interesting intersections.

We conclude with the following.

Theorem 2. *Given a pair R, B of plane straight-line graphs with n edges and vertices in total in both graphs, the RB-Preprocessing problem for R and B can be solved in $O(n \log n)$ time and $O(n)$ space, such that the resulting data structure answers each of the navigation queries in $O(1)$ time.*

Proof. The first phase of our procedure to build the data structure is an execution of the Mantler-Snoeyink algorithm, with the only difference that the bundle trees from the sweepline status are made partially persistent. The latter can be performed with amortized $O(1)$ time and space overhead per update step and a worst-case time cost of $O(1)$ per access step [10]. The total number of updates made to the sweepline status during the course of the Mantler-Snoeyink algorithm is $O(n)$ [16]. Thus, after the first phase is completed, we have the order of red segments and the persistent sweepline status. With this information, the life table can be built in $O(n)$ time and space: we fill the table with the horizontal red segments, and we access sequentially all the versions of the sweepline, retrieving the blue bundles and the information on their intersections with the red bundles, and drawing the vertical segments of the life table. Sweeping the life table with a horizontal line, and making the sweepline status partially persistent again costs $O(n \log n)$ time and $O(n)$ space.

For an edge $e = r_i$ of R, the version of the sweepline status structure at time i provides a balanced binary search tree T_e, required by the RB-Preprocessing problem, such that T_e can be navigated (but not modified) in the same way and with the same time complexity as the standard balanced binary search tree. Hence the *navigation queries* of the RB-Preprocessing problem can be answered in $O(1)$ each.

3 Applications

We proceed with more detail on the applications of our technique, which are listed in Section 1. Due to space constraints, we give here only the following results: red-blue batched search, the vertical distance problem, and a short version of the construction of the Hausdorff Voronoi diagram. The omitted parts can be found in the full version of this paper [13].

3.1 The red-blue batched search problem

Consider two sets, R (red) and B (blue), of line segments in $\mathbb{R}^2$, such that the segments in each set are pairwise interior-disjoint, and suppose that some of the red-blue intersections are *interesting*, and there is at most one interesting red-blue intersection per each segment. Let $\mathcal{O}$ be an oracle that, given an intersection point p of a red segment r and a blue segment b, determines to which side of p the interesting intersection on r lies.

Problem 2 (Red-blue batched search problem [9]). Given sets R, B and oracle $\mathcal{O}$, find all interesting intersections between the segments in R and the ones in B.

Dehne et al. [9] showed how to solve the red-blue batched search problem by using an augmentation of the hereditary segment tree data structure of Chazelle et al. [4]. Their solution requires $O(n \log^2 n)$ space and $O(n \log^3 n)$ queries to the oracle.

Our technique presented in Section 2.2 can be directly applied to solve the red-blue batched search problem with better time and space: We preprocess the sets R and B; after that for each red segment r we perform a binary search in the (implicit) tree storing the red-blue intersections along r. The search is guided by the oracle $\mathcal{O}$, and thus it requires $O(\log n)$ queries to the oracle. Since the number of red segments is $O(n)$, the total number of queries to the oracle required for searching all red edges is $O(n \log n)$. Theorem 2 implies the following.

Theorem 3. *The red-blue batched search problem for the sets R, B and the oracle $\mathcal{O}$ can be solved using $O(n)$ space and $O(n \log n)$ queries to the oracle, where n is the total number of segments in R and B.*

3.2 Vertical distance for a pair of 3D polyhedral terrains, one of which is convex

Let R and B be two polyhedral terrains of complexity n_R and n_B respectively, where terrain B is convex (that is, B is the upper envelope of a set of n_B planes in $\mathbb{R}^3$). Let $n = n_R + n_B$. We wish to determine the maximum vertical distance between R and B, i.e., the length of the longest vertical line segment connecting a point in R and a point in B.[8] As an illustration, the reader may imagine a (convex) approximate model of a mountain, and a need to compare it with the real mountain (of course, not necessarily convex) in order to estimate the quality of the approximation.

Since both surfaces R and B are composed of planar patches, the vertical distance between R and B is the vertical distance between these patches. The maximum vertical distance between R and B is thus attained either between a vertex of R and a facet of B (or vice versa), at infinity along an unbounded edge of one of the surfaces, or between an edge of R and an edge of B. In the second case the distance between R and B is $+\infty$. Both the first and the second case can be easily processed by point location queries of a (possibly infinite) vertex of one surface into the other one, which requires $O(n \log n)$ time in total. To deal with the last case one can preprocess the vertical projections of R and B following our technique, and perform the binary searches along each edge e of R for the intersection with an edge of B maximizing the vertical distance. Consider the cross-section of B by the vertical plane containing e. This is a convex monotone polygonal line. The sequence h_B of heights of its breakpoints is unimodal, and since all points of e lie on the same line, if we subtract from each member of h_B the height of the point in e lying on the same vertical line, the resulting sequence will still have one maximum. It then follows that given a point $p \in e$ vertically above/below an edge of B, in constant time we can find out in which direction this maximum lies, and this is exactly what the oracle for the binary search along e should do. Using Theorem 2, we conclude.

Theorem 4. *Given a pair of polyhedral terrains in 3D, where one of the terrains is convex, the maximum vertical distance between the terrains can be found in $O(n \log n)$ time and $O(n)$ space, where n is the total complexity of both terrains.*

Notice that by slightly changing the algorithm, we could be answering the minimum vertical distance, instead of the maximum one. In particular this gives an alternative $O(n \log n)$ algorithm to solve the shortest watchtower problem [21, 22].

3.3   Construction of the Hausdorff Voronoi diagram

Given a set of n distinct points in the plane, we partition this set, resulting in a family S of point clusters, where no two clusters share a point. Let the distance from a point $t \in \mathbb{R}^2$ to a cluster $P \in S$, denoted as $\mathsf{d}_f(t, P)$, be the maximum Euclidean distance from t to any point in P. The *Hausdorff Voronoi diagram* of S, denoted as $\mathsf{HVD}(S)$, is a subdivision of $\mathbb{R}^2$ into maximal regions such that every point within one region has the same nearest cluster according to distance $\mathsf{d}_f(\cdot, \cdot)$.

The diagram has worst-case combinatorial complexity $\Theta(n^2)$, and it can be constructed in optimal $O(n^2)$ time [11]. However, these bounds can be refined according to certain parameters of the family S. Two clusters are called *non-crossing* if their convex hulls intersect at most twice (i.e., their convex hulls are pseudocircles), and *crossing* otherwise. Below we consider separately the (simpler) case of non-crossing clusters, and the one of crossing clusters. The latter case subsumes the former one, and is given in detail in the full version [13].

---

[8] It may happen, that the distance keeps increasing as we move along some direction towards infinity. Then we say that the maximum vertical distance between R and B is $+\infty$.

Non-crossing clusters. If all clusters in S are pairwise non-crossing, the complexity of $\mathsf{HVD}(S)$ is $O(n)$. In this case the diagram can be constructed in expected $O(n \log^2 n)$ time and expected $O(n)$ space [5, 14]. The best deterministic algorithm to date requires $O(n \log^5 n)$ time and $O(n \log^2 n)$ space [9].[9] The latter algorithm follows the divide-and-conquer strategy. To merge two recursively computed diagrams, a bottleneck procedure is formulated as a red-blue batched segment search problem (see Section 3.1), where the two segment sets are the sets of edges of the two diagrams. The latter problem is then solved in $O(n \log^2 n)$ space and $O(n \log^3 n)$ queries to the oracle. The authors define an oracle to perform this search, which they assume can be implemented in $O(1)$ time. We were unable to reconstruct the claimed constant-time oracle, however we know how to implement it in $O(\log n)$ time per query. Theorem 3 implies an algorithm to construct the Hausdorff Voronoi diagram of a family of non-crossing clusters in $O(n \log^3 n)$ time and $O(n)$ space.[10] This result is subsumed by the one for arbitrary clusters, which is strictly more general (for a family of non-crossing clusters $m = 0$).

Arbitrary clusters. Consider the Hausdorff Voronoi diagram of a family S of arbitrary clusters, that could possibly cross. The essential parameter used to refine the quadratic bounds related to the diagram in that case, is the *number of crossings*,[11] denoted by m. The parameter m is bounded from above by half the number of intersections between the convex hulls of all pairs of crossing clusters. In the worst case $m = \Theta(n^2)$, however it is small in known practical applications, e.g., in VLSI CAD [19, 20]. The combinatorial complexity of the Hausdorff Voronoi diagram is shown to be $O(n + m)$ [19]. Apart from the $O(n^2)$ time algorithm mentioned above, there is a plane sweep [19] and a divide and-conquer [20] algorithm to construct the diagram. Both of them are sensitive to the parameter m, however their time complexity depends as well on some other parameters, which are unrelated to m. In particular, these algorithms may have $\Omega(n^2)$ time complexity even when clusters are non-crossing.

Our technique can be applied to reduce the time complexity of the divide-and-conquer construction of the Hausdorff Voronoi diagram of arbitrary clusters [20]. The resulting algorithm is the fastest to date deterministic algorithm for certain input families, where clusters may cross, but the number of crossings is small. See the detailed description of our algorithm in [13].

Acknowledgments. Research partially completed while J. I. was on sabbatical at the Algorithms Research Group of the Computer Science Dept. at ULB with support from a Fulbright Research Fellowship, F.R.S.-FNRS., and NSF grants CNS-1229185, CCF-1319648, and CCF-1533564. E. K. was supported by F.R.S.-FNRS, and by the SNF grant P2TIP2-168563 of the Early PostDoc Mobility program. S. L. is directeur de recherches du F.R.S.-FNRS.

---

[9] The time complexity claimed in Dehne et al. is $O(n \log^4 n)$, however we believe that in reality the described algorithm requires $O(n \log^5 n)$ time.

[10] Note that if it was possible to implement the oracle in $O(1)$ time, our algorithm would instantly be improved by a $O(\log n)$ factor. Thus in all cases our algorithm is faster than the previous one by a factor of $O(\log^2 n)$.

[11] See [19, 20] for the formal definition of the number of crossings.

References

1. Abellanas, M., Hurtado, F., Icking, C., Klein, R., Langetepe, E., Ma, L., Palop, B., Sacristán, V.: The farthest color Voronoi diagram and related problems. In: 17th Eur. Workshop on Comput. Geom. pp. 113–116 (2001), Tech. Rep. 002 2006, Univ. Bonn
2. Aronov, B., Bose, P., Demaine, E.D., Gudmundsson, J., Iacono, J., Langerman, S., Smid, M.: Data structures for halfplane proximity queries and incremental voronoi diagrams. In: 7th Latin American Symp. on Theoretical Informatics (LATIN). pp. 80–92. Springer (2006)
3. Chan, T.M.: A simple trapezoid sweep algorithm for reporting red/blue segment intersections. In: CCCG. pp. 263–268 (1994)
4. Chazelle, B., Edelsbrunner, H., Guibas, L.J., Sharir, M.: Algorithms for bichromatic line-segment problems and polyhedral terrains. Algorithmica 11(2), 116–132 (1994)
5. Cheilaris, P., Khramtcova, E., Langerman, S., Papadopoulou, E.: A randomized incremental algorithm for the Hausdorff Voronoi diagram of non-crossing clusters. Algorithmica 76(4), 935–960 (2016)
6. Cheong, O., Everett, H., Glisse, M., Gudmundsson, J., Hornus, S., Lazard, S., Lee, M., Na, H.S.: Farthest-polygon voronoi diagrams. Comput. Geom. 44(4), 234–247 (2011)
7. Claverol, M., Khramtcova, E., Papadopoulou, E., Saumell, M., Seara, C.: Stabbing circles for some sets of Delaunay segments. In: 32th Eur. Workshop on Comput. Geom. (EuroCG). pp. 139–143 (2016)
8. Claverol, M., Khramtcova, E., Papadopoulou, E., Saumell, M., Seara, C.: Stabbing circles for sets of segments in the plane. Algorithmica (2017), DOI 10.1007/s00453-017-0299-z
9. Dehne, F., Maheshwari, A., Taylor, R.: A coarse grained parallel algorithm for Hausdorff Voronoi diagrams. In: Int. Conf. on Parallel Processing (ICPP). pp. 497–504. IEEE (2006)
10. Driscoll, J.R., Sarnak, N., Sleator, D.D., Tarjan, R.E.: Making data structures persistent. In: 18th annual ACM Symp. on Theory of Computing. pp. 109–121. ACM (1986)
11. Edelsbrunner, H., Guibas, L.J., Sharir, M.: The upper envelope of piecewise linear functions: algorithms and applications. Discr. & Comput. Geom. 4(4), 311–336 (1989)
12. Guibas, L.J., Sharir, M., Sifrony, S.: On the general motion-planning problem with two degrees of freedom. Discr. & Comput. Geom. 4(5), 491–521 (1989)
13. Iacono, J., Khramtcova, E., Langerman, S.: Searching edges in the overlap of two plane graphs. ArXiv e-prints (2017), arXiv:1701.02229
14. Khramtcova, E., Papadopoulou, E.: Randomized incremental construction for the Hausdorff Voronoi diagram of point clusters. ArXiv e-prints (2016), arXiv:1612.01335
15. Mairson, H.G., Stolfi, J.: Reporting and counting intersections between two sets of line segments. In: Theoretical Foundations of Computer Graphics and CAD, pp. 307–325 (1988)
16. Mantler, A., Snoeyink, J.: Intersecting red and blue line segments in optimal time and precision. In: Discr. and Comput. Geom.: Japanese Conference, JCDCG 2000, Revised Papers. pp. 244–251 (2001)
17. Mount, D.M.: Storing the subdivision of a polyhedral surface. Discr. & Comput. Geom. 2(2), 153–174 (1987)
18. Palazzi, L., Snoeyink, J.: Counting and reporting red/blue segment intersections. CVGIP: Graphical Models and Image Processing 56(4), 304–310 (1994)
19. Papadopoulou, E.: The Hausdorff Voronoi diagram of point clusters in the plane. Algorithmica 40(2), 63–82 (2004)
20. Papadopoulou, E., Lee, D.T.: The Hausdorff Voronoi diagram of polygonal objects: A divide and conquer approach. Int. J. of Comput. Geom. & Appl. 14(06), 421–452 (2004)
21. Sharir, M.: The shortest watchtower and related problems for polyhedral terrains. Inf. Process. Lett. 29(5), 265–270 (1988)
22. Zhu, B.: Computing the shortest watchtower of a polyhedral terrain in $O(n \log n)$ time. Comput. Geom. 8(4), 181–193 (1997)

Improved Average Complexity for Comparison-Based Sorting

Kazuo Iwama[1] and Junichi Teruyama[2,3]

[1] Kyoto University, iwama@kuis.kyoto-u.ac.jp
[2] National Institute of Informatics,
[3] JST, ERATO, Kawarabayashi Large Graph Project, teruyama@nii.ac.jp

Abstract. This paper studies the average complexity on the number of comparisons for sorting algorithms. Its information-theoretic lower bound is $n \lg n - 1.4427n + O(\log n)$. For many efficient algorithms, the first $n \lg n$ term is easy to achieve and our focus is on the (negative) constant factor of the linear term. The current best value is -1.3999 for the MergeInsertion sort. Our new value is -1.4106, narrowing the gap by some 25%. An important building block of our algorithm is "two-element insertion," which inserts two numbers A and B, $A < B$, into a sorted sequence T. This insertion algorithm is still sufficiently simple for rigorous mathematical analysis and works well for a certain range of the length of T for which the simple binary insertion does not, thus allowing us to take a complementary approach with the binary insertion.

1 Introduction

A majority of existing sorting algorithms, including Bubble sort, Quick sort, Heap sort, Merge sort and Insertion sort, are so-called comparison-based sorts, in which our basic operation is a comparison of two input numbers. The complexity in terms of this measure, the number of comparisons needed to obtain a sorted sequence, is an obvious lower bound of the running time of the algorithm. Thus it has been a popular research topic in TCS to investigate its upper and lower bounds for several sorting algorithms. Note that any sorting algorithm for n elements can be described as a binary decision tree having $n!$ leaves corresponding to all different permutations of the n elements. The number of comparisons to obtain one of them is the number of nodes on the path from the root to the leaf corresponding to the sequence. Therefore we have an obvious lower bound, called an information-theoretic lower bound. Namely, any sorting algorithm needs

$$\lceil \lg n! \rceil \approx n \lg n - 1.4427n + O(\log n).$$

comparisons in the worst case.

Usually it is not very hard to obtain an upper bound of $n \lg n$. For instance, consider the BinaryInsertion sort that increases the length of the sorted sequence one by one using binary insertion. Obviously we have $n - 1$ steps and each of them consists of at most $\lceil \lg n \rceil$ comparisons (and much less for most of the steps).

© Springer International Publishing AG 2017
F. Ellen et al. (Eds.): WADS 2017, LNCS 10389, pp. 485–496, 2017.
DOI: 10.1007/978-3-319-62127-2_41

Thus our interest naturally comes to the constant factor for the linear term in n. Unfortunately, however, its analysis does not seem so easy and our knowledge is quite limited. For instance, it is at most -0.91 for Merge (and similar other) sort [6, 13] and the current best one is -1.32 for MergeInsertion sort obtained by Ford and Johnson more than five decades ago [3].

Our interest in this paper is the *average-case complexity* on the number of comparisons, which should be easier to obtain than the worst-case complexity. In fact we do have a number of better results; -1.26 for Merge sort [6], -1.38 for BinaryInsertion sort, and most recently -1.3999 for MergeInsertion sort [2]. Notice that 1.3999 is some 96.98% of 1.4427, but there still exists a gap and seeking the exact bound for this fundamental problem should be an important research goal.

Our Contribution We achieve 1.4034 by a new algorithm $(1,2)$INSERTION. Furthermore it is shown that the constant is improved to -1.4106 by combining the new algorithm with the MergeInsertion sort. Thus we have narrowed the previous gap between 1.3999 and 1.4427 by some 25%. Our new algorithm is based on binary insertion. Recall that the BinaryInsertion sort repeats a binary insertion of a new item into a sorted sequence of length $i - 1$ for $i = 2$ to n. Here the performance of binary insertion itself is optimal because it constitutes an optimal decision tree of height $\lceil \lg i \rceil$. However, if i is not a power of two, this tree is not completely balanced, i.e., there is a difference of one comparison due to the position of the inserted element. This small difference in each step accumulates during the repeated steps and finally creates a large imbalance. This is the reason for its relatively poor performance.

Our idea is to use a binary insertion if i is close to a power of two and to use what we call a "two-element merge," or 2MERGE otherwise. 2MERGE merges a two-element sequence (A, B), $A < B$, with a sorted sequence T of length $i - 2$ to obtain a sorted sequence of length i. We first insert A using a kind of binary search, meaning A is compared with an element in T whose position is approximately $1/3$ from the smallest. If A falls into the first third of T, then we use a standard (with a bit of care) binary search, called *right-heavy binary search* or *RHBS*. The key thing here is that the original "bad" i changes to a "good" i' in this binary insertion. If A falls into the right part of T, we simply recurse. Then, we insert B into T by using a standard binary search. Thus we can reduce the imbalance of each step of insertion, which contributes to the better bound for the whole sorting.

Due to [2], the performance of MergeInsertion differs a lot for different n and it hits a best peak when n is about one third from the previous power-of-two number, which achieves around -1.415. This is much better than our $(1,2)$INSERTION (but, unfortunately, it quickly gets worse as n leaves the best position and ends up with -1.3999 for a roughly power-of-two n). Thus here is a natural idea: For a given sequence X of length n that is bad for MergeInsertion, select the largest value n' that is less than n and is good for MergeInsertion. Then we use MergeInsertion sort for a length n' subsequence of X and insert the remaining elements using $(1,2)$INSERTION, which in fact gives us -1.4106.

Related Work The idea of inserting two elements into a sorted sequence is not new. [4] and [14] claimed two exactly optimal algorithms for such a merge operation in terms of the worst-case complexity [4] and in terms of the average-case complexity [14]. Unfortunately, both algorithms are a bit involved and their performance analysis did not give closed formulas for the complexity. Our 2MERGE is probably not exactly optimal, but is sufficiently simple for rigorous mathematical analysis.

The analysis of the BinaryInsertion sort by Edelkamp and Weiß [2] gives many hints to our new analysis. They show that the average number of comparisons is

$$\lceil \lg i \rceil + \mathcal{B}(i), \text{ where } \mathcal{B}(i) = 1 - \frac{2^{\lceil \lg i \rceil}}{i} \tag{1}$$

for a single insertion and is

$$\sum_{i=1}^{n} (\lceil \lg i \rceil + \mathcal{B}(i)) = n \lg n + \left(1 - \lg p_n - \frac{1 + \ln(4p_n)}{p_n}\right) \tag{2}$$

$$< n \lg n - 1.386n$$

for the entire BinaryInsertion sort, where $p_n = \frac{n}{2^{\lceil \lg n \rceil}}$ is a parameter indicating the deviation from a power of two. Edelkamp and Weiß [2] also includes a nice survey on this topic.

Although we have few results on the worst-case complexities for asymptotically large n, we do have a rather rich literature for small n's. For instance, the information-theoretic bound (actually its ceiling) cannot be achieved by any comparison-based sorting for $12 \leq n \leq 15$. The MergeInsertion sort achieves a matching upper bound for $1 \leq n \leq 15$, but $n = 16$ is still open, namely there is a gap of one between the lower and upper bounds (45 and 46, resp.) for the exact number of necessary comparisons. It is also known that MergeInsertion is not optimal for some n's, for instance, for $n = 47$. See [1, 3, 7–12, 15] for these results.

Notations and Assumptions. Our sorting algorithm takes a *sequence* of all different n *elements* as input. An *average complexity* (or simply *complexity*) of a sorting algorithm ALG is the expected number of comparisons ALG executes to sort each of $n!$ different sequences of length n. Note that the complexity of all sorting algorithms in this paper is written as $n \lg n + cn + O(\log n)$ for some negative constant c. It is important to mention that the value of c, that is our main issue, periodically changes depending on n usually and we are interested in its worst (largest) value for asymptotically large n, unless otherwise stated. We exploit the $O(\log n)$ term to make analysis simpler. In particular we assume, without loss of generality, that n is always even throughout this paper. Also, when summing up a cost function $f(i)$ for $i = 1$ to n, an $O(1/i)$ term in f is not important. For notation, we write $x = y \pm z$ if $|x - y| < z$, where z may be a big-O notation like $x = y \pm O(z)$. We may denote a sequence of one element (s_1) by simply s_1.

2   Our Algorithm and Its Analysis

See Algorithms 1, 2 and 3. The main algorithm is Algorithm 1. Note that Algorithm 2 is improved in the next section and Algorithm 1 is combined with MergeInsertion in Section 4. For a given sequence $S = (s_1, s_2, \ldots, s_{n-1}, s_n)$ with an even n, (1,2)INSERTION works in Round 0, Round 2, ... up to Round $n-2$. In Round 0, s_1 and s_2 are sorted by a single comparison to make a sorted sequence T_0 of length two. In Round i, s_{i+1} and s_{i+2} are inserted into T_{i-2} obtained in Round $i-2$ by using (i) a single call of 2MERGE or (ii) two calls of RHBS, depending on the value i. Recall that we wish to obtain the average complexity for all different $n!$ sequences, in other words, we wish to obtain the expected number of comparisons assuming that each S appears uniformly at random. It then turns out that we can also assume that the position of s_{i+1} (and that of s_{i+2} also) in each round is uniformly at random in the different $i+1$ positions of T_{i-2} that includes i elements. Thus the overall average complexity is a simple sum of the average complexity of each round.

We first make an analysis of 2MERGE. Note that 2MERGE uses RHBS which stands for Right-Heavy Binary Search. Note that the number, say q, of comparisons to insert A into a sequence $T = (t_1, \ldots t_i)$ is $q_0 = \lceil \lg(i+1) \rceil - 1$ or $q_0 + 1$ if we use the standard binary search. The feature of RHBS is that if $q = q_0 + 1$

Algorithm 1 (1,2)INSERTION(S)

Input: A (unsorted) sequence $S = (s_1, s_2, ..., s_n)$, where n is even.
Output: Sorted sequence
Step 1: If $n = 2$, then sort (s_1, s_2) with a single comparison.
Step 2: Sort $S' = (s_1, ..., s_{n-2})$ by (1,2)INSERTION to obtain T'.
Step 3: If $p_n \in [0.5511, 0.888]$ then insert s_{n-1} and s_n into T' by calling 2MERGE(s_{n-1}, s_n, T'). Otherwise insert s_{n-1} into T' by RHBS and then s_n by RHBS.

Algorithm 2 2MERGE(A, B, T)

Input: A and B are numbers and $T = (t_1, t_2, ..., t_{i-2})$ is a sorted sequence such that i is even and $i \geq 4$.
Output: Sorted sequence of length i.
Step 1. Compare A and B and swap them if $A > B$.
Step 2. Let $\alpha(r) = 1 - 2^{-r/2}$. For $r = 1, 2, \ldots$, up to $2 \lg i$, compare A with $t_{\lceil \alpha(r)i \rceil}$ and go to Step 3 if $A < t_{\lceil \alpha(r)i \rceil}$.
Step 3. Insert A to $(t_{\lceil \alpha(r-1)i \rceil} + 1, \ldots, t_{\lceil \alpha(r)i \rceil} - 1)$ using RHBS. Suppose that A falls between t_ℓ and $t_{\ell+1}$.
Step 4. Insert B to $(t_{\ell+1}, \ldots, t_{i-2})$ using RHBS.

Algorithm 3 RHBS(A, T)

Input: A is a number and $T = (t_1, t_2, ..., t_i)$ is a sorted sequence.
Output: Sorted sequence of length $i + 1$.
Step 1. If $i \leq 3 \times 2^{\lceil \lg(i+1) \rceil - 2} - 1$, then let set $d := 2^{\lceil \lg(i+1) \rceil - 2}$. Otherwise, let set $d := i - 2^{\lceil \lg(i+1) \rceil - 1} + 1$.
Step 2. Let $T_1 = (t_1, \ldots, t_{d-1})$ and $T_2 = (t_{d+1}, \ldots, t_i)$.
Step 3. Compare A with t_d. If $A < t_d$, return RHBS(A, T_1) $\circ t_d \circ T_2$. Otherwise, return $T_1 \circ t_d \circ$RHBS(A, T_2).

---

for some A, then $q = q_0 + 1$ for any A' such that $A' > A$, in other words, the number of comparisons is monotone. This is easily realized by selecting t_d (to be compared with A) in each recursion phase such that either the number of T's elements that is smaller than t_d or the number of those that is larger than t_d be (a power of two)-1. Suppose for instance $8 \le i \le 15$. Then if i is 11 or less, then the first comparison is with t_4 and if i is 12 or more, then the first comparison is with t_{i-7}. There would be no merit of this structure if the position of A is uniformly distributed. However, if small A's are more likely than large A's, there is an obvious advantage and that provides a real merit in 2MERGE. Notice that even if our improvement in each step is a small constant, something like 0.1, that constant significantly affect the value of our constant factor of the linear term.

In Step 2, we determine the range of the smaller element A. If the condition there ($A < t_{\lceil \alpha(r)i \rceil}$) is met for $r = 1$, then the range is $(t_1, \ldots, t_{\lceil (1-1/\sqrt{2})i \rceil - 1})$, where $(1 - 1/\sqrt{2}) \approx 0.2929$. In general, the range is $(t_{\lceil \alpha(r-1)i \rceil + 1}, \ldots, t_{\lceil \alpha(r)i \rceil - 1})$ for an integer $r \ge 1$, and we wish to compute the average complexity of Step 3, i.e., the average number of comparisons to insert A into this range. Here we have two technical issues: (i) We introduce a parameter w_r and let $w_r := (\sqrt{2} - 1)2^{-r/2}i$. Note that w_r is somehow related to the size of the above range but it may not be integral. The idea is that the complexity does not differ significantly if the size of the range differs by a small constant and approximating the size by w_r makes our job much easier. (ii) Although the positions of A and B are uniformly at random, we now know that $A < B$. Therefore the probability that A falls between $t_{\ell-1}$ and t_ℓ under the condition that $A < B$ is $(i - \ell)/\binom{i}{2}$. We also extend the definition of $p_x = \frac{x}{2^{\lceil \lg x \rceil}}$ for a real value x.

Lemma 1. *Suppose that A is to be inserted to $(t_{\lceil \alpha(r-1)i \rceil + 1}, \ldots, t_{\lceil \alpha(r)i \rceil - 1})$ for an even i. Then 2MERGE requires*

$$A(r) = \lceil \lg w_r \rceil + 7 - 4\sqrt{2} - \frac{10 - 6\sqrt{2}}{p_r} + \frac{3 - 2\sqrt{2}}{p_r^2} \pm O\left(\frac{2^{r/2}}{i}\right) \qquad (3)$$

comparisons on average at Step 3. Furthermore, the expected value of $A(r)$ is

$$\mathbf{Pr}[r = 1]A(1) + \mathbf{Pr}[r = 2]A(2) + \cdots = \lceil \lg i \rceil + \mathfrak{T}(i),$$

where

$$\mathfrak{T}(i) = 5 - 4\sqrt{2} - \frac{1}{p_i} + \frac{1}{6p_i^2} + \begin{cases} -\frac{1}{6p_i} - \frac{1}{16p_i^2} - \frac{2}{3} & p_i \in (1/2, \frac{1+\sqrt{2}}{4}], \\ -\frac{\sqrt{2}}{3p_i} - \frac{1}{3} & p_i \in (\frac{1+\sqrt{2}}{4}, \frac{2+\sqrt{2}}{4}], \\ -\frac{4}{3p_i} + \frac{1}{4p_i^2} + \frac{1}{3} & p_i \in (\frac{2+\sqrt{2}}{4}, 1]. \end{cases} \qquad (4)$$

See Section 2.1 for the proof. Now we are going to Step 4 to insert B and here is our analysis (see the full version of this paper [5] for its proof).

Lemma 2. *For an even i, 2MERGE requires*

$$\lceil \lg(i - 1) \rceil + 1 - \frac{2}{p_i} + \frac{1}{3p_i^2} + O(1/i)$$

comparisons on average at Step 4.

The entire complexity of 2MERGE is the sum of these two quantities in Lemmas 1 and 2 and another two values; (+1) for comparing A and B at Step 1 and the one for the expected number of comparisons in Step 2 that is $2 \pm O(1/i)$ (the precise analysis is shown in [5]). Thus the complexity of 2MERGE is

$$\lceil \lg i \rceil + \lceil \lg(i-1) \rceil + \mathcal{U}(i) + O(1/i)$$

where ($\mathcal{T}(i)$ is equation (4))

$$\mathcal{U}(i) = 1 + \mathcal{T}(i) - \frac{2}{p_{i-1}} + \frac{1}{3p_{i-1}^2}. \tag{5}$$

Since this is the complexity for inserting two elements, the complexity for a single insertion can be regarded as a half of it, or

$$\lceil \lg i \rceil + \mathcal{U}(i)/2 + O(1/i). \tag{6}$$

It then turns out that by comparing this value with (1) of the BinaryInsertion, 2MERGE is better than BinaryInsertion for $0.5511 < p_i < 0.888$. (Note that this range is obtained by a numerical calculation.) Thus we use 2MERGE for this range of p_i and RHBS for the other range. In summary our one step complexity is

$$\lceil \lg i \rceil + \begin{cases} \mathcal{B}(i) & p_i \in (1/2, 0.5511] \\ \mathcal{U}(i)/2 + O(1/i) & p_i \in (0.5511, 0.888] \\ \mathcal{B}(i) & p_i \in (0.888, 1] \end{cases}$$

By simple calculation, this is rewritten by

$$\lceil \lg i \rceil + \mathcal{D}(p_i) \tag{7}$$

where

$$\mathcal{D}(p_i) = \begin{cases} 1 - \frac{1}{p_i} & p_i \in (1/2, 0.5511], \\ \frac{25}{6} - 2\sqrt{2} - \frac{19}{12p_i} + \frac{7}{32p_i^2} & p_i \in \left(0.5511, \frac{1+\sqrt{2}}{4}\right], \\ \frac{13}{3} - 2\sqrt{2} - \frac{9+\sqrt{2}}{6p_i} + \frac{1}{4p_i^2} & p_i \in \left(\frac{1+\sqrt{2}}{4}, \frac{2+\sqrt{2}}{4}\right], \\ \frac{14}{3} - 2\sqrt{2} - \frac{13}{6p_i} + \frac{3}{8p_i^2} & p_i \in \left(\frac{2+\sqrt{2}}{4}, 0.888\right], \\ 1 - \frac{1}{p_i} & p_i \in (0.888, 1]. \end{cases} \tag{8}$$

Now by using the trapezoidal rule, we have

$$\sum_{i=1}^{n} \mathcal{D}(p_i) = 2^{\lceil \lg n \rceil} \times \left\{ \int_{1/2}^{1} \mathcal{D}(x)dx + \int_{1/2}^{p_n} \mathcal{D}(x)dx \right\} + O(\log n)$$

and the following theorem. We omit details of analyses, see [5].

Theorem 1. *The complexity of* (1,2)INSERTION *is at most* $n \lg n - 1.40118n$.

2.1   Proof of Lemma 1

We first prove formula (3). From the assumption, we call RHBS$(A, (t_{\ell_1+1}, \ldots, t_{\ell_2-1}))$, where $\ell_1 = \lceil (1 - 2^{-(r-1)/2})i \rceil$ and $\ell_2 = \lceil (1 - 2^{-r/2})i \rceil$. For an integer ℓ, let E_ℓ denote the event that A falls between $t_{\ell-1}$ and t_ℓ. Also F denotes the event that

A is inserted between t_{ℓ_1} and t_{ℓ_2}, namely $F = \bigcup_{\ell=\ell_1+1}^{\ell_2} E_\ell$. Let $w = \ell_2 - \ell_1$ and $z = 2i - \ell_1 - \ell_2 - 1$. Since $\Pr[E_\ell] = \frac{i-\ell}{\binom{i}{2}}$, we have

$$\Pr[F] = \sum_{\ell=\ell_1+1}^{\ell_2} \frac{i-\ell}{\binom{i}{2}} = \frac{w \cdot z}{2\binom{i}{2}} \quad \text{and} \quad \Pr[E_\ell \mid F] = \frac{\Pr[E_\ell]}{\Pr[F]} = \frac{2(i-\ell)}{w \cdot z}.$$

Let $k = 2^{\lceil \lg w \rceil} - w$. By its monotonicity, RHBS requires $\lceil \lg w \rceil - 1$ comparisons if $t_{\ell_1} < A < t_{\ell_1+k}$, and requires $\lceil \lg w \rceil$ comparisons otherwise. Therefore, the average number of comparisons is $\lceil \lg w \rceil - \sum_{\ell=\ell_1+1}^{\ell_1+k} \Pr[E_\ell \mid F]$, we need to calculate the summation $\sum_{\ell=\ell_1+1}^{\ell_1+k} \Pr[E_\ell \mid F] = \sum_{\ell=\ell_1+1}^{\ell_1+k} \frac{2(i-\ell)}{w \cdot z}$. Observing that $k/w = 1/p_w - 1$, we have

$$\sum_{\ell=\ell_1+1}^{\ell_1+k} \frac{2(i-\ell)}{w \cdot z} = \frac{k}{w} \cdot \frac{2i - 2\ell_1 - k - 1}{z}$$

$$= \frac{k}{w} \cdot \frac{z + w - k}{z} \quad (\because 2i - 2\ell_1 - 1 = w + z)$$

$$= \frac{k}{w} \cdot \left(1 + \frac{w-k}{w} \cdot \frac{w}{z} \right)$$

$$= \frac{1}{p_w} - 1 + \left(-2 + \frac{3}{p_w} - \frac{1}{p_w^2} \right) \cdot \frac{w}{z} \tag{9}$$

Since $\ell_1 = \lceil (1 - 2^{-(r-1)/2})i \rceil$ and $\ell_2 = \lceil (1 - 2^{-r/2})i \rceil$. we have $w = 2^{-r/2}(\sqrt{2} - 1)i \pm 1$ and $z = 2^{-r/2}(\sqrt{2}+1)i \pm 1$. Observe the value $\frac{w}{z}$ is close to $3 - 2\sqrt{2}$, in fact the difference is bounded as

$$\left| 3 - 2\sqrt{2} - \frac{w}{z} \right| < \frac{4 - 2\sqrt{2}}{z} < \frac{2^{r/2}}{i} \quad (\because r \le 2 \lg i).$$

Therefore, because $k/w = 1/p_w - 1$ and $-2 + \frac{3}{p_w} - \frac{1}{p_w^2} \le \frac{1}{4}$, (9) continues as

$$(*) = \frac{1}{p_w} - 1 + \left(-2 + \frac{3}{p_w} - \frac{1}{p_w^2} \right) \cdot \left(3 - 2\sqrt{2} \pm \frac{2^{r/2}}{i} \right)$$

$$= -7 + 4\sqrt{2} + \frac{10 - 6\sqrt{2}}{p_r} - \frac{3 - 2\sqrt{2}}{p_r^2} \pm \frac{2^{r/2}}{4i}$$

Thus, the average number of comparisons is

$$\lceil \lg w \rceil + 7 - 4\sqrt{2} - \frac{10 - 6\sqrt{2}}{p_w} + \frac{3 - 2\sqrt{2}}{p_w^2} \pm \frac{2^{r/2}}{4i}. \tag{10}$$

As mentioned before the statement of the lemma, we wish to replace $\lg w$ by $\lg w_r$, since there is no obvious way of treating the ceiling of the former that includes another ceilings for w. Now, recall that $w_r = 2^{-r/2}(\sqrt{2} - 1)i$ and $p_r = \frac{w_r}{2^{\lceil \lg w_r \rceil}}$. We show that it is possible to simply replace $\lceil \lg w \rceil$ by $\lceil \lg w_r \rceil$ almost as it is: (i) If $\lceil \lg w \rceil = \lceil \lg w_r \rceil$ holds, then because $|\frac{1}{p_w} - \frac{1}{p_r}| = O\left(\frac{2^{r/2}}{i} \right)$, it is enough to replace the last (error) term with $O\left(\frac{2^{r/2}}{i} \right)$. (ii) Otherwise suppose that $\lceil \lg w \rceil \ne \lceil \lg w_r \rceil$. Since $|w - w_r| < 1$ and w is an integer, w must be a power of two and $\lceil \lg w_r \rceil$ must be $\lceil \lg w \rceil + 1$. It then follows that $p_w = 1$ and we can write that $1/p_r = 2 - \epsilon$, where $|\epsilon| = \frac{2|w_r - w|}{w_r} < (2\sqrt{2} + 2) \cdot 2^{r/2}/i$. Substituting $p_w = 1$, (10) becomes

$$\lceil \lg w \rceil \pm \frac{2^{r/2}}{4i}.$$

Substituting $\lceil \lg w_r \rceil = \lceil \lg w \rceil + 1$ and $1/p_r = 2 - \epsilon$, (3) becomes

$$\lceil \lg w \rceil + (2\sqrt{2} - 2)\epsilon + (3 - 2\sqrt{2})\epsilon^2 \pm O\left(\frac{2^{r/2}}{i}\right) = \lceil \lg w \rceil \pm O\left(\frac{2^{r/2}}{i}\right).$$

Therefore (10) i.e., the value we want to obtain can be replaced by (3) with the error term. Thus the former part of lemma is proved.

For formula (4) we need to give the average values of $\lceil \lg w_r \rceil$, $1/p_r$ and $1/p_r^2$. Since $\lg i = \lceil \lg i \rceil + \lg p_i$ and $\lceil x \rceil = -\lfloor -x \rfloor$ for any value x, we have

$$\lceil \lg w_r \rceil = \left\lceil \lg i + \lg(\sqrt{2} - 1) - r/2 \right\rceil = \lceil \lg i \rceil - \left\lfloor r/2 - \lg(p_i(\sqrt{2} - 1)) \right\rfloor.$$

Also, for any value x and integer m, we have

$$\lfloor r/2 + x \rfloor = \begin{cases} \lfloor r/2 \rfloor + m & x \in [m, m + 1/2), \\ \lceil r/2 \rceil + m & x \in [m + 1/2, m + 1). \end{cases}$$

Let

$$c_r(p_i) = \left\lfloor r/2 - \lg(p_i(\sqrt{2} - 1)) \right\rfloor.$$

Then since $\lg(p_i(\sqrt{2} - 1)) \in (-2.5, -1)$, we have

$$c_r(p_i) := \begin{cases} \lfloor r/2 \rfloor + 2 & p_i \in (1/2, \frac{1+\sqrt{2}}{4}], \\ \lceil r/2 \rceil + 1 & p_i \in (\frac{1+\sqrt{2}}{4}, \frac{2+\sqrt{2}}{4}], \\ \lfloor r/2 \rfloor + 1 & p_i \in (\frac{2+\sqrt{2}}{4}, 1]. \end{cases}$$

We have the following lemma about the expected values of $\lceil r/2 \rceil$ and $\lfloor r/2 \rfloor$.

Lemma 3. $\mathbf{E}[\lfloor r/2 \rfloor] = 2/3 \pm O(1/i)$ and $\mathbf{E}[\lceil r/2 \rceil] = 4/3 \pm O(1/i)$.

This lemma implies

$$\mathbf{E}[\lceil \lg w_r \rceil] = \lceil \lg i \rceil \pm O(1/i) - \begin{cases} 8/3 & p_i \in (1/2, \frac{1+\sqrt{2}}{4}], \\ 7/3 & p_i \in (\frac{1+\sqrt{2}}{4}, \frac{2+\sqrt{2}}{4}], \\ 5/3 & p_i \in (\frac{2+\sqrt{2}}{4}, 1]. \end{cases}$$

Similarly, we can obtain the expected value of $1/p_r$ and $1/p_r^2$ as follows.

Lemma 4.

$$\mathbf{E}[1/p_r] = \begin{cases} \frac{3\sqrt{2}+5}{12p_i} & p_i \in (1/2, \frac{1+\sqrt{2}}{4}] \\ \frac{3+2\sqrt{2}}{6p_i} & p_i \in (\frac{1+\sqrt{2}}{4}, \frac{2+\sqrt{2}}{4}] \\ \frac{3\sqrt{2}+5}{6p_i} & p_i \in (\frac{2+\sqrt{2}}{4}, 1] \end{cases}, \quad \mathbf{E}[1/p_r^2] = \begin{cases} \frac{5(3+2\sqrt{2})}{48p_i^2} & p_i \in (1/2, \frac{1+\sqrt{2}}{4}] \\ \frac{3+2\sqrt{2}}{6p_i^2} & p_i \in (\frac{1+\sqrt{2}}{4}, \frac{2+\sqrt{2}}{4}] \\ \frac{5(3+2\sqrt{2})}{12p_i^2} & p_i \in (\frac{2+\sqrt{2}}{4}, 1] \end{cases}.$$

Adding all those values, we can obtain (4) and the lemma is proved. The proofs of Lemmas 3 and 4 are shown in [5].

3   Improvement of 2MERGE

As mentioned before, the value of $\alpha(r)$ is selected based on the observation that (1) the probability that A falls in the left part of T should be close to $1/2$ and (2) the length of the left part for $r = 1$ (which seems more important than other less happening cases for $r \geq 2$) should be close to a power of two. The previous selection is perfect in terms of (1) but is not in terms of (2) since $\alpha(r)$ does not depend on the length i of T. In this section, we put a priority to (2) by setting

$$\alpha(r, p_i) = \begin{cases} 1 - \frac{1}{2^{k-1}} + \frac{1}{p_i 2^{k+1}} & (r = 2k - 1 \text{ and } p_i \in (3/4, 1]), \\ 1 - \frac{1}{2^k} - \frac{1}{p_i 2^{k+2}} & (r = 2k - 1 \text{ and } p_i \in (1/2, 3/4]), \\ 1 - \frac{1}{2^k} & (r = 2k). \end{cases} \quad (11)$$

Note that it now depends on i and it turns out that if A falls into the left part of T for $r = 1$, then the length of the left part is exactly a power of two for any i when $p_i \in (3/4, 1]$. Note that for even r, $\alpha(r, p_i)$ is the same as the previous $\alpha(r)$.

We denote the modified 2MERGE as 2MERGE* and the whole sorting algorithm as (1,2)INSERTION*. (See Algorithm 4 and 5.) Our analysis, having two cases for $p_i \geq 3/4$ and $p_i < 3/4$, is more involved but we can obtain the average number of comparisons for a single step is

$$\lceil \lg i \rceil + \mathcal{B}(i) \pm O(1/i) + \begin{cases} \frac{1}{2} - \frac{3}{4p_i} + \frac{25}{96p_i^2} & p_i \in (1/2, 3/4), \\ 1 - \frac{3}{2p_i} + \frac{13}{24p_i^2} & p_i \in [3/4, 1]. \end{cases}$$

As with the previous section, comparing this value with (1), 2MERGE* is better than the binary insertion for $p_i \in \left[\frac{3}{4} - \frac{\sqrt{6}}{12}, \frac{3}{4} + \frac{\sqrt{3}}{12}\right]$. (Note that we did not use numerical analysis this time.) Then, one step complexity of (1,2)INSERTION*

Algorithm 4 (1,2)INSERTION*(S)

Input: A (unsorted) sequence $S = (s_1, s_2, ..., s_n)$, where n is even.
Output: Sorted sequence
Step 1: If $n = 2$, then sort (s_1, s_2) with a single comparison.
Step 2: Sort $S' = (s_1, ..., s_{n-2})$ by (1,2)INSERTION* to obtain T'.
Step 3: If $p_n \in \left[\frac{3}{4} - \frac{\sqrt{6}}{12}, \frac{3}{4} + \frac{\sqrt{3}}{12}\right]$ then insert s_{n-1} and s_n into T' by calling 2MERGE(s_{n-1}, s_n, T'). Otherwise insert s_{n-1} into T' by RHBS and then s_n by RHBS.

Algorithm 5 2MERGE*(A, B, T)

Input: A and B are numbers and $T = (t_1, t_2, ..., t_{i-2})$ is a sorted sequence such that i is even and $i \geq 4$.
Output: Sorted sequence of length i.
Step 1. Compare A and B and swap them if $A > B$.
Step 2. Define $\alpha(r, p_i)$ as Equation (11). For $r = 1, 2, \ldots$, up to $2 \lg i$, compare A with $t_{\lceil \alpha(r)i \rceil}$ and go to Step 3 if $A < t_{\lceil \alpha(r)i \rceil}$.
Step 3. Insert A to $(t_{\lceil \alpha(r-1)i \rceil} + 1, \ldots, t_{\lceil \alpha(r)i \rceil} - 1)$ using RHBS. Suppose that A falls between t_ℓ and $t_{\ell+1}$.
Step 4. Insert B to $(t_{\ell+1}, \ldots, t_{i-2})$ using RHBS.

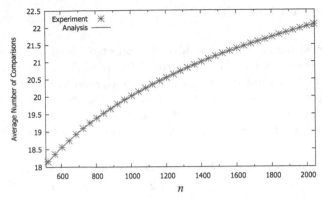

Fig. 1. The average number of comparisons of 2MERGE*: Experiment and analysis

is

$$\lceil \lg i \rceil + \mathcal{D}^*(p_i)$$

where

$$\mathcal{D}^*(p_i) = \begin{cases} 1 - \frac{1}{p_i} & (1/2, \frac{3}{4} - \frac{\sqrt{6}}{12}], \\ \frac{3}{2} - \frac{7}{4p_i} + \frac{25}{96p_i^2} & p_i \in (\frac{3}{4} - \frac{\sqrt{6}}{12}, 3/4], \\ 2 - \frac{5}{2p_i} + \frac{13}{24p_i^2} & p_i \in (3/4, \frac{3}{4} + \frac{\sqrt{3}}{12}], \\ 1 - \frac{1}{p_i} & p_i \in (\frac{3}{4} + \frac{\sqrt{3}}{12}, 1]. \end{cases}$$

Thus, we have

$$\sum_{i=1}^{n} \mathcal{D}^*(p_i) = 2^{\lceil \lg n \rceil} \times \left\{ \int_{1/2}^{1} \mathcal{D}^*(x)dx + \int_{1/2}^{p_n} \mathcal{D}^*(x)dx \right\} + O(\log n)$$

and can obtain the following theorem. The complete analysis for (1,2)INSERTION*
is given in the full version.

Theorem 2. *The complexity of* (1,2)INSERTION* *is at most* $n \lg n - 1.4034n$.

We conducted an experiment for 2MERGE*. We prepare sequences $N = (1, 2, \ldots, n)$ for n up to $2^{12} = 2046$. Then two elements I_1 and I_2 are selected
from N and they are inserted into $N - \{I_1, I_2\}$ using 2MERGE*. We take the
average for the number of comparisons for all possible pairs of I_1 and I_2. As one
can see the result matches the analysis very well. We also did a similar experiment for 2MERGE. The result is very close and the difference is not visible in
such a graph.

See Fig. 1, which illustrates our analysis and results of simulations. The
symbol '+' means the average number of comparisons of simulation for each n.
The line represents the value of analysis:

$$\lceil \lg i \rceil + \mathcal{B}(i) + \begin{cases} \frac{1}{2} - \frac{3}{4p_i} + \frac{25}{96p_i^2} & p_i \in (1/2, 3/4), \\ 1 - \frac{3}{2p_i} + \frac{13}{24p_i^2} & p_i \in [3/4, 1]. \end{cases}$$

4   Combination with MergeInsertion

See Fig. 2, which illustrates the performance of $(1,2)$INSERTION, $(1,2)$INSERTION*, and MergeInsertion [2] for the value of p_n. As one can see, MergeInsertion is way better than our algorithms in a certain range of p_n. In fact, due to [3, page 389], its best case happens for $n = \left\lceil \frac{2^k}{3} \right\rceil$ for an integer k, achieving a complexity of $n \lg n - (3 - \lg 3)n + O(\lg n) \approx n \lg n - 1.415n + O(\lg n)$. This best case can be easily included into our $(1,2)$INSERTION*, as follows (see Algorithm 6):

Suppose that our input satisfies $p_n \geq 2/3$. Then we select the largest k such that $n' := \left\lceil \frac{2^k}{3} \right\rceil \leq n$. Then we sort the first n' elements by MergeInsertion. After that the remaining elements are inserted by $(1,2)$INSERTION*. Since $n' = \frac{2n}{3p_n}$ as mentioned above, the complexity of MergeInsertion for that size is at most

$$n' \lg n' - (3 - \lg 3)n' = n' \lceil \lg n \rceil - \frac{4}{3p_n}n$$

and the additional comparisons in $(1,2)$INSERTION* cost is

$$\sum_{i=n'+1}^{n} \{\lceil \lg i \rceil + \mathcal{D}^*(i)\} = (n - n') \lceil \lg n \rceil + 2^{\lceil \lg n \rceil} \int_{2/3}^{p_n} \mathcal{D}^*(x)dx.$$

Summing up these two quantities, the complexity of COMBINATION is at most

$$n \lg n + \left\{ -\lg p_n - \frac{4}{3p_n} + \frac{1}{p_n} \int_{2/3}^{p_n} \mathcal{D}^*(x)dx \right\} n.$$

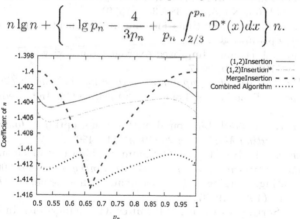

Fig. 2. Performance of the algorithms

Algorithm 6 COMBINATION(S)

Input: A (unsorted) sequence $S = (s_1, s_2, ..., s_n)$, where n is even.
Output: Sorted sequence
Step 1: If $p_n \geq 2/3$, then let $n' := \frac{2n}{3p_n}$. Otherwise, let $n' := \frac{n}{3p_n}$.
Step 2: Sort $S' = (s_1, ..., s_{n'})$ by the MergeInsertion sort to obtain T'.
Step 3: For $i = n' + 2, n' + 4, ..., n$, if $p_i \in \left[\frac{3}{4} - \frac{\sqrt{6}}{12}, \frac{3}{4} + \frac{\sqrt{3}}{12}\right]$, then insert s_{i-1} and s_i into T' by calling 2MERGE*(s_{i-1}, s_i, T'), otherwise insert s_{i-1} into T' by RHBS and then s_i by RHBS.

We can use exactly the same approach for the case that $p_n \leq 2/3$. It turns out however that the combined approach is worse than MergeInsertion itself for $0.638 \leq p_n \leq 2/3$. So it is better to use only MergeInsertion for this range. See Fig. 2 for the overall performance of the combined algorithm.

Theorem 3. *The complexity of* COMBINATION *is at most* $n \lg n - 1.41064n$.

5 Final Remarks

There is the wide agreement in the community that the information-theoretic lower bound $(= -1.4427)$ cannot be achieved by a specific sorting algorithm; to prove or disprove it is a big open question. Anyway, our upper bound for the average case seems quite close to the lower bound. So attacking the worst case using the ideas in this paper may be more promising.

References

1. Ayala-Rincón, M., De Abreu, B. T. and De Siqueira, J.: A variant of the Ford–Johnson algorithm that is more space efficient, *Information Processing Letters*, Vol. 102, No. 5, pp. 201–207 (2007).
2. S. Edelkamp and A. Weiß. QuickXsort: Efficient Sorting with n logn - 1.399n + o(n) Comparisons on Average. CSR 2014: pp. 139–152.
3. Ford, L. R. and Johnson, S. M.: A tournament problem, *The American Mathematical Monthly*, Vol. 66, No. 5, pp. 387–389 (1959).
4. F. K. Hwang and S. Lin, Optimal merging of 2 elements with n elements, Acta Informatica 1, pp.145–158, 1971.
5. K. Iwama and J. Teruyama, Improved Average Complexity for Comparison-Based Sorting, arXiv:1705.00849, 2017.
6. Knuth, D. E.: *The Art of Computer Programming, Volume 3: (2nd Ed.) Sorting and Searching*, Addison Wesley Longman Publishing Co., Inc., Redwood City, CA, USA (1998).
7. G.K. Manacher, The Ford–Johnson algorithm is not optimal, *Journal of the Association for Computing Machinery* 26 (1979) 441–456.
8. G.K. Manacher, T.D. Bui, T. Mai, Optimum combinations of sorting and merging, *Journal of the Association for Computing Machinery* 36 (1989) 290–334.
9. J. Schulte Mönting, Merging of 4 or 5 elements with n elements, *Theoretical Computer Science* 14 (1981) 19–37.
10. Peczarski, M.: Sorting 13 Elements Requires 34 Comparisons. In Proc. 10th Annual European Symposium on Algorithms. LNCS, vol. 2461, pp. 785–794. Springer (2002).
11. Peczarski, M.: New results in minimum-comparison sorting, *Algorithmica*, Vol. 40, No. 2, pp. 133–145 (2004).
12. Peczarski, M.: The Ford–Johnson algorithm still unbeaten for less than 47 elements, *Information processing letters*, Vol. 101, No. 3, pp. 126–128 (2007).
13. H. Steinhaus, Mathematical Snapshots, New Nork, 1950, pp. 37–40.
14. M. Thanh, V.S. Alagar, T. D. Bui, Optimal Expected-Time Algorithms for Merging. J. Algorithms 7(3): pp. 341–357, 1986.
15. Wells, M.: Applications of a Language for Computing in Combinatorics. In Proc. 1965 IFIP Congress, North-Holland, Amsterdam, pp. 497–498 (1966).

An EPTAS for Scheduling on Unrelated Machines of Few Different Types*

Klaus Jansen and Marten Maack

Christian-Albrechts-Universität zu Kiel, 24118 Kiel, Germany,
{kj,mmaa}@informatik.uni-kiel.de

Abstract. In the classical problem of scheduling on unrelated parallel machines, a set of jobs has to be assigned to a set of machines. The jobs have a processing time depending on the machine and the goal is to minimize the makespan, that is, the maximum machine load. It is well known that this problem is NP-hard and does not allow polynomial time approximation algorithms with approximation guarantees smaller than 1.5, unless P=NP. We consider the case that there is only a constant number K of machine types. Two machines have the same type, if all jobs have the same processing time for them. We present an efficient polynomial time approximation scheme (EPTAS) for this problem, that is, for any $\varepsilon > 0$ an assignment with makespan of length at most $(1+\varepsilon)$ times the optimum can be found in polynomial time in the input length and the exponent is independent of $1/\varepsilon$. In particular we achieve a running time of $2^{O(K \log(K)1/\varepsilon \log^4 1/\varepsilon)} + \text{poly}(|I|)$, where $|I|$ denotes the input length. Furthermore, we study the case where the minimum machine load has to be maximized and achieve a similar result.

1 Introduction

We consider the problem of scheduling jobs on unrelated parallel machines—or unrelated scheduling for short—in which a set $\mathcal{J}$ of n jobs has to be assigned to a set $\mathcal{M}$ of m machines. Each job j has a processing time p_{ij} for each machine i and the goal is to find a schedule $\sigma : \mathcal{J} \to \mathcal{M}$ minimizing the *makespan* $C_{\max}(\sigma) = \max_{i \in \mathcal{M}} \sum_{j \in \sigma^{-1}(i)} p_{ij}$, i.e., the maximum machine load. The problem is one of the classical scheduling problems studied in approximation. In 1990 Lenstra, Shmoys and Tardos [20] showed that there is no approximation algorithm with an approximation guarantee smaller than 1.5, unless P=NP. Moreover, they presented a 2-approximation and closing this gap is a rather famous open problem in scheduling theory and approximation (see e.g. [23]).

In particular we study the special case where there is only a constant number K of *machine types*. Two machines i and i' have the same type, if $p_{ij} = p_{i'j}$ holds for each job j. In many application scenarios this scenario is plausible, e.g. when considering computers which typically only have a very limited number of different

---

* This work was partially supported by the German Research Foundation (DFG) project JA 612/16-1.

F. Ellen et al. (Eds.): WADS 2017, LNCS 10389, pp. 497–508, 2017.
DOI: 10.1007/978-3-319-62127-2_42

types of processing units. We denote the processing time of a job j on a machine of type $t \in [K]$ by p_{tj} and assume that the input consist of the corresponding $K \times n$ processing time matrix together with machine multiplicities m_t for each type t, yielding $m = \sum_{t \in [K]} m_t$. Note that the case $K = 1$ is equivalent to the classical scheduling on identical machines.

We will also consider the reverse objective of maximizing the minimum machine load, i.e., $C_{\min}(\sigma) = \min_{i \in \mathcal{M}} \sum_{j \in \sigma^{-1}(i)} p_{ij}$. This problem is also known as max-min fair allocation or the Santa Claus problem. The intuition behind these names is that the jobs are interpreted as goods (e.g. presents), the machines as players (e.g. children), and the processing times as the values of the goods from the perspective of the different players. Finding an assignment that maximizes the minimum machine load, means therefore finding an allocation of the goods that is in some sense fair (making the least happy kid as happy as possible). We will refer to the problem as Santa Claus problem in the following, but otherwise will stick to the scheduling terminology.

We study approximation algorithms: Given an instance I of an optimization problem, an α-approximation A produces a solution in time poly($|I|$), where $|I|$ denotes the input length. For the objective function value $A(I)$ it is guaranteed that $A(I) \leq \alpha \mathrm{OPT}(I)$, in the case of an minimization problem, or $A(I) \geq (1/\alpha)\mathrm{OPT}(I)$, in the case of an maximization problem, where $\mathrm{OPT}(I)$ is the value of an optimal solution. We call α the *approximation guarantee* or *rate* of the algorithm. In some cases a polynomial time approximation scheme (PTAS) can be achieved, that is, for each $\varepsilon > 0$ an $(1 + \varepsilon)$-approximation. If for such a family of algorithms the running time can be bounded by $f(1/\varepsilon)$poly($|I|$) for some computable function f, the PTAS is called *efficient* (EPTAS), and if the running time is polynomial in both $1/\varepsilon$ and $|I|$ it is called *fully polynomial* (FPTAS).

Related work. It is well known that the unrelated scheduling problem admits an FPTAS in the case that the number of machines is considered constant [13] and we already mentioned the seminal work by Lenstra et al. [20]. Furthermore, the problem of unrelated scheduling with a constant number of machine types is strongly NP-hard, because it is a generalization of the strongly NP-hard problem of scheduling on identical parallel machines. Therefore an FPTAS can not be hoped for in this case. However, Bonifaci and Wiese [5] showed that there is a PTAS even for the more general vector scheduling case. However, in the case considered here, their algorithm has to solve m to the power of $\mathcal{O}(K(1/\varepsilon)^{1/\varepsilon \log 1/\varepsilon})$ linear programs. Gehrke et al. [10] presented a PTAS with an improved running time of $\mathcal{O}(Kn) + m^{\mathcal{O}(K/\varepsilon^2)}(\log(m)/\varepsilon)^{\mathcal{O}(K^2)}$ for unrelated scheduling with a constant number of machine types. On the other hand, Chen et al. [7] showed that there is no PTAS for scheduling on identical machines with running time $2^{(1/\varepsilon)^{1-\delta}}$ for any $\delta > 0$, unless the exponential time hypothesis fails. Furthermore, the case $K = 2$ has been studied: Imreh [14] designed heuristic algorithms with rates $2 + (m_1 - 1)/m_2$ and $4 - 2/m_1$, and Bleuse et al. [4] presented an algorithm with rate $4/3 + 3/m_2$ and moreover a (faster) 3/2-approximation, for the case that for each job the processing time on the second machine type is at most the one on the first. Moreover, Raravi and Nélis [22] designed a PTAS for the case with two machine types.

Interestingly, Goemans and Rothvoss [11] were able to show that unrelated scheduling is in P, if both the number of machine types and the number of job types is bounded by a constant. Job types are defined analogously to machine types, i.e., two jobs j, j' have the same type, if $p_{ij} = p_{ij'}$ for each machine i. In this case the matrix (p_{ij}) has only a constant number of distinct rows and columns. Note that already in the case we study, the rank of this matrix is constant. However the case of unrelated scheduling where the matrix (p_{ij}) has constant rank turns out to be much harder: Already for the case with rank 4 there is no approximation algorithm with rate smaller than 3/2 unless P=NP [8]. In a rather recent work, Knop and Koutecký [19] considered the number of machine types as a parameter from the perspective of fixed parameter tractability. They showed that unrelated scheduling is fixed parameter tractable for the parameters K and $\max p_{i,j}$, that is, there is an algorithm with running time $f(K, \max p_{i,j}) \text{poly}(|I|)$ for some computable function f that solves the problem to optimality.

For the case that the number of machines is constant, the Santa Claus problem behaves similar to the unrelated scheduling problem: there is an FPTAS that is implied by a result due to Woeginger [24]. In the general case however, so far no approximation algorithm with a constant approximation guarantee has been found. The results by Lenstra et al. [20] can be adapted to show that that there is no approximation algorithm with a rate smaller than 2, unless P=NP, and to get an algorithm that finds a solution with value at least $OPT(I) - \max p_{i,j}$, as was done by Bezkov and Dani [3]. Since $\max p_{i,j}$ could be bigger than $OPT(I)$, this does not provide an (multiplicative) approximation guarantee. Bezkov and Dani also presented a simple $(n-m+1)$-approximation and an improved approximation guarantee of $\mathcal{O}(\sqrt{n} \log^3 n)$ was achieved by Asadpour and Saberi [1]. The best rate so far is $O(n^\varepsilon)$ due to Bateni et al. [2] and Chakrabarty et al. [6], with a running time of $\mathcal{O}(n^{1/\varepsilon})$ for any $\varepsilon > 0$.

Results and Methodology. In this paper we show:

Theorem 1. *There is an EPTAS for both scheduling on unrelated parallel machines and the Santa Claus problem with a constant number of different machine types with running time* $2^{\mathcal{O}(K \log(K) 1/\varepsilon \log^4 1/\varepsilon)} + \text{poly}(|I|)$.

First we present a basic version of the EPTAS for unrelated scheduling with a running time doubly exponential in $1/\varepsilon$. For this EPTAS we use the dual approximation approach by Hochbaum and Shmoys [12] to get a guess T of the optimal makespan OPT. Then we further simplify the problem via geometric rounding of the processing times. Next we formulate a mixed integer linear program (MILP) with a constant number of integral variables that encodes a relaxed version of the problem. We solve it with the algorithm by Lenstra and Kannan [21, 18]. The fractional variables of the MILP have to be rounded and we achieve this with a cleverly designed flow network utilizing flow integrality and causing only a small error. With an additional error the obtained solution can be used to construct a schedule with makespan $(1 + \mathcal{O}(\varepsilon))T$. This procedure is described in detail in Section 2. Building upon the basic EPTAS we achieve the improved running time using techniques by Jansen [15] and by Jansen, Klein and Verschae [16]. The basic idea of these techniques is to make use of existential results about simple structured solutions of integer linear programs (ILPs). In particular these results can be used to guess the non-zero variables of the MILP, because they sufficiently limit the search space. We show how these techniques can be applied in our case in Section 3. Interestingly, our techniques can be adapted for the Santa Claus Problem, which typically has a worse approximation behaviour. We discuss the ideas needed for this in the last section of the paper. More details and omitted proofs can be found in the long version of the paper [17].

2 Basic EPTAS

In this chapter we describe a basic EPTAS for unrelated scheduling with a constant number of machine types, with a running time doubly exponential in $1/\varepsilon$. Wlog. we assume $\varepsilon < 1$. Furthermore $\log(\cdot)$ denotes the logarithm with basis 2 and for $k \in \mathbb{Z}_{\geq 0}$ we write $[k]$ for $\{1, \ldots, k\}$.

First, we simplify the problem via the classical dual approximation concept by Hochbaum and Shmoys [12]. In the simplified version of the problem a target makespan T is given and the goal is to either output a schedule with makespan at most $(1 + \alpha\varepsilon)T$ for some constant $\alpha \in \mathbb{Z}_{>0}$, or correctly report that there is no schedule with makespan T. We can use a polynomial time algorithm for this problem in the design of a PTAS in the following way. First we obtain an upper bound B for the optimal makespan OPT of the instance with $B \leq 2\text{OPT}$. This can be done using the 2-approximation by Lenstra et al. [20]. With binary search on the interval $[B/2, B]$ we can find in $\mathcal{O}(\log 1/\varepsilon)$ iterations a value T^* for which the mentioned algorithm is successful, while $T^* - \varepsilon B/2$ is rejected. We have $T^* - \varepsilon B/2 \leq \text{OPT}$ and therefore $T^* \leq (1 + \varepsilon)\text{OPT}$. Hence the schedule we obtained for the target makespan T^* has makespan at most $(1+\alpha\varepsilon)T^* \leq (1+\alpha\varepsilon)(1+\varepsilon)\text{OPT} = (1+\mathcal{O}(\varepsilon))\text{OPT}$. In the following we will always assume that a target makespan T is given. Next we present a brief overview of the algorithm for the simplified problem followed by a more detailed description and analysis.

Algorithm 2.

1. Simplify the input via geometric rounding with an error of εT.
2. Build the mixed integer linear program MILP($\bar{T}$) and solve it with the algorithm by Lenstra and Kannan ($\bar{T} = (1 + \varepsilon)T$).
3. If there is no solution, report that there is no solution with makespan T.
4. Generate an integral solution for MILP($\bar{T} + \varepsilon T + \varepsilon^2 T$) via a flow network utilizing flow integrality.
5. The integral solution is turned into a schedule with an additional error of $\varepsilon^2 T$ due to the small jobs.

Simplification of the Input. We construct a simplified instance $\bar{I}$ with modified processing times $\bar{p}_{tj}$. If a job j has a processing time bigger than T for a machine type $t \in [K]$ we set $\bar{p}_{tj} = \infty$. We call a job *big* (for machine type t), if $p_{tj} > \varepsilon^2 T$, and *small* otherwise. We perform a geometric rounding step for each job j with $p_{tj} < \infty$, that is, we set $\bar{p}_{tj} = (1+\varepsilon)^x \varepsilon^2 T$ with $x = \lceil \log_{1+\varepsilon}(p_{tj}/(\varepsilon^2 T)) \rceil$.

Lemma 1. *If there is a schedule with makespan at most T for I, the same schedule has makespan at most $(1 + \varepsilon)T$ for instance $\bar{I}$ and any schedule for instance $\bar{I}$ can be turned into a schedule for I without increase in the makespan.*

We will search for a schedule with makespan $\bar{T} = (1 + \varepsilon)T$ for $\bar{I}$.

We establish some notation for the rounded instance. For any rounded processing time p we denote the set of jobs j with $\bar{p}_{tj} = p$ by $J_t(p)$. Moreover, for each machine type t let S_t and B_t be the set of small and big rounded processing times. Obviously we have $|S_t| + |B_t| \leq n$. Furthermore $|B_t|$ is bounded by a constant: Let N be such that $(1 + \varepsilon)^N \varepsilon^2 T$ is the biggest rounded processing time for all machine type. Then we have $(1 + \varepsilon)^{N-1} \varepsilon^2 T \leq T$ and therefore $|B_t| \leq N \leq \log(1/\varepsilon^2)/\log(1 + \varepsilon) + 1 \leq 1/\varepsilon \log(1/\varepsilon^2) + 1$ (using $\varepsilon \leq 1$).

MILP. For any set of processing times P we call the P-indexed vectors of non-negative integers $\mathbb{Z}_{\geq 0}^P$ *configurations* (for P). The *size* size(C) of configuration C is given by $\sum_{p \in P} C_p p$. For each $t \in [K]$ we consider the set $C_t(\bar{T})$ of configurations C for the big processing times B_t and with size(C) $\leq \bar{T}$. Given a schedule σ, we say that a machine i of type t obeys a configuration C, if the number of big jobs with processing time p that σ assigns to i is exactly C_p for each $p \in B_t$. Since the processing times in B_t are bigger than $\varepsilon^2 T$ we have $\sum_{p \in B_t} C_p \leq 1/\varepsilon^2$ for each $C \in C_t(\bar{T})$. Therefore the number of distinct configurations in $C_t(\bar{T})$ can be bounded by $(1/\varepsilon^2 + 1)^N < 2^{\log(1/\varepsilon^2+1)1/\varepsilon \log(1/\varepsilon^2)+1} \in 2^{O(1/\varepsilon \log^2 1/\varepsilon)}$.

We define a mixed integer linear program MILP($\bar{T}$) in which configurations are assigned integrally and jobs are assigned fractionally to machine types. To this amount we introduce variables $z_{C,t} \in \mathbb{Z}_{\geq 0}$ for each machine type $t \in [K]$ and configuration $C \in C_t(\bar{T})$, and $x_{j,t} \geq 0$ for each machine type $t \in [K]$ and job $j \in J$. For $\bar{p}_{tj} = \infty$ we set $x_{j,t} = 0$. Besides this, the MILP has the following constraints:

$$\sum_{C \in C_t(\bar{T})} z_{C,t} = m_t \qquad \forall t \in [K] \quad (1)$$

$$\sum_{t \in [K]} x_{j,t} = 1 \qquad \forall j \in J \quad (2)$$

$$\sum_{j \in J_t(p)} x_{j,t} \leq \sum_{C \in C_t(\bar{T})} C_p z_{C,t} \quad \forall t \in [K], p \in B_t \quad (3)$$

$$\sum_{C \in C_t(\bar{T})} \text{size}(C) z_{C,t} + \sum_{p \in S_t} p \sum_{j \in J_t(p)} x_{j,t} \leq m_t \bar{T} \qquad \forall t \in [K] \quad (4)$$

With constraint (1) the number of chosen configurations for each machine type equals the number of machines of this type. Due to constraint (2) the variables

$x_{j,t}$ encode the fractional assignment of jobs to machine types. Moreover, for each machine type it is ensured with constraint (3) that the summed up number of big jobs of each size is at most the number of big jobs that are used in the chosen configurations for the respective machine type. Lastly, (4) guarantees that the overall processing time of the configurations and small jobs assigned to a machine type does not exceed the area $m_t \bar{T}$. It is easy to see that the MILP models a relaxed version of the problem:

Lemma 2. *If there is schedule with makespan $\bar{T}$ there is a feasible (integral) solution of MILP($\bar{T}$), and if there is a feasible integral solution for MILP($\bar{T}$) there is a schedule with makespan at most $\bar{T} + \varepsilon^2 T$.* $\qquad\square$

We have $K2^{\mathcal{O}(1/\varepsilon \log^2 1/\varepsilon)}$ integral variables, i.e., a constant number. Therefore MILP(T) can be solved in polynomial time, with the following classical result due to Lenstra [21] and Kannan [18]:

Theorem 3. *A mixed integer linear program with d integral variables and encoding size s can be solved in time $d^{\mathcal{O}(d)}\mathrm{poly}(s)$.*

Rounding. In this paragraph we describe how a feasible solution $(z_{C,t}, x_{j,t})$ for MILP($\bar{T}$) can be transformed into an integral feasible solution $(\bar{z}_{C,t}, \bar{x}_{j,t})$ of MILP($\bar{T} + \varepsilon T + \varepsilon^2 T$). This is achieved via a flow network utilizing flow integrality.

For any (small or big) processing time p let $\eta_{t,p} = \lceil \sum_{j \in J_t(p)} x_{j,t} \rceil$ be the rounded up (fractional) number of jobs with processing time p that are assigned to machine type t. Note that for big job sizes $p \in B_t$ we have $\eta_{t,p} \leq \sum_{C \in \mathcal{C}_t(\bar{T})} C_p z_{C,t}$ because of (3) and because the right hand side is an integer.

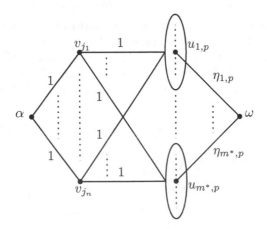

Fig. 1. A sketch of the flow network.

Now we describe the flow network $G = (V, E)$ with source α and sink ω. For each job $j \in \mathcal{J}$ there is a job node v_j and an edge (α, v_j) with capacity

1 connecting the source and the job node. Moreover, for each machine type t we have processing time nodes $u_{t,p}$ for each processing time $p \in B_t \cup S_t$. The processing time nodes are connected to the sink via edges $(u_{t,p}, w)$ with capacity $\eta_{t,p}$. Lastly, for each job j and machine type t with $\bar{p}_{t,j} < \infty$, we have an edge $(v_j, u_{t,\bar{p}_{t,j}})$ with capacity 1 connecting the job node with the corresponding processing time nodes. We outline the construction in Figure 1. Obviously we have $|V| \leq (K+1)n + 2$ and $|E| \leq (2K+1)n$.

Lemma 3. *G has a maximum flow with value n.*

Proof. Obviously n is an upper bound for the maximum flow, and the solution $(z_{C,t}, x_{j,t})$ for MILP$(\bar{T})$ can be used to design a flow f with value n, by setting $f((\alpha, v_j)) = 1$, $f((v_j, u_{t,\bar{p}_{t,j}})) = x_{j,t}$ and $f((u_{t,y}, w)) = \sum_{j \in J_t(y)} x_{j,t}$. □

Using the Ford-Fulkerson algorithm, an integral maximum flow f^* can be found in time $\mathcal{O}(|E|f^*) = \mathcal{O}(Kn^2)$. Due to flow conservation, for each job j there is exactly one machine type t^* such that $f((v_j, u_{t^*, y^*})) = 1$, and we set $\bar{x}_{j,t^*} = 1$ and $\bar{x}_{j,t} = 0$ for $t \neq t^*$. Moreover we set $\bar{z}_{C,t} = z_{C,t}$. Obviously $(\bar{z}_{C,t}, \bar{x}_{j,t})$ fulfils (1) and (2). Furthermore (3) is fulfilled, because of the capacities and because $\eta_{t,p} \leq \sum_{C \in \mathcal{C}_t(\bar{T})} C_p z_{C,t}$ for big job sizes p. Utilizing the geometric rounding and the convergence of the geometric series, as well as $\sum_{j \in J_t(p)} \bar{x}_{j,t} \leq \eta_{t,p} < \sum_{j \in J_t(p)} x_{j,t} + 1$, we get:

$$\sum_{p \in S_t} p \sum_{j \in J_t(p)} \bar{x}_{j,t} < \sum_{p \in S_t} p \sum_{j \in J_t(p)} x_{j,t} + \sum_{p \in S_t} p < \sum_{p \in S_t} p \sum_{j \in J_t(p)} x_{j,t} + \varepsilon^2 T \frac{1+\varepsilon}{\varepsilon}$$

Hence, we have $\sum_{C \in \mathcal{C}_t(T)} \text{size}(C) \bar{z}_{C,t} + \sum_{s \in S_t} \sum_{j \in J_{t,s}} p_{j,t} \bar{x}_{j,t} < m_t(\bar{T} + \varepsilon T + \varepsilon^2 T)$ and therefore (4) is fulfilled as well.

Analysis. The solution found for MILP$(\bar{T})$ can be turned into an integral solution for MILP$(\bar{T} + \varepsilon T + \varepsilon^2 T)$. This can easily be turned into a schedule with makespan $\bar{T} + \varepsilon T + \varepsilon^2 T + \varepsilon^2 T \leq (1 + 4\varepsilon)T$. It is easy to see that the running time of the algorithm by Lenstra and Kannan dominates the overall running time. Since MILP$(\bar{T})$ has $\mathcal{O}(K/\varepsilon \log 1/\varepsilon + n)$ many constraints, Kn fractional and $K2^{\mathcal{O}(1/\varepsilon \log^2 1/\varepsilon)}$ integral variables, the running time of the algorithm can be bounded by $2^{K2^{\mathcal{O}(1/\varepsilon \log^2 1/\varepsilon)}} \text{poly}(|I|)$.

3   Better running time

We improve the running time of the algorithm using techniques that utilize results concerning the existence of solutions for integer linear programs (ILPs) with a certain simple structure. In a first step we can reduce the running time to be only singly exponential in $1/\varepsilon$ with a technique by Jansen [15]. Then we further improve the running time to the one claimed in Theorem 1 with a very recent result by Jansen, Klein and Verschae [16]. Both techniques rely upon the following result about integer cones by Eisenbrandt and Shmonin [9].

Theorem 4. *Let $X \subset \mathbb{Z}^d$ be a finite set of integer vectors and $b \in \text{int-cone}(X) = \{\sum_{x \in X} \lambda_x x \mid \lambda_x \in \mathbb{Z}_{\geq 0}\}$. Then there is a subset $\tilde{X} \subseteq X$, such that $b \in \text{int-cone}(\tilde{X})$ and $|\tilde{X}| \leq 2d \log(4dM)$, with $M = \max_{x \in X} \|x\|_\infty$.*

For the first improvement of the running time this Theorem is used to show:

Corollary 1. MILP($\bar{T}$) *has a feasible solution where for each machine type $\mathcal{O}(1/\varepsilon \log^2 1/\varepsilon)$ of the corresponding integer variables are non-zero.*

We get the better running time by guessing the non-zero variables and removing all the others from the MILP. The number of possibilities of choosing $\mathcal{O}(1/\varepsilon \log^2 1/\varepsilon)$ elements out of a set of $2^{\mathcal{O}(1/\varepsilon \log^2 1/\varepsilon)}$ elements can be bounded by $2^{\mathcal{O}(1/\varepsilon^2 \log^4 1/\varepsilon)}$. Considering all the machine types we can bound the number of guesses by $2^{\mathcal{O}(K/\varepsilon^2 \log^4 1/\varepsilon)}$. The running time of the algorithm by Lenstra and Kannan with $\mathcal{O}(K/\varepsilon \log^2 1/\varepsilon)$ integer variables can be bounded by $2^{\mathcal{O}(K/\varepsilon \log K/\varepsilon \log^2 1/\varepsilon)} \text{poly}(|I|)$. This yields a running time of $2^{\mathcal{O}(K \log(K) 1/\varepsilon^2 \log^4 1/\varepsilon)} \text{poly}(|I|)$.

In the following we first proof Corollary 1 and then introduce the technique from [16] to further reduce the running time.

Proof of Corollary 1. We consider the so called *configuration ILP* for scheduling on identical machines. Let m' be a given number of machines, P be a set of processing times with multiplicities $k_p \in \mathbb{Z}_{>0}$ for each $p \in P$ and let $\mathcal{C} \subseteq \mathbb{Z}_{\geq 0}^P$ be some finite set of configurations for P. The configuration ILP for m', P, k, and $\mathcal{C}$ is given by:

$$\sum_{C \in \mathcal{C}} C_p y_C = k_p \qquad\qquad \forall p \in P \qquad\qquad (5)$$

$$\sum_{C \in \mathcal{C}} y_C = m' \qquad\qquad\qquad\qquad (6)$$

$$y_C \in \mathbb{Z}_{\geq 0} \qquad\qquad \forall C \in \mathcal{C} \qquad\qquad (7)$$

The default case that we will consider most of the time is that $\mathcal{C}$ is given by a target makespan T that upper bounds the size of the configurations.

Lets assume we had a feasible solution $(\tilde{z}_{C,t}, \tilde{x}_{j,t})$ for MILP($\bar{T}$). For $t \in [K]$ and $p \in B_t$ we set $\tilde{k}_{t,p} = \sum_{C \in \mathcal{C}_t(\bar{T})} C_p \tilde{z}_{C,t}$. We fix a machine type t. By setting $y_C = \tilde{z}_{C,t}$ we get a feasible solution for the configuration ILP given by m_t, B_t, $\tilde{k}_t$ and $\mathcal{C}_t(\bar{T})$. Theorem 4 can be used to show the existence of a solution for the ILP with only a few non-zero variables: Let X be the set of column vectors corresponding to the left hand side of the ILP and b be the vector corresponding to the right hand side. Then $b \in \text{int-cone}(X)$ holds and Theorem 4 yields that there is a subset $\tilde{X}$ of X with cardinality at most $2(|B_t|+1) \log(4(|B_t|+1)1/\varepsilon^2) \in \mathcal{O}(1/\varepsilon \log^2 1/\varepsilon)$ and $b \in \text{int-cone}(\tilde{X})$. Therefore there is a solution $(\breve{y}_C)$ for the ILP with $\mathcal{O}(1/\varepsilon \log^2 1/\varepsilon)$ many non-zero variables. If we set $\breve{z}_{C,t} = \breve{y}_C$ and $\breve{x}_{j,t} = \tilde{x}_{j,t}$ and perform corresponding steps for each machine type, we get a solution $(\breve{z}_{C,t}, \breve{x}_{j,t})$ that obviously satisfies constraints (1),(2) and (3) of MILP($\bar{T}$). The last constraint is also satisfied, because the number of covered big jobs of each size does not change and therefore the overall size of the configurations does not change either for each machine type. This completes the proof of Corollary 1.

Further Improvement of the Running Time. The main ingredient of the technique by Jansen et al. [16] is a result about the configuration ILP, for the case that there is a target makespan T' upper bounding the configuration sizes. Let $\mathcal{C}(T')$ be the set of configurations with size at most T'. We need some further notation. The *support* of any vector of numbers v is the set of indices with non-zero entries, i.e., $\text{supp}(v) = \{i \mid v_i \neq 0\}$. A configuration is called *simple*, if the size of its support is at most $\log(T' + 1)$, and *complex* otherwise. The set of complex configurations from $\mathcal{C}(T')$ is denoted by $\mathcal{C}^c(T')$.

Theorem 5. *Let the configuration ILP for m', P, k, and $\mathcal{C}(T')$ have a feasible solution and let both the makespan T' and the processing times from P be integral. Then there is a solution (y_C) for the ILP that satisfies the following conditions:*

1. $|\text{supp}(y|_{\mathcal{C}^c(T')})| \leq 2(|P| + 1) \log(4(|P| + 1)T')$ *and* $y_C \leq 1$ *for* $C \in \mathcal{C}^c(T')$.
2. $|\text{supp}(y)| \leq 4(|P| + 1) \log(4(|P| + 1)T')$.

We will call such a solution *thin*. Furthermore they argue:

Remark 1. *The total number of simple configurations is in* $2^{\mathcal{O}(\log^2(T') + \log^2(|P|))}$.

The better running time can be achieved by determining configurations that are equivalent to the complex configurations (via guessing and dynamic programming), guessing the support of the simple configurations, and solving the MILP with few integral variables. The approach is a direct adaptation of the one in [16] for our case. We now explain the additional steps of the modified algorithm in more detail and analyze its running time.

We have to ensure that the makespan and the processing times are integral and that the makespan is small. After the geometric rounding step we scale the makespan and the processing times, such that $T = 1/\varepsilon^3$ and $\bar{T} = (1 + \varepsilon)/\varepsilon^3$ holds and the processing times have the form $(1 + \varepsilon)^x \varepsilon^2 T = (1 + \varepsilon)^x/\varepsilon$. Next we apply a second rounding step for the big processing times, setting $\breve{p}_{t,j} = \lceil \bar{p}_{t,j} \rceil$ for $\bar{p}_{t,j} \in B_t$ and denote the set of these processing times by $\breve{B}_t$. Obviously we have $|\breve{B}_t| \leq |B_t| \leq 1/\varepsilon \log(1/\varepsilon^2) + 1$. We denote the corresponding instance by $\breve{I}$. Since for a schedule with makespan T for instance I there are at most $1/\varepsilon^2$ big jobs on any machine, we get:

Lemma 4. *If there is a schedule with makespan at most T for I, the same schedule has makespan at most $(1 + 2\varepsilon)T$ for instance $\breve{I}$ and any schedule for instance $\breve{I}$ can be turned into a schedule for I without increase in the makespan.*

We set $\breve{T} = (1 + 2\varepsilon)T$ and for each machine type t we consider the set of configurations $\mathcal{C}_t(\lfloor \breve{T} \rfloor)$ for $\breve{B}_t$ with size at most $\lfloor \breve{T} \rfloor$. Rounding down $\breve{T}$ ensures integrality and causes no problems, because all big processing times are integral. Furthermore let $\mathcal{C}_t^c(\lfloor \breve{T} \rfloor)$ and $\mathcal{C}_t^s(\lfloor \breve{T} \rfloor)$ be the subsets of complex and simple configurations. Due to Remark 1 we have:

$$|\mathcal{C}_t^s(\lfloor \breve{T} \rfloor)| \in 2^{\mathcal{O}(\log^2 \lfloor \breve{T} \rfloor + \log^2 |\breve{B}_t|)} = 2^{\mathcal{O}(\log^2 1/\varepsilon)} \tag{8}$$

Due to Theorem 5 (using the same considerations concerning configuration ILPs like in the last paragraph) we get that there is a solution $(\check{z}_C, \check{x}_{j,t})$ for MILP($\check{T}$) (adjusted to this case) that uses for each machine type t at most $4(|\check{B}_t|+1)\log(4(|\check{B}_t|+1)\lfloor\check{T}\rfloor) \in \mathcal{O}(1/\varepsilon\log^2 1/\varepsilon)$ many configurations from $\mathcal{C}_t(\lfloor\check{T}\rfloor)$. Moreover at most $2(|\check{B}_t|+1)\log(4(|\check{B}_t|+1)\lfloor\check{T}\rfloor) \in \mathcal{O}(1/\varepsilon\log^2 1/\varepsilon)$ complex configurations are used and each of them is used only once. Since each configuration corresponds to at most $1/\varepsilon^2$ jobs, there are at most $\mathcal{O}(1/\varepsilon^3\log^2 1/\varepsilon)$ many jobs for each type corresponding to complex configurations. Hence, we can determine the number of complex configurations m_t^c for machine type t along with the number of jobs $k_{t,p}^c$ with processing time p that are covered by a complex configuration for each $p \in \check{B}_t$ in $2^{\mathcal{O}(K/\varepsilon\log^2 1/\varepsilon)}$ steps via guessing. Now we can use a dynamic program to determine configurations (with multiplicities) that are equivalent to the complex configurations in the sense that their size is bounded by $\lfloor\check{T}\rfloor$, their summed up number is m_t^c and they cover exactly $k_{t,p}^c$ jobs with processing time p. The dynamic program iterates through $[m_t^c]$ determining $\check{B}_t$-indexed vectors y of non-negative integers with $y_p \leq k_{t,p}^c$. A vector y computed at step i encodes that y_p jobs of size p can be covered by i configurations from $\mathcal{C}_t(\lfloor\check{T}\rfloor)$. We denote the set of configurations the program computes with $\tilde{\mathcal{C}}_t$ and the multiplicities with $\tilde{z}_C$ for $C \in \tilde{\mathcal{C}}_t$. It is easy to see that the running time of such a program can be bounded by $\mathcal{O}(m_t^c(\prod_{p\in\check{B}_t}(k_{t,p}^c + 1))^2)$. Using $m_t^c \in \mathcal{O}(1/\varepsilon\log^2 1/\varepsilon)$ and $k_{t,p}^c \in \mathcal{O}(1/\varepsilon^3\log^2 1/\varepsilon)$ this yields a running time of $K2^{\mathcal{O}(1/\varepsilon\log^2 1/\varepsilon)}$, when considering all the machine types.

Having determined configurations that are equivalent to the complex configurations, we may just guess the simple configurations. For each machine type, there are at most $2^{\mathcal{O}(\log^2 1/\varepsilon)}$ simple configurations and the number of configurations we need is bounded by $\mathcal{O}(1/\varepsilon\log^2 1/\varepsilon)$. Therefore the number of needed guesses is bounded by $2^{\mathcal{O}(K/\varepsilon\log^4 1/\varepsilon)}$. Now we can solve a modified version of MILP($\check{T}$) in which z_C is fixed to $\tilde{z}_C$ for $C \in \tilde{\mathcal{C}}_t$ and only variables $z_{C'}$ corresponding to the guessed simple configurations are used. The running time for the algorithm by Lenstra and Kannan can again be bounded by $2^{\mathcal{O}(K\log K 1/\varepsilon\log^3 1/\varepsilon)}\mathrm{poly}(|I|)$. Thus we get an overall running time of $2^{\mathcal{O}(K\log K 1/\varepsilon\log^4 1/\varepsilon)}\mathrm{poly}(|I|)$. Considering the two cases $2^{\mathcal{O}(K\log K 1/\varepsilon\log^4 1/\varepsilon)} < \mathrm{poly}(|I|)$ and $2^{\mathcal{O}(K\log K 1/\varepsilon\log^4 1/\varepsilon)} \geq \mathrm{poly}(|I|)$ yields the claimed running time of of $2^{\mathcal{O}(K\log(K)1/\varepsilon\log^4 1/\varepsilon)} + \mathrm{poly}(|I|)$ completing the proof of the part of Theorem 1 concerning unrelated scheduling.

4 The Santa Claus Problem

Adapting the result for unrelated scheduling we achieve an EPTAS for the Santa Claus problem. It is based on the basic EPTAS together with the second running time improvement. In the following we briefly discuss the needed adjustments.

The dual approximation method can be applied in this case as well. However, since we have no approximation algorithm with a constant rate, the binary search is more expensive. For the simplification of the input it has to be taken into account that their may be huge jobs that are bigger than the optimal makespan, but otherwise it can be done similarly.

Moreover, in the Santa Claus problem it makes sense to use configurations of size bigger than $\check{T}$. Let $P = \lfloor \check{T} \rfloor + \max\{\check{p}_{t,j} \mid t \in [K], j \in \check{B}_t\}$. It suffices to consider configurations with size at most P and for each machine type t we denote the corresponding set of configurations by $\mathcal{C}_t(P)$. Again we can bound $\mathcal{C}_t(P)$ by $2^{\mathcal{O}(1/\epsilon \log^2 1/\epsilon)}$. The MILP has integral variables $z_{C,t}$ for each such configuration and fractional ones like before. The constraints (1) and (2) are adapted changing only the set of configurations and for constraint (3) additionally in this case the left-hand side has to be at least as big as the right hand side. The last constraint (4) has to be changed more. For this we partition $\mathcal{C}_t(P)$ into the set $\hat{\mathcal{C}}_t(P)$ of big configurations with size bigger than $\lfloor \check{T} \rfloor$ and the set $\check{\mathcal{C}}_t(P)$ of small configurations with size at most $\lfloor \check{T} \rfloor$. The changed constraint has the following form:

$$\sum_{C \in \check{\mathcal{C}}_t(P)} \text{size}(C) z_{C,t} + \sum_{p \in S_t} p \sum_{j \in J_t(p)} x_{j,t} \geq (m_t - \sum_{C \in \hat{\mathcal{C}}_t(P)} z_{C,t}) \check{T} \quad \forall t \in [K] \quad (9)$$

To solve the MILP with the claimed running time, some additional non-trivial considerations are needed that are omitted in this version of the paper.

Lastly, for the rounding of the MILP the flow network has to be changed as well, using lower and upper bounds for the flow.

Acknowledgements. We thank Florian Mai and Jannis Mell for helpful discussions on the problem.

References

1. Asadpour, A., Saberi, A.: An approximation algorithm for max-min fair allocation of indivisible goods. SIAM Journal on Computing 39(7), 2970–2989 (2010)
2. Bateni, M., Charikar, M., Guruswami, V.: Maxmin allocation via degree lower-bounded arborescences. In: Proceedings of the forty-first annual ACM symposium on Theory of computing. pp. 543–552. ACM (2009)
3. Bezáková, I., Dani, V.: Allocating indivisible goods. ACM SIGecom Exchanges 5(3), 11–18 (2005)
4. Bleuse, R., Kedad-Sidhoum, S., Monna, F., Mounié, G., Trystram, D.: Scheduling independent tasks on multi-cores with gpu accelerators. Concurrency and Computation: Practice and Experience 27(6), 1625–1638 (2015)
5. Bonifaci, V., Wiese, A.: Scheduling unrelated machines of few different types. arXiv preprint arXiv:1205.0974 (2012)
6. Chakrabarty, D., Chuzhoy, J., Khanna, S.: On allocating goods to maximize fairness. In: Foundations of Computer Science, 2009. FOCS'09. 50th Annual IEEE Symposium on. pp. 107–116. IEEE (2009)
7. Chen, L., Jansen, K., Zhang, G.: On the optimality of approximation schemes for the classical scheduling problem. In: Proceedings of the Twenty-Fifth Annual ACM-SIAM Symposium on Discrete Algorithms. pp. 657–668. SIAM (2014)
8. Chen, L., Ye, D., Zhang, G.: An improved lower bound for rank four scheduling. Operations Research Letters 42(5), 348–350 (2014)

9. Eisenbrand, F., Shmonin, G.: Carathéodory bounds for integer cones. Operations Research Letters 34(5), 564–568 (2006)
10. Gehrke, J.C., Jansen, K., Kraft, S.E., Schikowski, J.: A ptas for scheduling unrelated machines of few different types. In: International Conference on Current Trends in Theory and Practice of Informatics. pp. 290–301. Springer (2016)
11. Goemans, M.X., Rothvoß, T.: Polynomiality for bin packing with a constant number of item types. In: Proceedings of the Twenty-Fifth Annual ACM-SIAM Symposium on Discrete Algorithms. pp. 830–839. Society for Industrial and Applied Mathematics (2014)
12. Hochbaum, D.S., Shmoys, D.B.: Using dual approximation algorithms for scheduling problems theoretical and practical results. Journal of the ACM (JACM) 34(1), 144–162 (1987)
13. Horowitz, E., Sahni, S.: Exact and approximate algorithms for scheduling nonidentical processors. Journal of the ACM (JACM) 23(2), 317–327 (1976)
14. Imreh, C.: Scheduling problems on two sets of identical machines. Computing 70(4), 277–294 (2003)
15. Jansen, K.: An EPTAS for scheduling jobs on uniform processors: using an milp relaxation with a constant number of integral variables. SIAM Journal on Discrete Mathematics 24(2), 457–485 (2010)
16. Jansen, K., Klein, K., Verschae, J.: Closing the gap for makespan scheduling via sparsification techniques. In: 43rd International Colloquium on Automata, Languages, and Programming, ICALP 2016, July 11-15, 2016, Rome, Italy. pp. 72:1–72:13 (2016)
17. Jansen, K., Maack, M.: An EPTAS for scheduling on unrelated machines of few different types. arXiv preprint arXiv:1701.03263v1 (2017), https://arxiv.org/abs/1701.03263
18. Kannan, R.: Minkowski's convex body theorem and integer programming. Mathematics of operations research 12(3), 415–440 (1987)
19. Knop, D., Koutecký, M.: Scheduling meets n-fold integer programming. arXiv preprint arXiv:1603.02611 (2016)
20. Lenstra, J.K., Shmoys, D.B., Tardos, É.: Approximation algorithms for scheduling unrelated parallel machines. Mathematical programming 46(1-3), 259–271 (1990)
21. Lenstra Jr, H.W.: Integer programming with a fixed number of variables. Mathematics of operations research 8(4), 538–548 (1983)
22. Raravi, G., Nélis, V.: A ptas for assigning sporadic tasks on two-type heterogeneous multiprocessors. In: Real-Time Systems Symposium (RTSS), 2012 IEEE 33rd. pp. 117–126. IEEE (2012)
23. Williamson, D.P., Shmoys, D.B.: The design of approximation algorithms. Cambridge university press (2011)
24. Woeginger, G.J.: When does a dynamic programming formulation guarantee the existence of a fully polynomial time approximation scheme (FPTAS)? INFORMS Journal on Computing 12(1), 57–74 (2000)

A polynomial kernel for
Distance-Hereditary Vertex Deletion

Eun Jung Kim[1] and O-joung Kwon[2] *

[1] CNRS-Université Paris-Dauphine, Place du Marechal de Lattre de Tassigny, 75775
Paris cedex 16, France
[2] Logic and Semantics, TU Berlin, Berlin, Germany
eunjungkim78@gmail.com, ojoungkwon@gmail.com

Abstract. A graph is *distance-hereditary* if for any pair of vertices,
their distance in every connected induced subgraph containing both vertices is the same as their distance in the original graph. The DISTANCE-
HEREDITARY VERTEX DELETION problem asks, given a graph G on n
vertices and an integer k, whether there is a set S of at most k vertices
in G such that $G - S$ is distance-hereditary. This problem is important
due to its connection to the graph parameter rank-width [19]; distance-
hereditary graphs are exactly the graphs of rank-width at most 1. Eiben,
Ganian, and Kwon (MFCS' 16) proved that DISTANCE-HEREDITARY VER-
TEX DELETION can be solved in time $2^{\mathcal{O}(k)}n^{\mathcal{O}(1)}$, and asked whether it
admits a polynomial kernelization. We show that this problem admits a
polynomial kernel, answering this question positively. For this, we use a
similar idea for obtaining an approximate solution for CHORDAL VER-
TEX DELETION due to Jansen and Pilipczuk (SODA' 17) to obtain an
approximate solution with $\mathcal{O}(k^3 \log n)$ vertices when the problem is a
YES-instance, and we exploit the structure of split decompositions of
distance-hereditary graphs to reduce the total size.

1   Introduction

A graph is *distance-hereditary* if for every connected induced subgraph H and
two vertices u and v in H, the distance between u and v in H is the same as
their distance in G. A vertex subset X of a graph G is a *distance-hereditary mod-
ulator*, or a *DH-modulator* in short, if $G - X$ is a distance-hereditary graph. We
study the problem DISTANCE-HEREDITARY VERTEX DELETION (DH VERTEX
DELETION) which asks, given a graph G and an integer k, whether G contains
a DH-modulator of size at most k.

The graph modification problems, in which we want to transform a graph
to satisfy a certain property with as few graph modifications as possible, have
been extensively studied. For instance, the VERTEX COVER and FEEDBACK
VERTEX SET problems are graph modification problems where the target graphs

---

* O. Kwon is supported by the European Research Council (ERC) under the European
Union's Horizon 2020 research and innovation programme (ERC consolidator grant
DISTRUCT, agreement No. 648527).

© Springer International Publishing AG 2017   509
F. Ellen et al. (Eds.): WADS 2017, LNCS 10389, pp. 509–520, 2017.
DOI: 10.1007/978-3-319-62127-2_43

are edgeless graphs and forests, respectively. By the classic result of Lewis and Yannakakis [18], it is known that for all non-trivial hereditary properties that can be tested in polynomial time, the corresponding vertex deletion problems are NP-complete. Hence, the research effort has been directed toward designing algorithms such as approximation and parameterized algorithms.

When the target graph class $\mathcal{C}$ admits efficient recognition algorithms for some NP-hard problems, the graph modification problem related to such a class attracts more attention. Vertex deletion problems to classes of graphs of constant tree-width or constant tree-depth have attracted much attention in this context. TREE-WIDTH w VERTEX DELETION is proved to admit an FPT algorithm running in time $2^{\mathcal{O}(k)}n^{\mathcal{O}(1)}$ and a kernel with $\mathcal{O}(k^{g(w)})$ vertices for some function g [11,17]. Also, it was shown that TREE-DEPTH w VERTEX DELETION admits uniformly polynomial kernels with $\mathcal{O}(k^6)$ vertices, for every fixed w [12]. All these problems are categorized as vertex deletion problems for $\mathcal{F}$-minor free graphs in a general setting, when the set $\mathcal{F}$ contains at least one planar graph. However, $\mathcal{F}$-minor free graphs capture only sparse graphs in a sense that the number of edges of such a graph is bounded by a linear function on the number of its vertices. Thus these problems are not useful when dealing with very dense graphs.

Rank-width [19] and *clique-width* [5] are graph width parameters introduced for extending graph classes of bounded tree-width. Graphs of bounded rank-width represent graphs that can be recursively decomposed along vertex partitions (X, Y) where the number of neighborhood types between X and Y are small. Thus, graphs of constant rank-width may contain dense graphs; for instance, all complete graphs have rank-width at most 1. Courcelle, Makowski, and Rotics [4] proved that every MSO_1-expressible problem can be solved in polynomial time on graphs of bounded rank-width.

Motivated from TREE-WIDTH w VERTEX DELETION, Eiben, Ganian, and the second author [9] initiated study on vertex deletion problems to graphs of constant rank-width. The class of graphs of rank-width at most 1 is exactly the class of distance-hereditary graphs [19]. It was known that the vertex deletion problem for graphs of rank-width w can be solved in FPT time [16] using a meta-theorem [4]. Eiben et al. [9] devised the first elementary algorithm for this problem when $w = 1$, or equivalently DH VERTEX DELETION, that runs in time $2^{\mathcal{O}(k)}n^{\mathcal{O}(1)}$. Furthermore, they discussed that a DH-modulator of the size k can be used to obtain a $2^{\mathcal{O}(k)}n^{\mathcal{O}(1)}$-time algorithm for problems such as INDEPENDENT SET, VERTEX COVER, and 3-COLORING.

However, until now, it was not known whether DH VERTEX DELETION admits a polynomial kernel or not. A *kernelization* of a parameterized graph problem Π is a polynomial-time algorithm which, given an instance (G, k) of Π, outputs an equivalent instance (G', k') of Π with $|V(G')| + k' \leq h(k)$ for some computable function h. The resulting instance (G', k') of a kernelization is called a *kernel*, and in particular, when h is a polynomial function, Π is said to admit a *polynomial kernel*.

Our Contribution and Approach. Our main result is the following.

Theorem 1. DH VERTEX DELETION *admits a polynomial kernel.*

We introduce in Section 3 an approximate DH-modulator with $\mathcal{O}(k^3 \log n)$ vertices if the given instance is a YES-instance. An important observation here is that a distance-hereditary graph contains a complete bipartite subgraph (not necessarily induced) which is a balanced separator. Thus, if G admits a small DH-modulator, then there is balanced vertex separator $X \uplus K$ where X is small and K induces a complete bipartite subgraph. By recursively extracting such separators using an approximation algorithm for finding a balanced vertex separator [10], we will decompose the given graph into $D \uplus K_1 \uplus \cdots \uplus K_\ell \uplus X$, where $\ell = \mathcal{O}(k \log n)$, D is distance-hereditary, each K_i is a complete bipartite subgraph, $|X| = \mathcal{O}(k^2 \sqrt{\log k} \log n)$. In the next step, we argue that if a graph H is a disjoint union of a distance-hereditary graph and a complete bipartite graph and (H, k) is a YES-instance and satisfies a certain property, then in polynomial time, one can construct a DH-modulator of size $\mathcal{O}(k^2)$ for H (Proposition 2). Using this sub-algorithm ℓ times, we construct an approximate DH-modulator with $\mathcal{O}(k^3 \log n)$ vertices. This part follows a vein similar to the approach of Jansen and Pilipczuk [15] for CHORDAL VERTEX DELETION. Given a DH-modulator S of size $\mathcal{O}(k^3 \log n)$, we can obtain a new DH-modulator S' of size $\mathcal{O}(k^5 \log n)$ such that for every $v \in S'$, $G[(V(G) \setminus S') \cup \{v\}]$ is also distance-hereditary by adding $\mathcal{O}(k^2)$ vertices per each vertex in S. Such a DH-modulator is called a *good DH-modulator* and the details will be explained in Section 4.

The remaining part of the paper is contributed to reduce the number of vertices in $G - S'$. Two vertices v and w are twins if they have the same neighbors outside $\{v, w\}$. In Section 5, we present a reduction rule that bounds the size of each set of pairwise twins in $G - S'$. We give, in Section 6, a reduction rule that bounds the number of components of $G - S'$. Lastly in Section 7, we reduce the size of each component of $G - S'$ having at least 2 vertices. For the last part, we use split decompositions of distance-hereditary graphs. Briefly, a split decomposition displays a tree-like structure of a distance-hereditary graph in the form of a decomposition tree with bags for each nodes, such that each bag consists of a maximal set of pairwise twins (possibly with an extra vertex) in $G - S'$. We will provide a rule that bounds the number of bags in the decomposition tree, which results in bounding the size of each component.

2 Preliminaries

We follow [8] for basic graph terminology. A graph is *trivial* if it consists of a single vertex, and *non-trivial* otherwise. For two sets $A, B \subseteq V(G)$, we say A is *complete* to B if for every $v \in A$ and $w \in B$, v is adjacent to w. Two vertices v and w of a graph G are *twins* if they have the same neighbors in $G - \{v, w\}$. A vertex partition (A, B) of G is *split* if $N_G(A)$ is complete to $N_G(B)$.

A graph H is a *biclique* if there is a bipartition of $V(H)$ into non-empty sets $A \uplus B$ such that any two vertices $a \in A$ and $b \in B$ is adjacent. Notice that there may be edges among the vertices of A or B. For $K \subseteq V(G)$, we say that K is a biclique of G if $G[K]$ is a biclique. For a connected graph G, a vertex subset S

of G is called a *balanced vertex separator* of G if every component of $G - S$ has at most $\frac{2}{3}|V(G)|$ vertices. We allow $V(G)$ to be a balanced vertex separator of G. For a vertex subset S of G, a path is called an *S-path* if its end vertices are in S and all other internal vertices are in $V(G) \setminus S$.

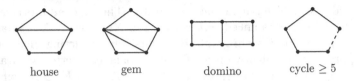

house   gem   domino   cycle ≥ 5

Fig. 1. The induced subgraph obstructions for distance-hereditary graphs.

A graph is called a *DH obstruction* if it is isomorphic to a gem, a house, a domino or an induced cycle of length at least 5, that are depicted in Figure 1. A DH obstruction is *small* if it has at most 6 vertices. Bandelt and Mulder [2] proved that a graph is distance-hereditary if and only if it has no induced subgraph isomorphic to one of DH obstructions. A DH-modulator S is *good* if $G[(V(G) \setminus S) \cup \{v\}]$ is distance-hereditary for every $v \in S$.

3   Approximation algorithm

We present a polynomial-time algorithm which constructs an approximate DH-modulator of G whenever (G, k) is a YES-instance.

Theorem 2. *There is a polynomial-time algorithm which, given a graph G and a positive integer k, either correctly reports that (G, k) is a NO-instance to DH VERTEX DELETION, or returns a DH-modulator of size $O(k^3 \log n)$.*

If G contains $k+1$ vertex-disjoint copies of small DH obstructions, then (G, k) is a NO-instance. We may assume a maximal packing of small DH obstructions has cardinality at most k. Since a maximal packing consists of at most $6k$ vertices, it is sufficient to prove Theorem 2 when G has no small DH obstruction.

We prove the following two propositions, implying Theorem 2 together.

Proposition 1. *There is a polynomial-time algorithm which, given an instance (G, k), either computes a decomposition $V(G) = D \uplus K_\ell \uplus \cdots \uplus K_1 \uplus X$ such that $G[D]$ is distance-hereditary, each K_i is a biclique, $|X| = O(k^2\sqrt{\log k} \log n)$ and $\ell = O(k \log n)$, or correctly reports that (G, k) is a NO-instance.*

When G does not contain any small DH obstructions, a linear program of DH VERTEX DELETION for G can be formulated as follows, where $x_v \geq 0$ for each $v \in V(G)$:

$$\min \sum_{v \in V(G)} x_v \text{ s.t} \sum_{v \in V(H)} x_v \geq 1 \quad \forall \text{ induced cycle } H \text{ of length at least 7}$$

A mapping $x = (x_v)_{v \in V(G)}$ from $V(G)$ to $\mathbb{R}$ is a *feasible fractional solution* to DH VERTEX DELETION for G if it is feasible to the above linear program for G. For a subgraph H of G, we write $x(H) := \sum_{v \in V(H)} x_v$ and $|x| := x(G)$.

Proposition 2. *Let* (G, k) *be an instance to* DH VERTEX DELETION *such that* G *has no small DH obstructions,* $V(G) = D \uplus K$ *where* $G[D]$ *is distance-hereditary, and* K *is a biclique. Let* x^* *be a feasible fractional solution to* DH VERTEX DELETION *for* G *such that* $x_v^* < \frac{1}{20}$, $\forall v \in V(G)$. *Given such* G *and* x^*, *one can in polynomial time find a DH-modulator* X *with* $O(|x^*|^2)$ *vertices.*

We first explain Proposition 1. First, we obtain an $O(n^3)$ bound on the number of maximal bicliques in a graph having no small DH obstructions. Secondly, we prove that every connected distance-hereditary graph on at least two vertices contains a balanced vertex separator that is a biclique. Combining these results, we can show the following.

Lemma 1. *Whenever* (G, k) *is a* YES-*instance and* G *has no small DH obstructions, one can in polynomial time find a balanced vertex separator* $K \uplus X$ *where* K *is a biclique or an empty set and* $|X| = O(k\sqrt{\log k})$.

Proof (Sketch of proof). Over all maximal bicliques K of G, we apply the $O(\sqrt{\log OPT})$-approximation algorithm for finding a balanced vertex separator in $G - K$, due to Feige et al [10]. One can observe that since (G, k) is a YES-instance, there is some set X of size at most k and a balanced vertex separator of K' of $G - X$ that is a biclique. Thus, a maximal biclique of G containing this K' is detected in the algorithm, and the approximation algorithm provides a set X' of size $O(k\sqrt{\log k})$ where $K' \uplus X'$ is a balanced vertex separator. □

We set $G_1 := G$, $K_0 = X_0 = \emptyset$, and at i-th recursive step, we apply Lemma 1 to a connected component G_i of $G - \bigcup_{j<i}(K_j \uplus X_j)$ which is not distance-hereditary and obtain a balanced vertex separator $K_i \uplus X_i$ of G_i. In the end, we obtain a decomposition $V(G) = D \uplus K_\ell \uplus \cdots \uplus K_1 \uplus X_\ell \uplus \cdots \uplus X_1$, where $G[D]$ is distance-hereditary, each K_i is a biclique or an empty set, and $|X_i| = O(k\sqrt{\log k})$. Since we only apply Lemma 1 to a component that is not distance-hereditary, if (G, k) is a YES-instance, then the size-k-modulator of G intersects every such component. By representing the recursive procedure as a collection of branching trees $\mathcal{T}$, we can show that $\ell = O(k \log n)$, as the maximum length of a root-to-leaf path in $\mathcal{T}$ is $O(\log n)$.

Now, we explain Proposition 2. Suppose G has no small DH obstructions and $V(G) = D \uplus K$ where $G[D]$ is distance-hereditary, and K is a biclique with a bipartition (A, B), and x^* is a feasible fractional solution to DH VERTEX DELETION such that $x_v^* < \frac{1}{20}$ for every $v \in V(G)$. We first observe that a new vector x' where $x'_v = 0$ if $v \in K$ and $x'_v = 2x_v^*$ if $v \in D$, is again a feasible fractional solution. For this, we show that every induced cycle H of length at least 7 in G satisfies that if $G[V(H) \cap K]$ has one component, then $|V(H) \cap K| \le 3$, and otherwise, H contains a K-path whose length is at least 3. In the former case, we have $x'(H) = x'(G[V(H) \cap D]) = 2x^*(G[V(H) \cap D]) \ge 1$ as $x^*(G[V(H) \cap K]) < 3 \cdot \frac{1}{20} < \frac{1}{2}$. In the latter case, the end vertices of the K-path P are contained in the same part of A or B, and it forms another DH obstruction H' with a vertex in the other part, where $G[V(H') \cap K]$ has one component. Thus, we have $x'(H) \ge x'(P) \ge x'(H') \ge 1$, as $x'_v = 0$ for $v \in K$.

We construct an instance $(G[D], \mathcal{T})$ of VERTEX MULTICUT with terminal pairs $\mathcal{T} := \{(s,t) \subseteq D \times D : \mathrm{dist}_{G[D],x'}(s,t) \geq 1\}$, where $\mathrm{dist}_{G[D],x'}(s,t)$ is the minimum $x'(P)$ over all (s,t)-paths P. Notice that for every terminal pair $(s,t) \in \mathcal{T}$, and for every (s,t)-path P in $G[D]$, we have $x'(P) \geq \mathrm{dist}_{G[D],x'}(s,t) \geq 1$, meaning that x' is a feasible fractional solution to VERTEX MULTICUT for the instance $(G[D], \mathcal{T})$. Using an approximation algorithm for VERTEX MULTICUT by Gupta [14], we can obtain a vertex set $X \subseteq D$ of size $O(|x'|^2)$ such that $G[D \setminus X]$ contains no (s,t)-path for every terminal pair $(s,t) \in \mathcal{T}$ in polynomial time. We prove, in the appendix, that the obtained set X is a DH-modulator.

Proof (of Theorem 2). It is sufficient to prove when G has no small DH obstructions. Let x^* be an optimal fractional solution to DH VERTEX DELETION for G. We may assume $|x^*| \leq k$, otherwise we report that (G,k) is a No-instance. Let $\tilde{X}$ be the set of all vertices v such that $x_v^* \geq \frac{1}{20}$. Observe that $|\tilde{X}| \leq 20k$ since otherwise, $|x^*| \geq \frac{1}{20}|\tilde{X}| > k$, a contradiction. Also x^* restricted to $V(G) \setminus \tilde{X}$ is a fractional feasible solution for $G - \tilde{X}$ such that $x_v^* < \frac{1}{20}$ for every v.

We compute a decomposition $V(G - \tilde{X}) = D \uplus \bigcup_{i=1}^{\ell} K_i \cup X$ as in Proposition 1, or correctly report (G,k) as a No-instance. Recall that $\ell = O(k \log n)$ and $|X| = O(k^2 \sqrt{\log k} \log n)$. Note that $V(G - (\tilde{X} \cup X)) = D \uplus \bigcup_{i=1}^{\ell} K_i$. From $i = 1$ up to ℓ, we want to obtain a DH-modulator S_i of G_i, where $G_1 := G[D \cup K_1]$ and for $i = 2, \ldots, \ell$, G_i is the subgraph of G induced by $(V(G_{i-1}) \setminus S_{i-1}) \cup K_i$. Note that $G_i - S_i$ is distance-hereditary and K_i is a blique. Hence, we can inductively apply the algorithm of Proposition 2 and obtain a DH-modulator S_i of size at most $O(|x^*|^2)$ of G_i. Especially, $G_\ell - S_\ell$ is distance-hereditary, implying that the set defined as $S := \tilde{X} \cup X \cup \bigcup_{i=1}^{\ell} S_i$ is a DH-modulator of G. From $|x^*| \leq k$, we have $|S_i| = O(k^2)$ for each i. It follows that $|S| = O(k^3 \log n)$. □

4   Good modulator

Theorem 3. *There is a polynomial-time algorithm which, given a graph G and a positive integer k, either correctly reports that (G,k) is a No-instance, or returns an equivalent instance (G', k') with a good DH-modulator of size $O(k^5 \log n)$.*

Proof (Sketch of Proof). If the algorithm of Theorem 2 reports that the instance is a No-instance, then we are done. Let S be a DH-modulator of size $\mathcal{O}(k^3 \log n)$ given by Theorem 2. Let $U := \emptyset$, and for $v \in S$, let $H_v := G[(V(G) \setminus S) \cup \{v\}]$. One can in polynomial time find either $k + 1$ small DH obstructions in H_v whose pairwise intersection is v, or a vertex set T_v of $V(G) \setminus S$ such that $|T_v| \leq 5k$ and $H_v - T_v$ has no small DH obstructions. In the former case, we add v to U.

Assume we obtain a vertex set T_v. Since $H_v - T_v$ has no small DH obstructions, every DH obstruction in $H_v - T_v$ is an induced cycle of length at least 7. We assert that either $H_v - T_v$ contains a vertex set X_v of size $\mathcal{O}(k^2)$ such that $H_v - (T_v \cup X_v)$ has no DH obstructions, or correctly reports that every DH-modulator of size at most k contains v.

We consider an instance $(H_v - (T_v \cup \{v\}), \mathcal{T})$ of VERTEX MULTICUT where $\mathcal{T} := \{(s,t) : s,t \in N_{H_v - T_v}(v), \mathrm{dist}_{H_v - (T_v \cup \{v\})}(s,t) \geq 3\}$. We can show that

$X \subseteq V(H_v) \setminus (T_v \cup \{v\})$ hits all induced cycles of $H_v - T_v$ of length at least 7 if and only if X is a vertex multicut for $(H_v - (T_v \cup \{v\}), \mathcal{T})$, because the restriction of an induced cycle of $H_v - T_v$ of length at least 7 is an induced path of length at least 3 between two neighbors of v in $H_v - T_v$, and the shortest path between those vertices and the induced path have the same length, as $H_v - (T_v \cup \{v\})$ is distance-hereditary. Let x^* be an optimal fractional solution to VERTEX MULTICUT, which can be efficiently found using the ellipsoid method and an algorithm for the (weighted) shortest path problem as a separation oracle. If $|x^*| \leq k$, then we can construct a multicut $X_v \subseteq V(H_v) \setminus (T_v \cup \{v\})$ of size $O(|x^*|^2) = O(k^2)$ using the approximation algorithm of Gupta [14]. If $|x^*| > k$, then any integral solution for $(H_v - (T_v \cup \{v\}), \mathcal{T})$ is larger than k, and any DH-modulator of size at most k must contain v. In this case, we add v to U.

We can confirm that $(G - U, k - |U|)$ is an instance equivalent to (G, k) and $S \cup (\bigcup_{v \in S \setminus U}(T_v \cup X_v))$ is a good DH-modulator for $G - U$.   □

5 Twin Reduction Rule

In a distance-hereditary graph, there may be a large set of pairwise twins. We introduce a reduction rule that bounds the size of a set of pairwise twins in $G - S$ by $\mathcal{O}(k^2|S|^3)$, where S is a DH-modulator (not necessarily good). The underlying observation is that it suffices to keep up to $k + 1$ vertices that are pairwise twins with respect to each subset of S of small size. For a subset $S' \subseteq S$, two vertices u and v in $V(G) \setminus S$ are S'-twins if u and v have the same neighbors in S'. It is not difficult to get an upper bound $\mathcal{O}(k|S|^5)$, by considering all subsets S' of S of size $\min\{|S|, 5\}$ and marking up to $k + 1$ S'-twins. To get a better bound, we proceed as follows.

Reduction Rule 1 *Let W be a set of pairwise twins in $G - S$, and let $m :=$ $\min\{|S|, 3\}$. (1) Over all subsets $S' \subseteq S$ of size m, we mark up to $k + 1$ pairwise S'-twins in W that are unmarked yet. (2) When $|S| \geq 4$, over all subsets $S' \subseteq S$ of size 4, if there is an unmarked vertex v of W such that $G[S' \cup \{v\}]$ is isomorphic to the house or the gem, then we mark up to $k + 1$ previously unmarked vertices in W including v that are pairwise S'-twins. (3) If there is an unmarked vertex v of W after finishing the marking procedure, we remove v from G.*

If (G, k) is irreducible with respect to Reduction Rule 1 and $|S| \geq 4$, then each set W of pairwise twins in $G - S$ contains $\mathcal{O}(k^2|S|^3)$ vertices, or (G, k) is a No-instance. This is because, if (G, k) is a YES-instance, then all chosen subsets S' in (2) can be covered by at most $4k$ vertices. If $|S| \leq 3$, then W has at most $8(k + 1)$ vertices. To see the safeness, suppose there is an unmarked vertex v of W after finishing the marking procedure. It is clear that if (G, k) is a YES-instance, then $(G - v, k)$ is a YES-instance. Suppose $G - v$ has a DH-modulator T of size at most k, and $G - T$ contains a DH obstruction F containing v. In case when $F - v$ is an induced path, let w, z be the end vertices of the path, and choose a set $S' \subseteq S$ of size 3 containing $\{w, z\} \cap S$. Since v is unmarked in Reduction Rule 1, there are $v_1, \ldots, v_{k+1} \in W \setminus \{v\}$ where $v_1, \ldots, v_{k+1}, v$ are pairwise S'-twins. Note that $V(F) \cap \{v_1, \ldots, v_{k+1}\} = \emptyset$ since no other vertex in F is adjacent to both w and z. Thus, there exists $v' \in \{v_1, \ldots, v_{k+1}\} \setminus T$ such that $G[V(F) \setminus \{v\} \cup \{v'\}]$ is a DH obstruction in $(G - v) - T$, contradiction.

If $|S \cap V(F)| \leq 3$, then we can proceed in the same way. We may assume $F - v$ is not an induced path and $|S \cap V(F)| \geq 4$. If F is the house or the gem, then we marked necessary vertices in (2), and thus we can proceed similarly. If F is the domino, then v should be a vertex of degree 2 in F. We can prove that the 3 vertices S' in $F - v$, two neighbors of v and the vertex farthest from v, satisfies that the existence of S'-twins with v is enough to get another DH obstruction.

6 The number of non-trivial components of $G - S$

We provide a reduction rule that bounds the number of non-trivial components of $G - S$, when S is a good DH-modulator. For $v \in S$ and a component C of $G - S$, let $N(v, C) := N_G(v) \cap V(C)$. We say that a pair (v, w) of vertices in S is a *witnessing pair* (for being non-split) for a component C of $G - S$ if $N(v, C) \neq \emptyset$, $N(w, C) \neq \emptyset$ and $N(v, C) \neq N(w, C)$. The following lemma is essential.

Lemma 2. *If $C_1, C_2, \ldots, C_m$ are distinct connected components of $G - S$ with $m \geq 2$ and $v_1, v_2, \ldots, v_m$ are distinct vertices of S $(v_{m+1} = v_1)$ such that for each $i \in \{1, \ldots, m\}$, (v_i, v_{i+1}) is a witnessing pair for C_i, then $G[\{v_1, v_2, \ldots, v_m\} \cup \bigcup_{i \in \{1, \ldots, m\}} V(C_i)]$ contains a DH obstruction.*

Lemma 2 for $m = 2$ observes that if a pair of vertices in S witnesses at least $k + 2$ non-trivial components in $G - S$, at least one of the pair must be contained in any size-k DH-modulator. Furthermore, keeping exactly $k + 2$ non-trivial components would suffice to impose this restriction. This suggests the following rule.

Reduction Rule 2 *For each pair of vertices v and w in S, we mark up to $k + 2$ non-trivial (previously unmarked) connected components C of $G - S$ such that (v, w) is a witnessing pair for C. If there is an unmarked non-trivial connected component C after the marking procedure, then we remove all edges in C.*

For the safeness of Reduction Rule 2, suppose there was an unmarked non-trivial connected component C after the marking procedure, and G' is the resulting graph. We mainly observe that if G' has a DH-modulator T of size at most k, then $(V(C) \setminus T, V(G) \setminus V(C) \setminus T)$ is a split in $G - T$. Otherwise, there are $v, w \in S$ and components $C_1, \ldots, C_{k+2}$ where (v, w) is a witnessing pair for $C, C_1, \ldots, C_{k+2}$ in G. Then there are 2 components among $C_1, \ldots, C_{k+2}$ that does not intersect T, and by Lemma 2, $G - T$ contains a DH obstruction, contradiction. Thus, if there is a DH obstruction H in $G - T$, then since $(V(C) \setminus T, V(G) \setminus V(C) \setminus T)$ is a split in $G - T$ and S is a good-modulator, we have $|V(H) \cap V(C)| \leq 1$. This implies that $G' - T$ also contains H, contradiction. We prove for the other direction in the similar way.

Proposition 3. *If (G, k) is irreducible with respect to Reduction Rule 2, then either the number of non-trivial components is $O(k^2 |S|)$ or it is a No-instance.*

Proof (of Proposition 3). Suppose (G, k) is a YES-instance. We define an auxiliary multigraph F on S such that for $v, w \in S$, the multiplicity of the edge vw equals the number of non-trivial components that are marked by the witness of (v, w) in Reduction Rule 2. It suffices to obtain a bound on the number of edges in F with the edge multiplicity taken into account.

Construct a maximal packing of 2-cycles in F and let $S_1 \subseteq S$ be the vertices contained in the packing. By Lemma 2, a packing of size $k + 1$ implies the existence of $k + 1$ vertex-disjoint DH obstructions. Therefore, $|S_1| \leq 2k$. Again, due to the assumption that (G, k) is a YES-instance, the subgraph $F - S_1$ does not have $k + 1$ vertex-disjoint cycles: otherwise, G contains $k + 1$ vertex-disjoint DH obstructions by Lemma 2. By the Erdős-Pósa property of cycles, there exists $S_2 \subseteq V(F) \setminus S_1$ hitting all cycles of $F - S_1$ with $|S_2| \leq rk \log k$ for some constant r. Now, the number of edges in F is at most $|S_1| \cdot |S|(k + 2) + |S_2| \cdot |S \setminus S_1| + (|S \setminus S_1 \setminus S_2|) = 2k(k + 2)|S| + rk \log k|S| + |S| \leq (7 + r)k^2|S|$. $\square$

7 The size of non-trivial components of $G - S$

It remains to bound the size of each non-trivial connected component of $G - S$. For this, we need to use split decompositions that present tree-like structure of distance-hereditary graphs. For the length constraint, we shortly define here with an example, and put the full description in the appendix (preliminary section).

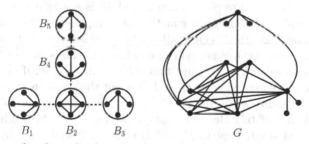

Fig. 2. An example of a split decomposition of a distance-hereditary graph. Dashed edges denote marked edges and each B_i denotes a bag.

A connected graph G is *prime* if $|V(G)| \geq 5$ and it has no split. A connected graph D with a distinguished set of cut edges $M(D)$ of D is called a *marked graph* if $M(D)$ forms a matching. An edge in $M(D)$ is a *marked edge*, and every other edge is an *unmarked edge*. A vertex incident with a marked edge is a *marked vertex*, and every other vertex is an *unmarked vertex*. Each component of $D - M(D)$ is a *bag* of D. See Figure 2 for an example. When G admits a split (A, B), we construct a marked graph D on the vertex set $V(G) \cup \{a', b'\}$ such that (1) $a'b'$ is a new marked edge, (2) there are no edges between A and B, (3) $\{a'\}$ is complete to $N_G(B)$, $\{b'\}$ is complete to $N_G(A)$, and (4) $G[A] = D[A]$ and $G[B] = D[B]$. The marked graph D is a *simple decomposition* of G. A *split decomposition* of a connected graph G is a marked graph D defined inductively to be either G or a marked graph defined from a split decomposition D' of G by replacing a bag B with its simple decomposition.

Cunningham and Edmonds [6] developed a canonical way to decompose a graph into a split decomposition. A split decomposition D of G is *canonical* if each bag of D is either a prime graph, a star, or a complete graph, and recomposing any marked edge of D violates this property. It is unique up to isomorphism [6] and can be computed in time $\mathcal{O}(|V(G)| + |E(G)|)$ [7]. In particular, Bouchet [3] proved that a graph is distance-hereditary if and only if every bag in its canonical split decomposition is a star or a complete graph.

Let D be the canonical split decomposition of a non-trivial component H of $G - S$. It is known that unmarked vertices in each bag of D consist of at most two twin sets in $G - S$. Thus, by Reduction Rule 1, it suffices to bound the number of bags of D. Since S is a good DH-modulator, for each $v \in S$, $G[V(H) \cup \{v\}]$ is distance-hereditary. Gioan and Paul [13, Theorem 3.4] described the way of extending D to the canonical split decomposition of $G[V(H) \cup \{v\}]$. In particular, there exists a bag or a marked edge that is modified when pushing v, and we can find this place in time $\mathcal{O}(|V(G)|)$. Such a bag or a marked edge is called *S-affected*. A bag B is a *branch bag* if $D - V(B)$ contains at least 3 connected components having at least two bags. For two adjacent bags B_1 and B_2, we denote by $e(B_1, B_2)$ the marked edge linking B_1 and B_2.

We first apply three reduction rules dealing with leaf bags. Firstly, we remove a vertex of degree 1 in G. Since any DH obstruction does not contain a vertex of degree 1, this is safe. Secondly, if there are a leaf bag B and its neighbor bag B' such that $B, B', e(B, B')$ are S-unaffected, and B' is a star bag whose center is adjacent to B, and $D - V(B')$ has exactly two components, then we remove all unmarked vertices of B'. In this case, all neighbors of a vertex in B' are contained in B whose unmarked vertices are pairwise twins in G. Thus, any DH obstruction does not contain a vertex of B', and this rule is safe. Lastly, if $A \subseteq V(H)$ is a maximal set of pairwise twins in G and flipping the adjacency between every two vertices of A reduces the number of bags, then we flip the adjacency. For instance, if B is a leaf bag that is a complete graph and its neighbor bag is a star bag whose leaf is adjacent to B, and B is S-unaffected, then by flipping the adjacency between two vertices in B, we can transform B into a star bag and merge with B'. This rule is also used in the FPT algorithm [9].

After applying those rules exhaustively, to find a reducible part, we color the bags of D with red and blue in the following way. (1) If a bag B is S-affected or incident with an S-affected edge, we color B with red. (2) If a bag B is adjacent to an S-affected leaf bag, we color B with red. (3) If B is a branch bag, then we color B with red. (4) All other bags are colored with blue. We can show that the number of red bags is at most $3|S|$, and for each bag B, there is at most one blue bag adjacent to B. We further show that the number of components of $D - \bigcup_{B \in \mathcal{R} \cup \mathcal{Q}} V(B)$ is at most $3|S|$, where $\mathcal{R}$ is the set of red bags, and $\mathcal{Q}$ is the set of blue leaf bags whose neighbor bags are red.

It remains to bound each connected component of $D - \bigcup_{B \in \mathcal{R} \cup \mathcal{Q}} V(B)$. Let D' be a connected component of $D - \bigcup_{B \in \mathcal{R} \cup \mathcal{Q}} V(B)$. As $\mathcal{R}$ contains all branch bags, D' contains no branch bags. Therefore, there is a sequence $B_1 - B_2 - \cdots - B_m$ of bags of D' that are not leaf bags, and all other bags are leaf bags adjacent to one of $B_1, \ldots, B_m$. For $1 \leq i_1 < i_2 < i_3 \leq m$, B_{i_2} is a (B_{i_1}, B_{i_3})-separator bag, if it is a star bag whose center is adjacent to neither B_{i_2-1} nor B_{i_2+1}. We first bound the number of (B_1, B_m)-separator bags, since if there are many, then we can merge two closest separator bags into one bag. This is similar to usual bypassing rule in FEEDBACK VERTEX SET. One interesting part is to bound the number of the sequence of consecutive bags that are not (B_1, B_m)-separator bags. We show that if there is such a sequence of length more than $5k + 11$, then we can always find a vertex that can be safely removed. In the end, we reduce the number of bags in each component of $D - \bigcup_{B \in \mathcal{R} \cup \mathcal{Q}} V(B)$ to $20k + 52$, and we conclude that the number of all bags is bounded by $3|S|(20k + 54)$ bags.

Proof (of Theorem 1). We first prove that given an instance (G, k) and a good DH-modulator S, one can output an equivalent instance of size $O(k^5|S|^5)$. We first apply Reduction Rule 2 to (G, k) with S. After that, $G - S$ has $O(k^2|S|)$ non-trivial components or we can correctly report that (G, k) is a NO-instance by Proposition 3. Next, we apply reduction rules for reducing the size of non-trivial components of $G-S$. We prove the safeness, polynomial-time applicability, and preserving the goodness of S in the appendix. In the end, the canonical split decomposition D of each non-trivial component of $G - S$ has at most $3|S|(20k+54)$ bags. Last, we apply Reduction Rule 1 exhaustively in polynomial time. This bounds the size of a twin set in $G - S$ by $O(k^2|S|^3)$. We note that the unmarked vertices of a bag form at most two twin sets. Therefore, the number of unmarked vertices in a bag is bounded by $O(k^2|S|^3)$. Especially, the same bounds apply to the number of trivial components in $G - S$ since they form an independent set in $G - S$. Combining the previous bounds altogether, we conclude that $V(G') = O(k^5|S|^5)$, when (G', k') is the resulting instance.

We may assume $n \leq 2^{ck}$ for some constant c. Recall that there is an algorithm for DH VERTEX DELETION running in time $2^{ck} n^{O(1)}$ by Eiben, Ganian, and Kwon [9]. If $n > 2^{ck}$, then the algorithm of [9] solves the instance (G, k) correctly in polynomial time, in which case we can output a trivial equivalent instance. By Theorem 3, we can obtain a good DH-modulator S of size $O(k^5 \log n) = O(k^6)$ in polynomial time or correctly report (G, k) as a NO-instance. The previous argument yields that in polynomial time, an equivalent instance (G', k') of size $O(k^{35})$ can be constructed. Now, applying Theorem 3 again[3] to (G', k'), we can either correctly conclude that (G', k'), and thus (G, k), is a NO-instance or output a good DH-modulator S' of size $O(k^5 \log k)$. Finally we obtain a kernel of size $O(k^{30} \log^5 k)$.

References

1. Agrawal, A., Lokshtanov, D., Misra, P., Saurabh, S., Zehavi, M.: Feedback vertex set inspired kernel for chordal vertex deletion. In: Proceedings of the Twenty-Eighth Annual ACM-SIAM Symposium on Discrete Algorithms, SODA 2017, Barcelona, Spain, Hotel Porta Fira, January 16-19. pp. 1383–1398 (2017)

---

[3] That applying kernelization twice can yield an improved bound was adequately observed in [1].

2. Bandelt, H.J., Mulder, H.M.: Distance-hereditary graphs. J. Combin. Theory Ser. B 41(2), 182–208 (1986)
3. Bouchet, A.: Transforming trees by successive local complementations. J. Graph Theory 12(2), 195–207 (1988)
4. Courcelle, B., Makowsky, J.A., Rotics, U.: Linear time solvable optimization problems on graphs of bounded clique-width. Theory Comput. Syst. 33(2), 125–150 (2000)
5. Courcelle, B., Olariu, S.: Upper bounds to the clique width of graphs. Discrete Appl. Math. 101(1-3), 77–114 (2000)
6. Cunningham, W.H., Edmonds, J.: A combinatorial decomposition theory. Canad. J. Math. 32(3), 734–765 (1980)
7. Dahlhaus, E.: Parallel algorithms for hierarchical clustering, and applications to split decomposition and parity graph recognition. Journal of Algorithms 36(2), 205–240 (2000)
8. Diestel, R.: Graph Theory. No. 173 in Graduate Texts in Mathematics, Springer, third edn. (2005)
9. Eiben, E., Ganian, R., Kwon, O.: A Single-Exponential Fixed-Parameter Algorithm for Distance-Hereditary Vertex Deletion. In: Faliszewski, P., Muscholl, A., Niedermeier, R. (eds.) 41st International Symposium on Mathematical Foundations of Computer Science, MFCS 2016. Leibniz International Proceedings in Informatics (LIPIcs), vol. 58, pp. 34:1–34:14. Schloss Dagstuhl–Leibniz-Zentrum fuer Informatik, Dagstuhl, Germany (2016)
10. Feige, U., Hajiaghayi, M., Lee, J.R.: Improved approximation algorithms for minimum weight vertex separators. SIAM J. Comput. 38(2), 629–657 (2008)
11. Fomin, F.V., Lokshtanov, D., Misra, N., Saurabh, S.: Planar F-Deletion: approximation, kernelization and optimal FPT algorithms. In: 2012 IEEE 53rd Annual Symposium on Foundations of Computer Science—FOCS 2012, pp. 470–479. IEEE Computer Soc., Los Alamitos, CA (2012)
12. Giannopoulou, A.C., Jansen, B.M.P., Lokshtanov, D., Saurabh, S.: Uniform kernelization complexity of hitting forbidden minors. In: Automata, languages, and programming. Part I, Lecture Notes in Comput. Sci., vol. 9134, pp. 629–641. Springer, Heidelberg (2015)
13. Gioan, E., Paul, C.: Split decomposition and graph-labelled trees: characterizations and fully dynamic algorithms for totally decomposable graphs. Discrete Appl. Math. 160(6), 708–733 (2012)
14. Gupta, A.: Improved results for directed multicut. In: Proceedings of the Fourteenth Annual ACM-SIAM Symposium on Discrete Algorithms, SODA 2003, January 12-14, 2003, Baltimore, Maryland, USA. pp. 454–455 (2003)
15. Jansen, B.M.P., Pilipczuk, M.: Approximation and kernelization for chordal vertex deletion. In: 28th Annual ACM-SIAM Symposium on Discrete Algorithms, SODA 2017, January 16-19, 2017, Barcelona, Spain. (2017), to appear
16. Kanté, M.M., Kim, E.J., Kwon, O., Paul, C.: An fpt algorithm and a polynomial kernel for linear rankwidth-1 vertex deletion. Algorithmica pp. 1–30 (2016)
17. Kim, E.J., Langer, A., Paul, C., Reidl, F., Rossmanith, P., Sau, I., Sikdar, S.: Linear kernels and single-exponential algorithms via protrusion decompositions. ACM Trans. Algorithms 12(2), Art. 21, 41 (2016)
18. Lewis, J.M., Yannakakis, M.: The node-deletion problem for hereditary properties is NP-complete. J. Comput. Syst. Sci. 20(2), 219–230 (1980)
19. Oum, S.: Rank-width and vertex-minors. J. Combin. Theory Ser. B 95(1), 79–100 (2005)

Busy Time Scheduling on a Bounded Number of Machines (Extended Abstract)[*]

Frederic Koehler[1] and Samir Khuller[2]

[1] Dept. of Mathematics, MIT, Cambridge MA 02139 USA
fkoehler@mit.edu
[2] Dept. of Computer Science, Univ. of Maryland, College Park MD 20742 USA
samir@cs.umd.edu

Abstract In this paper we consider a basic scheduling problem called the busy time scheduling problem - given n jobs, with release times r_j, deadlines d_j and processing times p_j, and m machines, where each machine can run up to g jobs concurrently, our goal is to find a schedule to minimize the sum of the "on" times for the machines. We develop the first correct constant factor online competitive algorithm for the case when g is unbounded, and give a $O(\log P)$ approximation for general g, where P is the ratio of maximum to minimum processing time. When g is bounded, all prior busy time approximation algorithms use an unbounded number of machines; note it is NP-hard just to test feasibility on fixed m machines. For this problem we give both offline and online (requiring "lookahead") algorithms, which are $O(1)$ competitive in busy time and $O(\log P)$ competitive in number of machines used.

1 Introduction

Scheduling jobs on multiple parallel machines has received extensive attention in the computer science and operations research communities for decades (see reference work [3]). For the most part, these studies have focused primarily on job-related metrics such as minimum makespan, total completion time, flow time, tardiness and maximum throughput. Our work is part of a line of recent results working towards a different goal: *energy efficiency*, in particular aiming to minimize the total time that a machine must be turned on, its *busy time* [4,12,8,11,15,5]. Equivalently, we seek to maximize the *average load* of machines while they are powered on, assuming we are free to turn machines off when they are idle. Note in this context we are concerned with *multi-processing* machines, as for machines which process only one job at a time the load is either 0 or 1 always. This measure has been studied in an effort to understand energy-related problems in cloud computing contexts; see e.g. [11,5,4] . The busy time metric also has connections to several key problems in optical network design, for example in minimizing the fiber costs of Optical Add Drop Multiplexers

[*] Full version: http://math.mit.edu/~fkoehler/busytime.pdf. Research supported by CCF 1217890 and CNS 1262805.

© Springer International Publishing AG 2017
F. Ellen et al. (Eds.): WADS 2017, LNCS 10389, pp. 521–532, 2017.
DOI: 10.1007/978-3-319-62127-2_44

(OADMs) [8], and the application of busy time models to optical network design has been extensively outlined in the literature [8,9,18,1].

Formally the problem is defined as follows: we are given a set of n jobs, and job j has a release time of r_j, a deadline d_j and a processing time of p_j (it is assumed $r_j + p_j \leq d_j$) and a collection of m multiprocessor machines with g processors each. The significance of processors sharing a machine is that they share busy time: the machine is on if a single processor on the machine is active. Each job j is assigned to the time window $[s_j, s_j + p_j)$ on some machine m_j. The assignment must satisfy the following constraints:

1. Start times respect job release times and deadlines, i.e., $[s_j, s_j + p_j) \subseteq [r_j, d_j)$.
2. At most g jobs are running at any time on any given machine. Formally, at any time t and on any machine m, $|\{j | t \in [s_j, s_j + p_j), m_j = m\}| \leq g$.

The busy time of a machine is the duration for which the machine is processing any non-zero number of jobs. The objective is to minimize the total sum of busy times of all the machines. Formally, the objective function is

$$\sum_{i=0}^{\infty} \mu \left(\bigcup_{j:m_j=i} [s_j, s_j + p_j) \right)$$

where μ measures the geometric length of a union of disjoint half intervals by summing their individual lengths; e.g. $\mu([1,2) \cup [3,4) \cup [3,5)) = 3$ i.e. μ is Lebesgue measure. *Note that this objective is constant if $g = 1$.*

All previous algorithms (described below) for busy time are forced to make the assumption that $m = \infty$, because the number of machines required by the schedules they generate can be as large as $\Omega(n)$, i.e. worst-possible. Our primary interest in this paper is in improving on this front. Thus our primary interest will really be in the *simultaneous optimization* problem of generating schedules whose performance is bounded in two objectives simultaneously: both the busy time and the number of machines required by the schedule. The best known approximation algorithms for each of these objectives separately is 3 [5] and $O(\sqrt{\log n / \log \log n})$ [6]. We conjecture that there exist schedules which achieve a $O(1)$ approximation in *both* objectives. However, as it stands the $O(1)$ machine minimization problem by itself remains a major open problem in combinatorial optimization, so such a result is out of reach for now. The main result of our paper will show that we can at least construct such a schedule under the assumption that $\log P$ is bounded by a constant, where $P = \max_{i,j} p_j / p_i$.

1.1 Related Work

Winkler and Zhang [18] first studied the interval job case of busy time scheduling, i.e. when $p_j = d_j - r_j$, and showed that even the special case when $g = 2$ is NP-hard. Their work was motivated by a problem in optical communication and assigning routing requests to wavelengths. Assuming that the number of machines available is unbounded, Alicherry and Bhatia [1], and independently

Kumar and Rudra [13], developed approximation algorithms with an approximation factor of 2 for the case of interval jobs. Being unaware of prior work on this problem, subsequently, Flammini et al [8] developed a very simple 4 approximation via a greedy algorithm for the interval job case.

The first constant factor approximation for the general problem, albeit on an unbounded number of machines, was given by Khandekar et al [11]. They first design a dynamic programming based algorithm for the case when $g = \infty$. This schedule is then used to fix the starting times of the jobs, and the resulting instance of interval jobs is scheduled by the greedy algorithm of Flammini et al [8]. Despite the "restriction" mapping, the approximation factor of 4 is unchanged. Since better approximation algorithms for the interval job case were available, it is natural to attempt to use those instead. Sadly, the restriction mapping can actually increase the cost of an optimal solution by a factor of 2, and so even if we use these algorithms we do not get a better bound than 4 (see [5] for a tight example). Chang et al [5] developed a 3-approximation algorithm by giving a new interval packing algorithm. We conjecture that the correct answer for this is 2, matching the interval job case.

Unfortunately, the number of machines used by all of these algorithms may be as large as $\Omega(n)$, even in the case when all jobs are equal length and released at time $r_j = 0$. This is because the $g = \infty$ reduction chooses start times oblivious to the true value of g. One may hope to resolve this problem from the other direction, by adapting an algorithm for minimizing the number of machines used. It is not difficult to get a $O(\log n)$ approximation algorithm for this problem via randomized LP rounding. The best known result is a $O(\sqrt{\log n / \log \log n})$ approximation algorithm by Chuzhoy et al [6] which uses a sophisticated recursive LP relaxation to the problem. Unfortunately, it appears to us quite difficult to adapt these LP rounding methods to account for the cost of the nonlinear busy time objective.

When $g < \infty$, very strong lower bounds for online minimization of busy time were given by Shalom et al [17]. They show that when $g < \infty$, no online algorithm can be better than g competitive algorithm against an online adaptive adversary. It should be noted that their general online model is harder than the one we consider; they have no notion of time, so in the online scenario they envision the algorithm must be able to deal with jobs arriving in arbitrary order. However, their proof of the lower bound does not need this additional power: it releases jobs in left-to-right order.

Some recent work [7,10] claims a 2-competitive online algorithm when $g = \infty$, but it is incorrect; see Fig. 1. *Independently and simultaneously to us*, Ren and Tang [16] recently studied the online problem when $g = \infty$ as well (see next section). They proved the same lower bound on the competitive ratio of this problem as we do and gave a slightly worse upper bound, $4 + 2\sqrt{2}$. They also analyzed a version of the problem where job lengths are unknown to the scheduler and proved a strong lower bound in this setting.

Figure 1: Counter-example to online algorithm of [7]. The optimal solution delays all the flexible unit jobs to the end and gets a busy time cost of $1 + g\delta$ rather than g. Setting $\delta = \frac{1}{g}$ gives the gap. The figure shows the schedule produced by the online algorithm with a cost of g.

1.2   Our Contributions

We divide the results into sections depending on the flexibility the algorithm has with m, the number of machines. We begin with the "classic" busy time model, where $m = \infty$.

- Our first result is an online 5-competitive algorithm for the busytime problem when machine capacity is unbounded $g = \infty$. In addition, we show that against an adaptive online adversary there is no online algorithm with competitive ratio less than $\varphi = (1 + \sqrt{5})/2$.
- The previous result is extended to the general busy time problem with $g < \infty$, and we get a competitive ratio of $O(\log P)$. No online algorithm for this problem was previously known. In the online setting with lookahead of $2p_{max}$ we can give a 12-competitive algorithm.

We then present our main results, concerned with simultaneous optimization of busytime and number of machines used:

- We present a constant-factor approximation algorithm for the busy time problem with fixed number of machines m, given the assumption of identical length jobs $p_j = p$.
- We give the first approximation algorithm for busy time scheduling with a non-trivial bound on the number of machines used. More precisely, for the simultaneous optimization problem we give a schedule which is $3 + \epsilon$-competitive on busy time and $O(\frac{\log P}{\log(1+\epsilon)})$ competitive on machine usage for $\epsilon < 1$.
- We give an online algorithm with $O(p_{max})$ lookahead in time, which remains $O(1)$-competitive for busy time and $O(\log P)$ competitive on machine usage.
- We also give *tradeoff lower bounds* which show the limits on the simultaneous optimizability of these objectives; if we optimize solely for one objective (e.g. machine usage), we may lose a factor of $\Omega(g)$ in the other (e.g. busy time).

1.3   Preliminaries

We recall the following fundamental scheduling lemma. The *interval graph* of a collection of half-intervals $\{[\alpha_i, \beta_i)\}_{i=1}^{n}$ is the graph with vertices the half-intervals, and an edge between two half-intervals I_1 and I_2 iff $I_1 \cap I_2 \neq \emptyset$. The interval graph is perfect, i.e.:

Proposition 1 *Given a collection of half-open intervals $\{[\alpha_i, \beta_i)\}_{i=1}^{n}$ there exists a k-coloring of the corresponding interval graph iff for all $t \in \mathbb{R}$,*

$$|\{i : [\alpha_i, \beta_i) \ni t\}| \leq k. \tag{1}$$

Proposition 2 *The following are lower bounds on the optimum busy time:*

1. *The optimal busy time for the same input instance with $g = \infty$.*
2. *The load bound $(1/g) \sum_{j=1}^{n} p_j$.*

We say a job is *available* at time t if $r_j \leq t$. It is often useful to refer to the latest start time $u_j = d_j - p$ of a job. An *interval job* is one with no choice in start time, i.e. j is an interval job when $d_j - r_j = p_j$. We define an algorithm to be a (r_1, r_2)-approximation if it generates a schedule using at most $r_1 m_{opt}$ machines and $r_2 \text{busy}_{OPT}$ busy time, where m_{opt} is the smallest number of machines for which a feasible schedule exists, and busy_{OPT} (or just OPT) is the minimum busy time on an unbounded number of machines.

2   Online Busy Time Scheduling with an Unbounded Capacity

2.1   The $g = \infty$ case, Upper Bound

We give a 5-competitive deterministic online algorithm for busy time scheduling when $g = \infty$. In this setting we may assign all jobs to a single machine so we assume w.l.o.g. $m = 1$. Informally, the algorithm is quite simple: everytime we hit a latest starting time u_j of an unscheduled job j, we activate the machine from time u_j to time $u_j + 2p_j$ and run all the jobs that fit in this window. To analyze this, we can pick an *arbitrary* optimal schedule, decompose its busy time into connected components, and then bound the cost of our schedule by charging the cost of running jobs to the connected components containing them.

In this section we will let T denote the *active time* of our machine; all jobs are processed during this active time, i.e. $\bigcup_j [s_j, s_j + p_j) \subset T$. We also maintain a set P of *primary jobs* but this is only for the purposes of the analysis. Note that at each time t the algorithm uses only information about jobs released by time t, so it is truly online.

Algorithm Doubler:

1. Let $P = \emptyset$. Let $T = \emptyset$.
2. For $t = 0$ to d_{max}:
   (a) Let U be the set of unscheduled, available jobs at time t.
   (b) Run every unscheduled job j s.t. $[t, t + p_j) \subset T$; remove j from U.
   (c) If $t = u_j$ for some $j \in U$, then pick such a j with p_j maximal and set $T = T \cup [t, t + 2p_j)$ (activating the machine from time t to $t + 2p_j$). Let $P = P \cup \{j\}$.

(d) Run[3] every unscheduled job j s.t. $[t, t + p_j) \subset T$; j is removed from U.

Suppose the algorithm fails to schedule a job j. Then at time u_j the job was available but was not scheduled; impossible because steps 2(c) ensures that $T \supset [u_j, u_j + p_j)$ and so step 2(d) would necessarily schedule it. Thus the algorithm schedules all jobs and, because we may trivially verify it respects r_j, d_j constraints, produces a valid schedule. Henceforth s_j refers to the start times chosen by algorithm Doubler; the following proposition is immediate.

Proposition 3 *Let T be the resulting active time and P the resulting set of primary jobs. Then $T = \bigcup_{j \in P} [s_j, s_j + 2p_j)$ and for every $j \in P$, $s_j = u_j$.*

Theorem 1. *Algorithm Doubler is 5-competitive.*

Proof. Fix an input instance (r_j, d_j, p_j) and an optimal offline schedule OPT with start times s_j^*. Let $T^* = \bigcup_j [s_j^*, s_j^* + p_j)$ so $\mu(T^*)$ is the busy time cost of OPT. Let P be the set of primary jobs. Let $P_1 \subset P$ consist of those jobs j in P with $[s_j, s_j + 2p_j) \subset T^*$ and $P_2 = P \setminus P_1$. By the Proposition,

$$\mu(T) = \mu\left(\bigcup_{j \in P} [s_j, s_j + 2p_j)\right) \leq \mu\left(\bigcup_{j \in P_1} [s_j, s_j + 2p_j)\right) + \mu\left(\bigcup_{j \in P_2} [s_j, s_j + 2p_j)\right)$$

$$\leq \mu(T^*) + \sum_{j \in P_2} 2p_j. \tag{2}$$

It remains to bound the cost incurred by jobs in P_2. Decompose T^* into connected components $\{C^i\}_{i=1}^k$ so $T^* = C^1 \cup \cdots \cup C^k$. The endpoints of C^i are $\inf C^i$ and $\sup C^i$. Let $J(C^i)$ be the set of jobs j with $[s_j^*, s_j^* + p_j) \subset C^i$. OPT schedules all jobs so $\bigcup_i J(C^i)$ is the set of all jobs, thus $\sum_{j \in P_2} 2p_j = \sum_{i=1}^k \sum_{j \in P_2 \cap J(C_i)} 2p_j$. We now claim that

$$\sum_{j \in P_2 \cap J(C_i)} p_j \leq 2\mu(C^i). \tag{3}$$

To show the claim, first we index so $\{e_j^i\}_{j=1}^{k_i'} = P_2 \cap J(C_i)$, where $k_i' = |P_2 \cap J(C_i)|$, and $(s(e_j^i))_{j=1}^{k_i'}$ is a monotonically increasing sequence.

   Observation: $r_{e_j^i} \leq s(e_1^i)$ *for all j.* Suppose for contradiction that $r_{e_j^i} > s(e_1^i)$ for some j. We know $[s^*(e_j^i), s^*(e_j^i) + p_{e_j^i}) \subset C^i$, hence $r_{e_j^i} + p_{e_j^i} \leq \sup C^i$. Because $[s(e_1^i), s(e_1^i) + 2p_1) \not\subset C^i$ we know that $s(e_1^i) + 2p_1 \geq \sup C_i \geq r_{e_j^i} + p_{e_j^i}$. Thus $[r_{e_j^i}, r_{e_j^i} + p_{e_j^i}] \subset [s(e_1^i), s(e_1^i) + 2p_1) \subset T$. We see then that at time $r_{e_j^i}$, step 2 (b) the algorithm must have scheduled job e_j^i. Thus $e_j^i \notin P \supset P_2 \cap J(C_i)$, which contradicts the definition of e_j^i. By contradiction $r_{e_j^i} \leq s(e_1^i)$ for all j.

---

[3] Step 2 (b) and step 2(d) are both necessary. Consider an interval job released at time 0 of length 2 and another at time 1 of length 4. Without step 2 (b) running at time 1, the machine will be turned on from time 5 to 9 unnecessarily.

Now it follows that $p_{e_j^i} > 2p_{e_{j-1}^i}$ (for $j \geq 2$): suppose otherwise, then because we know e_j^i was available at step 2 (c) at $t = s(e_{j-1}^i) \geq s(e_1^i) \geq r_{e_j^i}$, job e_j^i must have been scheduled at t with e_{j-1}^i and cannot have been added to P. By contradiction, $p_{e_j^i} > 2p_{e_{j-1}^i}$ hence by induction $p_{e_{k_i'}^i} > 2^{k_i'-j} p_{e_j^i}$. Now (3) follows: $\sum_{j=1}^{k_i'} p_{e_j^i} \leq \sum_{j=1}^{k_i'} 2^{j-k_i'} p_{e_{k_i'}^i} < p_{e_{k_i'}^i} \sum_{j'=0}^{\infty} 2^{-j'} = 2p_{e_{k_i'}^i} \leq 2\mu(\mathcal{C}^i)$. Thus $\sum_{j \in P_2} 2p_j \leq \sum_{i=1}^{k} 4\mu(\mathcal{C}^i) = 4\mu(T^*)$. Combining this with (2) proves the theorem.

Obviously we could have defined the above algorithm replacing 2 with any $\alpha > 1$, however $\alpha = 2$ minimizes $\alpha + \sum_{i=0}^{\infty} \alpha^{-i}$ and is thus optimal.

2.2 $g = \infty$, Online Lower Bounds

Proposition 4 *No online algorithm (without lookahead) against an online adaptive adversary has competitive ratio better than* $\varphi = \frac{1+\sqrt{5}}{2} \approx 1.618$.

Proof. Let $0 < \alpha < 1$ be a constant to be optimized later. Fix $1 > \epsilon > 0$ such that $\alpha = \epsilon k$ where $k \in \mathbb{Z}$. Here is the strategy for the adversary:

1. Release job A of length 1 available in $[0, 3)$.
2. Until job A is started, at each $t = n\epsilon$ for $n < k \in \mathbb{Z}$ release a single job of length ϵ available in $[t, t + \epsilon)$. (The ϵ jobs are interval jobs.)
3. If job A was started at $t - n\epsilon$, release a final job of length 1 available in $[2, 3)$.
4. Otherwise if job A is still not started at time $(k-1)\epsilon$, release no more jobs.

In the case corresponding to step (3), the online schedule has busy time $n\epsilon + 1 + 1$ whereas the optimal offline schedule, which runs job A at time 2, has busy time $(n+1)\epsilon + 1$. The ratio is thus $\frac{n\epsilon+2}{(n+1)\epsilon+1} \geq \frac{\alpha-\epsilon+2}{\alpha+1}$ because $f(x) = \frac{x-\epsilon+2}{x+1}$ is monotonically decreasing for $x > 0$. In the case corresponding to step (4), the online schedule has busy time at least $(k-1)\epsilon + 1 = \alpha - \epsilon + 1$ whereas the offline schedule has busy time 1. Thus the competitive ratio is at least $\min\left\{\frac{\alpha-\epsilon+2}{\alpha+1}, \alpha - \epsilon + 1\right\}$ and we may take the limit as $\epsilon \to 0$. The positive solution to $\frac{\alpha-2}{\alpha+1} = \alpha + 1$ is at $\alpha = \frac{\sqrt{5}-1}{2}$, and thus we get a lower bound of $\varphi = \frac{1+\sqrt{5}}{2}$.

A similar proof also gives a weaker lower bound when the algorithm is granted lookahead of $O(p_{max})$. Let $0 < \beta < 1$. Release job A with a very large availability span, and simultaneously release an interval job of length β, i.e. a job with $r_j = 0, p_j = \beta, d_j = \beta$. Without loss of generality the online algorithm either schedules job A at time 0 or chooses not to schedule job A until after time β. In the former case, release a job of length 1 at the very end of job A's availability window; in the latter case, release no more jobs. The lower bound on the competitive ratio now $\min\{\frac{1+\beta}{1}, \frac{2}{1+\beta}\}$, optimized at $\beta = \sqrt{2} - 1$, giving a ratio of $\sqrt{2}$.

Proposition 5 *An algorithm with lookahead a function of p_{max} has competitive ratio at least $\sqrt{2} \approx 1.414$.*

2.3   General Case, $g < \infty$

Combining with the bucketing algorithm given by Shalom et al [17] this gives a $O(\log \frac{p_{max}}{p_{min}})$-competitive online algorithm for busy time scheduling. More precisely, because the cost of their algorithm is bounded by 4 times the weight of the input jobs, and 1 times the $g = \infty$ lower bound, the approximation is $9 \log \frac{p_{max}}{p_{min}}$.

Running Algorithm Doubler offline and combining with the 3-approximation of Chang et al [5] gives a fast 7-approximation to the optimal busy time schedule. This is because the Greedy Tracking algorithm [5] takes as input a $g = \infty$ schedule using busytime T and outputs a schedule with cost at most $T + 2w(J)/g \leq T + 2OPT$ where $w(J)$ denotes the total processing time of all jobs. Since $T \leq 5OPT$ using our algorithm, the cost is bounded by $7OPT$.

If we are willing to grant the online algorithm a lookahead of $2p_{max}$ then we can get a constant factor online algorithm. We use our $g = \infty$ online algorithm to determine the time placement of jobs; this algorithm requires no lookahead so we now know the start time of jobs $2p_{max}$ ahead of the current time. We now run the offline machine-assignment algorithm in windows of the form $[kp_{max}, (k+2)p_{max})$ for $k \in \mathbb{N}$. We can bound the cost of even k by $5OPT + 2w(J_0)/g$ where $w(J_0)$ is the total processing time of jobs run in windows with even k; adding the matching term for odd k shows that this gives a fast $2 * 5 + 2 = 12$ approximation.

3   Offline Algorithm for Equal length jobs, Bounded Number of Machines

Although it is impossible in the general case (see Lower Bounds, Section 6), in the case of $p_j = p$ we are able to compute a schedule which is $(1, O(1))$-approximate, i.e. with the optimal number of machines and $O(1)$ busy time vs. the busy time optimum. Proposition 7 shows that a $(O(1), 1)$-approximation is impossible to achieve, even in this scenario. Our algorithm is two-step: it starts with a feasible schedule, and then uses a "pushing scanline" to push jobs together and reduce the busytime cost.

Algorithm Compact

1. Find the minimum number of machine required to feasibly schedule the jobs by binary search ($0 \leq m_{opt} \leq n$), using a feasibility algorithm for the problem with mg identical single-processor machines. Then construct a schedule on S on these jobs and m_{opt} machines that minimizes the sum of completion times, $\sum C_j$. A $O(n^2)$ time algorithm for these tasks is known [14].
2. Let s_j^0 be the start time of job j in S, and let $s_j := s_j^0$, $K := \emptyset$ and $P := \emptyset$.
3. For t from r_{min} to d_{max}: (*main loop*)
   (a) For every unscheduled job j, let $s_j := \max\{s_j^0, t\}$. Let U be the set of unscheduled jobs.
   (b) If $|\{j \in U : s_j \in [t, t + 2p]\}| \geq mg$, run each job j in this set at time s_j. Let $K := K \cup \{[t, t + 3p]\}$. We say these jobs were run in a *cluster*. Return to the main loop at $t := t + 2p$.

(c) Otherwise if $t = u_j$ for some unscheduled job j, run each job in the set $\{j \in U : s_j \in [t, t+p]\}$ at its start time s_j. Return to the main loop at $t := t + p$. Let $P := P \cup \{j\}$.

In step 3 it is necessary to consider only $t \in \{u_j, s_j - 2p\}$, so we can run this step in $O(n \log n)$ time.

Theorem 2. *Algorithm Compact is a 6-approximation for busy time, and generates a feasible schedule using m_{opt} machines.*

4   Offline Algorithm for Bounded Number of Machines

In this section we will use the fact that scheduling jobs on a minimum number of machines with integer release times and deadlines and with $p = p_j = 1$ is trivial offline. For a fixed m, it is well-known that an EDF (earliest-deadline first) schedule, i.e. one given by running at each time up to m of the jobs with earliest deadlines, gives a feasible schedule iff the input instance is feasible. Computing the minimum m can be done by binary search in $\log n$ steps.

We would like to describe some intuition before launching into the formal analysis. As before, we use something like a "pushing scanline" approach, moving jobs rightward from a "left-shifted" schedule and starting a group of jobs whenever a large number have been pushed together. To make this approach have bounded busy time usage, we first need to bucket jobs by similar lengths, but this alone cannot attain our desired performance ratio, because we may need for busy time purposes to group some long jobs with some short jobs. Therefore, in each bucket, when a job does not nicely group together with other jobs of the same length, we temporarily drop it. A second "clean-up pass" (step 3 below) runs the remaining jobs using an algorithm which has good busy-time performance but a priori unbounded machine usage. By arguing that we drop few jobs with overlapping availability times from each bucket, it is then possible to bound the machine usage. Below is our $(O(\log p_{max}/p_{min}), O(1))$-approximation algorithm for the general problem. Fix a constant $\alpha > 1$ to be optimized later.

1. Bucket jobs by processing time increasing exponentially by α, so the buckets contain jobs of processing time in the intervals $[p_{min}, \alpha p_{min}), [\alpha p_{min}, \alpha^2 p_{min}), \ldots,$ $[\alpha^{q-1} p_{min}, \alpha^q p_{min}]$ where $q = \left\lceil \log_\alpha \frac{p_{max}}{p_{min}} \right\rceil$.
2. For each bucket B_i
   (a) Let p be the supremum of the processing times of jobs in this bucket. We round job availability constraints down to multiples of p, so $r'_j = p\lfloor r_j/p \rfloor, u'_j = p\lfloor u_j/p \rfloor$, and $p'_j = p$. This is a unit job scheduling problem after we rescale by a factor of p.
   (b) We generate a left-shifted feasible schedule (referred to as the *initial schedule*) for the rounded (r'_j, d'_j, p'_j) instance using the minimum number of machines m. Let s^0_j be the start time of job j in this schedule.
   (c) Execute Algorithm RunHeavy.

(d) Let U_i' denote the set of jobs unscheduled in B_i after running Algorithm RunHeavy.

3. Now let U'' be the union of the U_i' for all buckets, and schedule the jobs in U'' by the 3-approximation of Chang et al [5] upon a new set of machines.

Algorithm RunHeavy

1. Let U initially be the set of all jobs in the bucket. Split machines into groups M_1 and M_0; we will require at most m machines in each group (see analysis).
2. For $t = kp$ from r_{min}' to u_{max}':
   (a) Let $J_t = \{j \in U : s_j^0 = t\}$. Let $k_1 = \lfloor |J_t|/g \rfloor$ and run k_1 groups of g jobs from this set on the group of machines $M_{k \bmod 2}$ with start times $s_j = \max(s_j^0, r_j)$. Remove these jobs from U.
   (b) Let $J_t' = \{j \in U : s_j^0 \le t \le u_j'\}$. Let $k_2 = \lfloor |J_t'|/g \rfloor$ jobs, and run k_2 groups of g jobs from this set on the group of machines $M_{k \bmod 2}$ with start times $s_j = \max(s_j^0, r_j)$. Remove these jobs from U.

Note in the loop in RunHeavy, we only need to do something when $t = s_j^0$ for some job j so the loop is really over polynomially many t.

Theorem 3. *The above algorithm generates a schedule feasible using $(2\alpha + 1)OPT$ busy time on $\lceil \log_\alpha p_{max}/p_{min} \rceil (2\lceil \alpha \rceil m_{opt} + 8)$ machines.*

5   Online Algorithm for Bounded Number of Machines

Since the formal details in this section are quite long, we give a brief summary of the main idea. In order to get an online algorithm, we still use the approach of the previous section, but interweave an agressive variant of Algorithm Doubler in order to pick start times for the "leftover" jobs which fit poorly into their buckets. In the previous section we could use that the $(r_j, d_j, p = p_j = 1)$ problem was exactly solvable offline; now, we must instead rely upon the online e-competitive online algorithm of [2] for this task. We also must use our $g = \infty$ algorithm in order to schedule the jobs in U'' online with bounded performance.

Theorem 4. *The online algorithm with lookahead $3p_{max}$ generates a schedule requiring at most $\lceil \log_\alpha p_{max}/p_{min} \rceil (16 + 4\lceil e\lceil \alpha \rceil m_{opt} \rceil)$ machines and $(20 + 42\alpha)$ busy time.*

6   Simultaneous Optimization of Busy Time and Machine Usage

6.1   Lower Bounds

Proposition 6 *For any input g, there exist input instances (with g processors per machine) where every machine-optimal schedule uses $(g-\epsilon)busy_{opt}$ busy time for ϵ arbitrary small.*

Proof. Fix $1 > \delta > 0$. Release g jobs of length 1 at time 0 with availability windows $[0, g)$. For $k = 0$ to $g - 1$, release $g - 1$ jobs of length δ with availability windows $[k, k + \delta)$, and $g - 1$ jobs of length δ with availability windows $[k + 1 - \delta, k + 1)$. The machine-optimal schedule runs all jobs on one machine, but due to the presence of the δ-jobs cannot overlap the execution of the long jobs, and thus has busy time cost g (see Fig. 2). The busy time optimal schedule runs the δ jobs on a separate machine and runs all of the long jobs in parallel, giving a busy time cost $1 + 2g\delta$. Thus the ratio is $\frac{g}{1+2g\delta}$ and taking δ sufficiently small gives the desired result.

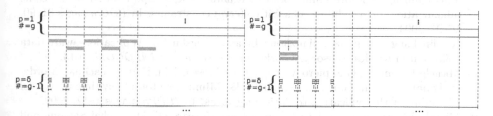

Figure 2: Illustrations of trade-off lower bounds. On one machine, the busy time is g. On two machines the busy time is $1 + 2g\delta$.

Proposition 7 *For any g, there exist input instances where every busy time optimal schedule uses gm_{opt} machines, even with the restriction $p_j = p$.*

Proof. We set $p = p_j = 1$ for all jobs. For $k = 0$ to $g - 1$, we release an interval job with availability window $[k/g, k/g+1)$, and we release $g(g-1)$ *unconstrained* jobs with availability windows $[0, 2g^2)$.

There exists a busy time optimal schedule using g machines, which runs $g - 1$ unconstrained jobs along with a single interval job together on a machine. Here the busy time cost equals the load bound exactly. There exists a feasible schedule using only 1 machine: for $k = 0$ to $g - 1$, on processor k of the machine it runs first the interval job followed by $g - 1$ unconstrained jobs, end-to-end. Thus $m_{opt} = 1$.

Now consider any schedule using fewer than g machines. By the pigeonhole principle, it must run two interval jobs on a single machine M. Let these jobs start at k_1/g and k_2/g respectively with $k_1 < k_2$; then the processor running the job at k_2/g must be idle in $[0, k_2/g) \supset [k_1/g, k_2/g)$. Since the load is positive but below g in this interval, the busy time exceeds the busy time lower bound, and so is greater than the cost of the busy time optimal schedule described earlier.

Acknowledgements: We are grateful to Chunxing Yin for extremely useful discussions.

References

1. Mansoor Alicherry and Randeep Bhatia. Line system design and a generalized coloring problem. In *ESA*, pages 19–30, 2003.
2. Nikhil Bansal, Tracy Kimbrel, and Kirk Pruhs. Speed scaling to manage energy and temperature. *J. ACM*, 54(1):3:1–3:39, March 2007.
3. Peter Brucker. *Scheduling algorithms*. Springer, 2007.
4. Jessica Chang, Harold Gabow, and Samir Khuller. A model for minimizing active processor time. *Algorithmica*, 70(3):368–405, November 2014.
5. Jessica Chang, Samir Khuller, and Koyel Mukherjee. Lp rounding and combinatorial algorithms for minimizing active and busy time. In *SPAA*, pages 118–127. ACM, 2014.
6. Julia Chuzhoy, Sudipto Guha, Sanjeev Khanna, and Joseph Seffi Naor. Machine minimization for scheduling jobs with interval constraints. In *FOCS*, pages 81–90. IEEE, 2004.
7. Xiaolin Fang, Hong Gao, Jianzhong Li, and Yingshu Li. Application-aware data collection in wireless sensor networks. In *Proceedings of INFOCOM*, 2013.
8. Michele Flammini, Gianpiero Monaco, Luca Moscardelli, Hadas Shachnai, Mordechai Shalom, Tami Tamir, and Shmuel Zaks. Minimizing total busy time in parallel scheduling with application to optical networks. In *IPDPS*, pages 1–12, 2009.
9. Michele Flammini, Gianpiero Monaco, Luca Moscardelli, Mordechai Shalom, and Shmuel Zaks. Approximating the traffic grooming problem with respect to adms and oadms. In *Proceedings of Euro-Par*, pages 920–929, 2008.
10. Chi Kit Ken Fong, Minming Li, Shi Li, Sheung-Hung Poon, Weiwei Wu, and Yingchao Zhao. Scheduling tasks to minimize active time on a processor with unlimited capacity. In *MAPSP*, 2015.
11. Rohit Khandekar, Baruch Schieber, Hadas Shachnai, and Tami Tamir. Minimizing busy time in multiple machine real-time scheduling. In *FSTTCS*, pages 169 – 180, 2010.
12. Frederic Koehler and Samir Khuller. Optimal batch schedules for parallel machines. In *WADS*, pages 475–486, 2013.
13. Vijay Kumar and Atri Rudra. Approximation algorithms for wavelength assignment. In *FSTTCS*, pages 152–163, 2005.
14. Alejandro López-Ortiz and Claude-Guy Quimper. A fast algorithm for multimachine scheduling problems with jobs of equal processing times. In *STACS*, pages 380–391, 2011.
15. George B. Mertzios, Mordechai Shalom, Ariella Voloshin, Prudence W.H. Wong, and Shmuel Zaks. Optimizing busy time on parallel machines. In *IPDPS*, pages 238–248, 2012.
16. Runtian Ren and Xueyan Tang. Online flexible job scheduling for minimum span. In *SPAA*, 2017. To Appear.
17. Mordechai Shalom, Ariella Voloshin, Prudence W.H. Wong, Fencol C.C. Yung, and Shmuel Zaks. Online optimization of busy time on parallel machines. *TAMC*, 7287:448–460, 2012.
18. Peter Winkler and Lisa Zhang. Wavelength assignment and generalized interval graph coloring. In *SODA*, pages 830 – 831, 2003.

Algorithms for Covering Multiple Barriers*

Shimin Li and Haitao Wang

Department of Computer Science
Utah State University, Logan, UT 84322, USA
shiminli@aggiemail.usu.edu,haitao.wang@usu.edu

Abstract. In this paper, we consider the problems for covering multiple intervals on a line. Given a set B of m line segments (called "barriers") on a horizontal line L and another set S of n horizontal line segments of the same length in the plane, we want to move all segments of S to L so that their union covers all barriers and the maximum movement of all segments of S is minimized. Previously, an $O(n^3 \log n)$-time algorithm was given for the problem but only for the special case $m = 1$. In this paper, we propose an $O(n^2 \log n \log \log n + nm \log m)$-time algorithm for any m, which improves the previous work even for $m = 1$. We then consider a line-constrained version of the problem in which the segments of S are all initially on the line L. Previously, an $O(n \log n)$-time algorithm was known for the case $m = 1$. We present an algorithm of $O((n + m) \log(n + m))$ time for any m. These problems may have applications in mobile sensor barrier coverage in wireless sensor networks.

1 Introduction

In this paper, we study algorithms for covering multiple barriers. These are basic geometric problems and have applications in barrier coverage of mobile sensors in wireless sensor networks. For convenience, in the following we introduce and discuss the problems from the mobile sensor barrier coverage point of view.

Let L be a line, say, the x-axis. Let $\mathcal{B}$ be a set of m pairwise disjoint segments, called *barriers*, sorted on L from left to right. Let S be a set of n sensors in the plane, and each sensor $s_i \in S$ is represented by a point (x_i, y_i). If a sensor is moved on L, it has a *sensing/covering range* of length r, i.e., if a sensor s is located at x on L, then all points of L in the interval $[x - r, x + r]$ are *covered* by s and the interval is called the *covering interval* of s. The problem is to move all sensors of S onto L such that each point of every barrier is covered by at least one sensor and the maximum movement of all sensors of S is minimized, i.e., the value $\max_{s_i \in S} \sqrt{(x_i - x_i')^2 + y_i^2}$ is minimized, where x_i' is the location of s_i on L in the solution (its y-coordinate is 0 since L is the x-axis). We call it *multiple-barrier* coverage problem, denoted by MBC.

We assume that covering range of the sensors is long enough so that a coverage of all barriers is always possible. Note that we can check whether a coverage is possible in $O(m + n)$ time by an easy greedy algorithm.

---

* This research was supported in part by NSF under Grant CCF-1317143.

© Springer International Publishing AG 2017
F. Ellen et al. (Eds.): WADS 2017, LNCS 10389, pp. 533–544, 2017.
DOI: 10.1007/978-3-319-62127-2_45

Previously, only the special case $m = 1$ was studied and the problem was solved in $O(n^3 \log n)$ time [10]. In this paper, we propose an $O(n^2 \log n \log \log n + nm \log m)$-time algorithm for any value m, which improves the algorithm in [10] by almost a linear factor even for the case $m = 1$.

We further consider a *line-constrained* version of the problem where all sensors of S are initially on L. Previously, only the special case $m = 1$ was studied and the problem was solved in $O(n \log n)$ time [3]. We present an $O((n + m) \log(n + m))$ time algorithm for any value m, and the running time matches that of the algorithm in [3] when $m = 1$.

1.1 Related Work

Sensors are basic units in wireless sensor networks. The advantage of allowing the sensors to be mobile increases monitoring capability compared to those static ones. One of the most important applications in mobile wireless sensor networks is to monitor a barrier to detect intruders in an attempt to cross a specific region. Barrier coverage [9, 10], which guarantees that every movement crossing a barrier of sensors will be detected, is known to be an appropriate model of coverage for such applications. Mobile sensors have limited battery power and therefore their movements should be as small as possible.

Dobrev et al. [7] studies several problems on covering multiple barriers in the plane. They showed that these problems are generally NP-hard when sensors have different ranges. They also proposed polygonal-time algorithms for some special cases, e.g., barriers are parallel or perpendicular to each other, and sensors have some constrained movements. In fact, if sensors have different ranges, by an easy reduction from the Partition Problem as in [7], we can show that our problem MBC is NP-hard even for the line-constrained version and $m = 2$.

Other previous work has been focused on the line-constrained problem with $m = 1$. Czyzowicz et al. [5] first gave an $O(n^2)$ time algorithm, and later, Chen et al. [3] solved the problem in $O(n \log n)$ time. If sensors have different ranges, Chen et al. [3] presented an $O(n^2 \log n)$ time algorithm. For the *weighted case* where sensors have weights such that the moving cost of a sensor is its moving distance times its weight, Wang and Zhang [15] gave an $O(n^2 \log n \log \log n)$ time algorithm for the case where sensors have the same range.

The *min-sum* version of the line-constrained problem with $m = 1$ has also been studied, which is to minimize the sum of the moving distances of all sensors. If sensors have different ranges, the problem is NP-hard [6]. Otherwise, Czyzowicz et al. [6] gave an $O(n^2)$ time algorithm, and Andrews and Wang [1] solved the problem in $O(n \log n)$ time. The *min-num* version of the problem was also studied, where the goal is to move the minimum number of sensors to form a barrier coverage. Mehrandish et al. [13, 14] proved that the problem is NP-hard if sensors have different ranges and gave polynomial time algorithms otherwise.

Bhattacharya et al. [2] studied a circular barrier coverage problem in which the barrier is a circle and the sensors are initially located inside the circle. The goal is to move sensors to the circle to form a regular n-gon (so as to cover the circle) such that the maximum sensor movement is minimized. An $O(n^{3.5} \log n)$-time algorithm was given in [2] and later Chen et al. [4] improved the algorithm to $O(n \log^3 n)$ time. The min-sum version of the problem was also studied [2, 4].

1.2 Our Approach

To solve MBC, one major difficulty is that we do not know the order of the sensors of S on L in an optimal solution. Therefore, our main effort is to find such an order. To this end, we first develop a *decision algorithm* that can determine whether $\lambda \geq \lambda^*$ for any value λ, where λ^* is the maximum sensor movement in an optimal solution. Our decision algorithm runs in $O(m + n \log n)$ time. Then, we solve the problem MBC by "parameterizing" the decision algorithm in a way similar in spirit to parametric search [12]. The high-level scheme of our algorithm is very similar to those in [3, 15], but many low-level computations are different.

The line-constrained problem is easier due to an *order preserving property*: there exists an optimal solution in which the order of the sensors is the same as in the input. This leads to a linear-time decision algorithm using the greedy strategy. Also based on this property, we can find a set Λ of $O(n^2 m)$ "candidate values" such that Λ contains λ^*. To avoid computing Λ explicitly, we implicitly organize the elements of Λ into $O(n)$ sorted arrays such that each array element can be found in $O(\log m)$ time. Finally, by applying the matrix search technique in [8], along with our linear-time decision algorithm, we compute λ^* in $O((n + m) \log(n+m))$ time. We should point out that implicitly organizing the elements of Λ into sorted arrays is the key and also the major difficulty for solving the problem, and our technique may be interesting in its own right.

The remaining paper is organized as follows. In Section 2, we introduce some notation. In Section 3, we present our algorithm for the line-constrained problem. In Section 4, we present our decision algorithm for MBC. Section 5 solves the problem MBC. Section 6 concludes the paper, with remarks that our techniques can be used to reduce the space complexities of the algorithms in [3, 15]. Due to the space limit, some proofs are omitted but can be found in the full paper [11].

2 Preliminaries

We denote the barriers of $\mathcal{B}$ by $B_1, B_2, \ldots, B_m$ sorted on L from left to right. For each B_i, let a_i and b_i denote the left and right endpoints of B_i, respectively. For ease of exposition, we make a general position assumption that $a_i \neq b_i$ for each B_i. The degenerated case can also be handled by our techniques, but the discussions would be more tedious.

For any point x on L (the x-axis), we also use x to denote its x-coordinate, and vice versa. We assume that the left endpoint of B_1 is at 0, i.e., $a_1 = 0$. Let β denote the right endpoint of B_m, i.e., $\beta = b_m$.

We denote the sensors of S by $s_1, s_2, \ldots, s_n$ sorted by their x-coordinates. For each sensor s_i located on a point x of L, $x - r$ and $x + r$ are the left and right endpoints of the covering interval of s_i, respectively, and we call them the *left and right extensions* of s_i, respectively.

Again, let λ^* be the maximum sensor movement in an optimal solution. Given any value λ, the *decision problem* is to determine whether $\lambda \geq \lambda^*$, or equivalently, whether we can move each sensor with distance at most λ such that all barriers can be covered. If yes, we say that λ is a *feasible value*. Thus, we also call it a *feasibility test* on λ.

3   The Line-Constrained Version of MBC

In this section, we present our algorithm for the line-constrained MBC. As in the special case $m = 1$ [5], a useful observation is that the following *order preserving* property holds: There exists an optimal solution in which the order of the sensors is the same as in the input. Due to this property, we have the following lemma.

Lemma 1. *Given any $\lambda > 0$, we can determine whether λ is a feasible value in $O(n + m)$ time.*

Let OPT be an optimal solution that preserves the order of the sensors. For each $i \in [1, n]$, let x_i' be the position of s_i in OPT. We say that a set of k sensors are in *attached positions* if the union of their covering intervals is a single interval of length equal to $2rk$. The following lemma is self-evident and is an extension of a similar observation for the case $m = 1$ in [5].

Lemma 2. *There exists a sequence of sensors $s_i, s_{i+1}, \ldots, s_j$ in attached positions in OPT such that one of the following three cases holds. (a) The sensor s_j is moved to the left by distance λ^* and $x_i' = a_k + r$ for some barrier B_k (i.e., the sensors from s_i to s_j together cover the interval $[a_k, a_k + 2r(j - i + 1)]$). (b) The sensor s_i is moved to the right by λ^* and $x_j' = b_k - r$ for some barrier B_k. (c) The sensor s_i is moved rightwards by λ^* and s_j is moved leftwards by λ^*.*

Cases (a) and (b) are symmetric in the above lemma. Let Λ_1 be the set of all possible distance values introduced by s_j in Case (a). Specifically, for any pair (i, j) with $1 \leq i \leq j \leq n$ and any barrier B_k with $1 \leq k \leq m$, define $\lambda(i, j, k) = x_j - (a_k + 2r(j - i) + r)$. Let Λ_1 consists of $\lambda(i, j, k)$ for all such triples (i, j, k). We define Λ_2 symmetrically be the set of all possible values introduced by s_i in Case (b). We define Λ_3 as the set consisting of the values $[x_j - x_i - 2r(j - i)]/2$ for all pairs (i, j) with $1 \leq i < j \leq n$. Clearly, $|\Lambda_3| = O(n^2)$ and both $|\Lambda_1|$ and $|\Lambda_2|$ are $O(mn^2)$. Let $\Lambda = \Lambda_1 \cup \Lambda_2 \cup \Lambda_3$.

By Lemma 2, λ^* is in Λ and is actually the smallest feasible value of Λ. Hence, we can first compute Λ and then find the smallest feasible value in Λ by using the decision algorithm. However, that would take $\Omega(mn^2)$ time. To reduce the time, we will not compute Λ explicitly, but implicitly organize the elements of Λ into certain sorted arrays and then apply the matrix search technique in [8]. Since we only need to deal with sorted arrays instead of more general matrices, we review the technique with respect to arrays in the following lemma.

Lemma 3. [8] *Given a set of N sorted arrays of size at most M each, we can compute the smallest feasible value of these arrays with $O(\log N + \log M)$ feasibility tests and the total time of the algorithm excluding the feasibility tests is $O(\tau \cdot N \cdot \log \frac{2M}{N})$, where τ is the time for evaluating each array element (i.e., the number of array elements that need to be evaluated is $O(N \cdot \log \frac{2M}{N})$).*

With Lemma 3, we can compute the smallest feasible values in the sets Λ_1, Λ_2, and Λ_3, respectively, and then return the smallest one as λ^*. For Λ_3, Chen et al. [3] (see Lemma 14) gave an approach to order in $O(n \log n)$ time the elements of Λ_3 into $O(n)$ sorted arrays of $O(n)$ elements each such that each array element can be obtained in $O(1)$ time. Consequently, by applying Lemma 3, the smallest feasible value of Λ_3 can be computed in $O((n + m) \log n)$ time.

For the other two sets Λ_1 and Λ_2, in the case $m = 1$, the elements of each set can be easily ordered into $O(n)$ sorted arrays of $O(n)$ elements each [3]. However, in our problem for general m, it becomes significantly more difficult to obtain a subquadratic-time algorithm. Indeed, this is the main challenge of our method. In what follows, our main effort is to prove Lemma 4.

Lemma 4. *For the set Λ_1, in $O(m \log m)$ time, we can implicitly form a set $\mathcal{A}$ of $O(n)$ sorted arrays of $O(m^2 n)$ elements each such that each array element can be computed in $O(\log m)$ time and every element of Λ_1 is contained in one of the arrays. The same applies to the set Λ_2.*

We note that our technique for Lemma 4 might be interesting in its own right and may find other applications as well. Before proving Lemma 4, we first prove the following theorem by using Lemma 4.

Theorem 1. *The line-constrained version of MBC can be solved in $O((n + m) \log(n + m))$ time.*

Proof. It is sufficient to compute λ^*, after which we can apply the decision algorithm on λ^* to obtain an optimal solution.

Let Λ'_1 denote the set of all elements in the arrays of $\mathcal{A}$ specified in Lemma 4. Define Λ'_2 similarly with respect to Λ_2. By Lemma 4, $\Lambda_1 \subseteq \Lambda'_1$ and $\Lambda_2 \subseteq \Lambda'_2$. Since $\lambda^* \in \Lambda_1 \cup \Lambda_2 \cup \Lambda_3$, we also have $\lambda^* \in \Lambda'_1 \cup \Lambda'_2 \cup \Lambda_3$. Hence, λ^* is the smallest feasible value in $\Lambda'_1 \cup \Lambda'_2 \cup \Lambda_3$. Let λ_1, λ_2, and λ_3 be the smallest feasible values in the sets Λ'_1, Λ'_2, and Λ_3, respectively. As discussed before, λ_3 can be computed in $O((n+m) \log n)$ time. By Lemma 4, applying the algorithm in Lemma 3 can compute both λ_1 and λ_2 in $O((n + m)(\log m + \log n))$ time. Note that $(n + m)(\log m + \log n) = \Theta((n + m) \log(n + m))$. $\square$

3.1 Proving Lemma 4

In this section, we prove Lemma 4. We will only prove the case for the set Λ_1, since the other case for Λ_2 is symmetric. Recall that $\Lambda_1 = \{\lambda(i, j, k) \mid 1 \leq i \leq j \leq n, 1 \leq k \leq m\}$, where $\lambda(i, j, k) = x_j - (a_k + 2r(j - i) + r)$.

For any j and k, let $A[j,k]$ denote the list $\lambda(i,j,k)$ for $i = 1, 2, \ldots, j$, which is sorted increasingly. Let $A[j]$ denote the union of the elements in $A[j,k]$ for all $k \in [1,m]$. Clearly, $\Lambda_1 = \bigcup_{j=1}^{n} A[j]$. In the following, we will organize the elements in each $A[j]$ into a sorted array $B[j]$ of size $O(nm^2)$ such that given any index t, the t-th element of $B[j]$ can be computed in $O(\log m)$ time, which will prove Lemma 4. Our technique replies on the following property: the difference of every two adjacent elements in each list $A[j,k]$ is the same, i.e., $2r$.

Notice that for any $k \in [1, m-1]$, the first (resp., last) element of $A[j,k]$ is larger than the first (resp., last) element of $A[j, k+1]$. Hence, the first element of $A[j,m]$, i.e., $\lambda(1,j,m)$, is the smallest element of $A[j]$ and the last element of $A[j,1]$, i.e., $\lambda(j,j,1)$, is the largest element of $A[j]$. Let $\lambda_{min}[j] = \lambda(1,j,m)$ and $\lambda_{max}[j] = \lambda(j,j,1)$.

For each $k \in [1,m]$, we extend the list $A[j,k]$ to a new sorted list $B[j,k]$ with the following property: (1) $A[j,k]$ is a sublist of $B[j,k]$; (2) the difference every two adjacent elements of $B[j,k]$ is $2r$; (3) the first element of $B[j,k]$ is in $[\lambda_{min}[j], \lambda_{min}[j] + 2r)$; (4) the last element of $B[j,k]$ is in $(\lambda_{max}[j] - 2r, \lambda_{max}[j]]$. Specifically, $B[j,k]$ is defined as follows. Note that $\lambda(1,j,k)$ and $\lambda(j,j,k)$ are the first and last elements of $A[j,k]$, respectively. We let $\lambda(1,j,k) - \lfloor \frac{\lambda(1,j,k) - \lambda_{min}[j]}{2r} \rfloor \cdot 2r$ and $\lambda(j,j,k) + \lfloor \frac{\lambda_{max}[j] - \lambda(j,j,k)}{2r} \rfloor \cdot 2r$ be the first and last elements of $B[j,k]$, respectively. Then, the h-th element of $B[j,k]$ is equal to $\lambda(1,j,k) - \lfloor \frac{\lambda(1,j,k) - \lambda_{min}[j]}{2r} \rfloor \cdot 2r + 2r \cdot (h-1)$ for any $h \in [1, \alpha[j]]$, where $\alpha[j] = 1 + \lceil \frac{\lambda_{max}[j] - \lambda_{min}[j]}{2r} \rceil$. Hence, $B[j,k]$ has $\alpha[j]$ elements. One can verify that $B[j,k]$ has the above four properties. Note that we can implicitly create the lists $B[j,k]$ in $O(1)$ time so that given any $k \in [1,m]$ and $h \in [1, \alpha[j]]$, we can obtain the h-th element of $B[j,k]$ in $O(1)$ time. Let $B[j]$ be the sorted list of all elements of $B[j,k]$ for all $1 \le k \le m$. Hence, $B[j]$ has $\alpha[j] \cdot m$ elements.

Let σ_j be the permutation of $1, 2, \ldots, m$ following the sorted order of the first elements of $B[j,k]$. For any $k \in [1,m]$, let $\sigma_j(k)$ be the k-th index in σ_j.

Lemma 5. *For any t with $1 \le t \le \alpha[j] \cdot m$, the t-th smallest element of $B[j]$ is the h_t-th element of the list $B[j, \sigma_j(k_t)]$, where $h_t = \lceil \frac{t}{m} \rceil$ and $k_t = t \mod m$.*

By Lemma 5, if σ_j is known, we can obtain the t-th smallest element of $B[j]$ in $O(1)$ time for any t. Computing σ_j can be done in $O(m \log m)$ time by sorting. If we do the sorting for every $j \in [1,n]$, then we wound need $O(nm \log m)$ time. Fortunately, Lemma 6 implies that we only need to do the sorting once.

Lemma 6. *The permutation σ_j is unique for all $j \in [1,n]$.*

In summary, after $O(m \log m)$ time preprocessing to compute σ_j for any j, we can form the arrays $B[j]$ for all $j \in [1,n]$ such that given any $j \in [1,n]$ and $t \in [1, \alpha[j] \cdot m]$, we can compute t-th smallest element of $B[j]$ in $O(1)$ time. However, we are not done yet, because we do not have a reasonable upper bound for $\alpha[j]$, which is equal to $1 + \lceil \frac{\lambda_{max}[j] - \lambda_{min}[j]}{2r} \rceil = 1 + \lceil \frac{\lambda(j,j,1) - \lambda(1,j,m)}{2r} \rceil = j + \lceil \frac{a_m - a_1}{2r} \rceil$. To address the issue, in the sequel, we will partition the indices $k \in [1,m]$ into groups and then apply our above approach to each group so that the corresponding $\alpha[j]$ values can be bounded, e.g., by $O(mn)$.

The Group Partition Technique We consider any index $j \in [1, m]$.

We partition the indices $1, 2, \ldots, m$ into groups each consisting of a sequence of consecutive indices, such that each group has an *intra-group overlapping property*: For any index k that is not the largest index in the group, the first element of $A[j, k]$ is smaller than or equal to the last element of $A[j, k + 1]$, i.e., $\lambda(1, j, k) \leq \lambda(j, j, k+1)$. Further, the groups have the following *inter-group non-overlapping property*: For the largest index k in a group that is not the last group, the first element of $A[j, k]$ is larger than the last element of $A[j, k + 1]$, i.e., $\lambda(1, j, k) > \lambda(j, j, k + 1)$.

We compute the groups in $O(m)$ time as follows. Initially, add 1 into the first group G_1. Let $k = 1$. While the first element of $A[j, k]$ is smaller than or equal to the last element of $A[j, k+1]$, we add $k+1$ into G_1 and reset $k = k + 1$. After the while loop, G_1 is computed. Then, starting from $k + 1$, we compute G_2 and so on until index m is included in the last group. Let $G_1, G_2, \ldots, G_l$ be the l groups we compute. Note that $l \leq m$.

Consider any group G_g with $1 \leq g \leq l$. We process the lists $A[j][k]$ for all $k \in G_g$ in the same way as discussed before. Specifically, for each $k \in G_g$, we create a new list $B[j][k]$ from $A[j][k]$. Based on the new lists in the group G_g, we form the sorted array $B_g[j]$ with a total of $|G_g| \cdot \alpha_g[j]$ elements, where $|G_g|$ is the number of indices of G_g and $\alpha_g[j]$ is corresponding $\alpha[j]$ value as defined before but only on the group G_g, i.e., if k_1 and k_2 are the smallest and largest indices of G_g respectively, then $\alpha_g[j] = 1 + \lceil \frac{\lambda(j,j,k_1) - \lambda(1,j,k_2)}{2r} \rceil$. Let $B[j]$ be the sorted list of all elements in the lists $B_g[j]$ for all groups. Due to the intra-group overlapping property of each group, it holds that $\alpha_g \leq |G_g| \cdot n$. Thus, the size of $B[j]$ is at most $\sum_{g=1}^{l} |G_g|^2 \cdot n$, which is at most $m^2 n$ since $\sum_{g=1}^{l} |G_g| = m$.

Suppose we want to find the t-th smallest element of $B[j]$. As preprocessing, we compute a sequence of values $\beta_g[j]$ for $g = 1, 2, \ldots, l$, where $\beta_g[j] = \sum_{g'=1}^{g} \alpha_{g'}[j] \cdot |G_{g'}|$, in $O(m)$ time. To compute the t-th smallest element of $B[j]$, we first do binary search on the sequence $\beta_1[j], \beta_2[j], \ldots, \beta_l[j]$ to find in $O(\log l)$ time the index g such that $t \in (\beta_{g-1}[j], \beta_g[j]]$. Due to the inter-group non-overlapping property of the groups, the t-th smallest element of $B[j]$ is the $(t - \beta_{g-1}[j])$-th element in the array $B_g[j]$, which can be found in $O(1)$ time. As $l \leq m$, the total time for computing the t-th smallest element of $B[j]$ is $O(\log m)$.

The above discussion is on any single index $j \in [1, n]$. With $O(m \log m)$ time preprocessing, given any t, we can find the t-th smallest value of $B[j]$ in $O(\log m)$ time. For all indices $j \in [1, n]$, it appears that we have to do the group partition for every $j \in [1, n]$, which would take quadratic time. To resolve the issue, we show that it suffices to only use the group partition based on $j = n$ for all other $j \in [1, n - 1]$. The details are given below.

Suppose from now on $G_1, G_2, \ldots, G_l$ are the groups computed as above with respect to $j = n$. We know that the inter-group non-overlapping property holds respect to the index n. The following lemma shows that the property also holds with respect to any other index $j \in [1, n - 1]$.

Lemma 7. *The inter-group non-overlapping property holds for any $j \in [1, n-1]$.*

Consider any G_g with $1 \leq g \leq l$ and any $j \in [1, n]$. For each $k \in G_g$, we create a new list $B[j][k]$ based on $A[j][k]$ in the same way as before. Based on the new lists, we form the sorted array $B_g[j]$ of $|G_g| \cdot \alpha_g[j]$ elements. We also define the value $\beta_g[j]$ in the same way as before. Lemma 8 shows that $\alpha_g[j]$ and $\beta_g[j]$ can be computed from $\alpha_g[n]$ and $\beta_g[n]$.

Lemma 8. *For any $j \in [1, n-1]$ and $g \in [1, l]$, $\alpha_g[j] = \alpha_g[n] - n + j$ and $\beta_g[j] = \beta_g[n] + \delta_g \cdot g \cdot (j - n)$, where $\delta_g = \sum_{g'=1}^{g} |G_{g'}|$.*

For each group G_g, we compute the permutation for the lists $B[n, k]$ for all k in the group. Computing the permutations for all groups takes $O(m \log m)$ time. Also as preprocessing, we first compute δ_g, $\alpha_g(n)$ and $\beta_g(n)$ for all $g \in [1, l]$ in $O(m)$ time. By Lemma 8, for any $j \in [1, n]$ and any $g \in [1, l]$, we can compute $\alpha_g[j]$ and $\beta_g[j]$ in $O(1)$ time. Because the lists $B[n, k]$ for all k in each group G_g have the intra-group overlapping property, it holds that $\alpha_g[n] \leq |G_g| \cdot n$. Hence, $\sum_{g=1}^{l} \alpha_g[n] \leq mn$. For any $j \in [1, n-1]$, by Lemma 8, $\alpha_g[j] < \alpha_g[n]$, and thus $\sum_{g=1}^{l} \alpha_g[j] \leq mn$. Note that $B[j]$ has at most $m^2 n$ elements.

For any $j \in [1, n]$ and any $t \in [1, \sum_{g=1}^{l} |G_g| \cdot \alpha_g[j]]$, to compute the t-th smallest element of $B[j]$, due to the inter-group non-overlapping property in Lemma 7, we can still use the previous binary search approach. As we can obtain each $\beta_g[j]$ for any $g \in [1, l]$ in $O(1)$ time by Lemma 8, we can still compute the t-th smallest element of $B[j]$ in $O(\log m)$ time. This proves Lemma 4.

4   The Decision Problem of MBC

In this section, we present an $O(m + n \log n)$-time algorithm for the decision problem of MBC: given any value $\lambda > 0$, determine whether $\lambda \geq \lambda^*$. Our algorithm for MBC in Section 5 will make use of this decision algorithm. The decision problem may have independent interest because in some applications each sensor has a limited energy λ and we want to know whether their energy is enough for them to move to cover all barriers.

Consider any value $\lambda > 0$. We assume $\lambda \geq \max_{1 \leq i \leq n} |y_i|$ since otherwise some sensor cannot reach L by moving λ (and thus λ is not feasible). For any sensor $s_i \in S$, define $x_i^r = x_i + \sqrt{\lambda^2 - y_i^2}$ and $x_i^l = x_i - \sqrt{\lambda^2 - y_i^2}$. Note that x_i^r and x_i^l are respectively the rightmost and leftmost points of L s_i can reach with respect to λ. We call x_i^r the *rightmost (resp., leftmost) λ-reachable location* of s_i on L. For any point x on L, we use $p^+(x)$ to denote a point x' such that $x' > x$ and x' is infinitesimally close to x. The high-level scheme of our algorithm is similar to that in [15]. Below we describe the algorithm.

We use a *configuration* to refer to a specification on where each sensor $s_i \in S$ is located. For example, in the *input configuration*, each s_i is at (x_i, y_i). We first move each sensor s_i to x_i^r on L. Let C_0 denote the resulting configuration. In C_0, each sensor s_i is not allowed to move rightwards but can move leftwards on L by a maximum distance $2\sqrt{\lambda^2 - y_i^2}$.

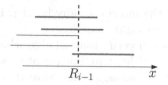 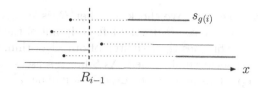

Fig. 1. Illustrating the set S_{i1}. The covering intervals of sensors are shown with segments (the red thick segments correspond to the sensors in S_{i1}). Every sensor in S_{i1} can be $s_{g(i)}$.

Fig. 2. Illustrating the set S_{i2}. The segments are the covering intervals of sensors. The red thick segments correspond to the sensors in S_{i2}. The four black points corresponding to the values $x_k^l - r$ of the four sensors x_k to the right of R_{i-1}. The sensor $s_{g(i)}$ is labeled.

If $\lambda \geq \lambda^*$, our algorithm will compute a subset of sensors with their new locations to cover all barriers of $\mathcal{B}$ and the maximum movement of each sensor of in the subset is at most λ.

For each step i with $i \geq 1$, let C_{i-1} be the configuration right before the i-th step. Our algorithm maintains the following *invariants*. (1) We have a subset of sensors $S_{i-1} = \{s_{g(1)}, s_{g(2)}, \ldots, s_{g(i-1)}\}$, where for each $1 \leq j \leq i - 1$, $g(j)$ is the index of the sensor $s_{g(j)}$ in S. (2) In C_{i-1}, each sensor s_k of S_{i-1} is at a new location $x_k' \in [x_k^l, x_k^r]$, and all other sensors are still in their locations of C_0. (3) A value R_{i-1} is maintained such that $0 \leq R_{i-1} < \beta$, R_{i-1} is on a barrier, every barrier point $x < R_{i-1}$ is covered by a sensor of S_{i-1} in C_{i-1}. (4) If R_{i-1} is not at the left endpoint of a barrier, then R_{i-1} is covered by a sensor of S_{i-1} in C_{i-1}. (5) The point $p^+(R_{i-1})$ is not covered by any sensor in S_{i-1}.

Initially when $i = 1$, we let $S_0 = \emptyset$ and $R_0 = 0$, and thus all algorithm invariants hold for C_0. The i-th step of the algorithm finds a sensor $s_{g(i)} \in S \setminus S_{i-1}$ and moves it to a new location $x_{g(i)}' \in [x_{g(i)}^l, x_{g(i)}^r]$ and thus obtains a new configuration C_i. The details are given below.

Define S_{i1} as the set of sensors that cover the point $p^+(R_{i-1})$ in C_{i-1}, i.e., $S_{i1} = \{s_k \mid x_k^r - r \leq R_{i-1} < x_k^r + r\}$. By the algorithm invariant (5), no sensor in S_{i-1} covers $p^+(R_{i-1})$. Thus, $S_{i1} \subseteq S \setminus S_{i-1}$. If $S_{i1} \neq \emptyset$, then we choose an *arbitrary* sensor in S_{i1} as $s_{g(i)}$ (e.g., see Fig. 1) and let $x_{g(i)}' = x_{g(i)}^r$. We then set $R_i = x_{g(i)}' + r$, i.e., R_i is at the right endpoint of the covering interval of $s_{g(i)}$. Note that C_i is C_{i-1} as $s_{g(i)}$ is not moved.

If $S_{i1} = \emptyset$, then we define $S_{i2} = \{s_k \mid x_k^l - r \leq R_{i-1} < x_k^r - r\}$ (i.e., S_{i2} consists of those sensors s_k that does not cover R_{i-1} when it is at x_k^r but is possible to do so when it is at some location in $[x_k^l, x_k^r]$). If $S_{i2} \neq \emptyset$, we choose the *leftmost* sensor of S_{i2} as $s_{g(i)}$ (e.g., see Fig. 2), and let $x_{g(i)}' = R_{i-1} + r$ (i.e., we move $s_{g(i)}$ to $x_{g(i)}'$ and thus obtain C_i). If $S_{i2} = \emptyset$, then we conclude that $\lambda < \lambda^*$ and terminate the algorithm.

Hence, if $S_{i1} = S_{i2} = \emptyset$, the algorithm will stop and report $\lambda < \lambda^*$. Otherwise, a sensor $s_{g(i)}$ is found from either S_{i1} or S_{i2}, and it is moved to $x_{g(i)}'$. In either case, $R_i = x_{g(i)}' + r$ and $S_i = S_{i-1} \cup \{s_{g(i)}\}$. If $R_i \geq \beta$, then we terminate the algorithm and report $\lambda \geq \lambda^*$. Otherwise, we further perform the following *jump-*

over procedure: We check whether R_i is located at the interior of any barrier; if not, then we set R_i to the left endpoint of the barrier right after R_i.

This finishes the i-th step of our algorithm. One can verify that all algorithm invariants are maintained. As S has n sensors, the algorithm will finish in at most n steps. This finishes the description of our algorithm. The algorithm correctness and implementation are omitted.

Theorem 2. *Given any value λ, we can determine whether $\lambda \geq \lambda^*$ in $O(m + n \log n)$ time.*

Our algorithm in Section 5 will perform feasibility tests multiple times, for which we have the following lemma.

Lemma 9. *Suppose the values x_i^r for all $i = 1, 2, \ldots, n$ are already sorted, we can determine whether $\lambda \geq \lambda^*$ in $O(m + n \log \log n)$ time for any λ.*

5 Solving the Problem MBC

To solve MBC, it suffices to compute λ^*. The high-level scheme of our algorithm is similar to that in [15], although some low-level computations are different.

We now use $x_i^r(\lambda)$ to refer to x_i^r for any λ, so that we consider $x_i^r(\lambda)$ as a function on $\lambda \in [0, \infty]$, which actually defines a half of the upper branch (on the right side of the y-axis) of a hyperbola. Let σ be the order of the values $x_i^r(\lambda^*)$ for all $i \in [1, n]$. To use Lemma 9, we first run a preprocessing step in Lemma 10.

Lemma 10. *With $O(n \log^3 n + m \log^2 n)$ time preprocessing, we can compute σ and an interval $(\lambda_1^*, \lambda_2^*]$ containing λ^* such that σ is also the order of the values $x_i^r(\lambda)$ for any $\lambda \in (\lambda_1^*, \lambda_2^*]$.*

Proof. To compute σ, we apply Megiddo's parametric search [12] to sort the values $x_i^r(\lambda^*)$ for $i \in [1, n]$, using the decision algorithm in Theorem 2. Indeed, recall that $x_i^r(\lambda) = x_i + \sqrt{\lambda^2 - y_i^2}$. Hence, as λ increases, $x_i^r(\lambda)$ is a (strictly) increasing function. For any two indices i and j, there is at most one root on $\lambda \in [0, \infty)$ for the equation: $x_i^r(\lambda) = x_j^r(\lambda)$. Therefore, we can apply Megiddo's parametric search [12] to do the sorting. The total time is $O((\tau + n) \log^2 n)$, where τ is the running time of the decision algorithm. By Theorem 2, $\tau = O(m + n \log n)$. Hence, the total time for computing σ is $O(m \log^2 n + n \log^3 n)$.

In addition, Megiddo's parametric search [12] will return an interval $(\lambda_1^*, \lambda_2^*]$ containing λ^* and σ is also the order of the values $x_i^r(\lambda)$ for any $\lambda \in (\lambda_1^*, \lambda_2^*]$. $\square$

As $\lambda^* \in (\lambda_1^*, \lambda_2^*]$, our subsequent feasible tests will be only on values $\lambda \in (\lambda_1^*, \lambda_2^*)$ because if $\lambda \leq \lambda_1^*$, then λ is not feasible and if $\lambda \geq \lambda_2^*$, then λ is feasible. Lemmas 9 and 10 together lead to the following result.

Lemma 11. *Each feasibility test can be done in $O(m + n \log \log n)$ time for any $\lambda \in (\lambda_1^*, \lambda_2^*)$.*

To compute λ^*, we "parameterize" our decision algorithm with λ as a parameter. Although we do not know λ^*, we execute the decision algorithm in such a way that it computes the same subset of sensors $s_{g(1)}, s_{g(2)}, \ldots$ as would be obtained if we ran the decision algorithm on $\lambda = \lambda^*$.

Recall that for any λ, step i of our decision algorithm computes the sensor $s_{g(i)}$, the set $S_i = \{s_{g(1)}, s_{g(2)}, \ldots, s_{g(i)}\}$, and the value R_i, and obtains the configuration C_i. In the following, we often consider λ as a variable rather than a fixed value. Thus, we will use $S_i(\lambda)$ (resp., $R_i(\lambda)$, $s_{g(i)}(\lambda)$, $C_i(\lambda)$, $x_i^r(\lambda)$) to refer to the corresponding S_i (resp., R_i, $s_{g(i)}$, C_i, x_i^r). Our algorithm has at most n steps. Consider a general i-th step for $i \geq 1$. Right before the step, we have an interval $(\lambda_{i-1}^1, \lambda_{i-1}^2]$ and a sensor set $S_{i-1}(\lambda)$, such that the following algorithm invariants hold.

1. $\lambda^* \in (\lambda_{i-1}^1, \lambda_{i-1}^2]$.
2. The set $S_{i-1}(\lambda)$ is the same (with the same order) for all $\lambda \in (\lambda_{i-1}^1, \lambda_{i-1}^2)$.
3. $R_{i-1}(\lambda)$ on $\lambda \in (\lambda_{i-1}^1, \lambda_{i-1}^2)$ is either constant or equal to $x_j + \sqrt{\lambda^2 - y_j^2} + c$ for some constant c and some sensor s_j with $1 \leq j \leq i - 1$, and $R_{i-1}(\lambda)$ is maintained by the algorithm.
4. $R_{i-1}(\lambda) < \beta$ for any $\lambda \in (\lambda_{i-1}^1, \lambda_{i-1}^2)$.

Initially when $i = 1$, we let $\lambda_0^1 = \lambda_1^*$ and $\lambda_0^2 = \lambda_2^*$. Since $S_0(\lambda) = \emptyset$ and $R_0(\lambda) = 0$ for any λ, by Lemma 10, all invariants hold for $i = 1$. In general, the i-th step will either compute λ^*, or obtain an interval $(\lambda_i^1, \lambda_i^2] \subseteq (\lambda_{i-1}^1, \lambda_{i-1}^2]$ and a sensor $s_{g(i)}(\lambda)$ with $S_i(\lambda) = S_{i-1}(\lambda) \cup \{s_{g(i)}(\lambda)\}$. The running time of the step is $O((m + n \log \log n)(\log n + \log m))$. The details are omitted.

The algorithm will compute λ^* after at most n steps. The total time is $O(n \cdot (m + n \log \log n) \cdot (\log m + \log n))$, which is bounded by $O(nm \log m + n^2 \log n \log \log n)$ as shown in Theorem 3. The space of the algorithm is $O(n)$.

Theorem 3. *The problem MBC can be solved in $O(nm \log m + n^2 \log n \log \log n)$ time and $O(n)$ space.*

6 Concluding Remarks

As mentioned before, the high-level scheme of our algorithm for MBC is similar to those in [3, 15]. However, a new technique we propose in this paper can help reduce the space complexities of the algorithms in [3, 15]. Specifically, Chen et al. [3] solved the line-constrained problem in $O(n^2 \log n)$ time and $O(n^2)$ space for the case where $m = 1$ and sensors have different ranges. Wang and Zhang [15] solved the line-constrained problem in $O(n^2 \log n \log \log n)$ time and $O(n^2)$ space for the case where $m = 1$, sensors have the same range, and sensors have weights. If we apply the similar preprocessing as in Lemma 10, then the space complexities of both algorithms [3, 15] can be reduced to $O(n)$ while the time complexities do not change asymptotically.

In addition, by slightly changing our algorithm for MBC, we can also solve the following problem variant: Find a subset S' of sensors of S to move them to

L to cover all barriers such that the maximum movement of all sensors of S' is minimized (and sensors of $S \setminus S'$ do not move). We omit the details.

References

1. Andrews, A., Wang, H.: Minimizing the aggregate movements for interval coverage. Algorithmica 78, 47–85 (2017)
2. Bhattacharya, B., Burmester, B., Hu, Y., Kranakis, E., Shi, Q., Wiese, A.: Optimal movement of mobile sensors for barrier coverage of a planar region. Theoretical Computer Science 410(52), 5515–5528 (2009)
3. Chen, D., Gu, Y., Li, J., Wang, H.: Algorithms on minimizing the maximum sensor movement for barrier coverage of a linear domain. Discrete and Computational Geometry 50, 374–408 (2013)
4. Chen, D., Tan, X., Wang, H., Wu, G.: Optimal point movement for covering circular regions. Algorithmica 72, 379–399 (2013)
5. Czyzowicz, J., Kranakis, E., Krizanc, D., Lambadaris, I., Narayanan, L., Opatrny, J., Stacho, L., Urrutia, J., Yazdani, M.: On minimizing the maximum sensor movement for barrier coverage of a line segment. In: Proc. of the 8th International Conference on Ad-Hoc, Mobile and Wireless Networks. pp. 194–212 (2009)
6. Czyzowicz, J., Kranakis, E., Krizanc, D., Lambadaris, I., Narayanan, L., Opatrny, J., Stacho, L., Urrutia, J., Yazdani, M.: On minimizing the sum of sensor movements for barrier coverage of a line segment. In: Proc. of the 9th International Conference on Ad-Hoc, Mobile and Wireless Networks. pp. 29–42 (2010)
7. Dobrev, S., Durocher, S., Eftekhari, M., Georgiou, K., Kranakis, E., Krizanc, D., Narayanan, L., Opatrny, J., Shende, S., Urrutia, J.: Complexity of barrier coverage with relocatable sensors in the plane. Theoretical Computer Science 579, 64–73 (2015)
8. Frederickson, G., Johnson, D.: Generalized selection and ranking: Sorted matrices. SIAM Journal on Computing 13(1), 14–30 (1984)
9. Kumar, S., Lai, T., Arora, A.: Barrier coverage with wireless sensors. In: Proc. of the 11th Annual International Conference on Mobile Computing and Networking (MobiCom). pp. 284–298 (2005)
10. Li, S., Shen, H.: Minimizing the maximum sensor movement for barrier coverage in the plane. In: Proc. of the 2015 IEEE Conference on Computer Communications (INFOCOM). pp. 244–252 (2015)
11. Li, S., Wang, H.: Algorithms for covering multiple barriers. arXiv:1704.06870 (2017)
12. Megiddo, N.: Applying parallel computation algorithms in the design of serial algorithms. Journal of the ACM 30(4), 852–865 (1983)
13. Mehrandish, M.: On Routing, Backbone Formation and Barrier Coverage in Wireless Ad Doc and Sensor Networks. Ph.D. thesis, Concordia University, Montreal, Quebec, Canada (2011)
14. Mehrandish, M., Narayanan, L., Opatrny, J.: Minimizing the number of sensors moved on line barriers. In: Proc. of IEEE Wireless Communications and Networking Conference (WCNC). pp. 653–658 (2011)
15. Wang, H., Zhang, X.: Minimizing the maximum moving cost of interval coverage. In: Proc. of the 26th International Symposium on Algorithms and Computation (ISAAC). pp. 188–198 (2015), full version to appear in *International Journal of Computational Geometry and Application (IJCGA)*

An Improved Algorithm for Diameter-Optimally Augmenting Paths in a Metric Space*

Haitao Wang

Department of Computer Science
Utah State University, Logan, UT 84322, USA
haitao.wang@usu.edu

Abstract. Let P be a path graph of n vertices embedded in a metric space. We consider the problem of adding a new edge to P such that the diameter of the resulting graph is minimized. Previously (in ICALP 2015) the problem was solved in $O(n \log^3 n)$ time. In this paper, based on new algorithmic techniques and observations, we present an $O(n \log n)$ time algorithm.

1 Introduction

Let P be a path graph of n vertices embedded in a metric space. We consider the problem of adding a new edge to P such that the diameter of the resulting graph is minimized. Let G be a graph and each edge has a non-negative length. The *length* of any path of G is the total length of all edges of the path. For any two vertices u and v of G, we use $d_G(u,v)$ to denote the length of the shortest path from u to v in G. The *diameter* of G is defined as $\max_{u,v \in G} d_G(u,v)$.

Let P be a path graph of n vertices $v_1, v_2, \ldots, v_n$ with an edge $e(v_{i-1}, v_i)$ connecting v_{i-1} and v_i for each $1 \le i \le n-1$. Let V be the vertex set of P. We assume $(V, |\cdot|)$ is a metric space and $|v_i v_j|$ is the distance of any two vertices v_i and v_j of V. Specifically, the following properties hold: (1) the triangle inequality: $|v_i v_k| + |v_k v_j| \ge |v_i v_j|$; (2) $|v_i v_j| = |v_j v_i| \ge 0$; (3) $|v_i v_j| = 0$ if $i = j$. In particular, for each edge $e(v_{i-1}, v_i)$ of P, its length is equal to $|v_{i-1} v_i|$. We assume that given any two vertices v_i and v_j of P, the distance $|v_i v_j|$ can be obtained in $O(1)$ time.

Our goal is to find a new edge e connecting two vertices of P and add e to P, such that the diameter of the resulting graph $P \cup \{e\}$ is minimized.

The problem has been studied before. Große et al. [11] solved the problem in $O(n \log^3 n)$ time. In this paper, we present a new algorithm that runs in $O(n \log n)$ time. As in [11], we refer to the problem as *the diameter-optimally augmenting path problem*, or DOAP for short.

1.1 Related Work

If the path P is in the Euclidean space $\mathbb{R}^d$ for a constant d, then Große et al. [11] also gave an $O(n + 1/\epsilon^3)$ time algorithm that can find a $(1 + \epsilon)$-approximation

---

* This research was supported in part by NSF under Grant CCF-1317143.

F. Ellen et al. (Eds.): WADS 2017, LNCS 10389, pp. 545–556, 2017.
DOI: 10.1007/978-3-319-62127-2_46

solution for DOAP, for any $\epsilon > 0$. If P is in the Euclidean plane $\mathbb{R}^2$, De Carufel et al. [4] gave a linear time algorithm for adding a new edge to P to minimize the *continuous diameter* (i.e., the diameter is defined with respect to all points of P, not only vertices). For a geometric tree T of n vertices embedded in the Euclidean plane, De Carufel et al. [5] gave an $O(n \log n)$-time algorithm for adding a new edge to T to minimize the *continuous* diameter. Unfortunately, both algorithms [4,5], which are particularly for the continuous diameter in the Euclidean plane, cannot be generalized to solve our problem for the "discrete" diameter in the more general metric space.

Some more general problems were also studied before, e.g., see [1, 3, 6, 7, 10, 13, 14, 16] and the references therein. Consider a general graph G in which edges have non-negative lengths. For an integer k, the goal of the general problem is to compute k new edges and add them to G such that the resulting graph has the minimum diameter. The problem is NP-hard [16] and some other variants are even W[2]-hard [7, 10]. Approximation results were given for the general problem and many of its variations, e.g., see [3, 7, 14]. The upper and lower bounds on the diameters of the augmented graphs were investigated, e.g., [1, 13].

Since the diameter is an important metric of network performance, which measures the worst-case cost between any two nodes of the network, as discussed in [3, 6], the problem of augmenting graphs for minimizing the diameter and its variations have many applications, such as in data networks, telephone networks, transportation networks, scheduling, etc. As an application of DOAP, consider the following example. Suppose there is a highway that connects several cities. In order to reduce the transportation time, we want to build a new highway connecting two cities such that the distance between the farthest two cities using both highways is minimized. Clearly, this is a problem instance of DOAP.

1.2 Our Approach

To tackle the problem DOAP, Große et al. [11] first gave an $O(n \log n)$ time algorithm for the *decision version* of the problem: Given any value λ, determine whether it is possible to add a new edge e into P such that the diameter of the resulting graph is at most λ. Then, by implementing the above decision algorithm in a parallel fashion and applying Megiddo's parametric search [15], they solved DOAP in $O(n \log^3 n)$ time [11]. For differentiation, we refer to the original problem DOAP as the *optimization problem*. Our improvement over the previous work [11] is twofold.

First, we solve the decision problem in $O(n)$ time. Our algorithm is based on the $O(n \log n)$ time algorithm in the previous work [11]. However, by discovering new observations on the problem structure and with the help of the range-minima data structure [2, 12], we avoid certain expensive operations and eventually achieve the $O(n)$ time complexity.

Second, comparing with the decision problem, our algorithm for the optimization problem is completely different from the previous work [11]. Let λ^* be the diameter of the resulting graph in an optimal solution. Instead of using the parametric search, we identify a set S of candidate values such that λ^* is in S and

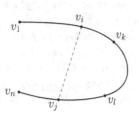

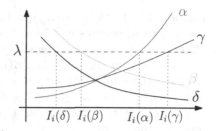

Fig. 1. Illustrating the resulting graph after a new edge $e(v_i, v_j)$ is added.

Fig. 2. Illustrating $f(i,j)$ as j changes in $[i, n]$ and $I_i(f)$ for $f \in \{\alpha, \beta, \gamma, \delta\}$.

then we search λ^* in S using our algorithm for the decision problem. However, computational difficulties arise for this approach due to that the set S is too large ($|S| = \Omega(n^2)$) and computing certain values of S is time-consuming (e.g., for certain values of S, computing each of them takes $O(n)$ time). To circumvent these difficulties, our algorithm has several steps. In each step, we shrink S significantly such that λ^* always remains in S. More importantly, each step will obtain certain information, based on which the next step can further reduce S. After several steps, the size of S is reduced to $O(n)$ and all the remaining values of S can be computed in $O(n \log n)$ time. At this point we can use our decision algorithm to find λ^* from S in additional $O(n \log n)$ time.

The remaining paper is organized as follows. In Section 2, we introduce some notation and observations. In Section 3, we present our algorithm for the decision problem. The optimization problem is solved in Section 4. Due to the space limit, some lemma proofs are omitted but can be found in the full paper [17].

2   Preliminaries

In this section, we introduce some notation and observations, some of which are from Große et al. [11].

For any two vertices v_i and v_j of P, we use $e(v_i, v_j)$ to denote the edge connecting v_i and v_j in the metric space. Hence, $e(v_i, v_j)$ is in P if $|i - j|$ is 1 or 0. The length of $e(v_i, v_j)$ is $|v_i v_j|$. For any i and j with $1 \le i \le j \le n$, we use $G(i, j)$ to denote the resulting graph by adding the edge $e(v_i, v_j)$ into P. If $i = j$ or $|i - j| = 1$, $G(i, j)$ is P. Let $D(i, j)$ denote the diameter of $G(i, j)$.

Our goal for the optimization problem DOAP is to find a pair of indices (i, j) with $1 \le i \le j \le n$ such that $D(i, j)$ is minimized. Let $\lambda^* = \min_{1 \le i \le j \le n} D(i, j)$, i.e., λ^* is the diameter in an optimal solution.

Given a value λ, the decision problem is to determine whether $\lambda \ge \lambda^*$, or in other words, determine whether there exist a pair (i, j) with $1 \le i \le j \le n$ such that $D(i, j) \le \lambda$. If yes, we say that λ is a *feasible* value.

Recall that for any graph G, $d_G(u, v)$ refers to the length of the shortest path between two vertices u and v in G.

Consider any pair of indices (i, j) with $1 \le i \le j \le n$. We define $\alpha(i, j)$, $\beta(i, j)$, $\gamma(i, j)$, and $\delta(i, j)$ as follows (refer to Fig. 1).

Definition 1. *1. Define $\alpha(i,j)$ to be the largest shortest path length in $G(i,j)$ from v_1 to all vertices v_k with $k \in [i,j]$, i.e., $\alpha(i,j) = \max_{i \leq k \leq j} d_{G(i,j)}(v_1, v_k)$.*
  2. Define $\beta(i,j)$ to be the largest shortest path length in $G(i,j)$ from v_n to all vertices v_k with $k \in [i,j]$, i.e., $\beta(i,j) = \max_{i \leq k \leq j} d_{G(i,j)}(v_k, v_n)$.
  3. Define $\gamma(i,j)$ to be the largest shortest path length in $G(i,j)$ from v_k to v_l for any k and l with $i \leq k \leq l \leq j$, i.e., $\gamma(i,j) = \max_{i \leq k \leq l \leq j} d_{G(i,j)}(v_k, v_l)$.
  4. Define $\delta(i,j)$ to be the shortest path length in $G(i,j)$ from v_1 to v_n, i.e., $\delta(i,j) = d_{G(i,j)}(v_1, v_n)$.

It can be verified (also shown in [11]) that Observation 1 holds.

Observation 1 ([11]) $D(i,j) = \max\{\alpha(i,j), \beta(i,j), \gamma(i,j), \delta(i,j)\}$.

Further, due to the triangle inequality of the metric space, the following monotonicity properties hold.

Observation 2 ([11])

1. *For any $1 \leq i \leq j \leq n-1$, $\alpha(i,j) \leq \alpha(i,j+1)$, $\beta(i,j) \geq \beta(i,j+1)$, $\gamma(i,j) \leq \gamma(i,j+1)$, and $\delta(i,j) \geq \delta(i,j+1)$.*
2. *For any $1 \leq i < j \leq n$, $\alpha(i,j) \leq \alpha(i+1,j)$, $\beta(i,j) \geq \beta(i+1,j)$, $\gamma(i,j) \geq \gamma(i+1,j)$, and $\delta(i,j) \leq \delta(i+1,j)$.*

For any pair (i,j) with $1 \leq i \leq j \leq n$, let $P(i,j)$ denote the subpath of P between v_i and v_j. Hence, $d_P(v_i, v_j)$ is the length of $P(i,j)$, i.e., $d_P(v_i, v_j) = \sum_{i \leq k \leq j-1} |v_k v_{k+1}|$ if $i < j$ and $d_P(v_i, v_j) = 0$ if $i = j$.

Our algorithms will need to compute $f(i,j)$ for each $f \in \{\alpha, \beta, \gamma, \delta\}$. Lemma 1 was already shown by Große et al. [11].

Lemma 1. ([11]) *With $O(n)$ time preprocessing, given any pair (i,j) with $1 \leq i \leq j \leq n$, we can compute $d_P(i,j)$ and $\delta(i,j)$ in $O(1)$ time, and compute $\alpha(i,j)$ and $\beta(i,j)$ in $O(\log n)$ time.*

For computing $\gamma(i,j)$, although one may be able to do so in $O(n)$ time, it is not clear to us how to make it in $O(\log n)$ time even with $O(n \log n)$ time preprocessing. As will be seen later, this is the major difficulty for solving the problem DOAP efficiently. We refer to it as the *γ-computation difficulty*. Our main effort will be to circumvent the difficulty by providing alternative and efficient solutions.

For any pair (i,j) with $1 \leq i \leq j \leq n$, we use $C(i,j)$ to denote the cycle $P(i,j) \cup e(v_i, v_j)$. Consider $d_{G(i,j)}(v_k, v_l)$ for any k and l with $i \leq k \leq l \leq j$. Notice that the shortest path from v_k to v_l in $C(i,j)$ is also a shortest path in $G(i,j)$. Hence, $d_{G(i,j)}(v_k, v_l) = d_{C(i,j)}(v_k, v_l)$. There are two paths in $C(i,j)$ from v_k to v_l: one is $P(k,l)$ and the other uses the edge $e(v_i, v_j)$. We use $d^1_{C(i,j)}(v_k, v_l)$ to denote the length of the latter path, i.e., $d^1_{C(i,j)}(v_k, v_l) = d_P(v_i, v_k) + |v_i v_j| + d_P(v_l, v_j)$. With this notation, we have $d_{C(i,j)}(v_k, v_l) = \min\{d_P(v_k, v_l), d^1_{C(i,j)}(v_k, v_l)\}$. According to the definition of $\gamma(i,j)$, we summarize our discussion in the following observation.

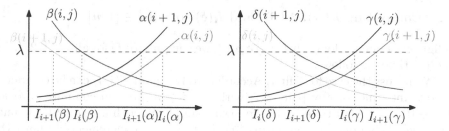

Fig. 3. Illustrating $f(i,j)$ and $f(i+1,j)$ as j changes and $I_i(f)$ and $I_{i+1}(f)$ for $f \in \{\alpha, \beta, \gamma, \delta\}$.

Observation 3 *For any pair (i,j) with $1 \leq i \leq j \leq n$, we have $\gamma(i,j) = \max_{i \leq k \leq l \leq j} d_{C(i,j)}(v_k, v_l)$, with $d_{C(i,j)}(v_k, v_l) = \min\{d_P(v_k, v_l), d^1_{C(i,j)}(v_k, v_l)\}$ and $d^1_{C(i,j)}(v_k, v_l) = d_P(v_i, v_k) + |v_i v_j| + d_P(v_l, v_j)$.*

In the following, to simplify the notation, when the context is clear, we will use index i to refer to vertex v_i. For example, $d_P(i,j)$ refers to $d_P(v_i, v_j)$.

3   The Decision Problem

In this section, we present our $O(n)$ time algorithm for the decision problem. For any value λ, our goal is to determine whether λ is feasible, i.e. whether $\lambda \geq \lambda^*$, or equivalently, whether there is a pair (i,j) with $1 \leq i \leq j \leq n$ such that $D(i,j) \leq \lambda$. If yes, our algorithm can also find such a *feasible edge* $e(i,j)$.

By Observation 1, $D(i,j) \leq \lambda$ if and only if $f(i,j) \leq \lambda$ for each $f \in \{\alpha, \beta, \gamma, \delta\}$. To determine whether λ is feasible, our algorithm will determine for each $i \in [1,n]$, whether there exists $j \in [i,n]$ such that $f(i,j) \leq \lambda$ for each $f \in \{\alpha, \beta, \gamma, \delta\}$.

For any fixed $i \in [1,n]$, we consider $\alpha(i,j)$, $\beta(i,j)$, $\gamma(i,j)$, and $\delta(i,j)$ as functions of $j \in [i,n]$. In light of Observation 2, $\alpha(i,j)$ and $\gamma(i,j)$ are monotonically increasing and $\beta(i,j)$ and $\delta(i,j)$ are monotonically decreasing (e.g., see Fig. 2). We define four indices $I_i(f)$ for $f \in \{\alpha, \beta, \gamma, \delta\}$ as follows. Refer to Fig. 2.

Definition 2. *Define $I_i(\alpha)$ to be the largest index $j \in [i,n]$ such that $\alpha(i,j) \leq \lambda$. Define $I_i(\gamma)$ similarly to $I_i(\alpha)$, i.e., $I_i(\gamma)$ is the largest index $j \in [i,n]$ such that $\gamma(i,j) \leq \lambda$. If $\beta(i,n) \leq \lambda$, then define $I_i(\beta)$ to be the smallest index $j \in [i,n]$ such that $\beta(i,j) \leq \lambda$; otherwise, let $I_i(\beta) = \infty$. Define $I_i(\delta)$ similarly to $I_i(\beta)$, i.e., if $\delta(i,n) \leq \lambda$, then $I_i(\delta)$ is the smallest index $j \in [i,n]$ such that $\delta(i,j) \leq \lambda$; otherwise, $I_i(\delta) = \infty$.*

Clearly, λ is feasible if and only if $[1, I_i(\alpha)] \cap [I_i(\beta), n] \cap [1, I_i(\gamma)] \cap [I_i(\delta), n] \neq \emptyset$ for some $i \in [1,n]$. By Observation 2, we have the following lemma.

Lemma 2. *For any $i \in [1, n-1]$, $I_i(\alpha) \geq I_{i+1}(\alpha)$, $I_i(\beta) \geq I_{i+1}(\beta)$, $I_i(\gamma) \leq I_{i+1}(\gamma)$, and $I_i(\delta) \leq I_{i+1}(\delta)$ (e.g., see Fig. 3).*

Proof. By Observation 2, $\alpha(i,j) \leq \alpha(i+1,j)$. This implies that $I_i(\alpha) \geq I_{i+1}(\alpha)$ by the their definitions. The other three cases for β, γ, and δ are similar.   □

3.1   Computing $I_i(\alpha)$, $I_i(\beta)$, and $I_i(\delta)$ for all $i \in [1, n]$

In light of Lemma 2, for each $f \in \{\alpha, \beta, \delta\}$, we compute $I_i(f)$ for all $i = 1, 2, \ldots, n$ in $O(n)$ time, as follows.

We discuss the case for δ first. According to Lemma 1, $\delta(i, j)$ can be computed in constant time for any pair (i, j) with $1 \leq i \leq n$. We can compute $I_i(\delta)$ for all $i \in [1, n]$ in $O(n)$ time by the following simple algorithm. We first compute $I_1(\delta)$, which is done by computing $\delta(1, j)$ from $j = 1$ incrementally until the first time $\delta(1, j) \leq \lambda$. Then, to compute $I_2(\delta)$, we compute $\delta(2, j)$ from $j = I_1(\delta)$ incrementally until the first time $\delta(2, j) \leq \lambda$. Next, we compute $I_i(\delta)$ for $i = 3, 4, \ldots, n$ in the same way. The total time is $O(n)$. The correctness is based on the monotonicity property of $I_i(\delta)$ in Lemma 2.

To compute $I_i(\alpha)$ or $I_i(\beta)$ for $i = 1, 2, \ldots, n$, using a similar approach as above, we can only have an $O(n \log n)$ time algorithm since computing each $\alpha(i, j)$ or $\beta(i, j)$ takes $O(\log n)$ time by Lemma 1. Lemma 3 gives an approach that only needs $O(n)$ time.

Lemma 3. $I_i(\alpha)$ and $I_i(\beta)$ for all $i = 1, 2, \ldots, n$ can be computed in $O(n)$ time.

Due to the γ-computation difficulty mentioned in Section 2, it is not clear to us whether it possible to compute $I_i(\gamma)$ for all $i = 1, \ldots, n$ in $O(n \log n)$ time. Recall that λ is feasible if and only if there exists an $i \in [1, n]$ such that $[1, I_i(\alpha)] \cap [I_i(\beta), n] \cap [1, I_i(\gamma)] \cap [I_i(\delta), n] \neq \emptyset$. Now that $I_i(f)$ for all $i \in [1, n]$ and $f \in \{\alpha, \beta, \delta\}$ have been computed but the $I_i(\gamma)$'s are not known, below we will use an "indirect" approach to determine whether the intersection of the above four intervals is empty for every $i \in [1, n]$.

3.2   Determining the Feasibility of λ

For each $i \in [1, n]$, define $Q_i = [1, I_i(\alpha)] \cap [I_i(\beta), n] \cap [1, I_i(\gamma)] \cap [I_i(\delta), n]$. Our goal is to determine whether Q_i is empty for each $i = 1, 2, \ldots, n$.

Consider any $i \in [1, n]$. Since $I_i(f)$ for each $f \in \{\alpha, \beta, \delta\}$ is known, we can determine the intersection $[1, I_i(\alpha)] \cap [I_i(\beta), n] \cap [I_i(\delta), n]$ in constant time. If the intersection is empty, then $Q_i = \emptyset$. In the following, we assume that the intersection is not empty, and let a_i be the smallest index in the intersection.

As in [11], an easy observation is that $Q_i \neq \emptyset$ if and only if $a_i \in [1, I_i(\gamma)]$. If $a_i \leq i$ (note that $a_i \leq i$ actually implies $a_i = i$ since $a_i \geq I_i(\beta) \geq i$), it is obviously true that $a_i \in [1, I_i(\gamma)]$ since $i \leq I_i(\gamma)$. Otherwise (i.e., $i < a_i$), according to the definition of $I_i(\gamma)$, $a_i \in [1, I_i(\gamma)]$ if and only if $\gamma(i, a_i) \leq \lambda$. Große et al. [11] gave an approach that can determine whether $\gamma(i, a_i) \leq \lambda$ in $O(\log n)$ time after $O(n \log n)$ time preprocessing. Below, by new observations and with the help of the range minima data structure [2, 12], we show that whether $\gamma(i, a_i) \leq \lambda$ can be determined in constant time after $O(n)$ time preprocessing.

For each $j \in [1, n]$, define g_j as the largest index k in $[j, n]$ such that $d_P(j, k) \leq \lambda$. Observe that $g_1 \leq g_2 \leq \cdots \leq g_n$.

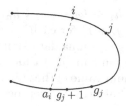

Fig. 4. Illustrating the graph $G(i, a_i)$ with $g_j + 1 \leq a_i$.

Consider any i and the corresponding a_i with $i < a_i$. Our goal is to determine whether $\gamma(i, a_i) \leq \lambda$. Since we are talking about $\gamma(i, a_i)$, we are essentially considering the graph $G(i, a_i)$. Recall that $C(i, a_i)$ is the cycle $P(i, a_i) \cup e(i, a_i)$. By Observation 3, $\gamma(i, a_i) = \max_{i \leq k \leq l \leq a_i} d_{C(i,a_i)}(k, l)$, and further, $d_{C(i,j)}(k, l) = \min\{d_P(k, l), d_{C(i,a_i)}^1(k, l)\}$ and $d_{C(i,a_i)}^1(k, l) = d_P(i, k) + |v_i v_{a_i}| + d_P(l, a_i)$.

For any $j \in [i, a_i - 1]$, if $g_j \leq a_i - 1$, then vertex $g_j + 1$ is in the cycle $C(i, a_i)$. By definition, $d_{C(i,a_i)}^1(j, g_j + 1) = d_P(i, j) + |v_i v_{a_i}| + d_P(g_j + 1, a_i)$. See Fig. 4.

Lemma 4. $\gamma(i, a_i) \leq \lambda$ if and only if for each $j \in [i, a_i - 1]$, either $g_j \geq a_i$ or $d_{C(i,a_i)}^1(j, g_j + 1) \leq \lambda$.

Recall that $g_1 \leq g_2 \leq \cdots \leq g_n$. For each $k \in [1, n]$, define h_k to be the smallest index j in $[1, k]$ with $g_j \geq k$. Observe that $h_1 \leq h_2 \leq \cdots \leq h_n$.

Note that if $i < h_{a_i}$, then for each $j \in [i, h_{a_i} - 1]$, $g_j < a_i$ and $g_j + 1 \leq a_i$. Due to Lemma 4, we further have the following lemma.

Lemma 5. $\gamma(i, a_i) \leq \lambda$ if and only if either $h_{a_i} \leq i$ or $d_{C(i,a_i)}^1(j, g_j + 1) \leq \lambda$ holds for each $j \in [i, h_{a_i} - 1]$.

Let $|C(i, a_i)|$ denote the total length of the cycle $C(i, a_i)$, i.e., $|C(i, a_i)| = d_P(i, a_i) + |v_i v_{a_i}|$. The following observation is crucial because it immediately leads to our algorithm in Lemma 6.

Observation 4 $\gamma(i, a_i) \leq \lambda$ if and only if either $\min_{j \in [i, h_{a_i} - 1]}\{d_P(j, g_j + 1)\} \geq |C(i, a_i)| - \lambda$ or $h_{a_i} \leq i$.

Proof. Suppose $h_{a_i} > i$. Then, for each $j \in [i, h_{a_i} - 1]$, $g_j < a_i$ and $g_j + 1 \leq a_i$. Note that $d_{C(i,a_i)}^1(j, g_j + 1) = |C(i, a_i)| - d_P(i, g_j + 1)$. Hence, $d_{C(i,a_i)}^1(j, g_j + 1) \leq \lambda$ is equivalent to $d_P(j, g_j + 1) \geq |C(i, a_i)| - \lambda$. Therefore, $d_{C(i,a_i)}^1(j, g_j + 1) \leq \lambda$ holds for each $j \in [i, h_{a_i} - 1]$ if and only if $\min_{j \in [i, h_{a_i} - 1]}\{d_P(j, g_j + 1)\} \geq |C(i, a_i)| - \lambda$.

By Lemma 5, the observation follows.   □

Lemma 6. With $O(n)$ time preprocessing, given any $i \in [1, n]$ and the corresponding a_i with $i < a_i$, whether $\gamma(i, a_i) \leq \lambda$ can be decided in constant time.

Proof. As preprocessing, we first compute g_j for all $j = 1, 2, \ldots, n$, which can be done in $O(n)$ time due to the monotonicity property $g_1 \leq g_2 \leq \cdots \leq g_n$. Then, we compute h_k for all $k = 1, 2, \ldots, n$, which can also be done in $O(n)$ time due to the monotonicity property $h_1 \leq h_2 \leq \ldots \leq h_n$. Next, we compute an array $B[1, \ldots, n]$ with $B[j] = d_P(j, g_j + 1)$ for each $j \in [1, n]$ (let $d_P(j, g_j + 1) = \infty$ if $g_j + 1 > n$). We build a range-minima data structure on B [2, 12]. The range minima data structure can be built in $O(n)$ time such that given any pair (i, j) with $1 \leq i \leq j \leq n$, the minimum value of the subarray $B[i \cdots j]$ can be returned in constant time [2, 12]. This finishes the preprocessing step, which takes $O(n)$ time in total.

Consider any i and the corresponding a_i with $i < a_i$. Our goal is to determine whether $\gamma(i, a_i) \leq \lambda$, which can be done in $O(1)$ time as follows.

According to Observation 4, $\gamma(i, a_i) \leq \lambda$ if and only if either $h_{a_i} \leq i$ or $\min_{j \in [i, h_{a_i} - 1]}\{d_P(j, g_j + 1)\} \geq |C(i, a_i)| - \lambda$. Since h_{a_i} has been computed in the preprocessing, we check whether $h_{a_i} \leq i$ is true. If yes, then we are done with the assertion that $\gamma(i, a_i) \leq \lambda$. Otherwise, we need to determine whether $\min_{j \in [i, h_{a_i} - 1]}\{d_P(j, g_j + 1)\} \geq |C(i, a_i)| - \lambda$ holds. To this end, we first compute $\min_{j \in [i, h_{a_i} - 1]}\{d_P(j, g_j + 1)\}$ in constant time by querying the range-minima data structure on B with $(i, h_{a_i} - 1)$. Note that $|C(i, a_i)|$ can be computed in constant time. Therefore, we can determine whether $\gamma(i, a_i) \leq \lambda$ in $O(1)$ time. □

With Lemma 6, the decision problem can be solved in $O(n)$ time. The proof of the following theorem summarizes our algorithm.

Theorem 1. *Given any λ, we can determine whether λ is feasible in $O(n)$ time, and if λ is feasible, a feasible edge can be found in $O(n)$ time.*

Proof. First, we do the preprocessing in Lemma 1 in $O(n)$ time. Then, for each $f \in \{\alpha, \beta, \delta\}$, we compute $I_i(f)$ for all $i = 1, 2, \ldots, n$, in $O(n)$ time. We also do the preprocessing in Lemma 6.

Next, for each $i \in [1, n]$, we do the following. Compute the intersection $[1, I_i(\alpha)] \cap [I_i(\beta), n] \cap [I_i(\delta), n]$ in constant time. If the intersection is empty, then we are done for this i. Otherwise, obtain the smallest index a_i in the above intersection. If $a_i \leq i$, then we stop the algorithm with the assertion that λ is feasible and report $e(i, a_i)$ as a feasible edge. Otherwise, we use Lemma 6 to determine whether $\gamma(i, a_i) \leq \lambda$ in constant time. If yes, we stop the algorithm with the assertion that λ is feasible and report $e(i, a_i)$ as a feasible edge. Otherwise, we proceed on $i + 1$.

If the algorithm does not stop after we check all $i \in [1, n]$, then we stop the algorithm with the assertion that λ is not feasible. Clearly, we spend $O(1)$ time on each i, and thus, the total time of the algorithm is $O(n)$. □

4 The Optimization Problem

In this section, we solve the optimization problem in $O(n \log n)$ time, by making use of our algorithm for the decision problem given in Section 3 (we will refer to it as the *decision algorithm*). We will focus on computing λ^*.

Notice that λ^* must be equal to the diameter $D(i, j)$ of $G(i, j)$ for some pair (i, j) with $1 \leq i \leq j \leq n$. Further, by Observation 1, λ^* is equal to $f(i, j)$ for some $f \in \{\alpha, \beta, \gamma, \delta\}$ and some pair (i, j) with $1 \leq i \leq j \leq n$.

For each $f \in \{\alpha, \beta, \gamma, \delta\}$, define $S_f = \{f(i,j) \mid 1 \leq i \leq j \leq n\}$. Let $S = \cup_{f \in \{\alpha, \beta, \gamma, \delta\}} S_f$. According to our discussion above, λ^* is in S. Further, note that λ^* is the smallest feasible value of S. We will not compute the entire set S since $|S| = \Omega(n^2)$. For each $f \in \{\alpha, \beta, \gamma, \delta\}$, let λ_f be the smallest feasible value in S_f. Hence, we have $\lambda^* = \min\{\lambda_\alpha, \lambda_\beta, \lambda_\gamma, \lambda_\delta\}$.

In the following, we first compute $\lambda_\alpha, \lambda_\beta, \lambda_\delta$ in $O(n \log n)$ time.

4.1 Computing $\lambda_\alpha, \lambda_\beta$, and λ_δ

For convenience, we begin with computing λ_β.

We define an $n \times n$ matrix $M[1 \cdots n; 1 \cdots n]$: For each $1 \leq i \leq n$ and $1 \leq j \leq n$, define $M[i,j] = \beta(i,j)$ if $j \geq i$ and $M[i,j] = \beta(i,i)$ otherwise. By Observation 2, the following lemma shows that M is a sorted matrix in the sense that each row is sorted in descending order from left to right and each column is sorted in descending order from top to bottom.

Lemma 7. *For each $1 \leq i \leq n$, $M[i,j] \geq M[i,j+1]$ for any $j \in [1, n-1]$; for each $1 \leq j \leq n$, $M[i,j] \geq M[i+1,j]$ for any $i \in [1, n-1]$.*

Note that each element of S_β is in M and vice versa. Since λ_β is the smallest feasible value of S_β, λ_β is also the smallest feasible value of M. We do not construct M explicitly. Rather, given any i and j, we can "evaluate" $M[i,j]$ in $O(\log n)$ time since $\beta(i,j)$ can be computed in $O(\log n)$ time if $i \leq j$ by Lemma 1. Using the sorted-matrix searching techniques [8,9], we can find λ_β in M by calling our decision algorithm $O(\log n)$ times and evaluating $O(n)$ elements of M. The total time on calling the decision algorithm is $O(n \log n)$ and the total time on evaluating matrix elements is also $O(n \log n)$. Hence, we can compute λ_β in $O(n \log n)$ time.

Computing λ_α and λ_δ can be done similarly in $O(n \log n)$ time, although the corresponding sorted matrices may be defined slightly differently. We omit the details. However, we cannot compute λ_γ in $O(n \log n)$ time in the above way, and again this is due to the λ-computation difficulty mentioned in Section 2. Note that having $\lambda_\alpha, \lambda_\beta$, and λ_δ essentially reduces our search space for λ^* from S to $S_\gamma \cup \{\lambda_\alpha, \lambda_\beta, \lambda_\delta\}$.

Let $\lambda_1 = \min\{\lambda_\alpha, \lambda_\beta, \lambda_\delta\}$. Thus, $\lambda^* = \min\{\lambda_1, \lambda_\gamma\}$. Hence, if $\lambda_\gamma \geq \lambda_1$, then $\lambda^* = \lambda_1$ and λ^* is computed. Otherwise (i.e., $\lambda_\gamma < \lambda_1$), it must be that $\lambda^* = \lambda_\gamma$ and we need to compute λ_γ. To compute λ_γ, again we cannot use the similar way as the above for computing λ_β. Instead, we use the following approach, whose success relies on the information implied by $\lambda_\gamma < \lambda_1$.

4.2 Computing λ^* in the Case $\lambda_\gamma < \lambda_1$

We assume $\lambda_\gamma < \lambda_1$. Hence, $\lambda^* = \lambda_\gamma$. Let $e(i^*, j^*)$ be the new edge added to P in an optimal solution. We also call $e(i^*, j^*)$ an *optimal edge*.

Since $\lambda^* = \lambda_\gamma < \lambda_1$, we have the following observation.

Observation 5 *If $\lambda_\gamma < \lambda_1$ and $e(i^*, j^*)$ is an optimal edge, then $\lambda^* = \gamma(i^*, j^*)$.*

For any $i \in [1, n]$, for each $f \in \{\alpha, \beta, \gamma, \delta\}$, with respect to λ_1, we define $I_i'(f)$ in a similar way to $I_i(f)$ defined in Section 3 with respect to λ except that we change "$\leq \lambda$" to "$< \lambda_1$". Specifically, define $I_i'(\alpha)$ to be the largest index $j \in [i, n]$ such that $\alpha(i, j) < \lambda_1$. $I_i'(\gamma)$ is defined similarly to $I_i'(\alpha)$. If $\beta(i, n) < \lambda_1$, then define $I_i'(\beta)$ to be the smallest index $j \in [i, n]$ such that $\beta(i, j) < \lambda_1$; otherwise $I_i'(\beta) = \infty$. $I_i'(\delta)$ is defined similarly to $I_i'(\beta)$. Note that similar monotonicity properties for $I_i'(f)$ with $f \in \{\alpha, \beta, \gamma, \delta\}$ to those in Lemma 2 also hold.

Recall that $e(i^*, j^*)$ is an optimal edge. An easy observation is that since λ_1 is strictly larger than λ^*, the intersection $[1, I_{i^*}'(\alpha)] \cap [I_{i^*}'(\beta), n] \cap [I_{i^*}'(\delta), n]$ cannot be empty. Let a_{i^*} be the smallest index in the above intersection. Note that $i^* \leq a_{i^*}$ since $i^* \leq I_{i^*}'(\beta) \leq a_{i^*}$. The following lemma shows that $e(i^*, a_{i^*})$ is actually an optimal edge.

Lemma 8. *If $\lambda_\gamma < \lambda_1$ and $e(i^*, j^*)$ is an optimal edge, then $j^* = a_{i^*}$.*

Lemma 8 is crucial because it suggests the following algorithm.

We first compute the indices $I_i'(\alpha), I_i'(\beta), I_i'(\delta)$ for $i = 1, \ldots, n$. This can be done in $O(n)$ time using the similar algorithms as those for computing $I_i(\alpha), I_i(\beta), I_i(\delta)$ in Section 3.1.

Next, for each $i \in [1, n]$, if $[1, I_i'(\alpha)] \cap [I_i'(\beta), n] \cap [I_i'(\delta), n] \neq \emptyset$, then we compute a_i, i.e., the smallest index in the above intersection. Let $\mathcal{I}$ be the set of indices i such that the above interval intersection for i is not empty. Lemma 8 leads to the following observation.

Observation 6 *If $\lambda_\gamma < \lambda_1$, then λ^* is the smallest feasible value of the set $\{\gamma(i, a_i) \mid i \in \mathcal{I}\}$.*

Proof. By Lemma 8, one of the edges of $\{e(i, a_i) \mid i \in \mathcal{I}\}$ is an optimal edge. By Observation 5, λ^* is in $\{\gamma(i, a_i) \mid i \in \mathcal{I}\}$. Thus, λ^* is the smallest feasible value in $\{\gamma(i, a_i) \mid i \in \mathcal{I}\}$. $\qquad\square$

Observation 6 essentially reduces the search space for λ^* to $\{\gamma(i, a_i) \mid i \in \mathcal{I}\}$, which has at most $O(n)$ values. It is tempting to first explicitly compute the set and then find λ^* from the set. However, again, due to the γ-computation difficulty, it is not clear to us how to compute the set in $O(n \log n)$ time. Alternatively, we use the following approach to compute λ^*.

4.3 Finding λ^* in the Set $\{\gamma(i, a_i) \mid i \in \mathcal{I}\}$

Recall that according to Observation 3, $\gamma(i, j) = \max_{i \leq k \leq l \leq j} d_{C(i,j)}(k, l)$, with $d_{C(i,j)}(k, l) = \min\{d_P(k, l), d_{C(i,j)}^1(k, l)\}$ and $d_{C(i,j)}^1(k, l) = d_P(i, k) + |v_i v_j| + d_P(l, j)$. Hence, $\gamma(i, j)$ is equal to $d_P(k, l)$ or $d_{C(i,j)}^1(k, l)$ for some $k \leq l$. Therefore, by Observation 6, there exists $i \in \mathcal{I}$ such that λ^* is equal to $d_P(k, l)$ or $d_{C(i,a_i)}^1(k, l)$ for some k and l with $i \leq k \leq l \leq a_i$.

Let $S_p = \{d_P(k, l) \mid 1 \leq k \leq l \leq n\}$ and $S_c = \{d_{C(i,a_i)}^1(k, l) \mid i \leq k \leq l \leq a_i, i \in \mathcal{I}\}$. Based on our above discussion, λ^* is in $S_p \cup S_c$. Further, λ^* is the smallest feasible value in $S_p \cup S_c$.

Let λ_p be the smallest feasible value of S_p and let λ_c be the smallest feasible value of S_c. Hence, $\lambda^* = \min\{\lambda_p, \lambda_c\}$. By using the technique of searching sorted-matrices [8,9], the following lemma computes λ_p in $O(n \log n)$ time.

Lemma 9. λ_p can be computed in $O(n \log n)$ time.

Recall that $\lambda^* = \min\{\lambda_p, \lambda_c\}$. In the case $\lambda_p \leq \lambda_c$, $\lambda^* = \lambda_p$ and we are done with computing λ^*. In the following, we assume $\lambda_p > \lambda_c$. Thus, $\lambda^* = \lambda_c$. With the help of the information implied by $\lambda_p > \lambda_c$, we will compute λ^* in $O(n \log n)$ time. The details are given below.

For any $j \in [1, n]$, let g'_j denote the largest index $k \in [j, n]$ such that the subpath length $d_P(j, k)$ is *strictly smaller* than λ_p. Note that the definition of g'_j is similar to g_j defined in Section 4.3 except that we change "$\leq \lambda$" to "$< \lambda_p$".

For each $k \in [1, n]$, let h'_k denote the smallest index $j \in [1, k]$ with $g'_j \geq k$. Let $\mathcal{I}'$ be the subset of $i \in \mathcal{I}'$ such that $i \leq h'_{a_i} - 1$. Hence, for each $i \in \mathcal{I}'$ and each $j \in [i, h'_{a_i} - 1]$, $g'_j < a_i$ and thus $g'_j + 1 \leq a_i$.

For each $i \in \mathcal{I}'$, define $d^1_{\max}(i, a_i) = \max_{j \in [i, h'_{a_i} - 1]} d^1_{C(i, a_i)}(j, g'_j + 1)$. The following lemma gives a way to determine λ^*.

Lemma 10. If $\lambda_\gamma < \lambda_1$ and $\lambda_c < \lambda_p$, then $\lambda^* = d^1_{\max}(i, a_i)$ for some $i \in \mathcal{I}'$.

By Lemma 10, in the case of $\lambda_c < \lambda_p$, $\lambda^* = \lambda_c$ is the smallest feasible value of $d^1_{\max}(i, a_i)$ for all $i \in \mathcal{I}'$. Note that the number of such values $d^1_{\max}(i, a_i)$ is $O(n)$. Hence, if we can compute $d^1_{\max}(i, a_i)$ for all $i \in \mathcal{I}'$, then λ^* can be easily found in additional $O(n \log n)$ time using our decision algorithm, e.g., by first sorting these values and then doing binary search.

Lemma 11 computes $d^1_{\max}(i, a_i)$ for all $i \in \mathcal{I}'$ in $O(n)$ time, with the help of the range-minima data structure [2,12].

Lemma 11. $d^1_{\max}(i, a_i)$ for all $i \in \mathcal{I}'$ can be computed in $O(n)$ time.

In summary, we can compute λ^* in $O(n \log n)$ time in the case $\lambda_\gamma < \lambda_1$ and $\lambda_c < \lambda_p$. Our overall algorithm for computing an optimal solution is summarized in the proof of Theorem 2.

Theorem 2. An optimal solution for the optimization problem can be found in $O(n \log n)$ time.

Proof. First, we compute λ_α, λ_β, and λ_δ, in $O(n \log n)$ time by using our decision algorithm and the sorted-matrix searching techniques. Then, we compute $\lambda_1 = \min\{\lambda_\alpha, \lambda_\beta, \lambda_\delta\}$.

Second, by using λ_1, we compute the indices $I'_i(\alpha)$, $I'_i(\beta)$, and $I'_i(\delta)$ for all $i = 1, 2, \ldots, n$. This can be done in $O(n)$ time. For each $i \in [1, n]$, if $[1, I'_i(\alpha)] \cap [I'_i(\beta), n] \cap [I'_i(\delta), n] \neq \emptyset$, we compute a_i (i.e., the smallest index in the above intersection) and add i to the set $\mathcal{I}$ (initially $\mathcal{I} = \emptyset$). Hence, all such a_i's and $\mathcal{I}$ can be computed in $O(n)$ time.

If $\mathcal{I} = \emptyset$, then we return λ_1 as λ^*.

If $\mathcal{I} \neq \emptyset$, then we compute λ_p in $O(n \log n)$ time by Lemma 9. We proceed to compute $d_{\max}^1(i, a_i)$ for all $i \in \mathcal{I}'$ by Lemma 11, and then find the smallest feasible value λ' in the set $\{d_{\max}^1(i, a_i) \mid i \in \mathcal{I}'\}$ in $O(n \log n)$ time. Finally, we return $\min\{\lambda_1, \lambda_p, \lambda'\}$ as λ^*.

The above computes λ^* in $O(n \log n)$ time. Applying $\lambda = \lambda^*$ on our decision algorithm can eventually find an optimal edge in additional $O(n)$ time. $\quad\square$

References

1. Alon, N., Gyárfás, A., Ruszinkó, M.: Decreasing the diameter of bounded degree graphs. Journal of Graph Theory 35, 161–172 (2000)
2. Bender, M., Farach-Colton, M.: The LCA problem revisited. In: Proc. of the 4th Latin American Symposium on Theoretical Informatics. pp. 88–94 (2000)
3. Bilò, D., Gualà, L., Proietti, G.: Improved approximability and non-approximability results for graph diameter decreasing problems. Theoretical Computer Science 417, 12–22 (2012)
4. Carufel, J.L.D., Grimm, C., Maheshwari, A., Smid, M.: Minimizing the continuous diameter when augmenting paths and cycles with shortcuts. In: Proc. of the 15th Scandinavian Workshop on Algorithm Theory (SWAT). pp. 27:1–27:14 (2016)
5. Carufel, J.L.D., Grimm, C., Schirra, S., Smid, M.: Minimizing the continuous diameter when augmenting a tree with a shortcut. arXiv:1612.01370 (2016)
6. Demaine, E., Zadimoghaddam, M.: Minimizing the diameter of a network using shortcut edges. In: Proc. of the 12th Scandinavian conference on Algorithm Theory (SWAT). pp. 420–431 (2010)
7. Frati, F., Gaspers, S., Gudmundsson, J., Mathieson, L.: Augmenting graphs to minimize the diameter. Algorithmica 72, 995–1010 (2015)
8. Frederickson, G., Johnson, D.: Generalized selection and ranking: Sorted matrices. SIAM Journal on Computing 13(1), 14–30 (1984)
9. Frederickson, G., Johnson, D.: Finding kth paths and p-centers by generating and searching good data structures. Journal of Algorithms 4(1), 61–80 (1983)
10. Gao, Y., Hare, D., Nastos, J.: The parametric complexity of graph diameter augmentation. Discrete Applied Mathematics 161, 1626–1631 (2013)
11. Große, U., Gudmundsson, J., Knauer, C., Smid, M., Stehn, F.: Fast algorithms for diameter-optimally augmenting paths. In: Proc. of the 42nd International Colloquium on Automata, Languages and Programming (ICALP). pp. 678–688 (2015)
12. Harel, D., Tarjan, R.: Fast algorithms for finding nearest common ancestors. SIAM Journal on Computing 13, 338–355 (1984)
13. Ishii, T.: Augmenting outerplanar graphs to meet diameter requirements. Journal of Graph Theory 74, 392–416 (2013)
14. Li, C.L., McCormick, S., Simchi-Levi, D.: On the minimum-cardinality-bounded-diameter and the bounded-cardinality-minimum-diameter edge addition problems. Operations Research Letters 11, 303–308 (1992)
15. Megiddo, N.: Applying parallel computation algorithms in the design of serial algorithms. Journal of the ACM 30(4), 852–865 (1983)
16. Schoone, A., Bodlaender, H., Leeuwen, J.V.: Diameter increase caused by edge deletion. Journal of Graph Theory 11, 409–427 (1997)
17. Wang, H.: An improved algorithm for diameter-optimally augmenting paths in a metric space. arXiv:1608.04456 (2016)

Covering Uncertain Points in a Tree*

Haitao Wang and Jingru Zhang

Department of Computer Science
Utah State University, Logan, UT 84322, USA
haitao.wang@usu.edu,jingruzhang@aggiemail.usu.edu

Abstract. We consider a coverage problem for uncertain points in a tree. Let T be a tree containing a set $\mathcal{P}$ of n (weighted) demand points, and the location of each demand point $P_i \in \mathcal{P}$ is uncertain but is known to be in one of m_i points on T each associated with a probability. Given a *covering range* λ, the problem is to find a minimum number of points (called *centers*) on T to build facilities for serving (or covering) these demand points in the sense that for each uncertain point $P_i \in \mathcal{P}$, the expected distance from P_i to at least one center is no more than λ. The problem has not been studied before. We present an $O(|T| + M \log^2 M)$ time algorithm, where $|T|$ is the number of vertices of T and M is the total number of locations of all uncertain points of $\mathcal{P}$, i.e., $M = \sum_{P_i \in \mathcal{P}} m_i$.

1 Introduction

Data uncertainty is very common in many applications, such as sensor databases, image resolution, and it is mainly due to measurement inaccuracy, sampling discrepancy, outdated data sources, resource limitation, etc. Problems on uncertain data have attracted considerable attention, e.g., [1–3, 9, 10, 15]. In this paper, we study a problem of covering uncertain points on a tree, defined as follows.

Let T be a tree. We consider each edge e of T as a line segment of a positive length so that we can talk about "points" on e. The distance of any two points p and q on T, denoted by $d(p, q)$, is defined as the sum of the lengths of all edges on the simple path from p to q in T. Let $\mathcal{P} = \{P_1, \ldots, P_n\}$ be a set of n uncertain (demand) points on T. Each $P_i \in \mathcal{P}$ has m_i possible locations on T, denoted by $\{p_{i1}, p_{i2}, \cdots, p_{im_i}\}$, and each location p_{ij} of P_i is associated with a probability $f_{ij} \geq 0$ for P_i appearing at p_{ij} (which is independent of other locations), with $\sum_{j=1}^{m_i} f_{ij} = 1$; e.g., see Fig. 1. In addition, each $P_i \in \mathcal{P}$ has a weight $w_i \geq 0$. For any point x on T, the (weighted) *expected distance* from x to P_i, denoted by $\mathsf{Ed}(x, P_i)$, is defined as $\mathsf{Ed}(x, P_i) = w_i \cdot \sum_{j=1}^{m_i} f_{ij} \cdot d(x, p_{ij})$.

Given a value $\lambda \geq 0$, called the *covering range*, we say that a point x on T *covers* an uncertain point P_i if $\mathsf{Ed}(x, P_i) \leq \lambda$. The *center-coverage problem* is to compute a minimum number of points on T, called *centers*, such that every uncertain point of $\mathcal{P}$ is covered by at least one center.

To the best of our knowledge, the problem has not been studied before. Let M denote the total number of locations of all uncertain points, i.e., $M = \sum_{i=1}^{n} m_i$.

---

* This research was supported in part by NSF under Grant CCF-1317143.

F. Ellen et al. (Eds.): WADS 2017, LNCS 10389, pp. 557–568, 2017.
DOI: 10.1007/978-3-319-62127-2_47

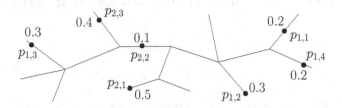

Fig. 1. Illustrating two uncertain points P_1 and P_2, where P_1 has four possible locations and P_2 has three possible locations. The numbers are the probabilities.

Let $|T|$ be the number of vertices of T. In this paper, we present an algorithm that solves the problem in $O(|T| + M \log^2 M)$ time, which is nearly linear as the input size of the problem is $\Theta(|T| + M)$.

As an application of our algorithm, we also solve a dual problem, called the *k-center* problem, which is to compute a number of k centers on T such that the covering range is minimized. The runtime of our algorithm is $O(|T| + n^2 \log n \log M + M \log^2 M \log n)$.

1.1 Related Work

Two models on uncertain data have been commonly considered: the *existential* model [3, 9, 10] and the *locational* model [1, 2, 15]. In the existential model an uncertain point has a specific location but its existence is uncertain while in the locational model an uncertain point always exists but its location is uncertain and follows a probability distribution function. Our problems belong to the locational model. In fact, the same problems under existential model are essentially the weighted case for "deterministic" points (i.e., each $P_i \in \mathcal{P}$ has a single "certain" location), and the center-coverage problem is solvable in linear time [11] and the k-center problem is solvable in $O(n \log^2 n)$ time [6, 14].

If T is a path, both the center-coverage problem and the k-center problem on uncertain points have been studied [17], but under a somewhat special problem setting where m_i is the same for all $1 \leq i \leq n$. The two problems were solved in $O(M + n \log k)$ and $O(M \log M + n \log k \log n)$ time, respectively. If T is tree, an $O(|T| + M)$ time algorithm was given in [16] for the one-center problem under the above special problem setting.

Recently Li and Huang [7] considered the same k-center problem under the same uncertain model as ours but in the Euclidean space, and they gave an approximation algorithm. Facility location problems in other uncertain models were also considered. Löffler and van Kreveld [12] gave algorithms for computing the smallest enclosing circle for imprecise points each of which is contained in a planar region (e.g., a circle or a square). Jørgenson et al. [8] studied the problem of computing the distribution of the radius of the smallest enclosing circle for uncertain points each of which has multiple locations in the plane. de Berg et al. [4] proposed algorithms for dynamically maintaining Euclidean 2-centers for a set of moving points in the plane (the moving points are considered uncertain).

1.2 Our Approach

For each uncertain point $P_i \in \mathcal{P}$, we find a point p_i^* on T that minimizes the expected distance $\mathsf{Ed}(p_i, P_i)$, and p_i^* is actually the weighted median of all locations of P_i. We observe that if we move a point x on T away from p_i^*, $\mathsf{Ed}(x, P_i)$ is monotonically increasing. We compute the medians p_i^* for all uncertain points in $O(M \log M)$ time. We show that there is an optimal solution in which all centers are in T_m, where T_m is the minimum subtree of T connecting/spanning all medians p_i^*. Next we find centers on T_m. For this, we propose a simple greedy algorithm, but the challenge is on developing efficient data structures to perform certain operations. We briefly discuss it below.

We pick an arbitrary vertex r of T_m as the root. Starting from the leaves, we consider the vertices of T_m in a bottom-up manner and place centers whenever we "have to". For example, consider a leaf v holding a median p_i^* and let u be the parent of v. If $\mathsf{Ed}(u, P_i) > \lambda$, then we have to place a center c on the edge $e(u, v)$ in order to cover P_i. The location of c is chosen to be at a point of $e(u, v)$ with $\mathsf{Ed}(c, P_i) = \lambda$ (i.e., on the one hand, c covers P_i, and on the other hand, c is close to u as much as possible in the hope of covering other uncertain points as many as possible). After c is placed, we find and remove all uncertain points that are covered by c. Performing this operation efficiently is a key difficulty for our approach. We solve the problem in an output-sensitive manner by proposing a dynamic data structure that also supports the remove operations. We also develop data structures for other operations needed in the algorithm. These data structures may be of independent interest.

For solving the k-center problem, by observations, we first identify a set of $O(n^2)$ "candidate" values such that the covering range in the optimal solution must be in the set. Subsequently, we use our algorithm for the center-coverage problem as a decision procedure to find the optimal covering range in the set.

We introduce some notations in Section 2. In Section 3, we describe our algorithmic scheme for the center-coverage problem, with details in the subsequent sections. Due to the space limit, many proofs and details, including our algorithm for the k-center problem, are omitted but can be found in the full paper [18].

2 Preliminaries

Note that the locations of the uncertain points of $\mathcal{P}$ may be in the interior of the edges of T. A *vertex-constrained case* happens if all locations of $\mathcal{P}$ are at vertices of T and each vertex of T holds at least one location of $\mathcal{P}$ (but the centers can still be in the interior of edges). As in [16], the general case can be reduced to the vertex-constrained case in $O(|T| + M)$ time (see the full paper for details). In the following, unless otherwise stated, we focus our discussion on the vertex-constrained case and assume that our problem on $\mathcal{P}$ and T is a vertex-constrained case.

For ease of exposition, we further make a general position assumption that every vertex of T has only one location of $\mathcal{P}$ (we explain in the full paper that our algorithm easily extends to the degenerate case). Under the assumption, $|T| = M \geq n$.

Let $e(u, v)$ denote the edge of T incident to two vertices u and v. For any two points p and q on T, denote by $\pi(p, q)$ the simple path from p to q on T.

Let π be any simple path on T and x be any point on π. For any location p_{ij} of an uncertain point P_i, the distance $d(x, p_{ij})$ is a convex (and piecewise linear) function as x changes on π [13]. As a sum of multiple convex functions, $\mathsf{Ed}(x, P_i)$ is also convex (and piecewise linear) on π, i.e., in general, as x moves on π, $\mathsf{Ed}(x, P_i)$ first monotonically decreases and then increases. In particular, for each edge e of T, $\mathsf{Ed}(x, P_i)$ is a linear function for $x \in e$.

For any subtree T' of T and any $P_i \in \mathcal{P}$, we call the sum of the probabilities of the locations of P_i in T' the *probability sum* of P_i in T'.

For each uncertain point P_i, let p_i^* be a point $x \in T$ that minimizes $\mathsf{Ed}(x, P_i)$. If we consider $w_i \cdot f_{ij}$ as the weight of p_{ij}, p_i^* is actually the *weighted median* of all points $p_{ij} \in P_i$. We call p_i^* the *median* of P_i. Although p_i^* may not be unique (e.g., when there is an edge e dividing T into two subtrees such that the probability sum of P_i in either subtree is exactly 0.5), P_i always has a median located at a vertex v of T, and we let p_i^* refer to such a vertex.

Recall that λ is the covering range of our problem. If $\mathsf{Ed}(p_i^*, P_i) > \lambda$ for some $i \in [1, n]$, then there is no solution for the problem since no point of T can cover P_i. Henceforth, we assume $\mathsf{Ed}(p_i^*, P_i) \leq \lambda$ for each $i \in [1, n]$.

3   The Algorithmic Scheme

In this section, we describe our algorithmic scheme for the center-coverage problem, and the implementation details will be presented later. We start with computing the medians p_i^* of all uncertain points of $\mathcal{P}$. We have the following lemma.

Lemma 1. *The medians p_i^* of all P_i of $\mathcal{P}$ can be computed in $O(M \log M)$ time.*

3.1   The Medians-Spanning Tree T_m

Denote by P^* the set of all medians p_i^*. Let T_m be the minimum connected subtree of T that spans/connects all medians. Note that each leaf of T_m must hold a median. We pick an arbitrary median as the root of T, denoted by r. The subtree T_m can be easily computed in $O(M)$ time by a post-order traversal on T (with respect to the root r), and we omit the details. The following lemma is based on the fact that $\mathsf{Ed}(x, P_i)$ is convex for x on any simple path of T and $\mathsf{Ed}(x, P_i)$ minimizes at $x = p_i^*$.

Lemma 2. *There exists an optimal solution for the center-coverage problem in which every center is on T_m.*

Due to Lemma 2, we will focus on finding centers on T_m. We also consider r as the root of T_m, and then we can talk about ancestors and descendants of the vertices in T_m. Note that for any two vertices u and v of T_m, $\pi(u, v)$ is in T_m.

We reindex all medians and the corresponding uncertain points so that the new indices will facilitate our algorithm, as follows. Starting from an arbitrary child of r in T_m, we traverse down the tree T_m by always following the leftmost child of the current node until we encounter a leaf, denoted by v^*. Starting from v^* (i.e., v^* is the first visited leaf), we perform a post-order traversal on T_m and reindex all medians of P^* such that $p_1^*, p_2^*, \ldots, p_n^*$ is the list of points of P^* visited in order in the above traversal. Recall that the root r contains a median, which is p_n^* after the reindexing. Accordingly, we also reindex all uncertain points of $\mathcal{P}$ and their corresponding locations on T, which can be done in $O(M)$ time. In the following paper, we will always use the new indices.

For each vertex v of T_m, we use $T_m(v)$ to represent the subtree of T_m rooted at v. The reason we do the above reindexing is that for any vertex v of T_m, the new indices of all medians in $T_m(v)$ must form a range $[i, j]$ for some $1 \le i \le j \le n$, and we use $R(v)$ to denote the range. It will be clear later that this property will facilitate our algorithm.

3.2 The Algorithm

Our algorithm for the center-coverage problem works as follows. Initially, all uncertain points are "active". During the algorithm, we will place centers on T_m, and once an uncertain point P_i is covered by a center, we will "deactivate" it (it then becomes "inactive"). The algorithm visits all vertices of T_m following the above post-order traversal of T_m starting from leaf v^*. Suppose v is currently being visited. Unless v is the root r, let u be the parent of v. Below we describe our algorithm for processing v. There are two cases depending on whether v is a leaf or an internal node, although the algorithm for them is essentially the same.

The Leaf Case If v is a leaf, then it holds a median p_i^*. If P_i is inactive, we do nothing; otherwise, we proceed as follows.

We compute a point c (called a *candidate center*) on the path $\pi(v, r)$ closest to r such that $\text{Ed}(c, P_i) \le \lambda$. Note that if we move a point x from v to r along $\pi(v, r)$, $\text{Ed}(x, P_i)$ is monotonically increasing. By the definition of c, if $\text{Ed}(r, P_i) \le \lambda$, then $c = r$; otherwise, $\text{Ed}(c, P_i) = \lambda$. If c is in $\pi(u, r)$, then we do nothing and finish processing v. Below we assume that c is not in $\pi(u, r)$ and thus is in $e(u, v) \setminus \{u\}$ (i.e., $c \in e(u, v)$ but $c \notin u$).

In order to cover P_i, by the definition of c, we must place a center in $e(u, v) \setminus \{u\}$. Our strategy is to place a center at c. Indeed, this is the best location for placing a center since it is the location that covers P_i and is closest to u (and thus is closest to every other active uncertain point). We use a *candidate-center-query* to compute c in $O(\log n)$ time, whose details will be discussed later. Next, we report all active uncertain points that are covered by c, by a *coverage-report-query* in output-sensitive $O(\log M \log n + k \log n)$ amortized time, where k is the number of uncertain points covered by c. The details for the operation will be discussed later. Further, we deactivate all these uncertain points. We will show that deactivating each uncertain point P_j can be done in $O(m_j \log M \log n)$ amortized time. This finishes processing v.

The Internal Node Case If v is an internal node, since we process the vertices of T_m following a post-order traversal, all descendants of v have already been processed. Our algorithm maintains an invariant that if the subtree $T_m(v)$ contains any active median p_i^* (i.e., P_i is active), then $\mathsf{Ed}(v, P_i) \leq \lambda$. When v is a leaf, this invariant trivially holds. Our way of processing a leaf discussed above also maintains this invariant.

To process v, we first check whether $T_m(v)$ has any active medians. This is done by a *range-status-query* in $O(\log n)$ time, whose details will be given later. If $T_m(v)$ does not have any active median, then we are done with processing v. Otherwise, by the algorithm invariant, for each active median p_i^* in $T_m(v)$, it holds that $\mathsf{Ed}(v, P_i) \leq \lambda$. If $v = r$, we place a center at v and finish the algorithm. Below, we assume $v \neq r$ and thus u is the parent of v.

We compute a point c on $\pi(v, r)$ closest to r such that $\mathsf{Ed}(c, P_i) \leq \lambda$ for all active medians $p_i^* \in T_m(v)$, and we call c the *candidate center*. By the definition of c, if $\mathsf{Ed}(r, P_i) \leq \lambda$ for all active medians $p_i^* \in T_m(v)$, then $c = r$; otherwise, $\mathsf{Ed}(c, P_i) = \lambda$ for some active median $p_i^* \in T_m(v)$. As in the leaf case, finding c is done in $O(\log n)$ time by a *candidate-center-query*. If c is on $\pi(u, r)$, then we finish processing v. Note that this implies $\mathsf{Ed}(u, P_i) \leq \lambda$ for each active median $p_i^* \in T_m(v)$, which maintains the algorithm invariant for u.

If $c \notin \pi(u, r)$, then $c \in e(u, v) \setminus \{u\}$. In this case, by the definition of c, we must place a center in $e(u, v) \setminus \{u\}$ to cover P_i. As discussed in the leaf case, the best location for placing a center is c and thus we place a center at c. Then, by using a coverage-report-query, we find all active uncertain points covered by c and deactivate them. Note that by the definition of c, c covers P_j for all medians $p_j^* \in T_m(v)$. This finishes processing v.

Once the root r is processed, the algorithm finishes.

3.3 The Time Complexity

To analyze the running time of the algorithm, it remains to discuss the three operations: range-status-queries, coverage-report-queries, and candidate-center-queries. For answering range-status-queries, we have the following Lemma 3.

Lemma 3. *We can build a data structure in $O(M)$ time that can answer each range-status-query in $O(\log n)$ time. Further, once an uncertain point is deactivated, we can remove it from the data structure in $O(\log n)$ time.*

For answering the coverage-report-queries and the candidate-center-queries, we have the following Lemmas 4 and 5. To prove them, we introduce a connector-bounded centroid decomposition on the tree T in Section 4. The proof of Lemma 4 is thus given in Section 5, and the proof of Lemma 5 is omitted. Using these results, we can eventually obtain Theorem 1.

Lemma 4. *We can build a data structure $\mathcal{A}_1$ in $O(M \log^2 M)$ time that can answer in $O(\log M \log n + k \log n)$ amortized time each coverage-report-query, i.e., given any point $x \in T$, report all active uncertain points covered by x, where k is the output size. Further, if an uncertain point P_i is deactivated, we can remove P_i from $\mathcal{A}_1$ in $O(m_i \cdot \log M \cdot \log n)$ amortized time.*

Lemma 5. *We can build a data structure $\mathcal{A}_2$ in $O(M \log M + n \log^2 M)$ time that can answer in $O(\log n)$ time each candidate-center-query, i.e., given any vertex $v \in T_m$, find the candidate center c for the active medians of $T_m(v)$. Further, if an uncertain point P_i is deactivated, we can remove P_i from $\mathcal{A}_2$ in $O(\log n)$ time.*

Theorem 1. *We can find a minimum number of centers on T to cover all uncertain points of $\mathcal{P}$ in $O(M \log^2 M)$ time.*

4   A Connector-Bounded Centroid Decomposition

In this section, we propose a tree decomposition of T, called a *connector-bounded centroid decomposition*, which will be repeatedly used later (e.g., for Lemmas 1, 4, 5). The decomposition is different from the centroid decompositions used before, e.g., [11, 14] and has certain properties that can facilitate our algorithms.

A vertex v of T is called a *centroid* if T can be represented as a union of two subtrees with v as their only common vertex and each subtree has at most $\frac{2}{3}$ of the vertices of T [11], and we say the two subtrees are *decomposed* by v. Such a centroid always exists and can be found in linear time [11]. For convenience, we consider v to be contained in only one subtree but an "open vertex" in the other subtree (thus, the location of $\mathcal{P}$ at v only belongs to one subtree).

Our decomposition of T corresponds to a *decomposition tree*, denoted by Υ and defined recursively as follows. Each internal node of Υ has two, three, or four children. The root of Υ corresponds to the entire tree T. Let v be a centroid of T, and let T_1 and T_2 be the subtrees of T decomposed by v. Note that T_1 and T_2 are disjoint since we consider v to be contained in only one of them. Further, we call v a *connector* in both T_1 and T_2. Correspondingly, in Υ, its root has two children corresponding to T_1 and T_2, respectively.

In general, consider a node μ of Υ. Let $T(\mu)$ represent the subtree of T corresponding to μ. We assume $T(\mu)$ has at most two connectors (initially this is true when μ is the root). We further decompose $T(\mu)$ into subtrees that correspond to the children of μ in Υ, as follows. Let v be the centroid of $T(\mu)$ and let $T_1(\mu)$ and $T_2(\mu)$ respectively be the two subtrees of $T(\mu)$ decomposed by v. We consider v as a *connector* in both $T_1(\mu)$ and $T_2(\mu)$.

If $T(\mu)$ has at most one connector, then each of $T_1(\mu)$ and $T_2(\mu)$ has at most two connectors. In this case, μ has two children corresponding to $T_1(\mu)$ and $T_2(\mu)$, respectively.

If $T(\mu)$ has two connectors but each of $T_1(\mu)$ and $T_2(\mu)$ still has at most two connectors (with v as a new connector), then μ has two children corresponding to $T_1(\mu)$ and $T_2(\mu)$, respectively. Otherwise, one of them, say, $T_2(\mu)$, has three connectors and the other $T_1(\mu)$ has only one connector (e.g., see Fig. 2). In this case, μ has a child in Υ corresponding to $T_1(\mu)$, and we further perform a *connector-reducing decomposition* on $T_2(\mu)$, as follows (this is the main difference between our decomposition and the traditional centroid decomposition used before [11, 14]). Depending on whether the three connectors of $T_2(\mu)$ are in a simple path, there are two cases.

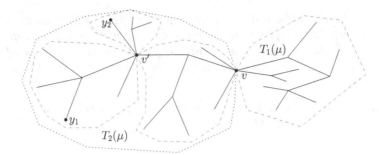

Fig. 2. Illustrating the decomposition of $T(\mu)$ into four subtrees in the (red) dashed cycles, where y_1 and y_2 are two connectors of $T(\mu)$. $T(\mu)$ is first decomposed into two subtrees $T_1(\mu)$ and $T_2(\mu)$. However, since $T_2(\mu)$ has three connectors, we further decompose it into three subtrees each of which has at most two connectors.

1. If they are in a simple path, without loss of generality, we assume v is the one between the other two connectors in the path. We decompose $T_2(\mu)$ into two subtrees at v such that they contain the two connectors respectively. In this way, each subtree contains at most two connectors. Correspondingly, μ has another two children corresponding the two subtrees of $T_2(\mu)$, and thus μ has three children in total.

2. Otherwise, there is a unique vertex v' in $T_2(\mu)$ that decomposes $T_2(\mu)$ into three subtrees that contain the three connectors respectively (e.g., see Fig. 2). Note that v' and the three subtrees can be easily found in linear time by traversing $T_2(\mu)$. Correspondingly, μ has another three children corresponding to the above three subtrees of $T_2(\mu)$, respectively, and thus μ has four children in total. Note that we consider v' as a connector in each of the above three subtrees. Thus, each subtree contains at most two connectors.

We continue the decomposition until each subtree $T(\mu)$ of $\mu \in \Upsilon$ becomes an edge $e(v_1, v_2)$ of T. According to our decomposition, both v_1 and v_2 are connectors of $T(\mu)$, but they may be only open vertices of $T(\mu)$. If both v_1 and v_2 are open vertices of $T(\mu)$, then we will not further decompose $T(\mu)$, so μ is a leaf of Υ. Otherwise, we further decompose $T(\mu)$ into an open edge and a closed vertex v_i if v_i is contained in $T(\mu)$ for each $i = 1, 2$. Correspondingly, μ has either two or three children that are leaves of Υ. In this way, for each leaf μ of Υ, $T(\mu)$ is either an open edge or a closed vertex of T. In the former case, $T(\mu)$ has two connectors that are its incident vertices, and in the latter case, $T(\mu)$ has one connector that is itself.

This finishes the decomposition of T. A major difference between our decomposition and the traditional decomposition [11, 14] is that each subtree in our decomposition has at most two connectors. As will be clear later, this property is crucial to guarantee the runtime of our algorithms.

Lemma 6. *The height of Υ is $O(\log M)$ and Υ has $O(M)$ nodes. The connector-bounded centroid decomposition of T can be computed in $O(M \log M)$ time.*

In the following paper, we assume that our decomposition of T and the decomposition tree Υ have been computed. We introduce some notation that will be used later. For each node μ of Υ, we use $T(\mu)$ to represent the subtree of T corresponding to μ. If y is a connector of $T(\mu)$, then we use $T(y, \mu)$ to represent the subtree of T consisting of all points q of $T \setminus T(\mu)$ such that $\pi(q, p)$ contains y for any point $p \in T(\mu)$ (i.e., $T(y, \mu)$ is the "outside world" connecting to $T(\mu)$ through y; e.g., see Fig. 3). By this definition, if y is the only connector of $T(\mu)$, then $T = T(\mu) \cup T(y, \mu)$; if $T(\mu)$ has two connectors y_1 and y_2, then $T = T(\mu) \cup T(y_1, \mu) \cup T(y_2, \mu)$.

5 The Data Structure $\mathcal{A}_1$

In this section, we present the data structure $\mathcal{A}_1$ for Lemma 4. The data structure $\mathcal{A}_1$ is for answering the coverage-report-queries, i.e., given any point $x \in T$, find all active uncertain points that are covered by x. Further, it also supports the operation of removing an uncertain point once it is deactivated.

Consider any node $\mu \in \Upsilon$. If μ is the root, let $L(\mu) = \emptyset$; otherwise, define $L(\mu)$ to be the sorted list of all indices $i \in [1, n]$ such that P_i does not have any locations in the subtree $T(\mu)$ but has at least one location in $T(\mu')$, where μ' is the parent of μ. Let y be any connector of $T(\mu)$. Let $L(y, \mu)$ be an index list the same as $L(\mu)$ and each index $i \subset L(y, \mu)$ is associated with two values: $F(i, y, \mu)$, which is the probability sum of P_i in the subtree $T(y, \mu)$, and $D(i, y, \mu)$, which is the expected distance from y to the locations of P_i in $T(y, \mu)$, i.e., $D(i, y, \mu) = w_i \cdot \sum_{p_{ij} \in T(y, \mu)} f_{ij} \cdot d(p_{ij}, y)$. We refer to $L(\mu)$ and $L(y, \mu)$ for each connector $y \in T(\mu)$ as the *information lists* of μ.

Lemma 7. *Suppose $L(\mu) \neq \emptyset$ and the information lists of μ are available. Let t_μ be the number of indices in $L(\mu)$. Then, we can build a data structure of $O(t_\mu)$ size in $O(|T(\mu)| + t_\mu \log t_\mu)$ time on $T(\mu)$, such that given any point $x \in T(\mu)$, we can report all indices i of $L(\mu)$ such that P_i is covered by x in $O(\log n + k \log n)$ amortized time, where k is the output size; further, if P_i is deactivated with $i \in L(\mu)$, then we can remove i from the data structure and all information lists of μ in $O(\log n)$ amortized time.*

Proof. As $L(\mu) \neq \emptyset$, μ is not the root. Thus, $T(\mu)$ has one or two connectors. We only discuss the most general case where $T(\mu)$ has two connectors since the other case is similar but easier. Let y_1 and y_2 denote the two connectors of $T(\mu)$, respectively. So the lists $L(y_1, \mu)$ and $L(y_2, \mu)$ are available. Note that for any two points p and q in $T(\mu)$, $\pi(p, q)$ is also in $T(\mu)$ since $T(\mu)$ is connected.

Consider any point $x \in T(\mu)$. Suppose we traverse on $T(\mu)$ from x to y_1, and let q_x be the first point on $\pi(y_1, y_2)$ we encounter (e.g. see Fig 4; so q_x is x if $x \in \pi(y_1, y_2)$). Let $a_x = d(x, q_x)$ and $b_x = d(q_x, y_1)$. Thus, $d(y_1, x) = a_x + b_x$ and $d(y_2, x) = a_x + d(y_1, y_2) - b_x$.

For any $i \in L(\mu)$, since P_i does not have any location in $T(\mu)$, we have $F(i, y_1, \mu) + F(i, y_2, \mu) = 1$. Note that $\mathrm{Ed}(x, P_i) = w_i \cdot \sum_{p_{ij} \in T} f_{ij} \cdot d(x, p_{ij}) = w_i \cdot \sum_{p_{ij} \in T(y_1, \mu)} f_{ij} \cdot d(x, p_{ij}) + w_i \cdot \sum_{p_{ij} \in T(y_2, \mu)} f_{ij} \cdot d(x, p_{ij})$. Further, $\sum_{p_{ij} \in T(y_1, \mu)} f_{ij} \cdot$

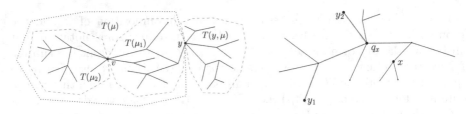

Fig. 3. Illustrating the subtrees $T(\mu_1), T(\mu_2),$ **Fig. 4.** Illustrating the definition of q_x and $T(y, \mu)$, where y is a connector of $T(\mu) =$ in the subtree $T(\mu)$ with two connectors $T(\mu_1) \cup T(\mu_2)$. Note that $T(y, \mu)$ is also y_1 and y_2. The path $\pi(y_1, y_2)$ is highlighted with thicker (red) segments.

$d(x, p_{ij}) = [F(i, y_1, \mu) \cdot (a_x + b_x) + D(i, y_1, \mu)]$ and $\sum_{p_{ij} \in T(y_2, \mu)} f_{ij} \cdot d(x, p_{ij}) = F(i, y_2, \mu) \cdot (a_x + d(y_1, y_2) - b_x) + D(i, y_2, \mu)$. Due to $F(i, y_1, \mu) + F(i, y_2, \mu) = 1$, we obtain $\mathsf{Ed}(x, P_i) = w_i \cdot [a_x + (F(i, y_1, \mu) - F(i, y_2, \mu)) \cdot b_x + D(i, y_1, \mu) + D(i, y_2, \mu) + F(i, y_2, \mu) \cdot d(y_1, y_2)]$.

Notice that for any $x \in T(\mu)$, all above values are constant except a_x and b_x. Therefore, if we consider a_x and b_x as two variables of x, $\mathsf{Ed}(x, P_i)$ is a linear function of them. In other words, $\mathsf{Ed}(x, P_i)$ defines a plane in $\mathbb{R}^3$, where the z-coordinates correspond to the values of $\mathsf{Ed}(x, P_i)$ and the x- and y-coordinates correspond to a_x and b_x respectively. In the following, we also use $\mathsf{Ed}(x, P_i)$ to refer to the plane defined by it in $\mathbb{R}^3$.

Remark. This nice property for calculating $\mathsf{Ed}(x, P_i)$ is due to that μ has at most two connectors. This is one reason our decomposition requires every subtree $T(\mu)$ to have at most two connectors.

Recall that x covers P_i if $\mathsf{Ed}(x, P_i) \leq \lambda$. Consider the plane $H_\lambda : z = \lambda$ in $\mathbb{R}^3$. In general the two planes $\mathsf{Ed}(x, P_i)$ and H_λ intersect at a line l_i and we let h_i represent the closed half-plane of H_λ bounded by l_i and above the plane $\mathsf{Ed}(x, P_i)$. Let x_λ be the point (a_x, b_x) in the plane H_λ. An easy observation is that $\mathsf{Ed}(x, P_i) \leq \lambda$ if and only if $x_\lambda \in h_i$. Further, we say that l_i is an *upper bounding line* of h_i if h_i is below l_i and a *lower bounding line* otherwise. Observe that if l_i is an upper bounding line, then $\mathsf{Ed}(x, P_i) \leq \lambda$ if and only if x_λ is below l_i; if l_i is a lower bounding line, then $\mathsf{Ed}(x, P_i) \leq \lambda$ if and only if x_λ is above l_i.

Given any query point $x \in T(\mu)$, our goal for answering the query is to find all indices $i \in L(\mu)$ such that P_i is covered by x. Based on the above discussions, we do the following preprocessing. After $d(y_1, y_2)$ is computed, by using the information lists of y_1 and y_2, we compute all functions $\mathsf{Ed}(x, P_i)$ for all $i \in L(\mu)$ in $O(t_\mu)$ time. Then, we obtain a set U of all upper bounding lines and a set of all lower bounding lines on the plane H_λ defined by $\mathsf{Ed}(x, P_i)$ for all $i \in L(\mu)$. In the following, we first discuss the upper bounding lines. Let S_U denote the indices $i \in L(\mu)$ such that P_i defines an upper bounding line in U.

Given any point $x \in T(\mu)$, we first compute a_x and b_x. This can be done in constant time after $O(|T(\mu)|)$ time preprocessing, as follows. In the preprocessing, for each vertex v of $T(\mu)$, we compute the vertex q_v (defined in the similar way as q_x with respect to x) as well as the two values a_v and b_v (defined similarly as

a_x and b_x, respectively). This can be easily done in $O(|T(\mu)|)$ time by traversing $T(\mu)$ and we omit the details. Given the point x, which is specified by an edge e containing x, let v be the incident vertex of e closer to y_1 and let δ be the length of e between v and x. Then, if e is on $\pi(y_1, y_2)$, we have $a_x = 0$ and $b_x = b_v + \delta$. Otherwise, $a_x = a_v + \delta$ and $b_x = b_v$.

After a_x and b_x are computed, the point $x_\lambda = (a_x, b_x)$ on the plane H_λ is also obtained. Then, according to our discussion, all uncertain points of S_U that are covered by x correspond to exactly those lines of U above x_λ. Finding the lines of U above x_λ is actually the dual problem of half-plane range reporting query in $\mathbb{R}^2$. By using the dynamic convex hull maintenance data structure of Brodal and Jacob [5], with $O(|U| \log |U|)$ time and $O(|U|)$ space preprocessing, for any point x_λ, we can easily report all lines of U above x_λ in $O(\log |U| + k \log |U|)$ amortized time (i.e., by repeating k deletions), where k is the output size, and deleting a line from U can be done in $O(\log |U|)$ amortized time. Clearly, $|U| \leq t_\mu$.

On the set of all lower bounding lines, we do the similar preprocessing, and the query algorithm is symmetric. Hence, the total preprocessing time is $O(|T(\mu)| + t_\mu \log t_\mu)$. Each query takes $O(\log t_\mu + k \log t_\mu)$ amortized time and each remove operation runs in $O(\log t_\mu)$ amortized time. As $t_\mu \leq n$, the lemma follows. □

The preprocessing algorithm for $\mathcal{A}_1$ consists of the following four steps. First, we compute the information lists for all nodes μ of Υ. Second, for each node $\mu \in \Upsilon$, we compute the data structure of Lemma 7. Third, for each $i \in [1, n]$, we compute a *node list* $L_\mu(i)$ containing all nodes $\mu \in \Upsilon$ such that $i \in L(\mu)$. Fourth, for each leaf μ of Υ, if $T(\mu)$ is a vertex v of T holding a location p_{ij}, then we maintain at μ the value $\mathsf{Ed}(v, P_i)$. Before giving the details of the above processing algorithm, we first discuss the algorithm for answering the coverage-report-queries by assuming that the preprocessing work has been done.

Given any point $x \in T$, we answer the coverage-report-query as follows. Note that x is in $T(\mu_x)$ for some leaf μ_x of Υ. For each node μ in the path of Υ from the root to μ_x, we apply the query algorithm in Lemma 7 to report all indices $i \in L(\mu)$ such that x covers P_i. In addition, if $T(\mu_x)$ is a vertex of T holding a location p_{ij} such that P_i is active, then we report i if $\mathsf{Ed}(v, P_i)$, which is maintained at v, is at most λ. The proof of the following lemma is omitted.

Lemma 8. *Our query algorithm correctly finds all active uncertain points that are covered by x in $O(k + \log M \log n)$ amortized time, where k is the output size.*

If an uncertain point P_i is deactivated, then we scan the node list $L_\mu(i)$ and for each node $\mu \in L_\mu(i)$, we remove i from the data structure by Lemma 7. The following lemma implies that the total time is $O(m_i \log M \log n)$.

Lemma 9. *For each $i \in [1, n]$, the number of nodes in $L_\mu(i)$ is $O(m_i \log M)$.*

Lemma 10. $\sum_{\mu \in \Upsilon} t_\mu = O(M \log M)$, *and the preprocessing time for constructing the data structure $\mathcal{A}_1$ except the second step is $O(M \log M)$.*

The proofs of Lemmas 9 and 10 are omitted. As $\sum_{\mu \in \Upsilon} t_\mu = O(M \log M)$, applying the preprocessing of Lemma 7 on all nodes of Υ takes $O(M \log^2 M)$

time and $O(M \log M)$ space in total. Hence, the total preprocessing time of $\mathcal{A}_1$ is $O(M \log^2 M)$ and the space is $O(M \log M)$. This proves Lemma 4.

References

1. Agarwal, P., Cheng, S.W., Tao, Y., Yi, K.: Indexing uncertain data. In: Proc. of the 28th Symposium on Principles of Database Systems (PODS). pp. 137–146 (2009)
2. Agarwal, P., Efrat, A., Sankararaman, S., Zhang, W.: Nearest-neighbor searching under uncertainty. In: Proc. of the 31st Symposium on Principles of Database Systems (PODS). pp. 225–236 (2012)
3. Agarwal, P., Har-Peled, S., Suri, S., Yıldız, H., Zhang, W.: Convex hulls under uncertainty. In: Proc. of the 22nd Annual European Symposium on Algorithms (ESA). pp. 37–48 (2014)
4. de Berg, M., Roeloffzen, M., Speckmann, B.: Kinetic 2-centers in the black-box model. In: Proc. of the 29th Annual Symposium on Computational Geometry (SoCG). pp. 145–154 (2013)
5. Brodal, G., Jacob, R.: Dynamic planar convex hull. In: Proc. of the 43rd IEEE Symposium on Foundations of Computer Science (FOCS). pp. 617–626 (2002)
6. Cole, R.: Slowing down sorting networks to obtain faster sorting algorithms. Journal of the ACM 34(1), 200–208 (1987)
7. Huang, L., Li, J.: Stochasitc k-center and j-flat-center problems. In: Proc. of the 28th Annual ACM-SIAM Symposium on Discrete Algorithms (SODA). pp. 110–129 (2017)
8. Jørgensen, A., Löffler, M., Phillips, J.: Geometric computations on indecisive points. In: Proc. of the 12nd Algorithms and Data Structures Symposium (WADS). pp. 536–547 (2011)
9. Kamousi, P., Chan, T., Suri, S.: Closest pair and the post office problem for stochastic points. In: Proc. of the 12nd International Workshop on Algorithms and Data Structures (WADS). pp. 548–559 (2011)
10. Kamousi, P., Chan, T., Suri, S.: Stochastic minimum spanning trees in Euclidean spaces. In: Proc. of the 27th Annual Symposium on Computational Geometry (SoCG). pp. 65–74 (2011)
11. Kariv, O., Hakimi, S.: An algorithmic approach to network location problems. I: The p-centers. SIAM J. on Applied Mathematics 37(3), 513–538 (1979)
12. Löffler, M., van Kreveld, M.: Largest bounding box, smallest diameter, and related problems on imprecise points. Computational Geometry: Theory and Applications 43(4), 419–433 (2010)
13. Megiddo, N.: Linear-time algorithms for linear programming in R^3 and related problems. SIAM Journal on Computing 12(4), 759–776 (1983)
14. Megiddo, N., Tamir, A.: New results on the complexity of p-centre problems. SIAM Journal on Computing 12(4), 751–758 (1983)
15. Tao, Y., Xiao, X., Cheng, R.: Range search on multidimensional uncertain data. ACM Transactions on Database Systems 32 (2007)
16. Wang, H., Zhang, J.: Computing the center of uncertain points on tree networks. Algorithmica 78(1), 232–254 (2017)
17. Wang, H., Zhang, J.: One-dimensional k-center on uncertain data. Theoretical Computer Science 602, 114–124 (2015)
18. Wang, H., Zhang, J.: Covering uncertain points in a tree. arXiv:1704.07497 (2017)

Stochastic Closest-pair Problem and Most-likely Nearest-neighbor Search in Tree Spaces

Jie Xue and Yuan Li (✉)

University of Minnesota — Twin Cities, Minneapolis MN 55455, USA
{xuexx193,lixx2100}@umn.edu

Abstract. Let $\mathcal{T}$ be a tree space represented by a weighted tree with t vertices, and S be a set of n stochastic points in $\mathcal{T}$, each of which has a fixed location with an independent existence probability. We investigate two fundamental problems under such a stochastic setting, the closest-pair problem and the nearest-neighbor search. For the former, we propose the first algorithm of computing the ℓ-threshold probability and the expectation of the closest-pair distance of a realization of S. For the latter, we study the k most-likely nearest-neighbor search (k-LNN) via a notion called the k most-likely Voronoi Diagram (k-LVD), where we show the combinatorial complexity of k-LVD is $O(nk)$ under two reasonable assumptions, leading to a logarithmic query time for k-LNN.

1 Introduction

In many real-world applications, due to the existence of noise or limitations of devices, the data obtained may be imprecise or not totally reliable. In this situation, the dataset may fail to capture well the features of the data. Motivated by this, the topic of uncertain data has received significant attention in the last few decades. Many classical problems have been investigated under uncertainty, including convex hull, minimum spanning tree, range search, linear separability, etc. [1,2,3,4,5,7,8,10,13,15,16]. Among these, there are two common models of uncertainty: existential uncertainty and locational uncertainty. In the former, each (stochastic) data point has a fixed location with an uncertain existence depicted by an independent existence probability, while in the latter the location of each point is uncertain and described as a distribution.

The closest-pair problem and nearest-neighbor search are two interrelated fundamental problems, which have numerous applications in various areas. The uncertain versions of both the problems have also been studied recently in [1,9,11,12,14]. Let S be a set of n stochastic points in some metric space $\mathcal{X}$. Concerning the closest pair problem, a basic question one may ask is how to compute elementary statistics about the stochastic closest-pair of S, e.g., the probability that the closest-pair distance of a realization of S is at least ℓ, the expected closest-pair distance, etc. Unfortunately, most problems of this kind have been shown to be NP-hard or #P-hard for general metrics, and some of them remain #P-hard even when $\mathcal{X} = \mathbb{R}^d$ for $d \geq 2$ [9,11]. Concerning the nearest-neighbor search, an important problem is the most-likely nearest-neighbor (LNN) search

© Springer International Publishing AG 2017
F. Ellen et al. (Eds.): WADS 2017, LNCS 10389, pp. 569–580, 2017.
DOI: 10.1007/978-3-319-62127-2_48

[14], which looks for the data point in S with the greatest probability of being the nearest-neighbor of a query point q. The LNN search introduces the concept of most-likely Voronoi diagram (LVD), which decomposes $\mathcal{X}$ into connected cells such that the query points in the same cell have the same LNN. However, as in [12,14], the bound of LVD in $\mathbb{R}^d$ is still high even on average. Due to the difficulties of both problems in general and Euclidean space, it is then natural to ask whether these problems are relatively easier in other metric spaces such as a *tree space*. Indeed, further exploring these problems in tree spaces will be helpful and interesting since any finite metrics (say a road network in practice) can be embedded on a tree space under some reasonable distortions [6].

With the above motivations, in this paper, we study the stochastic closest-pair (SCP) problem and k most-likely nearest-neighbor (k-LNN) search in tree spaces. A *tree space* $\mathcal{T}$ is represented by a positively-weighted tree T where the weight of each edge depicts its "length". Formally, $\mathcal{T}$ is the geometric realization of T, in which each edge weighted by w is isometric to the interval $[0, w]$. There is a natural metric over $\mathcal{T}$ which defines the distance $dist(x, y)$ as the length of the (unique) simple path between x and y in $\mathcal{T}$. See Fig. 1 for an example of tree space. Following [9,11,14], we study the problems under existential uncertainty: each stochastic point has a fixed location (in $\mathcal{T}$) associated with an (independent) existence probability. Due to limited space, the proofs of all lemmas and some theorems are omitted and can be found in the full version [17].

Our results. Let $\mathcal{T}$ be a tree space represented by a t-vertex weighted tree T, and S be the given set of n stochastic points in $\mathcal{T}$ each of which is associated with an existence probability. A *realization* of S refers to a random sample of S in which each point is sampled with its existence probability.

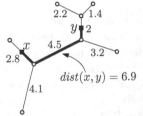

Fig. 1. A tree space and the unique simple path (in bold) between x and y

For the SCP problem, define $\kappa(S)$ as a random variable indicating the closest-pair distance of a realization of S. We first show that the ℓ-threshold probability of $\kappa(S)$ (i.e., the probability that $\kappa(S)$ is at least ℓ) can be computed in $O(t + n \log n + \min\{tn, n^2\})$ time for any given positive threshold ℓ. Based on this, we immediately obtain an $O(t + \min\{tn^3, n^4\})$-time algorithm for computing the expected closest-pair distance, i.e., the expectation of $\kappa(S)$. We then further show that one can approximate the expected closest-pair distance within a factor of $(1+\varepsilon)$ in $O(t+\varepsilon^{-1} \min\{tn^2, n^3\})$ time, by arguing that the expected closest-pair distance can be approximated via $O(\varepsilon^{-1}n)$ threshold probability queries.

For the LNN search, we first study the size of the the k-LVD $\Psi_{\mathcal{T}}^S$ of S on $\mathcal{T}$. A matching $O(n^2)$ upper bound for the worst-case size of $\Psi_{\mathcal{T}}^S$ is given. More interestingly, we show that (1) the worst-case size of $\Psi_{\mathcal{T}}^S$ is $O(kn)$, if the existence probabilities of the points in S are constant-far from 0; (2) the average-case size of $\Psi_{\mathcal{T}}^S$ is $O(kn)$, if the existence probabilities are i.i.d. random variables drawn from a fixed distribution. These results further imply the existence of an LVD data structure which answers k-LNN queries in $O(\log n + k)$ time using average-case $O(t + k^2 n)$ space, and worst-case $O(t + k^2 n)$ space if the existence probabilities of the points are constant-far from 0. Finally, we give an $O(t + n^2 \log n + n^2 k)$-time algorithm to construct such a data structure.

2 The stochastic closest-pair problem

Let $\mathcal{T}$ be a tree space represented by a t-vertex weighted tree T and $S = \{a_1, \ldots, a_n\} \subset \mathcal{T}$ be a set of stochastic points where a_i has an existence probability π_{a_i}. We use $\kappa(S)$ to denote the random variable indicating the closest-pair distance of a realization of S (if the realization is of size less than 2, we simply set its closest-pair distance to be 0).

2.1 Computing the threshold probability

We study the problem of computing the probability that $\kappa(S)$ is at least ℓ for a given threshold ℓ. We call this quantity the ℓ-threshold probability or simply threshold probability of $\kappa(S)$, and denote it by $C_{\geq \ell}(S)$. We show that $C_{\geq \ell}(S)$ can be computed in $O(t + n \log n + \min\{tn, n^2\})$ time. This result gives us an $O(t + n^2)$ upper bound for $t = \Omega(n)$ and an $O(n \log n + tn)$ bound for $t = O(n)$. In the rest of this section, we first present an $O(t + n^3)$-time algorithm for computing $C_{\geq \ell}(S)$, and then show how to improve it to achieve the desired bound. For simplicity of exposition, we assume $a_1, \ldots, a_n$ have distinct locations in $\mathcal{T}$.

An $O(t + n^3)$-time algorithm. In order to conveniently and efficiently handle the stochastic points in a tree space, we begin with a preprocessing step, which reduces the problem to a more regular setting.

Theorem 1. *Given $\mathcal{T}$ and S, one can compute in $O(t + n \log n)$ time a new tree space $\mathcal{T}' \subseteq \mathcal{T}$ represented by an $O(n)$-vertex weighted tree T' s.t. $S \subset \mathcal{T}'$ and every point in S is located at some vertex of T'. (See [17] for a proof.)*

By the above theorem, we use $O(t + n \log n)$ time to compute such a new tree space. Using this tree space as well as the $O(n)$-vertex tree representing it, the problem becomes more regular: every stochastic point in S is located at a vertex. We can further put the stochastic points in one-to-one correspondence with the vertices by adding dummy points with existence probability 0 to the "empty" vertices. In such a regular setting, we then consider how to compute the ℓ-threshold probability. For convenience, we still use T to denote the representation of the (new) tree space and $S = \{a_1, \ldots, a_n\}$ the stochastic dataset (though the actual size of S may be larger than n due to the additional dummy points, it is still bounded by $O(n)$). Since the vertices of T are now in one-to-one correspondence with the points in S, we also use a_i to denote the corresponding vertex of T.

As we are working on a tree space, a natural idea for solving the problem is to exploit the recursive structure of the tree and to compute $C_{\geq \ell}(S)$ in a recursive fashion. To this end, we need to define an important concept called *witness.* We make T rooted by setting a_1 as its root. The subtree rooted at a vertex x is denoted by T_x. Also, we use $V(T_x)$ to denote the set of the stochastic points lying in T_x, or equivalently, the set of the vertices of T_x. The notations $\bar{p}(x)$ and $ch(x)$ are used to denote the parent of x and the set of the children of x, respectively (for convenience we set $\bar{p}(a_1) = a_1$).

Definition 1. *Let $dep(a_i)$ be the depth of a_i in T, i.e., $dep(a_i) = dist(a_1, a_i)$. For any a_i and a_j, we define $a_i \prec a_j$ if $dep(a_i) < dep(a_j)$, or $dep(a_i) = dep(a_j)$ and $i < j$. Clearly, the relation $\prec$ is a strict total order over S (also, over the vertices of T). For any subset $S' \subseteq S$ and any vertex a_i of T, we define the* **witness** *of a_i with respect to S', denoted by $w(a_i, S')$, as the smallest vertex in $V(T_{a_i}) \cap S'$ under the $\prec$-order. If $V(T_{a_i}) \cap S' = \emptyset$, we say $w(a_i, S')$ is not defined.* See Fig. 2 for an illustration of witness. We say a subset $S' \subseteq S$ is *legal* if the closest-pair distance of S' is at least ℓ.

The following lemma allows us to verify the legality of a subset by using the witnesses, which will be used later.

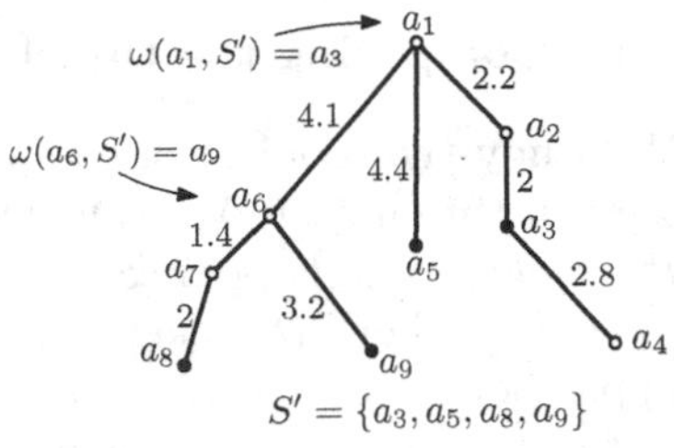

Lemma 1. *For any $S' \subseteq S$, we have S' is legal if and only if every point $a_i \in S' \backslash \{a_1\}$ satisfies one of the following three conditions:*
(1) $w(a_i, S')$ is not defined;
(2) $w(a_i, S') = w(\bar{p}(a_i), S')$;
(3) $dist(w(a_i, S'), w(\bar{p}(a_i), S')) \geq \ell$.
We say that S' is **locally legal** *at a_i whenever a_i satisfies one of the above conditions.*

Fig. 2. An illustration of witness

In order to compute $C_{\geq \ell}(S)$, we define, for all $x \in S$ and $y \in V(T_{\bar{p}(x)})$,

$$P_y(x) = \begin{cases} \Pr_{S' \subseteq_R V(T_x)} [S' \text{ is legal and } w(x, S') = y] & \text{if } y \in V(T_x), \\ \Pr_{S' \subseteq_R V(T_x)} [S' \cup \{y\} \text{ is legal and } w(\bar{p}(x), S' \cup \{y\}) = y] & \text{if } y \in V(T_{\bar{p}(x)}) \backslash V(T_x). \end{cases}$$

Here the notation $\subseteq_R$ means that the former is a realization of the latter, i.e., a random sample obtained by sampling each point with its existence probability. With the above, we immediately have that $C_{\geq \ell}(S) = \sum_{i=1}^{n} P_{a_i}(a_1) - P_0$, where P_0 is the probability that a realization of S contains exactly one point. We then show how $P_y(x)$ can be computed in a recursive way.

Lemma 2. *For $x \in S$ and $y \in V(T_x)$, we have that*

$$P_y(x) = Q \cdot \prod_{c \in ch(x)} P_y(c),$$

where $Q = \pi_x$ if $x = y$ and $Q = 1 - \pi_x$ if $x \neq y$.

Lemma 3. *For $x \in S$ and $y \in V(T_{\bar{p}(x)}) \backslash V(T_x)$, we have that*

$$P_y(x) = \prod_{a_i \in V(T_x)} (1 - \pi_{a_i}) + \sum_{z \in \Gamma} P_z(x),$$

where $\Gamma = \{z \in V(T_x) : y \prec z \text{ and } dist(z, y) \geq \ell\}$.

By the above two lemmas, the values of all $P_y(x)$ can be computed as follows. We enumerate $x \in S$ from the greatest to the smallest under $\prec$-order. For each x, we first compute all $P_y(x)$ for $y \in V(T_x)$ by applying Lemma 2. After this, we are able to compute all $P_y(x)$ for $y \in V(T_{\bar{p}(x)}) \backslash V(T_x)$ by applying Lemma 3. The entire process takes $O(n^3)$ time. Once we have the values of all $P_y(x)$, $C_{\geq \ell}(S)$ can be computed straightforwardly. Including the time for preprocessing, this gives us an $O(t + n^3)$-time algorithm for computing $C_{\geq \ell}(S)$.

In fact, we can further improve the runtime above to $O(t+n^2)$ by speeding up the computation of $P_y(x)$ for $y \in V(T_{\bar{p}(x)}) \backslash V(T_x)$ as they are the bottlenecks. In addition, if $t = O(n)$, we can even further reduce the runtime to $O(t + n \log n + \min(tn, n^2))$. Both optimizations are nontrivial and need new insights. However, due to the limited space, we leave these to [17] and conclude the following.

Theorem 2. *Given a weighted tree T with t vertices and a set S of n stochastic points in its tree space $\mathcal{T}$, one can compute the ℓ-threshold probability of the closest-pair distance of S, $C_{\geq \ell}(S)$, in $O(t + n \log n + \min\{tn, n^2\})$ time.*

2.2   Computing the expected closest-pair distance

Based on our algorithm for computing the threshold probability, we further study the problem of computing the expected closest-pair distance of S, i.e., the expectation of $\kappa(S)$. It is easy to see that our algorithm in Section 2.1 immediately gives us an $O(t+\min\{tn^3, n^4\})$-time algorithm to compute $\mathbf{E}[\kappa(S)]$. This is because the random variable $\kappa(S)$ has at most $\binom{n}{2}$ distinct possible values and hence we can compute $\mathbf{E}[\kappa(S)]$ via $O(n^2)$ threshold probability "queries" with various thresholds ℓ (note that after preprocessing our algorithm answers each threshold probability query in $O(\min\{tn, n^2\})$ time).

If we want to compute the exact value of $\mathbf{E}[\kappa(S)]$ (via threshold probability queries), $\Theta(n^2)$ queries are necessary in worst case. So it is natural to ask whether we can use less queries to approximate $\mathbf{E}[\kappa(S)]$. In the rest of this section, we show that one can use $O(\varepsilon^{-1}n)$ threshold probability queries to achieve a $(1 + \varepsilon)$-approximation for $\mathbf{E}[\kappa(S)]$, which in turn gives us an $O(t + \varepsilon^{-1}\min\{tn^2, n^3\})$-time approximation algorithm for computing $\mathbf{E}[\kappa(S)]$.

For simplicity of exposition, we assume that the stochastic points in S are now one-to-one corresponding to the vertices of T (this is what we have after preprocessing). We begin with a simple case, in which the *spread* of T, i.e., the ratio of the length of the longest edge to the length of the shortest edge is bounded by some polynomial of n. In this case, to approximate $\mathbf{E}[\kappa(S)]$ is fairly easy, and we only need $O(\varepsilon^{-1} \log n)$ threshold probability queries.

Definition 2. *For $\beta > \alpha > 0$ and $\tau > 1$, the (α, β, τ)-jump is defined as*

$$J = \{\alpha, \tau\alpha, \tau^2\alpha, \ldots, \tau^k\alpha, \beta\},$$

where $\tau^k \alpha < \beta$ and $\tau^{k+1}\alpha \geq \beta$.

Let $d_{\min}$ be the length of the shortest edge of T and $d_{\max}$ be the sum of the lengths of all edges of T. Also, let J be the $(d_{\min}, d_{\max}, 1+\varepsilon)$-jump. Suppose $J =$

$\{\ell_1, \ldots, \ell_{|J|}\}$. Then we do $|J|$ threshold probability queries using the thresholds $\ell_1, \ldots, \ell_{|J|}$, and compute

$$E = \sum_{i=1}^{|J|} C_{\geq \ell_i}(S) \cdot (\ell_i - \ell_{i-1})$$

as an approximation of $\mathbf{E}[\kappa(S)]$ (where $\ell_0 = 0$). Note that $|J| = O(\log_{1+\varepsilon} \frac{d_{\max}}{d_{\min}}) = O(\log_{1+\varepsilon} n) = O(\varepsilon^{-1} \log n)$. It is easy to verify that $E \leq \mathbf{E}[\kappa(S)] \leq (1 + \varepsilon)E$.

The problem becomes interesting when the spread of T is unbounded. In this case, although the above method still correctly approximates $\mathbf{E}[\kappa(S)]$, the number of the threshold probability queries is no longer well bounded. Imagine that the $O(n^2)$ possible values of $\kappa(S)$ are distributed as $\ell, (1+\varepsilon)\ell, (1+\varepsilon)^2\ell$, etc. Then the $(d_{\min}, d_{\max}, 1 + \varepsilon)$-jump J is of size $\Omega(n^2)$. Moreover, for guaranteeing the correctness, it seems that we cannot "skip" any element in J. However, as one will realize later, such an extreme situation can never happen. Recall that we are working on a weighted tree and the $O(n^2)$ possible values of $\kappa(S)$ are indeed the pairwise distances of the vertices of the tree. As such, these values are not arbitrary, and our insight here is to exploit the underlying properties of the distribution of these values.

Let $e_1, \ldots, e_{n-1}$ be the edges of T where e_i has the length (weight) w_i. Assume $w_1 \leq \cdots \leq w_{n-1}$. We define an index set $I = \left\{m : \sum_{i=1}^{m-1} w_i < w_m\right\}$. Suppose $I = \{m_1, \ldots, m_k\}$ where $m_1 < \cdots < m_k$. Note that $m_1 = 1$. For convenience, we set $m_{k+1} = n$. We design our threshold probability queries as follows. Let J_i be the $(w_{m_i}, s_i, 1 + \varepsilon)$-jump where $s_i = \sum_{j < m_{i+1}} w_j$, and $J = J_1 \cup \cdots \cup J_k$. Suppose $J = \{\ell_1, \ldots, \ell_{|J|}\}$ and set $\ell_0 = 0$. Similarly to the previous case, we do $|J|$ threshold probability queries using the thresholds $\ell_1, \ldots, \ell_{|J|}$, and compute

$$E = \sum_{i=1}^{|J|} C_{\geq \ell_i}(S) \cdot (\ell_i - \ell_{i-1})$$

as an approximation of $\mathbf{E}[\kappa(S)]$. We first verify the correctness, i.e., $E \leq \mathbf{E}[\kappa(S)] \leq (1 + \varepsilon)E$. The fact $E \leq \mathbf{E}[\kappa(S)]$ can be easily verified. To see the inequality $\mathbf{E}[\kappa(S)] \leq (1+\varepsilon)E$, we define a piecewise-constant function $h : \mathbb{R}^+ \cup \{0\} \to [0, 1]$ as

$$h(\ell) = \begin{cases} C_{\geq \ell_i}(S) & \text{if } (1+\varepsilon)\ell_i < \ell \leq (1+\varepsilon)\ell_{i+1}, \\ 0 & \text{if } \ell > (1+\varepsilon)l_{|J|}, \\ 1 & \text{if } \ell = 0. \end{cases}$$

Then it is clear that $(1+\varepsilon)E = \int_0^\infty h(\ell)d\ell$. We claim that $\int_0^\infty h(\ell)d\ell \geq \int_0^\infty C_{\geq \ell}(S)d\ell$, hence we have $\mathbf{E}[\kappa(S)] \leq (1 + \varepsilon)E$. Note that the jumps $J_1, \ldots, J_k$ are disjoint and each of them contains a consecutive portion of the sequence $\ell_1, \ldots, \ell_{|J|}$. Furthermore, if ℓ_i and ℓ_{i+1} belong to different jumps, then there is no possible value of $\kappa(S)$ within the range (ℓ_i, ℓ_{i+1}), i.e., $C_{\geq \ell}(S)$ is constant when $\ell \in [\ell_i, \ell_{i+1})$. With this observation, it is not difficult to verify that $h(\ell) \geq C_{\geq \ell}(S)$ for any $\ell \geq 0$. Consequently, we have $\mathbf{E}[\kappa(S)] \leq (1 + \varepsilon)E$, which implies the correctness of our method. Now the only thing remaining is to bound the number of the threshold probability queries, which we show in Lemma 4.

Lemma 4. *For each jump J_i, we have $|J_i| = O(\varepsilon^{-1}(m_{i+1} - m_i))$. As a result, the total number of the threshold probability queries, $|J|$, is $O(\varepsilon^{-1}n)$.*

Indeed, the above method can be extended to a much more general case, in which the stochastic dataset S is given in any metric space $\mathcal{X}$ (not necessarily a tree space). In this case, one can still define the threshold probability $C_{\geq \ell}(S)$ as well as the expected closest-pair distance $\mathbf{E}[\kappa(S)]$ in the same fashion. Our conclusion is the following.

Theorem 3. *Given a set S of n stochastic points in a metric space $\mathcal{X}$, one can $(1+\varepsilon)$-approximate the expected closest-pair distance of S, $\mathbf{E}[\kappa(S)]$, via $O(\varepsilon^{-1}n)$ threshold probability queries. (See [17] for a proof.)*

For the expected closest-pair distance in tree space, we can eventually conclude the following by plugging in our algorithm in Section 2.1 for computing $C_{\geq \ell}(S)$.

Corollary 1. *Given a tree space $\mathcal{T}$ represented by a weighted tree T with t vertices and a set S of n stochastic points in $\mathcal{T}$, one can compute a $(1+\varepsilon)$-approximation for the expected closest-pair distance of S, $\mathbf{E}[\kappa(S)]$, in $O(t + \varepsilon^{-1}\min\{tn^2, n^3\})$ time.*

3   The most-likely nearest-neighbor search problem

In this section, we study the k most-likely nearest-neighbor (k-LNN) search in a tree space. Again, let $\mathcal{T}$ be a tree space represented by a t-vertex weighted tree T and $S = \{a_1, \ldots, a_n\} \subset \mathcal{T}$ be the given stochastic dataset where the point a_i has an existence probability π_{a_i}. The k-LNN search problem can be defined as follows. Let $q \in \mathcal{T}$ be any point. For each $a_i \in S$, define $NNP_q(a_i)$ as the probability that the nearest-neighbor of q in a realization of S is a_i. Clearly, the nearest-neighbor of q in a realization is a_i iff a_i is in the realization and any point closer to q is not in the realization. Therefore, we have

$$NNP_q(a_i) = \pi_{a_i} \cdot \prod_{x \in \Gamma}(1 - \pi_x),$$

where $\Gamma = \{x \in S : dist(q, x) < dist(q, a_i)\}$. Given a query point $q \in \mathcal{T}$, the goal of the k-LNN search is to report the k-LNN of q, which is a k-sequence $(a_{i_1}, \ldots, a_{i_k})$ of points in S such that $NNP_q(a_{i_1}) \geq \cdots \geq NNP_q(a_{i_k}) \geq NNP_q(a_j)$ for all $j \notin \{i_1, \ldots, i_k\}$. For convenience, we assume $NNP_q(a_i) \neq NNP_q(a_j)$ for any $q \in \mathcal{T}$ and $a_i \neq a_j$ so that the k-LNN of any query point $q \in \mathcal{T}$ is uniquely defined.

A standard tool for nearest-neighbor search is the Voronoi diagram. In stochastic setting, we seek the most-likely Voronoi diagram (LVD), the concept of which is for the first time introduced in [14]. The k-LVD partitions the query space into connected cells such that points in the same cell have the same k-LNN. See [17] for an example (in color) of 1-LVD in a tree space.

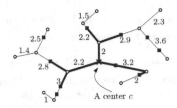

Fig. 3. A degree-3 center involving 5 points.

3.1 The size of the tree-space LVD

We use $\Psi_{\mathcal{T}}^S$ to denote the k-LVD of S on $\mathcal{T}$, i.e., the collection of the cells. Formally, $\Psi_{\mathcal{T}}^S$ can be defined as follows. For any k-sequence $\eta = (a_{i_1}, \ldots, a_{i_k})$, let Ψ_η be the set of the connected components of the subspace $\{q \in \mathcal{T} : \eta \text{ is the } k\text{-LNN of } q\}$. Then $\Psi_{\mathcal{T}}^S$ is the union of Ψ_η over all possible η. Clearly, the size of $\Psi_{\mathcal{T}}^S$ significantly influences the space efficiency of the LVD-based algorithm for k-LNN search. Let $m_{ij} \in \mathcal{T}$ be the "midpoint" of a_i and a_j, i.e., the midpoint of the path between a_i and a_j in $\mathcal{T}$. It is easy to see that the k-LNN only changes nearby these $\binom{n}{2}$ midpoints. However, this does not immediately imply that the size of $\Psi_{\mathcal{T}}^S$ is bounded by $O(n^2)$. The reason is that $O(n^2)$ points do not necessarily decompose $\mathcal{T}$ into $O(n^2)$ pieces (cells), unless these points are located only in the interiors of the edges. Note that throughout this section, we do not make any spatial assumption about the midpoints. In other words, it is allowed that different midpoints occupy the same location in $\mathcal{T}$, and some midpoints are located at the vertices of T. The reason why we allow this is explained in [17]. It is not surprising that even in such a general setting, the size of $\Psi_{\mathcal{T}}^S$ is still bounded by $O(n^2)$. We will see this later as a direct corollary of a technical result (Lemma 5).

Definition 3. *For any two midpoints m_{ij} and $m_{i'j'}$, we define $m_{ij} \equiv m_{i'j'}$ iff m_{ij} and $m_{i'j'}$ have the same location in $\mathcal{T}$ and $dist(a_i, m_{ij}) = dist(a_j, m_{ij}) = dist(a_{i'}, m_{i'j'}) = dist(a_{j'}, m_{i'j'})$. Clearly, $\equiv$ is an equivalence relation over the midpoints. We call the equivalence classes (under $\equiv$) **centers** of S and use $[m_{ij}]$ to denote the center that contains m_{ij}. A stochastic point $a_i \in S$ is said to be **involved** by a center c if $c = [m_{ij}]$ for some j. The **degree** of a center c, denoted by $deg(c)$, is defined as the number of the connected components of $\mathcal{T} \setminus \hat{c}$ that contain at least one point involved by c, where $\hat{c}$ denotes the point in $\mathcal{T}$ corresponding to c, and each such component is called a **branch** of c. A center c is said to be **critical** if $\hat{c}$ is not in the interior of any cell $C \in \Psi_{\mathcal{T}}^S$ and there exists at least one point involved by c that is in the k-LNN of $\hat{c}$. (See Fig. 3 for an intuitive illustration of a center.)*

Lemma 5. *Let Γ be the set of the critical centers and $\xi = \sum_{c \in \Gamma} deg(c)$. Then $|\Psi_{\mathcal{T}}^S| \leq \xi + 1$.*

The above lemma immediately gives us the $O(n^2)$ upper bound for the size of $\Psi_{\mathcal{T}}^S$. Indeed, a center c of S contains at least $\Omega(deg(c) \cdot m)$ midpoints, where m is the number of the points involved by c, so $\xi + 1$ is at most $O(n^2)$. Unfortunately, this upper bound is tight, following from the $\Omega(n^2)$ worst-case lower bound for the size of the 1-dim 1-LVD given by [14] (note that the 1-dim LVD is a special case of the tree-space LVD). Surprisingly, we show that, if we make reasonable assumptions for the existence probabilities of the stochastic points or consider the average case, the size of $\Psi_{\mathcal{T}}^S$ is significantly smaller. Our results are:

- If the existence probabilities of all points in S are *constant-far from* 0, i.e., there is a fixed constant $\varepsilon > 0$ such that $\pi_{a_i} \geq \varepsilon$ for all $a_i \in S$, then the size of the k-LVD $\Psi_{\mathcal{T}}^S$ is $O(kn)$. Note that this assumption about the existence probabilities is natural and reasonable. In applications, an extremely small existence probability means the data point is highly unreliable. Such a point can be considered as a noise and removed from the dataset.
- The average-case size of the k-LVD $\Psi_{\mathcal{T}}^S$ is $O(kn)$. For the average-case analysis we assume that the existence probabilities of the points in S are i.i.d. random variables drawn from any fixed distribution (e.g., the uniform distribution among $[0,1]$). In other words, we consider the expectation of $|\Psi_{\mathcal{T}}^S|$ when $\pi_{a_1}, \ldots, \pi_{a_n}$ are such random variables. The interesting point is that the $O(kn)$ upper bound is totally independent of the structure of $\mathcal{T}$ and the locations of the stochastic points. The randomness is only applied to the existence probabilities in our average-case analysis.

To prove these bounds requires new ideas. By Lemma 5, to bound the size of $\Psi_{\mathcal{T}}^S$, it suffices to bound the degree-sum of the critical centers. Intuitively, if a center c is far from the points it involves (compared with other points in S), then c is less likely to be critical, as the c-involved points are less likely to be in the k-LNN of $\hat{c}$. Along with this intuition, we define the following.

Definition 4. *For any center c, the **diameter** of c, denoted by $diam(c)$, is defined as the distance from $\hat{c}$ to the c-involved points. Let $A \subset \mathcal{T}$ be a finite set. We define the **depth** of c with respect to A as $dep_A(c) = |\{x \in A : dist(x,c) < diam(c)\}|$, i.e., the number of the points in A which are closer to c than the c-involved points.*

Our idea here is to first bound the "contribution" (degree-sum) of the "shallow" centers, and then further bound the degree-sum of the critical ones. Specifically, we investigate in Lemma 6 the degree-sum of the *d-shallow centers* of S, i.e., the centers of depth less than d with respect to S.

Lemma 6. *For $1 \leq d \leq n - 1$, the degree-sum of the d-shallow centers of S is at most $8dn$.*

Now we are ready to prove the $O(kn)$ bound for $|\Psi_{\mathcal{T}}^S|$ under the "constant-far from 0" assumption about the existence probabilities.

Lemma 7. *If the existence probabilities of the points in S are constant-far from 0, then a center of S is critical only if it is $O(k)$-shallow.*

Theorem 4. *If the existence probabilities of the points in S are constant-far from 0, then the size of the k-LVD $\Psi_{\mathcal{T}}^S$ is $O(kn)$.*

Proof. Suppose the existence probabilities $\pi_{a_1}, \ldots, \pi_{a_n}$ are constant-far from 0. Lemma 7 shows that all the critical centers of S are $O(k)$-shallow. By further applying Lemma 6, the degree-sum of the critical centers is $O(kn)$. Finally, by Lemma 5, the size of Ψ_T^S is $O(kn)$.   □

To prove the bound for the average-case size requires more efforts. Let f be a *fixed* probability distribution function whose support is in $(0, 1]$ and μ be the supremum of the support of f. Define two constants $\mu_0 = \mu/(1 + \mu)$ and $\lambda = 1 - \int_{-\infty}^{\mu_0} f(x)dx$. Clearly, if X is a random variable drawn from f, then $\lambda = \Pr[X > \mu_0]$. Note that λ is always positive by definition. The following lemma clarifies the meaning of μ_0.

Lemma 8. *Suppose $\pi_{a_1}, \ldots, \pi_{a_n}$ are i.i.d. random variables drawn from f. For any center c of S, the event "c is critical" does **not** happen if there are k (distinct) points $a_{i_1}, \ldots, a_{i_k}$ in S closer to $\hat{c}$ than the c-involved points such that $\pi_{a_{i_1}}, \ldots, \pi_{a_{i_k}}$ are all greater than μ_0.*

Theorem 5. *The average-case size of Ψ_T^S is $O(kn)$, given that the existence probabilities of the points in S are i.i.d. random variables drawn from a **fixed** distribution.*

Proof. Suppose the existence probabilities $\pi_{a_1}, \ldots, \pi_{a_n}$ are drawn independently from f. Lemma 8 implies that, if c is a center of S with $dep_S(c) = d \geq k$, then

$$\Pr[c \text{ is critical}] \leq u_d = \sum_{i=0}^{k-1} \binom{d}{i} \lambda^i (1 - \lambda)^{d-i}.$$

Then by applying Lemma 5, we have

$$\mathbf{E}[|\Psi_T^S|] \leq \sum_c \Pr[c \text{ is critical}] \cdot deg(c) \leq \sum_{c \in H_k} deg(c) + \sum_{d=k+1}^{n-1} \sum_{c \in H_d} (u_{d-1} - u_d) deg(c),$$

where H_d is the set of the d-shallow centers of S. Observe that

$$u_{d-1} - u_d = \binom{d-1}{k-1} \lambda^k (1 - \lambda)^{d-k}.$$

Based on this and Lemma 6, we further have

$$\mathbf{E}[|\Psi_T^S|] \leq 8kn + 8n \sum_{d=k+1}^{n-1} \binom{d-1}{k-1} \lambda^k (1 - \lambda)^{d-k} d.$$

Note that

$$\sum_{d=k+1}^{n-1} \binom{d-1}{k-1} \lambda^k (1 - \lambda)^{d-k} d = k \left(\frac{\lambda}{1 - \lambda} \right)^k \sum_{d=k+1}^{n-1} \binom{d}{k} (1 - \lambda)^d.$$

By an induction argument on k, it is not difficult to see that

$$\sum_{d=k+1}^{n-1} \binom{d}{k} (1 - \lambda)^d < \sum_{d=k}^{\infty} \binom{d}{k} (1 - \lambda)^d = \frac{(1 - \lambda)^k}{\lambda^{k+1}}.$$

Finally, by combining the inequalities, $\mathbf{E}[|\Psi_T^S|] \leq 8kn + \frac{8kn}{\lambda} = O(kn)$.   □

3.2 Constructing LVD and answering queries

In this section, we show how to construct the k-LVD $\Psi_{\mathcal{T}}^{S}$ and use it to answer k-LNN queries. Let $e_1, \ldots, e_{t-1}$ be the edges of T. Assume each edge e_i has a specified "start point" s_i (which is one of its two endpoints) and the query point q is specified via a pair (i, δ), meaning the point on e_i with distance δ to s_i.

We first explain the data structure used for storing the k-LVD $\Psi_{\mathcal{T}}^{S}$ and answering queries. The LVD data structure is simple. First, it contains $|\Psi_{\mathcal{T}}^{S}|$ arrays (called *answer arrays*) each of which stores the k-LNN answer of one cell of $\Psi_{\mathcal{T}}^{S}$. This part takes $O(k|\Psi_{\mathcal{T}}^{S}|)$ space. In addition to that, we also need to record the structure of $\Psi_{\mathcal{T}}^{S}$. For each edge e_i of T, we use a sorted list L_i to store the "cell-decomposition" of e_i. Specifically, the intersection of each cell $C \in \Psi_{\mathcal{T}}^{S}$ and e_i is an "interval" (may be empty). These intervals are stored in L_i in the order they appear on e_i. Note that this part takes $O(t + |\Psi_{\mathcal{T}}^{S}|)$ space. Indeed, if an edge is decomposed into p pieces (intervals) by $\Psi_{\mathcal{T}}^{S}$, then it at least entirely contains $(p-2)$ cells of $\Psi_{\mathcal{T}}^{S}$ (so we can charge these $(p-2)$ pieces to the corresponding cells and the remaining two pieces to the edge). Therefore, the total space of the LVD data structure is $O(t + k|\Psi_{\mathcal{T}}^{S}|)$. To answer a query $q = (i, \delta)$, we first do a binary search in the list L_i to know which cell q locates in, and then use the answer array corresponding to the cell to output the k-LNN of q directly. The query time is clearly $O(\log |\Psi_{\mathcal{T}}^{S}| + k)$.

Next, we consider the construction of the LVD data structure. The first step of the construction is to compute all the centers of S and sort the centers in the interior of each edge e in the order they appear on e. We are able to get this done in $O(t + n^2 \log n)$ time (see [17]). After the centers are computed and sorted, we begin to construct the LVD data structure. Choose a vertex v of T. Starting at v, we do a walk in $\mathcal{T}$ along with the edges of T. The walk visits each edge of T exactly twice and finally goes back to v; see Fig. 4. During the walk, we maintain a (balanced) binary search tree for $NNP_x(a_1), \ldots, NNP_x(a_n)$ w.r.t. the current location x. By exploiting this BST, we can work out the cell-decomposition of each edge e_i (i.e., the sorted list L_i) at the first time we visit e_i in the walk. Specifically, we track the k-LNN when walking along with e_i, which can be obtained by retrieving the k largest elements from the BST. Whenever the k-LNN changes, a new cell of $\Psi_{\mathcal{T}}^{S}$ is found, so we need to create a new answer array to store the k-LNN information. Also, we need to update the sorted list L_i. In this way, after we go through e_i (for the first time), L_i is cor-

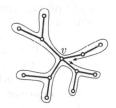

Fig. 4. A walk in tree visiting each edge exactly twice.

rectly computed. At the second time we visit e_i, we do nothing but maintain the binary search tree. When we finish the walk and go back to v, the construction of the LVD data structure is done. Clearly, in the process of the walk, we only need to maintain the binary search tree and retrieve the k-LNN when we arrive at (resp., leave from) a center of S from (resp., to) one of its branches. With a careful implementation and analysis (see [17]), we can complete the entire walk and hence the entire LVD structure in $O(t + n^2 \log n + n^2 k)$ time. Combined with the bounds in Section 3.1, we then have the following results.

Theorem 6. *Given a tree space $\mathcal{T}$ represented by a t-vertex weighted tree and a set S of n stochastic points in $\mathcal{T}$, one can construct in $O(t + n^2 \log n + n^2 k)$ time an LVD data structure to answer k-LNN queries in $O(\log n + k)$ time. The LVD data structure uses worst-case $O(t + kn^2)$ space and average-case $O(t + k^2 n)$ space. Furthermore, if the existence probabilities of the points in S are constant-far from 0, then the LVD data structure uses worst-case $O(t + k^2 n)$ space.*

References

1. Agarwal, P., Aronov, B., Har-Peled, S., Phillips, J., Yi, K., Zhang, W.: Nearest neighbor searching under uncertainty II. In: Proc. of the 32nd Sympos. on PODS. pp. 115–126. ACM (2013)
2. Agarwal, P., Cheng, S.W., Yi, K.: Range searching on uncertain data. ACM Transactions on Algorithms 8(4), 43 (2012)
3. Agarwal, P., Har-Peled, S., Suri, S., Yıldız, H., Zhang, W.: Convex hulls under uncertainty. In: Algorithms-ESA, pp. 37–48. Springer (2014)
4. Agarwal, P., Kumar, N., Sintos, S., Suri, S.: Range-max queries on uncertain data. In: Proc. of the 35th SIGMOD/PODS. pp. 465–476. ACM (2016)
5. Chen, J., Feng, L.: Efficient pruning algorithm for top-k ranking on dataset with value uncertainty. In: Proc. of the 22nd CIKM. pp. 2231–2236. ACM (2013)
6. Fakcharoenphol, J., Rao, S., Talwar, K.: Approximating metrics by tree metrics. ACM SIGACT News 35(2), 60–70 (2004)
7. Fink, M., Hershberger, J., Kumar, N., Suri, S.: Hyperplane separability and convexity of probabilistic point sets. In: Proc. of the 32nd SoCG. ACM (2016)
8. Ge, T., Zdonik, S., Madden, S.: Top-k queries on uncertain data: on score distribution and typical answers. In: Proc. of the 2009 SIGMOD. pp. 375–388. ACM (2009)
9. Huang, L., Li, J.: Approximating the expected values for combinatorial optimization problems over stochastic points. In: Intl. Colloquium on Automata, Languages, and Programming. pp. 910–921. Springer (2015)
10. Kamousi, P., Chan, T., Suri, S.: Stochastic minimum spanning trees in Euclidean spaces. In: Proc. of the 27th SoCG. pp. 65–74. ACM (2011)
11. Kamousi, P., Chan, T., Suri, S.: Closest pair and the post office problem for stochastic points. Computational Geometry 47(2), 214–223 (2014)
12. Kumar, N., Raichel, B., Suri, S., Verbeek, K.: Most likely Voronoi Diagrams in higher dimensions. In: LIPIcs-Leibniz International Proceedings in Informatics. vol. 65. Schloss Dagstuhl-Leibniz-Zentrum fuer Informatik (2016)
13. Löffler, M., van Kreveld, M.: Largest and smallest convex hulls for imprecise points. Algorithmica 56(2), 235–269 (2010)
14. Suri, S., Verbeek, K.: On the most likely Voronoi Diagram and nearest neighbor searching. In: ISAAC. pp. 338–350. Springer (2014)
15. Suri, S., Verbeek, K., Yıldız, H.: On the most likely convex hull of uncertain points. In: Algorithms-ESA, pp. 791–802. Springer (2013)
16. Xue, J., Li, Y., Janardan, R.: On the separability of stochastic geometric objects, with applications. In: Proc. of the 32nd SoCG. ACM (2016)
17. Xue, J., Li, Y.: Stochastic closest-pair problem and most-likely nearest-neighbor search in tree spaces. arXiv:1612.04890 (2016)

On the Expected Diameter, Width, and Complexity of a Stochastic Convex-hull

Jie Xue[⊠], Yuan Li, and Ravi Janardan

University of Minnesota — Twin Cities, Minneapolis MN 55455, USA
{xuexx193,lixx2100,janardan}@umn.edu

Abstract. We investigate several computational problems related to the stochastic convex hull (SCH). Given a stochastic dataset consisting of n points in $\mathbb{R}^d$ each of which has an existence probability, a SCH refers to the convex hull of a realization of the dataset, i.e., a random sample including each point with its existence probability. We are interested in computing certain expected statistics of a SCH, including diameter, width, and combinatorial complexity. For diameter, we establish the first deterministic 1.633-approximation algorithm with a time complexity polynomial in both n and d. For width, two approximation algorithms are provided: a deterministic $O(1)$-approximation running in $O(n^{d+1} \log n)$ time, and a fully polynomial-time randomized approximation scheme (FPRAS). For combinatorial complexity, we propose an exact $O(n^d)$-time algorithm. Our solutions exploit many geometric insights in Euclidean space, some of which might be of independent interest.

Keywords: uncertain data, expectation, diameter, width, combinatorial complexity

1 Introduction

The convex hull, which is one of the most fundamental structures in computational geometry, has a wide range of applications in various areas. Traditionally, the convex hull is studied on datasets whose information is known exactly. However, in many real-world applications, due to noise and limitation of devices, the data obtained may be imprecise or not totally reliable. In this situation, uncertain datasets (or stochastic datasets), in which the data points are allowed to have some uncertainty, can better model real data. In recent years, there have been several papers regarding the convex hull structure under uncertainty, known as *stochastic convex hull* (SCH) [3,10,11,13].

In this paper, we revisit several problems related to SCH under the well-known existential uncertainty model: each data point in the stochastic dataset has a certain (known) location in the space with an uncertain existence depicted by an associated (independent) existence probability. In real-world applications, the existence probabilities can be used to express the reliability or importance of the data points. Given a stochastic dataset $\mathcal{S}$ in $\mathbb{R}^d$ equipped with existential uncertainty, a SCH of $\mathcal{S}$ refers to the convex hull of a realization of $\mathcal{S}$, which can

© Springer International Publishing AG 2017
F. Ellen et al. (Eds.): WADS 2017, LNCS 10389, pp. 581–592, 2017.
DOI: 10.1007/978-3-319-62127-2_49

be regarded as a probabilistic polytope in $\mathbb{R}^d$. Our main focus is to compute the expected values of some basic statistics of a SCH, including diameter, width, and combinatorial complexity. (Formal definitions can be found in Sec. 1.1.) We give polynomial-time algorithms (both exact and approximate) for these problems.

Related work. Geometric computation on uncertain data has received considerable attentions in recent years. Many geometric problems have been studied under uncertainty (either existential uncertainty or locational uncertainty), e.g., nearest-neighbor search [1,12], minimum spanning trees [8], closest pair [5,9,16], range search [2], linear separability [4,14], dominance relation [15], etc. There have also been several papers concerning SCH [3,6,10,11,13]. We only summarize those that are strongly relevant to this paper. Li et al. [10] studied the expected computation of some basic statistics of a SCH in $\mathbb{R}^2$, e.g., area, perimeter, diameter (their results for diameter are summarized below), etc. The results in [10] are presented in a slightly different uncertainty model, but most of the algorithms also work under existential uncertainty. Huang et al. [6] studied ε-coresets of a stochastic dataset, which can be used to efficiently approximate the expected directional width of a SCH with respect to any given direction (see Sec. 1.1 for the definition of directional width). One should note that, although the diameter (resp., width) is defined as the largest (resp., smallest) directional width, the ε-coresets constructed in [6] cannot be used to approximate the expected diameter/width of a SCH, simply because the direction defining the diameter/width of a SCH varies from realization to realization. Specifically, the expected diameter of a SCH was investigated in some recent works. Huang and Li [5] provided an FPRAS for computing the expected farthest-pair distance of a stochastic dataset in a metric space (which works under both existential and locational uncertainty), which directly implies an FPRAS for computing the expected diameter of a SCH. However, an FPRAS can only obtain the desired approximation with high probability, and there seems no way to verify whether an answer obtained by the FPRAS is truly a good approximation. Li et al. [10] gave a deterministic $(2/\sqrt{3})$-approximation algorithm in $\mathbb{R}^2$, which is based on (exactly) computing the expected diameter of the stochastic smallest enclosing ball. Although [10] only considered the case in $\mathbb{R}^2$, the algorithm can be naturally extended to compute a $(\sqrt{2d}/\sqrt{d+1})$-approximation of the expected diameter of a SCH in $\mathbb{R}^d$. Nevertheless, the runtime of this algorithm grows exponentially as d increases, since computing the expected diameter of the stochastic smallest enclosing ball requires $n^{\Omega(d)}$ time [7]. The width and combinatorial complexity of a SCH have not yet been investigated previously, to our best knowledge.

Our results. The contributions of this paper are as follows:
- *Expected diameter.* We study the expected-diameter problem in $\mathbb{R}^d$ *without* assuming that d is a fixed. We establish an (n, d)-polynomial time (i.e., time polynomial in both the dataset-size n and the dimension d) 1.633-approximation algorithm for computing the expected diameter (Theorem 1).
- *Expected width.* We study the expected-width problem in $\mathbb{R}^d$ with a fixed dimension d. Two approximation algorithms are proposed for computing the expected width: a deterministic $O(1)$-approximation running in $O(n^{d+1} \log n)$

time (Theorem 2), and an FPRAS (Theorem 3).

• *Expected combinatorial complexity.* We study the expected-complexity problem in $\mathbb{R}^d$ with a fixed dimension d. We provide an exact algorithm for computing the expected combinatorial complexity of a SCH in $O(n^d)$ time.

Due to space limitations, some proofs and details are omitted in this paper. All missing proofs can be found in the full version [17].

1.1 Preliminaries

We give the formal definitions of some basic notions used in this paper. A *stochastic dataset* in $\mathbb{R}^d$ is a pair $\mathcal{S} = (S, \pi)$ where S is a set of points in $\mathbb{R}^d$ and $\pi : S \to (0, 1]$ specifies the existence probability of each point in S. A *realization* of $\mathcal{S}$ is a random sample $R \subseteq S$ where each point $a \in S$ is sampled with probability $\pi(a)$. For any subset $R \subseteq S$, we use $\Pr[R]$ to denote the probability that R occurs as a realization of $\mathcal{S}$, hence $\Pr[R] = \prod_{a \in R} \pi(a) \cdot \prod_{b \in S \setminus R}(1 - \pi(b))$. A *stochastic convex hull* (SCH) of $\mathcal{S}$ refers to the convex hull of a realization of $\mathcal{S}$, which can be regarded as a probabilistic polytope in $\mathbb{R}^d$.

Let P be a convex polytope in $\mathbb{R}^d$. The *combinatorial complexity* (or simply *complexity*) of P, denoted by $|P|$, is defined as the total number of the faces of P (the dimensions of the faces vary from 0 to $d - 1$). If $\mathbf{u}$ is a unit vector in $\mathbb{R}^d$, we define the *directional width* of P with respect to $\mathbf{u}$ as

$$\mathrm{wid}_{\mathbf{u}}(P) = \sup_{p,q \in P} (\langle \mathbf{u}, p \rangle - \langle \mathbf{u}, q \rangle),$$

where $\langle \cdot, \cdot \rangle$ denotes the inner product. Let U be the set of unit vectors in $\mathbb{R}^d$. Then the *diameter* of P is defined as $\mathrm{diam}(P) = \sup_{\mathbf{u} \in U} \mathrm{wid}_{\mathbf{u}}(P)$, and the *width* of P is defined as $\mathrm{wid}(P) = \inf_{\mathbf{u} \in U} \mathrm{wid}_{\mathbf{u}}(P)$. It is clear that the diameter of P is also the distance between the farthest-pair of points in P.

For two points $x = (x_1, \ldots, x_d)$ and $y = (y_1, \ldots, y_d)$ in $\mathbb{R}^d$, we define $x \prec y$ if the d-tuple $(x_1, \ldots, x_d)$ is smaller than the d-tuple $(y_1, \ldots, y_d)$ in lexicographic order. Then $\prec$ induces a (strict) total order on $\mathbb{R}^d$, called $\prec$-*order*.

The approximation algorithms in this paper use *relative* performance guarantees. Formally, a δ-approximation ($\delta \geq 1$) algorithm outputs an answer within the range $[res/\delta, res]$, where res is the exact answer of the problem.

2 Approximating the expected diameter

Given a stochastic dataset $\mathcal{S} = (S, \pi)$ in $\mathbb{R}^d$ (d is not assumed to be fixed) with $|S| = n$, we consider how to (approximately) compute the expected diameter of a SCH of $\mathcal{S}$, denoted by $\mathrm{diam}_{\mathcal{S}}$. Formally, $\mathrm{diam}_{\mathcal{S}} = \sum_{R \subseteq S} \Pr[R] \cdot \mathrm{diam}(\mathcal{CH}(R))$. Note that computing $\mathrm{diam}_{\mathcal{S}}$ exactly is #P-hard when d is not fixed [17].

2.1 The witness sequence

In order to approximate $\mathrm{diam}_{\mathcal{S}}$, we first introduce an important notion called *witness sequence*. Let P be a convex polytope in $\mathbb{R}^d$ with the vertex set V. For a point $x \in \mathbb{R}^d$, we define $\Phi_P(x)$ as the set of all points in P farthest from x. Formally, $\Phi_P(x) = \{y \in P : \operatorname{dist}(x,y) \geq \operatorname{dist}(x,y') \text{ for any } y' \in P\}$. Note that $\Phi_P(x) \subseteq V$, and in particular $\Phi_P(x)$ is finite. We have the following two observation about $\operatorname{diam}(P)$.

Lemma 1. *Let $x \in \mathbb{R}^d$ be a point. If there exist $p, q \in P$ such that $\operatorname{dist}(p,q) = \operatorname{diam}(P)$ and $\angle pxq = \theta > \pi/2$, then for any $y \in \Phi_P(x)$ and $z \in \Phi_P(y)$ we have*

$$\operatorname{dist}(y,z) \geq \operatorname{diam}(P)/(2\cos(\theta/4)).$$

Proof. Fixing $x \in \mathbb{R}^d$, and let $p, q \in P$ such that $\operatorname{dist}(p,q) = \operatorname{diam}(P)$ and $\angle pxq > \pi/2$. Also, let $y \in \Phi_P(x)$ be any point. Since $\operatorname{dist}(y,z) \geq \max\{\operatorname{dist}(y,p), \operatorname{dist}(y,q)\}$ for any $z \in \Phi_P(y)$, it suffices to show

$$\max\{\operatorname{dist}(y,p), \operatorname{dist}(y,q)\} \geq \operatorname{diam}(P)/(2\cos(\theta/4)).$$

Without loss of generality, we may assume $x = (0, \ldots, 0)$, $p = (\alpha, \beta, 0, \ldots, 0)$, $q = (\alpha, \gamma, 0, \ldots, 0)$, where $\alpha \geq 0$. Furthermore, we may also assume $\operatorname{dist}(x,y) = 1$, hence $\alpha^2 + \beta^2 \leq 1$ and $\alpha^2 + \gamma^2 \leq 1$. Since $\angle pxq > \pi/2$, we must have $\beta\gamma < 0$ (say $\beta > 0$ and $\gamma < 0$). We first claim that $\max\{\operatorname{dist}(y,p), \operatorname{dist}(y,q)\}$ is minimized when $y = (\sqrt{1 - (\beta + \gamma)^2/4}, \ (\beta + \gamma)/2, \ 0, \ldots, \ 0)$. Let y be the point with these coordinates, and $r = (r_1, \ldots, r_d)$ be another point satisfying $\operatorname{dist}(x,r) = 1$ (i.e., $\sum_{i=1}^d r_i^2 = 1$). First consider the case of $r_2 \leq (\beta + \gamma)/2$. In this case, we show $\operatorname{dist}(r,p) \geq \max\{\operatorname{dist}(y,p), \operatorname{dist}(y,q)\}$. Since $\operatorname{dist}(y,p) = \operatorname{dist}(y,q)$, it suffices to show $\operatorname{dist}(r,p) \geq \operatorname{dist}(y,p)$. We have $\operatorname{dist}^2(r,p) = 1 + \alpha^2 + \beta^2 - 2r_1\alpha - 2r_2\beta$ and $\operatorname{dist}^2(y,p) = 1 + \alpha^2 + \beta^2 - 2y_1\alpha - 2y_2\beta$, where y_1 and y_2 are the first two coordinates of y defined above. Now we only need to show $r_1\alpha + r_2\beta \leq y_1\alpha + y_2\beta$. Note that $r_1\alpha + r_2\beta \leq \alpha\sqrt{1 - r_2^2} + r_2\beta$ as $\alpha \geq 0$. Define vectors $\mathbf{v} = (\alpha, \beta)$, $\mathbf{u} = (\sqrt{1 - r_2^2}, r_2)$, $\mathbf{w} = (y_1, y_2)$. Since $\alpha \geq 0$, $y_1 > 0$, and $r_2 \leq y_2 < \beta$, the angle between $\mathbf{v}$ and $\mathbf{u}$ is greater than that between $\mathbf{v}$ and $\mathbf{w}$. Furthermore, $\|\mathbf{u}\|_2 = \|\mathbf{w}\|_2 = 1$. Therefore, $\alpha\sqrt{1 - r_2^2} + r_2\beta = \langle \mathbf{u}, \mathbf{v} \rangle \leq \langle \mathbf{w}, \mathbf{v} \rangle = y_1\alpha + y_2\beta$, which implies $r_1\alpha + r_2\beta \leq y_1\alpha + y_2\beta$. In the case $r_2 \geq (\beta + \gamma)/2$, symmetrically, we have $\operatorname{dist}(r,q) \geq \max\{\operatorname{dist}(y,p), \operatorname{dist}(y,q)\}$. Therefore, we know that $\max\{\operatorname{dist}(y,p), \operatorname{dist}(y,q)\}$ is minimized when y has the above coordinates. Moreover, in this situation, we have

$$\operatorname{dist}(y,p) = \operatorname{dist}(y,q) = \frac{\operatorname{dist}(p,q)}{2\sin(\angle pyq/2)} = \frac{\operatorname{diam}(P)}{2\sin(\angle pyq/2)}. \tag{1}$$

Next, we show that $\angle pyq \leq \pi - \theta/2$ where $\theta = \angle pxq$. Since $\operatorname{dist}(x,p) \leq \operatorname{dist}(x,y)$, $\angle xyp \leq \angle xpy$. Also, since $\operatorname{dist}(x,q) \leq \operatorname{dist}(x,y)$, $\angle xyq \leq \angle xqy$. It follows that $\angle pyq = \angle xyp + \angle xyq \leq \angle xpy + \angle xqy$. But $\angle pxq + \angle pyq + \angle xpy + \angle xqy = 2\pi$ and $\angle pxq = \theta$, which implies that $2\angle pyq \leq 2\pi - \theta$, as desired. Using Equation 1, we have that $\operatorname{dist}(y,p) \geq \operatorname{diam}(P)/(2\sin(\pi/2 - \theta/4))$, which completes the proof. $\square$

Lemma 2. *Let $v \in V$ be a vertex of P, and $u \in \Phi_P(v)$, $w \in \Phi_P(u)$ be two points. Suppose r is the ray with initial point u which goes through v, and x is the point on r which has distance $\operatorname{dist}(u,w)/2$ from u. Then if there exist $p, q \in P$ with $\operatorname{dist}(p,q) = \operatorname{diam}(P)$ and $\angle pxq = \theta$, we have*

$$\operatorname{dist}(u,w) \geq \min\left\{\operatorname{diam}(P), \frac{\operatorname{diam}(P)}{\sqrt{3}\sin(\theta/2)}\right\}.$$

Proof. Let B_v be the (closed) ball centered at u with radius $\mathrm{dist}(v, u)$, and B_u be the (closed) ball centered at u with radius $\mathrm{dist}(u, w)$. Then we have $P \subseteq B_u \cap B_v$, because $u \in \Phi_P(v)$ and $w \in \Phi_P(u)$. Now let r and x be the ray and the point defined in the lemma. Define v' as the point on r which has distance $\mathrm{dist}(u, w)$ from u, so x is the midpoint of the segment connecting v' and u. Set $B_{v'}$ to be the (closed) ball centered at v' with radius $\mathrm{dist}(u, w)$. Note that $B_v \subseteq B_{v'}$, since $\mathrm{rad}(B_{v'}) \geq \mathrm{rad}(B_v) + \mathrm{dist}(v, v')$ where $\mathrm{rad}(\cdot)$ denotes the radius of a ball. Therefore, $P \subseteq B_u \cap B_{v'}$. Next, we claim that $B_u \cap B_{v'} \subseteq B_x$, where B_x is the (closed) ball centered at x with radius $\sqrt{3} \cdot \mathrm{dist}(u, w)/2$. Suppose $y \in B_u \cap B_{v'}$ is a point, and assume $\mathrm{dist}(y, u) \geq \mathrm{dist}(y, v')$ without loss of generality (so $\angle yxu \geq \pi/2$). Define $\mu = \mathrm{dist}(u, x)$ and $\gamma = \mathrm{dist}(y, x)$. Then $\gamma = \mu \cdot \sin \angle yux / \sin \angle uyx$. Note that we have the restrictions $\angle yxu \geq \pi/2$ and $\mathrm{dist}(u, y) \leq \mathrm{dist}(u, v') = 2\mu$. Under these restrictions, it is easy to see that γ is maximized when $\mathrm{dist}(u, y) = 2\mu$ and $\angle yxu = \pi/2$. In this case, $\gamma = \sqrt{3}\mu = \mathrm{rad}(B_x)$. Consequently, $B_u \cap B_{v'} \subseteq B_x$, which in turn implies $P \subseteq B_x$. With this observation, we now show the inequality in the lemma. Let $p, q \in P \subseteq B_x$ be two points satisfying $\mathrm{dist}(p, q) = \mathrm{diam}(P)$ and $\angle pxq = \theta$. If $\mathrm{dist}(p, q) \leq \mathrm{dist}(u, w)$, we are done, so assume $\mathrm{dist}(p, q) > \mathrm{dist}(u, w)$. But both $\mathrm{dist}(x, p)$ and $\mathrm{dist}(x, q)$ are at most $\mathrm{rad}(B_x) = \sqrt{3} \cdot \mathrm{dist}(u, w)/2$. Therefore, θ is the largest angle of the triangle $\triangle pxy$. In this case, it is easy to see that $\mathrm{dist}(p, q)$ is maximized when $\mathrm{dist}(x, p) = \mathrm{dist}(x, q) = \mathrm{rad}(B_x)$. It follows that $\mathrm{dist}(p, q) \leq \sqrt{3} \sin(\theta/2) \cdot \mathrm{dist}(u, w)$, which completes the proof. $\square$

Lemma 2 tells us that for a vertex $v \in V$, the distance $\mathrm{dist}(u, w)$ where $u \in \Phi_P(v)$ and $w \in \Phi_P(u)$ gives a good approximation for $\mathrm{diam}(P)$ as long as there exists a pair $p, q \in P$ defining $\mathrm{diam}(P)$ with a "small" angle $\angle pxq$ (see the lemma for the definition of x). However, the approximation is not satisfactory when $\angle pxq$ is large. Fortunately, we have Lemma 1, which is helpful for this case. Indeed, in the case that $\angle pxq$ is large, if we further take $y \in \Phi_P(x)$ and $z \in \Phi_P(y)$, then Lemma 1 implies that $\mathrm{dist}(y, z)$ gives a good approximation for $\mathrm{diam}(P)$. Therefore, intuitively, by taking $\max\{\mathrm{dist}(u, w), \mathrm{dist}(y, z)\}$, we can well-approximate $\mathrm{diam}(P)$ no matter whether $\angle pxq$ is small or large. Formally, we conclude the following.

Corollary 1. *Let v, u, w, x be the points defined in Lemma 2. Also, let $y \in \Phi_P(x)$ and $z \in \Phi_P(y)$ be any two points. Set $\delta = 2\sqrt{2}/\sqrt{3}$. Then we have*

$$\mathrm{diam}(P)/\delta \leq \max\{\mathrm{dist}(u, w), \mathrm{dist}(y, z)\} \leq \mathrm{diam}(P).$$

Proof. It is clear that $\max\{\mathrm{dist}(u, w), \mathrm{dist}(y, z)\} \leq \mathrm{diam}(P)$, as $u, w, y, z \in P$. Let $p, q \in P$ such that $\mathrm{dist}(p, q) = \mathrm{diam}(P)$, and $\theta = \angle pxq$. If $\theta \leq \pi/2$, then Lemma 2 implies $\mathrm{dist}(u, w) \geq \mathrm{diam}(P)/(\sqrt{3}/\sqrt{2})$. So assume $\theta > \pi/2$. By Lemma 1, we have $\mathrm{dist}(y, z) \geq \mathrm{diam}(P)/(2\cos(\theta/4))$. Then by Lemma 2, we have $\mathrm{dist}(u, w) \geq \mathrm{diam}(P)/(\sqrt{3} \sin(\theta/2))$. Therefore,

$$\max\{\mathrm{dist}(u, w), \mathrm{dist}(y, z)\} \geq \frac{\mathrm{diam}(P)}{\min\{2\cos(\theta/4), \sqrt{3} \sin(\theta/2)\}}.$$

Note that for $\theta \in (\pi/2, \pi]$, $2\cos(\theta/4)$ is monotonically decreasing and $\sqrt{3}\sin(\theta/2)$ is monotonically increasing. Thus, the right-hand side of the above inequality is minimized when $2\cos(\theta/4) = \sqrt{3}\sin(\theta/2)$. We have this equality when $\sin(\theta/4) = 1/\sqrt{3}$. In this case, the r.h.s. is equal to $\mathrm{diam}(P)/\delta$. □

With the five points v, u, w, y, z (which are in fact the vertices of P) in hand, Corollary 1 allows us to approximate the diameter of P within a factor of $\delta = 2\sqrt{2}/\sqrt{3} \approx 1.633$. However, the choice of v, u, w, y, z is not unique in our above construction for a given P. For later use, we need to make it unique, which can be done by considering $\prec$-order (see Sec. 1.1). We define $v \in V$ as the largest vertex of P under $\prec$-order. Also, we require $u \in \Phi_P(v)$, $w \in \Phi_P(u)$, $y \in \Phi_P(x)$, $z \in \Phi_P(y)$ to be the largest under $\prec$-order. In this way, we obtain a uniquely defined 5-tuple (v, u, w, y, z) for the polytope P. We call this 5-tuple the *witness sequence* of P, denoted by $\mathrm{wit}(P)$. For a 5-tuple $\psi = (x_1, \ldots, x_5)$ of points in $\mathbb{R}^d$, define $\Lambda(\psi) = \max\{\mathrm{dist}(x_2, x_3), \mathrm{dist}(x_4, x_5)\}$. Then Corollary 1 implies

$$\mathrm{diam}(P)/\delta \le \Lambda(\mathrm{wit}(P)) \le \mathrm{diam}(P) \tag{2}$$

for any convex polytope P in $\mathbb{R}^d$.

2.2 An (n, d)-polynomial-time approximation algorithm

We now use the notion of witness sequence defined above to establish our approximation algorithm for computing $\mathrm{diam}_{\mathcal{S}}$. Given the stochastic dataset $\mathcal{S} = (S, \pi)$, we first do a preprocessing to sort all the points in S in $\prec$-order and compute the pair-wise distances of the points in S. This preprocessing can be done in $O(dn^2)$ time. To approximate $\mathrm{diam}_{\mathcal{S}}$, we define

$$\mathrm{diam}_{\mathcal{S}}^* = \sum_{R \subseteq S} \Pr[R] \cdot \Lambda(\mathrm{wit}(\mathcal{CH}(R))).$$

Inequality 2 implies $\mathrm{diam}_{\mathcal{S}}/\delta \le \mathrm{diam}_{\mathcal{S}}^* \le \mathrm{diam}_{\mathcal{S}}$. Thus, it now suffices to compute $\mathrm{diam}_{\mathcal{S}}^*$. Computing $\mathrm{diam}_{\mathcal{S}}^*$ by directly using the above formula takes exponential time, as S has 2^n subsets. However, since for any $R \subseteq S$ the witness sequence $\mathrm{wit}(\mathcal{CH}(R))$ must be a 5-tuple of points in S, we can also write

$$\mathrm{diam}_{\mathcal{S}}^* = \sum_{\psi \in \Psi_S} \Pr[\psi] \cdot \Lambda(\psi), \tag{3}$$

where Ψ_S is the set of all 5-tuples of points in S and $\Pr[\psi]$ is the probability that the witness sequence of a SCH of $\mathcal{S}$ is ψ. Note that $|\Psi_S| = O(n^5)$. Thus, we can efficiently compute $\mathrm{diam}_{\mathcal{S}}^*$ as long as $\Pr[\psi]$ and $\Lambda(\psi)$ can be computed efficiently for every $\psi \in \Psi_S$. Clearly, $\Lambda(\psi)$ can be directly computed in constant time (after our preprocessing). To compute $\Pr[\psi]$, suppose $\psi = (p_1, \ldots, p_5) \in \Psi_S$. It is easy to check that if $p_1 = p_2$, then either $\Pr[\psi] = 0$ or $\Lambda(\psi) = 0$. So we may assume $p_1 \ne p_2$. From the definition of witness sequence, we directly obtain the following criterion for checking if ψ is the witness sequence of a SCH of $\mathcal{S}$. We write $a \prec_b c$ for $a, b, c \in \mathbb{R}^d$ if $\mathrm{dist}(a, b) < \mathrm{dist}(c, b)$, or $\mathrm{dist}(a, b) = \mathrm{dist}(c, b)$ and $a \prec c$.

Lemma 3. *Let $\psi = (p_1, \ldots, p_5) \in \Psi_S$ with $p_1 \ne p_2$. Suppose r is the ray with initial point p_2 which goes through p_1, and x is the point on r which has distance* dist$(p_2, p_3)/2$ *from* p_2. *For a realization* R *of* S, *we have* $\psi = $ wit$(\mathcal{CH}(R))$ *iff* (1) R *contains* $p_1, \ldots, p_5$, *and* (2) R *does not contain any point* $a \in S$ *satisfying* $p_1 \prec a$ *or* $p_2 \prec_{p_1} a$ *or* $p_3 \prec_{p_2} a$ *or* $p_4 \prec_x a$ *or* $p_5 \prec_{p_4} a$.

By Lemma 3, it is quite easy to compute $\Pr[\psi]$ in linear time, just by multiplying the existence probabilities of the points in ψ and the non-existence probabilities of all the points which should not be included in R (according to the condition (2) in the lemma). Using Equation 5, we obtain an (n, d)-polynomial-time algorithm to compute diam$^*_S$. This algorithm runs in $O(n^6 + dn^2)$ time. But it is fairly easy to improve the runtime to $O(n^5 \log n + dn^2)$ by considering the 5-tuples in Ψ_S in a proper order (see [17] for details).

Theorem 1. *One can 1.633-approximate* diam$_S$ *in* (n, d)-*polynomial time. Specifically, the approximation can be done in* $O(n^5 \log n + dn^2)$ *time.*

Interestingly, our witness-sequence technique also gives an $O(dn)$-time 1.633-approximation algorithm for computing the diameter of the convex hull of a (non-stochastic) point-set S in $\mathbb{R}^d$, because wit$(\mathcal{CH}(S))$ can be computed in $O(dn)$ time. To our best knowledge, there has not been any linear-time algorithm which can achieve such an approximation factor when d is not fixed.

3   Approximating the expected width

Given a stochastic dataset $S = (S, \pi)$ in $\mathbb{R}^d$ (d is fixed) with $|S| = n$, we consider how to (approximately) compute the expected width of a SCH of S, denoted by wid$_S$. Formally, wid$_S = \sum_{R \subseteq S} \Pr[R] \cdot$ wid$(\mathcal{CH}(R))$.

3.1   The witness simplex

Recall that when solving the expected-diameter problem, we developed the notion of witness sequence, which well-captures the diameter of a polytope and satisfies (1) the total number of the possible witness sequences of a SCH is polynomial (though there are exponentially many realizations), (2) the probability of a sequence being the witness sequence of a SCH can be easily computed. We apply this basic idea again to the expected-width problem. To this end, we have to design some good "witness object" for width, which satisfies the above two conditions. The witness object to be defined is called *witness simplex*.

Let P be a convex polytope in $\mathbb{R}^d$ with wid$(P) > 0$, and V be the vertex set of P. We choose $d + 1$ vertices $v_0, \ldots, v_d \in V$ of P inductively as follows. Define $v_0 \in V$ as the largest vertex of P under $\prec$-order. Suppose $v_0, \ldots, v_i$ are already defined. Let E_i be the (unique) i-dim hyperplane in $\mathbb{R}^d$ through $v_0, \ldots, v_i$ (or the i-dim linear subspace of $\mathbb{R}^d$ spanned by $v_0, \ldots, v_i$). We then define $v_{i+1} \in V$ as the vertex of P which has the maximum distance to E_i, i.e., $v_{i+1} = \arg\max_{v \in V} $ dist(v, E_i). If there exist multiple vertices having maximum distance to E_i, we choose the largest one under $\prec$-order to be v_{i+1}. In this way, we obtain the vertices $v_0, \ldots, v_d$. The *witness simplex* Δ_P of P is defined as the

d-simplex with vertices $v_0, \ldots, v_d$. The (ordered) sequence $(v_0, \ldots, v_d)$ is said to be the *vertex list* of Δ_P. Note that the vertex list is determined by only Δ_P and independent of P. In other words, if we only know Δ_P without knowing the original polytope P, we can still recover the vertex list of Δ_P, just by ordering the $d + 1$ vertices of Δ_P into a sequence $(v_0, \ldots, v_d)$ such that v_0 is the largest under $\prec$-order, and each v_{i+1} is the one having the maximum distance to E_i (the linear subspace spanned by $v_0, \ldots, v_i$). A useful geometric property of the witness simplex Δ_P is that it well-captures the width of P.

Lemma 4. *Let P be a convex polytope in $\mathbb{R}^d$ with $\mathrm{wid}(P) > 0$, then we have $\mathrm{wid}(\Delta_P) = \Theta(\mathrm{wid}(P))$. The constant hidden in $\Theta(\cdot)$ could be exponential in d.*

Proof intuition. Due to space limitation, here we only give some intuition for proving the lemma. A detailed proof can be found in [17]. Note that $\mathrm{wid}(\Delta_P) \leq \mathrm{wid}(P)$ since $\Delta_P \subseteq P$. To see $\mathrm{wid}(\Delta_P) = \Omega(\mathrm{wid}(P))$, let $(v_0, \ldots, v_d)$ be the vertex list of Δ_P. Without loss of generality, we may assume that v_i is in the linear subspace of $\mathbb{R}^d$ spanned by the $x_1, \ldots, x_i$ axes. We only need to show $\mathrm{wid}_{\mathbf{u}}(\Delta_P) = \Omega(\mathrm{wid}_{\mathbf{u}}(P))$ for any unit vector $\mathbf{u} = (u_1, \ldots, u_d)$. If $|u_1| = \Omega(\mathrm{wid}_{\mathbf{u}}(P)/\mathrm{diam}(P))$, then $\mathrm{wid}_{\mathbf{u}}(\Delta_P) \geq |\langle \mathbf{u}, v_0 \rangle - \langle \mathbf{u}, v_1 \rangle| \geq |u_1| \cdot \mathrm{dist}(v_0, v_1)$. But $|u_1| \cdot \mathrm{dist}(v_0, v_1) \geq |u_1| \cdot \mathrm{diam}(P)/2 = \Omega(\mathrm{wid}_{\mathbf{u}}(P))$. On the other hand, $|u_1|$ is very "small", so essentially we have $\mathrm{wid}_{\mathbf{u}}(\Delta_P) \approx \mathrm{wid}_{\mathbf{u}'}(\Delta_P)$ and $\mathrm{wid}_{\mathbf{u}}(P) \approx \mathrm{wid}_{\mathbf{u}'}(P)$ where $\mathbf{u}' = (0, u_2, \ldots, u_d)$. Then it suffices to have $\mathrm{wid}_{\mathbf{u}'}(\Delta_P) = \Omega(\mathrm{wid}_{\mathbf{u}'}(P))$, which can be shown using an induction argument due to our recursive construction of the witness simplex Δ_P. □

3.2   An $O(1)$-approximation algorithm

With the notion of witness simplex in hand, we propose a $O(1)$-approximation algorithm for computing wid_S. The basic idea is similar to what we use for approximating diam_S. We define

$$\mathrm{wid}_S^* = \sum_{R \subseteq S} \Pr[R] \cdot \mathrm{wid}(\Delta_{\mathcal{CH}(R)}),$$

Lemma 4 implies $\mathrm{wid}_S^* = \Theta(\mathrm{wid}_S)$. Thus, in order to approximate wid_S within a constant factor, it suffices to compute wid_S^*. To compute wid_S^* by directly using the above formula takes exponential time, as S has 2^n subsets. However, since $\Delta_{\mathcal{CH}(R)}$ must be a d-simplex with vertices in S, wid_S^* can also be written as

$$\mathrm{wid}_S^* = \sum_{\Delta \in \Gamma_S^d} \Pr[\Delta] \cdot \mathrm{wid}(\Delta), \tag{4}$$

where Γ_S^d is the set of all d-simplices in $\mathbb{R}^d$ whose vertices are (distinct) points in S and $\Pr[\Delta]$ is the probability that the witness simplex of a SCH of S is Δ. Note that $|\Gamma_S^d| = O(n^{d+1})$, which is polynomial. So the above formula allows us to compute wid_S^* in polynomial time, as long as we are able to compute $\Pr[\Delta]$ efficiently for each $\Delta \in \Gamma_S^d$. Fixing $\Delta \in \Gamma_S^d$, we now investigate how to compute $\Pr[\Delta]$. As argued before, we can recover the vertex list $(v_0, \ldots, v_d)$ of Δ. By the construction of Δ, $v_0, \ldots, v_d$ are points in S. For $i \in \{0, \ldots, d-1\}$, we denote by E_i the i-dim hyperplane in $\mathbb{R}^d$ through $v_0, \ldots, v_i$. From the definition of witness simplex, we directly obtain the following criterion for checking if Δ is the witness simplex of a SCH of S. For a hyperplane H (of any dimension) in $\mathbb{R}^d$ and two points $a, b \in \mathbb{R}^d$, we write $a \prec_H b$ if $\text{dist}(a, H) < \text{dist}(b, H)$, or $\text{dist}(a, H) = \text{dist}(b, H)$ and $a \prec b$.

Lemma 5. *For a realization R of S, Δ is the witness simplex of $\mathcal{CH}(R)$ (i.e., $\Delta = \Delta_{\mathcal{CH}(R)}$) iff (1) R contains $v_0, \ldots, v_d$, and (2) R does not contain any point $a \in S$ satisfying $v_0 \prec a$ or $v_{i+1} \prec_{E_i} a$ for some $i \in \{0, \ldots, d-1\}$.*

Using the above lemma, we can straightforwardly compute $\Pr[\Delta]$ in linear time, just by multiplying the existence probabilities of $v_0, \ldots, v_d$ and the non-existence probabilities of all $a \in S$ which should not be included in R (according to the condition (2) in the lemma). Therefore, we obtain an $O(n^{d+2})$-time algorithm for computing wid_S^*. It is easy to improve the runtime to $O(n^{d+1} \log n)$ by considering the simplices in Γ_S^d in a proper order (see [17] for details).

Theorem 2. *One can $O(1)$-approximate wid_S in $O(n^{d+1} \log n)$ time. The constant approximation factor could be exponential in d.*

3.3 A fully polynomial-time randomized approximation scheme

In this section, we develop a fully polynomial-time randomized approximation scheme (FPRAS) for computing wid_S. An FPRAS should take S and a real number $\varepsilon > 0$ as input, and output a $(1+\varepsilon)$ approximation of wid_S in time polynomial in the size of S and $1/\varepsilon$ with probability at least $2/3$.

We first introduce some notations. As defined in the preceding section, Γ_S^d is the set of all d-simplices in $\mathbb{R}^d$ whose vertices are (distinct) points in S, and for each $\Delta \in \Gamma_S^d$ the notation $\Pr[\Delta]$ denotes the probability that the witness simplex of a SCH of S is Δ. Let R be a realization of S and $\Delta \in \Gamma_S^d$ be a simplex. From Lemma 5, we know that $\Delta = \Delta_{\mathcal{CH}(R)}$ iff R contains the vertices of Δ but does not contain some other points in S according to (2) in the lemma. We now use V_Δ to denote the set of the vertices of Δ, X_Δ to denote the set of the points in S that R must not contain if $\Delta = \Delta_{\mathcal{CH}(R)}$. Let $F_\Delta = S \setminus (V_\Delta \cup X_\Delta)$, which is the set of the points in S whose presence/absence in R does not influence whether $\Delta = \Delta_{\mathcal{CH}(R)}$. Define $\mathcal{F}_\Delta$ as the sub-dataset of S with the point-set F_Δ. Our FPRAS works as follows. First, for each $\Delta \in \Gamma_S^d$, we randomly generate $m = \gamma \log n/\varepsilon^2$ realizations of $\mathcal{F}_\Delta$, where γ is a large enough constant to be determined. Let $R_1^\Delta, \ldots, R_m^\Delta$ be the generated realizations of $\mathcal{F}_\Delta$, and set $T_i^\Delta = R_i^\Delta \cup V_\Delta$. Note that the witness simplex of $\mathcal{CH}(T_i^\Delta)$ is Δ by Lemma 5. We then compute

$$\text{wid}_S' = \sum_{\Delta \in \Gamma_S^d} \Pr[\Delta] \cdot \left(\sum_{i=1}^m \frac{\text{wid}(\mathcal{CH}(T_i^\Delta))}{m} \right), \qquad (5)$$

and output wid_S' as the approximation of wid_S.

Next, we discuss the choice of the constant γ and verify the correctness of our FPRAS. By Lemma 4, we can find positive constants k_1, k_2 such that

$k_1 \cdot \mathrm{wid}(\Delta_P) \leq \mathrm{wid}(P) \leq k_2 \cdot \mathrm{wid}(\Delta_P)$ for any convex polytope P in $\mathbb{R}^d$ with $\mathrm{wid}(P) > 0$. We set $\gamma = d(k_2/k_1)^2$. Using Hoeffding's inequality, it is easy to prove the following lemma, which guarantees the correctness of our FPRAS.

Lemma 6. $(1 - \varepsilon)\mathrm{wid}_{\mathcal{S}} \leq \mathrm{wid}'_{\mathcal{S}} \leq (1 + \varepsilon)\mathrm{wid}_{\mathcal{S}}$ with probability at least $2/3$.

Theorem 3. There exists an FPRAS for computing $\mathrm{wid}_{\mathcal{S}}$.

4   Computing the expected combinatorial complexity

Given a stochastic dataset $\mathcal{S} = (S, \pi)$ in $\mathbb{R}^d$ (d is fixed) with $|S| = n$, our goal in this section is to compute the expected complexity of a SCH of $\mathcal{S}$, denoted by $\mathrm{comp}_{\mathcal{S}}$. Formally, we have $\mathrm{comp}_{\mathcal{S}} = \sum_{R \subseteq S} |\mathcal{CH}(R)|$.

4.1   Reduction to SCH membership probability queries

Given a stochastic dataset $\mathcal{T}$ in $\mathbb{R}^d$ and a query point $q \in \mathbb{R}^d$, the SCH membership probability (of q with respect to $\mathcal{T}$) refers to the probability that q lies in a SCH of $\mathcal{T}$, which we denote by $\mathrm{mem}_{\mathcal{T}}(q)$. It is known that $\mathrm{mem}_{\mathcal{T}}(q)$ can be computed in $O(m^{d-1})$ time for $d \geq 3$ [4,14] and $O(m \log m)$ time for $d \in \{1, 2\}$ [3], where m is the number of the stochastic points in $\mathcal{T}$.

We reduce our problem of computing $\mathrm{comp}_{\mathcal{S}}$ to SCH membership probability queries. Let R be a realization of $\mathcal{S}$. It is clear that the faces of $\mathcal{CH}(R)$ must be simplices with vertices in S. Therefore, we can rewrite the formula for $\mathrm{comp}_{\mathcal{S}}$ as

$$\mathrm{comp}_{\mathcal{S}} = \sum_{R \subseteq S} \Pr[R] \cdot \left(\sum_{\Delta \in \Gamma_{\mathcal{S}}} \sigma(R, \Delta) \right) = \sum_{\Delta \in \Gamma_{\mathcal{S}}} F_{\Delta}, \tag{6}$$

where $\Gamma_{\mathcal{S}}$ is the set of all simplices (of dimension less than d) with vertices in S, σ is a indicating function such that $\sigma(R, \Delta) = 1$ if Δ is a face of $\mathcal{CH}(R)$ and $\sigma(R, \Delta) = 0$ otherwise, F_{Δ} is the probability that Δ is a face of a SCH of $\mathcal{S}$. We now show that for each $\Delta \in \Gamma_{\mathcal{S}}$, the computation of F_{Δ} can be reduced to a SCH membership probability query. Suppose Y is a set of m ($m \geq d+1$) points in $\mathbb{R}^d$ in general position. Let $y_0, \ldots, y_k \in Y$ be $k+1$ points where $0 \leq k \leq d-1$, and Δ be the k-simplex with vertices $y_0, \ldots, y_k$. Define vectors $\mathbf{u}_i = y_i - y_0$ for $i \in \{1, \ldots, k\}$. By the general position assumption, $\mathbf{u}_1, \ldots, \mathbf{u}_k$ generate a k-dim linear subspace H of $\mathbb{R}^d$. Set H^* to be the orthogonal complement of H in $\mathbb{R}^d$, which is by definition the $(d - k)$-dim linear subspace of $\mathbb{R}^d$ orthogonal to H. We then orthogonally project the points in Y to H^*, and denote the set of the projection images by Y^*. Note that $y_0, \ldots, y_k$ are clearly projected to the same point in H^*, say $\hat{y}$. We then have the following geometric observation.

Lemma 7. Δ is a face of $\mathcal{CH}(Y)$ iff $\hat{y}$ is a vertex of $\mathcal{CH}(Y^*)$ in H^*.

The above lemma allows us to reduce the computation of F_{Δ} for any $\Delta \in \Gamma_{\mathcal{S}}$ to a SCH membership query as follows. For each $i \in \{0, \ldots, d - 1\}$, let $\Gamma_{\mathcal{S}}^i \subseteq \Gamma_{\mathcal{S}}$ be the subset consisting of all i-simplices in $\Gamma_{\mathcal{S}}$ (then $\Gamma_{\mathcal{S}} = \bigcup_{i=0}^{d-1} \Gamma_{\mathcal{S}}^i$).

Suppose $\Delta \in \Gamma_S^k$ is a k-simplex with vertices $v_0, \ldots, v_k \in S$. As before, we define vectors $u_i = v_i - v_0$ for $i \in \{1, \ldots, k\}$. Then $u_1, \ldots, u_k$ generate a k-dim linear subspace H of $\mathbb{R}^d$, and set H^* to be the orthogonal complement of H in $\mathbb{R}^d$. Let $\rho : \mathbb{R}^d \to H^*$ be the orthogonal projection map. We define a multi-set $S' = \{\rho(a) : a \in S \setminus \{v_0, \ldots, v_k\}\}$ of points in H^*, which gives us a stochastic dataset $\mathcal{S}' = (S', \pi')$ in H^* where $\pi'(\rho(a)) = \pi(a)$. Set $q = \rho(v_0) = \cdots = \rho(v_k)$.

Corollary 2. $F_\Delta = \prod_{i=0}^{k} \pi(v_i) \cdot (1 - \mathrm{mem}_{\mathcal{S}'}(q))$.

Proof. Let R be a realization of $\mathcal{S}$. If Δ is a face of $\mathcal{CH}(R)$, then $v_0, \ldots, v_k$ must be contained in R. Furthermore, by Lemma 7, q must be a vertex of the projection image of $\mathcal{CH}(R)$ in H^*. By the general position assumption, this is equivalent to saying that q is outside the projection image of $\mathcal{CH}(R \setminus \{v_0, \ldots, v_k\})$. Conversely, if $v_0, \ldots, v_k$ are contained in R and q is outside the projection image of $\mathcal{CH}(R \setminus \{v_0, \ldots, v_k\})$, then Δ is a face of $\mathcal{CH}(R)$ by Lemma 7. The probability that R contains $v_0, \ldots, v_k$ is $\prod_{i=0}^{k} \pi(v_i)$, and the probability that q is outside the projection image of $\mathcal{CH}(R \setminus \{v_0, \ldots, v_k\})$ is $1 - \mathrm{mem}_{\mathcal{S}'}(q)$. These two events are clearly independent. Therefore, we have the formula in the corollary. $\square$

Since H^* is linearly homeomorphic to $\mathbb{R}^{d-k}$, computing $\mathrm{mem}_{\mathcal{S}'}(q)$ is nothing but answering a SCH membership probability query in $\mathbb{R}^{d-k}$. Therefore, using the algorithms for answering SCH membership probability queries [4,14], F_Δ can be computed in $O(n^{d-k-1})$ time if $k \in \{0, \ldots, d-3\}$. Note that $|\Gamma_S^k| = O(n^{k+1})$, so we can compute the sum $\sum_{i=0}^{d-3} \sum_{\Delta \in \Gamma_S^i} F_\Delta$ in $O(n^d)$ time. In order to further compute comp_S by Equation 6, we now only need to compute $\sum_{\Delta \in \Gamma_S^{d-1}} F_\Delta$ and $\sum_{\Delta \in \Gamma_S^{d-2}} F_\Delta$. But answering SCH membership probability queries in $\mathbb{R}^1$ and $\mathbb{R}^2$ requires $O(m \log m)$ time [3] (where m is the size of the given stochastic dataset). Thus, if we directly plug in the algorithm in [3], the computation task cannot be done in $O(n^d)$ time. The next section discusses how to handle this issue.

4.2   Handling $k = d - 2$ and $k = d - 1$

Set $\lambda_1 = \sum_{\Delta \in \Gamma_S^{d-1}} F_\Delta$ and $\lambda_2 = \sum_{\Delta \in \Gamma_S^{d-2}} F_\Delta$. Fix a point $o \in \mathbb{R}^d$ such that $S \cup \{o\}$ is in general position. For every hyperplane E with $o \notin E$, we denote by E^+ the connected component of $\mathbb{R}^d \setminus E$ containing o, and by E^- the other one. Define the S-*statistic* of E as a 3-tuple $\mathrm{stat}_S(E) = (p^+, p^-, A)$ where $p^+ = \prod_{a \in S \cap E^+} (1 - \pi(a))$, $p^- = \prod_{a \in S \cap E^-} (1 - \pi(a))$, $A = S \cap E$. Let $\mathcal{E}$ be the collection of the hyperplanes in $\mathbb{R}^d$ which go through exactly d points in S. Since $S \cup \{o\}$ is in general position, $\mathrm{stat}(E)$ is defined for every $E \in \mathcal{E}$. We say an algorithm computes the S-statistics for $\mathcal{E}$ if it outputs $\mathrm{stat}_S(E)$ for all $E \in \mathcal{E}$.

Lemma 8. *One can compute λ_1 and λ_2 in $O(t(n))$ time, provided an algorithm computing the S-statistics for $\mathcal{E}$ in $O(t(n))$ time.*

Proof sketch. We use the provided algorithm to compute the S-statistics for $\mathcal{E}$ in $O(t(n))$ time. With the statistics in hand, we can easily compute λ_1 in $O(n^d)$ time, because F_Δ for each $\Delta \in \Gamma_S^{d-1}$ can be computed in constant time via stat(E) (using Corollary 2), where $E \in \mathcal{E}$ is the hyperplane through the d vertices of Δ. To compute λ_2 in $O(n^d)$ time, we observe that as long as we have the statistics, the SCH membership query needed in Corollary 2 to compute F_Δ for each $\Delta \in \Gamma_S^{d-2}$ can be answered in linear time by using the witness-edge method in [3]. Note that $t(n) = \Omega(n^d)$, so the computation of λ_1 and λ_2 is done in $O(t(n))$ time. See [17] for a detailed proof. $\square$

It is implicitly known that one can compute the $\mathcal{S}$-statistics for $\mathcal{E}$ in $O(n^d)$ time [4,14]. We explicitly describe the algorithm in the full version [17]. With this algorithm in hand, Lemma 8 allows us to compute λ_1 and λ_2 in $O(n^d)$ time. Therefore, we can finally conclude the following.

Theorem 4. *One can compute the exact value of* $\mathrm{comp}_\mathcal{S}$ *in* $O(n^d)$ *time.*

References

1. Agarwal, P.K., Aronov, B., Har-Peled, S., Phillips, J.M., Yi, K., Zhang, W.: Nearest neighbor searching under uncertainty II. In: Proc. of the 32nd PODS. ACM (2013)
2. Agarwal, P.K., Cheng, S.W., Yi, K.: Range searching on uncertain data. ACM Transactions on Algorithms (TALG) 8(4), 43 (2012)
3. Agarwal, P.K., Har-Peled, S., Suri, S., Yıldız, H., Zhang, W.: Convex hulls under uncertainty. In: Algorithms-ESA 2014, pp. 37–48. Springer (2014)
4. Fink, M., Hershberger, J., Kumar, N., Suri, S.: Hyperplane separability and convexity of probabilistic point sets. In: Proc. of the 32nd SoCG. ACM (2016)
5. Huang, L., Li, J.: Approximating the expected values for combinatorial optimization problems over stochastic points. In: International Colloquium on Automata, Languages, and Programming. pp. 910–921. Springer (2015)
6. Huang, L., Li, J., Phillips, J.M., Wang, H.: ϵ-kernel coresets for stochastic points. arXiv preprint arXiv:1411.0194 (2014)
7. Jørgensen, A., Löffler, M., Phillips, J.: Geometric computations on indecisive points. In: WADS. pp. 536–547. Springer (2011)
8. Kamousi, P., Chan, T., Suri, S.: Stochastic minimum spanning trees in euclidean spaces. In: Proc. of the 27th SoCG. pp. 65–74. ACM (2011)
9. Kamousi, P., Chan, T.M., Suri, S.: Closest pair and the post office problem for stochastic points. Computational Geometry 47(2), 214–223 (2014)
10. Li, C., Fan, C., Luo, J., Zhong, F., Zhu, B.: Expected computations on color spanning sets. Journal of Combinatorial Optimization 29(3), 589–604 (2015)
11. Löffler, M., van Kreveld, M.: Largest and smallest convex hulls for imprecise points. Algorithmica 56(2), 235–269 (2010)
12. Suri, S., Verbeek, K.: On the most likely Voronoi Diagram and nearest neighbor searching. In: ISAAC. pp. 338–350. Springer (2014)
13. Suri, S., Verbeek, K., Yıldız, H.: On the most likely convex hull of uncertain points. In: Algorithms–ESA 2013, pp. 791–802. Springer (2013)
14. Xue, J., Li, Y., Janardan, R.: On the separability of stochastic geometric objects, with applications. In: Proc. of the 32nd SoCG. ACM (2016)
15. Xue, J., Li, Y.: Colored stochastic dominance problems. arXiv preprint arXiv:1612.06954 (2016)
16. Xue, J., Li, Y.: Stochastic closest-pair problem and most-likely nearest-neighbor search in tree spaces. arXiv preprint arXiv:1612.04890 (2016)
17. Xue, J., Li, Y., Janardan, R.: On the expected diameter, width, and complexity of a stochastic convex-hull. arXiv preprint arXiv:1704.07028 (2017)

Author Index

© Springer International Publishing AG 2017
F. Ellen et al. (Eds.): WADS 2017, LNCS 10389, pp. 593–594, 2017.
DOI: 10.1007/978-3-319-62127-2

Printed in the United States
By Bookmasters